Study Guide

to accompany

Life

The Science of Biology

Fifth Edition

Purves, Orians, Heller, and Sadava

Prepared by

Jon C. Glase
Cornell University

Jerry A. Waldvogel
Clemson University

SINAUER ASSOCIATES, INC.

W. H. FREEMAN AND COMPANY

THE COVER

Grizzly bears *(Ursus arctos)* take many years to mature reproductively, and
their cubs remain with their mothers for several years.
Photograph by Michio Hoshino/Minden Pictures.

Study Guide to accompany
LIFE: THE SCIENCE OF BIOLOGY, Fifth Edition

Address editorial correspondence to
Sinauer Associates, Inc.
23 Plumtree Road
Sunderland, Massachusetts 01375 U.S.A.
www.sinauer.com

Address orders to W. H. Freeman and Co.
Distribution Center, 4419 West 1980 South,
Salt Lake City, Utah 84104 U.S.A.
www.whfreeman.com

ISBN 0–7167–3221–1

Printed in U.S.A.

4 3

PREFACE

THE PURPOSE OF THIS STUDY GUIDE

A well written, current textbook is an essential element in the serious study of any complex subject such as biology. Our collective experience in teaching biology during the past 23 years suggests that many students can use their textbooks more effectively if a well conceived study guide is available to them. Our motivation in writing this guide is to help you learn about biology using the Fifth edition of the textbook *Life: The Science of Biology* (1998) by Purves, Orians, Heller, and Sadava.

To be effective, the study guide that accompanies a biology textbook should facilitate your learning process in two fundamental ways. First, it should help you *clarify* the important concepts and principles in each of the areas of biology. The textbook is designed to explain concepts and principles using a combination of descriptive text, illustrative artwork, and useful examples. Unfortunately, students sometimes lose sight of the major ideas and themes in biology because they fail to distinguish between the essential information and the supportive details. A major goal of our guide is to "unpack" the large quantity of information presented in the textbook, and to restate the most important ideas as a list of *Key Concepts* within each chapter. In writing these concept statements we have attempted to be concise and include only the essence of the idea being considered in the textbook. Sometimes a key concept item will include several closely related concepts which should be learned together. Bear in mind that this study guide is meant to be a *companion* to the textbook; the study guide's key concepts will only be meaningful after you have carefully studied how these ideas are developed in the corresponding chapter of the textbook.

A second major goal of our guide is to provide you with *feedback* about your level of understanding of the key concepts. For each chapter of the textbook, we have developed activities and questions that will test your mastery of the material covered. Some questions test your knowledge of terminology, others examine the relationships between key concepts, and some questions require you to think about biological concepts in a novel way, or perhaps use those concepts to solve a problem. The answers to all questions are presented in the *Answers and Explanations* sections of the guide, with each answer carefully explained and indexed to the key concept(s) related to that question. If you do not understand a question in the study guide, or our proposed answer and its explanation, then you may need to return to the textbook for further review.

HOW TO USE THIS STUDY GUIDE

We suggest that the best approach you can take to the study of biology is a structured approach. Treat the textbook as your primary information source and study it first. Scanning through assigned textbook pages will help to prepare you for lectures, laboratories, or discussion sections covering a specific topic. However, do not expect to completely comprehend all of the ideas you encounter during your first reading of the textbook. Rather, begin by scanning the chapter to understand how it is organized, see what the main topics are, and identify major themes. At the same time, read the corresponding portions of the study guide to further help you identify key concepts as they are presented in the textbook.

After you have attended the lecture, lab, or discussion, you will be in a much better position to clearly identify the material your instructor considers most important. To assist you in this process, we have included a set of *Learning Objectives* at the beginning of each chapter in the study guide. Your instructor may select a subset of objectives from this list, or even add to it. Careful study of the subject on your part should provide you with the necessary understanding to give a complete answer to each of these objectives. Check off each objective as you master it.

After identifying the relevant learning objectives for a specific topic, you should return to the textbook chapter and read the assignment in detail. During this in-depth study session, try to relate information in the textbook to the ideas presented in the study guide, your lecture notes, or laboratory exercises. We have divided each chapter of the study guide into several sections, each of which includes about as much textbook information as you are likely to assimilate during a short, productive study session before taking a short break.

Active involvement will promote your learning, so you should have paper available to write down important terms and to highlight sections of the textbook or study guide. Thoughtfully study the tables and figures included in each chapter. The artwork in your textbook has been carefully designed to help convey information that is difficult to explain using text alone; this includes conceptual relationships as well as the dynamic interactions that take place between the components of biological systems. Sometimes a single diagram can tie together and give meaning to several paragraphs of description. At all times, try to relate new information to what you already know. Use the index of your textbook wisely to locate previously covered information that may need to be reviewed as you take on new material. Remember that the study of biology is a cumulative process, and review is constantly needed as you progress to more complex, abstract material.

After you have actively studied part or all of a chapter in the textbook and the relevant sections of the study guide, you are ready to test your understanding by answering the accompanying study guide questions. We have provided space in the study guide so that you can write your responses to questions directly in the book, circle your answers to multiple choice questions, and make any sketches or calculations required in the various activities. If you encounter difficulty with a question, do not be tempted to short-circuit the learning process by looking at the *Answers and Explanations* section. Instead, you should review that section's key concepts and consult the textbook until you can make a reasonable choice from the possible answers given. Sometimes several choices will seem partially correct. In this case pick the *best*, most specific answer available. When you do make a choice, you should also be able to explain *why* the other choices are incorrect in terms of the key concepts presented in the study guide.

Since your mind works best in short, focused learning sessions, take a break after you have completed one section of the study guide. When you return to your study of biology, it is a good idea to quickly review the most recently studied section before proceeding to new material.

ADDITIONAL AIDS TO STUDYING BIOLOGY

Concept Mapping

It has been estimated that the number of new words encountered in a typical introductory biology course surpasses the vocabulary introduced during the first year of a foreign language course. Rote learning is a very inefficient approach for mastering this terminology, because the act of simply memorizing a new term without connecting it to the things you already know about the topic only promotes short-term learning. To take the analogy further, one does not learn a foreign language simply by memorizing the dictionary. The language can only be learned by understanding the concepts of grammar and syntax that allow you to manipulate the vocabulary in a meaningful fashion. Meaningful, long-term learning therefore occurs when you link new information to the knowledge you already possess. If you attempt to learn a concept meaningfully, then the concept will likely be remembered for a much longer time, and in a more readily accessible form. Educational psychologists have proposed several methods for promoting meaningful learning—one such method is called *concept mapping*.

A concept map is a diagram that shows the relationships between concepts within a topic, making it easier to relate new information to what you already know. As shown in the following generic example, concept maps are usually constructed from top to bottom. Each concept is placed in a box, with the most inclusive concepts near the top, and the more subordinate concepts nearer the bottom. The concept boxes are then connected by labeled lines that indicate the relationship between linked concepts.

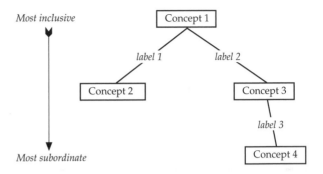

When constructing a concept map, it is best to select a fairly well-defined topic on which to focus. For example, the map at the top of the next page depicts information included under the heading *Atoms: The Constituents of Matter* found on pages 20–23 of the textbook, and described in key concepts 1–13 on pages 5–6 of the study guide. Notice that the concept map is simply a compact representation of the information found in the two book sources.

The real educational value of concept mapping comes from the actual construction of the map; you have to do it yourself! As you select concepts to include, you must define the relationships between these concepts and look for links to other concepts already present in the map. This makes the information meaningful for you. To make a map, select a group of key concepts that seem to be logically related. Terms in italics in the study guide are good candidates for inclusion as concepts in your map. Connect the concepts together with labels that make the relationships between concepts explicit. Since this is a dynamic learning process, it is best to do your map construction in pencil and to have an eraser handy—a map tends to change and evolve as your knowledge of the topic grows.

Taking Notes

Note taking is best viewed as yet another means of active learning. While this learning tool will most likely be applied to a lecture, laboratory, or discussion setting, it can also help in your efforts to understand the textbook and study guide materials. Like concept mapping, an emphasis should be placed on organizing the material in your notes in a way that has particular meaning for you. Your notes therefore should not simply be a written record of what the lecturer or textbook has to say about a topic. Instead, they should be a running account of your attempts to conceptually link the material you learned previously with the new material currently under study.

Although there is no one "right way" to take effective notes, following a few simple guidelines will increase the utility of your notes as a learning tool.

1. *Identify major themes* in the topic you are studying and use those themes as the section headings within your notes. In the case of lectures, this strategy will be greatly enhanced if you have already previewed the material being covered before coming to class.

2. *Record the key concepts* mentioned by the lecturer, but be sure to spend most of your time concentrating on what

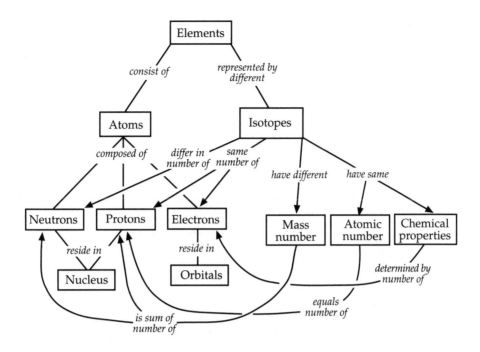

is being said about those concepts. One of the most important roles of a textbook is to provide the details that support the concepts presented in a lecture or laboratory setting. You should therefore rely on the book to provide much of the detail, rather than spending all of your time during lecture trying to write down things that are already available in your books and running the risk of missing most of what is being said. Think of the lecture as another way to prepare for the detailed reading of the textbook.

3. *Leave space in your notes* to include questions for the lecturer about the material just covered, to insert comments regarding relevant illustrations, and to fill in important or clarifying details as you read about the topic in the textbook. You should also strive to relate material from one section of a lecture to material in other sections, or to material presented in earlier lectures. If you get in the habit of asking "how" and "why" questions about the material you are currently studying, this will help you understand the connections between concepts and principles throughout the entire course.

4. *Develop a shorthand code* for common terms and phrases to increase your note-taking speed. Remember that your notes need not make sense to anyone but you, so feel free to use whatever abbreviations make sense in order to minimize the time spent writing while maximizing the time you devote to listening.

Preparing for and Taking Examinations

Unfortunately, examinations are a fact of life in the classroom. They come in many forms, ranging from oral examinations to essay and multiple choice formats. Although potentially stressful, examinations can also provide an excellent means of assessing your understanding of biology.

There is no better way to prepare for an exam than to practice answering exam questions. Your instructor may provide old examinations with which to practice, but even if this is not the case, you can still gain practice by creating your own exam questions. The idea here is not to "out-guess" the instructor, but rather to try and make connections between linked concepts by organizing ideas in a logical, related fashion.

Studying for an exam should be an ongoing process which begins immediately after the previous exam. Do not expect to master a complex subject such as biology by cramming for many hours just prior to a test. Careful study spread over a period of many days or weeks prior to the exam not only makes the task more manageable, but also fosters long-term understanding of concepts.

If your test will have essay questions, practice writing the essays as part of your studying. The mechanics of organizing an answer in your head and actually writing it down on paper will help promote conceptual understanding. During the exam, make a brief outline of your essay before writing the answer; this will help insure that you cover all aspects of the topic.

Many biology examinations, especially those given in large lecture classes, are of necessity done using a multiple choice format. Students are often frustrated by these examinations because they fail to apply their understanding of a topic when answering the question. Assuming that you have spent the time needed to master the conceptual basis of a topic, the following strategy will greatly improve your chances of successfully answering multiple choice questions:

1. *Read the question carefully*. It is impossible to answer a question correctly if you don't understand what's being asked, so take the time to thoroughly read the premise

statement of the question. For example, are you to choose the true or false statement, the exception or the generalization? If necessary, underline key words that help you focus on the main point of the question.

2. *Answer the question in your own way before looking at the answers.* After you understand the question, cover up the possible answers and attempt to think of all relevant concepts and examples related to that question. This puts the question in terms that you already understand, which makes it easier for you to evaluate the possible answers offered by your instructor.

3. *Systematically evaluate each possible answer.* Uncover the answers and carefully evaluate each possibility based on what you have just reviewed about the topic. Assuming that you really do understand the material, you will likely be surprised at how many of the items you thought of will be present in the possible answers.

4. *Don't overthink the solution.* If you have followed steps 1–3 above, then you should be able to identify one best answer without much effort. As long as you have carefully read the question and truly understand the material, more often than not, second-guessing will produce the wrong answer. If you do decide to change your answer, ask yourself why the alternative choice is a better one.

Post-Exam Review

You can often learn a great deal about biology, about your study habits, and about how to successfully take a test, from a thorough review of your exam. Go back and reconsider the questions you got wrong, trying to understand where your thinking went astray. Did you read the question carefully? Did you consider all possible answers? Did you simply spend too little time thinking carefully about each answer? Did you have an inaccurate understanding of the material? Were your notes accurate? Make it a point to return to the textbook or study guide and correct any faulty understanding you brought into the exam, since biology courses frequently build upon material presented during earlier portions of the course.

Seeking Help

The field of biology is vast. There will be an ever-growing body of information that builds as you progress through your biology course, and a time may come when you need assistance to clarify your understanding beyond that provided by this study guide. Realize that there is no dishonor in not understanding everything you read in a textbook or hear in lecture, and do not let yourself become overwhelmed or intimidated. If you are having trouble, please seek help from your instructor, tutors, fellow students, or other available resources.

ACKNOWLEDGEMENTS

We wish to acknowledge the assistance of several colleagues and the forbearance of our families and friends during the preparation of this study guide. Ideas for several of the questions included in this book came from examinations prepared for the introductory biology courses at Cornell University and Clemson University by Kraig Adler, Carl Hopkins, Antonie Blackler, and William Surver. We also thank the authors of *Life: The Science of Biology* for their careful reading of the study guide and for the many suggestions they provided during its development. Any remaining errors, mis-statements, or omissions are our own. Special thanks go out to the staff at Sinauer Associates for assistance during the developmental, editorial, and production phases of this book. We want to especially acknowledge the editorial assistance of Kathaleen Emerson.

The authors and publisher encourage and appreciate any suggestions that you may have for ways to make this study guide more useful. Please use the email addresses given below to provide us with any constructive feedback on the Guide or to ask for assistance. Good luck in your study of biology, and may you find it as interesting and exciting as we do!

Jon C. Glase
jcg6@cornell.edu
Ithaca, New York

Jerry A. Waldvogel
waldvoj@clemson.edu
Clemson, South Carolina

CONTENTS

1

An Evolutionary Framework for Biology

CHAPTER LEARNING OBJECTIVES—after studying this chapter you should be able to:

❑ Describe what a cell is, describe its characteristic features, and differentiate between prokaryotic and eukaryotic cells.

❑ Explain the importance of the changes in Earth's environment that were caused by the evolution of photosynthesis.

❑ Define the term "homeostasis," and explain the importance of homeostasis to metabolism.

❑ Explain what is meant when we say that an organism is adapted to its environment, and explain how adaptations arise.

❑ Describe a hypothesis for the origin of eukaryotic cells from prokaryotic cells.

❑ Describe what led to the evolution of multicellularity and the changes that multicellularity caused in sexual reproduction and regulated growth.

❑ Define and describe the importance of the process of speciation.

❑ Name and describe the distinguishing features of the six kingdoms of life.

❑ Describe the four stages of the hypothetico-deductive approach.

❑ Describe the relationship between a hypothesis and a null hypothesis and between a hypothesis and a theory.

❑ Describe the strengths and weaknesses of laboratory and field experimentation.

❑ Describe some of the arguments advanced by Buffon against the special creation of organisms.

❑ Explain the theory of evolutionary change developed by Lamarck.

❑ List the underlying ideas that Darwin used to develop his theory of evolution.

❑ Explain why variation within a species and the potential for exponential growth shown by many populations are important to Darwin's theory.

❑ Describe natural selection and how it differs from artificial selection.

❑ Describe the features and relationships of the following common levels in the hierarchical organization of biology: molecule, cell, tissue, organ, organism, species, population, community, biome, biosphere.

❑ Relate the concept of emergent properties to the hierarchical organization of biology.

EVOLUTIONARY MILESTONES • BIOLOGICAL DIVERSITY: DOMAINS AND KINGDOMS (pages 1–7)

Key Concepts

1. It is estimated that Earth formed about 4 billion years ago. The earliest fossils—remnants of past organisms—are dated from 3.8 billion years ago.

2. Life came from nonlife through interactions between small molecules. Complex organic molecules—lipids, carbohydrates, proteins, and nucleic acids—are formed only by living systems.

3. Interacting systems of molecules enclosed within a membrane became *cells*. Cells capture energy, control traffic through their membranes, and replicate themselves.

4. Several scientists were important in the development of the idea that the cell is the unit upon which all life is built.

Robert Hooke first observed dead cells. Antoni van Leeuwenhoek studied living unicellular organisms. Theodor Schwann asserted that all organisms consist of cells. Rudolf Virchow stated that all cells come from preexisting cells, and Louis Pasteur provided evidence supporting Virchow's hypothesis.

5. The first organisms to evolve consisted of prokaryotic cells and were unicellular. They were prevalent for more than 2 billion years and confined mostly to Earth's oceans.

6. The energy-consuming and energy-releasing chemical reactions occurring within living systems are called *metabolism*.

7. Simple chemical compounds available in prebiotic times were the major energy source of the first living cells. About 2.5 billion years ago, *photosynthesis* first appeared in certain prokaryotes that were probably similar to present-day cyanobacteria.

8. Oxygen gas (O_2), liberated as part of photosynthesis, accumulated in Earth's atmosphere and allowed the evolution of metabolic *aerobic* pathways. *Ozone* (O_3), a form of oxygen, accumulated in the atmosphere, forming a dense layer that filtered out damaging ultraviolet radiation.

9. The first cells divided asexually and the resulting progeny were identical to the parent cell. The evolution of sexual recombination—the combination of genetic material (genes) from two cells into a single cell—permitted reproduction with variation. Variable offspring are more likely to include individuals adapted to new environments.

10. *Eukaryotic cells* were formed about 1.5 billion years ago when certain prokaryotes were engulfed by other prokaryotes and began permanently living inside their host. The enclosed prokaryotes became the *organelles* of the host cell. Eukaryotic cells have a discrete *nucleus* and genetic material consisting of genes and chromosomes.

11. The evolution of *multicellular organisms* about 1 billion years ago allowed different cells to become specialized for certain functions, including sex. Sex and reproduction are linked in most multicellular organisms. In addition to simple cell division, or *mitosis*, *meiosis* evolved to provide for increased sexual recombination in many multicellular organisms.

12. With the evolution of multicellularity and aerobic respiration, organisms were able to adapt to terrestrial environments. Green plants first colonized the land, and they became a major food source for subsequent animal colonizers.

13. In response to changes in the external environment, organisms vary the rates of their chemical reactions in order to maintain relatively constant internal conditions that are suitable for metabolism. Maintenance of a steady internal state is called *homeostasis*. Homeostasis depends on specialized systems for responding to internal and external signals.

14. Through their metabolic activities, all living systems show *regulated growth*—a carefully controlled increase in the number of molecules of which they are composed.

15. *Speciation* is the process that creates genetically independent groups of organisms called *species*. Speciation has produced the great diversity of life present on Earth today.

16. In a common scheme for classifying organisms, all species are grouped into three *domains* and six *kingdoms*.

17. The domains *Archea* and *Bacteria* include all the single-celled *prokaryotic* organisms. Prokaryotic cells lack true nuclei and some other organelles. The Archea and Bacteria each consist of a single kingdom, the *Archaebacteria* and *Eubacteria*, respectively.

18. *Eukaryotic* cells were formed when certain prokaryotes began permanently living inside other prokaryotes. Members of the other four kingdoms constitute the domain Eukarya and consist of eukaryotic cells.

19. The kingdom *Protista* contains eukaryotic organisms, most of which are single-celled.

20. The kingdom *Fungi* includes multicellular, heterotrophic organisms called molds, mushrooms, yeasts, etc., that obtain their nutrients by *absorption*.

21. The kingdom *Plantae* includes multicellular, autotrophic organisms called plants that obtain their energy by *photosynthesis*.

22. The kingdom *Animalia* includes multicellular, heterotrophic organisms called animals that obtain their nutrients by *ingestion*.

23. The terms "ancestral" and "derived" are used to describe characteristics that appeared earlier or later in the evolution of a lineage of organisms. These terms are preferable to "primitive," "advanced," "lower," or "higher."

24. A theory to explain adaptation was first proposed by *Charles Darwin* and *Alfred Russel Wallace* in the 1850s.

Questions (for answers and explanations, see page 49)

1. Which of the following is *not* a characteristic of all living systems?
   a. Reproduction
   b. Meiosis
   c. Metabolism
   d. Cellularity
   e. Adaptation

2. Chronologically order the following events in Earth's history, where 1 = earliest and 6 = latest.

   _____ Eukaryotes appear

   _____ Aerobic respiration

   _____ Photosynthesis

   _____ Metabolism

   _____ Meiosis

   _____ Multicellularity

3. Which of the following kingdoms consists of organisms that are mostly unicellular and have eukaryotic cells?
   a. *Eubacteria*
   b. *Protista*
   c. *Fungi*
   d. *Plantae*
   e. *Animalia*

4. Which of the following kingdoms consists of organisms that are mostly multicellular, photosynthetic autotrophs?
   a. *Eubacteria*
   b. *Protista*
   c. *Fungi*
   d. *Plantae*
   e. *Animalia*

5. Which of the following kingdoms consists of organisms that are mostly multicellular, absorptive heterotrophs?
   a. *Archaebacteria*
   b. *Protista*
   c. *Fungi*
   d. *Plantae*
   e. *Animalia*

6. You discover a species of termite that has an animal-like protist called *Trichonympha* living within it. Inside the protists are bacteria that produce cellulase. This enzyme is used to digest the cellulose in wood, to the benefit of the

host protist and the termite. How many domains and king-doms are included in this association?

   a. 1 domain and 3 kingdoms
   b. 2 domains and 2 kingdoms
   c. 2 domains and 3 kingdoms
   d. 3 domains and 2 kingdoms
   e. 3 domains and 3 kingdoms

7. Describe a hypothesis for the evolution of the eukaryotic cell.

THE WORLD INTO WHICH DARWIN LED US • ASKING THE QUESTIONS "HOW?" AND "WHY?" • LIFE'S EMER-GENT PROPERTIES   (pages 8–14)

Key Concepts

1. *Buffon* advanced the following arguments against the spe-cial creation of organisms: all mammals share the same limb bones, although the bones are modified for the func-tions they perform, and some mammals have limbs with functionless toes.

2. *Jean Baptiste de Lamarck* argued that due to use or disuse structures become modified, and that these modifications are passed on to future generations.

3. Charles Darwin's theory of evolution (also independently developed by Alfred Wallace) is based on the ideas that Earth is very old, all organisms are related by common descent, diversity results from speciation, evolution pro-duces gradual changes in populations, and that evolution occurs by natural selection.

4. Evolution by natural selection is based on two observa-tions: considerable variation exists among individuals of the same species and most species have high reproductive rates.

5. The major inference of Darwin's theory is that slight varia-tion among individuals can lead to differential reproduc-tive success, in which some individuals leave more off-spring than other individuals.

6. The process by which some individuals leave more surviv-ing offspring than other individuals is called *natural selection*.

7. *Artificial selection* results from differential reproductive suc-cess as directed by humans and not the environment (as in natural selection).

8. Evolution by natural selection is not goal-directed. Organisms do not adapt to future environments.

9. *Science* is both a collection of discovered knowledge about the universe and a process that can be used to discover new knowledge.

10. To discover new truths, scientists employ a cyclical process called the *hypothetico-deductive method* consisting of five steps: (1) making *observations*, (2) asking questions, (3) forming a *hypothesis*, (4) using the hypothesis to make *predictions*, (5) testing the predictions.

11. A hypothesis is a tentative answer to a question that usual-ly predicts a specific effect. The corresponding *null hypothe-sis* states that the proposed effect is absent.

12. Assumptions and prior knowledge are usually involved in the formulation of a hypothesis.

13 In *experimentation,* extraneous factors are kept constant so that the effects of a single factor or group of related factors can be examined more clearly.

14. Greater control of environmental variables is possible in laboratory experimentation. Field experimentation is less controlled, but it may be easier to apply the results to nature.

15. A hypothesis cannot be proved or disproved with absolute certainty.

16. A *theory* is a hypothesis that has been extensively tested and supported.

17. The objects that biologists study, from molecules to ecosys-tems, can be organized into a hierarchy in which each level consists of the objects at the next lower level.

18. *Molecules* are the smallest objects studied by biologists. The most important large molecules found in living systems are *proteins, nucleic acids, lipids,* and *carbohydrates.*

19. Molecules aggregate to form complex structures, like mem-branes, that have unique properties of their own.

20. The *cell* is the simplest unit capable of independent exis-tence and reproduction. All organisms are cells or are com-posed of cells.

21. Structurally similar cells can associate to form *tissues,* and different tissues can aggregate to form *organs.*

22. The *organism* is the smallest unit of life. Organisms can be composed of a single cell *(unicellular)* or many cells *(multi-cellular).*

23. A group of structurally and functionally similar organisms that can interbreed with each other but not with other organisms is called a *species.*

24. Individuals of the same species living in a defined area are a *population.*

25. *An ecological community* consists of populations of different species living together in a particular environment.

26. Major types of ecological communities are *biomes.* All of the world's biomes form the *biosphere.*

27. *Emergent properties* are the new features of a system that result from the interaction of the system's components.

Questions (for answers and explanations, see page 49)

1. Which of the following was *not* a major hypothesis of the theory of evolution by natural selection advanced by Charles Darwin?

   a. Earth is ancient.
   b. Evolution proceeds by the accumulation of gradual changes within populations.
   c. Offspring inherit genes from their parents.
   d. Biological diversity is produced by speciation.
   e. All present-day organisms have a common ancestor in the past.

2. Describe natural selection and how it differs from artificial selection.

3. In reference to the studies done on palatability of cryptic and conspicuous caterpillars described in the text, classify each of the following as either an *observation, a hypothesis, a null hypothesis, a prediction, a result,* or an *assumption.*

   a. Birds are visually oriented predators: _____

   b. Conspicuous caterpillars tend to occur in groups: _____

   c. Bird 1 consumed five conspicuous caterpillars and one cryptic caterpillar: _____

   d. This species of conspicuous caterpillar should not be eaten by blue jays: _____

   e. Blue jays should eat equal numbers of conspicuous and cryptic caterpillars. _____

4. What do laboratory and field experiments have in common and how do they differ?

5. Which of the following is *not* a step in the hypothetico-deductive method of science?
   a. Stating hypotheses
   b. Making observations
   c. Establishing a conclusion with absolute certainty
   d. Deriving predictions from hypotheses
   e. Testing hypotheses

6. In the space provided before each of the following objects, provide a number that indicates the place of each object in a hierarchy from the least inclusive (1) to most inclusive (7) category. In the space following each object, specify its category.

   _____ A sperm   _____

   _____ A mouse   _____

   _____ Glucose   _____

   _____ A lung   _____

   _____ Earth   _____

   _____ A woodland   _____

   _____ *Peromyscus leucopus*   _____

7. Biological membranes are composed of molecules called proteins and phospholipids. Which of the following statements best describes an emergent property at the level of biological membranes?
   a. All cells are bounded by membranes.
   b. Membranes regulate what chemicals can pass into or out of the cell.
   c. Phospholipids are composed of phosphorus and lipids.
   d. Phospholipids and proteins are found in all cells.
   e. Cells contain membrane-bounded structures called organelles.

2

Small Molecules: Structure and Behavior

CHAPTER LEARNING OBJECTIVES—after studying this chapter you should be able to:

❑ Describe the structure of the atom, in terms of the locations of protons, electrons, and neutrons and the distribution of mass and charge.

❑ Explain the structure and organization of the periodic table of elements.

❑ Relate an element's atomic number, mass number, and chemical properties to the number of protons, electrons, and neutrons in its atoms.

❑ Write the symbols, atomic numbers, and mass numbers of the biologically important elements using the conventions described in the text.

❑ Differentiate between compounds and elemental substances.

❑ Explain how isotopes of an element differ in terms of atomic number and mass number and how an element's atomic weight is determined.

❑ Describe what radioisotopes are and their importance in biological research.

❑ Describe the distribution of electrons in an atom's orbitals and shells and how this distribution determines the chemical properties of atoms.

❑ Describe the octet rule and use it to predict chemical bond formation between different atoms.

❑ Explain the nature of the covalent bond and how the relative electronegativity of the atoms involved determines whether the bond will be polar or nonpolar.

❑ Use the covalent bonding capabilities of biologically important elements to determine the types of chemical bonds they will form.

❑ Understand how hydrogen bonds form, how they are represented, and their importance.

❑ Explain the similarities and differences between van der Waals and hydrophobic interactions and the biological importance of these two forces.

❑ Differentiate between molecules and ions and between covalent and ionic bonds.

❑ Relate molecular and structural formulas for biologically important compounds.

❑ Determine molecular weight and understand its relationship to molarity.

❑ Relate equations showing biologically important chemical reactions to changes in chemical bonding between reactants and products.

❑ Explain what calories, kilocalories, and joules are and how they are used to represent chemical bond energy.

❑ Describe how water's polar nature and its ability to form hydrogen bonds with other molecules affect its heat capacity, surface tension, cohesive strength, and heat of vaporization and result in its excellence as a solvent.

❑ Describe the biologically important properties of water.

❑ Understand the concept of pH, relate it to the $[H^+]$ and $[OH^-]$, and given the pH of an aqueous system, be able to determine if it is acidic, basic, or neutral.

❑ Identify an acid and a base given a reversible chemical reaction that affects the pH of an aqueous system.

❑ Explain what a buffer is and how a buffer can maintain a relatively constant pH within a buffering range.

❑ Recognize the distinguishing features of the following organic compounds: hydrocarbon, alcohol, aldehyde and ketone, organic acid, organic base, and amino acid.

❑ Recognize the distinguishing features of the following functional groups: hydroxyl, carbonyl, carboxyl, amino, phosphate, and sulfhydryl.

❑ Explain the terms "isomer" and "optical isomer" and be able to identify an asymmetric carbon atom.

ATOMS: THE CONSTITUENTS OF MATTER (pages 19–23)

Key Concepts

1. *Atoms* are composed of *protons*, *neutrons*, and *electrons*.

2. Protons and neutrons are found in the *nucleus* of the atom and represent most of the mass of the atom.

3. Electrons move in *orbitals* around the nucleus.

4. *Mass* is a measure of the amount of matter present. *Weight* is a measure of gravitational attraction for mass and differs for different planets.

5. Protons and neutrons have the same mass; each is equal to one *atomic mass unit (amu)* or *dalton*.

6. Protons have a unit charge of +1, electrons have a charge of −1 unit, and neutrons are electrically neutral.

7. Atoms are electrically neutral because they contain the same number of electrons and protons.

8. The more than 100 naturally-occurring elements can be arranged into a *periodic table* whose columns consist of elements with similar chemical properties but increasing mass.

9. *Elemental substances* are composed of only one type of atom. *Compounds* are substances consisting of two or more different atoms linked together by a chemical bond.

10. The *atomic number* of an element is the number of protons present in its atoms. The atomic number differs for each element.

11. The *mass number* is equal to the sum of the protons and neutrons in the nucleus of the atom; the mass of the electrons is negligible and is ignored.

12. Common notation for representing elements uses a one- or two-letter symbol with, optionally, the atomic number shown as a subscript and the mass number as a superscript preceding the symbol.

$$\begin{matrix} \text{mass number} \\ \text{atomic number} \end{matrix} \text{SYMBOL}$$

as in $^{12}_{6}C$ for carbon and $^{40}_{20}Ca$ for calcium.

13. Atoms with the same atomic number but a different mass number are *isotopes* of an element. Specifically, the isotopes of an element differ in the number of neutrons present, but have the same chemical properties.

14. Isotopes are usually shown with only the mass number preceding the symbol, as in ^{1}H, ^{2}H, ^{3}H for hydrogen's isotopes and ^{12}C, ^{13}C, and ^{14}C for carbon's isotopes.

15. The *atomic mass* or *weight* of an element is the average of the mass numbers for all isotopes of an element weighted according to the natural occurrence of each isotope.

16. *Radioisotopes* undergo radioactive decay to become another element by giving off energy or elementary particles. Radioisotopes are used as markers in biological research. Isotopes of this type include ^{3}H, ^{14}C, ^{32}P.

17. *Chemical reactions* occur when electrons are exchanged or shared differently between atoms.

18. An electron *orbital* is a region of space where an electron is located at least 90% of the time. A maximum of two electrons can occupy an orbital, and the two electrons spin in opposite directions.

19. Orbitals are grouped into *electron shells* that surround the nucleus and are designated by the letters K, L, M, N, O, P, and Q from the innermost to the outermost shell.

20. The K *shell* has a single orbital (*s* orbital) holding a maximum of two electrons.

21. The L *shell* has four orbitals, an *s* orbital and three *p* orbitals oriented in the *x*, *y*, and *z* axes relative to the center of the nucleus. The L shell accommodates a maximum of eight electrons (two per orbital).

22. Successive shells (M–Q) are located at increasing distance from the nucleus and hold differing numbers of orbitals.

23. The energy level of electrons increases with their distance from the nucleus. Therefore, electrons in outer shells have more energy than electrons in inner shells. Electron shells are filled from the innermost shell outward.

24. The outermost shell of an atom determines the chemical reactivity of the atom.

25. An atom with its outermost shell completely filled is most stable and, thus, least reactive with other atoms. The inert elements have completely filled shells and include helium (He), neon (Ne), and argon (Ar).

26. A *molecule* consists of two or more atoms linked together by chemical bonds (attractive forces between atoms).

27. The *octet rule* describes the tendency of many atoms in molecules to be stable when they have eight electrons in their outmost orbitals.

Questions (for answers and explanations, see page 49)

1. Which of the following statements about atomic structure is *not* true?
   a. Almost all the mass of an atom resides in the nucleus.
   b. Electrons are in orbitals that surround the nucleus.
   c. Protons and neutrons have approximately the same mass.
   d. All atoms contain electrons, protons, and neutrons.
   e. Protons and electrons have equal and opposite charge.

2. Which of the following statements about elements is *not* true?
   a. Elements with similar chemical properties are located in the same column of the periodic table.
   b. The mass number of an element equals the number of protons plus neutrons.
   c. The atomic number of an element equals the number of electrons or protons.
   d. All atoms of an element have the same atomic number.
   e. All atoms of an element have the same mass number.

3. The mass number minus the atomic number
   a. is equal to the number of electrons in the atom.
   b. is equal to the number of protons in the atom.
   c. is equal to the number of neutrons in the atom.
   d. is equal to the atomic weight of the atom.
   e. is the same for all isotopes of an element.

4. Sodium has an atomic number of 11. How many electrons are located in its M shell?
   a. 1
   b. 2
   c. 7
   d. 8
   e. 0

5. Neon has an atomic number of 10. It is chemically _____ and has ____ completely filled orbitals in its L shell.
   a. reactive, 2
   b. reactive, 3
   c. reactive, 4
   d. inert, 2
   e. inert, 3
   f. inert, 4

6. There are two isotopes of a hypothetical element X: ^{30}X and ^{32}X. If the atomic weight of X is 30.5, what is the natural proportion of ^{30}X to ^{32}X in the world?
   a. 100:1
   b. 50:50

c. 75:25
d. 25:75
e. 1:100

7. The element with which of the following atomic numbers would be *most* stable?
   a. 1
   b. 3
   c. 12
   d. 15
   e. 18

8. Given that the chemical properties of an element are determined by the number of electrons in its outer shell, which of the following pairs of elements would have the most similar chemical properties?
   a. Carbon ($_6$C) and nitrogen ($_7$N)
   b. Carbon ($_6$C) and oxygen ($_8$O)
   c. Helium ($_2$He) and fluorine ($_9$F)
   d. Fluorine ($_9$F) and chlorine ($_{17}$Cl)
   e. Chlorine ($_{17}$Cl) and sodium ($_{11}$Na)

Activities (for answers and explanations, see page 49)

- Write the symbols for the most common isotopes of the following elements, showing the atomic number and mass number, using the notation described in key concept 12.

  Hydrogen   **Carbon**

  Nitrogen   **Oxygen**

  Sulfur   **Phosphorus**

- Write the names and symbols for the less common, but biologically important, isotopes of hydrogen, carbon, and oxygen.

  Hydrogen

  Carbon

  Oxygen

- Write the numbers of electrons, protons, and neutrons found in atoms of each of the following elements or isotopes.

	Electrons	**Protons**	**Neutrons**
$_1^1$**H**			
$_6^{14}$**C**			
$_7^{14}$**N**			
$_{30}^{65}$**Zn**			

- In the following figures, place electrons in the appropriate shells based on the atomic numbers of the elements carbon ($_6$C), nitrogen ($_7$N), and sodium ($_{11}$Na).

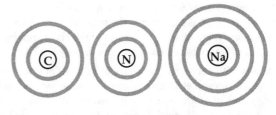

- In the following figures, place electrons in the appropriate shells based on the atomic numbers of the elements chlorine ($_{17}$Cl), argon ($_{18}$A), and oxygen ($_8$O).

CHEMICAL BONDS: LINKING ATOMS TOGETHER • EGGS BY THE DOZEN, MOLECULES BY THE MOLE (pages 23–30)

Key Concepts

1. In a molecule, atoms are held together by an attractive force called a *chemical bond*. Strong chemical bonds depend on the stability that atoms attain when their outermost electron orbitals are completely filled.

2. A *covalent chemical bond* results when two atoms share electrons in order to complete their outermost shell.

3. Atoms that complete their outer shells by sharing electrons are stable because the attraction of the shared electrons for the nuclei of the atoms exactly balances the mutual repulsion of the nuclei.

4. For an atom that can form covalent bonds, the number of vacant positions in its outer shell is the number of covalent bonds that it can form.

5. The *covalent bonding capabilities* of several biologically important elements include:

Hydrogen ($_1$H)	1
Oxygen ($_8$O)	2
Nitrogen ($_7$N)	3
Carbon ($_6$C)	4
Sulfur ($_{16}$S)	5 or 2

6. A single bond results if two atoms share a pair of electrons, a double bond results when four electrons are shared, and a triple bond results when six electrons are shared by two atoms.

7. Atoms with high affinity for electrons are described as having high *electronegativity*. Atoms with differing electronegativity form *polar covalent bonds* with partial electrical charges ($\delta-$ and $\delta+$) at opposite ends of the molecule.

8. A *hydrogen bond* results from the attraction of the positive pole of a covalently bound hydrogen in one molecule and the negative pole associated with an atom, usually oxygen or nitrogen, in the same or a different molecule. A hydrogen bond is about 1/20 as strong as the covalent bond between hydrogen and oxygen.

9. Hydrogen bonds form between water molecules, as each hydrogen atom is attracted to the negative pole of the oxygen atom in an adjacent water molecule. Hydrogen bonds are usually represented as a series of dots (–H····O–).

10. Atoms with very strong or weak affinity for electrons may interact to completely lose or gain electrons. *Ions* are atoms or groups of atoms that have gained or lost electrons and, as a result, are electrically charged. Positively charged ions are *cations*; negatively charged ions are *anions*.

11. Some covalently bonded molecules form ions when they dissolve in water. When a molecule ionizes it forms positively and negatively charged ions whose total charge is zero, as in: $HCl \rightarrow H^+ + Cl^-$.

12. Some ions have multiple charges, as in Ca^{2+}, Fe^{3+}, and Al^{3+}. Groups of atoms can be ions, as in NH_4^+, SO_4^{2-}, and PO_4^{3-}.

13. Oppositely charged ions can be bound together to form *stable ionic compounds* such as common salt (NaCl).

14. An *ionic bond* results from the electric attraction between oppositely charged ions. Ionic bonds are strong in solids because the ions are close together, but weak when the ions are dispersed in water. In biological systems, ionic bonds are less than one-tenth as strong as a covalent bond in which electrons are shared equally.

15. *Hydrocarbons* are compounds containing only hydrogen and carbon. Hydrocarbons are nonpolar molecules because carbon and hydrogen have nearly equal electronegativity

16. Random variations in the electron distributions of two nonpolar molecules create brief, weak attractions called *van der Waals attractions*. Van der Waals attractions are important in maintaining the three-dimensional shapes of macromolecules.

17. Substances with only nonpolar covalent bonds (*nonpolar substances*) are *hydrophobic*. Hydrophobic substances have no attraction for polar substances such as water.

18. The clumping of nonpolar substances in the presence of water is called *hydrophobic interaction*. The clumping minimizes the disruptive effects that the nonpolar molecules have on hydrogen bonding between water molecules.

19. A compound's *molecular formula* shows the symbols for each atom in a molecule followed by the numbers of each atom written as a subscript, as in H_2O. *Structural formulas* show how the atoms are linked together.

20. The *molecular weight* of a compound is the sum of the atomic weights of the atoms in the molecule.

21. A gram molecular weight of a substance (its molecular weight in grams) is called a *mole*. If one mole is dissolved in one liter of water, the resulting solution is a 1.0-*molar* solution (1 *M*).

22. A mole of any substance contains the same number of molecules, called *Avogadro's number*, or 6.023 X 10^{23} molecules per mole.

Questions (for answers and explanations, see page 50)

1. The molecular weight of glucose is 180. If you added 180 grams of glucose to 0.5 liter of water, what would be the molarity of the resulting solution?

a. 18
b. 1
c. 9
d. 0.5
e. 2

2. One milliliter (ml) of water weighs one gram (g), so one liter (l) weighs one kilogram (kg). Given that the molecular weight of water is 18, how many moles of water are there in one liter of pure water?

a. 1
b. 100
c. 56
d. 18
e. 1.8

3. If carbon can form the following stable molecule with Z, Z＝C＝Z, which of the following could be the atomic number of Z?

a. 2
b. 6
c. 8
d. 10
e. 11

4. Which of the following molecules would you expect to be *least* soluble in water?

a. C_2H_5OH
b. $H_3C—CH_3$
c. NH_4OH
d. CH_2NH_2COOH
e. NH_3

5. Which of the following molecules would you expect to form van der Waals or hydrophobic bonds with another molecule?

a. NH_3
b. C_2H_5OH
c. NH_4OH
d. CH_2NH_2COOH
e. $H_3C(CH_2)_4CH_3$

Activities (for answers and explanations, see page 50)

- In the following figures, place electrons in the appropriate locations based on the covalent bonding shown for each molecule.

O＝C＝O

O＝O

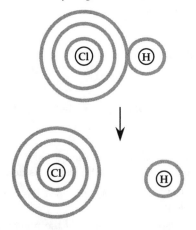

- Add electrons to the following diagram showing the ionization of hydrogen chloride: $HCL \rightarrow H^+ + Cl^-$.

CHEMICAL REACTIONS: ATOMS CHANGE PARTNERS • WATER: STRUCTURE AND PROPERTIES • ACIDS, BASES, AND THE pH SCALE   (pages 30–35)

Key Concepts

1. A *chemical reaction*, represented by an arrow ($\rightarrow$), occurs when atoms change their bonding with other atoms. Molecules to the left of the arrow are *reactants;* molecules to the right are *products.*

Reactants → Products

2. Chemical reactions usually result in the release or absorption of *energy,* where energy is defined as the capacity to do work or affect change.

3. Energy stored in chemical bonds is a form of *potential chemical energy.*

4. A *calorie (cal)* is the amount of heat required to raise 1 gram of water from 14.5°C to 15.5°C. A *kilocalorie (kcal)* equals 1,000 cal.

5. The *Joule (J)* is another commonly used unit of energy. One Joule equals 0.239 cal.

6. Calories and Joules can be used to represent changes in energy in any of its forms, such as heat, mechanical, electrical, or chemical.

7. The polarity of water explains its excellence as a solvent. More substances can dissolve in water than in any other solvent.

8. Most biologically important chemical reactions take place in water-based, or *aqueous,* solutions.

9. Water is a polar molecule because the shared electrons between the hydrogens and oxygen are more closely associated with the oxygen atom.

10. The four orbitals of water's outer electron shell are directed toward the vertices of a regular tetrahedron. The angles between the orbitals with bound hydrogens is about 104.5 degrees.

11. Ice floats because it is less dense than liquid water. This prevents bodies of water from freezing solid, and the ice serves as an insulating layer.

12. Because of all the hydrogen bonds that must be broken, a great deal of energy is required to melt ice.

13. *Heat capacity* is the amount of heat required to raise the temperature of a substance. Water's high heat capacity causes the temperature of bodies of water to remain relatively constant or to change only slowly.

14. Hydrogen bonding between water molecules explains the *cohesive strength* of liquid water.

15. Hydrogen bonding between water molecules is the basis for the high *surface tension* of liquid water.

16. Water has a high heat of vaporization. A significant quantity of heat is needed to change water from liquid to gas (*evaporation*). Thus, evaporation of sweat is an efficient cooling process for some organisms.

17. About 1 in every 550 million water molecules ionizes into a hydrogen ion (H^+) and a hydroxide ion (OH^-), $H_2O \rightarrow H^+ + OH^-$.

18. Pure water has equal concentrations of hydrogen and hydroxide ions and is called *neutral.* A solution with more hydrogen ions than hydroxide ions is called *acidic.* A solution with more hydroxide ions than hydrogen ions is called *basic.*

19. A compound that releases H^+ in water is called an *acid.* Carbonic acid is an example: $H_2CO_3 \rightarrow H^+ + HCO_3^-$. Adding an acid to water makes the solution *acidic.*

20. A compound that accepts H^+ is called a *base.* The bicarbonate ion is an example of a base: $HCO_3^- + H^+ \rightarrow H_2CO_3$. Adding a base to water makes the solution *basic* or *alkaline.*

21. Acids and bases occur in pairs. For example, H_2CO_3 is an acid; bicarbonate ion, HCO_3^-, is a base. $HCO_3^- + H^+ \rightleftharpoons H_2CO_3$ is an example of a *reversible reaction,* as indicated by double arrows.

22. The acidity or basicity of aqueous systems is expressed as *pH.* pH is the negative logarithm of the hydrogen ion concentration (pH = $-\log_{10}[H^+]$, where $[H^+]$ = molar concentration of hydrogen ions).

23. The pH scale is logarithmic. Each pH unit corresponds to a hydrogen ion concentration ten times greater than or less than the next nearest unit. Pure water has a pH of 7. Acidic solutions have pH values < 7; basic solutions have pH values > 7.

24. A *buffer* is a mixture of a weak acid (an acid that does not ionize completely) and its corresponding base.

25. Within a specific *buffering range,* buffers maintain a relatively constant pH by binding with or releasing H^+. If H^+ is added to a buffered system, the base binds with the additional H^+; if H^+ is removed from a buffered system, the weak acid ionizes to release additional H^+.

Activity (for answers and explanations, see page 51)

- Complete the missing parts of Table 1.

Table 1. The relationship between pH and the hydrogen ion concentration [H^+] and hydroxide ion concentration [OH^-] of water.

pH	[H^+]	[OH^-]
0	1.0	
1	0.1	
2	0.01	
3	0.001	
	10^{-4}	10^{-10}
	10^{-5}	10^{-9}
	10^{-6}	10^{-8}
	10^{-7}	10^{-7}
8		10^{-6}
9		10^{-5}
10		
11		
12		
13		
14		1.0

Questions (for answers and explanations, see page 51)

1. How many kilocalories are required to raise 250 grams of water from 14.5 to 15.5° C?
   a. 0.25
   b. 4
   c. 0.4
   d. 2.5
   e. 1

2. Which of the following shows the *correct* arrangement of two water molecules and the hydrogen bond(s) linking them?

   a. $H-O-H\text{······}H-O-H$ b.
   $$H-O-H$$
   $$H-O-H$$

   c.
   $$H-O-H$$
   $$H-O-H$$

   d. (structure showing H, O, H arrangement with $O\text{·····}H$)

   e.
   $$H-O-H$$
   $$H-O-H$$

3. Which of the following properties of water is *most* important in determining its excellence as a solvent?
   a. The pH of water is 7.
   b. Water has a high heat of evaporation.
   c. Water has a high surface tension.
   d. Water has a high heat capacity.
   e. The water molecule is polar.

4. In the reaction $CH_3COOH \rightleftharpoons H^+ + CH_3COO^-$, CH_3COO^- is an example of
   a. an acid.

   b. a base.
   c. a buffer.
   d. an ionic compound.
   e. a reversible reaction.

5. Cola has a pH of 3; blood plasma has a pH of 7. The hydrogen ion concentration of cola is ____ than the hydrogen ion concentration of blood plasma.
   a. four times greater
   b. four times lesser
   c. 400 times greater
   d. 400 times lesser
   e. 10,000 times greater

6. If solution A has a pH of 2 and solution B has a pH of 8, which of the following statements is true?
   a. A is basic and B is acidic.
   b. A is a base and B is an acid.
   c. A is acidic and B is basic.
   d. A is an acid and B is a base.
   e. A has a greater [OH^-] than B.

THE PROPERTIES OF MOLECULES (pages 35–39)

Key Concepts

1. *Organic molecules* contain carbon atoms and usually hydrogen and oxygen, but also nitrogen and phosphorus atoms.

2. Molecules vary in terms of their size (number of atoms), three-dimensional shape, and solubility.

3. Organic molecules also differ in terms of the functional groups they possess. A *functional group* is a specific grouping of atoms in larger molecules with particular reactive features.

4. A compound containing a *carboxyl* functional group (—COOH) is an example of an *organic acid*, since the carboxyl group can ionize to become —COO^- and release a hydrogen ion, as in $COOH \rightleftharpoons H^+ + COO^-$.

5. An *alcohol* results when a *hydroxyl* group (—OH) replaces one hydrogen in a hydrocarbon. The —OH end of an alcohol is polar and, as a result, some alcohols are soluble in water.

6. In addition to a hydroxyl group, *sugars* also have a *carbonyl* group (a carbon with a double-bonded oxygen). If the carbonyl group includes a hydrogen, it is an *aldehyde* sugar.

7. A compound containing an *amino* group (—NH_2) is an example of an *organic base* or amine, since the amino group can bind with a hydrogen ion to form the —NH_3^+ ion.

8. *Amino acids* have both an amino group and a carboxyl group attached to a central carbon called the *alpha (α) carbon*. Also attached to the α carbon are a hydrogen atom and a variable side chain designated R.

9. Each of the 20 amino acids found in proteins has a different side chain that determines its chemical characteristics.

10. Amino acids are simultaneously organic acids and bases. At normal pH values for cells, the carboxyl and amino groups of an amino acid are ionized and exist as COO^- and NH_3^+.

11. The *sulfhydryl* group, —SH, is an important constituent in the R groups of two amino acids. Two sulfhydryl groups react to form a disulfide bridge consisting of a S–S link between two different regions of a protein.

12. The *phosphate* group, —$OPO_3{}^{2-}$, is involved in energy transfer within cells and in the structure of DNA and RNA.

13. *Isomers* have the same chemical formula, but differ in the structural arrangement of their atoms. The structural differences can give isomers different chemical properties.

14. *Structural isomers* differ in how the atoms are joined together.

15. If a carbon atom is bound to four different atoms or groups it becomes an *asymmetric carbon*. The α carbon of an amino acid is an asymmetric carbon.

16. Each of the two different, mirror-image arrangements of the four groups around an asymmetric carbon is called an *optical isomer*. An asymmetric carbon's two optical isomers cannot be superimposed.

17. The two optical isomers of an amino acid are called the L- and D-forms. On Earth, only L-amino acids are commonly found in the proteins of living things.

Questions (for answers and explanations, see page 51)

1. Which of the following molecules is *not* polar?
   a. C_2H_5OH
   b. $H_3C—CH_3$
   c. H_2O
   d. CH_2NH_3COOH
   e. NH_3

2. Which of the following statements about optical isomers is *not* true?
   a. Optical isomers have the same molecular weight.
   b. Optical isomers must have an asymmetric carbon.
   c. Optical isomers only occur in the L-form.
   d. Optical isomers cannot be superimposed.
   e. Optical isomers must have identical groups attached to the asymmetric carbon.

3. Match the functional groups listed on the left with the correct structural formulas shown on the right.

_____	Amino	a.	—OH
_____	Phosphate	b.	—CHO
_____	Hydroxyl	c.	—SH
_____	Carboxyl	d.	—COOH
_____	Carbonyl	e.	—NH_2
_____	Sulfhydryl	f.	—$OPO_3{}^{2-}$

Activities (for answers and explanations, see page 51)

- Label each of the following as either an organic acid, an organic base, an amino acid, a hydrocarbon, or an alcohol.

   _____ $H_2C{=}CH_2$

   _____ $H_3C—CH_3$

   _____
$$H_2N—\underset{\underset{CH_2SH}{|}}{\overset{\overset{COOH}{|}}{C}}—H$$

   _____
$$H_2N—\overset{\overset{O}{\|}}{C}—NH_2$$

   _____ $H_3C—\overset{\overset{O}{\|}}{C}—CH_2—COOH$
   _____ $H_3C—CH_2—OH$

- On the following diagram, circle and label the amino group, carboxyl group, α carbon, and side chain of the amino acid serine.

$$H_2N—\underset{\underset{CH_2OH}{|}}{\overset{\overset{COOH}{|}}{C}}—H$$

Integrative Questions (for answers and explanations, see page 51)

1. Which of the following types of bonds is strongest?
   a. Hydrogen
   b. Covalent
   c. Van der Waals
   d. Hydrophobic
   e. Ionic

2. Consider the following element,

   $$_{20}^{40}X$$

   If an ion of this element has an electric charge of +2, how many electrons will there be in the outermost electron shell of the ion?
   a. 8
   b. 2
   c. 4
   d. 0
   e. 1

3. Consider a hypothetical ion in which the number of protons is X, the number of neutrons is Y, and the number of electrons is Z. The atomic number of the ion would be the same as
   a. X.
   b. Y.
   c. Z.
   d. X or Z.
   e. X + Y.

3

Macromolecules: Their Chemistry and Biology

CHAPTER LEARNING OBJECTIVES—after studying this chapter you should be able to:

❑ Recognize a fatty acid and determine if it is saturated, unsaturated, or polyunsaturated.

❑ Recognize a triglyceride and describe the effects that fatty acid characteristics have on the properties of triglycerides.

❑ Recognize a phospholipid and describe how its structure determines its function in biological membranes.

❑ Differentiate between the terms "hydrophilic" and "hydrophobic" and recognize hydrophilic and hydrophobic regions of macromolecules.

❑ Describe the characteristics and roles of carotenoids and steroids.

❑ Contrast condensation and hydrolysis reactions and how these reactions are involved in the assembly and disassembly of polymers.

❑ Explain the relationship between monosaccharides, disaccharides, oligosaccharides, and polysaccharides.

❑ Describe the transformation of glucose from straight chain to ring form, explain the numbering system used in referring to the carbons of a monosaccharide, and differentiate between α- and β-glucose.

❑ Recognize monosaccharides as pentoses or hexoses and list some of the common pentose and hexose isomers.

❑ Describe the formation of a disaccharide from two monosaccharides and differentiate between α and β linkages.

❑ Explain the principal features and biological functions of cellulose, starch, and glycogen.

❑ Explain the principal features and biological functions of some common sugar phosphates and amino sugars.

❑ Describe the classification of amino acids by the chemical properties of their R groups and the effect of R group properties on the location of specific amino acids within proteins.

❑ Identify the N-terminus and C-terminus of a polypeptide.

❑ Explain the formation of the following bonds within a polypeptide: peptide linkage, disulfide bridge, hydrogen bond.

❑ Determine the number of unique proteins that could be formed given a specific polypeptide length.

❑ Explain how protein primary structure determines higher levels of organization.

❑ Describe the two common types of protein secondary structure and give an example of each.

❑ Describe the organization of a protein such as hemoglobin relative to its primary, secondary, tertiary, and quaternary structure.

❑ Describe the main biological functions of proteins and how their primary structure is coded for by DNA.

❑ Explain the factors that can lead to denaturation of a protein and relate it to the process of renaturation.

❑ Explain what a chaperone protein is and its biological importance.

❑ Describe the structure of a nucleotide and a nucleoside and their polymerization to form DNA and RNA.

❑ Describe differences in the structures of DNA and RNA.

❑ Classify the five nitrogenous bases as purines or pyrimidines and indicate their occurrence in DNA and RNA.

❑ Explain the basis for complementary base pairing and its importance in the transfer of information from DNA to RNA and RNA to protein.

❑ Describe the structure of the following macromolecules: glycoprotein, glycolipid, lipoprotein, and nucleoprotein.

LIPIDS: WATER-INSOLUBLE MOLECULES • MACROMOLECULES: GIANT POLYMERS (pages 41–47)

Key Concepts

1. *Lipids* are water-insoluble hydrocarbons. Their insolubility results from the predominance of nonpolar bonds. Lipids can form large structures when they are aggregated by water molecules and held together by van der Waals forces.

2. *Fatty acids* have long hydrocarbon chains and a polar carboxyl group (—COOH).

3. *Saturated fatty acids* have no double bonds between carbon atoms in the hydrocarbon chain (they are saturated with hydrogen). Saturated fatty acids are rigid and straight.

4. *Unsaturated fatty acids* have some double bonds. Double bonds cause a kink in the molecule. *Polyunsaturated fatty acids* have more than one double bond.

5. *Triglycerides* (also called *simple lipids*) contain three fatty acid molecules chemically bonded to a molecule of *glycerol*. The carboxyl group of each fatty acid reacts with a hydroxyl group of the glycerol to form an ester linkage, as in:

$$-\overset{\overset{\displaystyle O}{\|}}{C}-OH + HO-C- \rightarrow -\overset{\overset{\displaystyle O}{\|}}{C}-O-C- + H_2O$$

6. The fatty acids in a triglyceride need not be the same in either the length of the chain or their saturated/unsaturated condition.

7. Triglycerides with long, saturated hydrocarbon chains are usually solid at room temperatures because their fatty acids pack tightly together. They are called *fats*. Triglycerides with short or unsaturated hydrocarbon chains are usually liquid at room temperature because their fatty acids are loosely packed due to the kinks. They are called *oils*.

8. Energy storage is a major function of cellular lipids, usually in the form of fat droplets.

9. Because of the natural tendency of lipids to form a layer in water, lipids are a major constituent of biological *membranes*. The movement of materials through the lipid-containing membranes of cells and organelles depends largely on their lipid solubility.

10. *Phospholipids* result when one of the fatty acids of a triglyceride is replaced by a phosphorus-containing compound. The phosphorus-containing end of the molecule is electrically charged (–) and, thus, *hydrophilic*. The hydrocarbon end of the molecule is *hydrophobic*.

11. Phospholipids are a major component of membranes and are oriented within the membrane to form a *bilayer*. In the bilayer, the phospholipids are arranged with the hydrophobic, hydrocarbon "tails" forming the center of the bilayer and the phosphorus-containing, hydrophilic "heads" to the outside of the bilayer.

12. The lipids called *carotenoids* are light-absorbing pigments found in both plants and animals. β-carotene and its breakdown product vitamin A are examples of important carotenoids.

13. *Steroids* are lipids with a specific four-ring structure. Many steroids, such as testosterone, estrogen, and the cortisones, are the chemical signaling molecules called hormones; other steroids, such as *cholesterol*, are important components of membranes.

14. Most lipids can be synthesized by animals. Humans cannot produce certain lipids (three types of unsaturated fatty acids and the vitamins A, D, E, and K), and these lipids must be obtained in our diet.

15. Macromolecules are *polymers* composed of simple subunits called *monomers*. Macromolecules have molecular weights in excess of 1,000 daltons.

16. Although triglycerides and phospholipids assemble into large structures in cells (fat droplets and membrane bilayers, respectively), lipids are not considered macromolecules.

17. The shapes and chemical properties of a macromolecule's monomers determine its functions. A large macromolecule may have many functional groups.

18. Polymerization involves *condensation* (also called *dehydration*) reactions of the type A—H + B—OH → A—B + H_2O, where A and B are the *reactant* monomers.

19. During a condensation reaction a water molecule is released for each subunit added to the polymer; the hydroxyl group (OH) and a hydrogen atom (H) are contributed by the reactants. Condensation reactions are energy-requiring reactions.

20. A *hydrolysis reaction* is an energy-yielding reaction that occurs when water is used to break the covalent bond linking monomers together.

Questions (for answers and explanations, see page 52)

1. Which of the following classes of lipids includes important light-capturing molecules?
   a. Phospholipids
   b. Carotenoids
   c. Fats
   d. Triglycerides
   e. Steroids

2. Which of the following pairs of lipid classes includes two molecules that are important components of biological membranes?
   a. Fats and oils
   b. Steroids and phospholipids
   c. Triglycerides and phospholipids
   d. Carotenoids and triglycerides
   e. Carotenoids and fatty acids

Consider the following triglycerides (*A* and *B*) in answering Questions 3 through 6.

$$CH_3(CH_2)_{16}-\overset{\overset{\displaystyle O}{\|}}{C}-O-CH_2$$
$$CH_3(CH_2)_{16}-\overset{\overset{\displaystyle O}{\|}}{C}-O-CH_2$$
$$CH_3(CH_2)_{16}-\overset{\overset{\displaystyle O}{\|}}{C}-O-CH_2$$

$$CH_3(CH_2)_3-\overset{\overset{\displaystyle O}{\|}}{C}-O-CH_2$$
$$CH_3-HC=CH-CH_2-\overset{\overset{\displaystyle O}{\|}}{C}-O-CH_2$$
$$CH_3-HC=CH-CH_2-\overset{\overset{\displaystyle O}{\|}}{C}-O-CH_2$$

A   *B*

3. In *B*, circle the remnant of the glycerol portion of the triglyceride.

4. Which triglyceride (*A* or *B*) is probably a solid at room temperature? _____ Explain your answer.

5. Which triglyceride (*A* or *B*) is probably derived from a plant? _____ Explain your answer.

6. How many water molecules result from the formation of triglyceride *B* from glycerol and three fatty acids? _____

7. What type of lipid is the following molecule?

a. Phospholipid
b. Carotenoid
c. Fat
d. Triglyceride
e. Steroid

Activities (for answers and explanations, see page 52)

• Circle and label the hydrophilic and the hydrophobic ends of the following phospholipid. Assuming that this phospholipid is part of a membrane, draw an arrow to indicate the direction to the outside of the membrane.

$$CH_3-CH_2-(CH_2)_{14}-CH_2-\overset{\overset{\displaystyle O}{\|}}{C}-O-CH_2$$

$$CH_3-CH_2-(CH_2)_6-CH_2-\overset{\overset{\displaystyle O}{\|}}{C}-O-CH$$

$$H_2-C-O-\overset{\overset{\displaystyle O}{\|}}{P}-O^-$$

• Circle the carboxylic end of the space-filling model of a fatty acid shown below.

How many carbons are included in this fatty acid? _____

Is this fatty acid saturated or unsaturated? _____

CARBOHYDRATES: SUGARS AND SUGAR POLYMERS (pages 47–51)

Key Concepts

1. Carbohydrates consist of *monosaccharides*, *disaccharides*, *oligosaccharides*, and *polysaccharides*. Monosaccharides are simple sugars; disaccharides result when two monosaccharides are linked together by condensation reactions; oligosaccharides are polymers of several simple sugars; polysaccharides are polymers of many simple sugars.

2. Most monosaccharides have the general formula $C_nH_{2n}O_n$. Because two hydrogens and an oxygen atom are lost in each of the condensation reactions that produces disaccharides, oligosaccharides, and polysaccharides, the structure of these carbohydrates differs from $C_nH_{2n}O_n$.

3. Plants produce *glucose*, $C_6H_{12}O_6$, through photosynthesis, and all cells utilize it as an energy source.

4. In an aqueous environment, glucose exists either as a straight chain or, more commonly, as a ring.

5. Two common isomers of the ring form of *glucose* (α- and β-glucose) differ in the placement of the —H and —OH groups on carbon 1.

6. All structural isomers of glucose have the same formula ($C_6H_{12}O_6$) and include the sugars *fructose, mannose,* and *galactose. Hexoses* include all six-carbon sugars.

7. *Ribose* and *deoxyribose* are five-carbon sugars (called *pentoses*) found in the backbone of RNA and DNA, respectively. Deoxyribose has one less oxygen atom than ribose.

8. Monosaccharides are covalently joined by condensation reactions that form *glycosidic linkages*, usually between carbons 1 and 4.

9. A *disaccharide* results when a condensation reaction joins two monosaccharides. If the monosaccharide providing carbon 1 of the linkage is an alpha sugar, then the linkage is an α linkage; if it is a beta sugar, then it is a β linkage.

10. *Maltose* results when carbon 1 of α-glucose is linked to another glucose. *Cellobiose* has the same formula as maltose ($C_{12}H_{22}O_{11}$), but involves a β-glucose linked to another glucose.

11. *Starch* is a polysaccharide that consists of hundreds to thousands of α-glucose molecules. Starch can be readily hydrolyzed to provide glucose for the cell.

12. *Cellulose* is a polysaccharide with only β linkages between glucose molecules. Unlike cellulose, starch can also form branches between carbon 6 of one glucose and carbon 1 of another. Cellulose is the main component of plant cell walls. Its β linkages are more resistant to breakage than the α linkages of starch.

13. *Glycogen* is identical to starch (α linkages between carbon 1 and 4 of glucose with some carbon 1 to 6 branches), except it is more branched.

14. Starch and glycogen are important energy storage molecules in plants and animals, respectively.

15. *Derivative carbohydrates* contain elements other than carbon, hydrogen, and oxygen. Some *sugar phosphates*, like fructose 1,6-bisphosphate, are important intermediates in photosynthesis and respiration. *Amino sugars* have an amino group in place of one —OH. *Galactosamine* is a major constituent of cartilage; *chitin* is derived from the amino sugar glucosamine.

Questions (for answers and explanations, see page 52)

1. Complete the following statements about the carbohydrate shown below.

   a. The carbohydrate shown is an example of a

   _____ .

   b. The two monosaccharides forming this carbohydrate are both _____ sugars.

   c. The two monosaccharides forming this carbohydrate are joined by a(n) _____ linkage.

   d. Prior to the condensation reaction that formed this carbohydrate, the two monosaccharides had the chemical formula _____ .

   e. Based on *d*, we can call these two monosaccharides_____ .

2. Which of the following statements correctly differentiates between starch and cellulose?
   a. Starch is composed of sucrose subunits, cellulose is formed from glucose subunits.
   b. Starch subunits are joined by β linkages; cellulose subunits are joined by α linkages.
   c. In addition to carbon-1 to carbon-4 linkages, starch also has carbon-6 to carbon-1 linkages; cellulose has only carbon-1 to carbon-4 linkages.
   d. Starch is found in animals, cellulose is found in plants.

3. Which of the following compounds is *not* classified as a derivative carbohydrate?
   a. Sugar phosphates
   b. Amino sugars
   c. Glucosamine
   d. Glycogen
   e. Fructose 1,6-bisphosphate

4. Circle all of the following statements that correctly compare ribose and deoxyribose sugar.
   a. They are both pentose sugars.
   b. They are isomers.
   c. Ribose has one more oxygen than deoxyribose.
   d. They are both found in DNA.
   e. They have the same number of carbons.

PROTEINS: AMAZING POLYMERS OF AMINO ACIDS (pages 51–58)

Key Concepts

1. Proteins are polymers of amino acids. The size of the chain varies in different proteins from about 50 amino acids to over 4600.

2. The functions of proteins include structural support, protection, transport, catalysis, regulation, and movement.

3. *Enzymes* are catalytic proteins. Most biologically important reactions are catalyzed by a unique enzyme.

4. *Prosthetic groups* are specific nonprotein substances (carbohydrates, lipids, metal ions, the iron-containing heme group) that associate with some proteins.

5. The twenty different *amino acids* are the monomers joined together to form proteins. The chemical properties of amino acids are determined by the nature of their side chains or *R groups* (*R* is for "residue").

6. Five amino acids have side chains with a net negative or positive charge. Water and oppositely charged ions are attracted to these amino acids.

7. Four amino acids have polar, but uncharged, R groups. They tend to form hydrogen bonds with water and other polar or charged molecules.

8. Eight amino acids have hydrophobic R groups composed predominantly of hydrocarbons. Hydrophobic R groups cluster together when in an aqueous environment.

9. *Cysteine* is a sulfur-containing amino acid. Two cysteines can form a covalent bond, called a *disulfide bridge*, between the sulfur atoms of their R groups.

10. Glycine can fit into tight corners within the interior of a protein because its side chain is a single hydrogen atom. Proline has a modified amino group as a side chain.

11. A *peptide linkage* results when a condensation reaction joins the carboxyl group of one amino acid to the amino group of a second amino acid.

12. *Polypeptides* are chains of many amino acids joined by peptide bonds. A polypeptide has an amino end (N terminus) and a carboxyl end (C terminus). *Proteins* can consist of one or more polypeptides.

13 Within a protein, hydrogen bonding occurs between nearby monomers because the $C=O$ oxygen has a partial negative charge and the N—H hydrogen has a partial positive charge.

14. The sequence of amino acids in a polypeptide is the *primary structure* of the protein. The *peptide backbone* of a polypeptide consists of the repeated sequence —N—C—C—.

15. Because each amino acid in a polypeptide can be one of twenty different types, the number of possible polypeptides with n amino acids is 20^n.

16. The sequence of monomers (called *nucleotides*) in DNA determines the primary structure of the protein coded for by the DNA. Within cells, protein synthesis always begins with the N terminus.

17. The sequence of amino acids in a protein (the primary structure) determines bonding potentials between different regions of the protein. Thus, the primary structure determines all higher-order structural levels.

18. Twisting and folding of polypeptide chains may result in consistent structural configurations collectively called *secondary structure*. Two major types of secondary structure are the α helix and the β pleated sheet. A protein may have one or more regions with a secondary structure, or none at all.

19. The *α helix* configuration results when a polypeptide twists so that hydrogen bonds can form between atoms in amino acids four monomers apart. A group of fibrous proteins called *keratins* show an α–helical secondary structure.

20. In the *β pleated sheet* configuration, several different polypeptides (or different regions of one polypeptide) are bonded together by hydrogen bonds to form a sheet with a zigzag, pleated appearance. This type of conformation occurs between separate polypeptide chains in the protein *silk*.

21. The overall, three-dimensional shape of the whole polypeptide is its *tertiary structure*. Tertiary structure includes any secondary structures such as α helix and β pleated sheet, but also unique bends and folds as dictated by the polypeptide's primary sequence.

22. Proteins that consist of two or more polypeptides have a higher-order structure called the *quaternary structure*. For example, *hemoglobin* consists of four similar polypeptide chains held together by a combination of ionic, hydrophobic, and hydrogen bonds. The relative arrangement of these subunits is hemoglobin's quaternary structure.

23. DNA specifies the primary structure of a polypeptide, which, in turn, specifies all higher levels of structure.

24. *Denaturation* is the loss of normal tertiary structure due to high temperature, inappropriate pH, or high concentrations of ions or polar substances. Proteins lose their biological function when denatured. Denaturation can sometimes be reversed; this is called *renaturation*.

25. A *chaperone protein* may attach to a newly forming protein and protect it from interactions with other molecules until it has assumed its normal tertiary structure. Chaperone proteins are usually nonspecific.

Activity (for answers and explanations, see page 52)

- In the following polypeptide, label the N terminus, the C terminus, and all the peptide linkages.

How many water molecules were released in the polymerization of this polypeptide? _____

Questions (for answers and explanations, see page 52)

1. Primary protein structure determines which of the following additional levels of protein organization?
   a. Secondary
   b. Tertiary
   c. Secondary and tertiary
   d. Quaternary
   e. Secondary, tertiary, and quaternary

2. Which of the following statements correctly categorizes the R groups (side chains) in the dipeptide shown next?

   a. Both R groups are hydrophobic.
   b. Both R groups are hydrophilic.
   c. The left R group is hydrophobic; the right is hydrophilic.
   d. The left R group is hydrophilic; the right is hydrophobic.
   e. None of these statements is correct.

3. How many different polypeptides *four* amino acids long could be made from *five* different kinds of amino acids?

_____

4. Which of the following types of protein structure is (are) maintained exclusively by hydrogen bonding?
   a. Primary structure
   b. Tertiary structure
   c. β pleated sheet
   d. α helix
   e. β pleated sheet and α helix

5. What is the *highest* structural organization found in *all* enzymes?
   a. Primary
   b. Secondary
   c. Tertiary
   d. Quaternary

NUCLEIC ACIDS: INFORMATIONAL MACROMOLECULES (pages 58–61)

Key Concepts

1. Nucleic acids are polymers of *nucleotides* involved in controlling the production of proteins. *Deoxyribonucleic acid (DNA)* codes for proteins; *ribonucleic acid (RNA)* transfers this information to protein-making machinery within the cell. During reproduction, information flows from DNA to DNA.

2. *Nucleotides* consist of a pentose sugar (deoxyribose in DNA, ribose in RNA), a phosphate group, and a nitrogenous base. The phosphate group is attached to carbon 5 of the sugar, the nitrogenous base to the sugar's carbon 1.

3. A *nucleoside* consists of a pentose sugar and a nitrogenous base without the phosphate group.

4. In a polynucleotide chain, the backbone is formed of alternating sugar and phosphate groups, with the nitrogenous bases attached to the sugars.

5. Nucleotides are joined together by *phosphodiester linkages* between the phosphate group of one nucleotide and carbon 3 of the next nucleotide.

6. DNA consists of two antiparallel chains of nucleotides bound together by hydrogen bonding between the nitrogenous bases on the different chains. The DNA molecule is twisted into a *double helix*.

7. The nature of the nitrogenous base determines the kind of nucleotide or nucleoside. Four bases are found in DNA: *adenine, cytosine, guanine,* and *thymine*. Adenine, cytosine, and guanine are also found in RNA, but instead of thymine, RNA has the base *uracil*.

8. Adenine and guanine are double-ringed molecules called *purines*; uracil, cytosine, and thymine are single-ringed molecules called *pyrimidines*.

9. There is just enough space between two polynucleotide chains for a pyrimidine to pair with a purine, so that all base pairs are the same size (three rings).

10. For optimum hydrogen bonding to occur, adenine can only pair with thymine (two hydrogen bonds) and cytosine can only pair with guanine (three hydrogen bonds). These hydrogen bonding restrictions are the basis for the *principle of complementary base pairing*.

11. Most RNA molecules are single-stranded, although complementary base pairing between different regions of an RNA occurs and determines the shape of the RNA.

12. In RNA, adenine pairs with uracil, but when RNA binds to DNA, adenine from RNA binds to the thymine from DNA.

13. The sequence of bases in one strand of DNA is the information that specifies the sequence of amino acids in proteins. This then specifies the information that is contained in the shape of proteins.

14. The base-pairing rules ensure that a DNA sequence specifies the RNA sequence and RNA transmits this information to the cytoplasm where it is used to manufacture proteins.

15. Comparing the base sequences of DNA from different species can provide information on their evolutionary relationships.

16. All three major classes of macromolecules can covalently bond together to produce *glycolipids, glycoproteins, lipoproteins,* and *nucleoproteins.*

17. Glycolipids usually have their carbohydrate portion directed to the outside of the cell membrane. Many nucleoproteins are involved with regulation of DNA. Lipoproteins move hydrophobic lipids through the hydrophilic tissues of the body.

Questions (for answers and explanations, see page 52)

1. Which of the following choices is to DNA and RNA what amino acids are to proteins?
   a. Sugar, phosphate, nitrogenous base
   b. Nucleosides
   c. Nitrogenous bases
   d. Nucleotides
   e. Adenine, cytosine, thymine, guanine, uracil

2. Which of the following choices is to DNA and RNA what amino acid R groups are to proteins?
   a. Sugar, phosphate, nitrogenous base
   b. Nucleosides
   c. Ribose or deoxyribose
   d. Nucleotides
   e. Adenine, cytosine, thymine, guanine, uracil

3. Which of the following is *not* a difference between RNA and DNA?
   a. Adenine pairs with different bases in RNA and DNA.
   b. Purines are only found in RNA; pyrimidines are restricted to DNA.

c. Only DNA is normally found in a double helix.
d. Ribose is found in RNA; deoxyribose is found in DNA.
e. DNA encodes the information for making proteins; RNA carries out these instructions.

Activities (for answers and explanations, see page 53)

- Use the base-pairing rules for DNA and RNA to label the complementary strand of RNA (right) to the single strand of DNA (left) shown below, where C = cytosine, G = guanine, A = adenine, T = thymine, and U = uracil.

- Circle and label an example of a *nucleotide* and a *nucleoside* in the figure showing the double-stranded DNA and RNA hybrid.

- Based on the orientation of the sugar molecules, label the four ends of the molecule as 3′ or 5′ in the figure showing the double-stranded DNA and RNA hybrid.

Integrative Questions (for answers and explanations, see page 53)

1. Which of the following did *not* result from polymerization?
   a. RNA
   b. An enzyme
   c. Cellulose
   d. Cholesterol
   e. Myoglobin

2. What would you expect to be *true* of the R groups of amino acids located on the surface of protein molecules found within the interior of biological membranes?
    a. The R groups should be hydrophobic.
    b. The R groups should be hydrophilic.
    c. The R groups should be polar.
    d. The R groups should be able to form disulfide.

3. Use numbers to represent the normal flow of information within a cell.

    _____   RNA nucleotide sequence

    _____   Primary structure of a protein

    _____   Folding of a protein into a three-dimensional
             shape

    _____   DNA nucleotide sequence

    _____   An enzyme catalyzes a specific reaction

4

The Organization of Cells

CHAPTER LEARNING OBJECTIVES—after studying this chapter you should be able to:

❑ State the two tenets of the cell theory.

❑ Describe the characteristic features of all cells.

❑ Explain the effect of the surface area–to–volume ratio on cell size.

❑ Describe the distinguishing features of prokaryotic and eukaryotic cells and know the kingdoms of the organisms composed of these two cell types.

❑ Describe prokaryotic cell structural features such as the nucleoid, cytoplasm, cytosol, ribosomes, cell walls, and several types of cell projections and membrane inclusions.

❑ Identify the major forms and shapes of bacteria.

❑ Explain the term "resolution" and know the relative resolutions of the light and electron microscopes.

❑ Distinguish between scanning and transmission electron microscopy.

❑ Name and describe the functions of all major eukaryotic organelles.

❑ Describe the characteristic features of the nucleus and the changes that occur in the nucleus as the cell prepares to divide.

❑ Describe the characteristic features of ribosomes.

❑ Describe the characteristic features of the mitochondrion.

❑ Describe the characteristic features of the chloroplast.

❑ Explain the endosymbiotic theory and describe the evidence supporting this theory.

❑ Name the components of the endomembrane system and explain their interacting functions.

❑ Describe differences in the activities of the rough and smooth endoplasmic reticulum.

❑ Describe the structure of the Golgi apparatus and the movement of vesicles through the Golgi.

❑ Describe the formation of lysosomes and their involvement with phagocytosis in intracellular digestion.

❑ Describe the characteristic features of the several common types of microbodies and vacuoles.

❑ Describe the structural features and characteristic uses of microfilaments, intermediate filaments, and microtubules in the cell cytoskeleton.

❑ Distinguish between cilia and flagella and explain their structural similarities and differences.

❑ Explain what a basal body is and its relationship to a centriole.

❑ Describe the relationship between cell walls and plasmo-desmata.

❑ Explain the techniques of cell fractionation and distinguish between differential and equilibrium centrifugation.

THE CELL: THE BASIC UNIT OF LIFE • PROKARYOTIC CELLS • MICROSCOPES: REVEALING THE SUBCELLULAR WORLD (pages 64–67)

Key Concepts

1. According to the *cell theory*, organisms are or consist of cells and all cells arise from other cells.

2. Cells are tiny, depend on external energy sources, selectively regulate exchange of material with the environment, and depend on information in their DNA to regulate their chemistry.

3. All cells have a *plasma membrane* responsible for regulating exchange of materials with the environment.

4. Most cells are small (volumes = 1 to 1000 μm^3) in order to maintain a large surface area–to–volume ratio. The cell's sur-

face area limits its rate of exchange of materials with the environment, but its volume determines its needs for materials.

5. *Prokaryotic cells* are characterized by the absence of a true nucleus and other membrane-bounded organelles. Organisms in the kingdoms Eubacteria and Archaebacteria consist of prokaryotic cells and are called *prokaryotes*.

6. *Eukaryotic cells* have true nuclei and membrane-bounded organelles. Organisms in the other four kingdoms are composed of eukaryotic cells and are called *eukaryotes*.

7. Prokaryotes are metabolically more diverse and tolerate a greater range of environmental conditions than eukaryotes. Most prokaryotic cells are smaller than eukaryotic cells.

8. Prokaryotes can occur as individual cells, short chains, or in colonies. Common shapes include the spherical, rod-shaped, and helical forms.

9. A prokaryotic cell has at least one *nucleoid* region containing DNA. The remainder of the prokaryotic cell is called the *cytoplasm*.

10. Cytoplasm includes both ions, small molecules, and macromolecules in solution (called the *cytosol*) as well as insoluble suspended particles, such as *ribosomes*.

11. Prokaryotic ribosomes consist of RNA and proteins. Ribosomes coordinate the synthesis of proteins within the cell.

12. Many prokaryotic cells also have an external *cell wall*. In bacteria the cell wall consists of *peptidoglycan* (a polymer of amino sugars), and some bacteria have an additional membrane called the *outer membrane*. An outer covering of polysaccharides called the *capsule* may protect the cell from dehydration or destruction by other cells.

13. In the photosynthetic prokaryotes, like the cyanobacteria, infolding of the plasma membrane produces an internal membrane system containing the molecules used in photosynthesis.

14. *Mesosomes* are other membrane systems involved in a variety of functions that also result from infolding of the plasma membrane.

15. Prokaryotic cell projections include *flagella* (sing. *flagellum*), whose function is propulsion, and *pili* (sing. *pilium*), used for adhesion to other cells.

16. *Resolution, or resolving power,* is the smallest distance between two objects so that they can be seen as separate entities. Objects closer than about 0.1 mm (100 μm) cannot be seen as separate entities by humans.

17. Microscopes are instruments designed to improve the resolution of the human eye. The *light microscope* uses glass lenses and visible light. It has a resolving power of about 200 nm (0.2 μm, 2×10^{-4} mm).

18. The *electron microscope* uses magnetic lenses and a beam of electrons. It provides a resolution of about 2 nm under ideal conditions. Images from an electron microscope can be used to expose photographic film; the resulting picture is called an *electron micrograph*.

19. In *transmission electron microscopy*, high-energy electrons pass through a thin section of material. *Scanning electron microscopy* provides a three-dimensional view of the cell by bouncing electrons off the exposed surface.

Questions (for answers and explanations, see page 53)

1. Which of the following features is *not* characteristic of both prokaryotic and eukaryotic cells?
   a. possession of a plasma membrane
   b. possession of a nucleoid region with DNA
   c. presence of ribosomes
   d. presence of cytoplasm
   e. dependence on external energy sources

2. Which of the following comparisons of prokaryotes and eukaryotes is *not* true?
   a. The cells of prokaryotes tend to be smaller than the cells of eukaryotes.
   b. Only the cells of eukaryotes have membrane-bound organelles, although the cells of some prokaryotes do have internal membrane systems.
   c. Eukaryotes are metabolically more diverse than prokaryotes.
   d. Prokaryotes can tolerate a greater range of environmental conditions than eukaryotes.
   e. Prokaryotes are more ancient than eukaryotes.

3. Scanning electron microscopy provides a resolution of about 10 nm. If the resolution of the unaided human eye is 0.1 mm, how much is our resolution increased by this technology?
   a. 10^6 times
   b. 10^5 times
   c. 10^4 times
   d. 10^3 times
   e. 10^2 times

4. Match the appropriate letters for characteristics listed on the right to the cell structures listed on the left.

d	Nucleoid	a. Fluid portion of cytoplasm
a	Cytosol	b. Locomotory structure
f	Capsule	c. Membranous structure
b	Flagella	d. Contains the cell's DNA
e	Pili	e. Cell adherence
c	Mesosome	f. Can have protective function

Activity (for answers and explanations, see page 53)

- Use the following terms to label the photomicrograph of the bacterial cell shown below: (A) peptidoglycan, (B) plasma membrane, (C) nucleoid, (D) capsule, (E) ribosomes, (F) cytoplasm.

EUKARYOTIC CELLS • ORGANELLES THAT PROCESS INFORMATION • ORGANELLES THAT PROCESS ENERGY (pages 71–80)

Key Concepts

1. Like prokaryotic cells, eukaryotic cells have a plasma membrane, cytosol, and ribosomes. An internal cytoskeleton maintains cell shape and moves materials.

2. The membrane-bounded systems that are unique to eukaryotic cells are included in the collection of intracellular structures called *organelles*. Although not membranous, ribosomes are also considered organelles.

3. Various membrane-bounded molecules carry out important functions found in eukaryotic cells, such as transport of materials between the cell and the environment and between different organelles, interactions between cells, and energy transformation.

4. In eukaryotic cells, the DNA resides in the *nucleus*, the largest organelle found within the cell. The nucleus is surrounded by a double phospholipid bilayer membrane called the *nuclear envelope*.

5. The nuclear envelope is perforated by *nuclear pores* formed in areas where the inner and outer membranes merge. Eight protein granules are arranged around each pore. RNA and some proteins can move through the nuclear pores.

6. The outer membrane of the nuclear envelope is continuous with the endoplasmic reticulum of the cytoplasm and, like the endoplasmic reticulum, the outer surface of this membrane may be embedded with *ribosomes*.

7. The complex of nuclear DNA and proteins found in eukaryotes is called *chromatin*. *Nucleoplasm* is comprised of water and dissolved substances. A meshwork of protein filaments called the *nuclear lamina* maintains the shape of the nucleus by interconnecting the chromatin with the nuclear envelope.

8. Chromatin usually exists as a diffuse collection of threads, but when the nucleus is about to divide, chromatin condenses to form *chromosomes*. Each chromosome includes a single DNA molecule.

9. The *nucleoli* (sing. *nucleolus*) are also found in the nucleus. Ribosomal RNA synthesis and assembly with the protein components of the ribosomes occur in the nucleoli.

10. In preparation for cell division, the nuclear envelope fragments into vesicles, the nuclear lamina depolymerizes, the chromatin condenses, and nucleoli disappear.

11. The ribosomes in both prokaryotic and eukaryotic cells are the sites of protein synthesis, where the hereditary information of DNA is translated into the primary structure of proteins.

12. Ribosomes in eukaryotic cells are found free in the cytoplasm, attached to the endoplasmic reticulum, and within mitochondria and plastids.

13. Prokaryotic and eukaryotic ribosomes are similar in design and function. Both consist of a large and a small subunit; each is composed of a unique type of RNA, called *ribosomal RNA*, and about 50 kinds of proteins.

14. *Cellular respiration*, the production of the short-term energy storage molecule *ATP* (adenosine triphosphate) from fuel molecules (such as glucose) and oxygen, occurs mainly in the eukaryotic organelle called the *mitochondrion*.

15. Energy transformations in prokaryotes are associated with enzymes embedded into the inner surface of the plasma membrane or extensions of it.

16. Mitochondria consist of a smooth outer membrane and a highly folded inner membrane. Folds of the inner membrane are called *cristae* and the region enclosed by the inner

membrane is called the *mitochondrial matrix*. The matrix includes ribosomes and DNA.

17. *Plastids* are another class of eukaryotic organelles with double membranes. All plastids are derived from *proplastids*; major types include the *chloroplast*, the *chromoplast*, and the *leucoplast*. Plastids are produced only by plants and certain protists.

18. Chloroplasts of flowering plants are surrounded by two smooth lipid bilayer membranes. Within the chloroplast are found *grana*, stacks of circular sacs called *thylakoids*. Each thylakoid has a single membrane in which arrays of chlorophyll and carotenoid molecules are embedded.

19. The fluid interior of the chloroplast is called the *stroma*. The stroma contains DNA and ribosomes.

20. Chromoplasts contain various types of carotenoids and give color to fruits, flowers, and vegetables. Leucoplasts store starch and lipids.

21. The *endosymbiotic theory* states that eukaryotic cells and their mitochondria and chloroplasts originated when certain prokaryotes began living within other prokaryotes as endosymbionts; aerobic endosymbionts became mitochondria, photosynthetic endosymbionts became chloroplasts.

22. Evidence for the endosymbiotic theory includes the presence of double membranes surrounding mitochondria and plastids, a similarity between present-day prokaryotic ribosomes and ribosomes found in eukaryotic organelles, the functional similarity of the prokaryotic plasma membrane and the inner mitochondrial membrane, and the existence of a number of examples of endosymbiosis among living species.

Questions (for answers and explanations, see page 53)

1. Which of the following statements is *not* true of the nucleus in eukaryotic cells?
   a. The nuclear envelope consists of a double phospholipid bilayer membrane.
   b. Cytoplasm and nucleoplasm are continuous through the nuclear pores.
   c. The nucleus contains DNA, but no RNA.
   d. Chromatin is attached to the inner nuclear membrane by the nuclear lamina.
   e. There are no mitochondria within the nucleus.

2. Which of the following does *not* occur to the nucleus of a eukaryotic cell when the cell prepares to divide?
   a. The nuclear pores seal.
   b. Chromatin condenses into chromosomes.
   c. The nuclear envelope fragments.
   d. The nuclear lamina depolymerizes.
   e. Nucleoli disappear.

3. Label the indicated structures in the following diagram of a eukaryotic organelle. This organelle is a **mitochondria**.

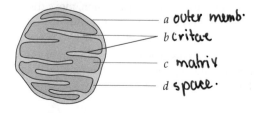

a outer memb.
b cristae
c matrix
d space.

4. According to the endosymbiotic theory, which structure(s) in the preceding diagram would have been derived from the original host prokaryote? **outer membrane**.

5. Match the appropriate letters for characteristics listed on the right to the chloroplast components listed on the left.

__c__	Stroma	a. In mitochondria, gives rise to cristae
__a__	Inner membrane	b. Chlorophyll located here
__d__	Granum	c. Similar to mitochondrial matrix
__b__	Thylakoid	d. Stack of membranous sacs

THE ENDOMEMBRANE SYSTEM • OTHER ORGANELLES (pages 80–85)

Key Concepts

1. The *endomembrane system* includes membranes of the endoplasmic reticulum, the Golgi apparatus, vesicles, lysosomes, and microbodies like peroxisomes.

2. The *endoplasmic reticulum* or *ER* is a system of interconnected tubules and sacs that branches throughout the cell. The ER is continuous with the outer nuclear membrane.

3. *Rough ER* has ribosomes attached to its outer membrane surface; *smooth ER* lacks ribosomes.

4. The ribosomes of the rough ER produce proteins for export from the cytosol to be used outside the cell or within the endomembrane system.

5. Proteins produced by rough ER ribosomes have a special sequence of amino acids called a *signal sequence* that directs them into the interior of the ER where they may be modified. For example, some gain carbohydrates and become glycoproteins.

6. Smooth ER is involved in the modification of proteins produced on the rough ER. Fatty acids, steroids, and phospholipids are also synthesized here.

7. Free ribosomes produce proteins that are used in the cytosol or within mitochondria or chloroplasts. These proteins lack signal sequences.

8. Cells that are specialized for protein synthesis, such as glandular cells, typically have an extensive endoplasmic reticulum.

9. The *Golgi apparatus* is a series of flattened sacs, with the sacs nearest the nucleus called the *cis* region. The sacs closest to the cell membrane are called the *trans* region. The middle sacs constitute the *medial* region.

10. The *trans* region of the Golgi apparatus receives *vesicles* containing protein from the ER. Small vesicles moving between the sacs of the medial region always move in a *cis*-to-*trans* direction.

11. Vesicles pinch off from the *trans* region, move through the cytosol, and merge with another membrane (either the plasma membrane or that of an organelle).

12. The Golgi apparatus chemically modifies proteins passing through it by adding specific "address tags" so the proteins are directed to the correct locations.

13. Lysosomes are vesicles formed, in part, by the Golgi apparatus. They have a single membrane and contain enzymes that can digest all major classes of macromolecules.

14. *Phagocytosis* results when a portion of the plasma membrane infolds to form a new vesicle, called a *phagosome*, containing material that had previously been outside the cell.

15. A primary lysosome, formed by the Golgi apparatus, becomes a secondary lysosome after it fuses with a phagosome. Secondary lysosomes then fuse with the plasma membrane to release waste products to the outside of the cell after digestion is complete.

16. *Microbodies* are small organelles that pinch off directly from the rough ER. They include *peroxisomes*, which protect the cell from reactions producing peroxides, and *glyoxysomes*, sites for conversion of lipids to carbohydrates in some plant cells.

17. *Vacuoles* are commonly found in protists and plant cells. The large vacuole in plant cells serves as a place for deposition of waste materials, and its turgidity gives nonwoody plants structural support. Some plant vacuoles contain pigments, such as the anthocyanins, that color petals and fruit.

18. Specialized vacuoles in protists include the *food vacuoles* formed by phagocytosis and the *contractile vacuoles* used to "bail out" excess water that diffuses into the cell.

Questions (for answers and explanations, see page 54)

1. Match the appropriate letters for characteristics listed on the right to the cell structures listed on the left.

__c__	Glyoxysome	a. Produces proteins used within cytosol
__d__	Peroxisome	b. Formed by phagocytosis
__b__	Food vacuole	c. Found only in plants
__a__	Free ribosome	d. Isolates dangerous reactions
__e__	Golgi apparatus	e. Endomembrane system component

2. Which of the following is the *correct* sequence of structures, tracing the path of a protein from its site of synthesis to its point of export from the cell?
   a. Mitochondrion, *cis* region of Golgi apparatus, *trans* region of Golgi apparatus, vesicle
   b. Smooth ER, *cis* region of Golgi apparatus, *trans* region of Golgi apparatus, vesicle
   c. Smooth ER, vesicle
   d. Rough ER, *cis* region of Golgi apparatus, *trans* region of Golgi apparatus, vesicle
   e. Rough ER, vesicle, *cis* region of Golgi apparatus, *trans* region of Golgi apparatus, vesicle

3. Which, if any, of the following is *not* a function of the endoplasmic reticulum?
   a. Synthesis of proteins exported from the cell
   b. Synthesis of proteins used in the cytosol
   c. Modification of proteins
   d. Lipid synthesis
   e. Transport of proteins

4. Which, if any, of the following is *not* a function of the types of vacuoles discussed in these sections?

a. Storage of cellular waste products
b. Providing turgidity to support the cell
c. Digestion of food
d. Attachment of signal sequence amino acids to proteins
e. Removal of excess water

Activity (for answers and explanations, see page 54)

- In the diagram of intracellular digestion that follows, use the letters shown to label the following structures: (A) plasma membrane, (B) *cis* region of Golgi, (C) medial region of Golgi, (D) *trans* region of Golgi, (E) primary lysosome, (F) secondary lysosome, (G) phagosome, (H) site of intracellular digestion, and (I) waste products.

THE CYTOSKELETON • EXTRACELLULAR STRUCTURES • CELLS VERSUS VIRUSES • ISOLATING ORGANELLES FOR STUDY   (pages 85–94)

Key Concepts

1. The *cytoskeleton* of the cell consists of microtubules, microfilaments, and intermediate filaments all located within the cytosol.

2. *Microfilaments* (actin filaments), arranged singly or in bundles, are involved in causing cell shape change, as in contraction of muscle cells; movement of organelles within the cell, as in *cytoplasmic streaming*; and the formation of cellular extensions, called *pseudopodia*, during amoeboid movement.

3. A microfilament is a double helical structure formed by two chains of the globular protein *G actin*. Microfilaments appear or disappear depending on whether the G actin chains are polymerized and depolymerized.

4. Actin-binding proteins interact with the microfilaments of the cytoskeleton to form networks directly below the plasma membrane that help to determine cell shape. Two important actin-binding proteins in red blood cells are *spectrin* and *ankyrin*.

5. *Intermediate filaments* consist of fibrous proteins of the keratin family and act like tiny ropes to anchor in place cellular structures, such as the nucleus.

6. *Microtubules* are long, hollow tubes formed by subunits composed of the protein *tubulin*, a dimer consisting of two polypeptide subunits (α-tubulin and β-tubulin). The circumference of the microtubule includes thirteen rows of tubulin dimers.

7. Microtubules change length by rapid addition or removal of tubulin subunits from one end (the + end) of the microtubule.

8. Microtubules are involved in the control of the construction of the cell wall in plants and are associated with the parts of the cell that are changing shape. Microtubules act as tracks along which motor proteins move protein-laden vesicles.

9. Microtubules are important components of the cellular organelles called *cilia* and *flagella*. Nine fused pairs of microtubules (called *doublets*) form the periphery of the cilium or flagellum and two unfused microtubules form the center (called a *9 + 2 arrangement*). A plasma membrane surrounds the cilium or flagellum.

10. Cilia are short and numerous; flagella are long and single or occur in pairs. Both move by bending that is caused by the sliding of enclosed microtubules. Both move the cell relative to the environment or move the aqueous environment relative to the cell.

11. *Dynein* molecules on one microtubule undergo a change in tertiary structure with energy from ATP and cause the movement of another microtubule to which they are bound.

12. Dynein and kinesin move cell organelles in opposite directions along microtubules; dynein toward the minus end and kinesin toward the plus end of the microtubule.

13. Prokaryotic flagella are much smaller and do not have the 9 + 2 arrangement of microtubules seen in the flagella of eukaryotes.

14. At the base of each eukaryotic flagellum or cilium is a structure called a *basal body*. The basal body gives rise to the cilium or flagellum and is structurally attached to it.

15. Basal bodies have nine sets of three microtubules. Two of each set are continuous with the paired peripheral microtubules found in the cilium or flagellum. Basal bodies lack central microtubules.

16. A *centriole* is structurally like an isolated basal body; it may organize the microtubules involved in cellular division.

17. Cells have a variety of structures that are *extracellular* (located outside of the plasma membrane), including the *peptidoglycan cell wall* in prokaryotic cells, the *cellulose cell wall* in plant cells, and the *extracellular matrix* in multicellular animals.

18. All plants have a semirigid cell wall composed of polysaccharides, mostly cellulose. Cell walls provide structural support and limit the uptake of water.

19. *Plasmodesmata* are the cytoplasmic connections between adjacent plant cells that extend through holes in their cell walls.

20. The extracellular matrix found in multicellular animals is composed of proteins such as collagen and of glycoproteins. The matrix is secreted by nearby cells and the extent of the matrix varies in different tissue types. Functions of the matrix include filtration, regulation of development, tissue regeneration, and communication between cells.

21. *Proteoglycan* is a very large molecule associated with some types of extracellular matrix. Composition of the matrix varies greatly and accounts for the differences between such tissues as bone, cartilage, and the *basal lamina.*

22. Viruses consist of nucleic acids and proteins; some are enveloped by a phospholipid membrane. Viruses are non-living parasites that are able to take over the metabolic machinery of living cells in order to make new viruses. They can not reproduce outside of their host.

23. *Cell fractionation* techniques include mechanically rupturing the cell to disperse its components, followed by *centrifugation* to separate the components into different classes based on their size and density.

24. In *differential centrifugation,* varying centrifuge rotor speeds (and, thus, centrifugal force) are used to separate the components of a mixture. After a spin at each speed, the *pellet* of particles that has settled is collected and the *supernatant* fluid is now spun at the next higher speed (and centrifugal force). The pellet of cell components can be resuspended and checked for purity with the electron microscope or enzyme assays.

25. In *equilibrium centrifugation* (also called *density gradient centrifugation*), the cell mixture is spun at a constant speed and allowed to move through a medium that differs in density from top to bottom. This causes components to reach an equilibrium position where their density equals that of the medium.

Questions (for answers and explanations, see page 54)

1. Match appropriate letters for the list on the right to the cytoskeletal elements listed on the left.

   c f   Microtubules   *a.* Actin

   a, d   Microfilaments   *b.* Fibrous protein

   b, e   Intermediate filaments   *c.* Tubulin

                                 d. Change cell shape

                                 e. Anchoring organelles

                                 f. Move cellular appendages

2. Which of the following statements about the cross section of a flagellum shown below is *not* true?

Flagellum

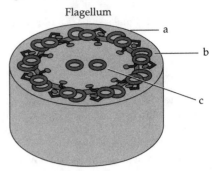

   a. Structure b is a doublet.

   b. Structure c is one of an unfused pair of microtubules.

   c. Structure a is the plasma membrane.

   d. The same arrangement of structures a, b, and c would be found in cilia of eukaryotic cells.

   e. The same arrangement of structures a, b, and c is found in the flagella of prokaryotic cells.

3. How would you modify the diagram shown in Question 2 to show the arrangement of microtubules seen in a *basal body*?

   a. Add a single microtubule to all pairs.

   b. Make all paired microtubules single.

   c. Remove the central pair and add a single microtubule to each of the doublets.

   d. Remove the central microtubule pair.

   e. Add a second plasma membrane.

4. Which of the following statements about equilibrium and differential centrifugation is true?

   a. Both equilibrium and differential centrifugation use a constant, high rotor speed.

   b. Both equilibrium and differential centrifugation use a centrifugation medium of fixed density.

   c. The rotor speed is varied in differential centrifugation.

   d. The density of the centrifugation medium varies in differential centrifugation.

   e. Differential centrifugation is used with intact cells; in equilibrium centrifugation, the cells must be ruptured.

Integrative Activity (for answers and explanations,
  see page 54)

On the transmission electron micrograph of an animal cell
shown below, label the following structures: (1) nuclear
envelope, (2) nuclear pore, (3) nucleolus, (4) mitochondrion, (5) rough endoplasmic reticulum, (6) cytoplasm, (7)
chromatin.

5

Cellular Membranes

CHAPTER LEARNING OBJECTIVES—after studying this chapter you should be able to:

❑ Describe the structure of a typical biological membrane and explain how that structure relates to the major roles of membranes.

❑ Describe the main types of lipids found in membranes and the functions they perform.

❑ Describe the main types of proteins found in membranes and the functions they perform.

❑ Differentiate between integral and peripheral proteins and explain what properties of a protein determine whether it will be integral or peripheral.

❑ Differentiate between glycoproteins and glycolipids and describe their primary functions.

❑ Differentiate between tight junctions, desmosomes, and gap junctions, and describe their primary functions.

❑ Describe the phenomenon of diffusion and the physical factors that affect the rate of diffusion.

❑ Differentiate between simple diffusion, facilitated diffusion, and active transport.

❑ With regard to a plot of a solute's transport rate against its concentration, explain how the concept of saturation can be used to differentiate between simple diffusion and facilitated diffusion.

❑ Describe the role of lipid solubility in determining a solute's transport rate.

❑ Differentiate between membrane transport proteins that are uniports and those that are coupled transport systems.

❑ Differentiate between coupled transport systems that are symports and those that are antiports.

❑ Describe the phenomenon of osmosis.

❑ Describe the concept of osmotic potential and relate it to the terms "isosmotic," "hyperosmotic," and "hypoosmotic"

❑ Explain the relative importance of osmotic potential and pressure potential in determining the osmotic movement of water in cells with walls and cells without walls.

❑ Differentiate between primary active transport and secondary active transport and give an example of each.

❑ Describe the importance of membranes to certain pathways of respiration and photosynthesis, to the establishment of electric and pH gradients, and to the cell's interaction with hormones, growth factors, antibodies, viruses, etc.

❑ Describe receptor-mediated endocytosis and the formation of a coated pit and coated vesicle.

❑ Explain what a ligand is and how ligands interact with membrane proteins.

❑ Discuss the importance of cell adhesion molecules to the formation of tissues.

❑ Describe the formation and assembly of the major membrane components and the typical flow of membranes within the cell.

MEMBRANE COMPOSITION AND STRUCTURE • ANIMAL CELL JUNCTIONS (pages 98–105)

Key Concepts

1. Biological *membranes* are flexible bilayers of lipids, principally *phospholipids*, in which are embedded protein molecules. A generalized description of biological membranes is called the *fluid mosaic model*.

2. The lipid bilayer is a barrier to movement of hydrophilic materials through the membrane and gives the membrane its structural integrity.

3. Biological membranes are *selectively permeable;* they regulate what materials can pass through them.

4. Some membrane proteins transport specific materials across the membrane; other proteins are receptors for chemicals outside the cell or act as membrane-bound enzymes.

5. Carbohydrates are usually found attached on the outside of the membrane and help the cell recognize environmental chemicals.

6. The membrane's *phospholipid bilayer* has a hydrophobic interior composed of phospholipid fatty acid chains, and a hydrophilic exterior where the polar phosphate groups of the phospholipids are located.

7. Phospholipid and protein molecules can move around within their layer of the membrane. Because of the hydrophobic membrane interior, phospholipids do not usually transfer from one side of the membrane to the other and the two sides of the membrane can differ in their phospholipid composition.

8. *Cholesterol* is an important lipid component of membranes, but its abundance can vary widely from cell to cell. Cholesterol influences the fluidity of membranes.

9. *Integral membrane proteins* extend into the lipid bilayer, with some actually extending from one side of the membrane to the other as *transmembrane proteins*. *Peripheral membrane proteins* are bound to the hydrophilic groups of the inner or outer membrane layer.

10. The presence of specific membrane-bound proteins gives different membranes their unique functional characteristics.

11. *Cell fusion experiments* involving cells from different species have shown that many proteins freely migrate within the membrane; other proteins seem to be anchored by elements of the cytoskeleton.

12. Membrane proteins, whether integral or peripheral, can vary in their occurrence on the two sides of a membrane or regionally within the cell, allowing the properties of the membrane to vary as well.

13. The tertiary structure of a membrane protein determines whether it will be integral or peripheral and, if integral, whether it will extend just into or entirely through the membrane.

14. Many proteins with extensive hydrophobic α-helical regions, but hydrophilic ends, extend completely through the phospholipid bilayer.

15. Carbohydrates bound to membrane lipids are called *glycolipids*. Glycolipids may be involved in cell–cell interactions.

16. *Glycoproteins* consist of a membrane protein and an attached oligosaccharide, usually 15 or fewer monosaccharides in length.

17. The great diversity of glycoproteins results from the specific sequence of the several kinds of monosaccharides, as well as the pattern of branching in the oligosaccharide chains.

18. Because of their role as extracellular recognition sites, all membrane carbohydrates (glycolipids and glycoproteins) are located on the outside of the plasma membrane.

19. In a tissue, cells can be joined to other cells or to the extracellular matrix by integral *cell adhesion proteins*. In some tissues, cells have special structures to limit movement of materials in the *intercellular space* between them.

20. Individual animal cells may be linked together by three types of *cell surface junctions*: tight junctions, desmosomes, and gap junctions.

21. *Tight junctions* result when adjacent cells share common membrane proteins that locally fuse the membranes of the cells and prevent extracellular material from "leaking" between the bound cells. Tight junctions are especially important in epithelial tissues lining such structures as the intestine.

22. Tight junctions also restrict movement of phospholipids and membrane proteins between different regions of the cell.

23. *Desmosomes* attach adjacent cells together via cell adhesion proteins between *plaques* located within the cytoplasm of the two cells. Plaques are anchored in the cytoplasm by keratin fibers. Cells in an epithelium are frequently bound by numerous desmosomes.

24. A *gap junction* of 2.7 nm between the plasma membranes of two cells results where intrinsic protein complexes called *connexons* are linked together. Materials can pass between the two cells through a strand of cytoplasm located within a central channel in the connexon.

Questions (for answers and explanations, see page 54)

1. Match the appropriate letters for the characteristics listed on the right with the membrane components listed on the left.

   _____ Lipid bilayer   *a.* Cellular communication

   _____ Membrane proteins *b.* Permeability barrier

   _____ Glycolipids   *c.* Membrane transport

   _____ Glycoproteins   *d.* Accelerate reactions

                            e. Molecule recognition

2. Which of the following is *not* a characteristic of the fluid mosaic model for biological membranes?
   a. Fluidity
   b. Components symmetrically distributed
   c. Membrane components can move about
   d. Structure reflects function
   e. Thickness varies

3. What would be the expected structural features of a transmembrane protein?
   a. Entirely hydrophobic
   b. Entirely hydrophilic
   c. Hydrophobic ends, hydrophilic middle
   d. Hydrophilic ends, hydrophobic middle
   e. Bound to carbohydrate (a glycoprotein)

4. Which of the following statements about membrane carbohydrates is *not* true?
   a. They are either glycoproteins or glycolipids.
   b. They are very diverse.
   c. They are attached to either side of the membrane.
   d. They are all peripheral to the membrane.
   e. They are not usually more than 15 monosaccharides in length.

5. Add the appropriate letters for characteristics listed on the right to the junction types listed on the left.

   _____ Desmosome   *a.* Joins cells together

   _____ Tight junction   *b.* Prevents leakage

   _____ Gap junction   *c.* "Spot weld"

                            d. Has connexons

                            e. Cell communication

                            f. Fused membranes

PASSIVE PROCESSES OF MEMBRANE TRANSPORT (pages 105–109)

Key Concepts

1. *Diffusion* is the random movement of molecules from regions of greater to lesser concentration. Diffusion results in a uniform distribution of molecules within a system at equilibrium.

2. The rate of diffusion increases with temperature and the steepness of the concentration gradient. Size and electric charge of the diffusing molecules also influence their rate of diffusion.

3. A membrane is *permeable* to a substance if the substance can pass through it, or *impermeable*, if it cannot.

4. In *simple diffusion* through a membrane, the substance must either pass directly through the lipid bilayer or through a protein *channel* in the membrane. Usually only small, non-polar molecules can move through the lipid bilayer; some ions diffuse through channels.

5. As a transport mechanism between cells, diffusion is only suitable for distances of one or two cells.

6. For substances that must move through the hydrophobic lipid bilayer, there is a direct relationship between their *lipid solubility* and their ability to pass through the membrane.

7. Ions like K^+ and Cl^-, which move through membranes at rates faster than their lipid solubilities warrant, do so via special channels located in membrane transport proteins.

8 Most biochemical molecules are too large or hydrophilic to pass through membranes by simple diffusion and depend on specific membrane proteins for transport.

9. *Membrane transport proteins* are integral proteins that span the membrane and are specialized, based on their tertiary structure, for transporting specific molecules or ions.

10. Membrane transport proteins can be *uniports* (transporting a single solute) or *coupled transport systems* (transporting two or more solutes). Coupled transport systems can be *symports* (transporting solutes in the same direction) or *antiports* (transporting solutes in opposite directions).

11. In *facilitated diffusion* (carrier-mediated diffusion), the molecule combines with a membrane-bound *carrier protein*, which allows it to pass through the membrane.

12. Facilitated diffusion carrier proteins allow solutes to move in both directions, with the net movement of solute being down the concentration gradient.

13. If we plot the rate of solute transport against solute concentration, the rate due to simple diffusion always increases directly with solute concentration, whereas in facilitated diffusion, the rate will eventually *saturate* when all of the transport proteins are fully engaged.

14. The *osmotic potential* of a solution is always negative, and its magnitude varies directly with the solute concentration of the solution. Pure water has an osmotic potential of zero.

15. *Osmosis* is the diffusion of water through a membrane. In animal cells, the net movement of water will be from the side with a less negative osmotic potential to the side with a more negative osmotic potential.

16. Solutions with equal osmotic potentials are *isosmotic*; a *hypoosmotic* solution has a less negative osmotic potential than another solution; a *hyperosmotic* solution has a more negative osmotic potential than another solution.

17. The osmotic movement of water in cells lacking walls (like those of animals) depends only on osmotic potential differences between the cells.

18. In cells with cell walls (those of bacteria, plants, fungi, and some protists), pressure can build up within the cell and

prevent further influx of water. Thus, in cells with walls, the net movement of water depends on the cell's osmotic potential and an opposing *pressure potential*.

Questions (for answers and explanations, see page 54)

1. Which of the following substances is most likely to permeate a membrane by diffusing *directly* through the lipid bilayer ?
   a. K^+
   b. Glucose
   c. Cl^-
   d. NH_3^+
   e. O_2

2. For a substance passing through a membrane by simple diffusion, select the curve in the following figure that best shows the relationship between its concentration outside the membrane and its rate of movement through the membrane.

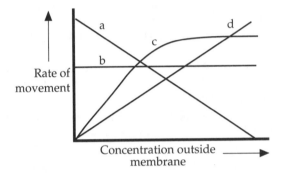

3. For a substance passing through a membrane by *facilitated diffusion*, which curve in the previous figure best shows the relationship between its concentration outside the membrane and its rate of movement through the membrane?

   Curve _____

4. An *animal cell* is put into a bathing solution and it gains additional water. Choose *all* of the following statements that are *true*.
   a. The cell was isosmotic to the bathing solution.
   b. The cell was hypoosmotic to the bathing solution.
   c. The cell was hyperosmotic to the bathing solution.
   d. The bathing solution was hypoosmotic to the cell.
   e. The bathing solution was hyperosmotic to the cell.

5. If the cell in Question 5 is a *plant cell*, which of the following statements is likely to be true at osmotic equilibrium?
   a. The cell has a greater pressure potential and a more negative osmotic potential than the bathing solution.
   b. The cell has a greater pressure potential and a less negative osmotic potential than the bathing solution.
   c. The cell has a lesser pressure potential and a more negative osmotic potential than the bathing solution.
   d. The cell has a lesser pressure potential and a less negative osmotic potential than the bathing solution.
   e. At equilibrium, the cell and bathing solution would be the same for these two measurements.

ACTIVE PROCESSES OF MEMBRANE TRANSPORT • MEMBRANES ARE NOT SIMPLY BARRIERS • MEMBRANES ARE DYNAMIC (pages 109–116)

Key Concepts

1. Whereas simple and facilitated diffusion can only move substances down their concentration gradients, *active transport* mechanisms expend energy to move substances from regions of lesser to greater concentration (up concentration gradients).

2. Active transport carrier proteins can only move their solutes in one direction, usually by expending energy to move solutes against their concentration gradients.

3. *Primary active transport* uses the energy released from the conversion of ATP to ADP to move ions across membranes against a concentration gradient.

4. In the *sodium–potassium pump*, an example of primary active transport found in all animal cells, an integral protein catalyzes the breakdown of ATP and couples the movement of two K^+ ions into the cell with the release of three Na^+ ions. Thus, the sodium–potassium pump is an antiport.

5. *Secondary active transport* uses the energy released when ions diffuse back down a concentration gradient established by primary active transport. The uptake of glucose, coupled to Na^+ ion movement back into the cell, is an example of secondary active transport involving a symport mechanism.

6. *Endocytosis* results when a portion of the plasma membrane infolds to form a new vesicle containing material that had previously been outside the cell.

7. *Pinocytosis* is a form of endocytosis in which only a small vesicle is formed, usually engulfing liquid; in *phagocytosis* a very large vesicle (called a food vacuole or phagosome) is formed, usually engulfing particulate material or another cell.

8. In *receptor-mediated endocytosis*, membrane-bound receptor proteins bind specific, extracellular macromolecules to form a *coated pit*, which invaginates to form a *coated vesicle*. Coated vesicles are stabilized by the fibrous proteins, such as *clathrin*.

9. Cholesterol, bound to a protein to form a lipoprotein such as *low-density lipoprotein (LDL)*, enters mammalian cells via receptor-mediated endocytosis.

10. *Exocytosis* results as material in a vesicle is moved to the outside of a cell when the vesicle fuses with the plasma membrane.

11. Substances that bind to membrane receptor proteins or carbohydrates are called *ligands*. Binding initiates a specific change in the cell or organelle. Examples of ligands include antibodies, many hormones, neurotransmitters, and growth factors.

12. Ligands have their effect on the cell by changing the tertiary structure of an integral membrane protein. The cytoplasmic portion of the integral protein then effects a change within the cell.

13. The ligand–receptor complex might also activate another membrane protein called a *G protein*. The G protein then dif-

fuses in the membrane and can interact with other molecules to produce diverse effects, such as increasing the rate of a chemical reaction, activating a transport protein, etc.

14. Important molecules in certain chemical pathways in cells (e.g., some steps of photosynthesis and respiration) are spatially arranged within membranes so that the reactions can take place in an orderly fashion.

15. By moving ions across membranes, certain membrane proteins can establish significant electric and pH differences across the membrane. Movement of charged particles back across the membrane can be used to perform work. Establishing a proton (H^+) gradient is an important function of both the chloroplast and mitochondrial membranes.

16. *Cell adhesion molecules* are membrane proteins that bind similar cells together to form *tissues*.

17. The phospholipid components of membranes are produced within the rough ER and then distributed to membranes throughout the cell.

18. Membrane proteins formed in ribosomes are deposited in the rough endoplasmic reticulum, and from there are moved to the Golgi apparatus. Carbohydrates may be added to these proteins within the ER or Golgi apparatus.

19. Golgi-derived vesicles move the completed membrane proteins to the plasma membrane where they are incorporated by exocytosis.

20. Formation and absorption of vesicles causes a steady flow of membranes from the ER to the Golgi apparatus (from *cis* to *trans* regions) to the plasma membrane.

21. Endocytosis moves membranes from the plasma membrane back into the cell interior.

Questions (for answers and explanations, see page 54)

1. Which of the following is *not* a ligand?
   a. Hormones such as insulin
   b. G protein
   c. Antibody
   d. Growth factor
   e. Neurotransmitter

2. Order the following terms in a way that depicts how membranes are likely to move through the cell: (1) ER, (2) nuclear envelope, (3) Golgi apparatus, (4) vesicle, (5) plasma membrane. [*Terms can be used twice.*]
   a. 1, 2, 3, 4, 5
   b. 1, 4, 2, 4, 3, 4, 5
   c. 2, 3, 1, 4, 5
   d. 2, 1, 4, 3, 4, 5
   e. 5, 4, 3, 4, 1, 2

3. Which of the following statements correctly characterizes the diagram shown at the top of page 30?
   a. Primary active transport involving a symport is shown on the right; secondary active transport involving an antiport is shown on the left.
   b. Primary active transport involving a symport is shown on the left; secondary active transport involving an antiport is shown on the right.
   c. Primary active transport involving an antiport is shown on the right; secondary active transport involving a symport is shown on the left.

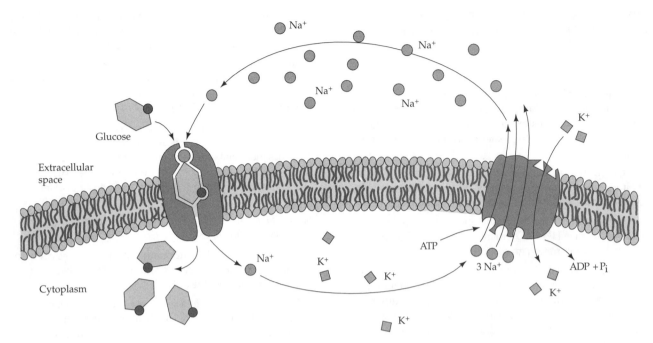

d. Primary active transport involving an antiport is shown on the left; secondary active transport involving a symport is shown on the right.

e. Both are examples of primary active transport; the antiport is on the right, the symport is on the left.

Integrative Questions/Activities (for answers and explanations, see page 55)

1. Circle *all* of the following activities in which cell membranes are normally involved.
   a. Communication between different cells
   b. Tissue formation
   c. Establishment of pH gradients
   d. Osmosis
   e. Nutrient procurement
   f. Waste removal
   g. Movement
   h. Control of chemical reactions

2. Label the diagram of the cell membrane shown below with the numbers accompanying the following labels: (1) hydrophilic portion of phospholipid, (2) hydrophobic portion of phospholipid, (3) intrinsic protein, (4) extrinsic protein, (5) glycolipid, (6) glycoprotein, (7) cytoplasm. How do you know which side of the membrane is the cell's interior?

- In the diagram to the right, add arrows to show membrane flow within cells.

  Label areas in this diagram where carbohydrates are added to newly synthesized membrane proteins.

6

Energy, Enzymes, and Metabolism

CHAPTER LEARNING OBJECTIVES—after studying this chapter you should be able to:

❏ Define "work" and state the first law of thermodynamics relative to your definition.

❏ Define "entropy" and "free energy" and state the second law of thermodynamics relative to your definitions.

❏ Explain how the equilibrium concentrations of reactants and products for a chemical reaction can be used to determine if a reaction is spontaneous or not.

❏ Relate the change in free energy (ΔG) of a chemical reaction to the equilibrium concentrations of reactants and products, and explain how this information can be used to determine if the reaction is endergonic or exergonic.

❏ Explain the equation relating ΔG to changes in heat and entropy of a reaction.

❏ Relate the structure of adenosine triphosphate (ATP) to its role as short-term energy storage molecule.

❏ Explain the role of a phosphorylated intermediate in a common two-step process whereby ATP drives an endergonic reaction.

❏ Explain the concept of activation energy in terms of the formation of transition species.

❏ Explain how a catalyst accelerates a chemical reaction.

❏ State the generalized formula for an enzyme-catalyzed reaction and define the terms "substrate" and "enzyme–substrate complex."

❏ Explain enzyme specificity, the interaction of enzyme and substrate at the enzyme's active site, and the concept of induced fit.

❏ Explain the effect of an enzyme on the ΔG, the equilibrium concentrations of substrate and product, and the activation energy of a reaction.

❏ Explain what an enzyme kinetic curve is and be able to draw a typical curve showing the effect of substrate concentration on reaction rate.

❏ Explain the role of cofactors in enzyme catalysis and differentiate between a prosthetic group and a coenzyme.

❏ Define what an inhibitor is and differentiate between irreversible, competitive, and noncompetitive inhibitors.

❏ Describe the structure of a simple allosteric enzyme and explain how its catalytic and regulatory subunits interact with substrates and effector molecules to determine the equilibrium between active and inactive forms of the enzyme.

❏ Explain how enzyme kinetic curves can be used to differentiate between competitive and noncompetitive inhibitors and between single-subunit and allosteric enzymes.

❏ Explain the concept of negative feedback and understand how negative feedback is accomplished in metabolic pathways.

❏ Explain the effects of pH and temperature on enzymes and what is meant by the "denaturation" of an enzyme.

❏ Define the terms "isozyme" and "ribozyme".

ENERGY CONVERSIONS: THE LAWS OF THERMO-DYNAMICS • CHEMICAL REACTIONS RELEASE OR TAKE UP ENERGY • ATP: TRANSFERRING ENERGY IN CELLS (pages 119–128)

Key Concepts

1. *Energy* is necessary to accomplish change, to do work. Stored energy or *potential energy*, due to position or state, can be converted to energy of motion, called *kinetic energy*.

2. Heat, light, electric and mechanical energy are forms of kinetic energy. The different forms of energy are interconvertible.

3. *Metabolism* is the total of all chemical reactions occurring within an organism. *Catabolic reactions* degrade complex molecules into simpler molecules; *anabolic reactions* build up complex molecules from simpler ones.

4. The *first law of thermodynamics* states that in a closed system, energy can neither be created nor destroyed.

5. *Entropy* is a measure of the disorder in a system.

6. *Free energy*, G, is energy that can be used to do work.

7. The relationship of free energy to changes in heat and entropy is $\Delta G = \Delta H - T\Delta S$, where ΔH is the change in heat, T is the absolute temperature, and ΔS is the change in entropy. Units for this equation are usually kilocalories (kcal).

8. According to the *second law of thermodynamics*, in any closed system free energy decreases and entropy increases.

9. *Chemical equilibrium* exists when the rate of the forward reaction equals the rate of the reverse reaction.

10. If at chemical equilibrium, the concentration of the products exceeds the concentration of the reactants, then the forward reaction dominates and the reaction is *spontaneous*. If

at chemical equilibrium, the concentration of the reactants exceeds the concentration of the products, then the reverse reaction dominates and the reaction is *nonspontaneous*.

11. The change in free energy of a reaction, ΔG, is directly related to the equilibrium concentrations of products and reactants: if the concentration of the products exceeds the concentration of the reactants, ΔG is negative and the reaction is *exergonic* (it releases energy); if the concentration of the reactants exceeds the concentration of the products, ΔG is positive and the reaction is *endergonic* (it absorbs free energy).

12. The free energy of a reaction is determined by the amount of heat it gives off or absorbs, and by differences in the orderliness of reactants and products.

13. *Adenosine triphosphate (ATP)* is a short-term energy storage molecule; carbohydrates and fats are long-term energy storage molecules.

14. The exergonic conversion of adenosine triphosphate (ATP) to adenosine diphosphate (ADP) plus inorganic phosphate (P_i) is a hydrolysis reaction with ΔG equal to about –12 kcal of free energy per mole of ATP under conditions typical of most cells.

15. In ATP, high-energy bonds result because several negative charges are held in close proximity in the last two phosphates. Hydrolyzing these phosphate groups reduces the free energy by spreading out the negative charges.

16. Exergonic reactions within the cell, such as those of cellular respiration, are coupled to the endergonic reaction that produces ATP,

$$ADP + P_i + \text{free energy} \rightarrow ATP + H_2O$$

The hydrolysis of ATP can then be used to drive a different endergonic reaction.

17. Frequently, ATP drives an endergonic reaction by first transferring a phosphate group to one of the reactants to form a higher energy, *phosphorylated intermediate*. The phosphorylated intermediate then interacts with the second reactant and releases the phosphate group. The net free energy change for both reactions is negative.

18. *Bioluminescence*, the production of light by an organism, is an example of a process that requires ATP.

Questions (for answers and explanations, see page 55)

1. Which of the following statements about spontaneous reactions is *not* true?
   a. In spontaneous reactions, the rate of the forward reaction exceeds the rate of the back reaction.
   b. Spontaneous reactions reach equilibrium quickly.
   c. The change in free energy, ΔG, of spontaneous reactions is negative.
   d. Spontaneous reactions release energy.
   e. In spontaneous reactions, the concentration of products at equilibrium is greater than the concentration of reactants.

2. Which expression in the equation for change in free energy, $\Delta G = \Delta H - T\Delta S$, reflects changes in the orderliness of reactants and products?
   a. ΔG
   b. ΔS
   c. T
   d. ΔH
   e. $T\Delta S$

3. When the reaction fructose 6-phosphate $\rightarrow$ glucose 6-phosphate reaches an equilibrium, the concentration of products is twice as great as the concentration of reactants. Which of the following conclusions about this reaction is true?
   a. The reaction is exergonic and ΔG is positive.
   b. The reaction is exergonic and ΔG is negative.
   c. The reaction is endergonic and ΔG is positive.
   d. The reaction is endergonic and ΔG is negative.
   e. None of these conclusions can be made based on a knowledge of the equilibrium concentration of reactants and products.

4. In the conversion of ATP to AMP, how many water molecules will be used as reactants or produced as products?
   a. Three will be used as reactants.
   b. Two will be used as reactants.
   c. One will be used as a reactant.
   d. Two will be produced as products.
   e. One will be produced as product.

Activity (for answers and explanations, see page 55)

- Label the adenine, ribose, and phosphate components of the ATP molecule shown below and indicate the portion of the molecule that constitutes adenosine, AMP, and ADP. Add the missing bonds using a straight line for a normal covalent bond and a wavy line for a high-energy covalent bond.

ENZYMES: BIOLOGICAL CATALYSTS • MOLECULAR STRUCTURE DETERMINES ENZYME FUNCTION (pages 128–134)

Key Concepts

1. Even if the free energy of the reactants is greater than the products (ΔG for the reaction is negative), some *activation energy* is usually required to start a reaction.

2. Activation energy helps change reactants into *transition-state species* before they can be converted to products. Transition-state species have greater free energy than either reactants or products.

3. A *catalyst* increases the rate of a reaction by decreasing the activation energy for the reaction.

4. Biological catalysts are globular proteins called *enzymes*. The names of many enzymes end with the suffix "-ase."

5. Reactants in enzyme-catalyzed reactions are called *substrates*.

6. The generalized form of an enzyme-catalyzed reaction is $E + S \rightleftharpoons E \bullet S \rightleftharpoons E + P$, where $E \bullet S$ is the *enzyme–substrate complex*.

7. Enzyme specificity results because the tertiary structure of the enzyme only allows chemically distinct types of substrates to attach to an area of the enzyme called its *active site*.

8. Substrate binding to the active site of the enzyme can involve hydrogen, hydrophobic, ionic, or covalent bonds.

9. ΔG and the equilibrium concentrations of reactants and products for the reaction remain unchanged by the enzyme; only the activation energy is lowered. Thus, enzymes speed up both forward and backward reactions equally.

10. At the active site the enzyme can chemically interact with the substrate in a variety of ways. Common modes of interaction are called *acid–base catalysis*, *covalent catalysis*, and *metal ion catalysis*. The enzyme must remain unchanged at the end of the reaction.

11. As with active transport and facilitated diffusion, enzyme kinetic curves plotting reaction rate against substrate concentration show *saturation* at high substrate levels. Substrate saturation occurs when all enzyme molecules are involved in the enzyme–substrate complex.

12. *Coupled reactions* occur on enzymes that have binding sites for each set of reactants. Typically, an endergonic reaction is coupled with an exergonic reaction so that the energy released by the exergonic reaction allows the endergonic reaction to proceed.

13. Coupling may also involve reactions that are part of the same pathway. Typically, if an exergonic reaction has as its reactants the products of an earlier endergonic reaction, it will speed up the endergonic reaction by preventing it from reaching equilibrium.

14. *X-ray crystallography* is used to obtain diffraction patterns from crystals of proteins and nucleic acids. Electron density maps can be produced from these patterns and analysis of the maps using computers permits one to deduce the structure of the molecule.

15. Typically, the enzyme undergoes a conformational change while in the enzyme–substrate complex. The change in shape, called an *induced fit* between substrate and enzyme, improves the conversion of reactant to product.

16. Many enzymes require nonprotein molecules called *cofactors* to function as catalysts. In addition to certain metal ions important in metal ion catalysis, cofactors also include *prosthetic groups* and *coenzymes*.

17. Metal ion prosthetic groups aid enzymes by binding the substrate or removing electrons from the substrate.

18. Coenzymes are complex organic compounds that are reversibly bound to the enzyme. In animals some coenzymes are derived from vitamins.

19. Coenzymes enhance the catalytic action of enzymes by either accepting or donating electrons, transferring phosphate groups, or altering the structure of the substrate in some other way.

Activities (for answers and explanations, see page 55)

- In the following graph, draw a curve showing the expected relationship between substrate concentration and the velocity of an enzyme-catalyzed reaction. Properly label the graph axes and mark areas on the curve where free enzyme molecules exist and where all enzyme is in the enzyme–substrate complex.

- Label the following graph to show the free energy change (ΔG) for the reaction, the activation energy without enzyme (E_a^{-e}), the activation energy with enzyme (E_a^{e}), the free energy associated with reactants (G_r), and the free energy associated with products (G_p).

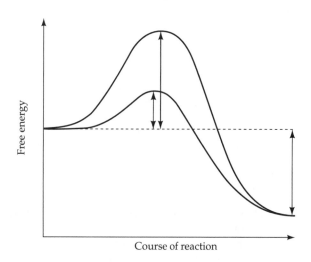

Course of reaction

Questions (for answers and explanations, see page 56)

1. Which of the following statements about the interaction of enzyme and substrate is *not* true?
   a. The conformation of the enzyme may change as it binds to substrate.
   b. The enzyme can only form hydrogen bonds with the substrate.
   c. The enzyme is unchanged at the end of the reaction.
   d. The substrate only interacts with a small portion of the enzyme molecule.
   e. The conformation of the enzyme determines its specificity for substrate.

2. Which of the following statements about enzyme-catalyzed reactions is *true*?
   a. Reactions are accelerated by changing ∆G for the reaction.
   b. Reactions are accelerated by changing the equilibrium point of the reaction.
   c. Reactions are accelerated by changing the activation energy of the reaction.
   d. Substrate can only become product through the action of the enzyme.
   e. Only the forward reaction is accelerated by the enzyme.

3. Which of the following statements about coupled reactions is *not* true?
   a. Usually an exergonic reaction is coupled to an endergonic reaction.
   b. Usually a single enzyme catalyzes both reactions.
   c. The product of one reaction is substrate for the other reaction.
   d. Energy released by the endergonic reaction allows the exergonic reaction to proceed.
   e. Active transport of molecules through membranes is an example of reaction coupling.

4. Which of the following choices correctly identifies the activation energy and ∆G for the reaction graphed below?

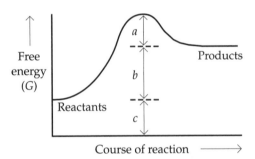

   a. Activation energy = *a*; ∆G = *a* + *b*
   b. Activation energy = *a*; ∆G =*b*
   c. Activation energy = *a* + *b*; ∆G = *a* + *b* + *c*
   d. Activation energy = *a*; ∆G = *a* + *b* + *c*
   e. Activation energy = *b*; ∆G = *b* + *c*

5. Which of the following conclusions is true for the reaction graphed above?
   a. The reaction is exergonic and spontaneous.
   b. The reaction is endergonic and spontaneous..
   c. The reaction is exergonic and nonspontaneous.
   d. The reaction is endergonic and nonspontaneous..
   e. More information is needed.

6. Which of the following statements about transition-state species is *not* true?
   a. The free energy of transition-state species is always greater than that of reactants.
   b. The free energy of transition-state species is always greater than that of products.
   c. The ∆G of the reaction is unaffected by the quantity of energy required to create transition-state species.
   d. Transition-state species are present in exergonic reactions, but not endergonic reactions.
   e. The activation energy equals the quantity of energy required to create transition-state species.

7. Which of the following statements about cofactors, prosthetic groups, and coenzymes is *not* true?
   a. Cofactors interact chemically only with the enzyme and not the substrate.
   b. Prosthetic groups and coenzymes are types of cofactors.
   c. Some coenzymes are free to move from enzyme to enzyme.
   d. Metal ions can function as cofactors.
   e. Some coenzymes are vitamins.

METABOLISM AND THE REGULATION OF ENZYMES (pages 134–139)

Key Concepts

1. Metabolic pathways consist of series of linked enzyme-catalyzed reactions, with the products of each serving as the substrate of the next reaction in the series.

2. Substances that interact with enzymes to reduce the catalytic activity of the enzyme are called *inhibitors*. Naturally occurring inhibitors control metabolism.

3. *Irreversible inhibitors* permanently modify the enzyme so that it can no longer act as a catalyst. The nerve gas DFP is an irreversible inhibitor of the enzyme acetylcholinesterase.

4. *Competitive inhibitors* resemble the enzyme's natural substrate and compete with it for the active site. Competitive inhibition can be overcome with additional substrate.

5. *Noncompetitive inhibitors* bind to the enzyme at a location other than the active site. Noncompetitive inhibitors change the conformation of the enzyme and alter the active site to make the enzyme less effective as a catalyst. Noncompetitive inhibition cannot be overcome with additional substrate.

6. Enzymes consisting of two or more subunits have a quaternary structure and may be subject to *allosteric control*.

7. *Allosteric enzymes* have two different types of subunits: a *catalytic subunit* with an active site and a *regulatory subunit* with an allosteric site. Substrate binds to the active site; *effector* molecules, either inhibitors or activators, bind to the allosteric site.

8. Allosteric enzymes can exist in two different forms, which are in equilibrium with each other: an *active form*, in which the active sites can bind substrate but the allosteric sites cannot bind an inhibitor, and an *inactive form*, in which the active sites cannot bind substrate but the allosteric sites can bind an inhibitor.

9. Addition of substrate shifts the equilibrium between the active and inactive forms so that more active forms exist and the reaction rate rapidly increases.

10. Addition of an effector that acts as an inhibitor shifts the equilibrium in favor of the inactive form, and the reaction rate rapidly decreases.

11. An *allosteric activator* is an effector that binds to the regulatory subunit of the active form and prevents it from converting to the inactive form.

12. Allosteric enzymes show more complex sigmoidal (S-shaped) kinetic curves when reaction rate is plotted against substrate concentration.

13. *Feedback inhibition* on metabolic pathways occurs when one or more products of the pathway are allosteric inhibitors of an enzyme catalyzing an earlier step in the pathway.

14. In a branched pathway, inhibition by a single end product of the *first committed step* in its portion of the pathway will not affect the other side of the pathway.

15. pH affects the ionization of carboxyl, amino, and other groups within the enzyme, which determines the shape of the enzyme and how it interacts with substrate.

16. Enzymes work best at specific pH values and may be *denatured* at extreme pH values.

17. Although enzyme activity increases with a moderate increase in temperature, at high temperatures enzymes become denatured.

18. *Isozymes* are enzymes that catalyze the same reaction but have slightly different structures; consequently, each isozyme requires different conditions (such as temperature) for optimum performance as a catalyst.

Questions (for answers and explanations, see page 56)

1. The following enzyme kinetics graph shows the results of running a series of enzyme-catalyzed reactions at different substrate concentrations, either without an inhibitor or with an inhibitor.

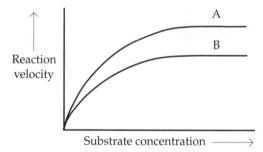

Which of the following statements is a *correct* interpretation of this graph?

a. Curve A is with the inhibitor, curve B is without the inhibitor, and the inhibitor probably resembles the normal substrate.

b. Curve B is with the inhibitor, curve A is without the inhibitor, and the inhibitor probably resembles the normal substrate.

c. Curve A is with the inhibitor, curve B is without the inhibitor, and the inhibitor probably does not resemble the normal substrate.

d. Curve B is with the inhibitor, curve A is without the inhibitor, and the inhibitor probably does not resemble the normal substrate.

e. Curve B is with the inhibitor, curve A is without the inhibitor, and the inhibitor probably resembles the enzyme.

2. Which of the following statements about an allosteric enzyme is *not* true?

a. At high substrate but low inhibitor concentrations, most of the enzyme is in the active form.

b. At high substrate but low inhibitor concentrations, most of the allosteric sites are empty.

c. At high inhibitor but low substrate concentrations, most of the enzyme is in the inactive form.

d. At high inhibitor but low substrate concentrations, most of the active sites are occupied.

e. At high inhibitor but low substrate concentrations, most of the allosteric sites are occupied.

3. Which of the following choices correctly shows the condition of the carboxyl and amino groups of an enzyme in an acidic environment?

a. —COOH and NH_2

b. —COOH and NH_3^+

c. —COO^- and NH_2

d. —COO^- and NH_3^+

4. In the following illustration, for substance G, the first committed step of the pathway is the

a. A to B step.

b. B to D step.

c. B to C step.

d. D to F step.

e. F to G step.

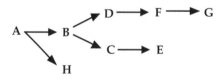

Activities (for answers and explanations, see page 56)

- In the diagram below showing allosteric regulation, label the following items: active site, allosteric site, substrate, inhibitor.

- Answer the following questions about the diagram shown below:

  a. If this enzyme catalyzed the first committed step in a metabolic pathway, what type of substance might act as an inhibitor?

  b. Which form of the enzyme (*a–f*) would be most prevalent if acted on by the type of substance described in *a* above? _____

  c. Could this enzyme be subject to concerted feedback inhibition? Explain your answer.

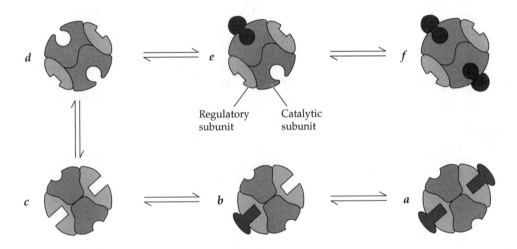

Regulatory subunit   Catalytic subunit

7

Cellular Pathways That Harvest Chemical Energy

OBTAINING ENERGY AND ELECTRONS FROM GLUCOSE
● AN OVERVIEW: RELEASING ENERGY FROM GLUCOSE
(pages 141–147)

Key Concepts

1. Energy flow in the biosphere is from the sun through photosynthetic autotrophs to heterotrophs and back to the environment as heat.

2. *Glycolysis, fermentation,* and *cellular respiration* are the major pathways for harvesting energy in cells.

3. The extraction of energy from food begins with glycolysis. Compounds that serve as food are converted to glucose or to intermediate molecules in the metabolism of glucose.

4. Under *aerobic* conditions (with oxygen present), cellular respiration follows glycolysis; under *anaerobic* conditions (with oxygen absent), fermentation follows glycolysis.

5. An *oxidation–reduction reaction*, or *redox reaction*, involves transfer of electrons or hydrogen atoms between chemical groups.

6. An atom or molecule is *reduced* when it gains an electron or hydrogen atom. An atom or molecule is *oxidized* when it loses an electron or hydrogen atom.

7. In a redox reaction, the reactant losing an electron or hydrogen atom is the *reducing agent*; the reactant accepting the electron or hydrogen atom is the *oxidizing agent*.

8. Redox reactions always involve both oxidation and reduction simultaneously; one reactant is reduced while the other reactant is oxidized.

9. The oxidative portion of a redox reaction is exergonic, with large negative ΔG values; the reductive portion of a redox reaction is endergonic, with large positive ΔG values. ΔG of the overall reaction will be negative if the reaction is spontaneous.

10. *Nicotinamide adenine dinucleotide (NAD)* is a reducing/oxidizing agent existing in two different forms that can bind or release hydrogens in the reaction

$$NAD^+ + 2H \rightarrow NADH + H^+$$

NAD^+ is the oxidizing agent; $NADH$ is the reducing agent.

11. Another important electron carrier in cells is *flavin adenine dinucleotide (FAD)*.

12. NAD and FAD transfer hydrogens and free energy during redox reactions.

13. Glycolysis is the oldest and most universal of the energy-releasing pathways; all organisms have the glycolytic pathway. The reactions of glycolysis occur in the cytosol.

14. Beginning with a molecule of glucose, the major products of glycolysis are a net yield of two ATP, pyruvate, and two electrons carried by NAD as $NADH + H^+$.

15. Cellular respiration operates when O_2 is available. It includes three pathways: *pyruvate oxidation*, the *citric acid cycle* (also called the Krebs or tricarboxylic acid cycle), and the *respiratory chain*.

16. In pyruvate oxidation, pyruvate is converted to acetate, with the loss of a carboxyl group ($—COO^-$) as CO_2 and two hydrogens. The acetate is activated by the addition of coenzyme A.

17. The reactions of the citric acid cycle, catalyzed by enzymes present in the mitochondrial matrix, move acetate through a series of reactions in which its carbons and hydrogens are removed. Products of the citric acid cycle are CO_2 and reduced NAD and FAD.

18. The respiratory chain transfers electrons from reduced NAD and FAD through a series of redox reactions that involve a chemiosmotic mechanism. Water is formed when oxygen serves as the final electron acceptor.

19. ATP production via the respiratory chain is called *oxidative phosphorylation*. Carriers of the respiratory chain are embedded in the folds of the inner mitochondrial membranes called the *cristae*.

20. Fermentation is an alternate pathway that can be used in the absence of oxygen for the oxidation of reduced NAD and other electron transfer agents

Questions *(for answers and explanations, see page 56)*

1. In eukaryotic cells with access to oxygen, the correct sequence showing movement of *carbon* atoms from glucose would include
   a. pyruvate, CO_2, citric acid cycle intermediates.
   b. glucose 6-phosphate, pyruvate, lactic acid.
   c. pyruvate, citric acid cycle intermediates, CO_2.
   d. glucose 6-phosphate, pyruvate, respiratory chain, CO_2.
   e. pyruvate, respiratory chain.

2. In eukaryotic cells with access to oxygen, the correct sequence showing movement of *hydrogen* atoms from glucose would include
   a. glucose 6-phosphate, coenzyme A, $NADH + H^+$, respiratory chain, H_2O.
   b. glucose 6-phosphate, $NADH + H^+$, respiratory chain, H_2O.

c. glucose 6-phosphate, pyruvate, $NADH + H^+$, citric acid cycle intermediates, respiratory chain, H_2O.
d. glucose 6-phosphate, $NADH + H^+$, H_2O, respiratory chain.
e. glucose 6-phosphate, pyruvate, citric acid cycle intermediates, respiratory chain, H_2O.

3. Which one of the following statements *incorrectly* characterizes the pair of coupled reactions shown below?

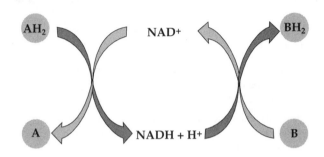

a. AH_2 is oxidized.
b. NAD^+ is a reducing agent.
c. The reaction producing BH_2 and NAD^+ is exergonic.
d. $NADH + H^+$ is a reducing agent.
e. B is an oxidizing agent.

GLYCOLYSIS: FROM GLUCOSE TO PYRUVATE • PYRUVATE OXIDATION • THE CITRIC ACID CYCLE • THE RESPIRATORY CHAIN: ELECTRONS, PROTON PUMPING, AND ATP   (pages 147–158)

Key Concepts

1. As a result of the ten reactions of glycolysis (which take place in the cytoplasm), one molecule of glucose is converted into two molecules of pyruvate. The net production of energy carriers is two $NADH + H^+$ and two ATP molecules per molecule of glucose.

2. The first five reactions of glycolysis are endergonic and result in the addition of two phosphates from ATP (reactions 1 and 3), the conversion of glucose to fructose (reaction 2), formation of two three-carbon sugars from fructose (reaction 4), and the production of two molecules of glyceraldehyde 3-phosphate (G3P) (reaction 5).

3. In reaction 6 of glycolysis, G3P undergoes three changes: it is oxidized with the addition of an oxygen atom, an inorganic phosphate group is added, and NAD removes two hydrogens to form two $NADH + H^+$ per molecule of G3P.

4. In both reactions 7 and 10 of glycolysis, one ATP is produced as phosphate groups are removed from three carbon compounds. Since two ATP were used in reactions 1 and 3, the net production of ATP is two per molecule of glucose.

5. *Substrate-level phosphorylation* produces ATP by the enzyme-catalyzed transfer of phosphate groups from donor molecules to ADP. In reaction 6, glyceraldehyde 3-phosphate is oxidized by NAD^+. The energy of oxidation is used to add a phosphate group to this compound to form 1,3-bisphosphoglycerate. In reaction 7, 1,3-bisphos-

phoglycerate donates its phosphate to ADP to form ATP through substrate-level phosphorylation.

6. The oxidation of pyruvate, a multistep process, is catalyzed by the *pyruvate dehydrogenase complex*. It involves the removal of CO_2 (decarboxylation) and hydrogens (dehydrogenation) by NAD. This in turn converts pyruvate to acetate, which then combines with *coenzyme A (CoA)* to form *acetyl coenzyme A (acetyl CoA)*.

7. The *citric acid cycle* is a cyclic series of reactions catalyzed by enzymes present within the matrix of the mitochondrion. Acetate, in the form of acetyl CoA, becomes oxidized by releasing its carbons as CO_2 and then passing its hydrogens on to NAD and FAD.

8. Of the four carriers involved in the citric acid cycle, three are NADH + H^+ and one is $FADH_2$.

9. Acetyl CoA enters the citric acid cycle by combining with *oxaloacetate* to form *citrate*, releasing coenzyme A unchanged (reaction 1).

10. Each of the following reactions in the cycle involves the reduction of NAD^+: isocitrate to α-ketoglutarate (reaction 3), α-ketoglutarate to succinyl CoA (reaction 4), and malate to oxaloacetate (reaction 8).

11. Decarboxylation occurs during reactions 3 and 4: isocitrate to α-ketoglutarate and α-ketoglutarate to succinyl CoA.

12. Reaction 5, succinyl CoA to succinate, produces guanosine triphosphate (GTP) via substrate-level phosphorylation. GTP is used to produce ATP.

13. Reaction 6, succinate to fumarate, also yields $FADH_2$ (the reduced form of FAD).

14. NADH + H^+ and $FADH_2$ pass their hydrogens to the *respiratory chain*, a series of oxidation–reduction agents built into the inner membrane of the mitochondrion.

15. The respiratory chain includes three large protein complexes (*NADH-Q reductase, cytochrome reductase, cytochrome oxidase*), a mobile carrier molecule (*ubiquinone*) that can move within the membrane, and an extrinsic protein (*cytochrome c*) lying in the space between the inner and outer membrane.

16. Cytochromes in the respiratory chain have heme groups with a central iron atom that is Fe^{3+} when oxidized or Fe^{2+} when reduced.

17. Each pair of hydrogens that is passed from NADH + H^+ to oxygen by way of the respiratory chain results in three new ATP molecules; each pair of hydrogens that is passed from $FADH_2$ to oxygen by way of the respiratory chain results in two new ATP molecules.

18. At various points within the respiratory chain, hydrogen atoms are separated into protons and electrons. Electrons continue to be moved by respiratory chain carriers; H^+ are deposited outside of the inner mitochondrial membrane.

19. The *chemiosmotic mechanism* for production of ATP depends on the proton gradient (a *proton-motive force*) set up between the inside and outside of the inner mitochondrial membrane.

20. ATP is formed when hydrogen ions pass through a channel-like *ATP synthase* protein as they diffuse from the outside to the inside of the inner membrane.

21. The final acceptor of electrons in the respiratory chain is oxygen; cytochrome oxidase passes two electrons to two protons and an oxygen atom to form H_2O.

22. The *chemiosmotic mechanism* is supported by the observation that mitochondria in equilibrium with a solution at pH 8 will rapidly produce ATP if transferred to a solution of pH 4 containing ADT and P_i.

23. *Respiratory uncouplers* are substances that uncouple ATP production from respiratory metabolism by allowing protons to diffuse back across the membrane discharging the proton-motive force.

Questions (for answers and explanations, see page 56)

Questions 1–4 are concerned with the ten reactions of glycolysis, which are summarized below.

(1) glucose + ATP → glucose 6-phosphate + ADP

(2) glucose 6-phosphate → fructose 6-phosphate

(3) fructose 6-phosphate + ATP → fructose 1,6-bisphosphate + ADP

(4) fructose 1,6-bisphosphate → glyceraldehyde 3-phosphate + dihydroxyacetone phosphate

(5) dihydroxyacetone phosphate → glyceraldehyde 3-phosphate

(6) 2 glyceraldehyde 3-phosphate + 2 P_i + 2 NAD → 2 1,3-bisphosphoglycerate + 2 NADH + H^+

(7) 2 bisphosphoglycerate + 2 ADP → 2 3-phosphoglycerate + 2 ATP

(8) 2 3-phosphoglycerate → 2 2-phosphoglycerate

(9) 2 2-phosphoglycerate → 2 phosphoenolpyruvate + 2 H_2O

(10) 2 phosphoenolpyruvate + 2 ADP → 2 pyruvate + 2 ATP

1. Of the reactions shown, list those that are exergonic: _____

2. Of the reactions shown, list those that involve oxidation–reduction agents: _____

3. Of the reactions shown, list those that are described in the text as "pump-priming": _____

4. In which reaction of glycolysis is a 6-carbon compound converted in to two 3-carbon compounds? _____

5. Which of the following are *not* products of the citric acid cycle?
   a. Coenzyme A
   b. CO_2
   c. NADH + H^+
   d. $FADH_2$
   e. ATP

6. Which of the following compounds are *not* cycled in the citric acid cycle?
   a. α-Ketoglutarate
   b. Citrate
   c. Acetyl CoA
   d. Oxaloacetate
   e. Fumarate

7. Which of the following is *not* a correct similarity between the reactions pyruvate → acetyl CoA and α-ketoglutarate → succinyl CoA?
   a. Both reactions are redox reactions.
   b. Both reactions yield NADH + H⁺.
   c. Both reactions yield CO_2.
   d. Both reactions are endergonic.
   e. The enzymes catalyzing both reactions are located on the inner mitochondrial membrane.

8. Which of the following *correctly* lists the first and last components in the respiratory chain?
   a. NADH-Q reductase, cytochrome oxidase
   b. Q (ubiquinone), cytochrome oxidase
   c. Succinate-Q reductase, cytochrome *c*
   d. Cytochrome reductase
   e. NADH-Q reductase, cytochrome reductase

9. Which of the following generalizations about the respiratory chain is *not* true?
   a. Except for Q (ubiquinone), all respiratory components are proteins.
   b. Some respiratory components move protons from the inside to the outside of the inner mitochondrial membrane.
   c. Some components of the respiratory chain are chemically similar to hemoglobin.
   d. The final acceptor of electrons and protons is oxygen.
   e. Because of where its hydrogens enter the respiratory chain, the carrier $FADH_2$ produces only 2 ATP.

Activities (for answers and explanations, see page 56)

- In the following transmission electron micrograph showing a portion of a mitochondrion and adjacent cytosol, add labeled arrows pointing to locations where (A) the reactions of glycolysis occur, (B) the pyruvate dehydrogenase complex and certain other enzymes of the citric acid cycle occur, (C) the remaining citric acid cycle reactions occur, and (D) the electron carriers of the respiratory chain are located. For the moment, ignore the arrows labeled 1 and 2.

- Complete the following sentence to make a *true* statement about the pH and ATP production in the areas of the mitochondrion indicated by arrows 1 and 2 in the previous transmission electron micrograph.

   The pH in area ____ is greater than area ____ , and ATP is formed in area ____ , but not in area ____ .

FERMENTATION: ATP FROM GLUCOSE WITHOUT O₂ • CONTRASTING ENERGY YIELDS • CONNECTIONS WITH OTHER PATHWAYS • REGULATING THE ENERGY PATHWAYS (pages 158–162)

Key Concepts

1. In *fermentation*, pyruvate produced by glycolysis is reduced by NADH + H⁺ to form either *lactic acid* or, via acetaldehyde, *ethanol*. The oxidized NAD⁺ produced by fermentation can be used again in glycolysis.

2. Some bacteria carry on cellular respiration but use oxidizing agents other than oxygen, (for example, nitrate, NO_3^-), as the final acceptor of electrons in the respiratory chain.

3. *Fermentation* yields a net of 2 ATP per molecule of glucose. Glycolysis followed by cellular respiration yields a net of 36 ATP per molecule of glucose.

4. Glycolysis produces 2 ATP and 2 NADH + H⁺. Movement of NADH + H⁺ into the mitochondrion uses ATP, so net production of ATP from glycolysis is 6 ATP (2 ATP per NADH + H⁺).

5. Pyruvic acid oxidation and the citric acid cycle produces 2 ATP directly, 8 NADH + H⁺, and 2 $FADH_2$, resulting in 30 ATP (3 ATP per NADH + H⁺ and 2 ATP per $FADH_2$).

6. After monosaccharides are converted into glucose, and polysaccharides such as starch and glycogen are hydrolyzed into glucose, they can enter glycolysis.

7. Upon digestion, fats become glycerol and fatty acids. Glycerol enters glycolysis after conversion to glyceraldehyde 3-phosphate and fatty acids enter the citric acid cycle after they are converted into acetyl CoA. When the cell has an adequate ATP supply, these reactions run in reverse to form fat.

8. Pyruvate and citric acid cycle intermediates can be used as building blocks for important amino acids.

9. In addition to its involvement in fat synthesis, acetyl CoA can also be used to produce pigments, plant growth substances, and steroid hormones.

10. Because aerobic respiration is 18 times more efficient than glycolysis and fermentation, much less glucose is needed to meet the energy requirements if the cell has access to oxygen.

11. That the rate of glycolysis in many cells increases when oxygen is not available and decreases when oxygen is available is called the *Pasteur effect*.

12. Regulation of the reactions of glycolysis, the citric acid cycle, and the respiratory chain depends on the *allosteric control* of the enzymes catalyzing these reactions.

13. In allosteric control, the intermediates of a biochemical pathway provide positive and negative feedback by inhibiting or enhancing the efficiency of enzymes catalyzing key reactions in the pathway.

14. Examples of negative feedback of respiratory pathways include inhibition of phosphofructokinase, the enzyme catalyzing the third reaction of glycolysis by fructose 1,6-bisphosphate (product of the reaction), and by citrate from the citric acid cycle.

15. Examples of positive feedback of respiratory pathways include the allosteric activation by acetyl CoA of the reaction that converts pyruvate directly into oxaloacetate and the stimulatory effect of citrate on the conversion of acetyl CoA into fatty acids.

16. Many reactions producing ATP and NADH + H$^+$ are inhibited by high concentrations of these molecules, but stimulated by high concentrations of ADP and NAD. For example, the rate of the third reaction of glycolysis is increased by ADP, but decreased by ATP. In the citric acid cycle, the rate of conversion of isocitrate to α-ketoglutarate is increased by ADP and NAD, but decreased by ATP and NADH + H$^+$.

17. ATP and NADH + H$^+$ are also allosteric inhibitors of the reaction that combines acetyl CoA with oxaloacetate to form citric acid.

Questions (for answers and explanations, see page 57)

(Note: in several of the following questions, you may wish to refer to the accompanying figure on page 42 showing feedback regulation of glycolysis and the citric acid cycle.)

1. Of the 36 ATP produced from the complete aerobic respiration of one molecule of glucose, how many result from the chemiosmotic mechanism?
   a. 24
   b. 28
   c. 30
   d. 32
   e. 36

2. Acetyl CoA is not usually involved in the synthesis of
   a. lactate.
   b. citrate.
   c. fatty acids.
   d. steroids.
   e. pigments.

3. Which of the following molecules will result in the greatest energy yield if linked to the respiratory chain?
   a. NAD$^+$ produced through lactic acid fermentation
   b. NADH + H$^+$ produced in glycolysis
   c. NADH + H$^+$ produced in the citric acid cycle
   d. FADH$_2$ produced in the citric acid cycle
   e. FADH$_2$ produced in glycolysis

4. Which one of the following mechanisms most directly causes the phenomenon called the Pasteur effect?
   a. Allosteric activation by acetyl CoA of the reaction converting pyruvate into oxaloacetate
   b. Allosteric inhibition by ATP and citrate of the enzyme (phosphofructokinase) catalyzing the reaction fructose 6-phosphate → fructose 1,6-bisphosphate
   c. Allosteric activation by citrate of fatty acid synthesis to form acetyl CoA
   d. Allosteric inhibition by ATP of the reaction isocitrate → α-Ketoglutarate
   e. Allosteric activation by NAD$^+$ of the reaction isocitrate
   f. α-Ketoglutarate

Activity (for answers and explanations, see page 57)

- In the following figure, circle allosteric control points (both positive and negative) that would be important in regulating glycolysis and respiration in a cell that had been undergoing fermentation but that is now starting to respire aerobically.

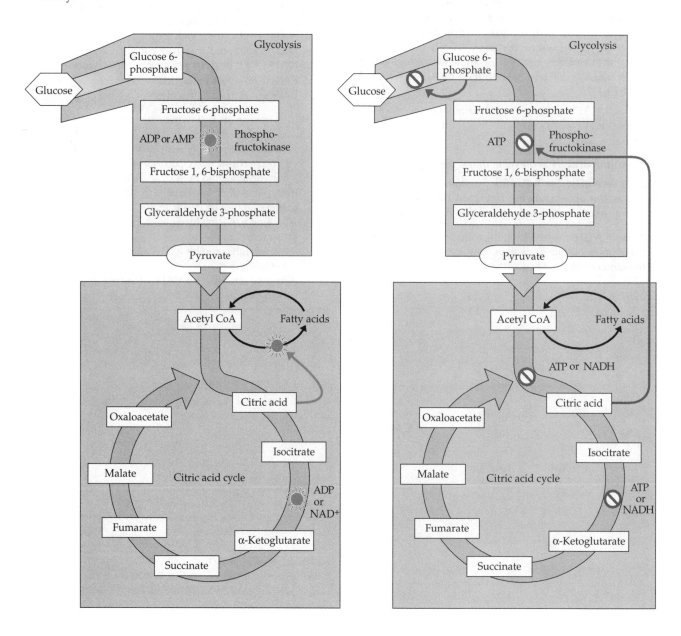

8

Photosynthesis: Energy from the Sun

IDENTIFYING PHOTOSYNTHETIC REACTANTS AND PRODUCTS • THE TWO PATHWAYS OF PHOTOSYNTHESIS: AN OVERVIEW • PROPERTIES OF LIGHT AND PIGMENTS (pages 165–172)

Key Concepts

1. *Autotrophs* are organisms that can use an energy source and simple raw materials to make reduced carbon compounds.

2. *Photosynthetic autotrophs* use the energy of visible light to reduce CO_2 with hydrogens supplied by water to produce carbohydrates, oxygen, and new water.

3. *Heterotrophs* are organisms that depend on preformed organic compounds to obtain needed free energy for life processes. Typically, heterotrophs consume autotrophs or other heterotrophs.

4. In studies using *heavy oxygen* (^{18}O), it was shown that in the overall reaction of photosynthesis the oxygen gas produced comes from water:

$$6\,CO_2 + 12\,H_2O \rightarrow C_6H_{12}O_6 + 6\,O_2 + 6\,H_2O$$

5. In higher land plants, water comes from the soil via the roots and is brought to the leaves; CO_2 enters the leaves through openings called *stomata* (sing. *stoma*). Oxygen produced in photosynthesis and water also exit through the stomata.

6. The oxidation–reduction agent used in photosynthesis is *nicotinamide adenine dinucleotide phosphate (NADP)*, which can exist in a reduced or oxidized form,

$$NADP^+ + 2H \rightarrow NADPH + H^+$$

7. Photosynthesis consists of two pathways: *photophosphorylation*, or the *light reactions*, and the *Calvin–Benson cycle*, or the *dark reactions*.

8. The two different systems of photophosphorylation are *noncyclic photophosphorylation* and *cyclic photophosphorylation*. Both use light energy to produce ATP; noncyclic photophosphorylation also produces NADPH + H^+. Both occur in molecular assemblies called *photosystems*.

9. The Calvin–Benson cycle fixes CO_2 and converts it to sugar using ATP and NADPH + H^+ produced in photophosphorylation.

10. In the dark, the Calvin–Benson cycle stops as soon as it depletes its supply of NADPH + H^+ and ATP; if CO_2 availability limits the Calvin–Benson cycle, photophosphorylation will stop when it runs out of $NADP^+$ and ADP. Thus, the two pathways are interdependent.

11. Light is a form of energy that can be described either as particles called *photons* or *quanta*, or in terms of its wave-like characteristics. Wavelength is the distance from peak to peak, and is usually measured in nanometers (nm).

12. Visible light (to humans) is *electromagnetic radiation* with wavelengths from 400 nm (violet) to 700 nm (red). Wavelengths from 100 to 400 nm are ultraviolet; over 700 nm, infrared and radio.

13. The frequency of light (v, Greek letter *nu*) is its speed (c) divided by its wavelength (λ, Greek letter *lambda*), or $v = c / \lambda$. The speed of light is a constant (3.0×10^{10} cm per sec, or 186,000 miles per sec). Frequency is sometimes expressed in hertz (cycles per second).

14. The energy (E) of a photon is directly proportional to its frequency (v), such that $E = hv$, where h is Planck's constant. Since $v = c / \lambda$ and $E = hc / \lambda$, light energy is directly dependent on wavelength, with light of shorter wavelength having more energy than light of longer wavelength.

15. Brightness, or light *intensity*, is a measure of the quantity of energy striking a defined surface area. It is expressed in energy units such as calories or as photons per centimeter per second.

16. When a molecule absorbs a photon, one of its electrons moves to a more outer orbital, raising the molecule from a *ground state* to an *excited state*. The difference between the two energy levels equals the energy of the absorbed photon.

17. *Pigments* are molecules that absorb light within the visible region of the electromagnetic spectrum. The colors of pigments are determined by the wavelengths not absorbed (i.e., the wavelengths transmitted or reflected).

18. Molecules can only absorb photons of particular energy levels as evidenced by the *absorption spectrum* of a molecule, in which we plot light absorption against wavelength. Light wavelengths in the blue and red regions are best absorbed by chlorophyll, as shown by its absorption spectrum.

19. An *action spectrum* is produced if we plot a particular light-dependent activity as a function of wavelength. The action spectrum for photosynthesis shows that light wavelengths in the blue and red regions of the spectrum are more effective than green, yellow, or orange light.

20. *Chlorophyll a* and *chlorophyll b* are two important photosynthetic pigments. Each has a complex ringed structure called a *chlorin*, with a central magnesium atom and a hydrocarbon chain. They differ only slightly in structure. Chlorophylls are found in plants, photosynthetic protists, and bacteria.

21. Photosynthetic *accessory pigments* include *carotenoids*, like β-carotene, and the *phycobilins*, like phycocyanin and phycoerythrin. Accessory pigments absorb mostly in the region between the blue and red peaks of chlorophyll *a* and *b*, and pass their energy on to chlorophyll *a*.

22. An energy-absorbing *antenna system* consists of chlorophyll *a* and chlorophyll *b* and accessory pigments. Excitation is passed between pigment molecules within the antenna and eventually ends up in the pigment molecule that absorbs light with the longest wavelength and lowest energy.

23. The molecule to which all excitation is funneled in an antenna is in the *reaction center* of the antenna. The reaction center in plants is always occupied by a special form of chlorophyll *a*. This molecule converts its excitation into chemical energy.

24. *Fluorescence* is the reemission of light as a boosted electron falls back to its original, lower-energy orbital. In photosynthesis, the excited electron is used in a redox reaction, passing on its energy to the molecule it reduces.

25. In photophosphorylation, the excited chlorophyll of the reaction center (Chl*) is oxidized by an oxidizing agent (A) in the following general reaction. The oxidized chlorophyll (Chl^+) has transferred an electron and its energy to A.

$$Chl^* + A \rightarrow Chl^+ + A^-$$

Questions (for answers and explanations, see page 57)

1. Which one of the following statements about photophosphorylation and the Calvin–Benson cycle is *not* true?
   a. The products of photophosphorylation are ATP, NADPH + H^+, and O_2.
   b. The products of the Calvin–Benson cycle are CO_2 and sugar.
   c. Photophosphorylation requires chlorophyll *a*.

d. ATP and NADPH + H$^+$ produced in photophosphorylation are used in the Calvin–Benson cycle.

e. Photophosphorylation is activated by light.

2. Examine the following absorption spectrum for an unknown pigment molecule.

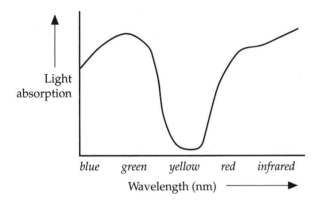

What color would this substance likely appear to you?

a. Blue

b. Green

c. Yellow

d. Blue-green

e. Infrared

3. Light beam A has a frequency of 1,000 hertz; light beam B has a frequency of 900 hertz. Which of the following comparisons is true?

a. Light beam A has a longer wavelength than B, and greater energy.

b. Light beam A has a longer wavelength than B, and lesser energy.

c. Light beam B has a longer wavelength than A, and greater energy.

d. Light beam B has a longer wavelength than A, and lesser energy.

e. More information is required to make these comparisons.

4. Explain what you would expect to happen if, in the following general reaction showing the reduction of oxidizing agent A by excited chlorophyll (Chl*), the amount of A becomes limiting?

$$Chl^* + A \rightarrow Chl^+ + A^-$$

PHOTOPHOSPHORYLATION AND REDUCTIONS • MAKING SUGAR FROM CO_2: THE CALVIN–BENSON CYCLE (pages 172–183)

Key Concepts

1. *Noncyclic photophosphorylation* involves two different systems of chlorophyll molecules called *photosystem I* and *photosystem II*.

2. Pairs of excited electrons, from photosystem I, and two hydrogen ions (H$^+$) are used to reduce NADP$^+$ to NADPH + H$^+$.

$$NADP^+ + 2\,e^- + 2\,H^+ \rightarrow NADPH + H^+$$

3. The reaction center of the photosystem I antenna is a chlorophyll *a* molecule called P$_{700}$ (its absorption peak is 700 nm). When it absorbs electrons it becomes excited P$_{700}$, or P$_{700}$*.

4. P$_{700}$* passes a pair of electrons to oxidized ferredoxin (Fd$_{ox}$), which becomes reduced ferredoxin (Fd$_{red}$). Fd$_{red}$ reduces NADP$^+$ to form NADPH + H$^+$.

5. Electrons from photosystem II pass through a series of electron carriers to oxidized chlorophylls of photosystem I. The sequence of carriers is pheophytin-I, plastoquinone (PQ), a cytochrome complex, and plastocyanin (PC).

6. The reaction center of the photosystem II antenna is a chlorophyll *a* molecule called P$_{680}$ (its absorption peak is 680 nm). When it absorbs electrons it becomes excited P$_{680}$, or P$_{680}$*.

7. Water ionizes to form H$^+$ and OH$^-$. Two OH$^-$ react to form O$_2$, two electrons, and two H$^+$. The electrons are used to replenish electrons in photosystem II.

8. The overall reaction for noncyclic photophosphorylation is H$_2$O + 4 photons + NADP$^+$ + ADP + P$_i$ → NADPH + H$^+$ + ATP + $^1/_2$O$_2$. Photosystems I and II each absorb two of the four photons. ATP production is via a chemiosmotic mechanism.

9. *Cyclic photophosphorylation* involves photosystem I. P$_{700}$ absorbs a photon and is raised to an excited state to become the reducing agent P$_{700}$*. Excited electrons from two different P$_{700}$* are used to reduce oxidized *ferredoxin* (Fd$_{ox}$) to form reduced ferredoxin (Fd$_{red}$),

$$2\,e^- + Fd_{ox} \rightarrow Fd_{red}$$

10. Pairs of electrons from Fd$_{red}$ are passed through a sequence of redox reactions involving the oxidation-reduction agents plastoquinone (PQ), a cytochrome complex (Cyt), and finally, plastocyanin (PC).

11. One of the reactions in the sequence is sufficiently exergonic to generate ATP from ADP, plus inorganic phosphate. Electrons are then returned by plastocyanin to oxidized P$_{700}$$^+$.

12. The electron carriers of photophosphorylation are located within the *thylakoid* membrane of the chloroplast.

13. Protons resulting from the ionization of water are pumped across the thylakoid membrane into the interior of the thylakoid. This creates a difference in pH and electric charge across the membrane—a *proton-motive force*.

14. *Chemiosmotic ATP formation* occurs when protons diffuse back out of the thylakoid into the stroma through proteins which are ATP synthases.

15. The model for chemiosmotic ATP formation during photophosphorylation was tested by varying the pH within the thylakoids of isolated chloroplasts. If the pH was lower inside the thylakoid than in the surrounding medium, ATP would be formed in the dark.

16. Noncyclic photophosphorylation evolved from cyclic photophosphorylation and photosystem I predates photosystem II.

17. Some photosynthetic bacteria have only cyclic photophosphorylation. Cyanobacteria, algae, and plants mostly use noncyclic photophosphorylation, but can switch to cyclic photophosphorylation when the ratio of NADPH + H⁺ to NADP⁺ in their chloroplasts is low.

18. Elucidation of the Calvin–Benson cycle depended on three techniques: labeling CO_2 with carbon-14 to get $^{14}CO_2$, *paper partition chromatography* for separating the intermediates in the pathway, and *autoradiography* for locating the colorless intermediates on the chromatogram.

19. In paper chromatography, movement of components in the mixture being separated depends on their relative affinity for the *stationary phase* (water bound to the paper) and the *mobile phase* (the solvent).

20. The distance moved by a substance in a chromatographic system divided by the distance moved by the solvent is called that substance's front ratio (R_F). The R_F value is constant for a substance under defined conditions.

21. In *two-dimensional chromatography*, a mixture is separated in one direction with one solvent system, the paper is rotated 90°, and the spots are run again with a different solvent system.

22. The method used by Calvin and Benson to determine the sequence of compounds in the Calvin–Benson cycle was to sample through time from a population of photosynthesizing algal cells exposed to $^{14}CO_2$ and identify the labeled organic compounds formed under different conditions.

23. Fixation of CO_2 by RuBP (ribulose bisphosphate) is catalyzed by the enzyme *RuBP carboxylase (rubisco)*. The resulting six-carbon compound immediately breaks down to form two molecules of 3PG (3-phosphoglycerate).

24. 3PG is reduced by NADPH + H⁺ to form a three-carbon sugar phosphate (G3P), in a reaction requiring ATP. There then follows a complex series of reactions that regenerates RuBP from G3P, while some G3P is directed into the synthesis of glucose and other organic compounds.

25. The Calvin–Benson cycle must "turn" six times to produce a molecule of glucose. This requires six CO_2, 12 NADPH + 12 H⁺, and 18 ATP from the light reactions.

26. Whereas the components of photophosphorylation are built into the thylakoid membrane, the enzymes of the Calvin–Benson cycle are found in the stroma of the chloroplast, except for rubisco, which is also part of the thylakoid membrane.

Questions (for answers and explanations, see page 57)

1. Select the correct path that an electron could take during the complete process of photosynthesis.
   a. CO_2 → ribulose bisphosphate → G3P → glucose
   b. H_2O → photosystem I → photosystem II → NADPH + H⁺ → G3P

c. photosystem II → H_2O → photosystem I → NADPH + H⁺ → glucose
   d. CO_2 → photosystem I → NADPH + H⁺ → ribulose bisphosphate → G3P
   e. H_2O → photosystem II → photosystem I → NADPH + H⁺ → G3P

2. Choose *all* of the following materials that are *not* cycled between the light reactions and the Calvin–Benson cycle.
   a. Water
   b. NADPH⁺
   c. ATP
   d. Oxygen
   e. NADPH + H⁺

3. The following graph shows changes in the concentration of substances A, B, and C during a "lollipop" experiment in which the CO_2 supply was *eliminated* at time 30.

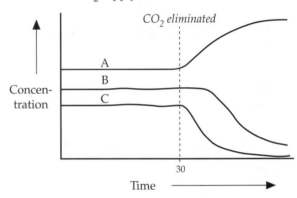

Based on these data, which of the following shows the correct pathway for the substances involved?
   a. A + CO_2 → B → C
   b. C + CO_2 → B → A
   c. A + CO_2 → C → B
   d. B + CO_2 → C → A
   e. A → B → C + CO_2

4. Two groups of chloroplasts (A and B) are kept in the dark and allowed to equilibrate with a solution of pH 6. Group A is then transferred to a solution of pH 7, and group B to a solution of pH 5. Which of the following would be true?
   a. Chloroplasts from both groups produce ATP.
   b. Chloroplasts from neither group produce ATP.
   c. Both groups produce ATP, but group A chloroplasts produce more ATP than group B chloroplasts.
   d. Both groups produce ATP, but group B chloroplasts produce more ATP than group A chloroplasts.
   e. Only group A chloroplasts produce ATP.

Activities (for answers and explanations, see page 57)

- A mixture of compounds was separated using two-dimensional chromatography, first with solvent A, then with solvent B. The following chromatogram results. Circle the substance that has the greatest total R_F for *both* solvents.

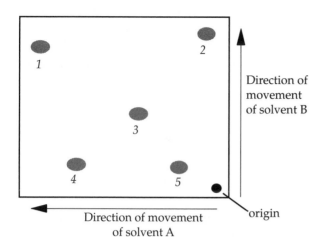

Direction of
movement
of solvent B

Direction of movement
of solvent A

origin

- In the following transmission electron micrograph showing a portion of a chloroplast and adjacent cytosol, label the arrows pointing to (A) the chloroplast outer membrane, (B) a thylakoid membrane, (C) the stroma, (D) the cytosol, (E) the location of photosystem I and photosystem II components, and (F) the location of the enzymes of carbon dioxide fixation.

PHOTORESPIRATION AND ITS EVOLUTIONARY CONSEQUENCES • PLANTS PERFORM CELLULAR RESPIRATION (pages 183–188)

Key Concepts

1. *Photorespiration* occurs when rubisco catalyzes the reaction of RuBP with oxygen to form, in part, the two-carbon compound *glycolate*.

2. Glycolate leaves the chloroplast and enters organelles called *microbodies* where it is further oxidized. The products of this pathway enter mitochondria, and eventually CO_2 is released. Photorespiration is a light-dependent process.

3. Photorespiration is a wasteful process because it uses RuBP, but produces no ATP. Photorespiration results because O_2 and CO_2 compete as substrates for rubisco.

4. C_4 *plants*, like sugar cane and corn, have an additional pathway for capturing CO_2 involving the enzyme *PEP carboxylase*, which has a much greater affinity for CO_2 than does rubisco.

5. PEP carboxylase catalyzes a reaction combining CO_2 with *phosphoenolpyruvate (PEP)* to form the four-carbon compound oxaloacetate. This pathway is called C_4 *photosynthesis* because the first stable compounds formed

have four carbons instead of three, as in the Calvin–Benson cycle, which is also called C_3 *photosynthesis*.

6. C_4 plants have a unique leaf anatomy, in which vascular bundles are surrounded by an outer *mesophyll layer* and an inner *bundle sheath layer*. The layers are composed of different types of cells, and each cell has a distinct type of chloroplast.

7. Cells of the mesophyll layer have the C_4 pathway only and produce oxaloacetate and other four-carbon acids. These four-carbon acids diffuse into the bundle sheath layer from the mesophyll.

8. Bundle sheath cells have the C_3 pathway only. The four-carbon acids from the mesophyll layer are decarboxylated to form CO_2 and three-carbon compounds that cycle back to the mesophyll cells. The CO_2 is incorporated into the C_3 pathway by rubisco.

9. The C_4 pathway and the unique anatomy of C_4 plants permit them to pump CO_2 from an area where it is in low concentration (the intercellular spaces) into the bundle sheath cells, thereby creating a high CO_2 concentration in these cells. This enables rubisco and the C_3 pathway to function normally and adapts C_4 plants for life in arid areas because they can keep their stomata closed more of the time.

10. Some plants, including many in the family Crassulaceae, keep their stomata closed all day to conserve water and carry on C_3 photosynthesis using CO_2 derived from four-carbon acids that were formed from CO_2 at night, using PEP carboxylase. This mode of carbon fixation is called *Crassulacean acid metabolism (CAM)*.

11. The C_4 pathway may be an adaptation for limiting photorespiration in arid environments, and C_4 plants show little photorespiration.

12. Photosynthetic autotrophs, such as the green plants, photosynthetic bacteria, and protists, cycle fixed carbon, CO_2, and O_2 with heterotrophic organisms.

13. Photosynthetic autotrophs simultaneously carry on photosynthesis and respiration and cycle fixed carbon, CO_2, and O_2 between chloroplasts and the sites of glycolysis and respiration in the cytosol and mitochondria.

Questions (for answers and explanations, see page 57)

1. Which one of the following reactions is *least likely* to occur within the cells of a C_4 plant?
   a. The Calvin–Benson cycle
   b. Photorespiration
   c. $CO_2 + RuBP \rightarrow 2$ 3PG
   d. Photophosphorylation
   e. $CO_2 + PEP \rightarrow$ oxaloacetate

2. Which one of the following categories does *not* differ between a C_3 and a C_4 plant?
   a. Initial CO_2 acceptor
   b. Extent of photorespiration
   c. Enzyme catalyzing reaction that fixes CO_2
   d. Presence of Calvin–Benson cycle
   e. Leaf anatomy

3. Which of the following features is *not* characteristic of some CAM plants?
   a. Stomata open only at night
   b. Trap CO_2 with PEP carboxylase
   c. PEP produced photosynthetically during day
   d. Calvin–Benson cycle during day only
   e. CO_2 captured by rubisco from four-carbon acids

4. Which of the following processes would *not* be inhibited if a plant cell were placed in the dark?
   a. Cellular respiration
   b. Cyclic photophosphorylation
   c. Noncyclic photophosphorylation
   d. Photorespiration
   e. The Calvin–Benson cycle

Integrative Questions (for answers and explanations, see page 57)

1. Which one of the following statements about NAD and NADPH is *not* true?
   a. Both compounds are oxidation–reduction agents.
   b. NAD participates in respiratory reactions; NADPH participates in photosynthetic reactions.
   c. NADPH is restricted to photosynthetic autotrophs; NAD is restricted to heterotrophs.

   d. Both compounds exist in two different molecular forms.
   e. Both compounds function as electron carriers.

2. Which of the following statements about chemiosmotic ATP production in chloroplasts and mitochondria is *not* true?
   a. Both depend on membrane proteins that are ATP synthases.
   b. In both, ATP is generated by passive diffusion of H^+.
   c. Both depend on a proton gradient across a membrane.
   d. Both depend on the impermeability of cell membranes to charged particles.
   e. The electron carriers are the same in both.

Integrative Activity (for answers and explanations, see page 58)

- In the diagram of noncyclic photophosphorylation shown below, label the appropriate parts with the following labels: (1) stroma, (2) thylakoid interior, (3) photosystem I, (4) photosystem II, (5) cytochrome complex, (6) ATP synthase, (7) NADP reductase, (8) location of dark reactions

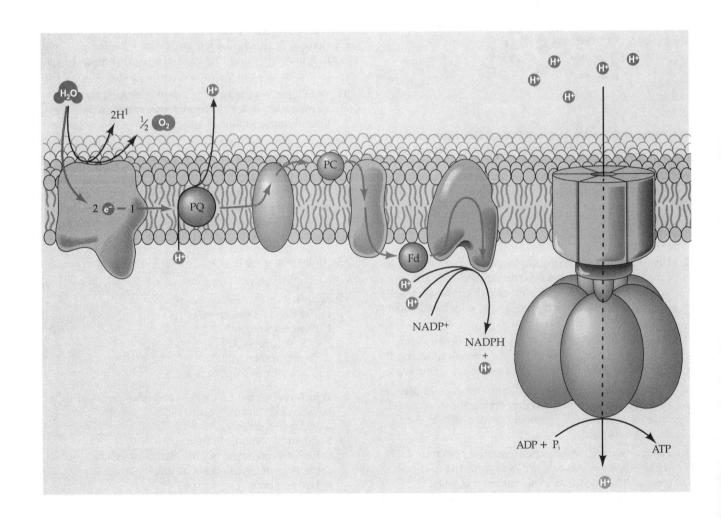

Answers and Explanations

Chapters 1–8

CHAPTER 1 – AN EVOLUTIONARY FRAMEWORK FOR BIOLOGY

Note: The page numbers listed with the section titles refer to the Study Guide; numbers in parentheses for each question refer to the relevant key concepts.

EVOLUTIONARY MILESTONES • BIOLOGICAL DIVERSITY: DOMAINS AND KINGDOMS (pages 1–3)

1. Choice *b*. Only a subset of all organisms have the type of cell division called meiosis. (3, 5–15)
2. (3, 5–15)

   4 Eukaryotes appear
   3 Aerobic respiration
   2 Photosynthesis
   1 Metabolism
   6 Meiosis
   5 Multicellularity

3. Choice *b*. The kingdom Protista includes organisms that are unicellular and have eukaryotic cells. (16–22)
4. Choice *d*. The kingdom Plantae includes organisms that are mostly multicellular, photosynthetic autotrophs. (16–22)
5. Choice *c*. The kingdom Fungi includes organisms that are mostly multicellular, absorptive heterotrophs. (16–22)
6. Choice *c*. The domains are Bacteria and Eukarya. The kingdoms are Eubacteria (cellulase-secreting bacterium), Protista (*Trichonymphia*), and Animalia (termite). (16–22)
7. Eukaryotic cells arose when prokaryotic cells that had been living inside other prokaryotic cells became permanent residents. (5, 10)

THE WORLD INTO WHICH DARWIN LED US • ASKING THE QUESTIONS "HOW?" AND "WHY?" • LIFE'S EMERGENT PROPERTIES (pages 3–4)

1. Choice *c*. Although Darwin recognized that offspring resemble their parents because they inherit characteristics from them, the concept of the gene was not developed until much later. (3–6)
2. Natural selection is differential reproductive success by individuals better adapted for their natural environment. Artificial selection is differential reproduction as directed by humans and not the natural environment. (6–8)
3. (9–16)
   a. Birds are visually oriented predators: *assumption*. This is not tested but is assumed true based on past studies.

   b. Conspicuous caterpillars tend to occur in groups: *observation*. Observations like this led to the development of the hypothesis that conspicuous caterpillars are not palatable.
   c. Bird 1 consumed five conspicuous caterpillars and one cryptic caterpillar: *result*. These are data collected in the study.
   d. This species of conspicuous caterpillar should not be eaten by blue jays: *prediction*. If the data show this to be true, the hypothesis is validated.
   e. Blue jays should eat equal numbers of conspicuous and cryptic caterpillars: *null hypothesis*. The null hypothesis always states that the factor being studied has no effect.

4. Both use an experimental approach in which extraneous factors are kept constant so that a single variable can be tested. It is easier to control extraneous factors in the laboratory than in the field, but the results of field experiments are easier to relate to nature than laboratory experiments. (13, 14)
5. Choice *c*. Absolute certainty is not possible in science. (10–16)
6. (17–26)

   2 A sperm .... **cell**
   4 A mouse .... **organism**
   1 Glucose .... **molecule**
   3 A lung .... **organ**
   7 Earth .... **biosphere**
   6 A woodland.................... **ecological community**
   5 *Peromyscus leucopus* .... **species**

7. Choice *b*. Emergent properties are the new features of a system that result from the interaction of the system's components. The molecules of which membranes are composed, such as proteins and phospholipids, result in the regulation that membranes exert over what enters or leaves cells. (27)

CHAPTER 2 – SMALL MOLECULES: STRUCTURE AND BEHAVIOR

Note: The page numbers listed with the section titles refer to the Study Guide; numbers in parentheses for each question refer to the relevant key concepts.

ATOMS: THE CONSTITUENTS OF MATTER (pages 5–7)

1. Choice *d*. Hydrogen, the most abundant element in the universe, consists of atoms with one proton, one electron, and no neutrons. (1–6)
2. Choice *e*. Isotopes of an element differ in the number of neutrons, so isotopes of an element have different mass numbers. (8, 10, 11, 13)

3. Choice *c*. Choices *a* and *b* are only true for elements in which the number of neutrons and protons is the same. (10, 11, 13, 15)

4. Choice *a*. Recall that the atomic number is equal to the number of electrons in the atom and that orbitals are filled from innermost to outermost, with the K shell holding a maximum of two electrons and the L shell holding a maximum of eight electrons. The electron distribution of sodium is 2 in the K shell, 8 in the L shell, and 1 in the M shell. (10, 18–24)

5. Choice *f*. Neon is inert with a completely filled outer (L) electron shell. Recall that the L shell can accommodate 8 electrons organized into four orbitals with 2 electrons per orbital, so all four orbitals are completely filled. (10, 18–25, 27)

6. Choice *c*. With mass numbers of ^{30}X and ^{32}X and an atomic weight of 30.5, the proportions of the two isotopes must be 75:25; $(30 \times .75) + (32 \times .25) = 30.5$. (11–15)

7. Choice *e*. The inert elements have their outermost shells completely filled and are least reactive. These include helium ($_2$He), neon ($_8$Ne), and argon ($_{18}$Ar). All other choices are elements with only partially filled outer shells. (10–12, 18, 27)

8. Choice *d*. Since chemical reactivity is determined by the number of electrons in the outermost shell, you must determine which elements have the same number of electrons in the outermost shell; chlorine and fluorine each have seven electrons in their outermost shell (10–12, 18, 27).

● (1–12)

Hydrogen $^1_1$H   Carbon $^{12}_6$C

Nitrogen $^{14}_7$N   Oxygen $^{16}_8$O

Sulfur $^{32}_{16}$S   Phosphorus $^{30}_{15}$P

● (1–12)
Deuterium $^2_1$H   Tritium $^3_1$H

Carbon-14 $^{14}_6$C

Oxygen-18 $^{18}_8$O

● (1–12)

	Electrons	Protons	Neutrons
$^1_1$H	1	1	0
$^{14}_6$C	6	6	8
$^{14}_7$N	7	7	7
$^{65}_{30}$Zn	30	30	35

● (2–7)

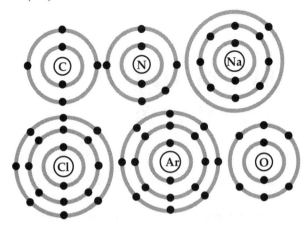

CHEMICAL BONDS: LINKING ATOMS TOGETHER ● EGGS BY THE DOZEN, MOLECULES BY THE MOLE (pages 7–9)

1. Choice *e*. One mole in one liter is a 1.0-molar solution. Adding one mole to 0.5 liter produces a 2.0-molar solution. (20–22)

2. Choice *c*. A 1.0-molar solution contains one gram molecular weight of a substance. The gram molecular weight of water is 18 grams. A liter contains 1,000g/18g/mole = 56 mole. (20–22)

3. Choice *c*. Since carbon is able to form a double bond with each Z atom, Z must need two electrons to complete its outer shell. An atomic number of 8 would result in 2 electrons in the K shell and 6 in the L shell. (3–15)

4. Choice *b*. Because water is polar, it cannot readily mix with or dissolve nonpolar substances. Only molecule *b* is nonpolar. (8, 9, 17)

5. Choice *e*. Van der Waals and hydrophobic bonds will only form between nonpolar molecules like the hydrocarbon shown in *e*. (8, 9, 18)

● (1–7)

• (11)

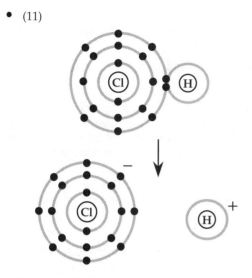

CHEMICAL REACTIONS: ATOMS CHANGE PARTNERS • WATER: STRUCTURE AND PROPERTIES • ACIDS, BASES, AND THE pH SCALE (pages 9–10)

• (22, 23)

pH	[H⁺]	[OH⁻]
0	1.0	10^{-14}
1	0.1	10^{-13}
2	0.01	10^{-12}
3	0.001	10^{-11}
4	10^{-4}	10^{-10}
5	10^{-5}	10^{-9}
6	10^{-6}	10^{-8}
7	10^{-7}	10^{-7}
8	10^{-8}	10^{-6}
9	10^{-9}	10^{-5}
10	10^{-10}	10^{-4}
11	10^{-11}	0.001
12	10^{-12}	0.01
13	10^{-13}	0.1
14	10^{-14}	1.0

1. Choice *a*. One kilocalorie is the heat needed to raise 1.0 kilogram (1,000 g) of water from 14.5°C to 15.5°C, so 0.25 kilocalorie would be needed to increase 250 g by the same amount. (4–6)
2. Choice *d*. Hydrogen bonding occurs between hydrogen and oxygen and the two hydrogens are oriented at an angle of 104.5 degrees relative to oxygen. (9, 10)
3. Choice *e*. The polarity of the water molecule allows it to bind to ions and polar molecules, preventing solute from clumping and coming out of solution. The properties listed in choices *b*, *c*, and *d* also result from the polar nature of the water molecule. (7, 11–16)
4. Choice *b*. Since CH_3COO^- combines with H^+ to form CH_2COOH, it is a base. (18–21, 24, 25)
5. Choice *e*. The pH scale is logarithmic, so each pH unit is ten times greater or lesser than the next nearest unit. Therefore, the [H⁺] of cola is $10 \times 10 \times 10 \times 10$ (10^4) greater than blood plasma! (22, 23)

6. Choice *c*. Recall that "acid" and "base" refer to the compounds or ions that are donating the H^+ and OH^-, while "acidic" and "basic" refer to solutions. (22, 23)

THE PROPERTIES OF MOLECULES (pages 10–11)

1. Choice *b*. The other molecules have hydrogens bound to atoms with greater electron affinity or have a structure in which the distribution of positive and negative poles is not balanced. (1–3)
2. Choice *c*. Optical isomers can occur in both L- and D-forms, although only the L-form of amino acids are found in biological systems. (15–17)
3. Amino: *e* (3–12)
   Phosphate: *f*
   Hydroxyl: *a*
   Carboxyl: *d*
   Carbonyl: *b*
   Sulfhydryl: *c*

• (3–12)

Hydrocarbon	$H_2C{=}CH_2$
Hydrocarbon	$H_3C{-}CH_3$
Amino acid	COOH | $H_2N{-}C{-}H$ | CH_2SH
Organic base	O || $H_2N{-}C{-}NH_2$
Organic acid	O || $H_3C{-}C{-}CH_2{-}COOH$
Alcohol	$H_3C{-}CH_2{-}OH$

• (7–9)

carboxyl
α-carbon
COOH
amino — $H_2N{-}C{-}H$
CH_2OH —— *side chain*

Integrative Questions

1. Choice *b*. Van der Waals and hydrophobic bonds are equal in strength and are the weakest of the bonds. Covalent bonds are strongest, while ionic and hydrogen bonds are about equal and intermediate in strength to the others.
2. Choice *a*. Element X, with an atomic number of 20, would normally have two electrons in its outer shell (2–8–8–2). When in the ionic form X^{2+}, it has lost two electrons and has a full outer shell with 8 electrons.
3. Choice *a*. In a neutral atom, the number of protons and electrons is the same and the atomic number is that number. Ions have gained or lost electrons, so the atomic number equals the number of protons.

CHAPTER 3 – MACROMOLECULES: THEIR CHEMISTRY AND BIOLOGY

Note: The page numbers listed with the section titles refer to the Study Guide; numbers in parentheses for each question refer to the relevant key concepts.

LIPIDS: WATER-INSOLUBLE MOLECULES • MACROMOLECULES: GIANT POLYMERS (pages 12–14)

1. Choice *b*. (7, 10, 12, 13)
2. Choice *b*. Steroids, especially cholesterol, and phospholipids are important lipid components of biological membranes. (7, 10, 12, 13)
3. (2–7)

4. Triglyceride *A* is probably solid at room temperature; its fatty acid chains are saturated (no double bonds) and relatively long, both characteristics of solid, animal-derived triglycerides. (2–7)
5. Triglyceride *B* is probably derived from a plant; its fatty acid chains are unsaturated (double bonds) and relatively short, both characteristics of liquid, plant-derived triglycerides. (2–7)
6. *Three* water molecules will result. A water molecule results for each of the three fatty acids added to glycerol by a condensation reaction. (18–20)
7. Choice *e*. The interlocking, four-ring structure is characteristic of steroids. (7, 10, 12, 13)

• (10, 11)

hydrophobic

$$CH_3-CH_2-(CH_2)_{14}-CH_2-\overset{\overset{O}{\|}}{C}-O-CH_2$$
$$CH_3-CH_2-(CH_2)_6-CH_2-\overset{\overset{O}{\|}}{C}-O-CH$$
$$H_2-C-O-\overset{\overset{O^-}{|}}{P}-O^-$$

hydrophilic

• (2–4)

18 carbons
unsaturated (actually polyunsaturated with two double bonds as indicated above by arrows)

CARBOHYDRATES: SUGARS AND SUGAR POLYMERS (pages 14–15)

1. disaccharide, hexose, α, $C_6H_{12}O_6$, isomers (1–10)
2. Choice *c*. Starch is joined by α linkages and cellulose is joined by β linkages, so choice *b* is false. In addition, starch can also form C-6 to C-1 branches between its glucose monomers. (11, 12)
3. Choice *d*. Derivative carbohydrates contain elements other than carbon, hydrogen, and oxygen; glycogen is a polymer of glucose. (15)
4. Choices *a, c, e*. (1–7)

PROTEINS: AMAZING POLYMERS OF AMINO ACIDS (pages 15–16)

• (11–16)

3 water molecules—one for each peptide bond formed.

1. Choice *e*. Because the sequence of amino acids in a protein determines the arrangement of R groups and bonding potentials between different regions of the protein, the primary structure determines all higher organizational levels. (17–23)
2. Choice *c*. The left R group is nonpolar; the right R group is polar. Nonpolar groups are hydrophobic; polar groups are hydrophilic. (5–10)
3. $m^n = 5^4 = 625$ different polypeptides, where n is the length of the chain and m is the number of different kinds of subunits. (15)
4. Choice *e*. Only these two examples of secondary structure result exclusively from hydrogen bonding. Primary structure depends on peptide bonds and tertiary structure is maintained by a variety of bond types. (11, 12, 17–21)
5. Choice *c*. Although some enzymes consist of several globular polypeptide subunits joined in a quaternary structure, the highest structural level seen in all enzymes is tertiary. (14–22)

NUCLEIC ACIDS: INFORMATIONAL MACROMOLECULES (pages 16–17)

1. Choice *d*. Nucleotides are the subunits of which DNA and RNA are composed, just as proteins are polymers of the subunits called amino acids. (1–10)
2. Choice *e*. Just as R groups give each amino acid its unique chemical features, the nitrogenous bases are unique for each nucleotide. (1–10)
3. Choice *b*. Purines and pyrimidines are found in both DNA and RNA. However, the pyrimidine uracil is restricted to RNA and the pyrimidine thymine is restricted to DNA. (1, 2, 11, 15)

- (2, 3, 10, 11)

Integrative Questions

1. Choice *d*. Lipids are a class of large molecules that are not polymers of simple subunits. Cholesterol is a lipid.

2. Choice *a*. Since the interior of membranes consists of the hydrophobic hydrocarbon tails of phospholipids, the R groups of amino acids located on the surface of protein molecules found within membranes would also be expected to be hydrophobic.

3.
   - *2* RNA nucleotide sequence
   - *3* Primary structure of a protein
   - *4* Folding of a protein into a three-dimensional shape
   - *1* DNA nucleotide sequence
   - *5* An enzyme catalyzes a specific reaction

CHAPTER 4 – THE ORGANIZATION OF THE CELLS
Note: The page numbers listed with the section titles refer to the Study Guide; numbers in parentheses for each ques-

---

THE CELL: THE BASIC UNIT OF LIFE • PROKARYOTIC CELLS • MICROSCOPES: REVEALING THE SUBCELLULAR WORLD (pages 19–20)

1. Choice *b*. The DNA of prokaryotes is located in a nucleoid region, but eukaryotes have true membrane-bound nuclei. (1–4, 9)

2. Choice *c*. Actually, prokaryotes are metabolically more diverse than eukaryotes. (5–7)

3. Choice *c*. Recall that 1 nm = 10^{-9} m or 10^{-6} mm, so 10 nm = 10^{-5} mm. If our unaided resolving power is 0.1 mm or 10^{-1} mm, then the increase in resolving power with the scanning electron microscope is $10^{-5} \div 10^{-1} = 10^{-4}$. (16, 17)

4. (9–15)
   Nucleoid: *d*
   Cytosol: *a*
   Capsule: *f*
   Flagella: *b*
   Pili: *e*
   Mesosome: *c*

- (9–15)

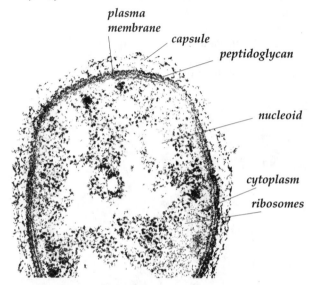

EUKARYOTIC CELLS • ORGANELLES THAT PROCESS INFORMATION • ORGANELLES THAT PROCESS ENERGY (pages 20–22)

1. Choice *c*. The nucleus contains both DNA and RNA. The RNA is in the form of ribosomal RNA and messenger RNA. (5–9)

2. Choice *a*. Since the pores are within the nuclear envelope, when the envelope disperses, the pores disappear. (7–10)

3. mitochondrion (16–20)

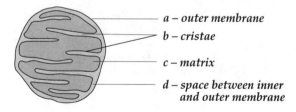

a – outer membrane
b – cristae
c – matrix
d – space between inner and outer membrane

4. Item *a*. The outer membrane of the organelle would have been derived from the plasma membrane of the host. (21, 22)

5. (18, 19)
   Stroma: *c*
   Inner membrane: *a*
   Granum: *d*
   Thylakoid: *b*

THE ENDOMEMBRANE SYSTEM • OTHER ORGANELLES (pages 22–23)

1. (7, 9, 16–18)
    Glyoxysome: *c*
    Peroxisome: *d*
    Food vacuole: *b*
    Free ribosome: *a*
    Golgi apparatus: **e**

2. Choice *e*. Exported proteins are formed by ribosomes attached to ER, are moved to the *cis* region of the Golgi apparatus by vesicles, depart from the *trans* region of the Golgi apparatus in other vesicles, and from there are transported to the outside of the cell. (1–11)

3. Choice *b*. Free ribosomes manufacture proteins for use within the cytosol. (2–6)

4. Choice *d*. Attachment of signal sequence amino acids to newly synthesized proteins prior to their entry into rough ER does not involve vacuoles. (17, 18)

*   (13–15)

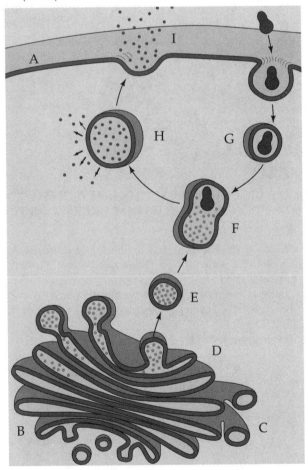

THE CYTOSKELETON • EXTRACELLULAR STRUCTURES • CELLS VERSUS VIRUSES • ISOLATING ORGANELLES FOR STUDY (pages 23–24)

1. (1–6)
    Microtubules: *c, f*
    Microfilaments: *a, d*
    Intermediate filaments: *b, e*

2. Choice *e*. Prokaryotic flagella do not have the 9+2 arrangement of microtubules. (9, 10, 13)

3. Choice *c*. Basal bodies lack central microtubules and have nine sets of three peripheral microtubules. (14–15)

4. Choice *c*. Differential refers to the variation in the rotor speed and, thus, centrifugal force. (23–25)

Integrative Activity

*   Please see page 75 in the textbook to determine the correct labeling of this photomicrograph.

CHAPTER 5 – CELLULAR MEMBRANES
Note: The page numbers listed with the section titles refer to the Study Guide; numbers in parentheses for each question refer to the relevant key concepts.

MEMBRANE COMPOSITION AND STRUCTURE • ANIMAL CELL JUNCTIONS (pages 26–27)

1. (1, 5–7, 17, 18)
    Lipid bilayer: *b*
    Membrane proteins: *a, c, d, e*
    Glycolipids: *a*
    Glycoproteins: *e*

2. Choice *b*. Membrane components, especially proteins, glycolipids, and glycoproteins, are distributed asymmetrically. (1–7)

3. Choice *d*. This type of tertiary structure allows intrinsic proteins to fit into the lipid bilayer, with its hydrophobic interior and hydrophilic exterior. (9, 12–14)

4. Choice *c*. Since most membrane carbohydrates are involved in interacting with the environment, most are attached to the outside of the membrane only. (15–18)

5. (20–24)
    Desmosome: *a, c*
    Tight junction: *a, b, f*
    Gap junction: *a, d, e*

PASSIVE PROCESSES OF MEMBRANE TRANSPORT (pages 27–28)

1. Choice *e*. Usually only small, nonpolar molecules can move through the lipid bilayer of the membrane by simple diffusion. Because a molecule of oxygen has a balanced electron distribution, it is nonpolar. (1–8)

2. Choice *d*. The rate of simple diffusion of a solute always increases directly with solute concentration. (11–13)

3. Curve *c*. In both facilitated diffusion and active transport, the rate of movement will eventually saturate when all of the transport proteins are fully engaged. (11–13)

4. Choices *c, d*. Because the cell absorbs water, we know that its osmotic potential was more negative than the osmotic potential of the environment. (11–18)

5. Choice *a*. Because the cell has absorbed some water and developed an opposing pressure potential, its osmotic potential will still be more negative than that of the environment. (11–18)

ACTIVE PROCESSES OF MEMBRANE TRANSPORT • MEMBRANES ARE NOT SIMPLY BARRIERS • MEMBRANES ARE DYNAMIC (pages 29–30)

1. Choice *b*. A ligand is an extracellular molecule that binds to a membrane receptor protein. A G protein is a membrane protein that can move within the membrane to effect change. (11–13)

2. Choice *d*. The typical flow of membrane within the cell is from nuclear envelope to rough ER, from rough ER to Golgi apparatus, and from Golgi apparatus to plasma membrane. Membrane movement between nuclear envelope, Golgi apparatus, and ER is via vesicles. (17–21)

3. Choice *c*. The sodium–potassium pump (see page 30 in the Study Guide) depends on ATP, so it carries out primary active transport, and since Na^+ and K^+ move in opposite directions, it is an antiport. Since Na^+ is passively diffusing down a concentration gradient established by active transport, and glucose is accompanying it (left), the process is secondary active transport involving a symport. (1–5)

Integrative Questions/Activities

1. Cell membranes are involved in all these activities!
2. Glycolipids and glycoproteins are on the external surface of the membrane.

- Carbohydrates are added to newly synthesized membrane proteins in the rough ER and the Golgi apparatus. (9–13)

CHAPTER 6 – ENERGY, ENZYMES, AND METABOLISM
Note: The page numbers listed with the section titles refer to the Study Guide; numbers in parentheses for each question refer to the relevant key concepts.

ENERGY CONVERSIONS: THE LAWS OF THERMODYNAMICS • CHEMICAL REACTIONS RELEASE OR TAKE UP ENERGY • ATP: TRANSFERRING ENERGY IN CELLS
(pages 31–32)

1. Choice *b*. The reaction rate is independent of whether the reaction is spontaneous or not. (9–12)
2. Choice *b*. The entropy (disorder) present in a system is expressed by *S*. (4–8)
3. Choice *b*. If the ratio of products to substrate is greater than 1, the reaction is exergonic and ΔG is negative. (9–12)
4. Choice *b*. One water molecule is used to break each bond. (13–17)

- (13–16)

ENZYMES: BIOLOGICAL CATALYSTS • MOLECULAR STRUCTURE DETERMINES ENZYME FUNCTION
(pages 32–34)

- (5–7, 11)

● (1, 3, 9)

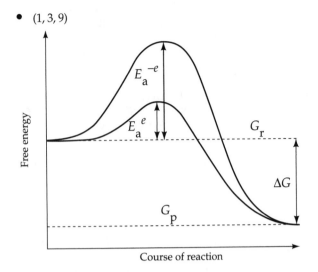

Course of reaction

1. Choice *b*. Substrate binding to an enzyme can involve all types of chemical bonds. (3–11, 15)
2. Choice *c*. Only the activation energy is lower. Also, in *d*, recall that a catalyst can only accelerate a reaction that would occur in the absence of the catalyst. (1–9)
3. Choice *d*. Energy released by the exergonic reaction allows the endergonic reaction to proceed. (12, 13)
4. Choice *b*. *b* is the difference between the free energy of reactants and products or ΔG. (1, 2, 9)
5. Choice *d*. If ΔG is positive, the reaction is endergonic and $K_{eq} < 1$. (1, 2, 9)
6. Choice *d*. Transition-state species are part of both forward and backward, exergonic and endergonic reactions. (1–4)
7. Choice *a*. Cofactors have their main effect by interacting chemically with substrate. (16–19)

METABOLISM AND THE REGULATION OF ENZYMES (pages 34–36)

1. Choice *d*. This is a graph typical of noncompetitive inhibition, with curve B resulting from reactions with the inhibitor. A noncompetitive inhibitor does not normally resemble the substrate. (2–5)
2. Choice *d*. At high inhibitor but low substrate concentrations, most of the allosteric sites are occupied, and consequently, most of the active sites are empty. (6–12)
3. Choice *b*. In an acidic environment, carboxyl groups would not ionize, but amino groups would ionize. (15–16)
4. Choice *b*. That step closes down only the branch leading to G. (13, 14)

● (6–12)

Allosteric site *Substrate*

Inhibitor

Active site

a. The end product of that branch of the pathway.
b. Form *a*
c. No. Concerted feedback inhibition would require an enzyme with at least two different allosteric sites—one for each of the two or more inhibitors.

CHAPTER 7 – CELLULAR PATHWAYS THAT HARVEST CHEMICAL ENERGY
Note: The page numbers listed with the section titles refer to the Study Guide; numbers in parentheses for each question refer to the relevant key concepts.

OBTAINING ENERGY AND ELECTRONS FROM GLUCOSE ● AN OVERVIEW: RELEASING ENERGY FROM GLUCOSE (pages 37–38)

1. Choice *c*. Carbons move from pyruvate into the citric acid cycle and are eventually released as CO_2. (4, 13–19)
2. Choice *b*. Hydrogens are moved from glycolysis intermediates, like glucose 6-phosphate, by NADH to the respiratory chain and eventually incorporated into water. (4, 13–19)
3. Choice *b*. NAD$^+$ is an *oxidizing* agent. See Figure 7.3 on page 144 in the textbook. (5–10)

GLYCOLYSIS: FROM GLUCOSE TO PYRUVATE ● PYRUVATE OXIDATION ● THE CITRIC ACID CYCLE ● THE RESPIRATORY CHAIN: ELECTRONS, PROTON PUMPING, AND ATP (pages 38–40)

1. *6, 7, 9, 10.* (1–5)
2. *6.* (1–5)
3. *1, 3, 6, 10.* In reactions 1 and 3, the ADP formed can be used elsewhere. (1–5)
4. *4.* (1–5)
5. Choice *a*. Coenzyme A is a permanent part of the citric acid cycle. (7–13)
6. Choice *c*. Acetyl coenzyme A is the compound derived from pyruvate that enters the cycle. (7–13)
7. Choice *d*. Both reactions are *exergonic*. (6–13)
8. Choice *a*. NADH + H$^+$ passes hydrogens to NADH-Q reductase and cytochrome oxidase passes electrons to molecular oxygen. (15–18)
9. Choice *b*. All carrier-based movement of protons is from the outside to the inside of the membrane. (15–21)

● (1, 6, 7, 14, 15)

● The pH in area *2* is greater than area *1*, and ATP is formed in area *1*, but not in area *2*. Hydrogen ions are moved from the inside to the outside of the membrane, so the pH is lower on the outside of the membrane; ATP is formed on the inside of the membrane as H$^+$ ions diffuse back. (19–22)

FERMENTATION: ATP FROM GLUCOSE WITHOUT O₂ • CONTRASTING ENERGY YIELDS • CONNECTIONS WITH OTHER PATHWAYS • REGULATING THE ENERGY PATHWAYS (page 40–42)

1. Choice *d*. The remaining four ATP are produced via substrate-level phosphorylation (a net of two during glycolysis and two from the citric acid cycle) and not via the chemiosmotic mechanism. (3–5)

2. Choice *a*. In certain organisms, lactate results from fermentation. (6–9)

3. Choice *c*. NADH + H⁺ produced in glycolysis results in only two ATP, as does FADH₂. (6–9)

4. Choice *b*. All other choices are valid allosteric control mechanisms but do not regulate the rate of glycolysis directly, which is what the Pasteur effect describes. (3–5)

• See Figure 7.19 on page 162 in the textbook. Because of the greater efficiency of aerobic respiration, less glucose needs to be processed, so the rate of glycolysis in the cell should decrease. Allosteric inhibition of the enzymes catalyzing the following reactions would accomplish this: glucose + ATP → glucose 6-phosphate + ADP by glucose 6-phosphate, fructose 6-phosphate + ATP → fructose 1,6-diphosphate + ADP by ATP and citrate. (10–17)

CHAPTER 8 – PHOTOSYNTHESIS: ENERGY FROM THE SUN

Note: The page numbers listed with the section titles refer to the Study Guide; numbers in parentheses for each ques-

IDENTIFYING PHOTOSYNTHETIC REACTANTS AND PRODUCTS • THE TWO PATHWAYS OF PHOTOSYNTHESIS: AN OVERVIEW • PROPERTIES OF LIGHT AND PIGMENTS (pages 43–45)

1. Choice *b*. CO₂ is a reactant in the Calvin–Benson cycle, not a product. All other statements are true. (7–10)

2. Choice *c*. The color of a pigment is indicated by the portion of the absorption spectrum where little light is absorbed. For the absorption spectrum shown, minimum absorbance occurs in the yellow portion of the spectrum. (11, 12, 18)

3. Choice *d*. Frequency and wavelength are inversely related and light of longer wavelength has less energy than light of shorter wavelength. (11–14)

4. *Fluorescence.* If reducing agent A were not available to accept the excited electron from Chl*, Chl* would release the electron's extra energy as a photon. (24, 25)

PHOTOPHOSPHORYLATION AND REDUCTIONS • MAKING SUGAR FROM CO₂: THE CALVIN–BENSON CYCLE (pages 45–47)

1. Choice *e*. Electrons move from water to photosystem II to photosystem I and are carried to G3P in the Calvin–Benson cycle by NADP + H⁺. (9–15, 22–25)

2. Choices *a* and *d*. Only NADP⁺, NADPH + H⁺, and ATP are cycled. Water and oxygen are reactants and products, respectively. (1–8, 22–25)

3. Choice *c*. Since the concentration of A increases when CO₂ is withheld, A must combine with CO₂. C is first to decrease, so A + CO₂ forms C first. B, which starts to decrease last, must be formed at a later step. (21)

4. Choice *e*. Recall that H⁺ ions are pumped into the interior of the chloroplast thylakoids and that ATP formation occurs when H⁺ ions diffuse back out of the thylakoids. In thylakoids of group A chloroplasts, H⁺ ions will be diffusing from inside to outside (the pH is greater outside, so the [H⁺] is less) and ATP *will* form; in group B thylakoids, H⁺ ion diffusion will be from outside to inside and *no* ATP will form. (13–15)

• Substance 1 moved the greatest distance with both solvents; thus, its total R_F is greatest. Substance 2 moved the greatest distance with solvent B, but little distance with solvent A. (21)

• (26)

PHOTORESPIRATION AND ITS EVOLUTIONARY CONSEQUENCES • PLANTS PERFORM CELLULAR RESPIRATION (pages 47–48)

1. Choice *b*. All of the reactions listed occur in a C₄ plant except for the reactions of photorespiration. Because of their unique anatomy and the presence of the enzyme PEP carboxylase, C₄ plants do not photorespire. (4–9)

2. Choice *d*. Both C₃ and C₄ plants have the Calvin–Benson cycle; C₄ plants have an additional CO₂ fixation pathway in their mesophyll cells called the C₄ cycle and the Calvin–Benson cycle occurs only in their bundle sheath cells. (1–9)

3. Choice *c*. In some plants in the family Crassulaceae, PEP is produced from CO₂ only at night using PEP carboxylase. The CO₂ is then passed on to the Calvin–Benson cycle during the day. (10)

4. Choice *a*. Cellular respiration continues in the dark and is only limited by glucose and oxygen. (12, 13)

Integrative Questions

1. Choice *c*. Since photosynthetic autotrophs have mitochondria and respire just like heterotrophs, plants use NAD as well as NADP.

2. Choice *e*. The electron carriers in the respiratory chain differ from those in photophosphorylation. All of the other statements are true.

Integrative Activity

- (See the figure below. Also, for more information see
  Figure 8.14 on page 177 in your text book)

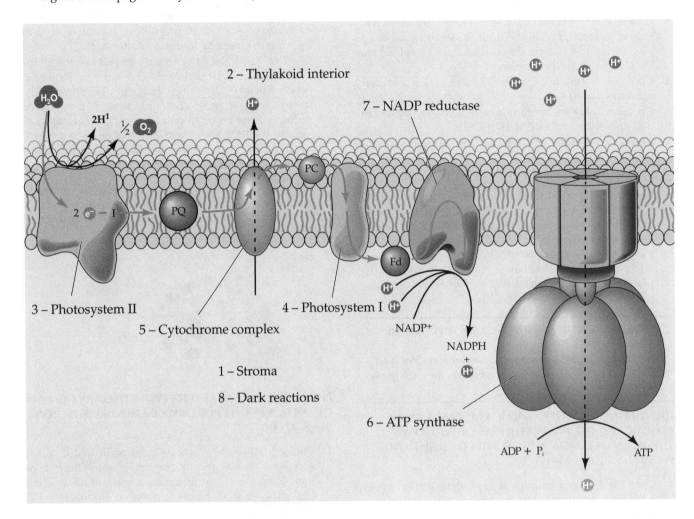

9

Chromosomes, the Cell Cycle, and Cell Division

CHAPTER LEARNING OBJECTIVES—after studying this chapter you should be able to:

❑ Identify the three basic events common to all types of cell division.

❑ Describe the fundamental structural and functional differences between prokaryotic and eukaryotic chromosomes.

❑ Compare and contrast prokaryotic fission with mitosis and meiosis in eukaryotes.

❑ State the fundamental genetic differences between mitosis and meiosis.

❑ Discuss why mitosis is useful for overall growth and repair while meiosis is necessary for sexual reproduction.

❑ Describe the component parts of a eukaryotic chromosome and indicate when they are present during the various stages of mitosis and meiosis.

❑ Diagram the generalized life cycle of a cell and discuss the events that occur in the S, G1, G2, and M phases.

❑ Explain the difference between interphase and cell division, indicate the relative length of time cells spend in these phases, and describe the molecular mechanisms that determine the transition from one phase to the next.

❑ Clearly differentiate between the terms chromatin, chromosome, chromatid, and centromere.

❑ Discuss the role of histone proteins in determining the shape of nucleosomes and chromosomes.

❑ Name the stages of mitosis and describe the condition of chromosomes and other important cellular structures during each stage.

❑ Identify the roles of mitotic spindles, centrosomes, asters, kinetochores, and microtubules in cell division.

❑ Explain the purpose of cytokinesis and describe how it differs between plant and animal cells.

❑ Differentiate between asexual and sexual reproduction, and discuss the relative genetic and evolutionary advantages or disadvantages of each.

❑ Associate the concepts of haploid and diploid with the terms gamete, zygote, egg, sperm, and spore. Be able to diagram how these terms and the concept of fertilization relate to one another in the life cyle of a sexually-reproducing organism.

❑ Explain what is meant by a karyotype and how karyotypes are determined.

❑ Describe the importance of the reduction division in meiosis.

❑ Identify the stages of meiosis and diagram the condition of chromosomes and other important cellular structures during each stage.

❑ Compare and contrast meiosis with mitosis, placing special emphasis on why meiosis has two sets of divisions while mitosis has only one.

❑ Discuss the genetic consequences of synapsis and crossing over in meiosis I.

❑ Explain what is meant by a cell's ploidy level, and describe how ploidy levels change during mitosis and meiosis.

❑ Explain how nondisjunction events that occur during meiosis can lead to the condition known as aneuploidy.

❑ Relate the terms trisomy and monosomy to chromosomal abnormalities, identify mechanisms for how these conditions can occur, and cite a specific example of trisomy in humans.

❑ Compare and contrast necrosis with apoptosis, and discuss the circumstances under which these mechanisms of cell death play a role in the lives of organisms.

SYSTEMS OF CELL DIVISION • INTERPHASE AND THE CONTROL OF CELL DIVISION • EUKARYOTIC CHROMOSOMES (pages 193–200)

Key Concepts

1. Nucleic acids contain the hereditary information that controls cell activity. When cells divide, they undergo three basic events: Their nucleic acids must be properly copied, the copies must be correctly allocated to new daughter cells, and the daughter cells must be successfully separated.

2. DNA is the primary nucleic acid that contains the genetic information of cells. It is a large, double-stranded molecule that combines with certain proteins to form structures called *chromosomes*. In eukaryotic cells, chromosomes reside in a membrane-bound organelle called the nucleus. In prokaryotes, the chromosome is found in a nonmembranous area called the nucleoid region.

3. Prokaryotic cells, such as the human intestinal bacterium *Escherichia coli*, have a single circular chromosome attached to the cell membrane during division.

4. Bacteria divide by *fission*, in which the circular chromosome is first copied and then attached to an adjacent part of the cell membrane. The new chromosome separates from its parent via cell elongation, at the end of which a new cell wall forms to complete the division process. Under ideal conditions the entire process takes only 30 minutes.

5. *Mitosis* is a type of eukaryotic cell division that produces genetically identical copies of the parent nucleus in two daughter cells. The overall growth and repair of organisms is achieved through mitosis.

6. *Meiosis* is a second type of eukaryotic cell division in which the amount of DNA is precisely reduced by one-half in the nuclei of four daughter cells. Meiosis promotes genetic variability between parents and offspring in sexually reproducing species.

7. The *cell cycle* consists of cell division (mitosis or *M phase*) and *interphase*, the period between successive cell divisions. Interphase typically comprises the vast majority of the cycle. Some cells never leave interphase, but most have at least some probability of dividing.

8. Interphase is divided into three subphases. *G1* is a gap period that follows immediately after mitosis; cells that will not divide again linger in this phase. *S* is a phase characterized by the synthesis of new DNA and proteins in preparation for nuclear division. *G2* represents a second gap period prior to the next mitosis.

9. The transitions from G1 to S and from G2 to M are regulated by the activity of an enzyme called *cyclin-dependent kinase* (Cdk). When activated, this enzyme phosphorylates amino acids in proteins that are essential for progress between the G, S, and M phases in the cell cycle. Cdk's are activated only when they associate with other proteins, called *cyclins*, to form a cyclin–Cdk complex.

10. Cyclin–Cdk complexes serve as inducers of mitotic activity by acting as *internal* controls for progress through the cell cycle. *External* controls can also affect the cell cycle. For example, growth factors derived from specialized cells such as platelets and white blood cells are activated when the body is damaged. These factors then promote cell division in relevant tissues to correct the damage.

11. DNA molecules replicate prior to cell division. After replication, each chromosome consists of two *chromatids* joined together by a *centromere*, which functions in chromosome movement.

12. Chromosomes contain five classes of *histone* proteins, which, combined with DNA, make up *chromatin*. A portion of an individual DNA molecule coils around a complex of four histone types. The histone–DNA complex is then locked together by the fifth histone class to form a spool, or *nucleosome*. Many repeating nucleosomes connected by a DNA thread yield a beaded chromatin chain.

13. During cell division, chromatin undergoes changes in form related to its function. While in interphase, chromatin exists as a long string spread throughout the nucleus. During mitosis and meiosis, the beaded chromatin becomes increasingly coiled and condensed into individual chromosomes. This condensed form facilitates chromosome movement.

Questions (for answers and explanations, see page 122)

1. If a cell has 44 chromosomes at the end of mitosis, how many will it have at the end of the next S phase?
   a. 4
   b. 11
   c. 22
   d. 44
   e. 88

2. Suppose you examine under the microscope a thin section of tissue removed from the actively growing portion of a plant shoot. Which of the following will be *true* of this section?
   a. Most of the cells will be in one of the three stages of interphase.
   b. All the cells will be undergoing a reduction in the amount of DNA present.
   c. The chromatin in most of the cells will already have condensed into clearly recognizable chromosomes.
   d. No histones or DNA will be present in the interphase cells.
   e. A large number of cells will have stopped dividing and will have become fixed in G2 of the cell cycle.

3. Most new histone molecules are manufactured during the _____ phase(s) of the cell cycle.
   a. G1
   b. S
   c. G2
   d. mitosis
   e. G1 and G2

4. Which of the following statements about chromosomes is *false*?
   a. Chromosomes observed just prior to nuclear division in mitosis will have two chromatids each.
   b. Chromosomes contain DNA and several types of histone proteins.
   c. The shape of chromosomes and chromatin stays remarkably constant during the cell cycle.
   d. New cells produced by meiosis have fewer chromosomes than the parent cell.
   e. The centromere connects chromatids within a chromosome.

5. During which of the following phases of the cell cycle is chromatin present in cells?
   a. G1
   b. S
   c. G2
   d. Mitosis
   e. All of the above

6. Which of the following statements about the structure or function of cyclin-dependent kinase (Cdk) in cells is *false*?
   a. Cdk is a type of protein with enzymatic properties.

b. Cdk's by themselves are able to phosphorylate proteins necessary for progress through the cell cycle.

c. Cyclin–Cdk complexes serve as internal controls of mitotic activity.

d. Cdk's become active only when associated with cyclin proteins.

e. Chromosome condensation is one cell division process initiated by a specific cyclin–Cdk complex.

7. Which of the following statements about platelet-derived growth factor and interleukins is *true*?
   a. They both represent external controls on cell division.
   b. They are both types of growth factors.
   c. They each stimulate appropriate cells to undergo mitosis.
   d. They both alter the cell cycle activity of specific body cells.
   e. All of the above

MITOSIS: DISTRIBUTING EXACT COPIES OF GENETIC INFORMATION • CYTOKINESIS: THE DIVISION OF THE CYTOPLASM (pages 200–203)

Key Concepts

1. Mitotic division yields two new nuclei that are genetically identical to the parent nucleus. Although mitosis is a continuous process, it can be divided into several phases.

2. *Interphase* is marked by the condensation of chromatin and the appearance of a pair of *centrosomes* near the nuclear envelope. In many organisms, each centrosome also contains a pair of *centrioles* oriented at right angles to one another. One centriole (the daughter) is smaller than the other (the parent).

3. During *prophase,* centrosomes move to opposite ends of the cell where they act as *mitotic centers,* organizing the microtubules necessary for chromosome movement.

4. Details of microtubule arrangement differ with cell type. All cells have *polar microtubules* running from one mitotic center to the midline of the cell; these form the two halves of the *spindle.* Some cells also have microtubules that radiate away from the spindle in patterns called *asters.*

5. Chromatin structure begins to change during prophase. Individual chromosomes condense, revealing that each chromosome consists of paired chromatids that are identical in genetic composition due to replication in the preceding S phase.

6. Late in prophase, *kinetochores* develop near each centromere. They produce *kinetochore microtubules,* which interact with polar microtubules in the spindle.

7. *Prometaphase* occurs when chromosomes actually begin moving. Kinetochore microtubules make contact with polar microtubules, and specialized motor protein enzymes attach at these contact points to provide energy for chromosome movement. Eventually the chromosomes align along the cell midplane, or *equatorial (metaphase) plate.* Disintegration of the nuclear envelope also occurs at this stage due to the action of a cyclin–Cdk complex.

8. *Metaphase* is a brief phase that commences when all kinetochores are aligned along the equatorial plate. Chromatin condensation is completed at this point, and the centromeres separate.

9. *Anaphase* begins with the separation of chromatids. The action of motor proteins and polar microtubules moves the sister chromatids, now called *daughter chromosomes,* toward opposite ends of the spindle. The new daughter chromosomes each contain one centromere and one double-stranded molecule of DNA.

10. Motor proteins cause the spindle poles to move apart late in anaphase, further separating the two sets of daughter chromosomes.

11. *Telophase* begins when daughter chromosomes stop moving as they reach the spindle poles. The chromatin uncoils to assume the dispersed state characteristic of interphase, and the nuclear envelope and nucleoli reform. This marks the end of mitosis.

12. *Cytokinesis* "packages" the new, genetically identical nuclei into unique cells by dividing the cell's cytoplasm. In animals, actin and myosin microfilaments contract, causing the center portion of the cell membrane to furrow and divide the cytoplasm.

13. The cell wall of plants prevents direct constriction of the plasma membrane. Instead, membrane vesicles group along the equatorial plate and, with the help of microtubules, form a cell plate that provides the foundation for a new cell wall.

14. Although genetically identical, the new cells may not share cytoplasmic structures equally. However, even if cytoplasm is divided unequally, a sufficient quantity of every organelle is passed on to each new cell to ensure proper metabolic function.

Questions (for answers and explanations, see page 122)

1. Polar microtubules are an important part of the spindle apparatus used in cell division. During which of the following phases of mitosis would you *not* expect to find well-formed polar microtubules?
   a. Interphase
   b. Prophase
   c. Prometaphase
   d. Metaphase
   e. Anaphase

2. During which of the following phases of mitosis do chromosomes contain distinct chromatids?
   a. G2 and prophase
   b. Prophase and prometaphase
   c. Prophase, prometaphase, and metaphase
   d. Metaphase and anaphase
   e. Anaphase and telophase

3. Daughter chromosomes first appear during _____ of mitosis.
   a. interphase
   b. prophase
   c. metaphase
   d. anaphase
   e. telophase

4. Match the letter of the appropriate phase(s) of mitosis with all relevant characteristics of cell division.

a. Interphase   _____ Chromosomes align on equatorial plate

b. Prophase   _____ Chromatin redisperses after cell division

c. Metaphase   _____ Centrosomes migrate to opposite ends of cell

d. Anaphase   _____ Double-stranded DNA present in cell

e. Telophase   _____ Centromeres present on chromosomes

                  _____ Cytokinesis occurs

5. Which one of the following statements about mitosis is *false*?

a. Cytokinesis in plants involves the production of a cell plate.

b. Chromatids separate during anaphase.

c. Anaphase represents the first time when chromosomes move during cell division.

d. The completion of telophase signals the end of mitosis.

e. Kinetochore microtubules interact with polar microtubules of the spindle.

6. In cells that have them, asters are usually present during _____ of mitosis.

a. prophase

b. prometaphase

c. metaphase

d. anaphase

e. all of the above

7. Which of the following would you expect to occur at the end of telophase for both a plant and an animal cell?

a. Separation of chromatids

b. Chromosome movement toward spindle poles

c. Cytokinesis

d. Movement of spindle poles

e. Polar microtubule formation

8. Which of the following would you expect to be absent from the newly formed daughter cells of an animal cell that has just completed cytokinesis?

a. Centrioles

b. Nuclei

c. Mitochondria

d. Cytoplasm

e. None of the above

Activity (for answers and explanations, see page 122)

- Consider the ovals shown next to represent an animal cell from a species with four chromosomes. Diagram what this cell and its chromosomes would look like during both metaphase and anaphase of mitosis.

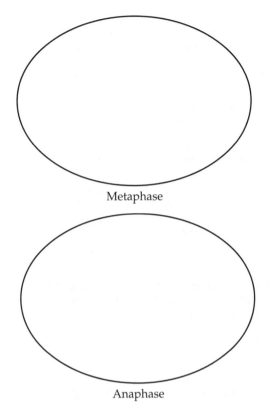

Metaphase

Anaphase

MITOSIS: ASEXUAL REPRODUCTION AND GENETIC CONSTANCY • MEIOSIS: SEXUAL REPRODUCTION AND DIVERSITY • MEIOSIS: A PAIR OF NUCLEAR DIVISIONS • MEIOTIC ERRORS: THE SOURCE OF CHROMOSOMAL DISORDERS • CELL DEATH   (pages 203–213)

Key Concepts

1. The formation of new individuals by mitotic cell division is called *asexual reproduction*. It is an easy, rapid, and effective way to reproduce.

2. A drawback to asexual reproduction is the lack of genetic diversity among the resulting offspring (i.e., all progeny are genetically identical, or *clones*). Genetic uniformity can be a disadvantage in changing environments.

3. *Sexual reproduction* promotes genetic diversity by recombining hereditary material from two genetically different individuals. Diversity promotes evolutionary opportunity through novel genetic combinations.

4. To reproduce sexually, specialized cells called *gametes* must be produced. Gametes are *haploid*, containing only one set of chromosomes, whereas body cells are typically *diploid*, containing two sets of chromosomes.

5. The number of chromosomes in each set is denoted by n; haploid cells thus have n chromosomes, diploid cells have $2n$ chromosomes. The haploid number of chromosomes for humans is 23 (i.e., 23 different types of chromosomes), and the diploid number 46 (i.e., two copies of each chromosome type).

6. Haploid female gametes (*eggs*) unite with haploid male gametes (*sperm*) during *fertilization* to form a diploid *zygote*. The zygote then undergoes repeated mitotic divisions to become a multicellular organism.

7. *Meiosis* is the cell division process that produces haploid gametes (animals) or *spores* (plants and some fungi). It involves the *reduction division* of a single diploid nucleus into four haploid nuclei.

8. Organisms alternate between haploid and diploid phases of their *life cycle* by successive meiosis and fertilization events. Many organisms have only a multicellular diploid phase in their life cycle, others have only a multicellular haploid phase. Some species, particularly plants, have both.

9. During metaphase of mitosis, it is possible to count the numbers and types of chromosomes present in a cell. Such *karyotypes* show that diploid cells have *homologous pairs* of chromosomes, one member of the pair coming from each parent.

10. Only a single member of each chromosome pair is present in a haploid cell. Fertilization restores the homologous pairs by combining two haploid sets of chromosomes in the diploid zygote. We can thus refer to cells by their *ploidy level*, which is usually either n or $2n$. However, in some cases *triploid (3n)*, *tetraploid (4n)*, or higher-order *polyploid* nuclei are formed.

11. The names of the phases of meiosis are the same as in mitosis; in meiosis, however, each phase occurs twice (i.e., there are two nuclear divisions in meiosis). There are also distinctly different chromosomal arrangements that occur during the phases of meiosis.

12. *Meiosis I* reduces the cell's ploidy level by one-half through separation of homologous chromosome pairs. *Meiosis II* separates sister chromatids in these haploid nuclei, ensuring that each haploid product has a full complement of chromosomes. In so doing, the overall process of meiosis promotes genetic diversity.

13. Homologous pairs of chromosomes are able to separate correctly because of an important pairing process known as *synapsis*, which lasts from prophase through the end of metaphase in meiosis I.

14. During *prophase I*, synapsed pairs of chromosomes form crosses among the arms of their chromatids. These chiasmata represent areas where genetic material is exchanged between chromatids. Since meiosis takes much longer to complete than does mitosis, synapsis and crossing-over of chromosomal material during meiosis I represent important ways in which genetic recombination can take place.

15. *Prometaphase I* occurs with the breakdown of the nuclear envelope, spindle formation, and kinetochore attachment to the polar microtubules. Unlike mitosis, only a single kinetochore is present in each chromosome.

16. During *metaphase I* the synapsed homologous pairs of chromosomes move to the equatorial plane. They separate (still with paired chromatids attached by the centromere) during *anaphase I*. Note how this process differs from chromosome alignment and separation during metaphase and anaphase of mitosis.

17. *Telophase I* follows anaphase I in some species, but not others, with features similar to those observed during telophase in mitosis.

18. *Interkinesis*, which is somewhat like interphase of mitosis, follows telophase I. However, no DNA synthesis occurs because the previously replicated chromosomes are still composed of two chromatids each.

19. Meiosis II resembles mitosis in that chromatids separate at their centromeres during metaphase to produce daughter chromosomes and a total of four haploid nuclei. Meiosis II differs from mitosis in that there is no additional DNA replication, sister chromatids may not be identical due to crossing over, and there is only the haploid number of chromosomes present.

20. Genetic diversity among the products of meiosis occurs for two important reasons. First, crossing over during synapsis in prophase I recombines genetic material. Second, chance alone determines homolog alignment in metaphase I, and thus in which cell a particular member of a chromosome pair will end up during anaphase I.

21. Division errors occasionally take place during meiosis. *Nondisjunction* occurs when a chromosome pair moves to the same pole during anaphase I or sister chromatids fail to separate during anaphase II. The resultant cells are aneuploid, having an abnormally low or high number of chromosomes.

22. *Trisomy* (three copies of the same chromosome in a cell) is one consequence of nondisjunction. In humans, trisomy of chromosome 21 leads to the developmental abnormality known as Down syndrome. Other aneuploid disorders in humans include Turner and Klinefelter syndromes (sex chromosome aneuploidy), and cri-du-chat (deletion of a portion of chromosome 5).

23. *Translocation* is another division error that produces aneuploidy. Large portions of one chromosome are moved to another, leading to the functional equivalent of a nondisjunction trisomy. Translocation of chromosome 21 in humans can also lead to Down syndrome.

24. Trisomies (or *monosomies*, where only one copy of a chromosome is present) are fairly common in humans. However, most are lethal to the fetus during early development and usually result in natural termination of the pregnancy before birth.

25. The ploidy of a cell places limits on whether meiosis can even occur. Because chromosomes must undergo synapsis, only cells with an even ploidy ($2n$, $4n$, etc.) can produce viable gametes or spores, which has important implications for the formation of new species through polyploidy. Mitosis is not similarly constrained by ploidy level.

26. One role of cell division is to replace existing cells that die. Cell death can occur by necrosis due to damage from poisoning or starvation. *Apoptosis*, or genetically-programmed cell death, is a natural mechanism that rids the body of cells that are no longer needed in the developmental process or that are too old to function properly.

Questions (for answers and explanations, see page 122)

1. Which of the following statements is *false?*
   a. Diploid cells contain two sets of chromosomes.
   b. Meiosis I represents a reduction division.
   c. Karyotypes indicate the numbers and kinds of chromosomes in a cell.
   d. Gametes are necessary for successful reproduction.
   e. Crossing over during meiosis is one factor contributing to increased genetic diversity.

2. A diploid plant cell has 36 chromosomes at the start of meiosis. How many *chromatids* does this cell have?
   a. 9
   b. 18
   c. 36
   d. 54
   e. 72

3. A diploid animal cell has 40 chromosomes at the start of meiosis. How many *homologous pairs of chromosomes* does this cell have?
   a. 10
   b. 20
   c. 40
   d. 60
   e. 80

4. A plant cell with a diploid number of 18 undergoes meiosis. How many *centromeres* does each of the four daughter cells have at the end of meiosis II?
   a. 9
   b. 18
   c. 27
   d. 36
   e. Cannot determine from information given

5. An important feature of metaphase in meiosis I is that
   a. chromatids separate to produce the final products of meiosis.
   b. kinetochores do not make attachments with the spindle.
   c. cytoplasmic division begins in preparation for chromosome separation.
   d. synapsed chromosome pairs line up along the equatorial plate.
   e. homologous pairs of chromosomes move to opposite poles of the spindle.

6. Synapsed chromosomes are present during which of the following phases of meiosis?
   a. Prophase I only
   b. Prophase I and II
   c. Prophase I through metaphase I
   d. Prophase II through metaphase II
   e. Metaphase II only

7. _____ can be the result of a nondisjunction event that occurs during meiosis.
   a. Haploidy
   b. Karyotypy
   c. Diploidy
   d. Interkinesis
   e. Aneuploidy

8. Which of the following is usually thought to result from an isolated error in cell division?
   a. Fertilization
   b. Translocation
   c. Necrosis
   d. Interkinesis
   e. Apoptosis

Activities (for answers and explanations, see page 123)

• Consider the oval below to represent an animal cell from a species whose diploid number equals four. Diagram how this cell and its chromosomes would appear during *metaphase I* of meiosis.

• Now diagram one of the immediate progeny of this same cell and show its chromosomes as they would appear during *metaphase II* of meiosis.

- Use the diagram below and the following labels to complete the generalized life cycle of an organism showing both multicellular haploid and diploid stages as well as sexual reproduction: *multicellular diploid organism, multicellular haploid organism, meiosis, fertilization, zygote, gametes, haploid cells, germ cells.*

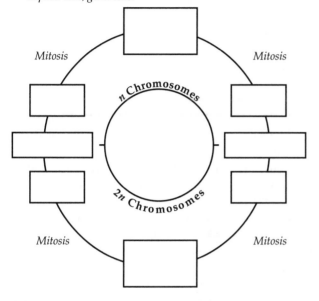

Integrative Questions (for answers and explanations, see page 123)

1. Anaphase of meiosis II is similar to anaphase of mitosis in that both involve
   a. reduction divisions.
   b. the production of aneuploid cells.
   c. separation of chromatids.
   d. condensation of chromatin.
   e. diploid cells.

2. You are given a microscope slide on which is mounted a cell from an organism whose diploid number is 18. This cell has 19 chromosomes, all of which appear to lack obvious chromatids. Which of the following is most likely a *true* statement about this cell?
   a. The cell is haploid and would probably have gone on to become a sperm cell.
   b. It had completed telophase of mitosis, and had probably suffered a nondisjunction event somewhere in its genetic past.
   c. If allowed to continue dividing, this cell would have undergone meiosis II.
   d. The cell had just completed interphase and was about to enter mitosis.
   e. It had just completed telophase of meiosis I, and was about to enter interkinesis.

3. All but one of the following would be present in animal cells isolated from both prophase of mitosis and metaphase of meiosis II. Select the *exception.*
   a. Kinetochore microtubules
   b. Polar microtubules
   c. Chromatids
   d. Diploid number of chromosomes
   e. Centromeres

10
Transmission Genetics: Mendel and Beyond

MENDEL'S WORK • MENDEL'S EXPERIMENTS AND LAWS OF INHERITANCE (pages 217–226)

Key Concepts

1. Gregor Mendel was the first to carefully quantify inheritance patterns in organisms (research published in 1866). However, his work on the genetics of garden peas went unnoticed for nearly 40 years.

2. Prior to Mendel, the botanist Kölreuter showed that cross-pollination yielded *hybrids*, which instead of being intermediate in form to their parents, often exhibited traits much like one or the other parent. *Reciprocal crosses*, where both possible mating combinations of parental sexes and traits are arranged, always gave identical results.

3. *Blending inheritance* was the model used to describe genetic transmission prior to Mendel's work. Like colors of paint, genetic traits were thought to mix irreversibly when combined through cross-breeding.

4. Mendel's approach identified several easily measured *heritable* (passed from parent to offspring) character traits in *true-breeding* pea plants (those whose progeny always resemble the parents). Suitable *characters* (e.g., flower color) were distinct and possessed contrasting *traits* (e.g., white and purple). Such careful attention to empirical quantification was uncommon in the science of Mendel's day.

5. The original male and female plants in Mendel's crosses are referred to as the parental generation *(P)*. Their immediate progeny belong to the first filial, or F_1, generation. The offspring of these individuals belong to the F_2 generation.

6. In Mendel's *monohybrid cross,* the parents differ for a single character only. The form of the trait for this character that shows up in the F_1 progeny is said to be *dominant* over the nonpresent, *recessive* form. When F_1 individuals from a monohybrid cross are allowed to interbreed, the "lost" recessive trait reappears in some of the F_2 individuals. The ratio of dominants to recessives is always approximately 3:1 in the F_2 generation.

7. Mendel's findings disproved the blending hypothesis of inheritance, suggesting instead that hereditary information travels in discrete units. This *particulate theory of inheritance* has since been verified by our understanding of *genes* and other important cellular structures in heredity.

8. The *genotype* of an individual represents its actual genetic makeup. Different forms of genes *(alleles)* are found at the gene's physical location, or *locus,* on the chromosome. Alleles combine in pairs to form the diploid genotype. The *phenotype* is the physical expression of an individual's genotype (e.g., purple flowers, wrinkled seeds, etc.).

9. The dominant allele for a gene is usually abbreviated with an uppercase letter, the recessive allele with a lowercase. *Homozygous* individuals have genotypes with two copies of the same allele, either dominant or recessive. *Heterozygous* individuals possess two different alleles for a particular gene.

10. For example, if the allele for spherical seeds in pea plants *(S)* is dominant to the allele for wrinkled seeds *(s),* both the homozygous dominant genotype *SS* and heterozygous *Ss* will produce plants with the same phenotype (spherical seeds), while the homozygous recessive genotype *ss* will yield a plant with wrinkled seeds.

11. When gametes are formed, only one allele for each trait in the parent's genotype is included in any given egg or sperm. Mendel's *law of segregation* recognizes that alleles separate during gamete formation, restoring the diploid genotype in the zygote at the time of fertilization.

12. The F_1 individuals from a monohybrid cross between true-breeding dominant and recessive parents are always heterozygotes. When the F_1 individuals form gametes, the dominant and recessive alleles segregate into separate gametes, which form with equal probability.

13. This known proportion of gamete types makes it possible to calculate the predicted outcomes of an F_1 cross. A Punnett square, in which gamete probabilities are plotted in tabular form to facilitate the determination of F_2 genotypes, is often used in this process.

14. The Punnett square method verifies the 3:1 ratio of dominant to recessive phenotypes found in the F_2 generation. This phenotypic ratio results because two of the three individuals with the dominant phenotype are heterozygous for the gene.

15. Individuals expressing a dominant phenotype can be either homozygous dominant or heterozygous. Which genotype they actually possess is determined using a *test cross*: the unknown dominant phenotype is crossed with a homozygous recessive individual. If all offspring have dominant phenotypes, then the test individual must be homozygous. If approximately half the offspring have recessive phenotypes, then the test individual is heterozygous.

16. *Dihybrid crosses* involve cases where individuals are mated with respect to two different characters (e.g., flower color and seed shape). Inheritance patterns from such crosses support Mendel's *law of independent assortment*, which states that alleles for different characters assort independently of one another during gamete formation.

17. As in a monohybrid cross, the F_1 progeny of a dihybrid cross between true-breeding parents will all have the same genotype and phenotype. In the F_2 generation, however, there are now nine unique genotypes and a phenotypic ratio of 9:3:3:1. Because they are genetically independent events, we can mathematically treat dihybrid crosses as the probabilistic interaction of two separate monohybrid crosses.

18. Some F_2 progeny of a dihybrid cross show phenotypes unlike either parent. These *recombinant phenotypes* result from the formation of novel genotypes due to the independent assortment of chromosomes during gamete formation.

19. Independent assortment of genes is not a universal hereditary law; it applies only to genes residing on different chromosomes, as was the case in all of Mendel's studies.

Activities (for answers and explanations, see page 123)

• Indicate how many different kinds of gametes can be made by individuals possessing the following diploid genotypes:

    AABBCCDDEE

    AaBbCcDdEe

    AABbCCDDEE

• Explain the fundamental differences between Mendel's laws of segregation and independent assortment.

• Complete a Punnett square for a dihybrid cross involving two individuals that are heterozygous for both gene loci.

Questions (for answers and explanations, see page 123)

1. An individual has the genotype *PpQQrr*. Which of the following is *true* about this individual?
   a. It is the result of a cross between two *PPqqRR* individuals.
   b. The *P* gene is dominant to both the *Q* and *R* genes.
   c. The individual is homozygous recessive for one character.
   d. It must be a fungus because the genotype is haploid.
   e. Gametes of the genotype *PQr* are not possible from this individual.

2. If all the F_2 progeny of a cross are homozygous recessive for the character in question, what does this tell you about the two individuals mated together in the original parental generation?
   a. They were also homozygous recessive for this character.
   b. One was homozygous recessive, the other homozygous dominant.
   c. Both were heterozygotes.
   d. Both were homozygous dominant.
   e. Neither was true-breeding.

3. Mendel's particulate theory of inheritance is supported by modern evidence about the genetic role of
   a. chromosomes.
   b. genes.
   c. alleles.
   d. gametes.
   e. all of the above

4. If allele *K* is dominant to *k*, then the offspring of a cross between *KK* and *kk* individuals will
   a. all be homozygous.
   b. display the same phenotype as the *KK* parent.
   c. display the same phenotype as the *kk* parent.
   d. have the same genotype as the *KK* parent.
   e. have the same genotype as the *kk* parent.

5. A cross between true-breeding short-haired and long-haired rats always produces short-haired individuals. If these F_1 animals are test crossed, half of the progeny are short-haired and the other half long-haired. This suggests that
   a. long hair is the dominant trait and that the F_1 progeny are all heterozygotes.
   b. long hair is the recessive trait and that the F_1 progeny are all homozygotes.
   c. short hair is the recessive trait and that the F_1 progeny are all homozygotes.
   d. short hair is the recessive trait and that the F_1 progeny are all heterozygotes.
   e. short hair is the dominant trait and that the F_1 progeny are all heterozygotes.

6. In cocker spaniels, black coat color *(B)* is dominant over red *(b)*, and solid color *(S)* is dominant over spotted *(s)*. If a red, spotted male is crossed with a black, solid female, and all the offspring from several litters express only the dominant traits, what is the most likely genotype for the female parent?

a. BBSS

b. BbSS

c. BbSs

d. BBSs

e. None of the above

7. In a species of ornamental fruit tree, curly leaves *(C)* are dominant to flat leaves *(c)*, and solid leaf color *(L)* is dominant to striped *(l)*. A cross between a tree that is heterozygous for curly, solid leaves and one that has flat, striped leaves will produce an F_1 generation with what percentage of individuals that are homozygous recessive for both characters?

a. 0

b. 25

c. 50

d. 75

e. 100

8. An individual with genotype *NnOO* is crossed with one having the genotype *nnOo*. How many different genotypes will be represented in the progeny of these parents?

a. One

b. Two

c. Three

d. Four

e. Cannot determine from information given

ALLELES AND THEIR INTERACTIONS • GENE INTERACTIONS • GENES AND CHROMOSOMES • CYTOPLASMIC INHERITANCE (pages 226–239)
Key Concepts

1. For certain genes, heterozygous individuals exhibit phenotypes intermediate to either parent. This arises because of *incomplete dominance*: the phenotypic effect of the single dominant allele in the heterozygote is qualitatively different from that of having two dominant alleles in a homozygote. Flower color in snapdragons is an example.

2. Incomplete dominance occurs more frequently than does complete dominance. It was fortuitous that all seven of Mendel's traits in peas showed complete dominance, making his analysis of inheritance patterns much easier.

3. Monohybrid crosses involving incomplete dominance produce phenotypic and genotypic ratios of 1:2:1 in the F_1 generation.

4. *Codominance* occurs when two alleles at a gene locus produce different phenotypic effects, both of which are visible in heterozygotes. The blood types produced by the ABO blood group system in humans are examples.

5. *Pleiotropic* alleles have phenotypic effects on more than one character. An example is the influence of a single gene on both coat coloration and crossed eyes in Siamese cats.

6. New alleles arise through *mutation*, the heritable change of nucleic acid material in the gene. The most common allelic form of the gene is called the *wild-type*; other forms are known as *mutant* alleles. A *polymorphic* gene locus is one where the wild-type allele is present less than 99% of the time.

7. *Multiple alleles* may exist at a single gene locus. An example is again the gene that regulates the ABO blood group system in humans. In this case, three different alleles interact to produce six genotypes and four phenotypes. The Rh factor in human blood also exhibits inheritance patterns consistent with multiple alleles.

8. In some dihybrid crosses, outcomes occur where homozygous recessive genotypes appear in the F_2 generation at levels in excess of those predicted by Mendel's laws. This is explained by *linkage*, where the genes involved occur on the same chromosome and do not assort independently during gamete formation.

9. *Epistasis* occurs when the action of one allelic form of a gene masks the expression of alleles at other gene loci. The epistatic gene usually affects production of substances manufactured early in the biochemical pathway that leads to the trait. An example is determination of coat color in mice.

10. Mendel studied characters showing *discontinuous* variation; they had discreet, contrasting forms. However, many heritable characters such as human height or skin pigmentation show *continuous* variation, which is under the combined control of many genes *(polygenes)*, often located on different chromosomes.

11. Genotype and environment also interact to affect an organism's phenotype. Nutrition, light, and temperature can all influence the translation of genotype into phenotype. Dark pigments in the fur of Siamese cats represent an example of temperature influences on phenotype.

12. Environmental factors can affect a genotype's *penetrance*, or the proportion of individuals in a population with a certain genotype that actually show the expected phenotype. The *expressivity* of a genotype is a measure of how much the phenotype emerges in any given individual with that genotype; it is also under environmental influence.

13. Differences between the influence of genetic and environmental factors are best studied using identical twins, where genetic variation is negligible and differences between individuals can usually only result from environmental influences.

14. Although Mendel's law of independent assortment was believed broadly applicable for many years, eventually it was found that this law could not explain the results of all dihybrid crosses. Some of these anomalous inheritance patterns were explained by T. H. Morgan using the concept of *linked* genes, where two gene loci are assumed to reside on the same chromosome. The full set of linked genes on a chromosome constitutes a *linkage group*.

15. Complete linkage is very rare. Chromatids often break and reseal during the formation of chiasmata in meiosis, allowing for *crossing over* and *recombination* of genetic material between homologous chromosome pairs.

16. By documenting phenotypic changes that occur through recombination relative to known markers on the chromosome, *recombination frequencies* (the number of recombinants divided by the total number of progeny) can be calculated.

17. Since the probability that crossing over will occur is directly proportional to the distance between linked genes, recombination frequencies can be used to generate a quantitative map of the relationships between gene loci on the chromosome. Distances are measured in *map units*, with one unit corresponding to a recombinant frequency of 0.01.

18. In Mendel's and Kölreuter's studies it did not matter which parental sex possessed a given trait; reciprocal crosses always yielded similar results. For many characters, however, parental origin has profound genetic consequences.

19. Most plants produce both male and female gametes on the same organism; they are *monoecious*. By contrast, most animals produce eggs and sperm inside separate individuals; they are *dioecious*.

20. Morphological sex in some dioecious species is determined by precise differences in the combination of *sex chromosomes* (X or Y), which may occur singly or in pairs. The sex chromosomes are typically present in unequal numbers. *Autosomes* represent all the other chromosomes and are present in equal numbers in both sexes.

21. In grasshoppers, the total number of X chromosomes determines sex. Females are XX, males are XO, where the O represents the absence of a second X chromosome.

22. In mammals, the presence or absence of the Y chromosome determines sex. XX individuals are female and XY individuals are male. Since only males produce Y-bearing gametes, it is the male who determines the sex of the offspring in mammals.

23. In humans, atypical combinations of the sex chromosomes (e.g. XO or XXY) produce males or females with physical abnormalities such as Turner and Klinefelter syndrome.

24. The actual gene for maleness in mammals is present on the Y chromosome. This *SRY* gene (sex-determining region on the Y) codes for a protein essential in *primary sex determination* (the type of gonads which develop). Without this protein, a developing embryo will become female, the "default" sex in mammals. *Secondary sex determination* (the nongonadal, outward manifestation of sex such as breasts and body hair) is determined by the actions of hormones coded for by genes other than *SRY*.

25. In the fruit fly *Drosophila melanogaster*, sex is determined by the ratio of the number of X chromosomes to autosome sets. In other insects such as honeybees, whether the individual is haploid or diploid determines its sex.

26. In birds, moths, and butterflies, the pattern of sex determination is the reverse of that in mammals. To avoid confusion, females in these species are designated ZW (equivalent to XY in mammals) and males ZZ (equivalent to XX). Females of these species are thus the ones who determine their progeny's genetic sex.

27. Important genes are located on the X and Y chromosomes. Depending on their sex, individuals may have only a single copy of each chromosome; thus they are *hemizygous* (having only one allele) for that character. Patterns of *sex-linked inheritance* differ markedly from those of autosomal inheritance, where two allelic copies are always present. Morgan's work on eye color in *Drosophila* provides an example.

28. *Carriers* of X-linked recessive traits (those who have one recessive allele masked by one dominant allele) will differentially pass the trait to their male and female offspring. In humans, where females are XX, a woman who carries an X-linked recessive allele in the heterozygous condition can pass it to either her sons or daughters, but it is only in her hemizygotic sons (XY) that the trait will actually be expressed. Hemophilia and red-green color blindness are examples.

29. Because of its small size, there are fewer genes on the Y chromosome than on the larger X chromosome. Nevertheless, some of these Y chromosome genes are quite important in determining primary male characteristics. In humans, Y-linked inheritance is obviously limited to males.

30. Since Mendelian ratios are statistical averages and not absolutes, it is essential to have large *sample sizes* to guard against the effects of chance deviations from predicted ratios. Slow-breeding species with long life spans (e.g., humans) therefore do not make good subjects for genetic study.

31. Mendelian inheritance is based on the precise allocation of genetic material from the nucleus during meiosis. However, some DNA is also passed between generations by non-Mendelian mechanisms. One mechanism is *maternal inheritance*, where cytoplasmic DNA from mitochondria or chloroplasts is provided to the zygote through the egg cytoplasm. Since cytoplasm does not divide precisely during cytokinesis, inheritance of cyto-

plasmic DNA is not as carefully controlled as with nuclear (chromosomal) DNA. Maternal inheritance produces the variegated color pattern of stems in the four-o'clock plant.

Activity (for answers and explanations, see page 124)

- For each combination of sex chromosomes and animal species listed below, indicate whether the individual would be male or female, and whether or not they would be reproductively normal.
  _____ ZZ chicken
  _____ X human
  _____ XXY human
  _____ XY fruit fly
  _____ XXY fruit fly
  _____ haploid honeybee

Questions (for answers and explanations, see page 124)

1. In cases of incomplete dominance, the phenotype associated with the homozygous recessive genotype is
   a. indistinguishable from the heterozygote phenotype.
   b. only compatible with the homozygous dominant genotype as a mating partner.
   c. different from both the homozygous dominant and heterozygous phenotypes.
   d. a very rare condition that occurs much less frequently than predicted by Mendelian ratios.
   e. unrecognizable because it looks just like the homozygous dominant phenotype.

2. Individuals in a species of deer with the genotype *BB* have dark brown skin, while those with the genotype *bb* have tan skin. Deer with the genotype *Bb* have mottled skin with mixes of brown and tan patches. This situation is probably an example of
   a. incomplete dominance.
   b. codominance.
   c. simple dominance.
   d. pleiotropy.
   e. hemizygosity.

3. The ABO blood group system in humans is controlled by
   a. multiple alleles.
   b. Rh factors.
   c. multiple genes.
   d. incomplete dominance.
   e. environmental influences on gene expression.

4. If a child belongs to blood type O, he or she could *not* possibly have been born to which of the following sets of parents?
   a. Type A mother and type B father
   b. Type O mother and type A father
   c. Type AB mother and type O father
   d. Type B mother and type O father
   e. Type O mother and type O father

5. A man with blood type A, one of whose parents was blood type O, has a child by a woman with blood type AB. What is the probability that their baby will have blood type O?
   a. 0%
   b. 25%
   c. 50%
   d. 75%
   e. 100%

6. In humans, hemizygous individuals for sex-linked characters are also known as
   a. males.
   b. females.
   c. heterozygotes.
   d. autosomal mutants.
   e. wild types.

7. Red–green color blindness is an X-linked recessive character in humans. A color-blind woman and a man with normal vision have a son. What is the probability that the son is color-blind?
   a. 0%
   b. 25%
   c. 50%
   d. 75%
   e. 100%

8. If a human female expresses an X-linked recessive trait, then she must have inherited the trait from
   a. her mother.
   b. her father.
   c. both parents.
   d. her maternal grandmother.
   e. through a mutation.

9. Data useful for constructing genetic maps come from which of the following sources?
   a. Measurements of the rate of mutation at specific gene loci
   b. Patterns of sex-linkage in the ABO blood group system of humans
   c. The speed at which wild-type alleles segregate during gamete formation
   d. Recombination frequencies that measure the amount of genetic change due to crossing over
   e. Counts of the numbers of carriers of X-linked recessive traits in the F_1 generation

10. Two genes regulate coat color in a breed of show dog: one for pigment production and one for pigment deposition. If either gene is homozygous recessive, the animal is an albino. This situation represents an example of which of the following genetic conditions?
    a. Epistasis
    b. Multiple alleles
    c. Discontinuous variation
    d. Expressivity
    e. Penetrance

11. Himalayan rabbits normally have an all-white coat except for black on the tail, legs, nose, and ears. If, however, you shave the hair off the back of one of these rabbits and put an ice bag on the skin, the new hair will grow in black. Which of the following represents a likely genetic explanation for this observation?
   a. Environmental influences on penetrance or expressivity
   b. Multiple allele interactions
   c. Masking of one gene by another in epistasis
   d. Simple dominance and recessiveness
   e. None of the above

12. Novel gene combinations produced by the events of meiosis are not a factor in maternal inheritance because
   a. genetic recombination is due entirely to the DNA found in the male sperm.
   b. Y-linked traits control gene recombination and are expressed only in males.
   c. maternally inherited DNA does not mutate, so change never occurs.
   d. errors that occur in replication of this DNA counteract all new gene combinations.
   e. maternally inherited DNA comes from mitochondria and chloroplasts and is not involved in meiosis.

13. Imagine a plant in which flower color is limited to three forms: red, white, and pink. By contrast, skin color in humans covers an enormous variety of colors and shades. This difference is best attributed to the fact that
   a. flower color is controlled by maternal inheritance and skin color is not.
   b. skin color is controlled by maternal inheritance and flower color is not.
   c. flower color involves only one gene whereas skin color involves polygenes.
   d. skin color involves only one gene whereas flower color involves polygenes.
   e. epistasis controls flower color and penetrance controls skin color.

Integrative Questions (for answers and explanations, see page 124)

1. If the linked gene sequence *ABCDEFGHIJK* becomes *ABHGFEDCIJK*, we would say that _____ had probably occurred.
   a. allele segregation
   b. an inversion mutation
   c. some polygenic event
   d. linkage reduction
   e. independent assortment

2. Which of the following statements is *not* true?
   a. Chromosomes have many genes, each of which may have multiple alleles.
   b. Incomplete dominance produces heterozygotes that are phenotypically different from either parent.
   c. The effects of one gene are masked by another in epistasis.
   d. The F_1 progeny of a cross between two homozygous dominant parents will all be heterozygotes.
   e. Wild-type alleles are the most common form of genes found under natural conditions.

3. Two separate crosses are performed using fruit flies. The first cross involves a dark-bodied male and a light-bodied female. The second involves a light-bodied male and a dark-bodied female. All of the offspring from these crosses are dark-bodied. This is an example of
   a. reciprocal crosses involving a likely autosomal character.
   b. reciprocal crosses involving a likely sex-linked character.
   c. test crosses involving a likely autosomal character.
   d. test crosses involving a likely sex-linked character.
   e. dihybrid crosses involving incomplete dominance.

4. Purple flowers are dominant to white flowers in pea plants, and tall plants are dominant to short plants. If a true-breeding, purple-flowered, tall plant is crossed with a true-breeding, white-flowered, short plant, what proportion of the offspring in the F_2 generation will be purple and short?
   a. 1/4
   b. 3/4
   c. 1/16
   d. 3/16
   e. 0/16

11

DNA and Its Role in Heredity

CHAPTER LEARNING OBJECTIVES—after studying this chapter you should be able to:

❑ Explain why, prior to the discovery of DNA as the genetic material, proteins were thought to be the most likely candidates for the hereditary molecule.

❑ Describe the molecular basis of bacterial transformation, and explain how the experiments of Griffith, Avery, MacLeod, and McCarty provided important clues in the search for a transforming principle.

❑ Discuss the reasoning behind, and the major conclusions from, Hershey and Chase's experiments on T2 bacteriophage, and explain why their results finally convinced the scientific community that DNA was the transforming principle.

❑ Discuss the importance of Franklin and Wilkins' work with X ray crystallography and its relationship to the study of DNA structure.

❑ Diagram the basic structure of a DNA nucleotide, and understand what is meant by a polymer of DNA nucleotides.

❑ Differentiate between the four purine and pyrimidine bases in DNA, and tell which specific bases pair with one another.

❑ Explain what is meant by complementary base pairing in DNA and how this concept relates to Chargaff's rules.

❑ Discuss how Watson and Crick used model building to finally determine the correct structure of DNA.

❑ Describe our current understanding of the physical structure of a DNA molecule.

❑ Correctly use the terms "double-stranded helix," "uniform diameter," "right-handed spiral," and "antiparallel strands" when describing a molecule of DNA.

❑ Explain what is meant by the 5' and 3' ends of a polynucleotide, and use these ideas to describe the direction in which new strands of DNA are formed.

❑ Differentiate between the three possible models for DNA replication, and discuss the evidence from Meselson and Stahl's experiments that led biologists to accept the semiconservative model.

❑ Describe in detail the molecular mechanism of DNA replication using terms such as "helicase," "polymerase," "replication origin," "replication fork," "primer," "lagging strand," "leading strand," "Okazaki fragments," "ligases," etc.

❑ Distinguish between the molecular mechanisms known as proofreading, mismatch repair, and excision repair that cells use to deal with errors in DNA replication.

❑ Discuss the energetics of DNA replication and repair, and identify sources of energy that cells have available for manipulating their DNA.

DNA: THE GENETIC MATERIAL • THE STRUCTURE OF DNA • DNA REPLICATION (pages 243–252)

Key Concepts

1. Early attempts to explain the chemical basis of heredity focused on proteins because of their chemical diversity, complexity, and in some cases great specificity. Nucleic acids were considered too "simple" to contain the necessary hereditary information.

2. In the early 1900s, circumstantial evidence began pointing to DNA as a possible hereditary molecule. It was found in the nucleus and chromosomes where genes are located and, when stained with Feulgen dye, appeared in species-specific amounts. Moreover, the quantity in diploid somatic cells was twice that found in haploid gametes.

3. In the 1920s, Griffith discovered the phenomenon of *transformation* in pneumococcus bacteria. His experiments with mice showed that a living R strain of nonvirulent bacteria could be transformed into a virulent S strain by exposure of the live R strain to heat-killed specimens of the S strain. This suggests that the live R strain had taken up some genetic information from the environment, presumably provided by the heat-killed S strain.

4. Shortly after Griffith's results were obtained, it was empirically demonstrated that bacterial transformation can occur in a test tube (i.e., without living mice) as well as in a cell-free extract of heat-killed S cells and living R cells. That is, the transformation process does not depend on the presence of living organisms or cells.

5. During the 1940s, Avery, MacLeod, and McCarty used Griffith's transformation technique along with highly purified cell extracts to clearly identify the *transforming principle* as DNA. However, their important work went unnoticed for several years because DNA was still believed to lack the complexity needed for the hereditary substance, and also because it was not known at the time whether bacteria actually possess genes.

6. In the 1950s, Hershey and Chase provided definitive evidence that DNA is the genetic material. They raised a T2 bacteriophage virus under two conditions: one with radioactive phosphorus (^{32}P), which labels the DNA, and the other with radioactive sulfur (^{35}S), which labels the virus's external proteins. Labeled bacteriophage were then allowed to infect normal bacteria, and the locations of radioactive markers were determined.

7. Only the ^{32}P was found inside the infected bacteria; most of the ^{35}S remained outside the cells. This finding that viruses transfer only their DNA from one generation to the next led to the conclusion that DNA (not protein) must be the hereditary material.

8. Although less clear-cut than Avery's experiment, Hershey and Chase's work finally convinced the scientific community that DNA was indeed the hereditary material. This happened partly because by 1950 mutations in viruses and bacteria, and thus the presence of genes, had also been discovered.

9. The next task was to determine the molecular structure of DNA. Precise X-ray crystallographic work by Franklin and Wilkins, coupled with earlier biochemical analyses, indicated that DNA is a long polymer of nucleotides. Each nucleotide consists of a deoxyribose sugar, a phosphate group, and a nitrogen-containing base.

10. There are four nucleotides of DNA, each differing only in their nitrogenous bases. Two are purines (*adenine* and *guanine*) and two are pyrimidines (*cytosine* and *thymine*).

11. In 1950, Chargaff noted that for any DNA sample the amount of adenine always equals the amount of thymine (A = T), and the amount of guanine always equals the amount of cytosine (G = C). Thus, the total amount of purines always equals the total amount of pyrimidines (A + G = T + C).

12. Watson and Crick used the X ray and chemical data in 1953 to build our current model of DNA structure. Their model portrays DNA as a *double-stranded helix* (cylindrical spiral) of *uniform diameter* that *twists to the right*. It contains two *antiparallel* chains of nucleotides (i.e., polynucleotides that run in opposite directions).

13. The "backbone" of each polynucleotide chain consists of repeated sugar–phosphate connections, with the nitrogenous bases oriented toward the center of the molecule. Hydrogen bonds form between the bases of one chain and those of the other. A system of *complementary base pairing* occurs in which G pairs only with C, and A only with T. The geometry of this bonding pattern produces a uniform width in DNA.

14. DNA is formed by joining the phosphate group attached to the 5' carbon of a new nucleotide with the 3' carbon of the preceding sugar molecule; the molecule is thus constructed in the 5' → 3' direction. This constraint on the formation of DNA explains why the two chains are antiparallel.

15. DNA carries hereditary information in a linear sequence of nucleotide bases. Because of the specific complementary base-pairing rules and double-stranded structure of the molecule, unwinding the helix provides an efficient mechanism for DNA replication. The helical structure also provides mechanisms that account for its overall function (genetic information is read linearly) and for gene mutation (mutations are changes in the sequence of bases).

16. In 1956, Kornberg showed that DNA will replicate itself in a cell-free extract in a test tube given an appropriate mixture of preformed DNA, precursors of the four specific nucleotide triphosphate (deoxy-ATP, -CTP, -GTP, and -TTP), and an enzyme called DNA polymerase. The pre-formed DNA serves as a template against which new DNA is formed using the precursors and enzyme.

17. Three possible DNA replication mechanisms can be conceived. *Conservative* replication would produce one new double-stranded DNA while retaining intact the old double-stranded molecule. *Dispersive* replication would require fragmentation of the old DNA, followed by its reassembly along with new DNA. All available evidence supports the third model of *semiconservative* replication (originally proposed by Watson & Crick), where each old strand serves as a template for the manufacture of one new strand.

18. In 1957, Meselson and Stahl verified the semiconservative model by raising bacteria on a medium enriched with "heavy" nitrogen (^{15}N). This forced the bacteria to manufacture all heavy DNA.

19. When bacteria that have grown for several generations on ^{15}N are transferred to a medium of ^{14}N, each parental strand of DNA is used as a template to manufacture a new double-stranded DNA molecule for the next generation. The result is that one strand of the new DNA molecule has nothing but heavy ^{15}N bases (old DNA), and the other strand contains only light ^{14}N bases (new DNA).

20. The difference in DNA types can be measured using density gradient centrifugation in a CsCl solution and interpreted using the semiconservative model. Three bands appear in the centrifuge tubes, corresponding to pure ^{15}N strands (heavy) in the parental generation, mixed ^{14}N and ^{15}N strands (intermediate) in the first generation, and pure ^{14}N strands (light) in all subsequent generations. With additional growth on a ^{14}N medium, the bacterial DNA becomes increasingly dominated by light ^{14}N strands.

Activities (for answers and explanations, see page 124)

- What will be the sequence of nucleotide bases in a new strand of DNA made from the DNA template strand shown below?

  5' – A T C C G T A A C G C A G G G C T T A – 3'

- Based on the results of Meselson and Stahl's experiments with heavy and light forms of DNA, explain why the following three density gradient centrifuge tubes and their contents provide support for the semiconservative model of DNA replication.

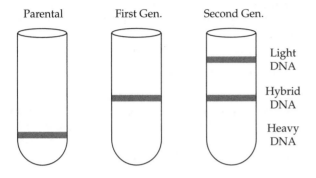

Parental   First Gen.   Second Gen.

Light DNA

Hybrid DNA

Heavy DNA

- Explain the conceptual relationship between Griffith's experiments on bacterial transformation and those of Hershey–Chase using T2 bacteriophage. What is the main conclusion regarding hereditary transfer that can be drawn from these two sets of experiments?

Questions (for answers and explanations, see page 125)

1. Which of the following represents a bond between a purine and a pyrimidine (in that order)?
   a. C – T
   b. G – A
   c. G – C
   d. T – A
   e. C – A

2. The Hershey–Chase experiments, in which T2 bacteriophage were formed in the presence of the radioactive precursors of potential hereditary macromolecules, demonstrated that
   a. protein, and not DNA, was the hereditary material in this virus.
   b. DNA, and not protein, was the hereditary material in this virus.
   c. DNA, and not protein, was the hereditary material in this bacterium.
   d. protein, and not DNA, was the hereditary material in this bacterium.
   e. none of the above

3. Thirty percent of the bases in a sample of DNA extracted from eukaryotic cells are adenine. What percentage of cytosine is present in this DNA?
   a. 10
   b. 20
   c. 30
   d. 40
   e. 50

4. In the Watson and Crick model of the DNA molecule
   a. neither nitrogen nor phosphorus is present.
   b. the width of the molecule changes regularly along its length.
   c. the amount of cytosine always equals the amount of thymine.
   d. a single helix of polynucleotides is held together exclusively by covalent bonds between bases.
   e. the double helix of polynucleotides is held together largely by hydrogen bonds between bases.

5. Which of the following was not a technique used to determine the structure of DNA?
   a. X ray crystallography
   b. Biochemical analysis of purine/pyrimidine ratios
   c. Radioactive labeling of sulfur and phosphorus
   d. Construction of three-dimensional molecular models
   e. Biochemical analysis of generalized nucleotide structure

6. The primary finding of Meselson and Stahl's experiments with isotopes of nitrogen was that
   a. DNA is the hereditary material.
   b. the semiconservative model of DNA replication is correct.
   c. the dispersive model of DNA replication is correct.
   d. Chargaff's rules of complementary base pairing are correct.
   e. purines and pyrimidines both possess nitrogen-containing bases.

7. Which of the following would not be found in the polynucleotide "backbone" of a DNA molecule?
   a. Carbon
   b. Deoxyribose sugar
   c. Nitrogen
   d. Phosphate
   e. Covalent bonds

THE MECHANISM OF DNA REPLICATION • DNA PROOFREADING AND REPAIR (pages 252–258)

Key Concepts

1. In addition to the necessary molecular precursors for DNA replication (i.e., DNA template, suitable polymerase, and appropriate nucleoside triphosphates), there must also be a source of chemical energy to drive this endergonic synthesis reaction.

2. DNA nucleotides are always added to the 3' end of the growing chain. A bond linking the terminal two phosphate groups on the 5' end of a nucleoside triphosphate is broken to provide energy used to link the new nucleotide to the existing chain. Additional free energy for this synthesis comes from breaking of the pyrophosphate (PP) ion released from the nucleoside triphosphate.

3. DNA replication is initiated when the double helix is unwound by enzymes called helicases. This action requires the expenditure of ATP energy to break the bonds between base pairs.

4. In prokaryotes, the unwinding of the DNA helix occurs at a specific location called the *origin of replication*. The exposed double strands of DNA form a bubble, or *replication fork*. Replication then proceeds in both directions around the exposed DNA strands of the circular chromosome until complete.

5. In the linear chromosomes of eukaryotes, multiple origins of replication and multiple replication forks occur. The y-shaped forks move along the length of the chromosome as new regions of DNA are added.

6. DNA replication requires the use of a DNA or RNA *primer*, a short sequence of bases to which DNA *polymerase* can attach new nucleotides that elongate the polynucleotide strand. Attachment of the new bases occurs at the 3′ end of the terminal nucleotide in the developing strand.

7. Enzymes called *primases*, which represent a portion of the aggregate polypeptides in a prokaryotic *primosome*, are used to make RNA primers for DNA replication. Once a primer is in place, *DNA polymerase III* catalyzes the elongation of new DNA strands at the replication forks.

8. The *leading strand* of newly synthesized daughter DNA is manufactured continuously along the 5′ → 3′ direction. However, the *lagging strand* produced from the other parent template cannot be manufactured in a continuous fashion because of the antiparallel nature of the parent strands.

9. Lagging strands are produced by working backwards (to preserve 5′ → 3′ synthesis) in relatively small DNA segments called *Okazaki fragments*. These fragments are made in the direction opposite that of the moving replication fork.

10. The finishing touches in replication of the lagging strand are provided by *DNA polymerase I*, which replaces the RNA primer with a segment of DNA between adjacent Okazaki fragments, and *DNA ligase*, which bonds the fragments together.

11. DNA replication is both rapid and precise. In excess of a thousand base pairs per second are added to developing bacterial DNA, with error rates of less than one base per billion or more base additions.

12. DNA polymerases, helicases, ligases, and other enzymes also carry out inspection functions to detect, excise, and replace mispaired bases. *Proofreading* allows DNA polymerase to notice mispaired bases and perform appropriate replacement during the actual synthesis of new DNA.

13. *Mismatch repair* occurs after new DNA has formed and during genetic recombination. The correct strand is repaired in part because the cytosine in new DNA has not yet been covalently methylated.

14. *Excision repair* involves the enzymatic inspection of DNA during a cell's lifetime to detect potentially dangerous base mispairings or chemical modifications. When detected, enzymes cut out the problem area so that new, healthy DNA can be synthesized to repair the damaged site.

15. The absence or malfunction of repair mechanisms greatly enhances an organism's susceptibility to permanent DNA damage. An example is the human skin disease xeroderma pigmentosum.

16. Although energetically expensive and sometimes even inefficient, DNA replication, inspection, and repair functions are so essential for proper cell health that energy is made available from numerous sources to insure proper DNA structure and activity.

Activity (for answers and explanations, see page 125)

- In the space below, diagram a replication fork as it would be seen in a replicating segment of DNA. In your diagram label the 5′ and 3′ ends of each parent strand and daughter strand. Indicate also which new strand is the leading and which the lagging strand of daughter DNA.

Questions (for answers and explanations, see page 125)

1. Which one of the following statements about DNA replication is *false*?
   a. Okazaki fragments are the initiators of continuous DNA synthesis along the leading strand.
   b. Replication forks represent areas of active DNA synthesis on the chromosome.
   c. Error rates for DNA replication are often less than one in every billion base pairings.
   d. Ligases and polymerases function in the vicinity of replication forks.
   e. Nucleic acid primers are essential for the process of DNA replication to begin.

2. DNA replication can be described as
   a. an overall endergonic synthesis reaction.
   b. an overall exergonic synthesis reaction.
   c. energetically easy to initiate and maintain.
   d. of little importance to the overall energy requirements of a cell.
   e. none of the above

3. In which of the following would you expect to find Okazaki fragments?
   a. DNA polymerase III
   b. DNA ligase
   c. Helicases
   d. Actively replicating DNA
   e. Primosomes

4. Which of the following most closely describes the number of replication forks that occur along a prokaryotic chromosome during DNA replication?
   a. 0
   b. 2

c. 20
d. 200
e. 2000

5. Consider the leading strand of a replicating segment of DNA in a bacterium. Which of the following would *not* be present at some time along this strand?
   a. Okazaki fragments
   b. DNA polymerase III
   c. RNA primer
   d. Replication fork
   e. Helicase enzyme

Integrative Questions (for answers and explanations, see page 125)

1. Consider the pneumococcus bacterium first studied by Griffith and later examined by Avery. Which of the following would likely be found in the DNA of this organism?
   a. Purines
   b. Guanine bases
   c. Phosphorus
   d. Nitrogen
   e. All of the above

2. _____ is (are) mainly responsible for holding pairs of complementary bases together in a stable molecule of DNA.
   a. Hydrogen bonds
   b. Covalent bonds
   c. Ligase
   d. Helicase
   e. Primase

3. The DNA in a cell of your body mutates as the result of exposure to some cancer-causing agent. A likely result of this mutation would be
   a. your immediate death.
   b. stimulation of DNA proofreading repair mechanisms.
   c. stimulation of DNA mismatch repair mechanisms.
   d. stimulation of DNA excision repair mechanisms.
   e. an overall lowering of energy use in the cell.

12

From DNA to Protein: Genotype to Phenotype

CHAPTER LEARNING OBJECTIVES—after studying this chapter you should be able to:

❑ Explain the basic sequence of molecular events that takes information coded in an organism's genes and turns it into a phenotype.

❑ Discuss the one-gene, one-polypeptide hypothesis and describe how Garrod's inheritance data regarding alkaptonuria in humans, as well as work by Beadle and Tatum on amino acid metabolism in the fungus *Neurospora*, provides support for this idea.

❑ Differentiate between auxotrophs and prototrophs, and explain their use in the study of molecular genetics.

❑ Explain the chemical similarities and differences between RNA and DNA, and describe how RNA base-pairing is different from that of DNA.

❑ State the central dogma of molecular biology, and explain how Crick contributed two key ideas to its development.

❑ Explain how the genetics of retroviruses necessitates some modification of the central dogma, and relate this to the importance of Temin and Baltimore's work on the role of reverse transcriptase in RNA viruses.

❑ Describe the roles of mRNA, tRNA, and rRNA in the overall process of protein synthesis.

❑ Compare and contrast in detail the mechanisms of transcription and translation with respect to the primary molecules involved, the sequence of molecular events in each mechanism, and where in the cell these events take place.

❑ Use the terms "RNA polymerase," "DNA template," and "antisense strand" when describing the process of transcription.

❑ Differentiate between the events of initiation, elongation, and termination for both transcription and translation.

❑ Define the term "codon" and explain how codons provide information for the construction of polypeptides.

❑ Differentiate between a degenerate genetic code and one that is ambiguous, and explain the importance of Nirenberg and Matthaei's work with artificial messengers in deciphering the genetic code.

❑ Use the terms "codon," "anticodon," "start codon," "stop codon," and "signal sequence" when describing stages in protein synthesis.

❑ Explain how tRNA is charged and how ribosomal proteins associate with various RNA molecules during translation.

❑ Discuss the role of ribozymes and polysomes in protein synthesis, and explain how antibiotics can affect the process of translation in prokaryotes.

❑ Describe how signal sequences are used to direct developing proteins towards specific cell regions in eukaryotes.

❑ Explain what biologists mean by the term "mutation," and discuss why mutations are important in the evolutionary process.

❑ Distinguish among the following types of mutations and describe their effects: point, chromosomal, silent, missense, nonsense, and frame-shift.

❑ Further distinguish among the following types of chromosomal mutations and describe their effects: deletions, duplications, inversions, and translocations.

❑ Describe molecular mechanisms by which mutagens can produce mutations and possibly act as carcinogens.

❑ Appreciate the rate at which natural mutations occur in organisms and how these mutations can influence evolution.

GENES AND THE SYNTHESIS OF POLYPEPTIDES • DNA, RNA, AND THE FLOW OF INFORMATION • TRANSCRIPTION: DNA-DIRECTED RNA SYNTHESIS • THE GENETIC CODE (pages 260–268)
Key Concepts

1. Phenotypic traits do not emerge directly from genes. Rather, genes code for the production of macromolecules necessary for the phenotype to appear. DNA "language" is first transcribed into RNA language, and then RNA information is subsequently translated into the amino acid sequence of a protein.

2. Garrod (1908) was the first to explore the mechanistic relationship between genes and phenotype in his study of the inherited disease alkaptonuria. He concluded that when a particular gene is homozygous recessive, a defective enzyme is produced that is incapable of

78

metabolizing the amino acid tyrosine. Unfortunately, his insight was too advanced for its time and remained unappreciated for over 30 years.

3. Beadle and Tatum extended Garrod's ideas in the 1940s using the fungus *Neurospora crassa*. Wild-type *prototrophs* of the fungus can be grown on a minimal medium, which provides only the most basic nutrients needed for successful growth. Mutant *auxotrophs*, produced by irradiation, require one supplemental nutrient in order to grow.

4. By producing a series of auxotroph mutants, Beadle and Tatum showed that each mutant lacked only one enzyme necessary for the synthesis of a required nutrient. Since they could produce many different kinds of autotrophs, this suggests that numerous genes are involved in the control of biosynthetic pathways.

5. Beadle and Tatum's findings spawned the *one-gene, one-polypeptide* hypothesis, which states that an individual gene controls the production of only one polypeptide. After its manufacture, the polypeptide usually functions in cell metabolism, often as an enzyme.

6. The *central dogma* of molecular biology, first proposed by Crick, states that DNA codes for the manufacture of RNA in a process called *transcription*. RNA in turn codes for the production of protein through a mechanism of *translation*. Since protein does not code for protein, RNA, or DNA, this system of transcription and translation represents a one-way flow of genetic information from DNA to protein.

7. Crick proposed that "adaptor" molecules permit RNA and the specific amino acid components of proteins to associate properly during translation. We call these adaptor molecules transfer RNA *(tRNA)*.

8. Crick, Brenner, and Jacob also suggested that DNA-based genetic information moves from the nucleus, where DNA is found, to the cytoplasm, where proteins are assembled, by way of messenger RNA *(mRNA)*.

9. Some viruses (which are noncellular) represent an exception to the central dogma in that RNA replaces DNA as their genetic material. *Retroviruses* carry a *reverse transcriptase* enzyme that transcribes RNA into DNA. The DNA is then replicated by normal host cell mechanisms to make new viruses. Although technically an exception to the central dogma, retroviruses still observe the one-way flow of genetic information.

10. The human immunodeficiency virus (HIV), which causes AIDS, is an example of a retrovirus. The AIDS drug AZT, currently one of the best treatments against this disease, works by inhibiting the action of reverse transcriptase and thus short-circuiting viral reproduction.

11. Before translation can occur, the DNA genetic message must first be transcribed into an mRNA message. The enzyme *RNA polymerase* binds to one side of the double-stranded DNA (the *template strand*) and begins assembling a continuous sequence of complementary ribonucleotides in the 5′ → 3′ direction. The other strand of DNA, called *antisense,* remains untranscribed. As the mRNA transcript is made, it peels away from the DNA template allowing the separated DNA strands to reassociate.

12. All forms of RNA are made by transcription. In addition to tRNA and mRNA, ribosomal RNA *(rRNA)* is coded for by DNA. rRNA is a major structural component of ribosomes, which are essential for the translation of mRNA into polypeptide chains.

13. Gene transcription is a three-step process, beginning with *initiation*, which occurs at well-defined *initiation sites* along the DNA molecule. Initiation sites contain special DNA sequences called promoters that allow RNA polymerase to bind tightly to the DNA and determine which strand will serve as the template. The relative affinity of a promoter for RNA polymerase determines the frequency with which a gene is transcribed.

14. Prokaryotes have a single type of RNA polymerase that serves for the transcription mRNA, tRNA, and rRNA. Eukaryotes have three different types of polymerase in addition to a collection of proteins that must first bind to the promoter before a polymerase can attach to the DNA. This added complexity provides a greater degree of transcriptional control in eukaryotes.

15. Once transcription has been initiated, the mRNA strand undergoes *elongation* in the 5′ → 3′ direction (i.e, antiparallel to the DNA strand) through the addition of new ribonucleotides. The large amount of energy needed for elongation comes from the removal and breakdown of the pyrophosphate group from each added nucleotide. Error rates in this process are on the order of one mistake per 100,000 bases, significantly higher than for DNA replication.

16. *Termination* of the developing mRNA strand occurs when specific *termination sites* are encountered along the DNA. The completed mRNA transcript may then simply fall away or be removed by a helper protein. Prokaryotes typically begin translation immediately after (if not before) the transcript is terminated. In eukaryotes, additional processing of the transcript often occurs before the start of translation.

17. mRNA transcripts can be viewed as a series of three-letter genetic "words," or *codons*, with nucleotide bases serving as the "letters."

18. The four possible nucleotides in RNA, taken three at a time, produce 64 possible codon combinations (e.g., UUU, UUC, UAC, etc.). Since there are only 20 amino acids to code for, we say that the genetic code is *degenerate*, meaning more than one codon codes for the same amino acid. This does not mean that the code is *ambiguous*, where one codon would code for multiple amino acids.

19. Several codons serve as translational punctuation marks. The *start codon* AUG acts as an initiation signal for transcription, while the three *stop codons* (UAA, UAG, and UGA) signal chain termination.

20. The genetic code is apparently universal among living organisms on this planet, and probably dates back to the very beginnings of organic evolution. The only known exceptions involve minor alterations in mitochondrial and chloroplast DNA, as well as a change in function for two stop codons in some protists. The functional significance of these alterations in coding remains unclear.

21. The first breakthrough in deciphering the genetic code came from the work of Nirenberg and Matthaei using simple, artificial polymers of mRNA. They found that poly-U, -C, and -A each produced proteins with only a single amino acid type present; that amino acid must therefore be what is called for in the codons UUU, CCC, and AAA. The remaining codons were later identified using various three-base mRNAs and determining which charged tRNA/amino acid combination bound to them.

Activities (for answers and explanations, see page 125)

• Using the universal genetic code depicted in Figure 12.5 on page 267 of your textbook, determine the amino acid sequences that would result from the translation of each of the following mRNA molecules:

Case 1: 5′ – A U G A C A U G U C G U G G G U G A – 3′

Case 2: 5′ – U G A A C A U G U C G U G G G C G U – 3′

Case 3: 3′ – G G U A C A U G A A C A G G G G U A – 5′

• Why must proteins be synthesized in a two-step process that first involves transcription, followed by translation?

Questions (for answers and explanations, see page 126)

1. In a species of *Neurospora*, it is known that ten genes have an important influence on the control of a biochemical pathway. How many auxotrophs related to this pathway can theoretically be made?
   a. One, since auxotrophs exhibit the most common genetic form of the trait.
   b. More than ten, since each gene that helps control the pathway could potentially mutate in many ways.
   c. Less than five, since two or more of these genes likely produce the same polypeptide product.
   d. None, since *Neurospora* is not a good organism for this kind of research.
   e. None, since autotroph strains cannot survive in the laboratory.

2. According to the one-gene, one-polypeptide hypothesis,
   a. all proteins are used as enzymes during DNA replication.
   b. each polypeptide codes for the production of a single nucleotide chain called a gene.
   c. genes can only be active in the presence of the polypeptide for whose production they code.
   d. each gene codes for production of a single polypeptide chain, which can often influence phenotype.
   e. the polypeptide made by the gene incorporates itself into the chromosome next to the gene locus.

3. Which of the following represents a proper hierarchical organization of genetic information?
   a. Gene – nucleotide – DNA – chromosome
   b. Nucleotide – DNA – gene – chromosome
   c. Gene – DNA – chromosome – nucleotide
   d. Nucleotide – gene – DNA – chromosome
   e. DNA – gene – nucleotide – chromosome

4. When interpreting the universal genetic code during translation, the codons being read to produce the correct sequence of amino acids for a protein are found in molecules of
   a. DNA.
   b. RNA polymerase.
   c. mRNA.
   d. tRNA.
   e. rRNA.

5. Which of the following is *true* about the DNA triplet code?
   a. Each DNA base codes for three proteins.
   b. Each gene codes for three proteins.
   c. Each triplet codes for three proteins.
   d. It takes three genes to code for one protein.
   e. Each amino acid in a protein is coded for by one set of three DNA bases.

6. A mutation occurs in a gene so that the first codon in the mRNA transcribed from this gene reads UAG. Which of the following is most likely to occur?
   a. Translation will proceed normally.
   b. Translation will be initiated, but will terminate after the first amino acid is added.
   c. Translation will take more energy than normal.
   d. The protein made will contain all arginine amino acids.
   e. No protein will be made because UAG is a stop codon.

7. Which of the following is *not* necessary for transcription to occur in a prokaryotic cell?
   a. DNA template strand
   b. Ribosome subunits
   c. RNA polymerase
   d. Promoter site
   e. Termination site

THE KEY PLAYERS IN TRANSLATION • TRANSLATION: RNA-DIRECTED POLYPEPTIDE SYNTHESIS • REGULATION OF TRANSLATION • MUTATIONS: HERITABLE CHANGES IN GENES (pages 268–279)

Key Concepts

1. Translation, like transcription, is a three-step process involving initiation, elongation, and termination. It can occur quite rapidly, with polypeptide sequences of several hundred amino acids being produced in a matter of minutes.

2. During translation, it is the job of tRNA to associate the proper amino acids with each codon in the mRNA message. tRNA molecules have a characteristic three-dimensional shape. Amino acids attach to one portion of the molecule. The *anticodon*, a 3-nucleotide sequence that is complementary to one particular mRNA codon, is found on another region of tRNA. Other portions of the tRNA molecule are involved with ribosome associations.

3. tRNA molecules become charged when one of a family of *activating enzymes* known as *aminoacyl-tRNA synthases* identifies an amino acid and attaches it to the correct tRNA. The energy needed to make the chemical bonds of this association comes from the splitting of ATP.

4. Ribosomes are the sites of protein assembly in both prokaryotic and eukaryotic cells. Each ribosome is capable of *self-assembly* and consists of one heavy and one light subunit. The subunits are made of several different rRNAs and many proteins arranged in a precise pattern held together by noncovalent bonds. Each ribosome has binding sites for two tRNAs and one mRNA molecule, and may participate in the synthesis of many different kinds of proteins.

5. Translation cannot begin without the formation of an *initiation complex*. This complex consists of the ribosomal light subunit bound to an appropriately-charged tRNA and the start codon of an mRNA message. The anticodon of a specific tRNA then forms hydrogen bonds with the initial mRNA codon. At this point, a heavy ribosomal subunit attaches to complete the functional ribosome complex. Proteins called *initiation factors* help direct this assembly process using GTP as an energy source.

6. Two RNA binding sites exist on each ribosome. The first mRNA codon and a charged tRNA bind to the P site (where the growing polypeptide forms). Other charged tRNAs bind to the A site (amino acid attachment location) in complementary accordance with the specific codons present in the mRNA sequence.

7. Once two tRNAs are in place on the ribosome, a peptide bond forms between the first and second amino acids. The first tRNA is then released from the P site. The second tRNA, bearing the newly-formed dipeptide, shifts position from the A to P site. A third charged tRNA then enters the vacated A site. Translation of the polypeptide continues in this manner until an mRNA stop codon appears in the A site, causing the binding of a *protein release factor* which catalyzes the addition of a water molecule instead of another amino acid to the end of the polypeptide.

8. Recent work (1992) by Noller and colleagues indicates that rRNA, rather than the large ribosomal subunit proteins, plays the critical role in catalyzing peptide bond formation on the ribosome. rRNA thus acts as a catalytic enzyme, or *ribozyme*. Certain antibiotics have their defensive effects on prokaryote invaders by inhibiting the action of catalytic ribosomal components, thus halting protein synthesis and killing the microbes.

9. Multiple ribosomes often move sequentially down the length of an mRNA thread during translation, trailing their growing polypeptide chains behind. These *polysomes* represent an efficient way of increasing the rate of synthesis in cells with high protein demands.

10. For proteins that remain suspended in the cytoplasm, translation takes place on "free" (unattached) ribosomes. Proteins intended for export from the cell, or for use in lysosomes and other membrane-bounded cell structures, are manufactured by ribosomes which become attached to the rough ER.

11. Regardless of the protein's final destination, all protein synthesis begins on free ribosomes. If specific amino acid chains known as *signal sequences* are present at the beginning of the polypeptide chain, then the protein is destined for export or incorporation into a membranous structure. To complete translation, these proteins and their ribosomes must attach to rough ER via *signal recognition particles* composed of protein and RNA. The signals act like a delivery coding system, directing a developing protein's cellular destination.

12. Errors in replication, transcription, or translation are infrequent, but do occur. Heritable changes in genetic material are called *mutations*. They result from alterations in the nucleotide sequence of DNA, and provide the ultimate source of genetic variation that fuels evolutionary change.

13. Genetic mutations are most easily detected if they have clear phenotypic effects, but many do not (e.g., primaquine sensitivity in humans). Some mutations are conditional and appear only under highly restrictive circumstances, such as with certain temperature-sensitive growth effects.

14. *Point mutations* affect the alleles of individual genes and may involve an alteration of only one nucleotide. *Chromosomal mutations* are more extensive, involving insertions, deletions, or inversions of many genes within entire chromosome segments.

15. Many point mutations are effectively *silent mutations* because the base substitution that occurs in DNA, and subsequently mRNA, is compensated for by the degeneracy of the genetic code. Silent mutations are common and account for much genetic diversity.

16. *Missense mutations* result from a single base substitution that causes only one codon to change; they do not alter the mRNA reading frame for translation. The result is a mutant protein with only one erroneous amino acid (e.g., human beta-globin produced by the sickle cell allele). Such proteins are usually still functional, and in some instances may even confer an evolutionary advantage on the bearer.

17. *Nonsense mutations* have more dramatic effects, since base substitution produces a chain-terminator codon that causes protein translation to stop prematurely. The resulting protein is typically nonfunctional.

18. *Frame-shift mutations* involve actual insertions or deletions of nucleotide bases. The effect is to alter the reading frame for all subsequent codons, causing dramatic changes in protein structure and thus phenotype. Such changes often prove lethal, or are otherwise evolutionarily maladaptive.

19. Chromosomal mutations can involve *deletions* (loss of genes), *duplications* (DNA copies on the same chromosome), *inversions* (reversed sequence of chromosome segment), or *translocations* (gene sequence retained but moved to another part of the chromosome, thus affecting synapsis during meiosis). All have dramatic effects on gene products or mechanisms of cell division, and thus are generally not beneficial to the mutant organism.

20. Genetic changes can be induced by *mutagens*: substances or treatments that cause alterations of nucleic acid sequences. Some mutagens are chemical analogs of nucleotide bases, such as 5-bromouracil, which is much more susceptible to base mispairings than is uracil. Other mutagens, such as substances present in polluted air or tobacco smoke, produce chemical changes in natural bases that lead to improper base pairing during DNA replication.

21. Many mutagens are also *carcinogens* (cancer-producing). These substances can be tested for using the *Ames test*, in which mutant bacteria are grown in the presence of suspected carcinogens. If these identifiable mutations are reversed, then the suspect chemical has a high probability of being mutagenic and potentially carcinogenic.

22. Mutation rates are typically less than 1 per 10^4 genes per DNA duplication (often as low as 1 per 10^9), and involve mostly point mutations. Many are harmful to the organism and are selected against.

23. Other mutations are neutral and have no selective effects. A very few mutations actually improve the organism's adaptation for its environment, and are thus favored evolutionarily. Regardless of their effect, mutations clearly provide the raw material for evolutionary change.

24. New genes may arise from chromosomal duplications that, over time, can subsequently mutate into a useful function that provides some new feature to the organism's phenotype. Because the original, unmutated genes still exist and are functional, species can "experiment" with mutations of the duplicate material and thus incur fewer ill effects.

Activity (for answer and explanation, see page 126)

- Consider the following sentence made entirely of three-letter words: THE CAT ATE HER NEW TOY. Now assume that these words represent DNA triplet bases, and that a deletion mutation occurs in this sequence which causes the loss of the letter "c" in the word cat. Rewrite the modified sentence below, still preserving the three-letter word format. Explain the consequences that this frame-shift mutation would have for the meaning of the message, both in terms of English language and the protein product of translation

Questions (for answers and explanations, see page 126)

1. Which of the following is *not* a product of translation?
   a. mRNA
   b. DNA polymerase
   c. Certain components of ribosomal subunits
   d. Signal sequence peptides
   e. Reverse transcriptase

2. Of the molecules listed below, which one is *not* directly involved in translation?
   a. DNA
   b. tRNA
   c. mRNA
   d. Ribosomes
   e. Enzymes

3. Transfer RNA is referred to as an adaptor molecule, in the sense that Francis Crick used this label, because
   a. tRNA adapts to changing cell environments by controlling the rate of mRNA production.
   b. tRNA connects the mRNA message with appropriate amino acids for protein synthesis.
   c. transcription cannot be accomplished without tRNA initiators binding to the DNA template.
   d. the shape of the other RNAs is based on the kind of tRNA present in the cell.
   e. none of the above

4. Which of the following best represents the sequence of events that takes place as a free amino acid is incorporated into a developing polypeptide chain?
   a. Free in cell cytoplasm – binds with mRNA on ribosome – attaches to tRNA – forms peptide bond
   b. Binds with mRNA on ribosome – free in cell cytoplasm – attaches to tRNA – forms peptide bond
   c. Free in cell cytoplasm – attaches to tRNA – binds with mRNA on ribosome – forms peptide bond
   d. Attaches to tRNA – forms peptide bond – binds with mRNA on ribosome – free in cell cytoplasm
   e. Attaches to tRNA – free in cell cytoplasm – forms peptide bond – binds with mRNA on ribosome

5. Digestive enzymes destined for use in the food vacuole of an amoeba are made
   a. primarily on ribosomes attached to the rough ER.
   b. exclusively on free ribosomes in the cytoplasm.
   c. by first translating the appropriate genes into mRNA and then transcribing this information into protein.
   d. using the enzyme reverse transcriptase.
   e. whenever polysomes are present within the cell.

6. Frame-shift mutations are particularly deleterious because
   a. entire segments of genetic material are lost from the chromosome.
   b. they can alter the kinds of amino acids coded for by large portions of a gene.
   c. they tend to affect only initiation codons, thus inhibiting the start of translation.
   d. they produce a situation where DNA can no longer replicate, leading to chromosomal abnormalities.
   e. they primarily affect genes that code proteins necessary in transcription.

7. Consider the short DNA sequence 3′ – C G A C A T A C
   G T G C – 5′. If this sequence suddenly mutated to
   become 3′ – C G A C A A A C G T G C – 5′, what kind of
   mutation would have to be involved?
   a. Insertion
   b. Deletion
   c. Inversion
   d. Substitution
   e. Chromosomal

***Integrative Questions (for answers and explanations,
see page 126)***

1. The energy used for the assembly of RNA and proteins
   comes mainly from the breaking of _____ bonds
   in high-energy molecules.
   a. hydrogen
   b. ionic
   c. hydrophobic
   d. covalent
   e. peptide

2. Suppose that proteins contained only 10 different amino
   acids, instead of the 20 known to exist. What effect
   could this difference in amino acid types possibly have
   had on the evolution of the genetic code?
   a. Codons might have had two bases instead of three.
   b. Codons would still need to have three bases.
   c. Codons might have had four bases instead of three.
   d. Protein synthesis would have been much costlier in
      energetic terms.
   e. There would probably have been no effect.

3. The elongation phase of transcription is similar to the
   elongation phase of translation in that
   a. energy is required to add new molecules.
   b. a polymer of some sort is produced.
   c. genetic information flows in one direction only.
   d. the process is organized and regulated in part by
      enzymes.
   e. all of the above

13

The Genetics of Viruses and Prokaryotes

CHAPTER LEARNING OBJECTIVES—after studying this chapter you should be able to:

❏ Provide several reasons that prokaryotes and viruses are good models for the study of molecular genetics.

❏ Describe the basic structure of a virus, explain why viruses are considered obligate intracellular parasites, and list the reasons viruses are not included within any of the six kingdoms of life.

❏ Discuss important historical advances in our understanding of viruses, including the work of Ivanovsky and Stanley.

❏ Discuss the difference between a virus and a virion, and explain why the molecular structure of viruses renders them immune to antibiotics.

❏ List the various criteria used to classify viruses, and describe the distinguishing characteristics of viruses that attack animals, plants, and bacteria.

❏ Discuss the difference between the lytic and lysogenic life cycles of a bacteriophage, and explain the role of a prophage in the lysogenic cycle.

❏ Distinguish between a virion and prion, and describe how they differ from viruses.

❏ Describe the differences between clonal reproduction, conjugation, transduction, and transformation in bacteria.

❏ Explain how bacterial colonies are produced under laboratory conditions, and describe how bacteriophages can be used to test for genetic mutations in bacteria.

❏ Discuss methods used to isolate and maintain mutant strains of bacteria, and describe how auxotrophic and prototrophic strains can aid the study of bacterial genetics.

❏ Define and use the terms "F factor," "F-pili," "Hfr mutant," and "F$^+$" and "F$^-$" mating types in bacteria.

❏ Explain how interrupting the mating of conjugating bacteria is used to map the bacterial chromosome.

❏ Compare and contrast the process of specialized transduction with that of generalized transduction.

❏ Define and discuss the following terms: "episome," "plasmid," "replicon," and "R factor."

❏ Explain what is meant by the terms "transposable element" and "transposon," and relate these to the function and evolution of plasmids, F factors, and Hfr mutants.

❏ Discuss the difference between inducible and constitutive enzymes, and explain the role of inducers in the function of such enzymes.

❏ Describe the roles of structural genes and regulatory genes in prokaryotic cell metabolism.

❏ Discuss the basic structure and function of an operon using the following terms: "promoter," "repressor," "corepressor," "operator," "regulator gene," and "inducer."

❏ Differentiate between inducible and repressible operon systems, and explain the molecular mechanisms that regulate the activity of each.

❏ Compare and contrast the molecular control mechanisms of the *lac* operon with those of the tryptophan operon in *E. coli*.

❏ Describe the control mechanism for an operon regulated by enhanced promotor efficiency, and compare this mechanism to those of repressible and inducible operon systems.

❏ Explain the differences between negative and positive transcriptional control of bacterial operons, and give examples of each.

USING PROKARYOTES AND VIRUSES FOR GENETIC EXPERIMENTS • VIRUSES: REPRODUCTION AND RECOMBINATION   (pages 282–290)

Key Concepts

1. Prokaryotic cells (bacteria) and viruses offer many advantages for the study of molecular genetics. They are small, easy to manipulate, and breed rapidly in large numbers. They are also haploid and thus easy to analyze phenotypically. Most importantly, they have less than one-thousandth as much DNA per cell as do humans and other eukaryotes.

2. Work by Ivanovsky in the 1890s first suggested the existence of viruses. He found that filtered extract of diseased tobacco leaves still carried the infectious agent, meaning that the agent had to be smaller than a bacterium. In 1935, Stanley succeeded in crystallizing viruses and showing that they consisted of nucleic acid and protein.

3. Viruses are acellular, do not metabolize energy, and are incapable of independent reproduction. They are thus considered *obligate intracellular parasites* and are extremely specific for their host cells. Viruses cannot be grouped with any other organisms in the six kingdoms of life.

4. Because they lack a distinct cell wall and do not have bacterial-like biochemistry, viruses are immune to the antibiotics commonly used against prokaryotes.

5. Outside of their host cells, viruses exist as individual particles called *virions*. A virion consists of a central core of either DNA or RNA surrounded by a protein capsid. Some viruses also possess structures such as lipid membranes acquired from an animal host, or specialized protein tails.

6. Viruses are commonly classified by their type of nucleic acid (DNA or RNA), whether the nucleic acid is single- or double-stranded, overall shape and symmetry (helical, icosahedral, or binal), the presence or absence of a membrane envelope, and capsid size.

7. *Bacteriophages* are viruses that parasitize bacteria. They often show a *lytic* life cycle in which new viruses are rapidly manufactured using the commandeered metabolic machinery of the host cell, leading to the eventual rupture of the host cell and release of new viruses.

8. Bacteriophages can also exhibit a *lysogenic* life cycle, where viral DNA known as a *prophage* becomes incorporated into the host cell's DNA for several generations. Environmental changes later induce the host to synthesize new viruses, which leads to cell lysis.

9. Bacteriophage viruses can undergo genetic recombination. They have a single, circular chromosome that, when injected into a host bacterium, may recombine with other phage chromosomes to produce novel phenotypes. Calculations of recombinant frequencies between two genetic markers can yield map distances and information on the arrangement of genes on the viral chromosome.

10. Animal viruses are very diverse. They include those responsible for the common cold, influenza, herpes, and HIV in humans. One group of arthropod-borne viruses, or *arboviruses*, can produce serious illnesses such as encephalitis in people but do not appear to affect the arthropod carrier.

11. Some animal viruses have RNA as their genetic material; these include the retroviruses like HIV, as well as others that do not carry out reverse transcription.

12. Animal viruses recognize their host cells by membrane interactions, and many emerge from the host with a membrane envelope surrounding their capsid. Animal viruses without membrane coats typically enter the host cell by endocytosis.

13. Plant viruses usually enter their hosts with the help of a vector (often an insect) that assists the virus in penetrating the plant cell wall. Mechanical damage to plant cells by wind or abrasion can also permit viral entry.

14. Plant viruses may be inherited vegetatively or sexually. Transmission of the virus from cell to cell is by way of the plasmodesmata that penetrate plant cell walls.

15. *Viroids* (pure viral nucleic acid) were first isolated in 1971 by Diener. They are circular, single-stranded RNA molecules that lack any kind of protein coating and are thus much smaller than regular viruses. How they cause disease is unknown.

16. Viroids have so far only been found in plants, and may be passed from plant to plant either horizontally (by physical contact) or vertically (inherited by offspring from their parents).

17. *Prions* are a class of protein fibrils that may cause some degenerative central nervous system diseases in mammals. They consist entirely of protein and contain no nucleic acids at all.

18. Although their mechanism of infection is still unknown (and in some instances doubted althogether), prions have been implicated in neural diseases such as scrapie, mad cow disease, kuru, and Creutzfeldt-Jakob disease.

Activity (for answer and explanation, see page 126)

- In the space below, list four basic criteria used to determine "life" in organisms. Then explain why viruses do not meet these basic criteria for life.

Questions (for answers and explanations, see page 126)

1. Which of the following is *not* a reason that prokaryotes and viruses make good subjects for the study of molecular genetics?
   a. Their DNA can mutate.
   b. They are small in size.
   c. They are easy to manipulate.
   d. They typically breed rapidly and in large numbers.
   e. They possess large amounts of DNA.

2. For which of the following viruses or virus-like entities is the mechanism of infection and disease production *not* linked to genetic information in the viral nucleic acid?
   a. Bacteriophages
   b. Animal viruses
   c. Plant viruses
   d. Prions
   e. Viroids

3. A physician collects a specimen of the agent that causes human influenza from a public drinking fountain. This sample is best described as a
   a. bacterium.
   b. viroid.
   c. virion.
   d. prion.
   e. bacteriophage.

4. An unknown virus is taken from a living cell and brought to a lab for analysis. It is found to contain DNA, protein, and a lipid membrane envelope. This virus most likely came from
   a. an animal.
   b. a plant.
   c. a bacterium.
   d. another virus.
   e. cannot determine from the information given

5. If the enzyme reverse transcriptase is present in a virus, then we know for sure that the virus
   a. is HIV.
   b. causes tobacco leaf disease.
   c. has DNA as its genetic material.
   d. has RNA as its genetic material.
   e. lacks a protein capsid.

6. In which of the following would you likely find a prophage?
   a. Lytic phase of the bacteriophage life cycle
   b. Lysogenic phase of the bacteriophage life cycle
   c. Lytic phase of the animal virus life cycle
   d. Lysogenic phase of the animal virus life cycle
   e. None of the above

PROKARYOTES: REPRODUCTION, MUTATION, AND RECOMBINATION • REGULATION OF GENE EXPRESSION IN PROKARYOTES (pages 290–303)

Key Concepts

1. Bacteria (and archaea) are living cells that metabolize and reproduce on their own. Their genetics thus differ from those of viruses and from the more complex eukaryotes.

2. Unlike eukaryotes, where each parent contributes equally to genetic recombination, prokaryote recombination involves an interaction between the parent cell genome and a small subset of genes from another cell. The smaller set of genes can enter the parent cell in three ways: conjugation, transformation, or transduction.

3. The bacterium *E. coli* is a representative prokaryote that can be easily grown in the laboratory on either a solid or liquid nutrient medium. Asexual reproduction in *E. coli* yields offspring that are exact genetic copies (i.e., clones) of the original parent cell. Cloning can be maintained indefinitely under favorable conditions, both in nature and the lab.

4. Bacterial mutation is demonstrated in experiments where bacteriophages are introduced into a culture dish supporting a well-developed *E. coli* colony (or "lawn"). Shortly after the virus is introduced, small circular openings called *plaques* appear in the lawn, indicating areas where viruses have killed bacteria.

5. In a few of these plaques, small colonies of virus-resistant bacteria arise due to the clonal growth of one mutant bacterium. If this mutant is grown alone on another culture dish, introduced bacteriophages will generate no plaques, indicating that the mutation for viral resistance is heritable.

6. Mutations may also occur in the bacteriophage, allowing it to better circumvent a host's defenses. This demonstrates the genetic basis for evolutionary change: mutations provide the raw material for phenotypic changes, and environmental selection pressures determine the viability of those changes.

7. The details of genetic recombination in bacteria were investigated by Lederberg and Tatum in the 1940s during studies using *E. coli* metabolic auxotrophs. They observed that when certain auxotrophic combinations were grown together, wild-type (prototrophic) colonies would occasionally appear.

8. One explanation for this outcome is that the prototrophs arise by spontaneous mutation. However, natural mutation is too rare an event to explain the high rate at which prototrophs were observed. A second explanation is that prototrophs are transformed bacteria that have been changed by a transforming principle secreted into the growth medium by one of the auxotrophic strains. This possibility was also ruled out.

9. A third, and correct, explanation is *conjugation*, a process in which bacteria pair and mix their genetic material. This type of genetic recombination produces some new bacteria that are wild-type for the mutant genes in their parents.

10. Conjugation in *E. coli* is a one-way process in which genetic material flows from the *donor* cell (male) to the *recipient* cell (female). During conjugation, the female acquires a small extrachromosomal piece of DNA called a *fertility factor (F)*, which transforms her into a male.

11. Male *E. coli* are thus F^+, and females are F^-. Genes on the F factor control the presence of hairlike, tubular projections, called *F-pili*, through which DNA is transferred between cells.

12. *Hfr* mutants are bacteria that act as high-frequency donors of genetic material. They yield a higher than normal percentage of recombinant offspring because their F factor is actually incorporated into the bacterial chromosome and can thus be transmitted more easily through conjugation.

13. Gene location on the bacterial chromosome can be mapped using the technique of *interrupted mating*. Conjugating bacteria are interrupted at different times during the mating process, permitting different amounts of DNA to be transferred between cells. By comparing offspring characteristics with the known duration of conjugation, a gene map can be constructed. The map reveals that the bacterial chromosome is circular.

14. Interrupted mating studies done in 1955 by Jacob and Wollman revealed that in Hfr males the F factor is actually incorporated into the chromosome, and that the insertion point of this factor varies with different Hfr strains.

15. The F factor also marks the point where the bacterial chromosome opens during conjugation. This opened end of the chromosome always leads the way into the female. The very end of the opened chromosome contains the portion of the F factor determining maleness.

16. Genetic transfer between Hfr and F⁻ cells begins when one strand of the Hfr chromosome is nicked and starts moving into the F⁻ cell through the F-pilus. This stimulates DNA replication of the remaining Hfr chromosome.

17. Once inside the F⁻ cell, the transferred Hfr DNA replicates and becomes double-stranded. It must then incorporate into the F⁻ cell's chromosome by crossing over in order for recombination to occur.

18. Bacteriophages can also act as vectors to move bacterial DNA between cells. In *specialized transduction*, only those bacterial genes closest to where the prophage actually attaches to the bacterial chromosome are transferred. Transducing phages cause infected bacteria to become lysogenic.

19. *Generalized transduction* occurs when some portion of the bacterial chromosome, minus the prophage, gets incorporated into the phage coat. When the resulting particle (which lacks phage genes and is not infectious) attaches to a bacterium, that cell gains a segment of bacterial DNA, which may recombine with its own chromosome.

20. Unlike specialized transduction, generalized transduction may move any portion of the bacterial chromosome. It is thus a useful tool for chromosomal mapping and genetic engineering.

21. F factors and viral prophages are *episomes*, nonessential genetic elements that can exist either independently of the host chromosome or incorporated into it. Neither arises from mutation; they are only acquired by infection.

22. *Plasmids* are another nonessential genetic element in bacteria that exist as self-replicating circles of DNA. They are not required for bacterial survival, nor are they ever incorporated into the bacterial chromosome.

23. Plasmids must exist as *replicons* to be maintained within a bacterial population. Each replicon has its own origin of DNA replication; it can thus divide independently and in synchrony with the bacterial chromosome.

24. *R factors* are an important class of plasmids that confer drug resistance in bacteria. Although they probably existed long before the advent of modern medicine, the increased drug resistance recently exhibited by microorganisms is likely due to R factor actitivty. Clinical and hospital environments, where antibiotics are used extensively, represent an intense area of selection favoring R factors.

25. Another agent for gene transport is a chromosomal segment or type of plasmid DNA known as a *transposable element*. These independently replicating, mobile elements carry the genes for their own insertion into DNA molecules, and often cause dramatic disruptions of normal gene activity.

26. Large transposable elements (more than 5,000 base pairs) are called *transposons*. Transposons may include additional genes besides the transposable elements themselves.

27. Plasmid evolution and F factor development in Hfr males are probably both the result of transposable elements. For example, the same drug-resistance genes found in transposable elements also exist in R factor plasmids, making it likely that R factors gained their resistance from transposons.

28. Cells do not require all of the substances they are capable of synthesizing at all times. Gene activity must therefore be regulated. Substances that cause enzyme synthesis are called *inducers*, and the enzymes made are known as *inducible enzymes*. If enzymes are made continuously at fairly constant rates, then they are known as *constitutive enzymes*.

29. The genetic blueprint for enzyme structure is contained in *structural genes*. When multiple enzymes are involved in the same biochemical pathway, the structural genes coding for those enzymes are usually adjacent to one another on the chromosome. These genes are often simultaneously transcribed to form an mRNA sequence that codes for the entire set of enzymes.

30. Some genes need to be transcribed more often than others. *Promoters* are segments of DNA on which RNA polymerase binds to begin the transcription process. Each set of structural genes has its own promoter whose binding efficiency for RNA polymerase controls the frequency with which those particular enzymes are produced.

31. Genes must be able to halt transcription when particular enzymes are no longer needed; that is, they must be able to negatively influence transcription rate. One strategy is to use a physical obstacle to block mRNA synthesis. A specific protein known as a *repressor* molecule does this by binding to a second site on the DNA called the *operator*. Such controllable units of transcription in bacteria are known as *operons*.

32. An operon consists of a repressor binding site (the operator), and one or more structural genes. Regulation of the operon occurs through control of repressor activity. When the repressor binds to the operator, transcription of the structural genes is halted. However, when the repressor binds to certain inducers, the repressor changes shape (i.e., becomes inactive) and transcription takes place.

33. The *lac* operon of *E. coli* is an example of an inducible operon. Lactose serves as the inducer; it combines with (inactivates) the repressor to permit mRNA transcription for the enzymes needed in lactose metabolism. If lactose concentration drops, the repressor is reactivated and blocks mRNA synthesis. Thus, by using the lactose stimulus itself as an inducer, the availability of enzymes for lactose breakdown is keyed directly to the need for those enzymes.

34. *Regulatory genes* code for repressor proteins; they may be located close to, or at some distance from, the operon they control.

35. In their unregulated condition, inducible operons are continuously "on." They make their structural gene products unless turned off by the repressor. Some regions in the operon make proteins whose only function is to control other genes, while other regions (operators and promoters) do not code for protein synthesis and may never be transcribed.

36. It is also useful for prokaryotes to switch off some operons in the presence of high concentrations of substances that are ordinarily manufactured by the cell. These *repressible* operons turn off mRNA synthesis in response to sudden increases in the concentration of important molecules. The tryptophan operon of *E. coli* is an example.

37. In repressible systems, the operon cannot be shut down unless the repressor first combines with a *corepressor*, which is often the stimulus substance itself (e.g., tryptophan) or some chemical analogue. Once combined, the repressor–corepressor complex blocks the operator to inhibit gene transcription.

38. Subtle but important differences exist between these two kinds of operons. Inducible operons use an inducer to inactivate the repressor. This frees up the operator, allowing the promoter to bind RNA polymerase and initiate transcription. Repressible systems use a corepressor to activate the repressor molecule. The activated repressor then binds to the operator and blocks transcription.

39. Some operons containing genes for catabolic enzymes are not regulated by being turned on or off. Rather, they adjust the efficiency of promoter function. In such systems, *CRP* (cAMP receptor protein) first forms a complex with cAMP. The CRP–cAMP complex then attaches near the promoter site to facilitate the binding of RNA polymerase. In this way polymerase binding and transcription rate are improved as much as 50-fold.

40. Because they prevent transcription via the action of repressor molecules, both inducible and repressible operon systems are examples of negative transcriptional control. By contrast, promoter operon systems are examples of positive transcriptional control because they actually enhance the rate of transcription.

Activities (for answers and explanations, see page 127)

- Consider a bacterial operon that controls the production of four enzymes needed in the synthesis of arginine. It is known from chromosome mapping studies that the structural genes for these enzymes are all adjacent to one another. Assuming that this is a repressible enzyme system, make a schematic diagram like the ones in your textbook that shows how the components of the arginine operon might be arranged on the chromosome, and briefly describe how this operon would function.

- Refer to the discussion of R factors and drug resistance on page 297 of your textbook. Propose a hypothesis to explain why R factors might be more widespread in certain microorganisms today than they were in the past.

Questions (for answers and explanations, see page 127)

1. In the studies by Lederberg and Tatum on sexual reproduction in bacteria, which of the following mechanisms was the basis for the appearance of prototrophs on the minimal medium plates?
   - *a* Transforming substances that convert one phenotype into another
   - *b.* Lysogenic induction
   - *c.* Spontaneous mutation of appropriate genes
   - *d.* DNA transfer via conjugation
   - *e.* Interrupted mating

2. Which of the following is *true* of bacterial conjugation?
   - *a.* Genetic exchange via conjugation always occurs from male to female mating types.
   - *b.* Mating types can change from one form to another via conjugation.
   - *c.* Conjugation by Hfr mutants tends to produce more recombinant offspring than non-Hfr matings.
   - *d.* After conjugation, the original F$^+$ cell in a mating of two *E. coli* remains F$^+$.
   - *e.* all of the above

3. Which of the following is *true* of generalized transduction in bacteria?
   - *a.* Bacterial DNA is incorporated into a phage coat.
   - *b.* Prophages play an active role in the process.
   - *c.* The infected cell is induced to become lysogenic.
   - *d.* Only highly restricted portions of bacterial chromosomes can be moved this way.
   - *e.* all of the above

4. Which of the following statements about Hfr bacteria is *false*?
   - *a.* They produce higher than normal numbers of recombinant offspring.
   - *b.* Their F factor is actually incorporated into the bacterial chromosome.
   - *c.* The F factor marks the spot where the chromosome opens during conjugation.

d. The leading end of the opened chromosome enters the female cell last.

e. Interrupted mating can be used to study the genetics of Hfr mutants.

5. Episomes and plasmids are important players in the molecular genetics of prokaryotes. Which of the following is *neither* an episome nor a plasmid?
a. F factor
b. R factor
c. F pilus
d. Prophage
e. All of the above are plasmids or episomes.

6. The reason that plasmids must be replicons in order to survive within a bacterial population is that
a. replicons allow the plasmid to integrate into the host chromosome prior to DNA replication.
b. plasmids use their replicons to extract themselves from the host chromosome after sexduction.
c. bacteria will reject the plasmid during conjugation unless the replicon is present.
d. the replicon allows the plasmid to divide independently of the host chromosome.
e. none of the above

7. Large segments of DNA that can make multiple copies of themselves and then insert these copies anywhere within the chromosomal material are known as
a. R factors.
b. transposons.
c. episomes.
d. regulator genes.
e. promoters.

8. The use of inducible enzymes is usually the cell's best option in situations where
a. the substance upon which the enzyme acts is always present.
b. constitutive enzymes can act as corepressors of the inducible enzyme.
c. the substance upon which the enzyme acts is present only on an irregular basis.
d. constitutive enzymes control the presence or absence of the inducible enzyme's target substance.
e. plasmids are needed to act on a continuously available nutrient that is required for cell growth.

9. In a repressible operon, the repressor molecule binds with the _____ before influencing _____ by interaction with the operator.
a. corepressor, gene transcription
b. regulator, protein translation
c inducer, conjugation
d. promoter, gene recombination
e. plasmid, episome transfer

10. With which of the following operon systems would you expect to find associated regulator genes?
a. Lac operon
b. Tryptophan operon
c. Inducible operon
d. Repressible operon
e. All of the above

11. The concentration of enzyme X in a bacterium is continuously maintained at a relatively constant level, regardless of changes in cell conditions. Enzyme X is best considered
a. an inducible enzyme.
b. a repressible enzyme.
c. a constitutive enzyme.
d. a repressor protein.
e. a corepressor protein.

12. In bacterial operons where the mechanism of regulation is by enhancement of promoter efficiency, which of the following is true?
a. The structural genes usually code for catabolic enzymes.
b. cAMP receptor protein is involved.
c. A CRP–cAMP complex is formed.
d. Transcription rates can sometimes be increased 50-fold.
e. All of the above

Integrative Questions (for answers and explanations, see page 127)

1. The patterns of mutations seen in bacteria and viruses demonstrate which of the following ideas regarding evolution?
a. Mutations almost never occur in eukaryotes.
b. Mutations are an important source of raw material for evolutionary change.
c. Mutations are limited to *E. coli* and bacteriophages.
d. Bacterial and viral mutations only occur via conjugation.
e. Mutations are limited exclusively to Hfr cells and bacteriophage.

2. Clear plaques appear in a cultured bacterial lawn when it is exposed to an appropriate bacteriophage. This occurs because
a. the viruses in the area of the plaques are killed by the bacteria, exposing the culture medium below.
b. the bacteria in the area of the plaques are killed by the virus, exposing the culture medium below.
c. bacteriophages secrete a clear fluid which covers up the bacterial lawn at the plaque locations.
d. antiviral drugs are made by the bacteria in plaque areas, thus killing the virus and clearing space.
e. None of the above

3. An interrupted mating experiment is performed using *E. coli.* Five phenotypic characters are monitored, and it is found that only those cells that engaged in conjugation for the longest time express character number five. What does this tell you about the gene for character five?
a. It is a recessive trait that is expressed in one-fifth of all offspring.
b. It controls the transcription of the other four genes.
c. It is situated furthest away from the F factor on the chromosome.
d. It regulates the ability of *E. coli* to engage in conjugation.
e. Nothing useful

14

The Eukaryotic Genome and Its Expression

CHAPTER LEARNING OBJECTIVES—after studying this chapter you should be able to:

❑ Compare and contrast the size and complexity of the eukaryotic and prokaryotic genomes.

❑ Describe the defining DNA sequences found in every chromosome.

❑ Describe the process of DNA hybridization, how it led to the discovery of noncoding and repetitive DNA in eukaryotes, and the function of a probe.

❑ Describe the three classes of DNA defined by the rate of reannealing and the characteristics of each class.

❑ Discuss the structure of the telomere, how its structure relates to the problem of chromosome clipping, and the function of the enzyme telomerase.

❑ Describe the importance of tranposable elements in the DNA of eukaryotes and to the endosymbiotic theory for the origin of mitochondria and chloroplasts.

❑ Describe what is meant by a gene family and how one might arise.

❑ Describe what is meant by a pseudogene and the two mechanisms that create pseudogenes.

❑ Explain the major steps in the processing of pre-RNA into mRNA molecules.

❑ Describe the principal features of the RNA splicing process in eukaryotes.

❑ Describe several examples of the transcriptional control of gene expression called gene inactivation, including those examples involving euchromatin and heterochromatin, and DNA methylation.

❑ Discuss the existence of the Barr body relative to the concepts of euchromatin and heterochromatin.

❑ Describe important examples of gene duplication as a mechanism for amplifying gene product.

❑ Discuss the involvement of steroid hormones in selective gene transcription.

❑ Describe differences in transcriptional control and the involvement of promoters in prokaryotes and eukaryotes.

❑ List the activities of the three types of RNA polymerase.

❑ Explain the involvement of the TATA box, transcription factors, and RNA polymerase II in forming a transcription complex.

❑ Provide some examples of translational control involving longevity of mRNAs and modification with the G cap.

❑ Provide some examples of posttranslational control.

THE EUKARYOTIC GENOME   (pages 306–313)

Key Concepts

1. The *genome* is the haploid DNA content of a cell; its size is usually expressed in base pairs (bp). The eukaryotic genome is larger and more complexly organized than the prokaryotic genome.

2. The majority of eukaryotic DNA does not code for protein. The noncoding DNA consists of repeated sequences, some with no known function. Coding DNA also includes interspersed sequences of noncoding DNA that must be removed.

3. Whereas most prokaryotes have a single circular piece of DNA, the eukaryotic genome consists of several molecules of DNA. Each molecule of DNA is compacted by proteins to form a structure called a chromosome.

4. Each chromosome consists of three defining DNA sequences: an origin of replication where DNA polymerase can attach, a centromere (essential for proper movement of chromosomes during mitosis and meiosis), and telomeres, so that replication can be completed at the end of the chromosome.

5. The nuclear envelope segregates DNA and its transcription into RNA in the nucleus from the translation of RNA into protein in the cytoplasm.

6. Eukaryotic genes contain stretches of noncoding DNA that must be removed from the initial transcript, called *pre-mRNA*, during a cutting and splicing step not found in prokaryotes. The processed RNA is much smaller than the pre-mRNA.

7. As in prokaryotes, eukaryotic genes have noncoding flanking *promoter* sequences, where RNA polymerase binds, and *terminator* sequences, for detachment of the RNA polymerase. Unlike prokaryotes, eukaryotic genes also have more distantly located *enhancer* and *silencer* sequences for controlling transcription.

8. *Nucleic acid hybridization* has been a powerful tool in the study of gene structure in eukaryotes. Nucleic acid hybridization depends on complementary base pairing.

9. When warmed, the two strands of a DNA molecule will separate as hydrogen bonds are broken to form single-stranded, *denatured* DNA. Cooling will allow complementary strands to *reanneal*.

10. If DNA is denatured and immobilized on filter paper, it can be probed with a second molecule of single-stranded DNA or RNA called the *probe*. The probe will bind to any complementary sequences in the target DNA. Probes can be labeled with radioactive or fluorescent nucleotides so that probe-target hybrids can be visualized.

11. Nucleic acid hybridization studies showed that not all eukaryotic DNA reanneals at the same rate, indicating that the number of copies of eukaryotic sequences in the *genome* varies widely.

12. The class of DNA that anneals most slowly consists of *single-copy sequences*, which include the genes coding for most enzymes and structural proteins. Some single copy sequences are noncoding and form long "spacers" between genes.

13. *Highly repetitive DNA* reanneals most quickly. These sequences are usually 5 to 50 base pairs long and are often repeated in tandem up to millions of times. This class of DNA may represent a third or more of the genome and much of it is located near the centromere, suggesting involvement in cell division. It is usually not transcribed.

14. *Moderately repetitive DNA* has intermediate reannealing rates and copy numbers of hundreds to thousands per genome. Included in this class of DNA are duplicate genes, such as those for tRNA and rRNA, the ends of eukaryotic chromosomes called *telomeres*, and *transposable elements*.

15. Telomeres are sequences of moderately repetitive DNA located at the ends of eukaryotic chromosomes. In humans, the repeated sequence is TTAGGG. Each time DNA is replicated, the lagging strand is unable to form the last Okazaki fragment and the chromosome loses a bit of double stranded DNA at its ends (it becomes *clipped*). The moderately repetitive telomeric DNA is a source of noncoding DNA to be used up as the cell divides.

16. Cells that constantly divide have an enzyme called *telomerase* that is bound to a short sequence of single-stranded RNA complementary to the lost repeat unit. The telomerase then adds the needed sequence to the end of the telomere, preventing clipping.

17. Some moderately repetitive DNA includes multiple copies of the genes that code for tRNA and rRNA, molecules that are needed in large supply.

18. *Transposable DNA* consists of moderately repetitive DNA that can move about within the genome. There are several distinct types of transposable elements: transposons, retrotransposons, viral retrotransposons, nonviral retrotransposons, LINEs, and SINEs.

19. Transposons have inverted repeat sequences at each end. This allows the transposon to be looped out of its location and moved to a new position in the genome.

20. By causing insertions within the coding regions of DNA, transposable elements are an important source of mutation in eukaryotes.

21. Some evidence suggests that transposable elements may have moved certain genes from the DNA of mitochondria and chloroplasts to the nuclear DNA where they now reside. The original genes in the mitochondria and chloroplasts may then have been lost.

Questions (for answers and explanations, see page 127)

1. Place numbers to the left of the following events to indicate their correct sequence in a typical application of nucleic acid hybridization.

    _____ Add the probe

    _____ Warm the nucleic acid mixture

    _____ Observe denaturation

    _____ Cool the nucleic acid mixture

    _____ Observe reannealing

    _____ Identify the probe–DNA hybrids

    _____ Immobilize DNA on nitrocellulose filter

2. Which of the following DNA types should have the *fastest* reannealing rate during nucleic acid hybridization?
    a. DNA coding for enzymes
    b. DNA coding for cellular structures
    c. DNA located near the centromeres
    d. DNA coding for ribosomal subunits (rRNA)
    e. DNA from the ends of chromosomes

3. Which one of the following features is *not* characteristic of telomeres?
    a. Found only in eukaryotes
    b. Located at the ends of chromosomes
    c. Consist of moderately repetitive DNA
    d. Restricted to the lagging strand
    e. Become clipped in absence of telomerase

4. Which one of the following statements about transposable elements is *not* true?
    a. Transposable elements employ a variety of mechanisms to move about in the genome.
    b. Transposable elements are an example of moderately repetitive DNA.
    c. Transposable elements may have been important in the endosymbiotic origin of mitochondria and chloroplasts.
    d. The gene products of many transposable elements have important cellular functions.
    e. Insertion of transposable elements are an important source of mutation for many genes.

5. Circle all of the following features that need to be present in any chromosome .
    a. Telomeres at each end
    b. A structural gene with introns and exons
    c. An origin of replication
    d. A centromere
    e. A transposable element

THE STRUCTURES OF PROTEIN-CODING GENES • RNA PROCESSING   (pages 313–318)

Key Concepts

1. The presence of noncoding DNA in eukaryotes was first discovered when cytoplasmic mRNA was allowed to hybridize with fragmented, denatured DNA from the same cell. Double-stranded DNA–mRNA coding sections were associated with thin, single-stranded DNA loops and thick, double-stranded DNA loops formed by the noncoding DNA.

2. The coding sections of a gene are called *exons*. The noncoding sections of a gene are called *introns*. Typical eukaryotic genes consist of several exons separated by intervening introns. A few prokaryotic genes also have introns.

3. Products of introns are edited out after transcription so that the base sequence of all the gene's exons determines the mature mRNA.

4. In some genes, such as the globin genes, exons code for the protein's different functional regions or domains.

5. Duplicate gene copies can be formed by several processes, such as an abnormal crossing over event and by retrotransposition.

6. Multiple copies of a gene can sustain different mutations, thereby producing a *gene family*. As long as one member of the gene family retains the sequence that codes for the proper protein, the other members may mutate. Members of a gene family may be clustered together on the same chromosome or may be on different chromosomes.

7. Human *hemoglobin* is a tetramer composed of four globin polypeptide subunits: two alpha (α) globins and two beta (β) globins. The α-globin family includes three functional genes; the β-globin family has five functional genes. Each family is on a different chromosome and in each the gene family members are organized into a cluster, separated by spacer DNA.

8. During development, different members of the globin families are expressed at different times and in different tissues, tailoring the form of hemoglobin to best suit the needs of the organism. For example, human fetal hemoglobin ($\alpha_2\gamma_2$) is able to bind oxygen more tightly than adult hemoglobin ($\alpha_2\beta_2$) so that oxygen is transferred from mother's blood to child's blood in the placenta.

9. Similarities in amino acid sequence indicate that the globin gene families descended from a common ancestral gene.

10. *Pseudogenes*, designated by the Greek letter psi (ψ), are nonfunctional genes that are not expressed; they are usually gene family members that have been extensively modified and may have inactive promoter regions or exons with nonsense mutations. Some pseudogenes may have been functional in the past, but their functions are now performed by other gene family members.

11. The primary RNA transcript, pre-RNA, contains the gene's exons and any introns that were present. The promoter and terminator are not transcribed and, thus, are not part of pre-RNA.

12. The first step in RNA processing is the addition of a *cap* consisting of a modified form of *guanosine triphosphate* to the 5′ end of the pre-RNA. The cap is later necessary for binding to a ribosome.

13. Next, a chain of 100 to 300 residues of adenylic acid (*poly A*) is added to the 3′ end of the pre-RNA. This tail of poly A may be important in movement of mRNA from the nucleus into the cytoplasm.

14. Removal of the introns and joining together of the exons, called *RNA splicing*, is the next step in RNA processing. In some genes, over 75 percent of the nucleotides are removed as introns are spliced out and exons joined together.

15. The boundaries between introns and exons consist of short sections of DNA called *consensus sequences* because the base sequences are very similar in many genes.

16. *Small ribonucleoprotein particles*, snRNPs or "snurps," contain RNA complementary to the consensus sequences. The RNA in one snRNP (U1) binds to the consensus 5′ exon–intron boundary; the RNA in another snRNP (U2) binds to the consensus 3′ exon–intron boundary. The RNAs of other snRNPs bind within the intron to form a large RNA–protein complex called a *spliceosome*.

17. snRNP molecules in the spliceosome interact to form a loop or *lariat* of the intron transcript. The exons at the base of the lariat are spliced together and the excised intron is later degraded in the nucleus. Splicing is ATP dependent.

18. The processed mRNA exits the nucleus through the nuclear pores. A nuclear pore receptor is specific to processed mRNA; pre-RNA must remain in the nucleus.

Questions (for answers and explanations, see page 127)

1. Which of the following statements about the genes in a gene family is *not* true?
   a. The number of pseudogenes in a family must be less than the number of functional genes.
   b. Gene families can be clustered together on the same part of a chromosome.
   c. Some members of a gene family may never be transcribed.
   d. Gene families can result from the action of certain types of transposable elements.
   e. Members of a gene family may be functional at certain times, but nonfunctional at other times.

2. Place numbers to the left of the following events to show the correct sequence during RNA processing in eukaryotes.

   _____ Removal of intron "lariat"

   _____ Cleavage of 5′ exon and intron

   _____ Cleavage of 3′ exon and splicing

   _____ Binding of snRNPs to consensus sequences in RNA

   _____ Degradation of intron

   _____ Addition of G cap

   _____ Addition of poly A tail

3. A gene that includes 1,200 nucleotides codes for a protein consisting of 250 amino acids. How many of the nucleotides reside in exons and how many reside in introns?

_____ Nucleotides in exons

_____ Nucleotides in introns

4. Select *all* of the following that are found in both a pre-RNA immediately after transcription and its mature mRNA.
   a. Introns
   b. Poly A tail
   c. Modified G cap
   d. Start and stop sequences
   e. Exons

Activities (for answers and explanations, see page 128)

- The following diagram shows a section of DNA with two exons and one intron during hybridization with complementary mRNA. Label (1) the coding DNA strand, (2) the noncoding DNA strand, (3) the mRNA, (4) exon 1, (5) exon 2, and (6) the intron.

- In the space provided below, diagram how the DNA–RNA shown above would appear if there were no intervening intron present.

TRANSCRIPTIONAL CONTROL • POSTTRANSCRIPTIONAL CONTROL • TRANSLATIONAL AND POSTTRANSLATIONAL CONTROL (pages 318–327)

Key Concepts

1. Multicellularity in eukaryotes led to division of labor among different cell types. Because genes must be turned on and off at appropriate times, regulation of gene expression is complex.

2. Regulation of gene expression occurs at various levels involving transcription, translation, and posttranslation.

3. Although all cells transcribe certain "housekeeping" genes, most genes are selectively transcribed in the cells of different tissues or at different developmental times.

4. Unlike the operons of prokaryotes, where related genes are transcribed as a unit, eukaryotic genes tend to be isolated. Steroid hormones, like estrogen, can simultaneously "turn on" several related genes, because they all share a common specific sequence near their promoters, called the *hormone-response element*.

5. Whereas transcriptional control in prokaryotes can be both positive and negative, regulation is almost always positive in eukaryotes—the genes are turned off until they are turned on by a protein.

6. Eukaryotes have three kinds of RNA polymerase: *RNA polymerase I* transcribes the DNA coding for rRNA; *RNA polymerase II* transcribes genes that produce mRNAs; *RNA polymerase III* transcribes genes that code for tRNAs and other small RNA molecules, such as the small nuclear RNAs U1 and U2.

7. *Promoters*—sequences of DNA to which RNA polymerase must bind to initiate transcription—are much more complex in eukaryotes than in prokaryotes. This allows more precise control over transcription.

8. Various regulatory protein molecules called *transcription factors* must bind to the *promoter region* of the gene before RNA polymerase II can attach and begin transcription. Each transcription factor binds to a unique sequence of nucleotides.

9. The *TATA box* is an eight–base-pair sequence of thymine and adenine pairs located about 25 base pairs before the starting point for transcription in many eukaryotic genes. The transcription factor TFIID binds to the TATA box. The TATA box helps to specify the exact location of the starting point for transcription.

10. After several additional transcription factors also bind to the promoter, the RNA polymerase II attaches to the growing transcription complex. After other factors are added, the complex is complete and transcription begins.

11. Although the TATA box is common to many genes and is recognized by transcription factors found in many cells, other genes have sequences only recognized by the transcription factors found in certain cell types.

12. *Regulators* are regions of DNA adjacent to the promoter of a gene that bind proteins responsible for activating the transcription complex.

13. *Enhancers* and *silencers* are DNA sequences located up to 20,000 base pairs away from the gene. Enhancers bind activator proteins that stimulate specific promoters and enhance transcription of the gene. Silencers bind repressor proteins that have the opposite effect. These distantly bound proteins may depend on DNA bending to have their effect on the promoter region.

14. In eukaryotes, coordination of related genes can be achieved if their regulators, enhancers, or silencers all share the same controlling sequences.

15. DNA-binding proteins have four different structural motifs that allow them to recognize and bind to specific DNA sequences. They are the helix-turn-helix motif, the zinc finger motif, the leucine zipper motif, and the helix-loop-helix motif.

16. During interphase, DNA can exist as uncoiled *euchromatin* or coiled *heterochromatin*. Because of the coiled nature of heterochromatin, DNA of this type cannot participate in transcription. The genes included in heterochromatin are said to be *inactivated*.

17. Heterochromatin at the chromosome level is shown by the presence of an *inactivated X chromosome* or *Barr body* in the cells of female mammals. One or the other of the two X chromosomes condenses into heterochromatin so that only a single active X chromosome remains in each cell.

18. *DNA methylation*, the addition of methyl groups to cytosine nucleotides, inactivates the affected genes. Most of the DNA of the inactivated X chromosome is methylated. The inactivated X chromosome does have one gene that is actively transcribed: *XIST*, for X inactivation specific tran-script. This gene's transcribed RNA binds to the X chromo-some and may lead to its inactivation.

19. In some cases, gene expression depends on rearrangement of the DNA, as in *mating type determination* in yeast, where the rearrangement of alleles between active and inactive loci determines the mating type of a cell.

20. DNA rearrangement is involved in producing the function-al genes that code for the highly variable antibody proteins.

21. *Gene amplification* is an important mechanism for increasing the amount of a needed gene product. A good example is amplification of the genes coding for rRNAs.

22. Many copies of the genes coding for rRNA are arranged in a series in egg cells to ensure adequate rRNA for the pro-duction of the ribosomes needed during the rapid develop-ment of the embryo. Electron micrographs of transcription in egg cells show the tandem arrangement of the copies of the rRNA genes. Each gene is separated from the next by a "spacer."

23. Alternate splicing of exons can result in different gene products. An example is the family of tropomyosins pro-duced in different tissues.

24. The stability and lifetime of mRNAs differ and this may play an important role in the development process.

25. Another translational control mechanism depends on the requirement that mRNA must be capped with a modified G unit before it can be translated. In some species, the G cap on a given variety of mRNA will not be modified until translation of that kind of mRNA is required. Once modi-fied, translation can begin.

26. Several mechanisms affecting translation of hemoglobin components are responsible for maintaining the proper balance of globin chains and the heme groups. For exam-ple, excess heme increases translation of the globin mRNAs by removing a block on the initiation of translation at the ribosome.

27. A common posttranslational control mechanism involves regulating the lifetime of the gene product (protein). A pro-tein can be targeted for degradation by attaching to it a small protein called *ubiquitin*. Chains of ubiquitin attract proteases, which then break down the protein.

Questions *(for answers and explanations, see page 128)*

1. Which of the following RNA polymerases transcribe genes that code for proteins?

a. RNA polymerase I
b. RNA polymerase II
c. RNA polymerase III
d. RNA polymerase I and RNA polymerase III
e. All three

2. The TATA box is
a. a transcription factor.
b. an enhancer.
c. a binding site for TFIID.
d. a leader sequence for a protein that must move through a membrane.
e. a growth factor.

3. How many Barr bodies would you observe in the cells of an individual with Klinefelter syndrome and the genotype XXXXY?
a. 0
b. 1
c. 2
d. 3
e. 4

4. Classify the following events associated with the control of gene expression as transcriptional, posttranscriptional, or posttranslational.

_____	Increasing the longevity of an mRNA
_____	Modification of the G cap
_____	Presence or absence of enhancers
_____	Methylation of a gene
_____	Presence or absence of the appropri-ate transcription factor
_____	Gene amplification through duplica-tion
_____	Attachment of ubiquitin to a protein
_____	Conversion of heterochromatin into euchromatin
_____	Modification of a gene product

Activity *(for answers and explanations, see page 128)*

- The following photomicrograph shows transcription of an rRNA gene in the nucleolus of an oocyte. Label (1) approximate location of an RNA polymerase protein, (2) the DNA molecule, (3) an RNA molecule, and (4) the location of a "spacer" region on the DNA. Use an arrow to show the direction of transcription.

  Is rRNA or mRNA being produced? _____

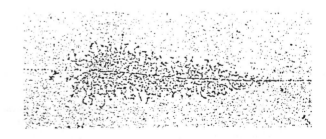

Integrative Activity (for answers and explanations,
   see page 128)

• Using the following descriptors, label the boxes next to the
   arrows in the diagram below showing the levels of control
   of gene expression in eukaryotes :

   Translational

   Processing

   Posttranslational

   Transcriptional

   Transport

   mRNA stability

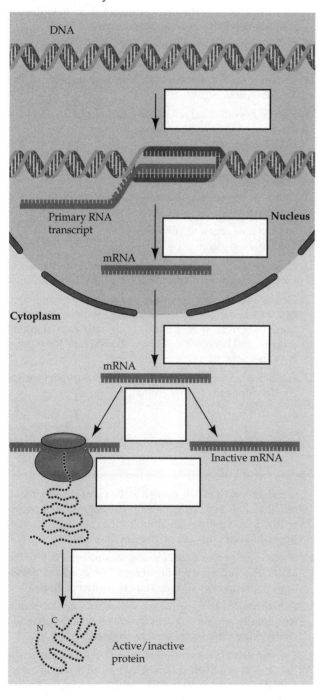

Integrative Questions (for answers and explanations,
   see page 128)

1. Select all of the following statements contrasting prokaryotes and eukaryotes that are *not* true.
   a. Eukaryotes lack operons.
   b. The proportion of total DNA that is noncoding in
      prokaryotes is about the same as in eukaryotes, even
      though the genome size is smaller in prokaryotes.
   c. In eukaryotes, hormones can trigger selective gene tran-
      scription.
   d. The promoter region of eukaryotes is more complex
      than the promoter region of prokaryotes.
   e. Eukaryotes have three types of RNA polymerases;
      prokaryotes have only one type.
   f. Transcription and translation are physically segregated
      in the eukaryotes.
   g. In most prokaryotes, RNA processing is not required.
   h. Prokaryotic chromosomes lack telomeres.
   i. The prokaryotic genome has more moderately repetitive
      DNA than the eukaryotic genome, but less highly repeti-
      tive DNA.

2. Label each of the following molecular components as consisting of nucleic acid (**n.a.**), protein (**p**), or both (**b**):

   _____   Promoter

   _____   Consensus sequence

   _____   Enhancer

   _____   Telomerase

   _____   Transposable elements

   _____   Poly A "tail"

   _____   snRNPs ("snurps")

   _____   U1 and U2

   _____   Hormone-response element

   _____   TATA box

   _____   Silencer

   _____   Activator and repressor

   _____   Transcription factor

   _____   Ubiquitin

   _____   RNA polymerase Type III

   _____   Heterochromatin

15

Development: Differential Gene Expression

CHAPTER LEARNING OBJECTIVES—after studying this chapter you should be able to:

❑ State a comprehensive definition of the term "development."

❑ Explain what is meant by growth and the roles of cell division and cell elongation in growth of plants and animals.

❑ Relate the following terms to each other: determination, differentiation, and totipotent.

❑ Define the terms morphogenesis, and pattern formation.

❑ Describe the significance of the nuclear transfer experiments performed by Robert Briggs and Thomas King and also by John Gurdon.

❑ Explain what complete metamorphosis is, using the terms "larva," "pupa," and "imaginal disc."

❑ Discuss the importance of the experiments involving the transplantation of imaginal discs in *Drosophila* fruit flies and the phenomenon of transdetermination.

❑ Explain the importance of studies of plant cell cloning to the idea that differentiation does not involve permanent changes in the genome.

❑ Discuss how DNA hybridization has been used in studies of differential gene expression.

❑ Explain the significance of the studies done by Sven Hörstadius on polarity in frog eggs and embryos.

❑ Describe the features of the nematode worm *Caenorhabditis elegans* that make it a model organism for developmental biologists.

❑ Describe the importance of segregation of cytoplasmic determinants in the zygote and embryo of *C. elegans*.

❑ Describe the inductive control of the development of the vertebrate eye.

❑ Discuss induction at the cellular level with respect to vulva formation in *C. elegans*.

❑ Explain how directed cell division and elongation in plants can be used to to achieve pattern formation in the absence of cell movement.

❑ Describe how positional information may determine vascular tissue differentiation in plants.

❑ Define apoptosis, and describe its importance in pattern formation.

❑ Explain how organ identity genes control flower development in some plants.

❑ Describe the work done by Lewis Wolpert on the role of morphogens in embryonic chick wing development.

❑ Describe the sequence of events that leads to pattern formation in *Drosophila* larvae, including the involvement of maternal effect, segmentation, gap, pair rule, segment polarity, and homeotic genes, and explain how these genes have their effects.

❑ Explain what the homeobox is and discuss its relevance to the development of many animals.

THE PROCESSES OF DEVELOPMENT • THE ROLE OF DIFFERENTIAL GENE EXPPRESSION IN DIFFERENTIATION (pages 330–336)

Key Concepts

1. *Development* is the progressive change in the structure and function of an individual as it passes through the stages of its life cycle.

2. The *embryo* stage begins with the zygote and continues until the organism becomes independent of the nutrients provided by the mother.

3. Development of an organism involves *growth,* an increase in size. Growth usually results from both mitotic *cell division* and *cell expansion.* In plant development, cell division and cell expansion usually occur concurrently; in animal development, cell expansion is sometimes delayed until a later stage in the embryo.

4. Somatic cells of multicellular organisms retain all of the DNA present in the fertilized egg. The events of development result from differential expression of genes.

5. *Differentiation* is the process whereby a cell becomes structurally and functionally specialized. Differentiation results from differential gene expression.

6. *Determination* is the commitment of a cell to a particular developmental path or *fate.* Once a cell's fate is determined, differentiation occurs and it becomes structurally and functionally specialized.

7. *Morphogenesis* is the progressive change in the shape of a multicellular body as caused by *pattern formation*, the organization of tissues to form composite structures. Differential cell division and expansion are important in the morphogenesis of plants; cell movements and programmed cell death are also important in animals.

8. Differential gene expression involves selective transcription of certain genes and the regulation of posttranscriptional events and translation.

9. Although the zygote is *totipotent*, able to differentiate into any cell type, the cellular descendants of the zygote become determined and differentiate into specific cell types. In most cases, differentiation does not result from irreversible genetic changes, such as loss of genes.

10. Robert Briggs and Thomas King transferred nuclei from the cells of frog embryos of various stages into enucleated eggs. If the nuclei were from cells of an early embryo, development of the activated donor egg was normal. Nuclei from the cells of older embryos did not usually develop normally.

11. John Gurdon, working with nuclei derived from older embryos and swimming tadpoles confirmed that the nuclei of differentiated cells remain totipotent and are still able to act like a zygote nucleus if placed into the proper environment (an enucleate egg).

12. Nuclear transplant experiments convinced biologists that differentiation does not result from the permanent inactivation or loss of genetic material.

13. In *complete metamorphosis*, an insect larva transforms into a *pupa*. After radical reorganization, the adult emerges from the pupa.

14. In a typical insect larva, clusters of undifferentiated cells called *imaginal discs* are responsible for transforming the larva into an adult during metamorphosis. Each disc will give rise to a distinct part of the adult insect.

15. An imaginal disc transplanted to a different site in another larva will form its distinctive organ wherever it is placed in the host insect. The cells of discs transplanted to adults continue to divide but give rise to undifferentiated cells. If such a disc is later transplanted to a larva it will usually form its distinctive body part.

16. Occasionally, imaginal discs transplanted from larvae to adults and back to larvae will *transdetermine*—they will give rise to unexpected organs—indicating that all genetic information is present in imaginal disc cells.

17. Plant cell cloning done with carrot root cells showed that a differentiated cell could be made to act like a fertilized egg and give rise to an adult plant if provided with the appropriate chemical signals.

18. Nucleic acid hybridization has shown that a probe for a gene that codes for a specialized protein normally is only found in a specific cell type, such as the α-globin gene in red blood cells, will also find its complement in a different cell type, such as a nerve cell. This indicates that differentiated cells do not lose genes.

19. A probe for a gene that codes for a specialized protein if applied to cellular mRNA's will normally only find its target in the cell type normally requiring that protein.

This indicates that although cells have all genes, only the appropriate genes are transcribed.

Questions (for answers and explanations, see page 128)

1. Select a sentence completion from the following choices to create a statement that is *not* true. When a cellular descendant of the zygote can no longer give rise to all the cells of a multicellular organism, we can say that
   a. it is transdetermined.
   b. it is no longer totipotent.
   c. it is determined.
   d. its developmental potential is reduced.
   e. its developmental fate is achieved.

2. In *Drosophila*, an imaginal disc that would normally form wings is transplanted into the head region of an adult. If these disc cells are later transplanted into a second larva, they will most likely give rise to
   a. a wing.
   b. a head.
   c. undifferentiated tissue.
   d. cells that die.
   e. a wing or head, depending on where in the larva they are placed.

3. The work of Briggs and King, and later John Gurdon, on the transplantation of nuclei into enucleated eggs in the frog showed that
   a. sperm are not necessary for development in the frog.
   b. while differentiation is reversible in plants, it is irreversible in animals.
   c. gene loss is not the cause of differentiation.
   d. the cells of the early frog embryo are fully determined.
   e. older nuclei are more likely to cause normal development in host eggs.

4. Suppose you have nucleic acid probes for each of the following sequences in the mouse:
   A. Myoglobin gene   B. Myoglobin mRNA
   C. Cytochrome *C* gene   D. Cytochrome *C* mRNA
   You apply the probe to cells from the mouse's salivary glands. Which probe(s) will hybridize?
   a. A, B, C, and D
   b. None
   c. Only A and C
   d. Only A, C, and D
   e. Only C and D

5. Which one of the following studies was *not* important in supporting the assertion that differentiation does not involve permanent changes in the genome?
   a. Nuclear transplant studies in frogs
   b. DNA hydridization studies
   c. *Drosophila* studies in which imaginal discs from larvae are transplanted into adult flies and give rise to undifferentiated tissue
   d. *Drosophila* imaginal disc studies in which transdetermination was observed
   e. Plant cell cloning studies

THE ROLE OF CYTOPLASMIC SEGREGATION IN CELL DETERMINATION • THE ROLE OF EMBRYONIC INDUCTION IN CELL DETERMINATION • THE ROLE OF PATTERN FORMATION IN ORGAN DEVELOPMENT (pages 337–344)

Key Concepts

1. There may be a *polarity* in the distribution of yolk and other substances in the egg, and after fertilization this polarity may change. Cytoplasmic differences in the cells of an embryo can result from polarity in the zygote.

2. Sven Hörstadius showed that if sea urchin embryos at the eight-cell stage were split in two, only those divisions that divided the embryo into right and left halves were able to produce normal, though dwarfed, larvae. Since nuclear transfer experiments have shown that the nuclei in these cells are still totipotent, these results must be due to cytoplasmic differences in the blastomeres.

3. Further studies by Hörstadius with sea urchin eggs showed that an unequal distribution of materials exists in the eggs, such that only eggs divided into halves with animal and vegetal portions have the potential to develop into normal larvae.

4. *Cytoplasmic determinants* are chemicals whose unequal distribution in the egg cytoplasm affects the course of development.

5. The worm *Caenorhabditis elegans* is a soil-dwelling nematode that has become a model organism for developmental biologists. Its entire genome is being sequenced and the developmental fate of each of its 959 cells has been mapped. Cells that will produce the gametes are called the *germ line*. *C. elegans* completes its development in about three and a half days

6. Cytoplasmic determinants, called *P granules*, specify which cells will eventually produce sperm and eggs in *C. elegans*. At each cell division during development, microfilaments are responsible for moving the granules into the correct cell.

7. *Induction* is the process in which one tissue causes an adjacent tissue to develop in a certain way. Generally a chemical signal called an *inducer* is released by the inducing tissue.

8. Formation of the lens of the eye is an example of induction. *Optic vesicles* from the forebrain induce the formation of a *lens placode* in overlying ectoderm. The lens placode invaginates, detaches from the ectoderm, and forms the lens. A signal released by the optic vesicle is essential for lens formation.

9. The lens also interacts with the optic vesicle so that the *optic cup* that forms from the vesicle is properly sized to fit the lens. The lens also induces ectoderm overlying it to become the *cornea* of the eye.

10. There is much yet to be learned about the chemical nature of *inducers*. Some are diffusible proteins, such as a growth factor discovered in frog embryos, but some inducers are insoluble, extracellular proteins, such as collagen.

11. One form of *C. elegans* is a hermaphrodite, having both sperm- and egg-producing organs.

12. Work on *C. elegans* has shown that development is controlled by chemical switches that determine which developmental path cells will take.

13. In the case of vulva formation in *C. elegans*, the primary switch is an inducer produced by the anchor cell. Cells receiving enough of this inducer become vulval precursors. Cells not receiving enough of the inducer become epidermis.

14. The vulval precursor that was nearest the anchor cell becomes the primary precursor and produces a second inducer. Only cells receiving enough of the secondary inducer become secondary precursors.

15. Inducers act by turning on specific genes in the cells receiving the inducers. Some inducers involved in differentiation of mesoderm in vertebrates have been identified as growth factors that cause a signal transduction cascade resulting in altered gene transcription.

16. Because plant cells are anchored in place by the cell wall, plant cells can not move. Orientation of cellulose microfibrils and cytoskeleton microtubules can control the direction of cell elongation. Microtubules forming a *preprophase band* determine the plane of cell division.

17. *Apoptosis*, programmed death of certain cells, is very important to the normal development of a multicellular organism. Apoptosis results from the expression of certain "death" genes.

18. Homologous genes for several of the death genes studied in *C. elegans*, *ced-3* and *ced-4*, and *ced-9*, an apoptosis inhibitor gene, have been found in humans.

19. Organ identity genes in plants, such as *Arabidopsis thaliana*, specify how the successive whorls of a developing flower—sepals, petals, stamens, and pistils—are organized.

20. *Pattern formation*—the organization of tissues to form composite structures such as organs—is regulated, in part, by *positional information*, the spatial relationship of one structure to another.

21. Positional information in plants seems to determine which cells will become water- and mineral-conducting tissue and which will become photosynthate-conducting tissue. Distance from the surface, and consequent O_2 and CO_2 gradients, may determine the course of differentiation.

22. *Lewis Wolpert* developed a theory of positional information based on his studies of limb bud formation in the chick, which postulated the existence of gradients of *morphogens*—diffusible chemicals that establish concentration gradients. Cells exposed to different concentrations of the morphogen develop along different lines.

23. The *zone of polarizing activity (ZPA)*, a region on the posterior margin of the wing bud, determines pattern formation along the anteroposterior axis. Grafting of an additional, donor ZPA to different positions on a wing bud leads to abnormal pattern formation, suggesting that distance from the ZPA determines what wing structures will develop.

24. A gene called *sonic hedgehog* appears to encode a protein that acts as a morphogen, forming a gradient in the wing bud. This gene is expressed only by cells in the ZPA.

25. *Retinoic acid*, a derivative of vitamin A, can create an effect like the ZPA in inducing a duplication of digits. Retinoic acid may cause cells in the ZPA to express *sonic hedgehog*.

Activity (for answers and explanations, see page 129)

• Beginning below with the simple diagram showing the initial stage in induction of the vertebrate eye, draw the stage where a complete lens has been formed. Label the following structures on these drawings: *optic vesicle*, *optic cup*, *lens placode*, and *cornea*.

Questions (for answers and explanations, see page 129)

1. Which of the following observations *does not* suggest that a polarity of cytoplasmic substances in eggs or embryos is important in normal development?
a. Bisection of the eight-celled sea urchin embryo only leads to normal development if each half receives cells from both animal and vegetal hemispheres.
b. The distribution of P granules in the eggs of *C. elegans* is initially uniform.
c. Sea urchin eggs divided into halves that receive only animal pole egg cytoplasm cannot undergo gastrulation.
d. Cytoplasmic determinants in *C. elegans* are only synthesized in the posterior end of the worm, where sex organs later develop.

2. Grafting an extra ZPA to the anterior margin of a wing bud would likely result in
a. a normal wing.
b. mirror-image duplication of the wing with two complete sets of wing bones.
c. mirror-image duplication of the wing with two sets of lower wing bones.
d. mirror-image duplication of the wing with two sets of upper wing bones.
e. an abnormal wing with no wing bones.

3. Which of the following statements about vulva development in the roundworm *C. elegans* is *not* true?
a. The anchor cell produces the primary inducer.
b. The activated gene in the epidermal precursor cells is is different from activated genes in primary and sec-

ondary precursors.
c. Only primary precursor cells produce the secondary inducer.
d. Proximity to the anchor cell determines the fate of cells relative to vulva formation.
e. If the anchor cell is destroyed, a neighboring cell will produce the primary precursor.

4. Which of the following statements about pattern formation in organ development is *not* true?
a. Humans and the roundworm *C. elegans* share similar genes that control apoptosis during development.
b. Organ identity genes are important in determining whether a cell becomes specialized for conducting water or the products of photosynthesis.
c. Positional information influences pattern formation in both plants and animals.
d. Cell movement is more important in pattern formation in animals than plants.
e. Preferential cell elongation in plants is determined by the orientation of cellulose microtubules.

THE ROLE OF DIFFERENTIAL GENE EXPRESSION IN ESTABLISHING BODY SEGMENTATION   (pages 344–348)

Key Concepts

1. Research on *Drosophila melanogaster* larvae has provided important insights into the process of pattern formation. The *Drosophila* adult consists of 13 segments—one head segment, three thoracic segments, eight abdominal segments, and a genital segment.

2. Development of the *Drosophila* larva depends on the sequential activation of key genes. The genes that produce the cytoplasmic determinants are active first and establish the anteroposterior and dorsoventral axes of the larva.

3. Polarity in the egg of *Drosophila melanogaster* is based on the distribution of more than a dozen morphogens, some mRNAs and some proteins. These chemicals are produced by specific *maternal-effect genes* in the supporting, nurse cells around the developing egg within the mother.

4. Cytoskeleton microtubules help to distribute the morphogens within the egg, often in an asymmetric manner. Their distribution determines the dorsoventral (top–bottom) and the anteroposterior (head–rear) axes of the embryo.

5. Female *Drosophila* with mutant alleles for any of the maternal effect genes produce abnormal larvae. For example, females homozygous for the *bicoid* allele produce larvae with no head and no thorax. Females homozygous for the *nanos* allele produce larvae with missing abdominal segments.

6. If eggs from females homozygous for the *bicoid* or *nanos* genes receive cytoplasm from the region of a normal egg where the mutation has its effect, they will develop normally. Also, removal of cytoplasm from specific regions of a normal egg can produce larvae similar in appearance to a *bicoid* or *nanos* larva.

7. Three classes of *segmentation genes* in *Drosophila* larva are activated by the maternal effect genes. First the *gap genes* (organizing large areas along the anteroposterior axis) become active, then the *pair rule genes* (dividing the embryo into two-segment-long units), and last, the *segment polarity genes* (determining the boundaries and orientation of the segments). Each class of segmentation genes determines finer and finer details of the larval segmentation pattern.

8. *Homeotic genes* specify what specific structures will develop on each segment. *Homeotic mutations* result in the development of an inappropriate body part on a segment. The *antennapedia* mutation causes legs to grow in place of antennae; the *opthalmoptera* mutation results in wings where eyes should be.

9. Loci of the homeotic genes fall into several tight clusters. The best known cluster is the *bithorax complex*— eight or more genes that control development of the abdomen and posterior thorax of the fly. The *antennapedia complex* controls the development of the head and anterior thorax.

10. Mutations of one or more genes in the bithorax and antennapedia complexes produce larvae with the normal number of segments, since the segmentation genes determine the number and polarity of segments, but the way the segments differentiate is abnormal. For example, some mutations of the bithorax complex produce adults with the third thoracic segment like the second— a fly with two pairs of wings, but no halteres.

11. Recombinant DNA techniques have shown that a sequence of DNA called the *homeobox* is common to the homeotic and segmentation genes and to some genes of all animals with segmentation, including humans.

12. The homeobox, a sequence of DNA about 180 base pairs in length, codes for a 60–amino acid region, called the *homeodomain*, that is part of the protein produced by the gene in which the homeobox resides. Proteins containing this 60–amino acid sequence return to the nucleus and bind to DNA, regulating the transcription of other genes.

13. The presence of the homeobox suggests that both the bithorax and antennapedia complexes evolved from an ancestral gene that gave rise to what is now a widespread developmental control system in many organisms.

14. Differences in the concentration of mRNAs for different morphogens in eggs of *Drosophila* produce morphogen concentration gradients in the larvae. Concentration of the *bicoid* protein morphogen is greatest in the anterior end, lowest in the posterior end; the *nanos* protein concentration gradient runs in the opposite direction.

15. Bicoid protein is a transcription factor for gap genes; nanos protein affects translation of gap genes. Different gap genes are enhanced or inhibited by these proteins depending on where they are located within the interacting concentration gradients of nanos and bicoid.

16. The gap genes code for transcription factors that control the pair rule genes. The pair rule genes code for transcription factors that control the segment polarity genes. The interacting gradients of the various transcription factors set up the segmentation pattern of the larva within the multicellular cell that is yet to begin dividing.

17. Different homeotic genes are expressed in different regions of the body, and these six genes are arranged in the same order on a single chromosome. Each homeotic gene has a homeobox and codes for a transcription factor.

18. Both mice and humans have homeobox genes arranged in clusters on different chromosomes. Within each cluster, the genes are arranged in the same sequence as the anterior-posterior order of their expression.

19. The appearance of homeotic genes in diverse organisms argues for the presence of a single ancestral gene that was the evolutionary progenitor for a developmental control system common to many organisms with segmented body plans.

Questions (for answers and explanations, see page 129)

1. Pattern formation in insects depends on the sequential activation of key groups of genes. Select the gene group that is normally *first* to be activated in the fruit fly *Drosophila melanogaster*.
   a. Gap genes
   b. Segment polarity genes
   c. Homeotic genes
   d. Segmentation genes
   e Pair rule genes

2. Select the gene group in *D. melanogaster* that causes differentiation of the segments?
   a. Gap genes
   b. Segment polarity genes
   c. Homeotic genes
   d. Segmentation genes
   e Pair rule genes

3. The homeobox is
   a. a gene involved in pattern formation in *D. melanogaster*.
   b. part of a protein that remains in the nucleus and regulates transcription of other genes.
   c. a DNA sequence common to the genes of many animals with a segmented body plan.
   d. important in determining pattern formation in the chick wing bud.
   e. a group of genes subject to homeotic mutations.

4. If an egg from a homozygous *bicoid* female is incubated with cytoplasm from the anterior region of the egg of a homozygous *nanos* female, the resulting embryo will likely
   a. be normal.
   b. be *bicoid*.
   c. be *nanos*.
   d. be both *bicoid* and *nanos*.
   e. die.

5. Complete the following statements about pattern formation in *Drosophila melanogaster*.

Within the *Drosophila* larva, the concentration gradients of the morphogens _____ and _____ are established by the unequal distribution of their mRNAs within the egg.

The proteins that are coded for by the segmentation and homeotic genes are _____.

The _____ genes are arranged on their chromosome in the same order as the order of their expression from anterior to posterior in the larva.

Integrative Question (for answers and explanations, see page 129)

- Match the organisms on the right with the developmental descriptors or scientists on the left.

_____ Imaginal discs	*a.* Chick
_____ L. Wolpert	*b.* Sea urchin
_____ Organ identity genes	*c.* Frog
_____ P granules	*d.* Fruit fly
_____ R. Briggs, T. King	*e.* *Arabadopsis*
_____ S. Hörstadius	*f.* *C. elegans*

16

Recombinant DNA and Biotechnology

CHAPTER LEARNING OBJECTIVES—after studying this chapter you should be able to:

❏ Describe what is meant by the terms "recombinant DNA" and "transgenic."

❏ Explain what restriction endonucleases are, what functions they perform in the bacteria that produce them, and how a bacterium is protected from the action of its own restriction endonucleases.

❏ Differentiate between restriction endonucleases that produce sticky-ended and blunt-ended fragments, and discuss the utility of each in recombinant DNA studies.

❏ Explain the principles underlying electrophoresis and the importance of size standards to this technique.

❏ Describe what a cloning vector is, and list the essential features of a cloning vector.

❏ Describe the steps involved in using restriction endonucleases and DNA ligase to insert foreign DNA into a plasmid.

❏ Evaluate the relative importance of plasmids and viruses as cloning vectors.

❏ Discuss the various methods that have been used to insert DNA into cells, specifically transformation, transfection, use of microprojectiles, and electroporation.

❏ Explain why baker's yeast has become the most commonly used eukaryotic host cell for recombinant DNA studies.

❏ Describe the *Ti* plasmid in *Agrobacterium tumefaciens* and how it is used as a cloning vector in plant biotechnology.

❏ Describe how plasmids with antibiotic resistance genes are used to select for transgenic cells.

❏ Explain how the *lac* operon has been used to select for transgenic cells.

❏ Explain what a reporter gene is and describe how it is used.

❏ Explain how genomic libraries and cDNA libraries differ and describe the special uses of each.

❏ Discuss how synthetic genes and probes are produced and describe their uses.

❏ Explain the DNA sequencing technique involving ddNTPs with attached fluorescent dyes.

❏ Explain the steps involved in the polymerase chain reaction, the principles underlying it, and its main uses.

❏ Contrast plant agricultural biotechnology with traditional plant breeding and give some examples of biotechnology applications in agriculture.

❏ Explain what an expression vector is and describe how it is used.

❏ Describe several examples of medically useful products made through biotechnology.

❏ Describe the basis for the production of a DNA fingerprint and discuss its principal uses.

CLEAVING AND REJOINING DNA • CLONING GENES (pages 351–359)

Key Concepts

1. *Recombinant DNA* results when DNA from different sources (different species or natural versus synthetic) are combined into a single molecule. Biotechnology is contributing to basic research in many areas and has practical applications in medicine, agriculture, environmental science, and so forth.

2. Recombinant DNA technology uses naturally occurring enzymes that cleave, replicate, transcribe, and repair DNA to sequence, recombine, locate, and identify DNA fragments in the laboratory. The complementary base pairing rules underlie most of the recombinant DNA techniques.

3. *Restriction endonucleases* are enzymes produced by bacteria to protect themselves against attack by *bacteriophages*. Each different enzyme recognizes and binds to a specific sequence of bases in DNA. The enzyme cleaves the DNA within this *recognition* or *restriction site*. Many recognition sites are palindromic.

4. Once the polynucleotide chains have been broken by the restriction endonuclease, only the hydrogen bonds between the bases hold the DNA pieces together. Warm temperatures disrupt these bonds and the pieces separate.

5. Restriction endonucleases are named for the bacterial species in which they were first identified. *Eco*RI cuts within the sequence G–A–A–T–T–C; it was first discovered in *E. coli*. Several hundred different restriction endonucleases are now available to molecular biologists for cleaving DNA.

6. The bacterium is protected from attack by its own endonucleases because methyl groups have been added to bases within the recognition sites of its DNA by specific *methylase* enzymes.

7. In *electrophoresis*, negatively charged DNA fragments migrate through a porous gel in response to an electrical field. The rate of fragment movement is proportional to the size of the fragments; smaller fragments move faster than larger fragments.

8. Several samples are run in different *lanes* on an electrophoresis gel. *Size standards* (DNA fragments of known size) are also usually run so that the sizes of all fragments can be estimated. To visualize the resulting bands, they are stained with a dye that fluoresces in ultraviolet light.

9. To locate a specific DNA sequence, DNA fragments are first separated by electrophoresis. A basic pH is used to denature the double-stranded DNA into single strands and a nylon membrane is used to blot the gel. A DNA *probe* with a base sequence complementary to a "target" DNA sequence is poured over the membrane and base pairing between the probe and its complementary DNA locates the fragments with the target sequence.

10. Some restriction endonucleases cut the two DNA chains between different pairs of bases within the recognition site to produce *sticky ends*; blunt ends are made by enzymes that cut between the opposite pairs.

11. Sticky ends can be rejoined by hydrogen bonding between complementary bases, but the enzyme *DNA ligase* must join the polynucleotide chains together to effect a permanent union of the sticky ends.

12. A *transgenic* cell or organism has a genome that includes some recombinant DNA.

13. Cloning DNA involves use of host cells to produce many copies of a particular gene. Cloning involves choosing a host, getting DNA into the host, and using genetic markers so that transgenic cells can be identified.

14. Bacteria make good hosts, but because they lack the ability to remove introns and do not have many of the needed enzymes for modifying proteins, bacteria are not ideal for cloning eukaryotic DNA. Yeast and plant cells are better hosts for cloning eukaryotic DNA.

15. Foreign DNA must be part of a replicating unit—a *replicon*—to be successfully cloned. A replicating unit has an origin of replication for DNA polymerase. It can be a host chromosome or a piece of carrier DNA called a *vector*.

16. Vectors are used to move foreign DNA, either synthetic or natural, into a cell, thereby making the cell transgenic. Vectors must have an origin of replication so it can replicate with the host chromosome.

17. A cloning vector should have a single recognition site for the restriction endonuclease so that the plasmid is opened at a single point for insertion of the gene.

18. A cloning vector should carry genes conferring properties on the host cell that will make it easier to identify the transgenic cells. Antibiotic resistance genes are particularly popular because antibiotics can be used to select for cells with resistance.

19. Plasmids are small, circular bacterial chromosomes. Many plasmids can replicate independently of the main chromosome, amplifying the foreign DNA they carry. One limitation to the use of plasmid vectors is that the DNA insert cannot be more than about 5000 base pairs.

20. A piece of foreign DNA can be inserted into a *plasmid* if both DNAs have recognition sites for the same restriction endonuclease. DNA ligase then joins the complementary sticky ends of the two DNA sources to create a plasmid containing the foreign DNA.

21. Viruses are now commonly used cloning vectors. An advantage of some viruses is that they naturally infect host cells and insert their genes into their chromosomes.

22. Baker's yeast, *Saccharomyces cerevisiae*, has become a model organism for recombinant DNA studies of eukaryotic genes. The *yeast artificial chromosome (YAC)* was created to include an origin of replication, a centromere, telomeres, selectable genetic markers, and recognition sites for commonly used restriction endonucleases.

23. A popular cloning vector for plant biotechnology is the *Ti* (tumor inducing) *plasmid* in the bacterium *Agrobacterium tumefaciens*, an organism that causes *crown gall disease* in many plants. Diseased plants have large tumors.

24. The uptake of foreign DNA by a cell *(transformation)* can be encouraged by a high calcium ion concentration, which neutralizes the negative charges of the DNA and the plasma membrane.

25. Genes to be cloned are inserted into a *transposon* in the Ti plasmid, called *T-DNA*. Transformation is used to insert the plasmid into *A. tumefaciens* and the T-DNA transposon transfers copies of itself and any other genes present to the chromosomes of an infected plant. Tissue culture can be used to grow whole transgenic plants from tumor cells.

26. *Transfection* is the uptake of foreign DNA by a eukaryotic cell. Enzymatic removal of cell walls to create *protoplasts* may be necessary when transfecting plant or fungal cells.

27. The so-called "DNA particle gun" fires microprojectiles—tiny, DNA-coated tungsten particles—at plant cells.

28. Other methods for inserting DNA into cells include electroporation (use of pulses of high-voltage current) or direct injection of DNA into individual cells.

29. A common approach to identifying cells with introduced DNA is to use plasmids with two antibiotic resistance genes, one of which includes a recognition site of a restriction endonuclease. Cells that did not get any plasmids or that picked up only the foreign DNA will be killed by the antibiotics. Cells with the plasmid not containing the foreign DNA will survive in the presence of both antibiotics, and cells containing a recombinant plasmid will be killed by the antibiotic whose resistance gene was inactivated by insertion of the foreign DNA.

30. Another method for selecting transgenic cells utilizes a plasmid vector with an antibiotic resistance gene and a restriction site within the *lac* operon gene. Cells without plasmids will not grow in the presence of the antibiotic. Cells with plasmids will, but cells containing the plasmid with the foreign DNA inserted into the operon will not have a functional operon, whereas nontransgenic cells will.

31. A *reporter gene* is any gene that is used to identify the presence of a vector in eukaryotic cells. The gene coding for green fluorescent protein is an example.

Activities (for answers and explanations, see page 129)

- The following diagram shows the recognition sequences for the two restriction endonucleases **(a)** *Hpa*I and **(b)** *Hind*III. Complete the complementary ends for each fragment.

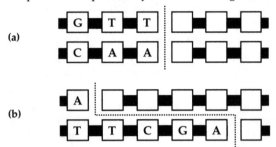

(a)

(b)

- Which of the two restriction endonucleases (*Hpa*I or *Hind*III) would be best for inserting foreign DNA into a plasmid? Explain your answer.

Questions (for answers and explanations, see page 129)

1. Which of the following statements about restriction endonucleases is *not* true?
   a. On the average, longer restriction endonuclease recognition sites (more base pairs) would be less common in DNA than shorter recognition sites (fewer base pairs).
   b. Methylated recognition sites are not recognized by restriction endonucleases.
   c. Restriction endonucleases catalyze the breakage of hydrogen bonds between the nitrogenous bases within the recognition site.
   d. Restriction endonucleases are named after the bacterial species in which they were first discovered.
   e. The blunt-ended fragments produced by some restriction endonucleases can rejoin with any other blunt-ended fragment.

2. Which of the following is *not* an important characteristic of a plasmid cloning vector?
   a. It should be a replicon.
   b. It should have many recognition sites for the restriction endonuclease used to isolate the foreign DNA.
   c. It should have a gene for antibiotic resistance.
   d. It should have a single origin of replication.
   e. It should have a gene that makes it possible to identify transgenic cells easily.

3. Which of the following techniques is *not* commonly used to create a transgenic cell?
   a. Transformation
   b. Lysing a protoplast in the presence of the vector
   c. Infection by a viral cloning vector

   d. Use of microprojectiles
   e. Use of electroporation

4. Order the following steps in the gene cloning process.
   _____ Encourage growth of the transgenic cells
   _____ Place host cells and DNA into calcium chloride solution
   _____ Treat with ligase
   _____ Create fragments with restriction endonucleases
   _____ Transform cells
   _____ Use selection plating with antibiotics

5. The plasmid *p*BR322 has genes for resistance to both ampicillin (*amp*^r^) and tetracycline (*tet*^r^). A *Bam*HI recognition site lies within the *tet*^r^ gene. Foreign DNA and the plasmid are restricted with *Bam*HI, and DNA ligase is added. Wild-type *E. coli*, which is sensitive to both antibiotics, is transformed with the ligation mixture. Cells are spread onto plates containing media of three types: no antibiotics, ampicillin-only, tetracycline-only. Add the labels that follow to the appropriate rows in column one of the next table indicating the type of DNA taken by *E. coli*:

A = No DNA taken up   **B** = Foreign DNA only
C = Nonrecombinant pBR322 **D** = recombinant pBR322

DNA taken up by wild-type *E. coli*	Growth (+) or no growth (−) on plates containing…		
	No antibiotics	Tetracycline	Ampicillin
	+	−	−
	+	−	+
	+	−	−
	+	+	+

6. Which of the following is *not* a reason baker's yeast (*Saccharomyces cerevisiae*) is a popular organism for recombinant DNA studies?
   a. It is a eukaryote.
   b. It can accommodate yeast artificial chromosomes (YAC).
   c. It completes its life cycle in 2 to 8 hours.
   d. It is multicellular.
   e. It has a small genome size.

7. Which of the following characteristics of the bacterium *Agrobacterium tumefaciens* is *not* a reason it is much used in plant recombinant DNA research?
   a. It only infects a few plants.
   b. It has a large plasmid called Ti.
   c. Its plasmid contains the transposon T-DNA.
   d. The plasmid can serve as a cloning vector.
   e. Transformation can be used to insert the plasmid into the plant.

8. Five DNA samples (a through e) were treated with a restriction endonuclease, and the resulting fragments separated in the following electrophoresis gel. Which of the DNA samples *both* contained the greatest number of recognition sites for this enzyme and produced the smallest fragments?

SOURCES OF GENES FOR CLONING • SOME ADDITIONAL TOOLS FOR DNA MANIPULATION • BIOTECHNOLOGY: APPLICATIONS OF DNA MANIPULATION (pages 359–370)

Key Concepts

1. A *genomic library* is the fragmented genome of an organism inserted into a cloning vector. First a DNA source, such as a species' genome, is fragmented with restriction endonucleases. The collection of fragments is then inserted into plasmids or bacteriophage DNA and cloned into bacteria. Each clone contains a single fragment from the library.

2. Gene libraries include all of the DNA, including introns, promoters, and regulatory sequences like the TATA box.

3. *Complementary DNA*, called *cDNA*, is made from RNA, usually mRNA, with the enzyme *reverse transcriptase*. First a string of T residues (*oligo dT*) is allowed to bind to the poly A tail on the 3′ end of mRNA. The oligo dT serves as a primer for reverse transcriptase.

4. If DNA precursors (2′-deoxyribonucleoside triphosphates) are available, reverse transcriptase produces a strand of cDNA complementary to the mRNA. The pH is then adjusted to favor separation of the mRNA–cDNA chain and degradation of the mRNA.

5. cDNA is made double-stranded using *DNA polymerase* and can then be incorporated into bacteriophage vectors to produce a *cDNA library* of the source DNA.

6. Because a cDNA library contains only *expressed DNA* (DNA representing mRNAs), comparing cDNA from different cells or tissues can provide insight into which genes are on or off during the differentiation process

7. Automated DNA synthesis equipment can be used to produce *synthetic DNA* for short to medium length genes. If the amino acid sequence of the desired protein is known, the base sequence of an artificial gene can be determined and initiation and termination codons added. Appropriate restriction endonuclease recognition sequences must also be added to the ends of the gene so it can be spliced into the cloning vector.

8. Synthetic DNA technology has allowed scientists to create specific mutations in genes and observe the effect of these mutations. This approach has been useful for testing hypotheses about gene action and also in the design of specific drugs.

9. Computational biology makes use of computers to analyze the three-dimensional shapes of molecules. One goal is to construct genes that code for polypeptides that act as inhibitors of disease-causing enzymes.

10. The *DNA base sequencing* technique makes use of 2′,3′-dideoxyribonucleoside triphosphates (*ddNTPs*) in addition to the normal DNA precursors 2′-deoxyribonucleoside triphosphates (*dNTPs*). Each of the four ddNTPs is bound to a fluorescent molecule that emits a different colored light when excited by laser light.

11. The single-stranded DNA to be sequenced is mixed with DNA polymerase, a primer, the four dNTPs (dATP, dGTP, dCTP, and dTTP), and a small amount of each of the ddNTPs (ddATP, ddGTP, ddCTP, and ddTTP). Whenever a ddNTP is added to the growing complementary strand, the chain is terminated.

12. After synthesis is complete, the reaction mixture is separated into fragments on an electrophoresis gel. By progressing from the shortest to the longest fragments, the color of each band identifies the base at that position in the newly synthesized DNA. The complementary base pairing rules then allow us to deduce the sequence of bases in the original DNA.

13. DNA sequence data is now routinely stored in computer databases that can be searched for known sequences of promoters, start-stop codons, consensus sequences, and so forth. Or DNA sequences from different sources can be compared to find homologous genes such as the homeotic genes.

14. The *polymerase chain reaction* (PCR) makes it possible to produce billions of copies of a gene in a cell-free environment in a few hours. First, two probes of several dozen bases in length must be made that are complementary to sequences at the 3′ ends bounding the DNA chain to be cloned. The probes serve as primers to initiate synthesis of strands complementary to the target DNA.

15. By alternately heating and cooling the DNA in the presence of DNA polymerase, additional primer, and DNA precursors, DNA strands separate and hybridize with the primer, and synthesis of new strands continues. The amount of DNA doubles with each cycle.

16. Because of the high temperatures needed to denature the DNA (over 90°C), a temperature-resistant DNA polymerase from the bacterium *Thermus aquaticus* is used in the polymerase chain reaction.

17. The polymerase chain reaction is used to amplify DNA collected from crime scenes, in genetic screening, and in the study of prehistoric DNA.

18. mRNA must remain single-stranded to be translated. *Antisense RNA* has a base sequence that is complementary to a specific mRNA. When it binds to its mRNA, it blocks translation. This is useful for preventing the expression of a target gene. Antisense RNA can be added directly or a gene coding for the antisense RNA can be inserted into a vector. *Ribozymes* are sometimes added to the antisense RNA to ensure destruction of the mRNA.

19. The field of biotechnology includes any use of microbial, plant, and animal cells to benefit humans. Modern biotechnology relies heavily on recombinant DNA technology.

20. *Expression vectors* have the necessary additional sequences to insure that the inserted gene is expressed, including promoters, transcription terminator, and ribosome binding sites for prokaryotic hosts, and a poly A addition site, transcription factor sites, and enhancers for eukaryotic hosts. Occasionally a genetic mechanism to turn gene expression on and off or a means to package and export the gene product will be added to the expression vector.

21. Medically useful products made through biotechnology include *tissue plasminogen activator (TPA)*, an enzyme that catalyzes conversion of plasminogen to plasmin. Plasmin dissolves blood clots and is used to remove clots in heart attack and stroke victums. cDNA made from human TPA mRNA was inserted into an expression vector and cloned in *E. coli.* Other examples include production of erythropoietin (EPO) and human insulin.

22. *Traditional plant breeding* involves the selection and breeding of strains of plants with desirable phenotypes. The limitations of traditional plant breeding are that it is space and time intensive, the entire genome of the parents is passed on to the progeny, and the availability of natural genetic strains is limited.

23. *Plant agricultural biotechnology* is much more economical of time and space, only the genes of interest need to be transferred from one plant to another, and genes can be transferred from any other living cell.

24. Agricultural applications of biotechnology include the production of transgenic plants that can produce proteins toxic to many pest insect larvae. The genes producing these proteins come from bacteria such as *Bacillus thuringiensis.* They were extensively modified and inserted into the Ti plasmid. Other examples include transgenic tomatoes whose fruit ripens slowly and transgenic crop plants that are resistant to the herbicide glyphosate.

25. DNA fingerprinting refers to the use of very diverse regions of a genome as a way to characterize the genetic uniqueness of individuals. In humans, an important system for DNA fingerprinting is the moderately repeated DNA sequences that occur in tandem throughout the genome.

26. To do a DNA fingerprint, primers are needed so that the region of interest can be amplified, typically with PCR. Then, the amplified DNA is cut with appropriate restriction endonucleases, and the fragments are separated with electrophoresis. Individuals differing in the number of tandem repeats will have different size fragments. By looking at eight or more different tandem repeat regions, each individual will have a unique pattern.

27. DNA fingerprinting has had many diverse applications, such as forensic analysis and selective breeding of endangered species and crop plants.

Questions (for answers and explanations, see page 130)

1. Which of the following statements characterizing gene and cDNA libraries is *not* true?
   a. Gene libraries can include the entire genome of a species.
   b. Gene libraries are especially useful for determining which genes are active during differentiation.
   c. Gene libraries can be used to study regulatory sequences like the TATA box.
   d. cDNA libraries contain only expressed DNA.
   e. cDNA libraries do not include introns.

2. Order the following steps in the process of creating a cDNA library.

   _____ Add oligo dT primer to mRNA

   _____ Denature DNA–RNA hybrid molecules

   _____ Insert cDNA into DNA vector

   _____ Add DNA polymerase

   _____ Degrade mRNA

   _____ Add reverse transcriptase

3. Match the molecular techniques and tools on the left with the appropriate characteristics on the right.

   _____ DNA fingerprinting   a. Makes cDNA from mRNA

   _____ Antisense RNA   b. Requires primers

   _____ Expression vector   c. Allows gene expression

   _____ Reverse transcriptase   d. Used in electrophoresis

   _____ Size standards   e. Variable tandem repeats

   _____ DNA base sequencing   f. Uses ddNTPs

   _____ PCR   g. Prevents translation

4. The first step in the use of the polymerase chain reaction to amplify a gene is to
   a. make cDNA for the gene of interest.
   b. produce probes that can serve as primers on the borders of the gene.
   c. add DNA polymerase.
   d. sequence the gene using the Sanger method.
   e. add reverse transcriptase.

5. Assume that it requires 5 minutes for one cycle of the polymerase chain reaction. If you start with one copy of a gene, theoretically how many minutes will be required to produce a thousand copies?
   a. 9
   b. 20
   c. 30
   d. 50
   e. 60

6. What are some of the major limitations of traditional plant breeding and how is recombinant DNA technology helping to address these limitations?

Activity (for answers and explanations, see page 130)

- The following diagram shows an electrophoresis gel produced by application of the Sanger sequencing method to a fragment of DNA 12 base pairs in length. The DNA was divided into four samples, and the specific ddNTP (A, T, C, G) added to each sample is shown across the top of the gel.

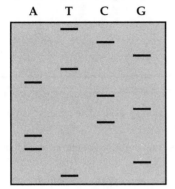

What is the sequence in the newly synthesized strand?

What is the sequence in the original DNA fragment?

Integrative Questions (for answers and explanations, see page 130)

1. The enzyme needed to make cDNA from mRNA is
   a. DNA ligase.
   b. DNA polymerase.
   c. reverse transcriptase.
   d. a restriction endonuclease.
   e. RNA polymerase.

2. The enzyme that is used to combine nucleic acid restriction fragments is
   a. DNA ligase.
   b. DNA polymerase.
   c. reverse transcriptase.
   d. a restriction endonuclease.
   e. RNA polymerase.

3. Select all of the following techniques that depend on the rules of complementary base pairing.
   a. Estimating DNA fragment sizes with gel electrophoresis
   b. Production of a DNA probe
   c. Production of a cDNA library
   d. Polymerase chain reaction
   e. DNA sequencing
   f. Production of antisense RNA

17

Molecular Biology and Medicine

CHAPTER LEARNING OBJECTIVES—After studying this chapter you should be able to:

❑ Explain the mode of inheritance and general features of the following genetic diseases: phenylketonuria, sickle-cell anemia, familial hypercholesterolemia, cystic fibrosis, Duchenne muscular dystrophy, osteogenesis imperfecta, fragile-X syndrome, hemophilia, and Huntington's disease.

❑ Describe how studies of the causes of some human genetic diseases support the ideas of one gene → one enzyme and the one mutation → one amino acid change.

❑ Compare two common strategies that have been used for isolating and describing human disease-causing mutations.

❑ Explain the technique of positional cloning and why it is especially useful if the gene to be isolated produces no detectable protein.

❑ Describe what a restriction fragment length polymorphism (RFLP) is and how RFLP mapping is used to locate human genes.

❑ Discuss the differences in the sizes of human gene mutations and give examples of disease-causing genes.

❑ Explain why mutational "hot spots" are sometimes associated with areas of the DNA where cytosines have been methylated.

❑ Explain what a triplet repeat is and describe the characteristics of triplet repeats seen in genetic diseases such as fragile-X syndrome.

❑ Describe the objectives of genetic screening and its application in the disease phenylketonuria.

❑ Explain the advantages of DNA testing in genetic screening and the features of two common approaches to DNA testing: allele-specific cleavage and allele-specific oligonucleotide hybridization.

❑ Explain why cancer is considered a genetic disease that is not inherited.

❑ Use the following terms to discuss the general characteristics of cancer: "metastasis," "benign tumor," "malignant tumor," "carcinoma," "sarcoma," "lymphoma," and "leukemia."

❑ Describe the characteristics of carcinogens and list common types.

❑ Explain the involvement of proto-oncogenes and tumor-suppressor genes in the regulation of cell growth.

❑ Describe the principal mechanisms that can disrupt normal cell growth regulation and cause cancer.

❑ Discuss the involvement of tumor-suppressor genes, such as *p53*, in the development of cancer.

❑ Explain the "two-hit" hypothesis for the onset of cancer.

❑ Describe the three approaches to treating genetic disorders: modifying the mutant phenotype, using metabolic inhibitors, and supplying the missing protein.

❑ Discuss the advantages and disadvantages of gene therapy using somatic cells such as white blood cells or stem cells.

❑ Explain the overall objectives of the human genome project (HGP) and the major methods being used.

❑ Describe the stepwise approach used in sequencing a genome, starting with a collection of chromosomes to the sequencing of ordered, 500-base-pair pieces.

❑ Discuss the currrent achievements of the HGP and some ethical concerns.

PROTEIN AS PHENOTYPE • MUTATIONS AND HUMAN DISEASES (pages 373–383)

Key Concepts

1. Mutations in genes are often expressed phenotypically as abnormal protein, including enzymes and carrier, receptor, transport, structural, and blood clotting proteins.

2. Individuals homozygous for the recessive allele causing *phenylketonuria (PKU)* cannot convert phenylalanine into tyrosine because they lack a functional version of the enzyme phenylalanine hydroxylase. Phenylalanine accumulates and is converted into phenylpyruvic acid, which is excreted in the urine. Frequently, the abnormal enzyme differs by only a single amino acid (tryptophan instead of alanine at position 408).

3. Because tyrosine is needed to produce the pigment melanin, many individuals with phenylketonuria have light colored hair and skin. The cause of mental retardation in PKU is unknown.

4. *Alkaptonuria* is another genetic disease due to a block in the metabolism of an amino acid. In this disease the

amino acid is tyrosine and the block, due to the absence of a required enzyme, results in an accumulation of homogentisic acid, which turns the urine black.

5. A protein is *polymorphic* if there are at least two molecular variants and the least common form occurs a minimum of 1 percent of the time.

6. *Sickle-cell anemia*, common in populations derived from equatorial Africa and the Mediterranean region, is characterized by abnormal hemoglobin and defective red blood cells in individuals homozygous for the recessive, autosomal allele.

7. Sickle-cell anemia is caused by a point mutation that substitutes glutamic acid for valine at position 6 of the 141 amino acids in the β chain. Because glutamic acid is charged, whereas valine is not, this change has a significant effect on the tertiary structure of hemoglobin.

8. Many other single amino acid substitutions in the β chain have been observed. Many of these hemoglobin variants do not have altered function. Only hemoglobins S, C, and E lead to severe anemia.

9. *Familial hypercholesterolemia (FH)* results from a mutation in a gene that codes for a receptor protein in liver cells that binds lipoprotein (specifically low-density lipoprotein), the form of cholesterol in the blood. Because cholesterol cannot be transported into liver cells, it cannot be converted into useful forms and accumulates on the walls of blood vessels, potentially leading to heart attack or stroke.

10. *Cystic fibrosis* results from a point mutation and a single amino acid change in a gene that is responsible for production of a transmembrane chloride channel protein called the chloride transporter. As chloride is moved to the outside of the cell, water osmotically follows, and the cells are covered with a moist mucus. Afflicted individuals cannot export chloride and produce thick extracellular mucus in the lungs and other organs.

11. *Duchenne muscular dystrophy* results from a mutant allele that incorrectly codes a muscle cell membrane protein called dystrophin. Progressive muscular deterioration usually leads to death in the twenties. *Osteogenesis imperfecta* is also a disease resulting from a mutation in a gene coding for a structural protein, in this case collagen, essential for bone formation.

12. The vast majority of human diseases are *multifactorial*—influenced by many genes interacting with the environment. Most people have at least some diseases that are genetically influenced, but their expression is directly affected by the environment.

13. PKU, sickle-cell anemia, and cystic fibrosis are examples of human diseases that are caused by autosomal recessive alleles. Typically, parents are both heterozygous carriers and, on average, only one-quarter of their children show the disease.

14. Heterozygotes for an autosomal recessive allele only have one functional allele and thus can only produce one-half the normal gene product. If the body can compensate for this deficiency, the disease is inherited as an autosomal recessive.

15. Diseases such as familial hypercholesterolemia and osteogenesis imperfecta are inhereted as autosoml dominants because having only one-half of the normal gene product is not enough to produce a disease-free state.

16. Diseases determined by X-linked genes are passed from fathers to daughters and mothers to sons. In diseases inherited as recessives, such as the most common form of *hemophilia* and *Duchenne muscular dystrophy*, the sons will be afflicted, the daughters will be carriers.

17. Chromosomal abnormalities, such as additional or too few chromosomes *(aneuploidy), deletion* of a chromosome piece, or *translocation* of a piece to a different chromosome, cause some human diseases. Many spontaneous abortions are associated with chromosomal abnormalities. *Fragile-X syndome*, an inherited mental retardation, is associated with a visible constriction of an abnormal X chromosome.

18. If the protein product of a gene is known, its mRNA can be used to produce a cDNA copy. The cDNA can then be used to probe a gene library. Once the gene is found, DNA sequencing is used to compare the normal allele with mutant alleles from afflicted individuals. These methods were used to identify the sickle-cell anemia mutation.

19. Duchenne's muscular distrophy was identified as an X-linked recessive trait when several males with the disease were found to have a small deletion in their X chromosome. The disease can be caused by several different X chromosome deletions or point mutations in intron splice (consensus) sequences.

20. *Positional cloning* allows isolation of disease genes even if the protein product is unknown and no chromosomal abnormalities can be observed. Positional cloning uses the principle of linkage to locate the disease gene relative to well-known molecular markers that result from differences in the occurrence of recognition sites of restriction endonuclease enzymes within the genome.

21 Polymorphic restriction endonuclease recognition sites create differences in band patterns called a *restriction fragment length polymorphism (RFLP)*. RFLP band patterns are inherited in a Mendelian fashion and can be followed through a pedigree.

22. If a disease-causing gene and a particular RFLP are always associated in afflicted individuals, then the gene must be near the restriction site producing the RFLP. Positional cloning has helped to isolate the genes for cystic fibrosis, fragile-X syndrome, Huntington disease, and hereditary breast cancer.

23. DNA mutational "hot spots" include regions where cytosine has been methylated to 5-methylcytosine. A repair system exists to fix a cytosine that has lost an amino group and become uracil (G-C → G-U), since uracil is not appropriate for DNA. However, when 5-methyl-cytosine loses an amino group it becomes thymine (G-C → G-T) and the repair system misses this change since thymine is appropriate for DNA. Another system that deals with mispaired bases (G-T) sometimes converts the guanine to adenine, and sometimes the thymine is converted to cytosine.

24. The observation that within a family fragile-X syndrome was more severe in successive generations suggested that the defective allele was changing over time. The increase in severity is caused by expanding *triplet repeats*.

25. A triplet repeat of the bases CGG occurs within the fragile-X gene. CGG repeats of 6 to 54 are seen in the normal allele. Carriers of the disease without symptoms have a "premutated" allele with 50 to 200 repeats. Mentally retarded individuals have an allele with 200 to 1,300 repeats.

26. Triplet repeats are common in the normal alleles of many genes of humans and other animals. Triplet repeats can be found both within and outside the protein-coding regions of genes.

27. Some other human genes represented by a normal allele with few repeats, a premutation allele with more repeats, and a disease-causing allele with the most repeats include myotonic dystrophy (the repeat unit is CTG) and Huntington's disease (the repeat unit is CAG).

28. Mutations causing diseases in humans are both of the "micro" and "macro" types. Point mutations affecting single amino acids can cause disease if they alter protein function, typically by modifying the protein's tertiary structure. Examples include sickle-cell anemia and PKU. Macro alterations include chromosomal additions, deletions, and translocations and expanding triplet repeats, as in Duchenne's muscular dystrohy and fragile-X syndome.

Activity (for answer and explanation, see page 130)

• Sickle-cell anemia results from a single point mutation in the β-globin gene. This point mutation also eliminates a recognition site for the restriction endonuclease *Mst*II. Thus, whereas *Mst*II cuts the normal β-globin gene at three sites, producing restriction fragments of 1150 and 200 base pairs (bp), *Mst*II cuts the mutated β-globin gene at only two sites, producing a large restriction fragment of 1350 bp. The following pedigree shows the incidence of sickle-cell anemia in a family. Also shown are RFLP patterns for all individuals, obtained using a probe for the β-globin gene. Using symbols described in Figure 17.6 on page 381 in the textbook, complete the pedigree to indicate unaffected individuals, affected individuals, and carriers of the disease. Add the inferred genotypes below each symbol, where *A* is the normal allele and *S* is the mutant allele.

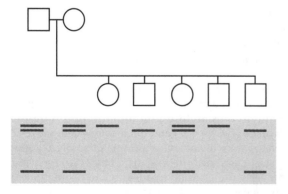

Questions (for answers and explanations, see page 130)

1. Which of the following statements about triplet repeats is *not* true?
   a. Triplet repeats are not restricted to disease genes.
   b. The mechanism of expansion of triplet repeats is not known.
   c. Triplet repeats only occur in the mutant alleles of a gene, not the normal alleles.
   d. The number of triplet repeats is intermediate in the premutation condition.
   e. Triplet repeats can occur within or outside the protein-coding region of the gene.

2. Which of the following statements does *not* apply to the genetic disease sickle-cell anemia?
   a. It is caused by a recessive, autosomal allele.
   b. It is one of several different mutations that affect the functioning of β-globin.
   c. It results from a point mutation.
   d. The mutant allele causes production of abnormal hemoglobin.
   e. The onset of the disease occurs in postreproductive years.

3. Match the following genetic diseases on the left with all of the appropriate characteristics on the right.

   ___ Cystic fibrosis   a. Affects hair/skin color

   ___ Osteogenesis imperfecta   b. Common in tropical areas

   ___ Phenylketonuria   c. Blood-clotting disorder

   ___ Sickle-cell anemia   d. Expanded triplet repeats

   ___ Duchenne's muscular   e. Chloride transporter
   dystrophy

   ___ Fragile-X syndrome   f. Dominant trait

   ___ Hemophilia A   g. X-linked allele

4. Select the following condition that is *not* important in utilizing a restriction fragment length polymorphism (RFLP) as a marker for a genetic disease.
   a. The availability of many pedigrees of families afflicted with the disease
   b. Establishing that the RFLP is tightly linked to the disease gene
   c. The presence of a restriction site within the normal allele or the mutant allele, but not both
   d. Identifying the protein produced by the gene
   e. Probes for the RFLP used

5. Explain why regions of DNA in which cytosines have been methylated are 50 percent more likely to undergo mutation than unmethylated regions.

DETECTING HUMAN GENETIC VARIATIONS • CANCER: A DISEASE OF GENETIC CHANGES (pages 383–390)

Key Concepts

1. The objective of *genetic screening* is to identify people who have, or are likely to develop, a specific disease.

2. Genetic screening tests are based on detecting chromosome abnormalities, observing certain proteins, verifying the presence or absence of biochemical reactions, or examining the DNA directly

3. Bacterial growth in the presence of blood phenylalanine is the basis for the *Guthrie test*, used in genetic screening for PKU. Screening infants for PKU is required by law in many countries.

4. DNA testing eliminates the need to observe gene expression. Observing gene expression can be difficult if it is restricted to certain cell types, under cellular control, or compensated for in the heterozygous state.

5. *Polar body diagnosis* uses PCR to amplify the DNA in the polar body produced after the first meiotic division. If the mother was heterozygous for a disease like cystic fibrosis, the polar body or the associated egg will have the mutant allele. If the polar body has the mutant allele, the egg can be used in *in vitro* fertilization.

6. DNA analysis of fetal cells can involve *chorionic villus sampling* on about week 10, or *amniocentesis* from week 13 to 17. DNA testing using PCR amplification of DNA can also be done on newborns and adults.

7. The form of genetic analysis called *allele-specific cleavage* depends on differences in the nucleotide sequences of the normal and mutant alleles that determine the presence of a restriction endonuclease recognition sequence. This creates differences in band fragment lengths that can be detected with electrophoresis.

8. In an easier, more direct form of genetic analysis called *allele-specific oligonucleotide hybridization*, a probe is synthesized that will hybridize with a section of the wild-type or mutant allele. Hybridization can be visualized in several ways.

9. *Cancer* is a category of over 200 different diseases that usually result from genetic alteration of an individual's somatic cells. As such, cancer is an example of a genetic disease that is not inherited.

10. When normal cells are transformed into cancerous cells they lose control of their rate of division and they spread to other parts of the body. Spreading of cancer is called *metastasis*.

11. Rapidly dividing cells can form cell masses called *tumors*. Noncancerous tumors, called *benign tumors*, do not metastasize, but remain localized. Cancerous tumors, called *malignant tumors*, undergo metastasis.

12. After spreading to surrounding tissue, cancerous cells can then either enter the lymphatic system or the bloodstream.

13. *Carcinomas*, such as lung, breast, colon, and liver cancer, are cancers of skin or the gut lining and account for the majority of all cancers. *Sarcomas* are cancers of bone, blood vessels, and muscle. *Lymphomas* are cancers of the white blood cells and *leukemias* affect the red blood cells.

14. About 10 percent of all human cancers are caused by viruses. Disease-causing viruses that are associated with cancers in humans include the hepatitis B virus, the Epstein-Barr virus, a variant of the herpes simplex virus, and papilloma viruses.

15. Cancer-causing agents, called *carcinogens*, include chemicals in the diet, high energy radiation (ultraviolet light, X rays), tobacco smoke, and so forth. The best way to reduce your risk of developing cancer is to avoid exposure to carcinogens.

16. Cancer may be a disease of old age because it depends on the accumulation of mutations within cells. Carcinogens react directly or indirectly with DNA and cause mutations, especially of genes involved in controlling normal growth.

17. *Proto-oncogenes* code for growth factors that simulate mitosis, growth factor receptors, parts of signal transduction pathways, transcription factors for genes causing cell division, and proteins that control apoptosis, or programmed cell death. Proto-oncogenes are essential for normal growth.

18. *Oncogenes* are the cancer-producing alleles that arise by mutation from proto-oncogene. Oncogenes lead to uncontrolled cell division. Cancer can result from the mutation of a proto-oncogene into an oncogene.

19. Tumor-suppressor genes encode proteins that inhibit cell division. For example, a common tumor-suppressor gene, *p53*, codes for a protein that arrests cells during G1. Mutations of *p53* are common in many human cancers.

20. Transformation of a normal cell into a cancerous cell may require mutation of several tumor-suppressor genes and also conversion of a proto-oncogene into an oncogene.

21. Individuals who inherit a mutated form of a tumor suppressor gene are likely to develop cancer early in life because they only need one additional mutational inactivation of the same gene. Individuals with two normal alleles require two mutational events.

22. Oligonucleotide-specific probes are now used to identify tumors and to determine if an individual has a mutated form of a tumor supressor gene.

Questions (for answers and explanations, see page 131)

1. Match the terms on the left with the appropriate characteristics on the right.

___ Benign	*a.* Cancer-causing agents
___ Metastasis	*b.* Lung cancer
___ Apoptosis	*c.* Nonmalignant
___ Carcinogens	*d.* Programmed cell death
___ Carcinoma	*e.* Spreading of cancerous cells
___ Leukemia	*f.* Cancer of blood cells

2. Which one of the following statements about two common methods of genetic analysis is *not* true?
   a. Allele-specific cleavage depends on the presence of restriction recognition sites within the gene of interest.
   b. Allele-specific cleavage is similar to RFLP analysis.
   c. Allele-specific oligonucleotide hybridization utilizes a probe that must be synthesized.
   d. Both methods of genetic analysis allow only the identification of individuals who are carriers for the disease.
   e. Both methods of genetic analysis can benefit from PCR amplification of DNA.

3. Which one of the following statements about oncogenes, proto-oncogenes, and tumor supressor genes is *true*?
   a. Given enough time, proto-oncogene always become oncogenes.
   b. A mutation can convert a proto-oncogene into a tumor supressor gene.
   c. The "two-hit" hypothesis relates only to tumor supressor genes.
   d. Inactivation of a proto-oncogene would cause cancer.
   e. Activation of some proto-oncogenes increases apoptosis.

4. A woman who is a carrier for cystic fibrosis is attempting to have a baby who will not express the disease or be a carrier. Her husband has both normal alleles. Polar body diagnosis shows that the polar body carries the normal allele. What should she do?

5. What is the connection between the issue of abortion and the technology of genetic screening?

6. Discuss the "two hit" hypothesis for cancer and explain why most forms of cancer are diseases of old age.

TREATING GENETIC DISEASES • THE HUMAN GENOME PROJECT   (pages 391–397)

Key Concepts

1. Modification of the mutant phenotype is a basic approach to treating genetic disorders. In PKU, this approach involves restricting the substrate (phenylalanine) for the secondary pathway activated by the inability to metabolize phenylalanine. People with PKU are put on a diet low in phenylalanine.

2. Another approach to modifying the mutant phenotype involves use of metabolic inhibitors, especially important in cancer therapy.

3. Supplying the missing protein is a third way to modify the mutant phenotype. Treatment of hemophilia A involves administering human clotting protein, usually produced through recombinant DNA technology.

4. *Gene therapy* utilizes DNA technlogy to insert a new gene into an individual who has inherited a mutant, disease-causing gene and to cause the new gene to be expressed in the host. DNA uptake may involve use of viral vectors, Ca^{2+} complexes, or liposomes.

5. *Ex vivo* gene therapy involves removal of cells from the body, insertion of the new gene into the cells, and return of the cells to the patient. *In vivo* gene therapy uses a vector to accomplish the insertion of the new gene into cells within the patient's body.

6. In the most effective form of gene therapy, the gene is inserted into reproductive cells that continue to divide and produce a line of cells with the new gene.

7. The goal of the human genome project (HGP) is to determine the nucleotide sequence of the human genome. Three intermediate goals are to develop methods to accomplish the primary goal, to test these methods by sequencing the genomes of several key organisms, and to study the ethical issues that the HGP will raise.

8. Each human chromosome can be isolated and cut into smaller pieces of about 500 base pairs, which can then be sequenced. *Mapped markers* are required to determine how to put the sequenced pieces back together.

9. Mapped markers include the *bands* that are visible when chromosomes are stained with certain dyes. In *fluorescence in situ hybridization (FISH)*, denatured metaphase or interphase chromosomes are exposed to a probe specific for a gene. A fluorescence microscope can be used to isolate the probe relative to known band locations.

10. Restriction endonuclease recognition sites are markers that can be used in *physical mapping*. Restriction maps express in base pairs the locations of the restriction sites of various enzymes relative to each other.

11. A piece of band, such as a band region from a chromosome, can be cut with a restriction endonuclease that has a recognition sequence with as many as 8 to 12 base pairs. This will generate relatively few, fairly large fragments that can then be cloned into yeast artificial chromosome (YAC) vectors to create a small gene library.

12. *Sequence tagged sites (STS)* are stretches of DNA, 60 to 100 base pairs in length, whose sequences are known and whose location has been mapped to a specific chromosome. Approximately 20,000 STSs, each about 100,000 base pairs from the next, have been mapped on human chromosomes.

13. DNA fragments from a YAC library produced with different restriction enzymes can be pieced together by examining them for the presence of the same STS. Common STS sites show where the fragments overlap. The YAC fragments can then be cut into 500 base pair pieces and sequenced.

14. *Genetic maps* are based on linkage between known genes and specific RFLPs. Side-by-side repeats are also useful as genetic linkage markers. Genetic maps complement physical maps and provide useful information for the positional cloning of disease genes.

15. Genomic sequencing has been completed for 141 viruses, 51 organelles, two bacteria, one archaean, and yeast. Many yeast genes have homologous sequences in humans.

16. The approach that is being taken in human DNA mapping has been to focus on the protein-coding genes first by sequencing cDNA produced from mRNA. *Expressed sequence tags* are being used as mapped markers (STSs) to locate the sequences of complete protein-coding genes. Some 30,000 human genes have be isolated and sequenced with this cDNA-STS approach.

17. Databases of polymorphic gene sequences from different people can be compared with disease databases to look for correlations between certain gene variants and specific diseases. Concerns about how this information will be used are growing.

Questions (for answers and explanations, see page 131)

1. Which of the following steps involved in mapping an entire genome would be third in sequence?
 a. Cut band fragment using a rectriction endonuclease specific for about 8 to 12 base pairs.
 b. Cut YAC fragments into 500 base pair pieces and sequence.
 c. Use FISH to localize DNA sequences.
 d. Clone into yeast artificial chromosome vectors.
 e. Determine overlap of different libraries for STSs.

2. Which of the following genetic diseases is/are being treated by use of metabolic inhibitors?
 a. PKU
 b. Various cancers
 c. Hemophilia A
 d. Familial hypercholesterolemia
 e. Sickle-cell anemia

3. Which of the following is not an example of a mapped marker?
 a. STS
 b. FISH
 c. Chromosome band
 d. Short side-by-side repeats
 e. RFLP

4. Initially, the human genome project has focused on which of the following approaches to mapping human DNA?
 a. Restriction mapping
 b. cDNA-STS
 c. Locating disease-causing genes only
 d. Constructing genetic maps
 e. RFLP mapping

5. How is information on common human genetic polymorphisms collected as part of the human genome project being used to identify genes that cause or influence disease?

18

Natural Defenses against Disease

CHAPTER LEARNING OBJECTIVES—after studying this chapter you should be able to:

❏ Describe the most important nonspecific defenses that organisms use against pathogens.

❏ Explain how the complement system functions and how it is activated.

❏ Differentiate between the terms "blood," "lymph," "plasma," and "serum."

❏ List the major types of white blood cells and describe the general roles that each type plays in the nonspecific or specific immune response.

❏ Characterize the features of the cellular immune response and the humoral immune response.

❏ Differentiate between the terms "antigen," "antibody," "immunoglobulin," and "antigenic determinant."

❏ Characterize the roles of macrophages, T lymphocytes, and B lymphocytes in the cellular and humoral immune responses.

❏ Describe the basis for immunological memory and its involvement in the process of immunization.

❏ Describe the clonal selection theory.

❏ Explain the roles that clonal deletion and clonal anergy may play in preventing the production of antiself antibodies.

❏ Explain what immunological tolerance is and how it normally develops.

❏ Describe the events that occur during plasma cell development.

❏ List the steps in the production of monoclonal antibodies and describe the major uses of this technology.

❏ Describe the structure of an immunoglobulin and identify the functions of its variable and constant regions.

❏ List the major classes of immunoglobulins and describe their functions.

❏ List the classes of the major histocompatibility complex (MHC) proteins and describe the functions of each class.

❏ Characterize the structure of the T-cell receptor and how it is able to bind to both processed antigen and MHC proteins.

❏ List the major antigen-presenting cells and describe how they present processed antigen.

❏ List and describe the roles of the major types of T cells.

❏ Describe the normal flow of events that occurs in the activation and effector phases of the humoral response.

❏ Describe the normal flow of events that occurs in the activation and effector phases of the cellular response.

❏ Describe the genetic mechanisms involved in the production of antibody diversity.

❏ Explain the basis for the phenomenon known as class switching.

❏ Describe the underlying genetic mechanisms controlling production of antibodies, MHC proteins, and T cell receptors and similarities that exist in the genes coding for these molecules.

❏ Describe the involvement of the various interleukins in the cellular and humoral immune responses.

❏ List some autoimmune diseases and immune deficiency disorders.

❏ Describe the life cycle of HIV.

❏ Describe the normal course of AIDS, the groups most at risk, what conditions cause its spread to uninfected individuals, and the prospects for a cure or development of a vaccine.

DEFENSIVE CELLS AND PROTEINS • NONSPECIFIC DEFENSES AGAINST PATHOGENS • SPECIFIC DEFENSES: THE IMMUNE RESPONSE   (pages 400–411)

Key Concepts

1. *Pathogens* are disease-producing organisms, such as a bacterium, virus, protist, fungus, or animal parasite.

2. Major functions of the vertebrate immune system are to

recognize self from nonself materials and to selectively eliminate and remember foreign invaders.

3. The cells of the immune system are distributed throughout the body but move mainly in *blood* and *lymph*. Blood consists of a cellular component, including red blood cells and white blood cells called *leukocytes*, and a fluid component called *plasma*. Plasma consists of water and simple and complex molecules; when blood clots, the liquid remaining is *serum*.

4 Lymph, a blood filtrate including water, solutes, and leukocytes, accumulates in extracellular spaces. It is returned to blood vessels near the heart by lymph vessels. *Lymph nodes* filter the lymph and contain a variety of white blood cells.

5. White blood cells (leukocytes) are large, colorless cells with nuclei. They move by extending pseudopods and can squeeze between the cells that form capillaries. Two kinds of white blood cells are *lymphocytes* and *phagocytes*.

6. Lymphocytes are the major cells of the immune system; they include *B cells* and *T cells*. B and T cells originate in the bone marrow. T cell precursors migrate to the *thymus gland,* where they develop into mature T cells. After B lymphocytes leave the bone marrow, they disperse thoughout the circulatory and lymph vessels.

7. Phagocytes are part of the nonspecific defense system, responsible for engulfing and digesting nonself material. The phagocytes called *macrophages* also present digested nonself molecules to the T cells. This activates a specific immune response.

8. Cells of the immune system interact using a variety of special proteins, including receptors, surface markers, signaling molecules, and toxins. B cells have *antibodies* on their cell membranes. Antibodies bind selectively to nonself substances. T cells have membrane-bound *T cell receptors*. *Cytokines* constitute a variety of signaling chemicals used to coordinate the cells of the immune system.

9. Both the physical integrity of unbroken skin and the *normal flora* of bacteria and fungi that live on the skin protect against invasion by a pathogen to deeper layers. Frequently, the normal flora outcompete pathogens for nutrients or produce substances that are toxic to them.

10. The *mucous membranes* covering surfaces of the visual, digestive, respiratory, and urogenital systems provide several lines of defense against penetration by pathogens, including secretion of *lysozyme* (in tears, saliva, and nasal secretions), entrapment in mucus, removal by beating cilia, and the sneeze and cough reflexes. Low pH in the stomach and bile salts in the small intestine further discourage pathogens.

11. Twenty common antimicrobial blood proteins make up the nonspecific defense system called the *complement system*.

12. Once certain antibodies have attached to an antigen, the first complement protein binds to the foreign cell surface. A cascade of reactions involving other complement proteins leads to the development of a doughnutlike protein complex that causes the membrane to become leaky.

13. Two other complement-mediated responses are the promotion of endocytosis by attaching to and marking microorganisms for easier recognition by phagocytes and the attraction of phagocytes to sites of infection.

14. *Interferons* are a class of antimicrobial molecules that provide generalized resistance to a variety of viruses, such as the influenza virus. Interferons are glycoproteins of about 160 amino acid subunits; they bind to cell surfaces and inhibit viral replication.

15. Nonspecific defenses against pathogens that have penetrated the host include ingestion via endocytosis by phagocytes. Three important types of phagocytes are short-lived *neutrophils, monocytes* (precursors of macrophages), and eosinophils that defend against worm parasites.

16. Other white blood cells called *natural killer cells* destroy cells infected with certain viruses and some tumor cells.

17. The *inflammation response* includes a localized redness, heat, and swelling. It results from dilation of blood vessels, leakage of blood plasma (including complement proteins), and invasion by phagocytes. Increased heat may also help to destroy pathogens. The inflammation response is initiated by the release of *histamine* by damaged cells, especially *mast cells* and *basophils*.

18. An *antigen* is a foreign cell or macromolecule that bears one or more specific sites, called *antigenic determinants*, that the immune systems attacks. A large antigen may contain several antigenic determinants.

19. Two forms of immune response are recognized: the *cellular immune response* is carried out by specialized cells against fungi, foreign tissue, multicellular parasites, and cells that are infected with viruses; the *humoral immune response* is carried out by protein molecules and is directed against bacteria and free viruses.

20. The humoral immune response is based on the secretion of *antibodies* by B cells that have differentiated into *plasma cells*. Antibodies are highly specific; they recognize and bind to a single antigenic determinant. Antibodies can circulate freely within the blood and lymph or may be part of a B cell membrane.

21. *T lymphocytes*, the cells responsible for the cellular immune system, have *T cell receptors* on their surfaces. T cell receptors are glycoproteins that recognize and bind to antigenic determinants just like antibodies. T cells attack either foreign cells or altered self cells. Important types of T cells include helper T cells (T_H cells), and cytotoxic T cells (T_C).

22. According to the *clonal selection theory*, the body contains an enormous variety of B cells and each B cell produces a single antibody. An antigen selects for a particular preexisting lymphocyte that is specific to it. The activated lymphocyte (either a T or a B cell) then gives rise to a large clone of cells that are all able to recognize the antigen.

23. Activated lymphocytes produce two types of progeny: *effector cells* and *memory cells*. Effector cells are relatively short-lived, antibody-secreting B cells or T cells with receptors that can bind antigenic determinants. Memory cells may circulate in the body for decades and respond rapidly upon a second exposure to the same antigen.

24. During the first encounter with an antigen (the *primary immune response*), there is a lag period of several days before the antibody and T cell numbers increase. *Immunological memory* refers to the more rapid and effective response to second and subsequent exposure to an

antigen. During the *secondary immune response*, the immune system remembers a previously encountered antigen, and production of antibodies and T cells specific to it are enhanced.

25. The presence of memory cells confers *natural immunity* to previously encountered pathogens. The success of *immunization (artificial immunity)* depends on immunological memory. *Vaccination* involves the injection of a small inoculum of a pathogen, a bacterium or virus, usually rendered harmless by heat treatment, or foreign protein from a pathogen. Subsequent exposure will elicit a massive response from the immune system.

26. The *clonal deletion theory* states that immature "antiself" lymphocytes are eliminated by programmed cell death (apoptosis) when they encounter their antiself antigens and that no clones of antiself lymphocytes ever develop. *Clonal anergy* is also important in preventing T cells from initiating clonal expansion when their receptor binds to an antigenic complex on a body cell. Activation of the T cell requires a costimulatory signal, a protein called CD28, that is normally present on the body cell presenting the antigen, but absent or blocked if the antigen is self.

27 Based on the discovery that some nonidentical twin cattle had blood cells of two different types, it was suggested that the blood may have been exchanged between the twins before the immune system had differentiated self from nonself and the blood cells from the other twin became recognized as self. This phenomenon of *immunological tolerance* was verified by injecting highly inbred, newborn mice of one strain with lymphoid cells from another strain. Later, only the injected mice would accept skin grafts from the other strain as self tissue; uninjected controls rejected skin grafts as nonself.

Questions (for answers and explanations, see page 131)

1. Which one of the following is *not* a major function of the immune system?
   a. Recognition of self
   b. Production of the inflammation response
   c. Selective elimination of pathogens
   d. Immunological memory
   e. Recognition of antigens

2. Match the names on the left with the most appropriate characteristics on the right.

_____ Plasma	a.	Foreign protein or cell
_____ White blood cells	b.	Blood minus blood cells
_____ Lymph	c.	Produced by plasma cell
_____ Lymphocytes	d.	Includes phagocytes
_____ Antigen	e.	B and T cells
_____ Antibody	f.	Blood filtrate with white blood cells

3. Which of the following does *not* characterize the humoral immune response?
   a. Directed against circulating viruses and bacteria

b. Based on activities of phagocytes
c. Recognition of antigens is by specific antibodies
d. Effector molecules bind to antigenic determinants
e. Second exposure to antigen is more rapid and effective

4. Which of the following statements is *not* true?
   a. Antigens can lead to the production of more than one antibody.
   b. Antigens serve as templates for the production of antibodies specific to them.
   c. Only plasma cells make antibodies.
   d. B and T cells both originate in bone marrow.
   e. A protein molecule can have more than one antigenic determinant.

5. Examine the figure shown below and answer the following questions:

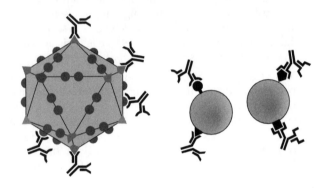

How many antigens are shown? _____

How many different antibodies are present? _____

How many different antigenic determinants are shown?

_____

6. Which of the following statements is *not* in agreement with the clonal selection theory?
   a. Antigens with more than one antigenic determinant can select for plasma cells each coding for more than one antibody.
   b. An antigen activates a preexisting lymphocyte with receptors for it.
   c. Exposure to an antigen leads to the production of both memory and effector cells.
   d. Exposure to an antigen leads to the proliferation of specific T cells and B cells.
   e. An antigen may lead to the production of more than one clone of B cells.

7. Which of the following statements about memory or effector cells is *not* true?
   a. Only effector cells carry out the attack on the antigen.
   b. Only memory cells are long-lived.
   c. Only effector cells are specific to a single antigenic determinant.
   d. Only memory cells can divide to produce both memory and effector cells.
   e. Effector cells include both B cell and T cell descendants.

8. Which of the following statements about immunological tolerance is *not* true?
   a. "Forbidden" antiself lymphocyte clones are eliminated by apoptosis.
   b. The development of immunological tolerance is independent of the age of the animal.
   c. T cells with receptors complexed to antigens on self cells show clonal anergy.
   d. Continued exposure to nonself antigens is necessary to maintain immunological tolerance.
   e. The failure of immunological tolerance may lead to an autoimmune disease.

9. The complement system is involved in all but one of the following activities. Select the *exception.*
   a. Elimination of foreign cells
   b. Attraction of phagocytes to sites of infection
   c. Formation of a lytic complex
   d. Recognition of specific pathogens
   e. Increasing the effectiveness of endocytosis

Activity (for answers and explanations, see page 131)

- Removal of the thymus gland in young mice usually results in most of the mice losing weight and dying. Mice that are thymectomized but maintained in a germ-free environment survive. Explain these observations.

B CELLS: THE HUMORAL IMMUNE RESPONSE • T CELLS: THE CELLULAR IMMUNE RESPONSE   (pages 411–419)

Key Concepts

1. B cell activation begins when the antigenic determinant of an antigen binds to the antibody carried on the surface of a particular B cell. A helper T (T_H) cell induces the B cell to divide and form a line of plasma cells and memory cells.

2. Plasma cell development involves an increase in the number of ribosomes and endoplasmic reticulum. The cell is now ready to synthesize antibodies identical to those on the surface of the parent B cell.

3. Antibodies are also called *immunoglobulins (Igs)*. The structure of the most common immunoglobulin is a tetramer, consisting of four polypeptides. Two identical light chains and two identical heavy chains are held together by disulfide bonds. Each of the halves of the Y-shaped molecule consists of one light and one heavy chain.

4. Each chain includes a *constant region,* in which the amino acid sequence does not differ much from one species to the next, and a *variable region,* in which the amino acid sequence does differ.

5. The variable regions of heavy and light chains are unique for each immunoglobulin. Each arm of the immunoglobulin molecule combines with a specific antigenic determinant.

6. Since the two halves of an immunoglobulin are identical, each can bind to an identical antigenic determinant, leading to the formation of large antibody–antigen complexes.

7. The constant regions of the molecule determine if it will be incorporated into a cell membrane or be secreted into the bloodstream. The constant regions of the heavy chains differ in the five classes of immunoglobulins.

8. *Immunoglobulin M (IgM)* is the first immunoglobulin produced by a plasma cell; the other four types arise by *class switching.* IgM consists of five immunoglobulin subunits and is the immunoglobulin produced in the primary immune response. It activates the complement system and promotes phagocytosis of antibody-coated cells. It has a μ heavy chain.

9. *IgG* (γ heavy chain) molecules consist of single subunits and are produced during the secondary immune response. Like IgM, IgG with bound antigens promotes phagocytosis by attaching to macrophages; they also activate the complement system.

10. The major function of *IgD* (δ heavy chain) is as a membrane receptor on B cells.

11. *IgE* (ε heavy chain) is the most common immunoglobulin in allergic reactions (such as hives, hay fever, eczema, and asthma) and in inflammation. IgE bound to antigenic determinants on allergens attaches to receptor sites on *mast cells* and the mast cell–IgE–allergen complex stimulates the release of *histamine.* IgE also helps fight worm parasites, like those causing *schistosomiasis.*

12. *IgA* (α heavy chain) is transported across epithelial membranes and added to body secretions such as tears, saliva, milk, and gastric secretions. It is a monomer when in plasma, a polymer in body secretions, where it binds to complex antigens.

13. Most antigens contain many different antigenic determinants. Lymphocytes produce large quantities of a single antibody but cannot grow in tissue culture. Fusing single lymphocytes with myelomas (cancerous plasma cells) produces hybrid cells called *hybridomas.* Hybridomas grow well in tissue culture and produce large quantities of a single antibody called a *monoclonal antibody.*

14. The steps in the production of monoclonal antibodies are the injection of an animal with an antigen, isolation of lymphocytes from the animal's lymphoid tissue (like the spleen), fusion of lymphocytes with myeloma cells to produce hybridomas, growth of clones of hybridomas in tissue culture, and selection of the cell line producing the desired antibody.

15. Hybridomas grown in tissue culture secrete large quantities of a specific antibody. Hybridomas can be stored by freezing or injecting into animals to give rise to tumors that produce a monoclonal antibody.

16. Monoclonal antibodies are important in studies of antibody chemistry and cell membranes, and for tissue typing and transplants. *Immunoassays* rely on the great specificity of antibodies for their target molecules. Medical applications include *passive immunization* (inoculation with antibody instead of antigen) and detection of specific cancers. Cancer treatments include the attachment of poisons to monoclonal antibodies specific for certain tumors.

17. Various types of T cells are the effectors of the cellular immune response but also coordinate the humoral immune response. T cells attack self cells that have become abnormal because of a mutation or a virus.

18. T cell receptors are membrane-bound glycoproteins composed of two polypeptide chains (α and β) with variable and constant regions. T cell receptor genes are similar to the immunoglobulin genes.

19. T cells activated by antigen produce two kinds of effector cells, *cytotoxic T cells* (T_C) and *helper T cells* (T_H). A given clone of T cells produces T_C and T_H cells with receptors that only respond to the same antigenic determinant.

20. T_C cells eliminate virus-infected cells by causing the infected cell to lyse. T_H cells are involved in both cellular and humoral responses.

21. In mice, the genes determining the major histocompatability complex (MHC) proteins are located in a tight cluster of loci. Because the large number of possible alleles for each gene generates a huge number of possible genotypes, each individual of a species can differ in its cell surface MHC proteins. MHC loci fall into three classes. A similar system in humans is the human leukocyte antigens (HLA).

22. Class I MHC loci code for proteins present on the surface of every nucleated cell of the body. Virus-infected cells present processed antigen on class I MHC proteins.

23. Class II MHC loci code for proteins found only on the surface of B cells, T cells, and macrophages. Interaction between these cells depends on recognition of self-class II MHC proteins and underlies the ability of these cells to recognize self versus nonself cells.

24. Class III MHC loci code for some of the complement proteins involved in the antibody-mediated lysis of foreign cells.

25. T cell receptors recognize processed antigen fragments that are attached to class II MHC proteins on the surface of macrophages. One region of the T cell receptor recognizes the antigen, another region recognizes the MHC protein.

26. Cells that can take up antigens, degrade them in their lysosomes, and present the antigen fragments as *processed antigen* in association with their MHC proteins are called *antigen-presenting cells*. Antigen-presenting cells include macrophages, B cells, and virus-infected cells of any type.

27. *Interleukins* are cytokines passed among macrophages, T cells, and B cells. Immunosuppressant drugs commonly used during tissue transplants, like cyclosporin, function by inhibiting interleukin synthesis.

28. The *humoral activation phase* begins when a macrophage presents processed antigen in association with class II MHC proteins anchored to its cell surface. Helper T cells with T cell receptors that recognize the processed antigen and the class II MHC protein bind to them and are activated to proliferate by interleukin-1 released by the macrophage.

29. The *humoral effector phase* begins when a B cell ingests an antigen that has bound to one of its membrane-anchored IgM receptors. The B cell displays processed antigen in association with its class II MHC proteins. When a T_H cell recognizes and binds to the processed antigen and the class II MHC protein, it releases interleukins that stimulate the B cell to produce a clone of plasma cells that secrete antibody specific to the antigen.

30. The *cellular activation phase* begins when a virus-infected cell presents processed antigen in association with membrane-bound class I MHC proteins. Cytotoxic T cells with appropriate T cell receptors recognize the processed antigen and the class I MHC protein and bind to the complex. The bound T_C cell is activated to proliferate to form a clone.

31. The *cellular effector phase* begins when an activated T_C cell recognizes processed antigen bound to class I MHC protein on a virus-infected cell. It binds to the antigen–MHC complex and releases chemicals (perforin) that cause lysis of the infected cell.

32. Base sequence similarities among the genes coding for MHC proteins, antibodies, and T cell receptors suggest that all three systems may have evolved from an ancestral defense system.

33. MHC proteins of one individual can act as antigens to another individual during organ transplant. Unless transplant is done shortly after birth (before the immune system has matured), the MHC compatibility of host and donor must be carefully matched or immunosuppressant drugs used to inhibit B and T cells.

Questions (for answers and explanations, see page 131)

1. Which of the following statements about the structure of an immunoglobulin molecule is *not* true?
   a. The constant regions of heavy chains are identical for a given molecule.
   b. The variable regions of light chains differ for a given molecule.
   c. The two antigen-binding sites are identical for a given molecule.
   d. The constant regions of light chains are the same for the same type of immunoglobulin.
   e. Precipitation of antigen would not occur if each immunoglobulin had only one antigen-binding site.

2. Select all the letters for characteristics listed on the right that apply to the immunoglobulin types listed on the left.

   _____ IgG   a. Primary immune response

   _____ IgM   b. Secondary immune response

   _____ IgA   c. Activates the complement system

   _____ IgD   d. Membrane receptor on B cells

   _____ IgE   e. Binds to mast cells

                      f. Found in tears and saliva

                      g. Combats schistosomiasis

                      h. Can bind ten antigenic determinants

3. In which of the following activities are major histocompatibility complex proteins *not* involved?
   a. Binding of T_H cells to macrophages
   b. Binding of antigenic determinant to B cells
   c. Complement-induced lysis of foreign cells
   d. Binding of T_C cells to virus-infected cells
   e. Transplant rejection

4. To reduce the risk of rejection of a tissue transplant, the immune system can be suppressed by administration of drugs that inhibit production of interleukin-2 (IL-2). Which one of the following activities would *not* be affected by this drug?
   a. Secretion of antibodies by B cells
   b. Activation of T cells by macrophages
   c. Rapid growth of T cell clones
   d. Activation of natural killer cells
   e. Rapid growth of B cell clones

5. Which of the following does *not* occur as a plasma cell matures from a B cell?
   a. Increase in the number of ribosomes
   b. Synthesis of antibodies specific to the antigenic determinant-binding receptors on the surface of the parent B cell
   c. Increase in the amount of ER
   d. Development of an extensive Golgi complex
   e. Loss of the nucleus and ability to divide mitotically

6. Which of the following steps is *not* normally a part of the production of monoclonal antibody?
   a. Isolation of the antigens from the immunized animal
   b. Immunization with the antigen
   c. Tissue culture
   d. Cell fusion
   e. Formation of hybridomas

Activities*(for answers and explanations, see page 132)*

- The following figure shows the effector phase of the humoral immune response. Label all important structures.

- The following figure shows the effector phase of the cellular immune response. Label all important structures.

- In the following diagram of an immunoglobulin, label the light and heavy chains and the variable and constant regions of each chain. Circle areas on the molecule where antigen can bind.

THE GENETIC BASIS OF ANTIBODY DIVERSITY • THE EVOLUTION OF ANIMAL DEFENSE SYSTEMS • DISORDERS OF THE IMMUNE SYSTEM   (pages 419–426)

Key Concepts

1. The genes specifying the four polypeptides of an immunoglobulin (two identical light and two identical heavy chains) are assembled from hundreds of separate DNA segments during the maturation of a B cell.

2. DNA segments coding for immunoglobulin light and heavy chains are on separate chromosomes. The light-chain variable region is determined by two families of DNA segments, the kappa and the lambda families. The heavy-chain variable region is determined by three families of DNA segments. Since each family may have several hundred different forms and the heavy and light chains are independently combined to form a single immunoglobulin, many millions of different immunoglobulins are possible.

3. In addition to the random DNA rearrangements, the process that joins the segments together is imprecise and this creates more variety. Terminal transferase also creates insertion mutations and there is a relatively high rate of mutation in the immunoglobulin genes.

4. The heavy-chain variable region has three families of DNA fragments. A mouse has over 100 different *V* (variable) fragments, 30 *D* (diversity) fragments, and 6 *J* (joining) fragments. Since each variable region is formed by one *V*, one *D*, and one *J* fragment, $100 \times 30 \times 6 = 18,000$ different heavy-chain variable regions can be produced before mutation further increases the diversity.

5. Assembly of the segments is accomplished with the deletion of the unused *V*, *D*, and *J* segments and the joining of the three used segments. The variable region of the heavy chain is then joined to one of eight *C* (constant) segments representing the constant region of the heavy chain. The other constant region segments are not deleted. The functional heavy-chain gene consists of one *V*, one *D*, one *J*, and one *C* region.

6. After transcription, editing of the primary RNA transcript of the functional gene removes any products of introns or *J* segments remaining between the selected J segment and the constant region.

7. Initially all B cells produce IgM molecules because the gene coding for that type of heavy-chain constant region is adjacent to the variable-region gene. *Class switching* occurs if a later deletion removes constant-region segments and juxtaposes a different segment with the variable-region gene. Class switching causes a plasma cell to produce antibodies that still recognize the same antigen, but have a different function.

8. T cells induce class switching by secretion of specific cytokines. The type of cytokine determines the class to which the B cell will switch.

9. T cell receptor genes are composed of *V*, *D*, and *J* regions and are formed by the same joining process seen in B cells.

10. Some protists use phagocytosis defensively, as do invertebrates, and invertebrates recognize self, but lack immunological memory. Invertebrates do not produce immunoglobulins, lymphocytes, or complement proteins, but they do have cytokine-like proteins.

11. *Allergies* represent an overresponse of the immune system to an antigen, such as pollen or bee sting toxin. Immediate hypersensitivity is an allergic reaction initiated when IgE binds directly to the allergen. In delayed hypersensitivity, the allergin is processed by antigen-presenting cells and a T cell response ensues.

12. *Autoimmune diseases* (including rheumatic fever, rheumatoid arthritis, ulcerative colitis, myasthenia gravis, and perhaps multiple sclerosis) result when the immune recognition of self fails and one or more "forbidden clones" of B or T cells proliferate and attack components of the body.

13. Examples of *immune deficiency disorders* include cases where B cells never form or cannot produce plasma cells. In both cases, antibodies are never produced.

14. In *acquired immune deficiency syndrome*, or *AIDS*, the immune system is devastated because the virus that causes AIDS, *HIV* (human immunodeficiency virus), attacks the T_H cells. The T_H cells play a key role in both the cellular and humoral immune responses.

15. After a brief peak in HIV concentration in the body immediately after infection, an immune attack against it is mounted and most of the viruses are destroyed. The remaining viruses infect T_H and other cells. After a delay lasting as much as several years, T_H cell numbers drop and the individual becomes susceptible to opportunistic diseases and rare cancers.

16. HIV is a *retrovirus* that inserts a reverse transcript of its RNA genetic material into the host genome. The central core of the virus includes a protein coat (p24 capsule protein) enclosing two identical RNA molecules and enzymes including the *reverse transcriptase*, used to make DNA from the RNA genome, integrase, which catalyzes the insertion of the newly made DNA into the host chromosome, and protease, responsible for formation of viral proteins. An envelope surrounds the core.

17. The HIV envelope, derived from the host cell plasma membrane, is studded with protein (gp120) that can couple with a membrane protein called *CD4* that is found only on T_H cells. After coupling, the viral core is released into the cell, reverse transcriptase makes double-stranded DNA from the RNA genetic material, and the viral DNA is inserted into a host chromosome.

18. The integrated viral DNA may remain latent within the host chromosome for a decade or more. Upon activation, viral DNA is transcribed into HIV RNA, translation produces viral enzymes and structural proteins, and assembly of new virus takes place. As the virus buds from the infected cell, it acquires an envelope of modified cell membrane. During rapid growth, lysis of the T_H cell occurs.

19. The high rate of mutation of the virus, its rapid integration into T_H cells, and the long life of infected cells are all problems that face the development of a vaccine.

20. Several drugs used alone or in combination may control replication of the virus. AZT, ddI, and ddC inhibit reverse transcriptase. Protease inhibitors block the production of viral proteins.

Activities (for answers and explanations, see page 132)

- Explain why the high mutation rate of the HIV gene coding for the antigenic properties of the envelope protein is complicating efforts to develop a vaccine for AIDS.

- List ways that HIV-I can and cannot be transmitted.

Questions (for answers and explanations, see page 132)

1. You discover a simple animal in which the variable regions of *both* the heavy and light immunoglobulin chains are assembled from three families of DNA segments and each family is represented by only three different segments. How many different antigenic determinants could this genetic system recognize?
   a. 9
   b. 18
   c. 27
   d. 729
   e. 10,512

2. Which of the following statements about the determination of class in immunoglobulins is *not* true?
   a. The DNA segment for the constant-region gene that is adjacent to the assembled variable-region gene for the heavy chain determines the immunoglobulin class.
   b. Class switching produces an immunoglobulin with a different class, but the same antigenic specificity.
   c. Class switching can only occur once for a given immunoglobulin.
   d. All immunoglobulins begin as members of the IgM class.
   e. Deletion of DNA is the mechanism underlying class switching.

3. What do immunoglobulins, T cell receptors, and MHC proteins have in common?
   a. The base sequences of their genes are similar.
   b. They are restricted to cells of the immune system.
   c. They can all be membrane-bound proteins.
   d. Their genes are assembled through the same V–D–J joining process.

4. Which of the following statements contrasting T cell receptors and immunoglobulins is *not* true?
   a. Unlike immunoglobulins, T cell receptors can only recognize processed antigen.
   b. The genetic system underlying T cell receptor diversity is different from the system specifying immunoglobulins.
   c. T cell receptors are more specific than immunoglobulins; they have a dual specificity for antigen and MHC protein.
   d. Both T cell receptors and immunoglobulins consist of multiple polypeptide chains held together with disulfide chains.
   e. Like immunoglobulins, T cell receptors have constant regions and variable regions.

5. Select all of the following that are *not* autoimmune diseases.
   a. AIDS
   b. Rheumatic fever
   c. Allergies
   d. Rheumatoid arthritis
   e. Multiple sclerosis

Integrative Questions (for answers and explanations, see page 132)

1. Characterize each of the following as part of A) the non-specific defense system or B) the immune system.

   _____ Natural killer cells

   _____ Production of mucus

   _____ Macrophages

   _____ Secretion of lysozyme

   _____ Inflammation

   _____ T_H cells

   _____ T_S cells

   _____ Low pH in stomach

   _____ Plasma cells

   _____ Interleukins

   _____ Complement proteins

   _____ IgM

   _____ Interferons

   _____ Normal flora

2. Number the following events in the humoral immune response to show their correct sequence.

   _____ Binding of T_H cell to processed antigen-class II MHC on B cell

   _____ Proliferation of T_H cell clone

   _____ Proliferation of effector and memory cells

   _____ Binding of antigen to a surface receptor on B cell

   _____ Stimulation of B cell by interleukin-2

   _____ Maturation of plasma cells

   _____ Binding of T_H cell to processed antigen-class II MHC on macrophage

Answers and Explanations

Chapters 9–18

CHAPTER 9 – CHROMOSOMES, THE CELL CYCLE, AND CELL DIVISION

Note: The page numbers listed with the section titles refer to the Study Guide; numbers in parentheses for each question refer to the relevant key concepts.

SYSTEMS OF CELL DIVISION • INTERPHASE AND THE CONTROL OF CELL DIVISION • EUKARYOTIC CHROMOSOMES (pages 59–61)

1. Choice *d*. The number of chromosomes does not change between the end of mitosis and the beginning of the next cell division. During the S phase, new DNA and chromatids are synthesized, but no new chromosomes appear. (7, 11)

2. Choice *a*. Since most of the cell cycle involves the various stages of interphase, most cells from this tissue sample will be in interphase. (7, 12)

3. Choice *b*. Although other kinds of proteins are synthesized throughout interphase, histones and DNA are manufactured during the S phase. (8, 12)

4. Choice *c*. Chromatin and chromosome forms change dramatically during the cell cycle. (5–6, 11, 13)

5. Choice *e*. Chromatin is present at all times during the cell cycle; only its form changes. (7–8, 12–13)

6. Choice *b*. Cdk's by themselves are unable to phosphorylate proteins; they must first combine with cyclin to form cdk-cyclin complexes which are the structures that actually have enzyme capability. (9–10)

7. Choice *e*. (9–10)

MITOSIS: DISTRIBUTING EXACT COPIES OF GENETIC INFORMATION • CYTOKINESIS: THE DIVISION OF THE CYTOPLASM (pages 61–62)

1. Choice *a*. Polar microtubules are present whenever any trace of the spindle is present, which includes all phases except interphase. (3–11)

2. Choice *c*. Chromatids are present from prophase until they separate at the beginning of anaphase. (3–9)

3. Choice *d*. (9)

4. *c*: Chromosomes align on equatorial plate (7–8)
   e: Chromatin redisperses after cell division (11)
   b: Centrosomes migrate to opposite ends of cell (2, 10)
   a–e: Double-stranded DNA present in cell (2–11)
   b–e: Centromeres present on chromosomes (3–11)
   a: Cytokinesis occurs (12)

5. Choice *c*. Chromosomes begin moving during prometaphase. (7)

6. Choice *e* (9, 10)

7. Choice *c*. Cytokinesis occurs after telophase for both plants and animals. (11–13)

8. Choice *e*. In mitosis, sufficient quantities of all important cell structures are passed along to each daughter cell, although not necessarily in equal proportions. (14)

- (6–10)

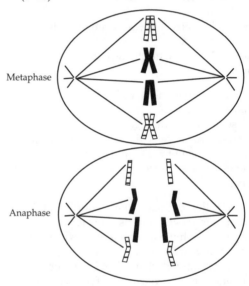

Metaphase

Anaphase

MITOSIS: ASEXUAL REPRODUCTION AND GENETIC CONSTANCY • MEIOSIS: SEXUAL REPRODUCTION AND DIVERSITY • MEIOSIS: A PAIR OF NUCLEAR DIVISIONS • MEIOTIC ERRORS: THE SOURCE OF CHROMOSOMAL DISORDERS • CELL DEATH (pages 62–65)

1. Choice *d*. Reproduction can involve mitosis only (i.e., it can be asexual). Gametes (or spores) are formed through meiosis and are used in sexual reproduction. (1–4, 7)

2. Choice *e*. The cell is diploid at this stage, so it has all of its chromosomes. Since the chromosomes have yet to divide, there are twice as many chromatids as chromosomes. (12–13)

3. Choice *b*. This cell is diploid, which by definition means that the number of homologous pairs of chromosomes equals one-half the diploid number. (9, 12–13)

4. Choice *a*. Cells are haploid at the end of meiosis II. The number of centromeres equals the number of chromosomes. (12, 19)

5. Choice *d*. (16)

6. Choice *c*. Homologous chromosomes remain paired from interphase through metaphase of meiosis I. (13–16)

7. Choice *e*. Nondisjunction causes an improper number of chromosomes to end up in daughter cells, a condition known as aneuploidy. (21)

8. Choice *b*. Translocation is a replication/division error where large segments of one chromosome are incorrectly moved onto another. (23)

- (19)

Metaphase I

- (19)

Metaphase II

- (3–6, 8)

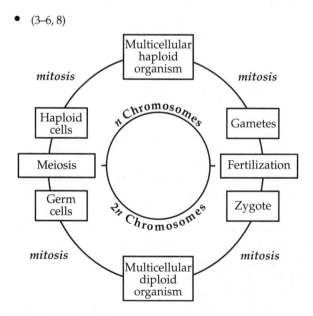

Integrative Questions

1. Choice *c*. In anaphase of mitosis, the chromatids of diploid cells separate. In anaphase of meiosis II, the chromatids of haploid cells separate.

2. Choice *b*. Since the cell has 19 chromosomes, one more than the normal diploid number, we assume it has just completed telophase and cytokinesis of mitosis. The one extra chromosome probably resulted from a nondisjunc-

tion event.

3. Choice *d*. Only cells from prophase of mitosis are diploid; those from metaphase of meiosis II would be haploid.

CHAPTER 10 – TRANSMISSION GENETICS: MENDEL AND BEYOND

Note: The page numbers listed with the section titles refer to the Study Guide; numbers in parentheses for each question refer to the relevant key concepts.

MENDEL'S WORK • MENDEL'S EXPERIMENTS AND LAWS OF INHERITANCE (pages 66–69)

- All five genes are homozygous dominant in the first genotype, so only **one** kind of gamete can be made (*ABCDE*). In the second case, all five genes are heterozygous, which means that there are two alleles for each gene, and thus a total of 32 (*i.e.*, 2^5) different allele combinations possible in the gametes. The third genotype is heterozygous for only one gene locus, so only **two** (2^1) kinds of gametes are possible (*ABCDE* and *AbCDE*). (8–11)

- Mendel's law of segregation states that alleles at the same diploid gene locus separate from one another during gamete formation. The law of independent assortment states that alleles from one gene assort into gametes independently of all other genes, unless the genes are members of a linkage group (i.e., on the same chromosome). Segregation thus concerns the separation of alleles for the same gene during meiosis, while assortment has to do with the interaction between alleles from genes on different chromosomes. (11, 16)

- We are told that both individuals are heterozygotes, which we can symbolize as *AaBb*. To set up the Punnett square, the four possible gametes that each individual can make are assigned to rows (sperm) and columns (eggs). The F_1 genotypes are then assembled in each cell of the square. (9, 11, 13–14, 16–17)

1. Choice *c*. This individual could not have come from *PPqqRR* parents because the allele combinations are inappropriate. Moreover, there is no evidence that *P* is dominant over the other genes. Finally, this diploid genotype could indeed make a *PQr* gamete. (9–11)

2. Choice *a*. The way to maintain absolutely true-breeding characteristics throughout several generations is to mate only those individuals that are homozygous for the same form of the desired trait. Thus, all parents in this lineage had to be homozygous recessive. (4, 12–14)

3. Choice *e*. The functions of all these components in the hereditary transfer system are consistent with Mendel's particulate theory. (7)

4. Choice *b*. All offspring from this cross will be heterozygotes *(Kk)* and will show the dominant phenotype. (5, 9–10)

5. Choice *e*. True breeding indicates that the parents are both homozygotes, probably with short hair being dominant. Since the test cross (mating with homozygous recessive individual) produces one-half dominant phenotypes, the F_1 individuals must be heterozygotes. (5, 10–12, 15)

6. Choice **a**. We know that the male has to be homozygous recessive *(bbss)*. The female must have at least one dominant allele for each trait (*B—S—*). In essence, this is simply a test cross. Since all the subsequent offspring show only the dominant phenotype, the female's genotype is probably *BBSS*. (5, 10, 15, 16)

7. Choice **b**. The cross is *CcLl × ccll*. Four F$_1$ genotypes will result, of which 25% will be *ccll*. (9, 13, 16, 17)

8. Choice **d**. Each parent can make two kinds of gametes (*NO* or *nO* for one, and *nO* or *no* for the other), which can then combine to form four different genotypes in the offspring. (9, 13, 16, 17)

ALLELES AND THEIR INTERACTIONS • GENE INTERACTIONS • GENES AND CHROMOSOMES • CYTOPLASMIC INHERITANCE (pages 69–72)

● (21–23, 25–26)
normal male: ZZ chicken
abnormal female: X human
abnormal male: XXY human
normal male: XY fruit fly
normal female: XXY fruit fly
normal male: haploid honeybee

1. Choice **c**. Incomplete dominance is characterized by a unique phenotype for all three genotypes. (1)

2. Choice **b**. In codominance, neither trait is recessive nor dominant. Both alleles have equal but different effects on the phenotype of the heterozygote. (4)

3. Choice **a**. There is only one gene locus for the ABO component of human blood type. However, there are multiple alleles for this gene, which combine to produce five different genotypes. (7)

4. Choice **c**. A type O child must have the genotype *ii*, meaning that each parent needs at least one *i* allele to contribute. A woman with blood type AB cannot make eggs with the *i* allele. (7)

5. Choice **a**. Again, the mother is type AB and so cannot make eggs with the *i* allele. Even though the father has the *i* allele to contribute (by way of his mother), none of his sons by this woman can have type O blood. (7)

6. Choice **a**. Human males have an XY combination of sex chromosomes, whereas females are XX. Males are thus hemizygous for traits on either of the sex chromosomes. (22, 27)

7. Choice **e**. Since women carry two alleles for X-linked traits, this woman must be homozygous recessive because she expresses the color-blind phenotype. The man is hemizygous, so he must have a single copy of the dominant allele to produce his normal phenotype. But the father will always donate a Y chromosome when producing a son, so the son's X chromosome must come from his homozygous mother, who has only the recessive allele to contribute. All sons from this couple will therefore be color-blind. (27–28)

8. Choice **c**. For females to express an X-linked recessive phenotype they must have the homozygous recessive genotype. Since one X chromosome in a female comes from each of her parents, each parent must have contributed to their daughter's homozygous condition. (22, 28)

9. Choice **d**. Useful gene mapping information must have something to say about the relative distances between genes. Of the things listed here, only recombination frequencies provide such information. (15–17)

10. Choice **a**. This is an example of epistasis, where the effect of one gene masks the other. If no pigment is produced, it doesn't matter that the alleles for proper pigment deposition are present. Likewise, the animal will be colorless even if pigment is made, as long as the recessive alleles for deposition are present. (1)

11. Choice **a**. The products manufactured by the gene that controls pigment deposition are temperature sensitive, showing how environment can influence penetrance or expressivity of a phenotype. (11–12)

12. Choice **e**. Maternal inheritance refers to the passage of extranuclear DNA from mother to offspring by way of cytoplasmic sources, usually the mitochondria or chloroplasts. The DNA in these structures does not undergo meiosis during cell division, and so genetic recombination is not an issue. (31)

13. Choice **c**. Skin color is a likely polygenically controlled trait in humans, whereas flower color in our hypothetical plant involves only a single gene showing incomplete dominance. (10)

Integrative Questions

1. Choice **b**. The sequence of genes from C through H has been inverted within the linkage group.

2. Choice **d**. This statement is false because homozygous dominants can only produce gametes containing the dominant allele, which cannot then combine with one another to form heterozygous offspring.

3. Choice **a**. These crosses are reciprocals of one another with respect to both sex and trait. Since the offspring are the same regardless of which parent possesses which trait (i.e., the trait is gender-independent), the inheritance pattern is likely to be autosomal.

4. Choice **d**. The information about both parents being true-breeding tells us that they are homozygous for their respective traits. Independent assortment allows us to treat this dihybrid cross as two monohybrid crosses occurring simultaneously. Thus, the chance of getting purple flowers in the F$_2$ generation is 3/4 and the chance of getting short plants is 1/4. The chance of both phenotypes occurring together is the product of the two independent probabilities, or 3/16.

CHAPTER 11 – DNA AND ITS ROLE IN HEREDITY

Note: The page numbers listed with the section titles refer to the Study Guide; numbers in parentheses for each question refer to the relevant key concepts.

DNA: THE GENETIC MATERIAL • THE STRUCTURE OF DNA • DNA REPLICATION (pages 73–75)

● (10, 13–14)
(parent strand)
5′ – A T C C G T A A C G C A G G G C T T A – 3′
3′ – T A G G C A T T G C G T C C C G A A T – 5′
(daughter strand)

- If parental DNA is made exclusively with ^{15}N, as was the case in Meselson and Stahl's experiments, it can only form one heavy band during centrifugation. The first generation of DNA raised on ^{14}N will, however, form a different centrifugation band due to its hybrid DNA, which is made from a ^{15}N template strand and a ^{14}N replicate strand. In the second and subsequent generations, the hybrid band and a new light DNA band will appear. Some new DNA will still be made using the old ^{15}N templates (hybrid band), but most new DNA will now be made by ^{14}N templates (light band). (17–20)

- Griffith's experiments with pneumococci showed that bacterial transformation occurs, but did not provide an actual mechanism for this process. Hershey and Chase used radioactive-specific labeling of nucleic acids and proteins to conclusively demonstrate that DNA is the transforming principal. The main conclusion from the combined experiments is that DNA is the hereditary material that ultimately controls the characteristics of organisms. (3–8)

1. Choice *c*. Guanine and cytosine are the only combination shown that fits both requirements. (10)

2. Choice *b*. These experiments provided definitive evidence that DNA, not protein, was the hereditary material for T2 bacteriophage (a type of virus). (6–7)

3. Choice *b*. If 30 percent of the bases are adenine, then according to Chargaff's rules 30 percent must also be thymine, accounting for a total of 60 percent of the DNA in this particular sample. That leaves 40 percent to be divided equally among the guanine and cytosine, or 20 percent for each. (10–11)

4. Choice *e*. The Watson–Crick model contains nitrogen in the bases and phosphorous in the phosphate groups. It has a constant cylindrical shape due to the double helix, and the amount of any given purine always equals the amount of its complementary pyrimidine. The double helix is held together mainly by hydrogen bonds between the bases of opposing strands. (12–13)

5. Choice *c*. Although radioactive labeling of phosphorus and sulfur in the Hershey–Chase experiments showed that DNA is the hereditary material, it did not provide information about the actual structure of DNA. (6–9, 12)

6. Choice *b*. The use of isotopes of nitrogen in these DNA replication experiments produced characteristic centrifuge gel banding patterns that indicated the semiconservative model was correct. (17–20)

7. Choice *c*. Nitrogen is part of the purine and pyrimidine bases of DNA nucleotides, and does not occur in the sugar-phosphate "backbone" of the molecule. However, all the other elements or bond types listed do. (9, 10, 13)

THE MECHANISM OF DNA REPLICATION • DNA PROOF-READING AND REPAIR (pages 75–77)

- Your diagram should resemble the segment of replicating DNA shown in Figure 11.16 of the textbook. (2, 4, 8–9)

1. Choice *a*. Okazaki fragments form in 100 to 200 nucleotide segments along the lagging strand of DNA, and are then connected together to make a functional strand. They cannot form continuous strands of DNA because the lagging strand has the wrong sugar–phosphate backbone orientation. (8–10)

2. Choice *a*. The addition of new nucleotides during DNA replication is a type of synthesis reaction, and considerable free energy must be added to the developing molecule to make the reaction take place (i.e., it is endergonic, not exergonic). The reaction is neither easy to maintain nor energetically inconsequential to the cells where it occurs. (1–2, 16)

3. Choice *d*. Okazaki fragments are formed as part of the lagging strand during DNA replication. The other molecules listed are used to help make DNA replication happen. (3, 6–7, 9–10)

4. Choice *b*. Two replication forks, one moving in each direction around the circular chromosome, occur during DNA replication in prokaryotes. (4)

5. Choice *a*. Okazaki fragments are formed on the lagging strand of replicating DNA. (9)

Integrative Questions

1. Choice *e*. All four of these elements or molecules are common components of bacterial DNA.

2. Choice *a*. While covalent bonds are what hold together the sugars and phosphates of the DNA backbone, hydrogen bonds hold the nucleotide bases of complementary DNA strands together. The use of these weaker hydrogen bonds makes the unwinding of double-stranded DNA energetically reasonable during replication while still providing stability to the DNA molecule.

3. Choice *d*. Excision repair mechanisms are what remove functional DNA that has been damaged due to chemical modification. These mechanisms make it possible for the cell to replace the damaged DNA with a newly synthesized sequence of nucleotides.

CHAPTER 12 – FROM DNA TO PROTEIN: GENOTYPE TO PHENOTYPE
Note: The page numbers listed with the section titles refer to the Study Guide; numbers in parentheses for each question refer to the relevant key concepts.

GENES AND THE SYNTHESIS OF POLYPEPTIDES • DNA, RNA, AND THE FLOW OF INORMATION • TRANSCRIPTION: DNA-DIRECTED RNA SYNTHESIS • THE GENETIC CODE (pages 78–80)

- (17–19)
  Case 1: *methionine – threonine – cysteine – arginine – glycine*
  Case 2: *nothing*, because there is no initiator codon to start translation
  Case 3: *methionine – alanine – threonine – serine – threonine – tryptophan*
  (Note: during translation mRNA, codons are read in the 5' → 3' direction)

- Genetic information is inherited in the form of DNA (or RNA in the case of retroviruses). Because this material remains in the nucleus, a messenger nucleic acid (mRNA) must first be transcribed to get the genetic message out into the cytoplasm where protein synthesis occurs. Once in the cytoplasm, a second process occurs as the mRNA message is translated into the amino acid language of proteins. (1, 6)

1. Choice *b.* Genes code for the production of a single polypeptide, so any one gene can control only one part of a pathway. Thus, assuming that these are the only genes involved, it is theoretically possible to have as few as ten different auxotroph strains, although each gene can mutate in many different ways. (3–5)

2. Choice *d.* The one-gene, one-polypeptide theory says that any given gene can produce only a single polypeptide, which then usually plays a role in shaping the organism's phenotype. (1, 5)

3. Choice *b.* Genetic information starts with nucleotides (the "letters" in the triplet codons). The nucleotides are organized into DNA segments, which are grouped into genes. Many genes are contained on each chromosome. (1, 15, 17)

4. Choice *c.* The universal genetic code as it relates to translation is based on the nucleotide triplets found in mRNA, the blueprint for protein manufacture. (17)

5. Choice *e.* One gene leads to one polypeptide, with the triplet bases in the DNA sequence corresponding to specific amino acids in the polypeptide. (1, 6, 17)

6. Choice *e.* UAG is a chain terminator codon, and so no protein would be manufactured from this mutant mRNA. (19)

7. Choice *b.* Ribosomes play an important role in the translation of mRNA into protein, but are not involved in the process of transcribing mRNA from DNA. (11, 13, 16)

THE KEY PLAYERS IN TRANSLATION • TRANSLATION: RNA-DIRECTED POLYPEPTIDE SYNTHESIS • REGULATION OF TRANSLATION • MUTATIONS: HERITABLE CHANGES IN GENES   (pages 80–83)

• The mutated "sentence" would now read THE ATA TEH ERN EWT OY. In other words, complete nonsense as far as English language is concerned! Likewise, a frame-shift mutation in DNA and mRNA leads to a highly distorted sequence of amino acids in the protein that is produced by translation. This almost always renders the protein nonfunctional. (12, 18)

1. Choice *a.* mRNA, not being a protein, is not a product of translation. Rather, it is produced by transcription of a DNA sequence. All the other choices are proteins. (1–2)

2. Choice *a.* Translation involves turning an mRNA message into a polypeptide. DNA does not play a direct role in this process, although it does in transcription. (1–10)

3. Choice *b.* Adaptors allow things to coexist or work together. tRNA is an adaptor that connects the nucleic acid language of DNA and mRNA with the amino acid language of proteins. (2–3)

4. Choice *c.* Free amino acids in the cell are attached to an appropriate tRNA through the action of activating enzymes. The charged tRNA then combines with mRNA on the ribosome and forms a dipeptide bond with an adjacent amino acid in the growing polypeptide sequence. (3–7)

5. Choice *a.* Digestive enzymes would most likely be packaged in membranous vesicles to avoid self-digestion. This suggests that these particular proteins would be manufactured on ribosomes attached to the rough ER. (10–11)

6. Choice *b.* Frame-shift mutations can radically alter the sequence of amino acids coded for by DNA and mRNA, and thus have dramatic effects on the characteristics of proteins translated from this altered genetic information. (18)

7. Choice *d.* Careful examination of the two sequences reveals a substitution of the sixth nucleotide base, thymine, with an adenine. No other detectable mutations have occurred. (14–17)

Integrative Questions

1. Choice *d.* The release of covalently bonded phosphate groups in ATP and triphosphate nucletides typically provides the energy needed to synthesize RNA and protein polymers.

2. Choice *a.* With only ten amino acids to code for, a doublet-base code using four nucleotides (i.e., 16 possible two-base combinations) might have sufficed to provide all the necessary information for the genetic code, including start and stop codons. Clearly, such an abbreviated code would not have anywhere near the degree of degeneracy that the three-base codon system that codes for the known twenty amino acids now displays.

3. Choice *e.* Despite the fact that very different kinds of products are being formed (RNA vs. protein), there are still striking biochemical, energetic, and information flow similarities between the elongation phases of both translation and transcription.

CHAPTER 13 – THE GENETICS OF VIRUSES AND PROKARYOTES
Note: The page numbers listed with the section titles refer to the Study Guide; numbers in parentheses for each question refer to the relevant key concepts.

USING PROKARYOTES AND VIRUSES FOR GENETIC EXPERIMENTS • VIRUSES: REPRODUCTION AND RECOMBINATION   (pages 84–86)

• Four obvious criteria for "life" are (1) being made of cells, (2) having metabolic activity, (3) being able to self-replicate, and (4) reacting to environmental stimuli. Viruses are acellular, do not metabolize because they lack internal biochemical mechanisms for such reactions, can only replicate by commandeering a host cell's metabolic machinery, and outside of their hosts are quite inert with respect to environmental stimuli. (3)

1. Choice *e.* Bacteria and viruses have relatively small amounts of DNA compared with eukaryotes. (1)

2. Choice *d.* Prions completely lack nucleic acid and therefore cannot code for their mechanism of infection or disease through DNA or RNA. (17)

3. Choice *c.* Outside of host cells viruses exist as independent particles called virions. (5)

4. Choice *a.* The presence of a lipid envelope on this virus suggests that its host is an animal cell. (10–12)

5. Choice *d.* Reverse transcriptase is the enzyme used to convert the RNA of retroviruses into host cell DNA. (11)

6. Choice *b.* Prophages are the form of that bacteriophage DNA assumes when it is incorporated into the host cell's DNA for several generations before new viruses are actually manufactured. (8)

PROKARYOTES: REPRODUCTION, MUTATION, AND RECOMBINATION • REGULATION OF GENE EXPRESSION IN PROKARYOTES (pages 86–89)

- In this operon, an inactive repressor (R) cannot normally block the transcription of the structural genes. All four enzymes needed for the synthesis of arginine are therefore made continuously. Arginine acts as the corepressor (co-R) in this system. When present it binds with R to form an active R/co-R complex that settles down at the operator (O) to block transcription of the structural genes. When arginine is removed, the repression ends and the genes are once again transcribed. (31–37)

- The most likely hypothesis is that as the use of new antibiotics increased with the advent of modern medicine, bacteria were subjected to intense selection pressures. Mutant forms that were resistant to these drugs appeared in the bacterial population, and the genes for drug resistance were transmitted via R factors. Since drug resistance was selectively advantageous, the frequency of R factors increased in these strains of bacteria. (24, 27)

1. Choice *d*. All other mechanisms could be ruled out. (7–9)
2. Choice *e*. All of these statements are true. (10–12)
3. Choice *a*. General transduction occurs when bacterial DNA is inserted into a phage coat, which can associate with another bacterium and thus pass along the bacterial DNA without the need for viral infection. (18–20)
4. Choice *d*. The opened end of the broken chromosome always enters the female cell first for Hfr cells. (12, 14–15)
5. Choice *c*. The F pilus is a hairlike, tubular projection through which genetic material is exchanged during conjugation. All others in the list are indeed plasmids and episomes. (10–11, 21–22)
6. Choice *d*. Replicons are necessary for independent plasmid replication, since the plasmid remains distinct from the host cell chromosome and must replicate in synchrony with the cell in order to remain in the bacterial population. (22–23)
7. Choice *b*. See the definition of a transposon. (26)
8. Choice *c*. Inducible enzymes have the advantage of only being made when the substance on which they are intended to act is actually present. This mechanism is particularly efficient when the substance is available to the cell only irregularly. (28, 33)
9. Choice *a*. In repressible operon systems, the repressor is normally incapable of preventing gene transcription, which goes on routinely in the absence of an external stimulus. The repressor becomes active when combined with a corepressor. The complex then proceeds to block transcription. (31, 36–37)
10. Choice *e*. All four types (of which *a* and *b* are specific examples of *c* and *d*) contain regulatory genes that code for repressor molecules. (28–40)
11. Choice *c*. See the definition of a constitutive enzyme. (28)
12. Choice *e*. All of these structures, events, or features occur in the regulatory mechanism of enhanced promoter efficiency. (39–40)

Integrative Questions

1. Choice *b*. Advantageous mutations can be naturally selected and result in heritable phenotypic changes in organisms and viruses. Thus, mutations are the raw material of evolution.
2. Choice *b*. The lawn in a bacterial culture represents large numbers of individual bacteria. When phages are applied, they kill all (or virtually all) of the bacteria near them, thus exposing the medium below.
3. Choice *c*. The F factor is where the circular chromosome breaks during conjugation. The DNA replicate formed starts entering the recipient cell in a linear sequence beginning with the F factor. If character five only appears in cells that mated the longest, this means that this character must be located farthest away from the F factor on the chromosome.

CHAPTER 14 – THE EUKARYOTIC GENOME AND ITS EXPRESSION
Note: The page numbers listed with the section titles refer to the Study Guide; numbers in parentheses for each question refer to the relevant key concepts.

THE EUKARYOTIC GENOME (pages 90–91)

1. (8–10)
   4: Add the probe
   1: Warm the nucleic acid mixture
   2: Observe denaturation
   5: Cool the nucleic acid mixture
   6: Observe reannealing
   7: Identify the probe-DNA hybrids
   3: Immobilize nucleic acids on nitrocellulose filter
2. Choice *c*. The DNA type that is most abundant will reanneal most rapidly. This includes the type known as highly repetitive DNA, typically located near the centromeres. (11–14)
3. Choice *d*. The telomeres are moderately repetitive DNA that compose the ends of chromosomes. Therefore, the repeat sequence is present in both the leading *and* the lagging strands. (15, 16)
4. Choice *d*. Although the gene products of some transposable elements have been characterized, their functions are unknown. (18–21)
5. Choices *a, c, d*. (3, 4)

THE STRUCTURES OF PROTEIN-CODING GENES • RNA PROCESSING (pages 92–93)

1. Choice *a*. There is no evident restriction on the relative numbers of functional genes and pseudogenes in a family. (5–10)
2. (11–18)
   6: Removal of intron "lariat"
   4: Cleavage of 5′ exon and intron
   5: Cleavage of 3′ exon and splicing
   3: Binding of snRNPs to consensus sequences in RNA
   7: Degradation of intron
   1: Addition of G cap
   2: Addition of poly A tail

3. (1–3)

750 nucleotides in exons

450 nucleotides in introns

Since each amino acid in the protein is coded for by three nucleotides, 750 nucleotides are included in the exons and the remaining 450 nucleotides must be in introns.

4. Choices *d, e*. The start and stop sequences and exons are found in both preRNA immediately after transcription and its mature mRNA. The introns are removed and a modified G cap and a poly A tail are added during processing. (11–18)

● (1–3)

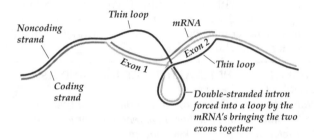

Thin loop

Noncoding strand

mRNA

Exon 1

Exon 2

Thin loop

Coding strand

Double-stranded intron forced into a loop by the mRNA's bringing the two exons together

● (1–3)

mRNA

Displaced DNA strand

TRANSCRIPTIONAL CONTROL ● POST-TRANSCRIPTIONAL CONTROL ● TRANSLATIONAL AND POSTTRANSLATIONAL CONTROL   (pages 93–95)

1. Choice *b*. Only RNA polymerase II catalyzes transcription of genes that code proteins; RNA polymerase I and III catalyze transcription of genes coding for either rRNA (I) or tRNAs (III). (6)

2. Choice *c*. The TATA box is a binding site on DNA for the transcription factor TFIID. (8–11)

3. Choice *d*. Since each cell needs one functional X chromosome, and no more, three of the X chromosomes in each cell of this individual will be inactivated, so three Barr bodies will be evident. (17, 18)

4. (2–27)

posttranscriptional	Increasing the longevity of the mRNA coding for a gene product
posttranscriptional	Modification of the G cap
transcriptional	Presence or absence of enhancers
transcriptional	Methylation of the gene
transcriptional	Presence or absence of the appropriate transcription factor
transcriptional	Gene amplification through duplication
posttranslational	Attachment of ubiquitin to a protein
transcriptional	Conversion of heterochromatin into euchromatin
posttranslational	Modification of a gene product

● *rRNA*. The nucleolus represents the chromosomal region that specializes in production of the RNA subunits for ribosomes, so rRNA is being produced. (21, 22)

4

3

1

2

Integrative Activity

● See Figure 14.15 on page 320 in the textbook for the correct labeling of this figure.

Integrative Questions

1. *b, i*

2. *n.a.* Promoter

n.a. Consensus sequence

n.a. Enhancer

p. Telomerase

n.a. Transposable elements

n.a. Poly A "tail"

b. snRNPs ("snurps")

n.a. U1 and U2

n.a. Hormone-response element

n.a. TATA box

n.a. Silencer

p. Activator and repressor

p. Transcription factor

p. Ubiquitin

p. RNA polymerase Type III

b. Heterochromatin

CHAPTER 15 – DEVELOPMENT: DIFFERENTIAL GENE EXPRESSION

Note: The page numbers listed with the section titles refer to the Study Guide; numbers in parentheses for each question refer to the relevant key concepts.

THE PROCESSES OF DEVELOPMENT ● THE ROLE OF DIFFERENTIAL GENE EXPPRESSION IN DIFFERENTIATION (pages 96–97))

1. Choice *c*. At this point, the developmental potential of the cell is less than the zygote; it is no longer totipotent. (5, 6, 9, 16)

2. Choice *a*. Normally, imaginal discs retain their ability to produce the distinctive structures they would have produced if undisturbed. (13–16)

3. Choice *c*. Their studies convincingly demonstrated that differentiation is not caused by loss or permanent inactivation of the genetic material. (10–12)

4. Choice *d*. The salivary gland cell will have both myoglobin and cytochrome c genes, but will only transcribe the cytochrome *c* gene (cytochrome *c* is needed by all cells for cellular respiration). The salivary gland cell will, therefore, not have myoglobin mRNA. (18, 19)

5. Choice *c*. Only the *Drosophila* imaginal disc studies in which transdetermination was demonstrated directly supported the hypothesis that differentiation does not involve permanent changes in the genome. (14–16)

THE ROLE OF CYTOPLASMIC SEGREGATION IN CELL DETERMINATION • THE ROLE OF EMBRYONIC INDUCTION IN CELL DETERMINATION • THE ROLE OF PATTERN FORMATION IN ORGAN DEVELOPMENT (pages 98–99)

- (7–9)

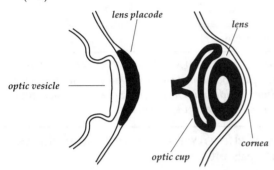

1. Choice *d*. Cytoplasmic determinants, called *P granules*, in *C. elegans* are uniformly distributed in the egg, but differential movement of granules during cell division leads to their segregation in the posterior end of the worm. (1–6)

2. Choice *d*. A ZPA grafted to the anterior margin should produce a mirror-image duplication of the entire wing. (22–25)

3. Choice *e*. If the anchor cell is destroyed, no vulva will be formed. Only the anchor cell can produce the primary inducer. (11–14)

4. Choice *b*. Organ identity genes have been implicated in specifying the successive whorls of a flower. Gradients of CO_2 and O_2, set up by distance from the epidermis, seem to influence whether cells become xylem or phloem. (19–21)

THE ROLE OF DIFFERENTIAL GENE EXPRESSION IN ESTABLISHING BODY SEGMENTATION (pages 99–101)

1. Choice *a*. The gap genes are activated first. They organize large areas along the anteroposterior axis. (7–10)

2. Choice *c*. The homeotic genes are activated last and they determine what specific structures will develop on each segment. (7–10)

3. Choice *c*. The homeobox is a sequence of DNA common to the genes of a host of organisms with segmented body plans. Choice *b* is the definition of the homeodomain, the polypeptide segment coded for by the homeobox. (11, 12)

4. Choice *a*. The *nanos* female is mutant for a gene coding for a cytoplasmic determinant affecting the abdominal segments of the larva. Because it has a normal allele for the *bicoid* gene, cytoplasm from the anterior region of its egg would have the normal cytoplasmic determinant that the *bicoid* female cannot produce. Consequently, this cytoplasm should permit normal development of the *bicoid* female egg. (3–6)

5. Within the *Drosophila* larva, the concentration gradients of the morphogens **bicoid protein** and **nanos protein** are established by the unequal distribution of their mRNAs within the egg. (14–17)

The proteins that are coded for by the segmentation and homeotic genes are *transcription factors*.

The *homeotic* genes are arranged on their chromosome in the same order as the order of their expression from anterior to posterior in the larva.

Integrative Question

- *d* Imaginal discs
  a L. Wolpert
  e Organ Identity genes
  f P granules
  c R. Briggs, T. King
  b S. Hörstadius

CHAPTER 16 – RECOMBINANT DNA AND BIOTECHNOLOGY
Note: The page numbers listed with the section titles refer to the Study Guide; numbers in parentheses for each question refer to the relevant key concepts.

CLEAVING AND REJOINING DNA • CLONING GENES (pages 102–105)

- Frequently the recognition sequences for restriction endonucleases are six base pairs long and palindromic, that is, the two sequences are the same when read in the opposite direction. (3, 10, 11)

- Because *Hpa*I produces a blunt-ended cut, its fragments will rejoin with any other fragment. *Hind*III produces so-called "sticky-ended" fragments that can only rejoin with a fragment that has a complementary end. This allows precise control over which fragments will produce recombinant molecules.

1. Choice *c*. Restriction endonucleases break the phosphodiester bonds within the backbone of the DNA molecule; hydrogen bonds between the bases are not broken, but cannot persist at warm temperatures. Choice *a* is true, because the distribution of the four bases in DNA is random, so longer recognition sequences will be less common than shorter sequences. (3–6, 10, 11)

2. Choice *b*. The ideal plasmid would have a single recognition site to open the plasmid for insertion of the new DNA. (13–20)

3. Choice *b*. Lysing the protoplast would kill the cell. All other techniques are used to insert foreign DNA into cells. (24–28)

4. (10–12, 18–20, 29, 30)
   5: Encourage growth of the transgenic cells
   3: Place host cells and DNA into calcium chloride solution
   2: Treat with ligase
   1: Create fragments with restriction endonucleases
   4: Transform cells
   6: Use selection plating with antibiotics

5. (29)

DNA taken up by wild-type *E. coli*	Growth (+) or no growth (−) on plates containing…		
	No antibiotics	Tetracycline	Ampicillin
A *or* **B**	+	−	−
D	+	−	+
A *or* **B**	+	−	−
C	+	+	+

See Figure 16.7 on page 359 in the textbook for more information.

6. Choice *d.* Yeast is a *unicellular* fungus. (22)

7. Choice *e.* The tumors have no role in the insertion of the plasmid into the plant. Normally, when the T DNA is replaced, the plasmid no longer produces tumors. (25)

8. Choice *e.* The sample with the greatest number of recognition sites would produce the largest number of fragments. Choices *b* and *e* both have three fragments. Since DNA prepared for electrophoresis is negatively charged, it will move toward the positive pole and the smallest fragments will move the greatest distance. (7, 11)

SOURCES OF GENES FOR CLONING • SOME ADDITIONAL TOOLS FOR DNA MANIPULATION • BIOTECHNOLOGY: APPLICATIONS OF DNA MANIPULATION (pages 105–107)

1. Choice *b.* cDNA libraries would be most useful for determining which genes are active during differentiation because they are normally obtained from mRNA. (1–6)

2. (3–6)
1: Add oligo dT primer to mRNA
3: Denature DNA–RNA hybrid molecules
6: Insert cDNA into DNA vector
5: Add DNA polymerase
4: Degrade mRNA
2: Add reverse transcriptase

3. (1–27)
e: DNA fingerprinting
g Antisense RNA
c: Expression vectors
a: Reverse transcriptase
d: Size standards
f: DNA base sequencing
b: PCR

4. Choice *b.* Probes with base sequences complementary to the sequences at the boundaries of the gene must be constructed to act as primers for the DNA polymerase. (14–17)

5. Choice *d.* Since the number of DNA molecules doubles with each cycle, it will require ten 5-minute cycles, or 50 minutes, to get 1024 copies. (14–17)

6. Traditional plant breeding is a slow process because it depends on the completion of a plant's normal life cycle. Also, seed production from the selected strains requires a significant land area. Since recombinant DNA techniques are directed at the cellular level, the time and space constraints are reduced considerably. A traditional genetic cross between two plants involves all the the genes, not just the genes of interest, and only plants that are capable of interbreeding can be used. Recombinant DNA technology allows transfer of individual genes, not whole genomes, and gene transfer can involve any organism, not just those that are closely related. (22–24)

- (10–12)
  New strand: T–C–G–T–A–C–G–C–A–A–G–T
  Original strand: A–G–C–A–T–G–C–G–T–T–C–A

Integrative Questions

1. Choice *c.*

2. Choice *a.*

3. Choices *b, c, d, e, f.*

CHAPTER 17–MOLECULAR BIOLOGY AND MEDICINE
Note: The page numbers listed with the section titles refer to the Study Guide; numbers in parentheses for each question refer to the relevant key concepts.

PROTEIN AS PHENOTYPE • MUTATIONS AND HUMAN DISEASES (pages 108–110)

- (21–22)

Remember that individuals have two chromosomes and thus will show bands for both alleles. Carriers would have an RFLP with three bands (1350, 1150, and 200 bp).

1. Choice *c.* Many non-disease-causing genes have triple repeats. For example, in the normal allele associated with fragile-X syndrome, the triplet CGG is repeated 6 to 54 times. All other statements are true. (24–27)

2. Choice *e.* Onset of the disease is early in life. (6–9)

3. (2–11, 13–17, 24–27)

e	Cystic fibrosis
f	Osteogenesis imperfecta
a	Phenylketonuria
b	Sickle-cell anemia
g	Duchenne's muscular dystrophy
d	Fragile-X syndrome
c	Hemophilia A

4. Choice *d.* All of these are important for use of a RFLP as a genetic marker, except identifying the protein produced by the gene. This may be good to know, but does not relate to use of the RFLP. (21–22)

5. Methylated cytosine becomes 5-methylcytosine. If this molecule loses an amino group it becomes thymine, which is now mispaired with the guanine that had been paired to the original cytosine. A repair system that deals with mispaired bases 50 percent of the time replaces the thymine with cytosine, but the other 50 percent of the time it replaces the guanine with adenine, rendering a mutation in the DNA. (23)

DETECTING HUMAN GENETIC VARIATIONS • CANCER: A DISEASE OF GENETIC CHANGES (pages 111–112)

1. (9–15)
   - *c:* Benign
   - *e:* Metastasis
   - *d:* Apoptosis
   - *a:* Carcinogens
   - *b:* Carcinoma
   - *f:* Leukemia

2. Choice *d*. Both methods reveal differences in the DNA and would show the condition for both alleles of a gene, so differentiating between all three genotypes is possible (7, 8).

3. Choice *c*. (17, 18)

4. If the woman is a carrier, either the polar body or the egg will have the mutant allele for cystic fibrosis. Polar diagnosis reveals that the polar body has the normal allele, so the egg must have the disease allele. If the objective is to have a baby who is not a carrier for the disease, this egg cannot be used in *in vitro* fertilization.

5. Genetic screening can be used to detect defects in the fetus and abortion may be considered if gene therapy is not available to correct the problem. (21–22)

6. Since tumor supressor genes inhibit cell division, both alleles of a gene must be mutated ("hit") before uncontrolled cell division can begin in a cell. This is referred to as the "two-hit" hypothesis. Conversion of a proto-oncogene into an oncogene can be accomplished by a single mutation. Since conversion of a normal cell into a cancerous cell requires mutation of at least two tumor-suppressor alleles and also conversion of a proto-oncogene into an oncogene. The accumulation of mutations requires time. (7, 12)

TREATING GENETIC DISEASES • THE HUMAN GENOME PROJECT (pages 112–113)

1. Choice *d*. The correct sequence is $c \rightarrow a \rightarrow d \rightarrow e \rightarrow b$. (8–13)

2. Choices *b*, *d*. Chemotherapy agents are usually metabolic inhibitors targeted for actively dividing cells. (1–3)

3. Choice *b*. FISH is an acronym for fluorescenece in situ hybridization. It is a method for visualizing a probe that is hybridized to a denatured chromosome. (15–17)

4. Choice *b*. The intention is to initially identify protein-coding genes, so cDNA is being mapped relative to expressed sequence tags, a type of STS. (16)

5. Databases containing information on polymorphic gene sequences in humans are being compared with disease databases for the same groups of people. Correlations are revealing which genes have influences on which diseases.

CHAPTER 18 – NATURAL DEFENSES AGAINST DISEASE
Note: The page numbers listed with the section titles refer to the Study Guide; numbers in parentheses for each ques-

DEFENSIVE CELLS AND PROTEINS • NONSPECIFIC DEFENSES AGAINST PATHOGENS • SPECIFIC DEFENSES: THE IMMUNE RESPONSE (pages 114–117)

1. Choice *b*. The inflammation response is a nonspecific response that is primarily hormonal in nature. (2, 17)

2. (3–8, 18)
   - *b*: Plasma
   - *d*: White Blood cells
   - *f*: Lymph
   - *e*: Lymphocytes
   - *a*: Antigen
   - *c*: Antibody

3. Choice *b*. Although the T cells facilitate the humoral immune response, the B cells that have differentiated into plasma cells secrete the antibodies. (19, 20)

4. Choice *b*. It is now known that plasma cells specific for a huge diversity of antibodies preexist in the body and are selected for by the antigen. In choice *a*, an antigen with more than one antigenic determinant can lead to the production of more than one antibody, as stated in *e*. (6, 18, 22)

5. The figure shows *three* different antigens. The antigens share *four* different antigenic determinants. Each antigenic determinant has its own antibody, so there are *four* different antibodies. (18)

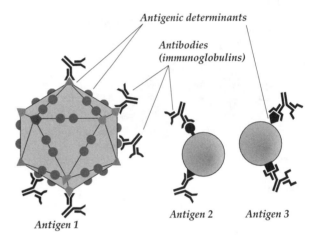

Antigenic determinants

Antibodies (immunoglobulins)

Antigen 1   *Antigen 2*   *Antigen 3*

6. Choice *a*. A fundamental postulate of the clonal selection theory is that a plasma cell can make only one immunoglobulin. In choice *b*, an antigen with more than one antigenic determinant can lead to the production of more than one clone of B cells. (18–23)

7. Choice *c*. Both effector cells *and* memory cells must be able to recognize the antigen. (23, 24)

8. Choice *b*. Normally, the development of immunological tolerance occurs in a very young animal, although continued exposure to self-antigens is required throughout the life of the animal. (26–27)

9. Choice *d*. The complement system is part of the nonspecific defenses of the body; it attacks any cell marked by agents of the immune system. (11–13)

• Since T cells mature in the thymus gland, removing this gland would seriously cripple the immune system. Only animals kept in isolation from pathogens could survive. (5–8)

B CELLS: THE HUMORAL IMMUNE RESPONSE • T CELLS: THE CELLULAR IMMUNE RESPONSE (pages 117–119)

1. Choice *b*. For a given molecule, the two light chains are identical in amino acid sequence, as are the two heavy chains (the two halves of the "Y" are identical), so their variable and constant regions are also identical. (3–7)

2. (8–12)

    b, c: IgG

    a, c, h: IgM

    f : IgA

    d: IgD

    e, g: IgE

3. Choice *b*. Antigenic determinants are bound by the membrane-embedded immunoglobulins of the B cell. In choice *c*, remember that some of the complement proteins involved in cell lysis are class III MHC proteins.(21–26, 28–31)

4. Choice *b*. Activation of the T cell by the macrophage is caused by interleukin-1 released by the macrophage. (17–31)

5. Choice *e*. Plasma cells have nuclei and are capable of continued cell division. (1, 2)

6. Choice *a*. *Lymphoid tissue*, not antigens, is isolated from the animal immunized to the antigen for which the monoclonal antibodies are desired. (15, 16)

- (28, 29)

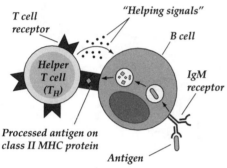

"Helping signals"

T cell receptor

B cell

Helper T cell (T$_H$)

IgM receptor

Processed antigen on class II MHC protein

Antigen

- (30, 31)

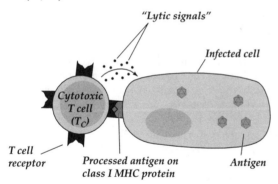

"Lytic signals"

Infected cell

Cytotoxic T cell (T$_C$)

T cell receptor

Processed antigen on class I MHC protein

Antigen

- (3–7)

Variable region

Light chain

Constant region

Heavy chain

THE GENETIC BASIS OF ANTIBODY DIVERSITY • THE EVOLUTION OF ANIMAL DEFENSE SYSTEMS• DISORDERS OF THE IMMUNE SYSTEM (pages 119–121)

- Developing a vaccine means isolating an antigen, like an HIV envelope protein, and exposing animals to it so that they can develop immunoglobulins specific to the antigen. If the antigenic determinants of the envelope proteins are constantly changed by mutation of the gene that codes for them, then previously vaccinated individuals will have immunoglobulins unable to recognize the changed antigen. (19)

- The HIV virus is transmitted by exchange of blood (transfusion, contaminated hypodermic needles), various forms of sexual intercourse, and through the placenta to the fetus. HIV cannot be transmitted by insect vectors or by kissing and casual contact. (14, 15)

1. Choice *d*. Since the variable regions of both chains will be assembled by joining only one of the three different segments of each family, there will be $3 \times 3 \times 3$ or 27 different variable regions for both light and heavy chains. Since the heavy and light chains are independently assembled, there will be 27×27 or 729 different combinations, each able to specify a different antigenic determinant. (1–4)

2. Choice *c*. An antibody always starts in the IgM class and class-switches to other classes. Class switching can be repeated if there are remaining constant region genes present on the chromosome. (7, 8)

3. Choice *a*. They are all proteins (T cell receptors are glycoproteins) and the genes that code for them have similar base sequences, suggesting evolution from a common ancestral gene. Choice *b* does not apply to MHC proteins (they are found on all body cells), choice *c* does not apply to immunoglobulins (immunoglobulins can occur free in the blood/lymph), and choice *d* does not apply to the genetic determination of MHC proteins (large numbers of multiple alleles produce their diversity). (1, 2, 9)

4. Choice *b*. T cell receptor genes are composed of V, D, and J regions and are formed by the same joining process seen in B cells. (8, 9)

5. Choices *a, c*. AIDS is an immune deficiency disease and allergies involve overreaction of the immune system to a harmless antigen. (11–14)

Integrative Questions

1. *A*: Natural killer cells

    A: Production of mucus

    A: Macrophages

    A: Secretion of lysozyme

    A: Inflammation

    B: T$_H$ cells

    B: T$_S$ cells

    A: Low pH in stomach

    B: Plasma cells

    B: Interleukins

    A: Complement proteins

    B: IgM

    A: Interferons

    A: Normal flora

2. Note: Step 3, binding of antigen to a surface receptor on B cell could have occurred earlier.

  4: Binding of T_H cell to processed antigen-class II MHC on B cell

  2: Proliferation of T_H cell clone

  6: Proliferation of effector and memory cells

  3: Binding of antigen to a surface receptor on B cell

  5: Stimulation of B cell by interleukin-2

  7: Maturation of plasma cells

  1: Binding of T_H cell to processed antigen-class II MHC on macrophage

19

The History of Life on Earth

CHAPTER LEARNING OBJECTIVES—after studying this chapter you should be able to:

❏ Differentiate among the geological terms "eons," "eras," "periods," and "epochs."

❏ Describe how radioactive isotopes can be used to age fossils and rocks, and give details of this technique as it relates to the use of ^{14}C.

❏ Describe changes that have occurred in the composition of Earth's atmosphere and how these changes affected the course of evolution.

❏ Explain the concept of continental drift, and describe how it has affected life on Earth.

❏ Describe how climate change can influence the patterns of evolution.

❏ Discuss the evidence supporting the hypothesis that the great Cretaceous mass extinction was caused by Earth's collision with a large meteorite.

❏ Discuss how and where fossils are formed, and why such a small percentage of the organisms that ever lived are known from the fossil record.

❏ Explain the various problems associated with interpreting the fossil record.

❏ Differentiate between the terms "flora," "fauna," and "biota," and explain what is meant by "provincialization of biota."

❏ Name the major eras of life on Earth, their historical times in millions of years, and the main evolutionary events that occurred in each.

❏ Discuss how and when the supercontinent Pangaea formed, and how it split to form Laurasia and Gondwanaland.

❏ Explain how Laurasia and Gondwanaland eventually split to form the modern continents.

❏ Differentiate between background extinction and mass extinction, and give the geologic times of several important mass extinction events.

❏ Explain our current understanding of the rate at which evolutionary change occurs, and describe the ways in which such change can be measured.

❏ Provide possible explanations for mass extinctions and the subsequent radiative explosions of new organisms.

DETERMINING HOW EARTH HAS CHANGED • UNIDIRECTIONAL CHANGES IN EARTH'S ATMOSPHERE • PROCESSES OF MAJOR CHANGE ON EARTH • THE FOSSIL RECORD OF LIFE (pages 433–439)

Key Concepts

1. Earth's history can be divided into four *eons*: the Hadean, the Archean, the Proterozoic, and the Phanerozoic. These eons are further subdivided into *eras* and *periods*. The boundaries of the geological time periods are based on major changes in the fossil record.

2. One of the most remarkable achievements of modern science has been to develop technologies that accurately date fossils and rocks. Radioactive decay of isotopes is used for this purpose.

3. Isotopes decay at regular rates, so during any given time a fixed fraction of the material will change from one form to another. Isotopes are commonly categorized

by their half-lives, the period of time it takes for an amount of radioactive material to be reduced by one-half. Many kinds of radioactive isotopes are used to date fossils, including phosphorus, tritium, potassium, uranium, and carbon.

4. The isotope ^{14}C can be used to determine the age of organisms that died within the past 15,000 years. The age of these fossils is then used to infer the age of the surrounding rock.

5. In living organisms, ^{14}C converts to ^{12}C at the same rate as ^{12}C changes to ^{14}C in the atmosphere. Because carbon is continuously exchanged between the atmosphere and organisms, new ^{14}C is constantly acquired and a steady state is reached. But as soon as an organism dies, this exchange ceases and the total amount of ^{14}C in the organism's remains starts to decline. By measuring the ratio of ^{14}C to ^{12}C, the time since the organism died can be calculated.

6. Earth's early atmosphere had very little oxygen. A unidirectional increase in atmospheric O_2 concentration began when certain sulfur bacteria, and the cyanobacteria that descended from them, evolved a form of photosynthesis that liberated oxygen from water.

7. Increased atmospheric O_2 levels allowed the evolution of larger unicellular organisms, first prokaryotes, then eukaryotes. About 700 to 570 mya, the atmospheric O_2 levels became great enough to enable multicellularity to evolve.

8. Earth's crust is made of a number of thick, solid plates that float atop a liquid mantle. Movement of these plates is caused by sea floor spreading. Fault lines and mountain ranges are produced where moving plates intersect.

9. Movement of these crustal plates, called *continental drift*, has had profound influences on climate, sea level, and the distribution of organisms.

10. Climate also changed as the continents shifted positions. During much of Earth's history, climates were considerably warmer than today. Other periods were colder and were accompanied by extensive glacial ice sheets. Most climate changes were gradual, but a number were quite rapid, occurring during only a few hundred years. These appear as "instantaneous" events in the geologic record, and have often been responsible for the sudden extinctions of organisms.

11. Volcanism is frequent in areas where continental plates collide. The ash discharged by volcanoes can reduce penetration of sunlight to Earth's surface, lower global temperatures, and trigger glaciation.

12. The mass extinction at the end of the Cretaceous period may have been caused by a collision with a large meteorite. Evidence for this includes the presence of an iridium-rich layer in rocks from this period and the discovery of a large impact crater near the Yucatán Peninsula.

13. Fossilization is the critical process that leaves a biological record of ancient times. The process occurs only under favorable conditions, which include low oxygen concentrations and rapid sedimentation due to the action of wind or water.

14. Although we have no direct way of knowing the total number of species that have ever lived on Earth, the number of species that have so far been identified in the fossil record (~300,000) certainly represents only a tiny fraction of those that existed.

15. Estimates of the number of species that have ever lived, based on the current *biota* (present species in all kingdoms) and the rate of turnover in the fossil record, lead to the conclusion that less than 2 percent of all species are known from fossils.

16. Marine animals with hard skeletons, such as mollusks, often fossilize well. Nearly two-thirds of all fossil species are from nine phyla of hard-shelled organisms. By contrast, soft-bodied organisms, many of which also do not live in the best habitats for fossil formation, are not particularly well represented in the fossil record.

17. Another potential problem with interpreting the fossil record is that most fossils come from limited geographic areas. Often information from widely separated geographic locations is necessary to decide when and where organisms first appeared during evolution.

18. For example, we know that horses evolved slowly over millions of years in North America, but periodically crossed the Bering land bridge into Asia. If only the Asian fossils were available, one could erroneously conclude that horses evolved very rapidly on several different occasions in Asia.

Questions (for answers and explanations, see page 162)

1. The half-life of an isotope is best defined as the
   a. time a fixed fraction of isotope material will take to change from one form to another.
   b. amount of time it takes for an herbivore to eat half of this material in its food.
   c. age over which the isotope is useful for dating rocks.
   d. ratio of one isotope species to another in a sample of organic matter.
   e. none of the above

2. Earthquakes and mountain ranges are ultimately the result of
   a. plates in Earth's crust that move against one another on top of a liquid mantle.
   b. climate changes and the movement of glacial ice sheets.
   c. the rising and falling of shallow oceans.
   d. leftover vibrations and debris from ancient collisions with an asteroid or meteor.
   e. the breakup of Laurasia and Gondwanaland.

3. Which of the following would likely be the *best* setting for fossilization to occur?
   a. The surf zone along a sand beach
   b. A shallow, cool swamp with good deposition rates of mud sediments
   c. The bottom of a hot, dry cave with no running water
   d. A fast running mountain stream
   e. Dry desert areas with rapidly shifting sand dunes

4. Despite being incomplete as a whole, the fossil record is rather detailed for
   a. soft-bodied insects.
   b. cnidarians and sponges.
   c. most terrestrial animals.
   d. hard-shelled mollusks.
   e. all of the above

5. The best estimate for the total number of species that have ever lived on Earth is approximately
   a. two hundred thousand.
   b. half a million.
   c. one million.
   d. five million.
   e. more than fifteen million.

LIFE IN THE REMOTE PAST (pages 439–445)

Key Concepts

1. During the late Cambrian period (550 mya), six continents were spread across Earth's surface at equatorial latitudes. Some 300 million years later, during the Permian period, the continents united to form a single large continent called Pangaea. Shallow seas formed and receded several times during the Paleozoic era, alternately covering and exposing substantial portions of Pangaea.

2. A process known as *continental drift* both formed and initiated the breakup of Pangaea during the Permian. As Pangaea broke up, two smaller continents were formed. Laurasia included all of the northern land masses, and Gondwanaland the southern ones.

3. During the mid-Cretaceous period, widening of oceanic channels resulted in the breakup of Gondwanaland into South America, Africa, and a third landmass that would ultimately divide during the late Cretaceous to yield Australia, Antarctica, and India. The Himalayas were formed when India collided with the Asian continent.

4. Over the next 70 to 100 million years the continents continued to separate, reaching roughly their current positions by the mid-Tertiary period. At several times, land bridges of various sizes and durations connected formerly isolated land masses. Some, such as the Isthmus of Panama, still exist today.

5. The fossil record is sketchy prior to the Cambrian period. Nevertheless, we know that some forms of protists, fungi, and animals, but not plants, already existed by that time.

6. Biologists use the word *fauna* to refer to the animals living in a given area or time, and *flora* to describe the plant life. The word *biota* refers to all organisms in all kingdoms living an an area.

7. The best known pre-Cambrian animals are a group of soft-bodied invertebrates known as the *Ediacaran fauna*. Many of these organisms were members of lineages that lack surviving descendants.

8. During the early Cambrian (600 mya), a rich fauna had already developed, including early forms of annelids, arthropods, echinoderms, and cnidarians.

9. By 540 mya (mid-Cambrian) all current animal phyla existed, along with species from ten now-extinct phyla. Many of these animals had hard external skeletons, suggesting a strong evolutionary influence of predation at that time.

10. The Ordovician period (500 to 440 mya) witnessed a great increase of evolutionary lineages. Many filter-feeding marine animals flourished, and on land primitive ancestors of the club mosses and horsetails appeared. This period ended when sea levels dropped, driving many groups to extinction.

11. The Silurian period (440 to 400 mya) saw the continuation of abundant marine communities and increased colonization of the land. Simple marine vertebrates first appeared, as did the initial groups of terrestrial arthropods, the millipedes and scorpions.

12. Another major radiation occurred during the Devonian period (400 to 345 mya). Cephalopods, corals, and land plants in particular proliferated. The evolution of jaws in fishes and the loss of heavy armor plating were major milestones in the evolution of vertebrates. Gymnosperms appeared toward the end of the Devonian, along with the first insects.

13. The Carboniferous period (345 to 290 mya) is named for the extensive terrestrial forests of the time, which gave rise to modern coal deposits. The rocks associated with these deposits are rich in fossils that document a great increase in the diversity of terrestrial animals. Amphibians, some of them quite large, were the dominant land vertebrates of the time.

14. Giant insects dominated invertebrate life on land in the Permian period (245 to 220 mya), and reptiles replaced amphibians as the dominant vertebrates. Aquatic habitats experienced an explosion in the number of bony fishes.

15. The largest of all mass extinctions took place at the end of the Permian when 95 percent of all invertebrate species disappeared. Abundant forms such as the trilobites vanished, and substantial reductions in species diversity for many other groups also occurred. On land, ginkgos, cycads, and conifers replaced the great carboniferous trees. The most probable cause for the Permian mass extinction was a radical shift in climate as Pangaea coalesced.

16. Earth's biota, which had previously been rather homogeneous, became increasingly *provincialized* during the Mesozoic era as the continents spread further apart and acquired their own unique geographic and climatic characteristics. As a result, distinctive terrestrial floras and faunas arose around the world.

17. During the Triassic period (245 to 195 mya), the number of species of corals, mollusks, and other aquatic invertebrate groups increased dramatically. On land, the main event was the rise in prominence of reptiles. The first mammals also appeared late during this period.

18. Another mass extinction occurred at the end of the Triassic and the beginning of the Jurassic period (194 to 138 mya). Many groups underwent major radiations in the wake of this extinction, including several lineages of shelled aquatic invertebrates. Bony fishes came to dominate the seas, and the modern amphibians appeared. The terrestrial dinosaurs became abundant, as well as aquatic and aerial forms. One lineage of dinosaurs gave rise to the first birds in the mid-Jurassic.

19. The Cretaceous period (138 to 66 mya) witnessed the continued diversification of dinosaurs and increased provincialization of terrestrial biotas. Most modern mammal groups appeared at this time, as did the first flowering plants. Coincident with the emergence of angiosperms was a major radiation of their insect pollinators. The Cretaceous ended with another mass extinction, probably caused by Earth's collision with a large meteorite.

20. The most recent 65 million years in Earth's history is called the Cenozoic era, and it is here that mammals have come to dominate the terrestrial vertebrate fauna.

Because it is the most recent time, the Cenozoic is rich in fossils. This allows paleontologists to more finely divide the Cenozoic into epochs as well as periods.

21. During the Tertiary period (65 to 2 mya) angiosperm diversification continued. Herbaceous forms of flowering plants appeared in response to cooler climates. Most of the modern groups of invertebrates were present by the beginning of the Tertiary, and the remaining modern vertebrate groups appeared as well.

22. Birds in particular radiated dramatically during the Tertiary. Mammalian lineages generally increased in body size, and great herds of grazing animals appeared in response to the herbaceous plants that evolved in the drier climates. The number of species of bats, rodents, and primates also increased dramatically.

23. The Quaternary period represents the latest geologic time slice in evolutionary history. The Pleistocene epoch commenced about 2 mya and was highlighted by many glacial episodes, the most recent occurring less than 10,000 years ago. The climatic changes associated with these events produced few extinctions, but did dramatically alter the range of many organisms.

24. The majority of human evolution has taken place during the Quaternary. From their beginnings it appears that humans have been, and continue to be, an important factor in determining the evolutionary fate of other organisms.

Questions (for answers and explanations, see page 162)

1. Although the continents drifted slowly apart, creating large oceans in the process, terrestrial animals could still sometimes get from one major land mass to another by means of
   a. glacial ice sheets.
   b. deep ocean trenches.
   c. high mountain ranges.
   d. fault lines.
   e. narrow land bridges.

2. The Himalayas formed when
   a. the Indian subcontinent collided with Asia.
   b. the Bering land bridge sank below the ocean's surface.
   c. Gondwanaland separated from Laurasia.
   d. sea levels fell in the Cambrian period.
   e. none of the above

3. When biologists say that Earth's biota became provincialized, they mean that
   a. mass extinctions took place.
   b. continental drift did not occur.
   c. distinctive floras and faunas arose on different continents.
   d. reductions in species diversity took place.
   e. all of the above

4. All but one of the following occurred during the Ordovician period. Select the *exception*.
   a. Marine filter feeders flourished.
   b. Club mosses appeared on land.
   c. The number of classes and orders increased.

   d. Modern mammals appeared.
   e. Many groups became extinct at the end of the period.

5. Which of the following events are coincident with the rapid diversification of the angiosperms during the Cretaceous?
   a. A sudden decrease in species diversity of mollusks
   b. A sudden increase in species diversity of corals
   c. A sudden increase in species diversity of amphibians
   d. A sudden decrease in species diversity of humans
   e. A sudden increase in species diversity of insects

6. Most of human evolution has occurred during the
   a. Paleozoic era.
   b. Devonian period.
   c. Quaternary period.
   d. Carboniferous period.
   e. Mesozoic era.

7. The most recent mass extinction event that occurred *prior* to the evolution of humans took place approximately
   a. 10 mya.
   b. 65 mya.
   c. 220 mya.
   d. 400 mya.
   e. none of the above

8. The main reason(s) the Cenozoic era can be more finely divided into geologic epochs and periods is that
   a. it is more recent.
   b. there are relatively more Cenozoic fossils present than in other eras.
   c. geologic processes have not had time to destroy these fossils.
   d. Cenozoic rock layers tend to be closer to the surface.
   e. all of the above

PATTERNS OF EVOLUTIONARY CHANGE • THE SPEED OF EVOLUTIONARY CHANGE • THE FUTURE OF EVOLUTION   (pages 445–449)

Key Concepts

1. The number of major new features, called *evolutionary innovations*, that arose in the history of life on Earth is rather small. Most new features are modifications of preexisting structures.

2. Slight changes in developmental rates can produce major changes in the shapes of structures. Many evolutionary changes resulted from genetically based variation in growth rates.

3. Large numbers of new evolutionary lineages emerged on three separate occasions in Earth's history: during the Cambrian, the Permian, and the Triassic.

4. The Cambrian explosion produced all of the major groups of organisms, along with other phyla that later became extinct. Major evolutionary innovations were favored because there were few species and many unoccupied ecological niches.

5. The Paleozoic and Triassic explosions greatly increased the number of species but no new phyla evolved. The new species had modifications of the body plans that were already present.

6. Within many lineages, there has been a trend toward an increase in size. Large size helps an organism dominate other organisms, but on average larger organisms survive for less time than smaller organisms.

7. There are numerous examples in the history of life on Earth of predators causing major evolutionary changes in prey or prey extinction.

8. Rates of evolutionary change are often difficult to measure due to the incompleteness of the fossil record and the fact that only the hard parts of an organism's body are preserved. Many characteristics are therefore unavailable for analysis.

9. Nevertheless, the fossil record does indicate that frequent long periods of evolutionary *stasis*, during which rates of morphological change are slow, have occurred in many lineages. These periods are separated by occasional episodes of more rapid evolution.

10. Evolutionary rates have varied dramatically among lineages. Some lineages have undergone rapid changes in form while others have remained unchanged over equivalent timespans.

11. Organisms that have changed very little through evolutionary time are often found in environments that have also remained unchanged for millenia. Modern examples of such "living fossils" are horseshoe crabs and chambered nautiluses.

12. Alternatively, if the physical or biological environment changes suddenly, then morphological evolution can be rapid. An example comes from work on a small, typically marine fish called the three-spined stickleback, which has undergone morphological changes in pelvic girdle structure and spine size upon invading freshwater habitats. Current population information and known predator pressures can be combined with fossil evidence to help understand the rapid and repeated evolution of differences in stickleback morphology.

13. Although extinction is the norm in evolution (more than 99 percent of all species that ever lived are now extinct), our understanding of the causes for extinction is still limited. It is known, however, that rates of extinction have not been constant throughout evolutionary history.

14. Background extinction is the rate at which species disappear under "normal" conditions, whereas mass extinction is characterized by a much higher rate of disappearance than normal.

15. There have been numerous mass extinctions on Earth. The earliest, at the end of the Cambrian, eliminated nearly half of the known animal families. The most recent, at the end of the Cretaceous, claimed the dinosaurs as well as many marine groups.

16. The primary effect of mass extinctions is to radically alter the flora and fauna of the next period. Those traits that enhance survival during background extinction periods may not insure survival during a mass extinction event.

17. During some extinctions, terrestrial organisms are particularly hard hit, while during others, aquatic organisms suffer most. Still other extinctions have been simultaneous in both environments.

18. Mass extinctions do not appear to have a single cause. Some were probably caused by natural events such as climate change, mountain building, volcanic activity, and changes in sea level. Others may have been caused by events of an external origin, such as collision with a large meteorite.

19. Burgeoning human populations are a major agent of evolutionary change today. In addition to loss of many large vertebrate species by the direct action of humans, humans are now indirectly eliminating many small species by our effect on global ecosystems and through spread of foreign species around the world.

Questions (for answers and explanations, see page 162)

1. One of the main factors that distinguishes the Cambrian explosion from all others is that
   a. evolutionarily, it was the most recent explosion.
   b. many novel lineages (particularly phyla) appeared at this time, but not during other explosions.
   c. it was the time when the dinosaurs became extinct.
   d. there was a dramatic drop in species diversity, especially among marine organisms.
   e. it was the time when the mammals first appeared.

2. During which one of the following geologic times did the most new kinds of body plans appear?
   a. Paleozoic
   b. Triassic
   c. Jurassic
   d. Cambrian
   e. Cretaceous

3. The event that precipitated the sudden disappearance of the dinosaurs some 55 mya may have been the result of
   a. Earth's collision with a large meteorite.
   b. slow climate changes due to planetary cooling.
   c. competition from better adapted organisms.
   d. the rise of birds and mammals.
   e. all of the above

4. Which of the following statements about evolution is *not* true?
   a. Rates of evolutionary change are known to have varied throughout Earth's history.
   b. Background extinction has been occurring ever since the appearance of the first species.
   c. Marine mollusks provide a good example of the increase in species diversity through time.
   d. There is no evidence in the fossil record for periods of rapid evolutionary change.
   e. Humans appear to have been influencing the evolution of other animals for at least 2 million years.

*Integrative Questions (for answers and explanations,
    see page 162)*

1. Which of the following groups of organisms has undergone an overall decrease in body size during evolution?
   a. Mammals
   b. Fish
   c. Mollusks
   d. Plants
   e. None of the above

2. Which of the following statements about patterns or
   processes in the evolution of life is *not* true?
   a. ^{14}C can be used to date the ages of dinosaur bones.
   b. Mass extinctions have occurred numerous times in
        Earth's history.
   c. The supercontinent Pangaea formed during the
        Permian.
   d. Simple marine vertebrates first appeared in the
        Silurian.
   e. Rates of evolutionary change can be rapid during
        times of dramatic change in physical environments.

3. Which of the following pairs of organisms were *not* present on Earth in their living forms at the same time?
   a. Humans and angiosperms
   b. Dinosaurs and bony fish
   c. Amphibians and birds
   d. Gymnosperms and insects
   e. Humans and trilobites

20
The Mechanisms of Evolution

VARIATION IN POPULATIONS • PRESERVING GENETIC VARIABILITY: THE HARDY–WEINBERG RULE (pages 453–458)

Key Concepts

1. A *population* is a geographically isolated group of organisms of the same species.

2. A population undergoes *biological evolution* when it changes genetically over time.

3. Genetic changes in a population over a small number of generations constitute *microevolution. Macroevolution* is genetic changes in a population over hundreds, thousands, or more years.

4. *Genotype* is the genetic constitution of an individual relative to a *heritable* trait. Genetic change in populations occurs when individuals with different genotypes within the population survive and reproduce at different rates.

5. *Phenotype* is the observable expression of a genotype. Agents of evolutionary change act on genetically based phenotypic variation of population members.

6. The *gene pool* of a population consists of all the alleles for all the genes present in the species.

7. The environment can influence the expression of the genotype—that is, some portion of phenotypic variation is environmentally based.

8. Most populations possess a high level of genetic variation among their members.

9. Plant and animal breeders have produced new varieties of organisms with desirable traits by artificially selecting from the high levels of genetic variation found in most species.

10. Artificial selection of populations of *Drosophila* fruit flies with low and high bristle number has shown that enough genetic variation exists to produce individuals with bristle numbers outside the range found in the original population.

11. A *deme* (or *Mendelian population*) is a local population of potentially interbreeding individuals.

12. Genetic variation is determined by the number of different alleles for the genes of a species found in population members.

13. *Allele frequencies* are the proportions of the total number of alleles for a gene represented by each different allele. Allele frequencies range from 0 to 1.0.

14. The *genetic structure* of a population is determined by the genotype frequencies and allele frequencies in the population.

15. Populations can have the same allele frequencies for a gene (the same gene pool), but because the alleles can be distributed differently in the genotypes, they can have different genetic structures.

16. *Genetic equilibrium* within a population occurs when allele and genotype frequencies do not change over time.

17. *G. H. Hardy* and *W. Weinberg* independently showed in 1908 that under certain conditions a population will remain in genetic equilibrium.

18. Important Hardy–Weinberg conditions are that the population must be large, all individuals must be equal in terms of reproduction and survival, and mating must be random (independent of genotype).

19. The *Hardy–Weinberg formula, $p^2 + 2pq + q^2$*, where p is the frequency of the first allele and q is the frequency of the second allele, applies the principle of independent probabilities to determine the expected genotype frequencies. The expected frequencies of the two homozygous genotypes are given by p^2 and q^2, and $2pq$ is the expected frequency of the heterozygous genotype.

20. A population under Hardy–Weinberg conditions will show an equilibrium between allele and genotype frequencies.

21. Very few real populations meet the Hardy–Weinberg conditions.

22. The main use of the Hardy–Weinberg rule is to determine if evolutionary change is occurring; if a population is evolving, its genotype frequencies differ from those predicted by the Hardy–Weinberg rule.

Questions (for answers and explanations, see page 162)

1. Evolution occurs at the level of
   a. the individual genotype.
   b. the individual phenotype.
   c. environmentally based phenotypic variation.
   d. the population.
   e. the species.

2. Which choice would form a statement about demes that is *not* true? A deme (Mendelian population) must
   a. consist of members of the same species.
   b. be geographically isolated from other populations.
   c. have members that are capable of interbreeding.
   d. show genetic variation.
   e. have a gene pool.

3. In comparing two populations of the same species, the population with the greatest genetic variation would have (select all correct answers):

   a. the greatest number of genes.
   b. the greatest number of alleles per gene.
   c. the greatest number of population members.
   d. the greatest number of different genotypes.
   e. the largest gene pool.

4. The Hardy–Weinberg expression is an expansion of a binomial or

$$(p + q)^2 = p^2 + 2pq + q^2$$

In this equation, the left-hand side represents _____ frequencies, and the right-hand side represents _____ frequencies.

5. In a population with 600 members, the numbers of individuals of three different genotypes are A1A1 = 350, A1A2 = 100, and A2A2 = 150.

   What are the *genotype frequencies* in this population?

   A1A1 =

   A1A2 =

   A2A2 =

   What are the *allele frequencies* in this population?

   A1 =

   A2 =

   What would the expected genotype frequencies be if this population were in genetic equilibrium?

   A1A1 =

   A1A2 =

   A2A2 =

   Is this population in genetic equilibrium? Explain.

6. The ability to taste the chemical PTC (phenylthiocarbamide) is determined in humans by a dominant allele T, with tasters having the genotypes *Tt* or *TT* and nontasters, *tt*. If you discover that 36% of the members of a population cannot taste PTC, then according to the Hardy–Weinberg rule, the frequency of the *T* allele should be
   a. 0.2.
   b. 0.4.
   c. 0.6.
   d. 0.64.
   e. 0.8.

7. A gene in humans has two alleles, *M* and *N*, that code for different surface proteins on red blood cells. If you know that the frequency of allele *M* is 0.3, according to the Hardy–Weinberg rule, the frequency of the genotype *MN* in the population should be
   a. 0.21.
   b. 0.3.
   c. 0.42.
   d. 0.49.
   e. 0.7.

8. If the frequency of allele *b* in a gene pool is 0.2, according to the Hardy–Weinberg rule, the expected frequency of the genotype *bbb* in a triploid (3*n*) plant species would be
   a. 0.008.
   b. 0.04.
   c. 0.08.
   d. 0.2.
   e. 0.4.

9. Discuss the main use of the Hardy–Weinberg rule in evolutionary biology.

CHANGING THE GENETIC STRUCTURE OF POPULATIONS • MAINTAINING GENETIC VARIATION • SHORT-TERM VERSUS LONG-TERM EVOLUTION (pages 458–467)

Key Concepts

1. *Evolutionary agents* are forces that alter the genetic structure of populations. They can change both allele and genotype frequencies.

2. Major evolutionary agents are mutation, gene flow, genetic drift, nonrandom mating, and natural selection.

3. Although mutation rates are low (usually one mutation in a million zygotes per generation), mutation creates genetic variation by producing new alleles, and also replaces alleles that are being eliminated by other evolutionary agents.

4. Although mutation is present in all populations, its rate is so low that mutation per se disturbs Hardy–Weinberg genetic equilibrium very little.

5. *Gene flow* results when individuals migrate from one population to another and breed in the new location.

6. Movement of individuals into a population (*immigration*) adds alleles to the population gene pool; movement of individuals out of a population (*emigration*) removes alleles from the pool. Both events can change the genetic structure of the population.

7. A population undergoes *genetic drift* when allele frequencies change randomly in a population. Genetic drift is most prevalent in small populations.

8. Because genetic drift is random, deleterious alleles may become more frequent in the gene pool, or advantageous alleles may be entirely lost.

9. A population *bottleneck* is a form of genetic drift brought on when a population is reduced to a very small size by some natural or human-produced catastrophe.

10. If a small number of individuals (founders) that are not genetically representative of the larger population from which they come founds a new population, the new population may have allele frequencies that are quite different from that of the parent population. This is called the *founder effect.*

11. The most common type of nonrandom mating is *assortative mating.* In assortative mating, individuals select mates that are similar in phenotype. Assortative mating is common in vertebrates.

12. *Self-fertilization (selfing)*, common in many plants, is also a form of nonrandom mating.

13. Assortative mating and self-fertilization both lead to more homozygous genotypes than are predicted by the Hardy–Weinberg rule, but usually do not alter the expected allele frequencies.

14. *Sexual selection* is a form of nonrandom mating in which some individuals mate more than others. This can lead to changes in both genotype and allele frequencies.

15. *Natural selection* occurs when individuals differing in genotype reproduce and survive differentially. Natural selection leads to populations whose members are better adapted to the environment.

16. For traits determined by multiple genes, a bell-shaped curve usually results if you plot some measurement of the trait on the y-axis and the proportion of the population with that measurement on the x-axis.

17. *Stabilizing selection* results if individuals with extreme forms of a trait contribute fewer offspring to the next generation than individuals that are intermediate for the trait. Stabilizing selection leads to reduced genetic variation.

18. Enhanced reproduction and survival of individuals from one of the two ends of the distribution curve can result in *directional selection.*

19. *Disruptive selection* occurs if individuals from both extremes of the distribution are reproductively superior to the intermediate forms.

20. An African finch called the black-bellied seed cracker (*Pyrenestes ostrinus*) is a dramatic example of disruptive selection for individuals with small and large bills.

21. *Sexual recombination* increases the *evolutionary potential* of populations by producing new combinations of alleles.

22. One approach to the study of the adaptive nature of natural selection is to systematically modify the phenotype of individuals and observe the effect of these modifications on reproduction and survival.

23. In the comparative approach to the study of natural selection, one determines if patterns of interspecific adaptations agree with those predicted by the evolutionary hypothesis being tested.

24. Natural selection is a cumulative process occurring over many generations. It is a nonrandom process acting on genetic variation produced by random mutatons.

25. The reproductive contribution of a genotype or phenotype to the next generation, relative to all other genotypes or phenotypes, is its *fitness.*

26. *Individual fitness* results from direct reproduction by an individual.

27. *Kin selection* results from behavior of an individual that promotes the survival and reproduction of relatives who are genetically similar.

28. The *inclusive fitness* of an individual is the sum of its individual fitness and any kin selection component that may result due to the organism's behavior.

29. Individual fitness dominates inclusive fitness in solitary species. Kin selection may be important in highly social species.

30. Genetic drift, stabilizing selection, and directional selection tend to reduce genetic variation within populations.

31. Some genetic variation is not adaptive, but neutral. On the molecular level, any change that does not influence the functioning of a molecule may be neutral.

32. Some genetic variation in populations is correlated with geography.

33. *Clines* are gradual geographic changes in phenotypes and underlying genotypes within a species.

34. *Step clines* are created when geographically based phenotypic variation changes abruptly across some environmental gradient.

35. *Polymorphism* is the presence of two or more phenotypes for a given characteristic within a population, at frequencies greater than those caused only by mutation pressure.

36. *Balanced polymorphism* exists in a population when two or more alleles exist at stable frequencies within the population

37. Preservation of genetic variation within a population can involve natural selection favoring different traits in different areas, or can involve natural selection favoring different traits in the same area, but at different times.

38. Macroevolutionary change may be dependent on events that occur infrequently and that are not likely to be observed in short-term microevolutionary studies.

Questions (for answers and explanations, see page 163)

1. A small, isolated population would most likely be subject over time to
   a. assortative mating.
   b. a founder effect.
   c. genetic drift.
   d. gene flow.
   e. natural selection.

2. Allele frequencies for genes determining traits such as blood groups are *least* likely to be changed by
   a. assortative mating.
   b. the founder effect.
   c. genetic drift.
   d. gene flow.
   e. mutation.

3. Select *all* of the following evolutionary agents that produce *nonrandom* changes in the genetic structure of a population.
   a. Self-fertilization
   b. The founder effect
   c. Bottlenecks
   d. Mutation
   e. Natural selection

4. The following graph shows the range of variation among population members for a trait determined by multiple genes.

If this population is subject to *stabilizing selection* for several generations, which of the distributions (*a-d*) is most likely to result?

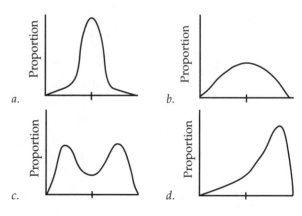

5. The following data were collected in a study in which dark and light moths of the peppered moth *(Biston betularia)* were released and later recaptured in several different areas *(a, b, c, and d)*. Based on these data, in which area is the *light* phenotype most advantageous?

Area	Moth Type	Released	Recaptured
a	dark	125	25
	light	200	40
b	dark	50	20
	light	200	40
c	dark	250	40
	light	425	125
d	dark	125	75
	light	200	25

6. Peppered moths are subject to predation by birds who eat them as the moths rest on bark surfaces during the day. Assuming that moths that are more cryptic in coloration (look more like their background) will be preyed on less, what can you say about the color of bark surfaces in area *d*?

7. A bird species in which females help sisters feed their nestlings is an example of
   a. kin selection.
   b. fitness.
   c. inclusive fitness.
   d. artificial selection.
   e. nonrandom mating.

8. Describe the relationship between inclusive fitness, individual fitness, and kin selection.

9. In areas of Africa where malaria is prevalent, many human populations exist in which the frequencies of the phenotypes normal, sickle-cell trait, and severe sickle-cell anemia are constant. This is an example of
   a. the founder effect.
   b. balanced polymorphism.
   c. transitional polymorphism.
   d. a cline.
   e. mutation pressure.

10. Which of the following statements about clinal variation is *not* true?
   a. Species showing clinal variation usually are widely distributed.
   b. Clinal variation usually involves external morphology.
   c. Some important environmental factor usually varies geographically over the range of the species.
   d. Clinal variation can be gradual or occur in steps.
   e. Clinal variation is the result of natural selection.

21

Species and Their Formation

CHAPTER LEARNING OBJECTIVES—after studying this chapter you should be able to:

❏ Define what a biological species is and explain each criterion of the definition.

❏ Differentiate between anagenesis and cladogenesis, or speciation.

❏ Discuss the sympatric mode of speciation caused by polyploidy and the groups of organisms that are most subject to it.

❏ Explain how sympatric speciation can occur in animals through nonpolyploid means and give an example.

❏ Explain the allopatric, or geographic, mode of speciation and illustrate its features by referring to the evolution of Darwin's finches.

❏ Discuss the parapatric mode of speciation, the necessary conditions that promote it, and the groups of organisms that are most subject to it.

❏ Describe what is meant by "reproductive isolating mechanisms" and differentiate between those that are prezygotic and postzygotic in their action.

❏ Describe the several possible outcomes that can result when two populations that have been genetically isolated from each other come into contact again.

❏ Discuss what is known about the genetic similarities of sympatric species based on studies of the Hawaiian *Drosophila*.

❏ Explain the influence of behavioral complexity on the time required for speciation.

❏ Explain what is known about the influence of ecology, especially diet, on the time required for speciation, based on studies of African mammals.

❏ Describe what is meant by an "evolutionary radiation" and explain why evolutionary radiations are most prominent on island archipelagos.

❏ Explain what is meant by the term "endemic" and explain the observation that many endemic species live on islands.

❏ Explain what is known about evolutionary radiations on islands based on studies of Hawaiian silverswords and tarweeds.

❏ Discuss the view that most population genetic change takes place during periods of speciation that are followed by intervals of genetic stability called stasis.

WHAT ARE SPECIES? • HOW DO NEW SPECIES ARISE? (pages 471–477)

Key Concepts

1. The biological species concept, developed by *Ernst Mayr* in 1940, defines a *species* as groups of natural populations of actually or potentially interbreeding organisms that are reproductively isolated from other such groups. The groups are collections of local populations, or demes.

2. Because organisms of the same species share a common gene pool, species are *independent evolutionary units*.

3. The potential to interbreed means that species members would produce viable offspring if a natural barrier that prevents them from doing so were removed—it excludes organisms that are able to interbreed only in captivity.

4. *Anagenesis* results when a single lineage changes through time to the extent that its descendants are given a different species name.

5. *Cladogenesis (speciation)* is the temporal process whereby one species gives rise to two species.

6. Separation of the gene pool of the ancestral species into two isolated pools is a necessary part of all speciation modes. Evolutionary agents can cause changes in allele and genotype frequencies in the isolated gene pools.

7. *Sympatric speciation* occurs when a population's gene pool becomes divided without geographic isolation to form two reproductively isolated subpopulations.

8. Among plants, and some animals, sympatric speciation results from *polyploidy*—an increase in the chromosome (ploidy or n) number.

9. Polyploidy arises in two different ways. In *autopolyploidy*, duplication of the chromosomes of a single species occurs, usually due to nondisjunction of chromosomes during meiosis. *Allopolyploids* result from mating between different species, whose chromosomes are so different that they do not pair properly during meiosis.

10. If polyploids can mate with other polyploids or are self-fertile, polyploidy can create a new species in one generation.

11. Animal species that arose by polyploidy are either *parthenogenetic* (self-fertilizing), or initially depend on mating with siblings to maintain the new species.

12. Sympatric speciation among animals (especially insects) occurs when individuals within the same area become specialized for specific resources; for example, herbivores or parasites specializing in different species of host.

13. Sympatric speciation is common among plants via polyploidy, but uncommon in animals.

14. *Allopatric speciation* (also called *geographic speciation*) begins with the establishment of two reproductively isolated populations from a formerly continuously breeding population.

15. Geographic speciation may result from the imposition of a physical barrier that divides a species into two separate gene pools. If environmental conditions are different, natural selection may operate differently in the two pools.

16. Geographic speciation may also result when a small number of founding individuals disperse across an already existing barrier.

17. In general, the longer reproductive isolation has existed within a group of species, the greater the degree of genetic divergence.

18. The 14 species of *Galapagos*, or *Darwin's finches*, arose by allopatric speciation from a single ancestral species that arrived from the South American mainland.

19. The scale of the barrier required to create reproductive isolation varies with the mobility and lifestyle of the organism.

20. The area required for speciation also varies with the mobility of organisms; birds need more space than small mammals, and animals like land snails can form new species in adjacent valleys.

21. When speciation occurs without geographic isolation due to an environmental discontinuity, it is called *parapatric speciation*.

22. Parapatric speciation is most common in plants adapted for different soils. Because their hybrids are less fit, reproductive isolating mechanisms may quickly evolve.

Questions (for answers and explanations, see page 163)

1. It is difficult to apply the biological species concept to groups of organisms that (select all correct answers):
   a. are asexual.
   b. are widespread.
   c. produce hybrids only in captivity.
   d. exist only in the fossil record.
   e. show little morphological diversity.

2. Which of the following statements about geographic speciation is *not* true?
   a. Geographic speciation is also called allopatric speciation.
   b. Geographic speciation only occurs in species that are widely distributed.
   c. Geographic speciation always involves a physical barrier that interrupts gene flow.

   d. Geographic speciation can sometimes involve chance events.
   e. Geographic speciation can sometimes involve small populations.

3. You look at several polymorphic loci in land snails removed from the areas A through E of four adjacent city blocks, as shown in the following map.

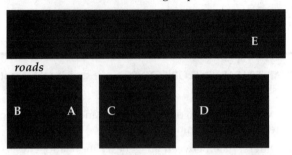

Snails from area A should be genetically most like individuals from area _____ and least like individuals from area _____ .

4. The activities of a mining company result in deposition of a new soil type within the range of a widespread plant species. Which of the following phenomena is likely to occur as a result of this?
   a. Geographic speciation
   b. Sympatric speciation
   c. Parapatric speciation
   d. Allopatric speciation
   e. Anagenesis

5. By means of sympatric speciation due to nondisjunction of chromosomes during meiosis, how many different tetraploid (4*n*) species could be formed from three diploid (2*n*) species?
   a. 0
   b. 3
   c. 4
   d. 6
   e. 9

6. Which type of speciation is most common among flowering plants?
   a. Geographic speciation
   b. Sympatric speciation
   c. Parapatric speciation
   d. Allopatric speciation
   e. Anagenesis

7. Match the evolutionary concepts on the left with the most appropriate descriptions on the right.
   _____ Allopatric speciation *a.* Associated with the fossil record
   _____ Anagenesis *b.* Sometimes seen in specialized parasites
   _____ Sympatric speciation *c.* Can be associated with soil differences
   _____ Parapatric speciation *d.* Most common speciation mode

8. Discuss the conditions on the Galapagos Islands that led to the evolution of the birds known as Darwin's finches.

REPRODUCTIVE ISOLATING MECHANISMS • VARIATION IN SPECIATION RATES • EVOLUTIONARY RADIATIONS: HIGH SPECIATION AND LOW EXTINCTION • SPECIATION AND EVOLUTIONARY CHANGE • THE SIGNIFICANCE OF SPECIATION (pages 477–483)

Key Concepts

1. *Reproductive isolating mechanisms* are organismal traits that prevent hybrid formation between individuals in two populations.

2. Isolating mechanisms are adaptations that prevent interbreeding between genetically different individuals in cases where hybrids would be less fit.

3. Differences in breeding seasons, courtship behavior, and reproductive physiology (including differences in courtship pheromones) are examples of *prezygotic isolating mechanisms* that reduce the chances of producing hybrid zygotes.

4. *Postzygotic isolating mechanisms* reduce the number of offspring produced by matings between genetically dissimilar individuals because the hybrids are less viable or sterile.

5. More effective prezygotic barriers may be selected for if postzygotic mechanisms fail to prevent the production of less viable hybrids.

6. If few changes have accumulated in the gene pools of isolated populations, when the barrier is removed interbreeding may occur and the two populations may again become one.

7. Stable *hybrid zones* can form between two species, and hybrids may be restricted to these areas. Hybrid zones sometimes occur in areas where the environment is intermediate for some important habitat variable and hybrids are only well adapted in that area.

8. Interbreeding between snow geese and blue geese and the formation of a hybrid zone in the Canadian Arctic presents biologists with an opportunity to study the development of isolating mechanisms.

9. Studies on species of Hawaiian *Drosophila* show that sympatric species—species living in the same area—can be morphologically distinct, yet genetically very similar.

10. There is a wide range in the time required for speciation: some plants can speciate in one season by polyploidy, but American and European sycamores can still interbreed after 20 million years of geographic isolation.

11. Areas with more species tend to have higher rates of speciation than areas with fewer species.

12. Increased behavioral complexity, especially in the degree of mate selection, tends to reduce the time required for speciation.

13. Studies of large hoofed mammals in Africa show that the diet of these animals is correlated with the rate of speciation: lineages of grazers and browsers show higher speciation rates than omnivores and anteaters.

14. We expect organisms with ecological roles that cause their populations to fragment and be more sensitive to environmental change to show the highest speciation rates.

15. Species with short generation times speciate more readily because they can diverge and evolve isolating mechanism more rapidly than species with long generation times.

16. *Evolutionary radiations* occur when a lineage of organisms with a low extinction rate gives rise to a large number of new species.

17. Because islands offer geographic isolation and ecological opportunities due to reduced competition with other organisms, the best examples of evolutionary radiations have occurred on islands.

18. The Hawaiian islands are the most isolated archipelago in the world. The islands also vary in age. Over 90 percent of all plant species on the Hawaiian islands are *endemic*—found nowhere else.

19. The group of Hawaiian plants called the *silverswords* and *tarweeds* provide an example of an evolutionary radiation. The 28 species of silverswords are much more diverse and widespread than are their mainland relatives.

20. The silversword radiation shows that major diversification can be accomplished by allele differences in a small number of genes. The silversword radiation occurred because the original ancestors encountered few competing species and could diversify ecologically.

21. The evolution of novel adaptations can also lead to an evolutionary radiation. This is illustrated by the radiation of American blackbird species with the evolution of the morphological adaptations that underly the feeding behavior called *gaping*.

22. It is theorized that major population genetic changes take place at the time of speciation due to the presence of small, founding populations and intense natural selection. Long intervals of *stasis* characterized by genetic stability often separate periods of active speciation.

23. This pattern of evolution, with short, rapid periods of speciation followed by long periods of little change *(stasis)*, is called *punctuated equilibrium*.

Questions (for answers and explanations, see page 164)

1. Classify each of the following as an example of a **pre**zygotic isolating mechanism, a **post**zygotic isolating mechanism, or **not** an isolating mechanism.

   _____   Nonoverlapping species ranges

   _____   Nonoverlapping breeding seasons

   _____   Production of sterile offspring

   _____   Differences in courtship pheromones

   _____   Incompatible gametes

   _____   Poorly adapted offspring

2. Which of the following factors would *not* be expected to increase the rate of speciation in a group of organisms?
   a. A species range consisting of fragmented populations
   b. A diet consisting of food items whose abundance varies widely
   c. High birth rates
   d. Increased behavioral complexity
   e. A high degree of mate selection

3. Which of the following is *not* a suggested reason for the evolutionary radiation of silverswords and tarweeds on the Hawaiian archipelago?
   a. Water is an effective barrier for many organisms.
   b. Because islands are small compared with mainland areas, you would expect more species to develop there.
   c. Frequently, competition is reduced on islands.
   d. The Hawaiian islands differ in age from west to east.
   e. More ecological opportunities exist on islands.

4. Studies of species of Hawaiian *Drosophila* show that
   a. sympatric speciation via polyploidy is common in insects.
   b. few genes need be involved in establishing reproductive isolation.
   c. matings between siblings can sometimes lead to new species.
   d. Species with nonoverlapping ranges can be morphologically very similar.
   e. A geographic barrier is not always necessary for the establishment of reproductive isolation.

5. Explain what is known about the influence of ecology, especially diet, on the time required for speciation, based on studies of African mammals.

6. Discuss what is known about the evolutionary radiations on islands, based on studies of Hawaiian silverswords and tarweeds.

Integrative Questions (for answers and explanations, see page 164)

1. Select *all* of the following statements about speciation that are *not* true.
   a. A small founding population can be involved in speciation.
   b. The time required for speciation can vary for different types of organisms.
   c. Speciation always involves interruption of gene flow between different groups of organisms.
   d. The rate of speciation can vary for different groups of organisms.
   e. Speciation can occur in a single generation.
   f. Only plants can speciate by polyploidy.
   g. Speciation is a temporal process.

2. Select all of the following observations suggesting that two overlapping populations that had been geographically isolated have *not* diverged into distinct species.
   a. A stable hybrid zone exists where their ranges overlap.
   b. Matings between members of the two populations produce viable hybrids.
   c. Electrophoresis shows that the two populations share a common gene pool.
   d. Interbreeding is common between members of the two populations.
   e. Distinct prezygotic isolating mechanisms have evolved.

3. If you examine geographic variation in a widely distributed species, you would expect to see the least morphological variation in
   a. mountainous areas.
   b. island archipelagos.
   c. environmentally homogeneous regions.
   d. environmentally heterogeneous regions.
   e. areas with numerous large rivers.

22

Constructing and Using Phylogenies

CHAPTER LEARNING OBJECTIVES—after studying this chapter you should be able to:

❏ Explain what systematics is and discuss the major objectives of this discipline of biology.

❏ Explain what taxonomy is and discuss its relationship to systematics.

❏ Explain some of the past and present uses of biological classification schemes.

❏ Discuss the limitations of using common names to identify organisms.

❏ Employ the conventions used in the Linnaean nomenclature system for constructing binomial names.

❏ Identify the major hierarchical taxa within the Linnaean classification system.

❏ Explain what cladistic systematics is and discuss its major objectives.

❏ Define and relate the following terms used in cladistic systematics: "clade," "cladogram," "ancestral traits," and "derived traits."

❏ Differentiate between homologous traits and homoplastic traits.

❏ Explain the roles of convergent evolution in creating homoplasy.

❏ Differentiate between general homologous traits and special homologous traits.

❏ Discuss what an outgroup is and explain the relationship of the outgroup to general and special homologous traits.

❏ Explain the provisional assumptions employed in the construction of a cladogram.

❏ Construct and interpret cladograms using the conventions presented in the textbook.

❏ Define what a fossil is and explain the major role of the fossil record in cladistic systematics.

❏ Discuss the application of the principle of parsimony in cladistic systematics.

❏ Discuss the strengths and limitations of using developmental traits to infer phylogenies.

❏ Discuss how biochemical differences in proteins and nucleic acids can be used to infer phylogenies.

❏ Explain why DNA from eukaryotic organelles has been especially useful in systematics.

❏ Discuss some of the features of chloroplast DNA and give an example of its use in inferring phylogenies.

❏ Explain the DNA hybridization technique and the rationale for its use in systematics.

❏ Differentiate between the terms "monophyletic," "paraphyletic," and "polyphyletic" and explain their application in evaluating phylogenies.

❏ Describe use of the term "domain" in biological classifications and list the three domains of the phylogenetic tree of life.

THE HIERARCHICAL CLASSIFICATION OF SPECIES
(pages 487–489)

Key Concepts

1. A *taxon* (plural, *taxa*) is a group of organisms treated as a unit within a classification system.

2. *Systematics* is the scientific study of the evolution of biological diversity.

3. *Taxonomy* is a branch of systematics specifically concerned with classifying organisms.

4. The goal of systematics is to devise taxonomic classification schemes that express the evolutionary relationships—the *phylogeny*—among organisms.

5. Another objective of biological classification is to provide a system for naming different species in a uniform, unambiguous way.

6. Common names for organisms often vary regionally and often do not reflect evolutionary relationships.

7. The *Linnaean system*, originated by *Carolus Linnaeus* (1707 to 1778), forms the basis of modern classification systems.

8. In the Linnaean system, each species is assigned a two-part, or *binomial*, name composed of a genus name and a species name. This two-name system is called *binomial nomenclature*.

9. A *genus* (plural, *genera*) is a group of related species. The species name is unique within the genus.

10. In some cases, the name of the scientist who named the organism is included after the binomial, as in *Homo sapiens* Linnaeus.

11. Both genus and species names are italicized; the genus name is capitalized, the species name is not.

12. To refer to more than one species within a genus, follow the genus name with "spp."; to refer to an unknown species within a known genus, follow the genus name with "sp." Within a paragraph, after the genus name is given once, it can be abbreviated to the first letter of the genus name followed by a period in further references.

13. Taxonomic classification schemes are hierarchical; each higher taxon includes all lower taxa, and organisms in higher taxa have fewer characteristics in common than do those in lower taxa.

14. The number of species included in a taxon depends on their relative degrees of similarity and dissimilarity.

15. Higher taxa contain organisms that presumably shared a common ancestor in the more distant past than organisms in lower taxonomic groups, although this information is usually not available.

16. Related genera are grouped into *families*. Family names are based on a name of one of the included genera and end in "*-idae*" for animals or "*-aceae*" for plants.

17. Related families are grouped into *orders*, orders are grouped into *classes*, related classes are grouped into either *phyla* (animals and protists), or *divisions* (plants, bacteria, and fungi), and phyla and divisions are grouped into *kingdoms*.

Questions (for answers and explanations, see page 164)

1. The *most* important attribute of a biological classification scheme is that it
   a. allows a more systematic procedure for naming organisms.
   b. reflects the evolutionary relationships among organisms.
   c. helps us remember organisms and their traits.
   d. improves our ability to make predictions about the morphology and behavior of organisms.
   e. avoids the ambiguity created by using common names.

2. Suppose you are writing a scientific paper about a unicellular green alga called *Chlamydomonas reinhardi*. What would be the proper way to refer to this species after you had used the full binomial earlier in the same paragraph?
   a. *Chlamydomonas reinhardi*
   b. *Chlamydomonas* sp.
   c. *Chlamydomonas* spp.
   d. *C. reinhardi*.
   e. *C.* sp.

3. Organisms in a higher taxon are _____ similar, usually have diverged from a common ancestor _____ recently, and include _____ species than organisms in a lower, included taxon.
   a. less, less, fewer
   b. less, more, fewer
   c. less, less, more
   d. more, more, fewer
   e. more, more, more

4. Number, from lower to higher, the following taxonomic categories as you would use them to classify a new species of salamander. If a listed taxon should not be used, leave it blank.

   _____ Order

   _____ Species

   _____ Family

   _____ Kingdom

   _____ Division

   _____ Genus

   _____ Phylum

   _____ Class

INFERRING PHYLOGENIES (pages 489–495)

Key Concepts

1. *Fossils* are any evidence of organisms that lived in the past, including remains of the organism or its impressions, such as footprints. The fossil record helps reveal ancestral traits.

2. *Cladistics* attempts to identify evolutionary lineages of organisms.

3. In cladistics, all of the species derived from a common ancestor form a *clade,* and members of a clade are all descended from a common ancestor.

4. A *cladogram* is an evolutionary tree showing the relationships among different clades. Contemporary organisms are shown across the top of the cladogram; the common ancestor of all the clades being represented is at the base of the cladogram.

5. A phylogenetic tree is a cladogram containing additional information, such as the proposed dates when common ancestors occurred in the past.

6. *Ancestral traits* are those traits that are shared by a group of species and that were also present in their common ancestor.

7. A *derived trait* has been modified from an ancestral trait due to evolution within a group.

8. Cladograms are based on the identification of ancestral and derived traits.

9. Structures derived from a common ancestor are called *homologous* traits. Homologous traits come from ancestral traits (i.e., they are derived traits), but may be modified by divergence.

10. *General homologous traits* are shared by many organisms because they appeared only once and in the distant past.

11. *Special homologous traits* are shared by only a few organisms because they appeared only recently.

12. *Homoplasy* is the presence of similar structures in different lineages that were not present in the ancestor common to the lineages. Homoplasy can result from convergent evolution.

13. In *convergent evolution*, formerly dissimilar structures become more alike because the different lineages are subject to similar selection pressures. Structures created by

convergent evolution that perform similar functions are *analogous* to one another.

14. Constructing cladograms that accurately reflect the phylogeny of a lineage requires distinguishing ancestral from derived traits and homologies from homoplasies.

15. An *outgroup* is a taxon that is closely related to the group whose phylogeny is being constructed. General homologous traits are found within some species of the group being studied and within the outgroup; special homologies are only found within members of the studied group.

16. A method for devising cladograms developed by Willi Hennig provisionally assumes that a trait shared by two species is homologous unless other evidence indicates that the trait is homoplastic.

17. Other assumptions provisionally used in constructing cladograms are that derived traits appear only once, and that no derived traits are lost during the evolution of the lineage.

18. Using Hennig's system, species are clustered according to the number of special homologous traits they share. If numerous traits are considered, most will be homologous if the species share a recent common ancestor.

19. The number of derived traits that two species share is a reflection of how recently they had a common ancestor.

20. In constructing a cladogram, contradictory data may suggest that derived traits may have appeared multiple times or been lost in the evolution of the lineage.

21. *Parsimony* is employed in arranging the taxa in a cladogram so that the number of changes in traits is minimized. Parsimony is used because multiple appearances of derived traits is relatively rare.

Activities (for answers and explanations, see page 164)

- Based on the following cladogram showing the phylogeny of five species *(A through E)* relative to five traits *(1 through 5)*, fill in the table shown below using 1 to indicate the presence of a derived trait, and 0 to indicate the presence of an ancestral trait.

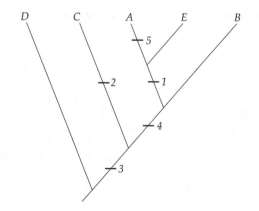

SPECIES	TRAIT 1	2	3	4	5
A					
B					
C					
D					
E					

- In the cladogram shown in activity 1, which species is considered the outgroup? _____ Explain your answer.

- Based on the following table showing the ancestral and derived traits of five species *(A through E)*, construct a cladogram that represents the phylogeny of this group using conventions presented in the textbook.

SPECIES	TRAIT 1	2	3	4	5
A	1	1	1	0	0
B	0	0	0	0	0
C	1	1	0	1	0
D	0	1	0	0	0
E	1	1	0	1	1

The ancestral state of each trait is indicated by 0 and the derived state is indicated by 1.

- Discuss the application of parsimony in the construction of cladograms.

Questions (for answers and explanations, see page 165)

1. Which of the following would *not* be expected to result in homoplasy?
   a. Convergent evolution
   b. The independent evolution of similar structures in different lineages
   c. The evolution of analogous traits
   d. The inheritance of ancestral traits
   e. Selection for traits that perform similar functions

2. Which of the following choices completes a statement about clades that is *not* true? A clade
    a. includes all organisms in one branch of a cladogram.
    b. is the basic taxon of cladistic systematics.
    c. includes only living species.
    d. shares a common ancestor.
    e. shares homologous structures only.

3. Choose a letter to make an *incorrect* statement. In a cladogram, the outgroup would
    a. have a branch point nearest the bottom.
    b. have all ancestral traits.
    c. not share any derived traits with other species.
    d. be determined based on evidence from the fossil record.
    e. share the least recent common ancestor with the group.

4. Which of the following is *not* an assumption used in the construction of cladograms?
    a. Shared traits are provisionally considered to be homologous.
    b. Special homologous traits are not found within the outgroup.
    c. General homologous traits are only found within the outgroup.
    d. Derived traits appear only once in the evolution of the lineage.
    e. Once derived traits appear, they can not disappear.

CONSTRUCTING PHYLOGENIES • BIOLOGICAL CLASSIFICATION AND EVOLUTIONARY RELATIONSHIPS • THE FUTURE OF SYSTEMATICS • THE THREE DOMAINS AND THE SIX KINGDOMS   (pages 495–500)

Key Concepts

1. Because the sizes and shapes of body parts, called *gross morphology*, are more likely to become part of the fossil record and are easy to measure and quantify, they are widely used by systematists.

2. Developmental stages of organisms may provide useful evidence about their phylogenies, as shown by the similarities between sea squirts and vertebrates.

3. Biochemical differences in molecules of organisms, especially DNA, RNA, and proteins, is valuable information for systematists.

4. Amino acid sequencing techniques allow direct comparison of the primary structures of homologous proteins, such as cytochrome *C*, in different species. Amino acid differences in the primary structure reflect genetic differences between the taxa.

5. A comparison of the *base sequences* of nucleic acids is also an important method for determining the evolutionary relationships of different species.

6. DNA from the organelles of eukaryotes, such as chloroplast DNA (cpDNA) and mitochondrial DNA (mtDNA), has been especially useful for systematists because it is evolutionarily more stable than nuclear DNA.

7. cpDNA is a circular, double-stranded molecule with a size of 151 kilobases. It includes about 100 genes that code for transfer RNAs, ribosomal subunits, and some proteins involved in photosynthesis.

8. *DNA hybridization* results when DNAs from two different species that have been denatured into single-stranded molecules by heating are cooled and allowed to form interspecific double helices.

9. DNA hybridization can be used to compare DNAs even if their exact sequences are not known.

10. Nucleic acid hybridization studies have helped settle debates about the phylogenies of such organisms as humans, chimpanzees, and the giant panda.

11. Classification systems should be judged based on how well they achieve their objectives; reflecting evolutionary history is the major objective of today's biological classification systems.

12. Some classification schemes used in the past were concerned with the usefulness of organisms for humans or with revealing divine laws.

13. Biological classification systems help us place organisms into easily remembered groups that share common characteristics.

14. Biological classification systems allow us to make predictions about the characteristics of new organisms once they have been placed within a group.

15. Because of the difficulty of objectively selecting the "best" traits for classifying a group, numerical taxonomists used as many traits as possible, but weighed them all equally. Computer programs developed by numerical taxonomists are widely used today.

16. Since a clade contains only descendants from a common ancestor, a clade is *monophyletic*. If a taxon consists of members with more than one common ancestor, the taxon is said to be *polyphyletic*. A taxon with some, but not all descendants is *paraphyletic*.

17. Polyphyletic taxa are not considered valid by most taxonomists, but paraphyletic groups are tolerated as a way to emphasize groups that have undergone rapid evolution, and to maintain stability in the classification system.

18 *Domains* are the most inclusive taxa of the phylogenetic tree of life. Domains are groupings of kingdoms. *Bacteria* and the *Archaea* are the prokaryotic domains; the *Eukarya* includes all of the eukaryotic kingdoms.

Questions (for answers and explanations, see page 165)

1. Which of the following statements about traits used in constructing phylogenies is *not* true?
    a. Gross morphology is the only type of systematic evidence obtainable from the fossil record.
    b. DNA hybridization between two different species reflects the degree of their relatedness even if the precise sequences being compared are unknown.
    c. DNA from organelles, such as mitochondria, is useful to study because it is evolutionarily more stable than nuclear DNA.
    d. The presence of a notochord in sea squirt larvae reveals their close relationship with other chordates.
    e. Larval characteristics can be highly modified by natural selection.

2. DNA hybridization has shown that the percentage difference in human and chimpanzee DNA is
   a. less than 1 percent.
   b. approximately 1–2 percent.
   c. approximately 10–20 percent.
   d. approximately 40–50 percent.
   e. greater than 50 percent.

3. Which of the following statements about chloroplast DNA (cpDNA) is *not* true?
   a. cpDNA is very similar in most flowering plant species.
   b. cpDNA is a circular, double-stranded molecule.
   c. cpDNA includes about 100 genes.
   d. cpDNA and mitochondrial DNA (mtDNA) are similar in many features.
   e. Studies show that cpDNA is subject to a high degree of mutation.

4. Discuss the implications the following statement has for systematics: "DNA is the genetic material for all eukaryotes."

5. Why are paraphyletic clades acceptable to many systematists, whereas polyphyletic clades are not?

23

Molecular Evolution

DETERMINING THE STRUCTURE OF MOLECULES • HOW MOLECULAR FUNCTIONS CHANGE (pages 504–510)

Key Concepts

1. The objectives of the study of *molecular evolution* are to reveal the patterns of evolutionary change at the molecular level and to understand the major processes underlying those changes.

2. Early studies of insulin from different mammals showed that most amino acid substitutions were restricted to positions that did not affect insulin's biological function.

3. The hypothesis of *neutral evolution* is that most nucleotide substitutions are neutral. Neutral substitutions are those causing changes in molecules that do not effect the functioning of the molecules.

4. The rate of accumulation of neutral substitutions is not influenced by natural selection, but is determined by the mutation rate.

5. The *polymerase chain reaction (PCR)* is important to the study of molecular evolution because it is used to amplify DNA samples for nucleotide sequencing.

6. *Cytochrome c* is an important protein component of the electron transport chain of mitochondria. The amino acid sequences of this protein have been determined for more than 100 species of eukaryotes.

7. Many of the positions in the cytochrome *c* molecule are occupied by the same amino acids in all species. These positions are important in determining the overall shape and functioning of the molecule. For example, positions 14, 17, 18, and 80 are invariant in all species. These residues interact with the iron-containing heme group and are essential for enzymatic functioning.

8. Amino acids that *do* vary in cytochrome *c* molecules from different species may either be functionally less adaptive or selectively neutral.

9. The functionally important regions of molecules can be discovered by determining which regions *do not* vary from species to species.

10. In addition to the mutational substitutions of nucleotides, the genomes of organisms can be changed by horizontal transfer between different species. Examples include bacterial transformation by plasmids and episomes and transduction by viruses.

11. The three globulin gene families (myoglobin family, alpha globulin family, and beta globulin family) all evolved as *gene duplications* of a single ancestral gene.

12. Based on amino acid differences between the genes, the myoglobin family diverged from the other globulins about 500 mya. The split between the alpha and beta globulin families occurred about 450 mya.

13. After gene duplication has occurred, the genes and the proteins they code can evolve independently. Newly duplicated genes can evolve entirely different functions. For example, myoglobin, a monomer consisting of a single myoglobin chain, has evolved a higher affinity for oxygen than hemoglobin. Hemoglobin, a tetramer that consists of two alpha chains and two beta chains, has become a molecule with multiple functions, including transport of four oxygen molecules, transfer of H^+ and CO_2, and the ability to respond to blood pH.

14. Gene duplication can also result in nonfunctional DNA, called *pseudogenes*. The production of gene families and pseudogenes by the gene duplication mechanism increases genomic complexity.

15. In humans, the rRNA genes, tRNA genes, lactate dehydrogenase, and pyruvate kinase are examples of gene families that arose from gene duplication.

16. The *homeotic gene complex*, which includes a region called the *homeobox* region, is a group of genes that controls development in a wide variety of organisms. All homeotic genes have a common evolutionary origin and are homologous.

17. *Lysozyme*, an enzyme produced by virtually all animals to kill invading bacteria, has evolved to serve a different function in a group of unrelated animals (ruminants, and leaf-eating monkeys, such as the langurs) with a unique mode of digestion called *foregut fermentation*. The amino acid substitutions that have independently occurred in the lysozymes of these animals allow the enzyme to function in the acidic environment of the digestive tract and in the presence of digestive enzymes.

18. A comparison of the lysozymes of foregut fermenters with those of other animals shows that amino acid substitutions occurred at about twice the rate in the former group. These rapid, convergent changes in the lysozymes of the foregut fermenters were directed by natural selection.

19. A leaf-eating bird called the hoatzin also uses a lysozyme very similar to that of the foregut fermenters in an enlarged crop that serves as a fermentation chamber.

Questions (for answers and explanations, see page 165)

1. Select the item below that is *not* involved in causing evolutionary change at the molecular level.
   a. Mutation
   b. Transposable elements
   c. Gene duplication
   d. Plasmids and episomes
   e. All of the above are involved.

2. Which of the following is *not* a true statement about gene families?
   a. Gene families evolve via gene duplication.
   b. Pseudogenes are quickly removed from gene families by deletion.
   c. After duplication, gene families can evolve different functions.
   d. Members of a gene family can include several functional genes.
   e. Examples of gene families include the homeotic gene complex in *Drosophila* and the alpha hemoglobin family in humans.

3. Which of the following is *not* a valid conclusion based on studies of the structure of proteins, such as insulin and cytochrome *c*, in different species?
   a. Most amino acid substitutions are restricted to positions not affecting the protein's function.
   b. The rate of accumulation of neutral substitutions is based on the mutation rate.
   c. Most nucleotide substitutions result in neutral changes in the amino acid sequence of the protein.
   d. Fewer differences were observed in the amino acid sequences of a protein if the organismal sources of the proteins were closely related.
   e. Functionally important regions of a protein can be discovered by identifying the regions with the most amino acid substitutions.

4. Select the correct statement about the enzyme lysozyme.
   a. Lysozyme is uncommon, except in animals known as the foregut fermenters.
   b. A small group of closely related mammals have evolved a special form of lysozyme that functions in digestion.
   c. The lysozymes found in the foregut fermenters resulted from convergent evolution.
   d. Lysozyme could not have evolved a secondary function as it did if it had been an enzyme with a vital primary function.
   e. A higher mutation rate in the foregut fermenters allow their lysozymes to evolve rapidly.

5. Discuss the importance of the polymerase chain reaction (PCR) to studies of molecular evolution.

GENOME SIZE: SURPRISING VARIABILITY • MOLECULAR CLOCKS • USING MOLECULES TO INFER PHYLOGENIES • MOLECULES AND HUMAN EVOLUTION (pages 510–515)

Key Concepts

1. Genome size, as measured by total DNA, does not correlate well with measures of organismal complexity. However, the amount of DNA that codes for functional RNA or protein does vary with the overall complexity of the organism.

2. Eukaryotes have more coding DNA than prokaryotes, multicellular organisms have more coding DNA than unicellular organisms, and vertebrates have more coding DNA than invertebrates.

3. Many proteins, such as cytochrome *c*, show a linear relationship between amino acid substitutions and the presumed time since divergence of the taxon providing the protein. The specific relationship observed defines the *molecular clock* of the protein.

4. Different molecular clocks tick at different rates. Molecules with great functional constraints, such as rRNA, have clocks that tick very slowly. Molecules with more flexible functions, such as lysozyme, have clocks that tick at a faster rate.

5. *Neutral changes* tend to accumulate at a constant rate. During periods of evolution, *adaptive changes* in a molecule's structure can change rapidly. These factors affect the suitablity of molecular clocks for dating lineage separations.

6. Molecules that have changed slowly are most useful for inferring phylogenies for taxa that have diverged long ago. Ribosomal RNA (rRNA) is a molecule that has changed very little because of strong functional constraints. rRNA can be used to study the evolutonary relationships of all extant organisms.

7. Molecules that have changed rapidly are useful for studying organisms that have diverged recently. Lysozyme and mitochrondial DNA (mtDNA) are examples.

8. After the molecule to be compared has been selected, the sequences (either amino acids or nucleotide pairs) from the different species are determined and compared. From this comparison, a *similarity matrix* is constructed. This matrix displays the numbers of similarities and differences for each pair of sequences.

9. A *molecular phylogeny* or *gene tree* is a visual representation of the similarities and differences shown in the similarity matrix. Typically, the branches in the tree represent the major changes that occurred in the evolution of the molecule.

10. To avoid being misled by convergence of characters (homoplasy), many characters, molecular, morphological, and fossil, should be studied.

11. 16S rRNA was used for identifying the major branches (domains) of life's phylogenetic tree. These molecules are found in all organisms and the 16S rRNA molecular clock ticks very slowly.

12. DNA recovered from extinct organisms is now commonly used to study phylogenetic relationships between extinct and extant organisms. Cloning extinct organisms is not possible because the fossil DNA is degraded, we have no way to piece together the many small amplified fragments, and the proper regulation of gene expression during development is unknown.

13. Studies of mitochondrial DNA (mtDNA) from humans has provided support for the hypothesis that all modern human populations share a recent mitochondrial ancestor and that the single transition from *Homo erectus* to *H. sapiens* occurred in Africa about 200 thousand years ago.

14. mtDNA is useful for studying the recent evolution of closely related species because it accumulates mutations rapidly and it is maternally inherited. Male lineages can be studied with DNA of the Y chromosome since it is paternally inherited.

Questions (for answers and explanations, see page 165)

1. Order the following organisms in terms of the expected total amount of coding DNA in their genomes, with 1 = least and 5 = most.

   _____ Insect

   _____ Bird

   _____ Bacterium

   _____ Single-celled eukaryote

   _____ Round worm

2. Which of the following statements about molecular clocks is *not* true?
   a. The molecular clock of cytochrome *c* ticks more slowly than that of rRNA.
   b. A molecular clock that mostly reflects neutral changes is best for dating lineage separations in the distant past.
   c. The tick rate of a clock for a molecule with great functional constraints would be very slow.
   d. The tick rate of a molecular clock that mostly reflects neutral changes is determined by the mutation rate.
   e. The tick rate of a molecular clock that mostly reflects neutral changes is directly related to the slope of the line showing the relationship between the time of divergence of different taxa and amino acid sequence differences for the protein.

3. Which of the following molecules would be best to use in a study of the domains of the phylogenetic tree of life?
   a. Lysozyme
   b. Cytochrome *c*
   c. Mitochrondrial DNA
   d. 16S rRNA
   e. Myoglobin

4. What is the most important conclusion about our origins that has been obtained by a study of the mitochondrial DNA (mtDNA) from over 100 ethnically distinct modern human groups.

5. Based on your study of this chapter, what assurances could you offer a local PTA that is concerned that some "mad scientist" might set up a Jurassic Park in their neighborhood?

6. You determine the amino acid sequence for an eight-residue section of a small protein from five different species. (1–5)

Position	1	2	3	4	5	6	7	8
Species 1:	Arg	Cys	Leu	Leu	Ser	Thr	Asn	Met
Species 2:	Arg	Cys	Phe	Leu	Leu	Ser	Thr	Asn
Species 3:	Arg	His	Leu	Leu	Ser	Thr	Asn	Met
Species 4:	Arg	Cys	Leu	Ser	Ser	Thr	Asn	Met
Species 5:	Arg	His	Leu	Leu	Ser	Gln	Asn	Met

Complete the following similarity matrix from these sequences:

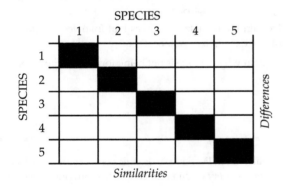

Which species differ because of an amino acid insertion or deletion?

Which species differ due to amino acid substitutions?

Complete the following gene tree by adding the species numbers across the top.

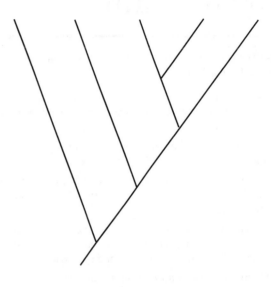

24

The Origin of Life on Earth

CHAPTER LEARNING OBJECTIVES—after studying this chapter you should be able to:

❑ Describe the work of Pasteur on spontaneous generation and explain why his studies were more convincing than earlier experiments.

❑ Explain what is known about Earth's early atmosphere and how it differs from the present-day atmosphere.

❑ Describe the work done by Stanley Miller on prebiotic synthesis of molecules, and discuss the importance of these studies.

❑ Explain how polymerization of simple organic molecules could have produced larger, more complex molecules, such as pyrimidine and purine bases and monosaccharides.

❑ Discuss the importance of phosphorylation to the formation of the first nucleic acids.

❑ Describe what a ribozyme is and give several examples of ribozymes.

❑ Explain the evidence for the notion that the first genetic code was based on RNA.

❑ Describe how the evolution of RNA molecules might have improved the catalytic efficiency of ribozymes.

❑ Discuss the properties of coacervates that suggest that they may have evolved into cells.

❑ Explain the conditions that would have favored the replacement of RNA by DNA as the molecule of heredity.

❑ Describe the insights into the metabolism of early organisms provided by a study of the metabolism of living bacteria.

❑ Discuss the characteristic features of the anaerobic photosynthetic bacteria, the purple and green sulfur bacteria, and the nonsulfur bacteria.

❑ Discuss the characteristic features of the first aerobic photosynthetic organisms.

❑ Explain the planetary changes that followed the liberation of free oxygen by cyanobacteria.

❑ Explain why new life is not being formed on Earth today.

IS LIFE FORMING FROM NONLIFE TODAY? • HOW DID LIFE EVOLVE FROM NONLIFE?   (pages 518–523)

Key Concepts

1. Studies done by *Louis Pasteur* in 1862 convinced most scientists that organisms can arise only from other organisms.

2. The universe began with a "big bang" about 10 to 20 billion years ago and has been expanding since that time.

3. Our solar system coalesced from cold dust particles in a small region of the Milky Way galaxy about 5 billion years ago.

4. Earth's early atmosphere probably consisted mostly of carbon dioxide (CO_2), water vapor (H_2O), hydrogen (H_2), nitrogen (N_2), ammonia (NH_3), and methane (CH_4).

5. Unlike the present-day atmosphere, the early atmosphere was a *reducing* atmosphere, because all of the original free oxygen had combined with other elements.

6. In 1951, *Stanley Miller* demonstrated that amino acids and other organic compounds formed if a reducing atmosphere of hydrogen, ammonia, methane, and water vapor was exposed to electric discharge within a spark chamber.

7. In Miller's studies, simple organic compounds were formed under a variety of different conditions, provided no oxygen was present and an energy source was available.

8. Prebiotically formed organic compounds accumulated in Earth's oceans and reached high concentrations in drying pools.

9. *Polymerization* of small organic molecules (monomers) via condensation reactions can form larger, more complex molecules (polymers). Polymerization of hydrogen cyanide and other molecules can form the nucleic acid bases. Polymerization of formaldehyde can generate sugars.

10. Phosphorylation (the addition of a phosphate group) produces stable yet reactive polymers. Nucleic acids may have formed as phosphorylated polymers of nucleic acid bases and three- and four-carbon sugars.

11. A central problem that slowed research on the origin of life was the need to explain how catalysis was provided to allow replication of early nucleic acids before protein synthesis had evolved.

12. Recent studies of the protist *Tetrahymena thermophilia* have shown that RNA can catalyze the excision and splicing of its own intron. Also, in the *Tetrahymena* ribosome, RNA rather than ribosomal protein is the catalyst of protein synthesis.

13. *Ribozymes* are RNAs with catalytic activity. Ribozymes have been discovered in many organisms.

14. In the late 1970's, *Manfred Eigen* showed that RNA is auto-catalytic—it could catalyze the polymerization of new RNA. Other studies revealed that some RNA can also catalyze the processing of pre-tRNA.

15. Many scientists now believe that the first genetic code was based on RNA that catalyzed its own replication as well as other reactions. RNA's enzymatic functions were assumed by proteins after the evolution of protein synthesis.

16. Evolution of RNAs that were more efficient catalysts for replication involved competition between different RNAs for RNA monomers. RNAs with base sequences that allow the molecule to fold into shapes that are more catalytically efficient could be favored by natural selection if they also unfold and replicate.

17. Ribozymes have "evolved" from random-sequence RNA under laboratory conditions.

18. The Russian scientist *Alexander Oparin* believed that the first cells evolved from *coacervates*—drops of protein and polysaccharide with some water that form spontaneously in an aqueous solution with low concentrations of these substances.

19. Coacervates can become coated with a membranelike lipid layer and will differentially exchange materials with the aqueous environment.

20. Coacervates with enzyme molecules can absorb substrate and release products. Oparin was able to create coacervates with chlorophyll that could absorb oxidized dye from the environment, reduce it using light energy, and release the reduced dye.

21. Competition for limited organic molecules might favor coacervates with better controlled chemical reactions, perhaps involving enzymes to make them more stable.

22. DNA is most stable in the hydrophobic environment provided within cells. DNA probably evolved as the molecule of inheritance after RNA-based life became cellular. RNAs then assumed their roles in translation of genetic information into proteins.

Questions (for answers and explanations, see page 166)

1. Which of the following gases was probably *least* abundant in Earth's early atmosphere? __O_2__
Which of the following gases was probably *most* abundant in Earth's early atmosphere? __CO_2__

CO_2 CO N_2 H_2 NH_3 H_2S O_2

2. Which of the following statements about experiments to synthesize organic compounds under simulated early-Earth conditions is *not* true?
  a. Organic compounds were formed in atmospheres containing different proportions of methane, ammonia, and hydrogen.
  b. Organic compounds were formed in atmospheres containing a variety of gases, provided oxygen was absent.
  c. A variety of energy sources could be used to promote the synthesis of organic compounds.

  d. A variety of organic compounds were formed in these experiments, including the amino acids found in most organisms.
  e. These studies showed that thermal energy released by deep sea vents was probably the most important energy source for prebiotic synthesis of organic compounds.

3. Which of the following molecules was probably formed *last* during the synthesis of prelife biochemicals?
  a. Hydrogen cyanide
  b. Methane
  c. Amino acids
  d. Tetramers of HCN
  e. Adenine

4. Which one of the following statements about the earliest catalysts is most probably *not* true?
  a. The first catalysts were nonproteins and lacked the specificity shown by enzymes.
  b. The earliest catalysts were nonprotein cofactors.
  c. The earliest catalysts were restricted to membranes.
  d. RNA had important catalytic functions before the evolution of enzymes.
  e. DNA had important catalytic functions before the evolution of enzymes.

5. Which of the following statements about the importance of coacervates in the evolution of cells is *not* true?
  a. Under appropriate conditions, coacervates form spontaneously.
  b. Early coacervates developed a system of nucleic acid replication after they became true cells.
  c. Oparin was able to create coacervates that showed some of the properties of chloroplasts.
  d. Coacervates can modify the chemistry of the solution in which they reside.
  e. Coacervates are clumps of protein and polysaccharide in an aqueous solution.

6. Which of the following was *not* presented in the text as evidence that catalysis by RNA preceded catalysis by enzymes?
  a. RNA can excise and splice its own intron.
  b. RNA is only stable in hydrophobic environments.
  c. Ribozymes have been discovered in many different organisms.
  d. RNA can catalyze the polymerization of RNA.
  e. Ribozymes have "evolved" from random-sequence RNA under laboratory conditions.

EARLY ANAEROBIC CELLS • EUKARYOTIC CELLS: INCREASING COMPLEXITY • MATTER, TIME, AND INEVITABLE LIFE (pages 523–525)

1. For a billion years after life evolved, Earth had a reducing atmosphere without oxygen.

2. Like some present-day bacteria, the earliest organisms were probably *obligate anaerobes*, obtaining energy in an oxygen-free environment. The earliest photosynthesizers were also obligate anaerobes and did not generate oxygen.

3. Present-day bacteria may provide clues about the stages in the evolution of photosynthetic pathways in early autotrophs.

4. Present-day anaerobic photosynthesizers include the *green sulfur bacteria*, the *purple sulfur bacteria*, and the *purple nonsulfur bacteria*.

5. The anaerobic photosynthesizers contain a type of chlorophyll called *bacteriochlorophyll*. They also contain red and yellow carotenoids that assist chlorophyll in capturing light energy.

6. All anaerobic photosynthesizers have their photosynthetic pigments and electron transport components embedded in membrane complexes.

7. Unlike plants and the cyanobacteria, which use water as the source of hydrogen to reduce CO_2 into carbohydrates and liberate oxygen, the anaerobic photosynthesizers use other compounds as hydrogen donors.

8. The sulfur bacteria, both green and purple, use H_2S and release sulfur; the purple nonsulfur bacteria use a variety of organic compounds as hydrogen donors, including ethanol, lactic acid, pyruvic acid, and molecular hydrogen.

9. Anaerobic bacteria were key organisms in Earth's early history, and today they continue to be important in cycling many important elements.

10. About one billion years ago, certain sulfur bacteria evolved into *cyanobacteria* that released oxygen as a product of photosynthesis, as evidenced by the first appearance of oxidized minerals in rocks from that period.

11. The remains of the first aerobic photosynthetic bacteria as fossil *stromatolites* show that these organisms became abundant.

12. As molecular oxygen (O_2) accumulated in the atmosphere, its presence favored those species that could tolerate and use it. Aerobic metabolism became dominant because it was more efficient.

13. Two major planetary changes caused by early photosynthesizers were 1) the increase in atmospheric oxygen and 2) the decrease in atmospheric CO_2 and the deposition of CO_2 in marine sediments.

14. As the oxygen concentration of the atmosphere increased, cells could become larger. Some larger cells used smaller cells as food.

15. Some engulfed cells survived and became the chloroplasts and mitochrondria of the first eukaryotic cells, through endosymbiosis.

16. As the amount of DNA needed to encode the cell's metabolic information increased, the single DNA molecule was replaced by multiple, independently replicating chromosomes.

17. Originally, all cells were haploid and divided by fusion after chromosome duplication. Chromosome duplication without division may have produced the first diploid cells.

18. Two copies of each chromosomes enable DNA repair if a chromosome is broken or suffers a point mutation. Many haploid unicellular organisms produce diploid cells when environmentally stressed.

19. Diploid cells can also be produced by fusion of haploid cells—sexual recombination. Repair of damaged chromosomes may have been the initial benefit of sexual recombination. Later, sexual recombination became important for gene exchange and production of variable offspring.

20. Given the early conditions on Earth, the origin of life was highly probable.

21. The prevalence of oxygen on present-day Earth prevents the accumulation of the simple organic molecules that would be necessary for the origin of new life.

Questions (for answers and explanations, see page 166)

1. The earliest organisms were probably
   a. chemosynthetic.
   b. photosynthetic.
   c. obligate anaerobes.
   d. obligate aerobes.
   e. like present-day cyanobacteria.

2. Select *all* of the following ways that present-day cyanobacteria and anaerobic photosynthesizers (purple and green sulfur bacteria and nonsulfur bacteria) are similar.
   a. Type of chlorophyll used
   b. Source for hydrogens used in photosynthesis
   c. Use of CO_2 as carbon source for photosynthesis
   d. Production of O_2 as a waste product of photosynthesis
   e. Energy source for photosynthesis

3. The earliest photosynthetic organisms were probably *least* like present-day
   a. eukaryotic photosynthesizers.
   b. cyanobacteria.
   c. purple sulfur bacteria.
   d. purple nonsulfur bacteria.
   e. green sulfur bacteria.

4. Which of the following is *not* a planetary effect caused by the evolution of photosynthesis and increased oxygen production?
   a. Development of an oxidizing atmosphere
   b. Reduction in amount of ultraviolet light reaching the Earth's surface
   c. Increase in the atmospheric CO_2 concentration
   d. Predominance of aerobic organisms
   e. Increased deposition of oxidized minerals in the ocean

5. Which of the following statements about the anaerobic photosynthesizers (purple and green sulfur bacteria and nonsulfur bacteria) is *not* true?
   a. All use a form of bacteriochlorophyll.
   b. All have carotenoid pigments that assist chlorophyll in light capture.
   c. All use inorganic compounds as hydrogen donors for reducing CO_2.
   d. All have their photosynthetic pigments embedded in thylakoid membrane systems.
   e. None liberate oxygen.

6. Discuss why life does not evolve from nonliving matter on Earth today.

Integrative Question (for answer and explanation, see page 166)

1. Place numbers next to the following events to reflect the sequence of their occurrence in Earth's history.

   4   Coacervates become cells

   6   Photosynthesis without oxygen production appears

   5   DNA assumes central role in genetic code

   7   Cyanobacteria appear

   2   Polymerization of monomers

   3   Ribozymes appear

   1   Organic compounds formed in a reducing atmosphere

Answers and Explanations

Chapters 19–24

CHAPTER 19 – THE HISTORY OF LIFE ON EARTH

Note: The page numbers listed with the section titles refer to the Study Guide; numbers in parentheses for each question refer to the relevant key concepts.

DETERMINING HOW EARTH HAS CHANGED • UNIDIRECTIONAL CHANGES IN EARTH'S ATMOSPHERE • PROCESSES OF MAJOR CHANGE ON EARTH • THE FOSSIL RECORD OF LIFE (pages 134–135)

1. Choice *a*. Radioactive decay is measured as the time it takes for one-half the amount of a substance to spontaneously convert into another substance. (3)
2. Choice *a*. As the crustal plates move past each other, the friction they cause generates earthquakes and pushes rocks upward to make mountains. (8, 9, 11)
3. Choice *b*. Fossilization occurs best in areas of low oxygen concentration and rapid sedimentation, where scavengers cannot destroy the body. (13)
4. Choice *d*. Hard-shelled animals such as mollusks are good candidates for fossilization because their shells can withstand decay long enough to become buried. Many also tend to live in quiet, shallow waters. (16).
5. Choice *e*. If about 300,000 species are currently known as fossils, and only 2 percent of all species become fossilized, then at least 15 million species must have existed throughout evolutionary history. (14, 15)

LIFE IN THE REMOTE PAST (pages 136–137)

1. Choice *e*. Although continental drift caused land masses to separate by forming large oceans, periodic land bridges allowed animals to overcome this separation. (4)
2. Choice *a*. The forces produced when India "rammed" into Asia caused uplifting that resulted in the tallest mountain range on Earth. (3)
3. Choice *c*. Coming from the same root word as "province," which is a specific area within a country or state possessing unique characteristics, provincialized biotas are geographically unique assemblages of organisms produced by a variety of physical and biological isolating factors. (16)
4. Choice *d*. The modern mammals did not appear until the Tertiary, more than 350 million years after the Ordovician. (10, 21, 22)
5. Choice *e*. An integral part of the success of angiosperms has been their coevolution with insect pollinators. (19)
6. Choice *c*. The genus *Homo* first appeared about 2 mya, at the start of the Quaternary. (23)

7. Choice *b*. The mass extinction at the end of the Cretaceous, 65 mya, was the most recent before the evolution of humans. (15, 18, 19)
8. Choice *e*. (20)

PATTERNS OF EVOLUTIONARY CHANGE • THE SPEED OF EVOLUTIONARY CHANGE • THE FUTURE OF EVOLUTION (pages 137–139)

1. Choice *b*. The Cambrian explosion occurred when many open habitats were available for exploitation. Thus, the largest numbers of new phyla were produced then, with later explosions causing increases in diversity only within already existing phyla. (3, 4)
2. Choice *d*. For the reasons outlined in the previous answer, the Cambrian is the time of choice. (1–4)
3. Choice *a*. Although there is still much debate, many paleontologists believe that an asteroid's collision with Earth so altered conditions that the dinosaurs rapidly succumbed during what we know as the great Cretaceous mass extinction. (15–18)
4. Choice *d*. Evolutionary rates apparently have changed through time. Long periods of stasis may be punctuated by periods of rapid change. (8–12)

Integrative Questions

1. Choice *e*. Looking at evolution from an overall viewpoint, all groups of organisms, with the exception of insects, have increased in size.
2. Choice *a*. Carbon dating is generally only reliable for fossils less than 15,000 years old.
3. Choice *e*. The trilobites disappeared in the Permian, over 200 million years before humans evolved.

CHAPTER 20 – THE MECHANISMS OF EVOLUTION

Note: The page numbers listed with the section titles refer to the Study Guide; numbers in parentheses for each question refer to the relevant key concepts.

VARIATION IN POPULATIONS • PRESERVING GENETIC VARIABILITY: THE HARDY–WEINBERG RULE (pages 140–142)

1. Choice *d*. Evolution is defined as changes in the genetic structure of a population over time, so evolution occurs at the level of the population. (1–6)
2. Choice *d*. Although most populations show genetic variation, this condition is not part of the definition of a deme. (6–11)

3. Choices **b** and **d**. Genetic variation has to do with the number of different alleles per gene, which would also result in the greatest number of different genotypes. Regarding choice *a*, recall that all species members have the same number of genes. Population size, per se, has little to do with genetic variation, so choices *c* and *e* are also incorrect. (1, 6, 12–15)

4. *allele, genotype* (19)

5. (18–20)

   A1A1 = 350/600 = 0.58
   A1A2 = 100/600 = 0.17
   A2A2 = 150/600 = 0.25

   A1 = (350 + 50)/600 = 0.67
   A2 = (150 + 50)/600 = 0.33

   A1A1 = p^2 = $(0.67)^2$ = 0.45
   A1A2 = $2pq$ = 2(0.67)(0.33) = 0.44
   A2A2 = q^2 = $(0.33)^2$ = 0.11

   No. The observed genotype frequencies differ from those predicted by the Hardy–Weinberg rule using the observed allele frequencies for *p* and *q*. The population is *not* in genetic equilibrium.

6. Choice **b**. Recall that q^2 = the frequency of the *tt* genotype, so $\sqrt{q^2}$ = the frequency of the *t* allele or $\sqrt{0.36}$ = 0.6, and 1 − 0.6 = 0.4 is the frequency of the *T* allele. (18–20)

7. Choice **c**. Since *p* = 0.3 and, therefore, *q* = 1 − 0.3 = 0.7, the frequency of the *MN* genotype is $2pq$ or 2(0.3)(0.7) = 0.42. (18–20)

8. Choice **a**. The probability of one allele *b* in a genotype is equal to its frequency, or 0.2, so the probability of 3 *b* alleles in a genotype (*bbb*) would be $(0.2)^3$ or 0.008. (18–20)

9. Most populations do not meet the Hardy–Weinberg conditions, and the main use of the rule is to determine if evolutionary change is occurring within a population. (21, 22)

CHANGING THE GENETIC STRUCTURE OF POPULA-TIONS • MAINTAINING GENETIC VARIATION • SHORT-TERM VERSUS LONG-TERM EVOLUTION (pages 142–144)

1. Choice **c**. Genetic drift is always most prevalent in small populations. The founder effect may be important in the act of creating a small population, but after it is established, random variation due to small size is called genetic drift. (5–8)

2. Choice **a**. Since assortative mating involves mate selection based on phenotype, a mostly invisible phenotype like blood group is not likely to be involved. (1–13)

3. Choices **a** and **e**. Self-fertilization leads to increased numbers of homozygous individuals and natural selection is also nonrandom. (13–15)

4. Distribution **a**. Stabilizing selection results when individuals that are intermediate in phenotype make a larger contribution to future generations than individuals of more extreme phenotype. This leads to reduced variation for the trait and causes the curve to be higher and narrower. Curve *b* shows greater variation, curve *c* would result from disruptive selection, and curve *d*, from directional selection. (16–19)

5. Area **c**. Relatively more light moths are recaptured in this area than dark moths, indicating that this phenotype has greater fitness. (15)

6. Dark moths seem to survive best in area *d*. If the increased survival of the moths is due to cryptic coloration then the bark surfaces in that area should also be dark. (15)

7. Choice **a**. Helping raise young that are related through a sister contributes to the bird's inclusive fitness through kin selection. (25–29)

8. Inclusive fitness is a measure of the reproductive success of an individual relative to all other individuals. Inclusive fitness results from an individual's direct reproduction (individual fitness) and any behavior shown by the individual that promotes reproduction and survival of relatives (kin selection). (25–29)

9. Choice **b**. Balanced polymorphism is the existence of two or more alleles at stable frequencies within the population over time. If phenotypes are constant through time, then the underlying alleles will also be constant. (35, 36)

10. Choice **b**. As shown by the example in the text of clinal variation in cyanide production in the European clover *Trifolium repens*, clines can involve any type of phenotypic feature of the organism, including biochemical characteristics. (32–34)

CHAPTER 21 – SPECIES AND THEIR FORMATION

Note: The page numbers listed with the section titles refer to the Study Guide; numbers in parentheses for each question refer to the relevant key concepts.

WHAT ARE SPECIES? • HOW DO NEW SPECIES ARISE? (pages 145–147)

1. Choices **a, d**. The key criterion of a biological species is that its members be reproductively isolated from other such groups. This criterion is impossible to evaluate in asexual and fossil species. (1–6)

2. Choice **b**. A wide distribution is not requisite for geographic speciation. (16–20)

3. Snails from area A would be genetically most like individuals from area **B** and least like individuals from area **D**. Snails from area A are in the same block of habitat with area B snails, but must cross two barriers (roads) to interbreed with area D snails. They need only cross one road to interbreed with area C or E snails. (19, 20)

4. Choice **c**. Differences in soil type and other environmental discontinuities frequently lead to parapatric speciation. (21, 22)

5. Choice **b**. Nondisjunction of homologous chromosomes during meiosis I can lead to a doubling of the chromosome number within a single species; polyploid individuals would have cells with four copies of each chromosome. Each of the three species could form a different tetraploid species. (9–13)

6. Choice **b**. Sympatric speciation is most common among flowering plants. It has been estimated that over one-half of all flowering plant species are polyploid. (9–13)

7. (7–22)

   d: Allopatric speciation
   a: Anagenesis
   b: Sympatric speciation
   c: Parapatric speciation

8. The relatively great distance between the Galapagos Islands and the South American mainland and also between each of the islands in the archipelago insured that once emigrants had arrived on an island, they would be genetically isolated for a substantial period of time. Also, because the islands differ greatly in climate and vegetation, the resident birds were subject to different selection pressures. This in combination with reduced gene flow between the islands led to a rapid evolutionary radiation of finches. (18, 19)

REPRODUCTIVE ISOLATING MECHANISMS • VARIATION IN SPECIATION RATES • EVOLUTIONARY RADIATIONS: HIGH SPECIATION AND LOW EXTINCTION • SPECIATION AND EVOLUTIONARY CHANGE • THE SIGNIFICANCE OF SPECIATION (pages 147–148)

1. (1–5)

not	Nonoverlapping species ranges
pre	Nonoverlapping breeding seasons
post	Production of sterile offspring
pre	Differences in courtship pheromones
pre	Incompatible gametes
post	Poorly adapted offspring

2. Choice *c*. Birth rates *per se* do not seem to affect the rate of speciation in organisms. All other factors have been shown to increase speciation rates in the lineages of some organisms. (10–15)

3. Choice *b*. Actually, biogeographers have found that larger islands tend to have more species than smaller islands, so you might expect the reverse effect. (17–19)

4. Choice *b*. Studies of the Hawaiian *Drosophila* show that these species have only slight genetic differences, despite being morphologically diverse. (9)

5. Studies of the African savanna mammals show that grazers and browsers have undergone more rapid speciation than omnivores and anteaters. Climate change has a more direct effect on the availability of food for grazers and browsers, causing frequent fragmentation of populations of these species. Climate change influences the food supply of omnivores and anteaters less, and their population ranges are more continuous. Thus, diet indirectly affects the degree of gene flow in these mammals, and this results in different speciation rates. (13)

6. Studies of the silverswords and tarweeds of the Hawaiian archipelago show that taxa that evolve on islands frequently show great morphological diversity because of the reduced competition that emigrants encounter on islands. Thus, Hawaiian silverswords and tarweeds have evolved tree- and shrublike species because there were few resident tree and shrub species there with which they had to compete. (19, 20)

Integrative Questions

1. Choice *f*. There is evidence that some parthenogenetic animals have undergone sympatric speciation by polyploidy. All other statements are true.

2. Choice *c*. If the two populations share a common gene pool, they are still a single species. Limited interbreeding, production of viable hybrids, and establishment of a hybrid zone do not necessarily mean that speciation is not complete.

3. Choice *c*. Areas where environmental change is least rapid (environmentally homogeneous areas) and gene flow greatest are areas where the least geographic (clinal) variation would be expected.

CHAPTER 22 – CONSTRUCTING AND USING PHYLOGENIES

Note: The page numbers listed with the section titles refer to the Study Guide; numbers in parentheses for each question refer to the relevant key concepts.

THE HIERARCHICAL CLASSIFICATION OF SPECIES (pages 149–150)

1. Choice *b*. Although all of the statements listed are important attributes of biological classification schemes, the most important attribute is that biological classification reflects evolutionary relationships. (2–5)

2. Choice *d*. After a scientific name is referenced once in a paragraph, the genus is abbreviated, but the species name is given in full. (7–12)

3. Choice *c*. Organisms in a higher taxon are *less* similar, have diverged from a common ancestor *less* recently, and include *more* species than organisms in a lower, included taxon. (15)

4. (8, 9, 16, 17)

4	Order
1	Species
3	Family
7	Kingdom
	Division (plants, fungi, bacteria only)
2	Genus
6	Phylum
5	Class

INFERRING PHYLOGENIES (pages 150–152)

- (4–12)

			TRAIT		
SPECIES	1	2	3	4	5
A	1	0	1	1	1
B	0	0	1	1	0
C	0	1	1	0	0
D	0	0	0	0	0
E	1	0	1	1	0

- Species *D*. The outgroup has the ancestral form for all traits. (5)

- (4–12)

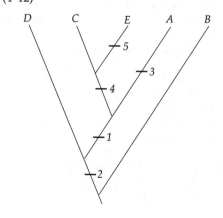

• In the construction of a cladogram, you initially assume that derived traits appear only once and never disappear. Given a set of traits for a group of species, sometimes these restrictions must be relaxed in order to produce a cladogram for the group. Parsimony involves arranging the species so that you minimize the number of required reversals and multiple origins. Generally, the simplest explanation is most likely to be correct. (16–21)

1. Choice *d*. Homoplasy is the appearance of similar structures in different lineages that were not present in the common ancestor. (16–21)

2. Choice *c*. A clade can include living species or species known only from the fossil record. (3–5)

3. Choice *d*. The outgroup includes the taxon with ancestral traits for those traits being considered (although some of the ancestral traits could have been lost). The fossil record is used for deciding what are ancestral traits, but it is not used directly in the selection of the outgroup. (15)

4. Choice *c*. General homologous traits are found in *both* the outgroup and the group being classified. (9–15)

CONSTRUCTING PHYLOGENIES • BIOLOGICAL CLASSIFICATION AND EVOLUTIONARY RELATIONSHIPS • THE FUTURE OF SYSTEMATICS • THE THREE DOMAINS AND THE SIX KINGDOMS (pages 152–153)

1. Choice *a*. With the availability of polymerase chain reaction technology, DNA from fossils can now be amplified, sequenced, and compared. All other statements are true. (1–10)

2. Choice *b*. See Table 21.3 on page 475 in the textbook. (8–10)

3. Choice *e*. Actually, cpDNA is a very stable molecule that has changed very little in flowering plants. This has made the changes that have occurred very useful systematic data. (6, 7)

4. Some of the implications of this statement are that DNA evolved as the genetic material before eukaryotes had diverged from prokaryotes, that DNA is an ancestral and general homologous trait, and that all surviving eukaryotes use DNA as the genetic material. (11–14)

5. Although clades are by definition monophyletic, that is consisting of only descendants of a common ancestor, a clade with some but not all descendants, a paraphyletic clade, is tolerated as a way to emphasize groups that have undergone rapid evolution. For example, the birds are put in their own Class, Aves, even though most biologists agree that the birds share a common ancestor with the reptiles. This makes the Class Reptilia paraphyletic. Polyphyletic clades are not acceptable, because they include members with more than one ancestor. This is antithetic to the purpose of constructing phylogenies. (16, 17)

CHAPTER 23 – MOLECULAR EVOLUTION

Note: The page numbers listed with the section titles refer to the Study Guide; numbers in parentheses for each question refer to the relevant key concepts.

DETERMINING THE STRUCTURE OF MOLECULES • HOW MOLECULAR FUNCTIONS CHANGE (pages 154–155)

1. Choice *e*. Mutation is the ultimate source of changes in nucleotide sequences. Gene duplication is an important mechanism for producing gene families. Plasmids and episomes function in "horizontal" movement of genes bewteen different bacterial species, as in Yops, mentioned in the introduction to this chapter. All other statements are true. (4, 10, 11)

2. Choice *b*. Pseudogenes (nonfunctional DNA) *may* be removed via deletion, but numerous pseudogenes persist in many gene families. (11–16)

3. Choice *e*. Functionally important regions of a protein can be discovered by identifying the regions with the *fewest* amino acid substitutions. Since the sequences in these regions have been optimized by natural selection to accomplish the function of the molecule, random (neutral) changes in these areas are not tolerated. (2–4, 6–9)

4. Choice *c*. Since the animals that evolved a similar lysozyme do not share a recent common ancestor, the mechanism involved is convergent evolution. All other statements are false. (17–19)

5. Since molecular evolution depends on obtaining fairly large amounts of pure DNA for sequencing purposes, the polymerase chain reaction is commonly used to amplify and purify DNA. This is especially important in studying DNA from fossil organisms where the amount available may be very small and degraded. (5)

GENOME SIZE: SURPRISING VARIABILITY • MOLECULAR CLOCKS • USING MOLECULES TO INFER PHYLOGENIES • MOLECULES AND HUMAN EVOLUTION (pages 155–157)

1. Recall that there is a rough relationship between the amount of *coding* DNA and some measure of organismal complexity. (1, 2)

   4 Insect
   5 Bird
   1 Bacterium
   2 Single-celled eukaryote
   3 Round worm

2. Choice *a*. The molecular clock of cytochrome *c* would tick more rapidly (i.e., the rate of neutral changes per unit time would be greater) than the clock of rRNA. rRNA has changed very little because it is a molecule with great functional constraints. (3–7)

3. Choice *d*. Unlike the other molecules, 16S rRNA is a molecule that is found in all living organisms. This molecule, as stated in 2 above, has also evolved very slowly. (6, 7, 11)

4. The major important consclusion is that all humans share a recent common mitochondrial ancestor about 200 thousand years ago. This supports the "out of Africa" hypothesis that our species evolved in Africa and dispersed from there to the rest of the world.

5. The main assurance is that although fossil dinosaur DNA can be obtained, it is very fragmented. Putting all the fragments together in the correct sequence to reconstruct the entire genome of a dinosaur is next to impossible. Controlling the expression of the genome during development is another insurmountable obstacle to cloning dinosaurs.

6. (8, 9)

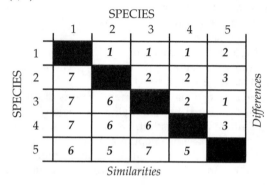

SPECIES

	1	2	3	4	5
1		1	1	1	2
2	7		2	2	3
3	7	6		2	1
4	7	6	6		3
5	6	5	7	5	

Similarities

Species *2* shows an amino acid insertion (Phe) at position 3. There are no deletions.

Species *3*, *4*, and *5* all show amino acid substitutions. Species *1* shows no changes.

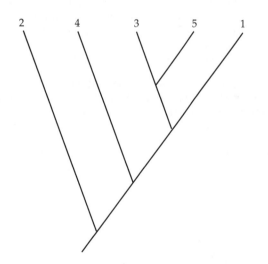

CHAPTER 24 – THE ORIGIN OF LIFE ON EARTH

Note: The page numbers listed with the section titles refer to the Study Guide; numbers in parentheses for each question refer to the relevant key concepts.

IS LIFE FORMING FROM NONLIFE TODAY? ● HOW DID LIFE EVOLVE FROM NONLIFE? (pages 158–159)

1. Oxygen (O_2) was probably the least abundant and carbon dioxide (CO_2) was probably the most abundant gas in Earth's early atmosphere. (4, 5)

2. Choice *e*. Although it is hypothesized that thermal energy may have been a more important energy source than ultraviolet light and electric discharge (because of their damaging effects on life), these studies did not deal with this issue directly. (6, 8)

3. Choice *e*. Hydrogen cyanide (HCN) and methane were constituents of the early atmosphere. Some amino acids were formed directly in Miller's experiments and presumably in the early atmosphere. The synthesis of adenine is more complex, requiring the polymerization of HCN and rearrangement of its HCN tetramers. (6–10)

4. Choice *e*. Unlike RNA, there is no evidence that DNA has ever had a catalytic function. (11–17)

5. Choice *b*. The theory explaining the evolution of the first cells from coacervate droplets assumes that replication via nucleic acids appeared later. (18–22)

6. Choice *b*. DNA is only stable in hydrophobic environments. (12–17, 22)

EARLY ANAEROBIC CELLS ● EUKARYOTIC CELLS: INCREASING COMPLEXITY ● MATTER, TIME, AND INEVITABLE LIFE (pages 159–161)

1. Choice *c*. Very little oxygen was present in the atmosphere of the early Earth, and until organisms evolved adaptations to use it, oxygen was poisonous . (1, 2)

2. Choices *c* and *e*. Except for the use of CO_2 as a carbon source and sunlight as an energy source, these groups differ in all other ways listed. (4–10)

3. Choice *a*. The earliest photosynthetic organisms were prokaryotic and, therefore, most like the bacteria listed in choices *b–e* and least like eukaryotic photosynthesizers. (4–10)

4. Choice *c*. With the evolution of noncyclic photophosphorylation, photosynthetic organisms became very successful, resulting in a *reduction* in CO_2 within the atmosphere as more became fixed in organic form or deposited in ocean bottoms. (10–14)

5. Choice *c*. The purple nonsulfur bacteria use a variety of organic compounds as hydrogen donors. (4, 5, 8)

6. The present-day oxidative atmosphere prevents the accumulation of abiotically formed simple organic compounds. Even if these materials could accumulate, they would be consumed by living organisms. (20, 21)

Integrative Question

1. *4* Coacervates become cells
   6 Photosynthesis without oxygen production appears
   5 DNA assumes central role in genetic code
   7 Cyanobacteria appear
   2 Polymerization of monomers
   3 Ribozymes appear
   1 Organic compounds formed in a reducing atmosphere

25

Bacteria and Archaea: The Prokaryotic Domains

CHAPTER LEARNING OBJECTIVES—after studying this chapter you should be able to:

❏ Name the three domains of living things, and describe the fundamental evolutionary relationships among them.

❏ Provide the biological rationale for separating prokaryotes into two distinct domains.

❏ Describe the fundamental differences in structure and function between prokaryotic and eukaryotic cells.

❏ Explain the difference between "bacillus," "coccus," and "spiral" cell shapes in prokaryotes.

❏ Discuss the various methods of locomotion used by prokaryotes, and explain the structural and functional differences between prokaryotic and eukaryotic flagella.

❏ Explain the importance of peptidoglycan as a taxonomic character in prokaryotic classification.

❏ Explain the Gram-staining method, describe the difference between gram-positive and gram-negative bacteria, and discuss how this technique can be used as a medical diagnostic tool.

❏ Describe the process of asexual or vegetative reproduction in prokaryotes, and discuss how the terms "fission," "transformation," "conjugation," and "transduction" are related to this process.

❏ Discuss the metabolic characteristics of the prokaryotic groups designated as obligate aerobes, obligate anaerobes, and facultative anaerobes.

❏ Explain the differences between the following modes of prokaryotic nutrition: photoautotrophy, photoheterotrophy, chemoautotrophy, and chemoheterotrophy.

❏ Describe the importance of sulfur and nitrogen metabolism for some groups of prokaryotes, and distinguish between the terms "nitrification," "denitrification," and "nitrogen fixation."

❏ Name the principal photosynthetic pigments used by photoautotrophic bacteria, and explain how prokaryotic photosynthesis has influenced the environment over the course of evolution.

❏ Discuss examples of how bacteria form important functional associations with eukaryotic organisms.

❏ List Koch's postulates and explain their use in associating a disease with a particular microoogranism.

❏ Differentiate between the concepts of invasiveness and toxigenicity when discussing bacterial infection, and explain the difference between bacterial endo- and exotoxins.

❏ Explain why the biological characteristics of rRNA make it a good tool for classifying prokaryotes, and discuss what is meant by a "signature sequence" of rRNA.

❏ Discuss the evolutionary reasons why prokaryotic phylogenies as determined by molecular means do not necessarily match affinities based on the phenotypic characteristics of these organisms.

❏ Discuss the unique biological characteristics that separate the kingdoms Archaebacteria and Eubacteria.

❏ Describe the diversity of the kingdom Eubacteria by listing important physiological and ecological features of the following groups: thermophiles, cyanobacteria, gliding bacteria, spirochetes, gram-negative rods, chlamydias, gram-positive rods, actinomycetes, and mycoplasmas.

❏ Describe the diversity of the kingdom Archaebacteria by listing important physiological and ecological features of the following groups: hyperthermophiles, methanogens, extreme halophiles, and thermoplasmas.

❏ Explain the role of "heterocysts," "axial filaments," "endospores," "ester and ether linkages in membrane lipids," and "bacteriorhodopsin" in the biology of prokaryotes.

WHY THREE DOMAINS? • GENERAL BIOLOGY OF THE PROKARYOTES • PROKARYOTES IN THEIR ENVIRONMENTS (pages 532–540)

Key Concepts

1. The term *domain* refers to a taxonomic grouping of organisms above the level of the kingdom. Three domains of living things are now recognized: Bacteria, Archaea, and Eukarya.

2. The Bacteria and Archaea domains are comprised exclusively of prokaryotic organisms. Each contains one kingdom, the Eubacteria and Archaebacteria, respectively. The Eukarya domain includes all of the eukaryotic kingdoms (Protista, Fungi, Plantae, and Animalia).

3. Available evidence suggests that Eukarya and Archaea are the more closely related of the three domains, sharing a common evolutionary ancestor about 1.8 bya. The common ancestor between these domains and the Bacteria dates back to about 2 bya. Since the oldest prokaryote fossils are about 3.5 billion years old, prokaryotes have a long history in the evolution of life on Earth.

4. The two kingdoms of prokaryotes include an incredibly abundant, diverse, and highly successful array of microorganisms. They display a great array of metabolic lifestyles, occupy a wide variety of habitats, and survive in more extreme conditions than any other taxonomic group.

5. One traditional method for classifying prokaryotes is by their shape. Three types are recognized: spherical (*coccus*), rod (*bacillus*), and spiral forms. Each may exist individually or in two- and three-dimensional arrays of adhering cells.

6. All bacteria show prokaryotic cell structure. They differ from eukaryotes in terms of the molecular organization of their genetic material and its replication mechanisms. Prokaryotes also lack a cytoskeleton and the typical membrane-bounded cytoplasmic organelles of eukaryotes. Some bacteria do, however, show an elaborate internal membrane structure used in respiration and photosynthesis.

7. A number of prokaryotes are motile. Many of these move by means of flagella, which may occur singly or in tufts along the cell's exterior. The structure of prokaryotic flagella differs significantly from that of eukaryotic flagella. Bacterial flagella consist of a single fiber of the protein flagellin, which rotates about its base, rather than beating from side to side as in eukaryotes. Other prokaryotes move by means of internal fibrils, gliding mechanisms, or by adjusting gas vesicles.

8. The chemical structure of the cell wall differs between the Eubacteria and Archaebacteria. The polymerized amino sugar peptidoglycan is unique to the cell walls of Eubacteria. The Archaebacteria lack any trace of peptidoglycan, but tend to contain significant amounts of glycoproteins in their cell walls.

9. The amount of peptidoglycan in the cell walls of Eubacteria influences their response to the chemical dye known as *Gram stain*. Those with large amounts of peptidoglycan readily take up this stain and appear purple under the microscope; they are called *gram-positive*. Other bacterial cell walls contain much less peptidoglycan, and thus do not absorb the stain. These are called *gram-negative*, and appear pink or red after counterstaining.

10. The Gram stain response of Eubacteria represents a useful diagnostic tool in medicine, since chemical treatments against bacterial infection are often aimed at destroying the cell wall of the pathogen.

11. Most bacteria reproduce asexually by fission, a form of *vegetative* reproduction. However, processes such as transformation, conjugation, and transduction do allow for exchange of genetic material between individuals.

12. The replication time of bacteria can be extremely rapid. Under optimal conditions it can be as short as 10 to 20 minutes, but is more commonly on the order of a few hours. In rare instances, replication times of several decades to more than a century can occur.

13. Some bacteria are *obligate anaerobes* and can only survive in oxygen-free environments. Others require oxygen for survival and are designated *obligate aerobes*. Bacteria that are able to shift their metabolism between aerobic and anaerobic mechanisms are called *facultative anaerobes*.

14. Four main categories of nutrition are recognized in bacteria: photoautotrophy, photoheterotrophy, chemoautotrophy, and chemoheterotrophy.

15. *Photoautotrophs* use light energy to fix CO_2 into organic carbon compounds. Two types of photosynthesis are known to occur in prokaryotes. The cyanobacteria use *chlorophyll a* as their key photosynthetic pigment in the process of noncyclic photophosphorylation, which produces oxygen as a by-product. They are thus similar to photosynthetic eukaryotes. Other bacterial species use the pigment *bacteriochlorophyll* in conjunction with hydrogen sulfide as an electron donor; they generate pure sulfur instead of oxygen as a by-product. Because of differences in the absorption spectra between these two photopigments, different species of bacteria can grow in diverse habitats spanning a broad range of light quality and intensity.

16. *Photoheterotrophs* also use light energy, but obtain carbon from organic compounds produced by other organisms instead of inorganic sources. Included are species such as the purple nonsulfur bacteria, which typically "feed" on alcohols, fatty acids, and carbohydrates.

17. *Chemoautotrophs* gather energy by oxidizing inorganic substances and then use it to reduce CO_2. Included are species that oxidize ammonia and nitrite ions (the nitrifiers important to plants), as well as those capable of oxidizing hydrogen gas, hydrogen sulfide, and sulfur. Oceanic ecosystems based on chemoautotrophy exist at great depths (~2,500 m) where no light reaches. Here, chemoautotrophic bacteria oxidize hydrogen sulfide emanating from thermal vents in the ocean floor to obtain energy that they use to fix carbon, thus creating food for other heterotrophic organisms.

18. *Chemoheterotrophs* obtain both energy and carbon from one or more preexisting organic compounds. The vast majority of bacteria are chemoheterotrophic, as are all archaea and most of the nonplant eukarya.

19. Some prokaryotes perform respiratory electron transport in the absence of oxygen by using inorganic ions such as nitrate, nitrite, or sulfate to accept electrons. *Nitrogen fixers* are the bacteria that convert atmospheric nitrogen gas into chemical forms usable by other cells. The process of nitrification is a vital function performed only by certain bacteria, many of which live in symbiotic association with organisms that require this nitrogen.

20. *Nitrifiers* are soil bacteria that oxidize ammonia into nitrate, a chemical form more easily taken up from the soil by plants. *Denitrifiers* are the bacteria that return nongaseous nitrogen back to the atmosphere to complete the nitrogen cycle. Comparable reactions occur for sulfur-using bacteria.

21. Prokaryotes play a critical role, either directly or indirectly, in the lives of all other organisms. Of particular importance are the nitrogen fixers and denitrifiers. Other prokaryotes are equally important in driving the sulfur cycle.

22. Cyanobacteria played a crucial historical role in shaping the nature of Earth's atmosphere. Through the process of photoautotrophy, and over many millions of years, these bacteria converted the global environment from an anaerobic to aerobic one, making possible the emergence of cellular respiration and eukaryotic life.

23. Many important relationships exist between prokaryotes and eukaryotes. These include the fact that some eukaryotic organelles (e.g., mitochondria and chloroplasts) are evolutionarily descended from free-living prokaryotes, that nitrogen-fixing bacteria form permanent cooperative associations with the plants for whom they provide nitrogen, and that animals often harbor large populations of prokaryotes in their digestive tracts to assist in the breakdown of cellulose or to provide metabolic products such as vitamins to supplement their nutrition.

24. Although most prokaryotes play very beneficial roles such as aiding animal digestion, processing nitrogen, acting as agents of decomposition in the environment, and assisting humans in the production of foods and medicines, a small minority do act as agents of disease in other organisms.

25. *Koch's postulates* state that, to definitively associate a disease with some particular microorganism
   a. the microorganism must always be found in diseased individuals;
   b. microorganisms taken from the host must be grown in pure culture;
   c. samples of this culture must produce disease when injected into a new, healthy host; and
   d. a newly infected host must yield a new pure culture of the suspected microorganism.

26. The consequences of infection for a host depend on factors related to a pathogen's *invasiveness* (ability to multiply within the host) as well as its *toxigenicity* (ability to produce chemical substances injurious to the host). Some bacterial pathogens have low invasiveness but high toxigenicity (e.g., the diphtheria bacterium), while others have high invasiveness and low toxigenicity (e.g., the anthrax bacterium).

27. Bacterial toxins are of two types. *Endotoxins* are lipopolysaccharides produced by lysing gram-negative bacteria; they are rarely fatal but often produce discomfort in the form of fever, vomiting, and diarrhea. Examples are the toxins that cause certain forms of food poisoning. *Exotoxins* are highly toxic proteins produced by living, multiplying bacteria; they do not typically cause fever, but are often fatal. Examples are the toxins that cause tetanus, botulism, and the plague.

Activity *(for answers and explanations, see page 221)*

- Briefly explain how each of the following is related to the general biology of prokaryotes.

   1. Invasiveness and toxigenicity:

   2. Facultative anaerobism:

   3. Chemoautotrophy:

   4. Differential uptake of Gram stain:

   5. Coccus, bacillus, or spiral:

Questions *(for answers and explanations, see page 221)*

1. Which of the following is *not* a necessary condition to satisfy Koch's postulates?
   a. It must be possible to grow a pure culture of the isolated microorganism.
   b. Cultured samples must be able to produce disease symptoms in newly infected host.
   c. Isolated microorganisms must respond positively to gram-staining procedures.
   d. Newly infected hosts must yield the same suspect microorganism in pure culture.
   e. The suspected microorganism must always be found in diseased hosts.

2. After using the Gram-staining procedure on a sample of bacteria from your toothbrush, you observe a cluster of tiny, rod-shaped purple cells under the microscope. These are most likely
   a. gram-positive bacilli.
   b. gram-negative bacilli.
   c. gram-positive cocci.
   d. gram-negative cocci.
   e. spiral bacteria insensitive to Gram stain.

3. An important distinction between eukaryotic and prokaryotic flagella is that
   a. the overall design of eukaryotic and prokaryotic flagella is quite different.
   b. prokaryotic flagella consist of a single fiber of flagellin, while eukaryotic flagella are made of other proteins.
   c. the mechanism of motion in the two types of flagella is completely different.
   d. All of the above

4. Which of the following statements about bacterial metabolism is *false*?
   a. Cyanobacteria utilize bacteriochlorophyll as their key photosynthetic pigment.
   b. Bacteria that use chlorophyll *a* for photosynthesis also produce oxygen as a by-product.
   c. Chemosynthetic bacteria sometimes form the basis of deep-water oceanic food chains.
   d. In chemoautotrophic bacteria, energy to fix carbon comes from oxidizing inorganic substances.
   e. Most bacteria are chemoheterotrophic.

5. Match the following categories of bacterial nutrition with the correct biochemical mechanism for that category.

   _b_ Photoautotrophy   a. Energy and carbon from preexisting organic compounds

   _d_ Photoheterotrophy   b. Light used as energy source, CO_2 is the carbon source

   _c_ Chemoautotrophy   c. Carbon comes from CO_2, energy from oxidizing inorganic materials

   _a_ Chemoheterotrophy   d. Preformed organics from other organisms, energy from light

6. Which of the following statements about the structure of prokaryotic and eukaryotic cells is *false*?
   a. The molecular organization and shape of the chromosomal material in eukaryotes is different from that of prokaryotes.
   b. Unlike eukaryotes, prokaryotes lack a membrane-bounded nucleus.
   c. While there are extensive internal membrane systems within eukaryotes, no such membranes exist in prokaryotes.
   d. Some prokaryotes have gas vesicles that assist in cell movement.
   e. Photosynthesis based on the pigment chlorophyll *a* occurs in both eukaryotes and prokaryotes.

7. A new species of prokaryote is discovered that converts nongaseous nitrogen to nitrogen gas. This cell would best be described as a
   a. nitrogen fixer.
   b. nitrifier.
   c. denitrifier.
   d. producer of endotoxins.
   e. producer of exotoxins.

8. A particular bacterium is pathogenic. When it invades a host organism, it multiplies slowly, but produces very potent toxins. This bacterium is best described as having
   a. high invasiveness and low toxigenicity.
   b. high invasiveness and high toxigenicity.
   c. low invasiveness and high toxigenicity.
   d. low invasiveness and low toxigenicity.
   e. none of the above

PROKARYOTE PHYLOGENY AND DIVERSITY • THE BACTERIA • THE ARCHAEA (pages 540–549)

Key Concepts

1. A truly evolutionarily-based classification of prokaryotes has, until recently, been problematic. Analysis of ribosomal RNA now provides a more reliable measure of evolutionary distance among prokaryotes. rRNA is notable as a taxonomic tool because it a) is evolutionarily ancient, b) is present and plays a role in translation for all species, and c) has evolved slowly so that base sequence similarities between species (called *signature sequences*) are easily found.

2. The importance of rRNA signature sequences notwithstanding, other factors in the biology of prokaryotes provide confounding sources of variation for the determination of bacterial phylogenies. Transformation, transduction, and conjugation all contribute to this variability.

3. Clearly the most important source of variation in prokaryotes is genetic mutation of the haploid bacterial genome, which can lead to significant phenotypic changes not reflected in the more evolutionarily conservative rRNA measures. Such changes are intensified by the rapid multiplication rate of most bacteria, so that new phenotypic characteristics can appear and disappear very quickly within species.

4. The great majority of prokaryotes are members of the domain Bacteria, kingdom Eubacteria. Although 12 distinct phyla can be identified through molecular analysis, complicated phenotypic variation in metabolism and ecology is widespread among these phyla. Thus, functional traits that cross phyletic boundaries are often better suited to describe the bacteria.

5. Proteobacteria (commonly known as the Purple Bacteria) is by far the phylum with the largest number of species. They are an extremely diverse group, especially with respect to mode of nutrition. Other bacteria include the thermophiles, cyanobacteria, gliding bacteria, spirochetes, gram-negative rods, chlamydias, gram-positive bacteria, actinomycetes, and mycoplasmas.

6. The *thermophiles* are thought to represent one of the earliest branches in bacterial evolution. Their name denotes the ability to withstand extreme heat, a trait that developed in the hot environment of early Earth, and one that is shared with most of the Archaea.

7. The *cyanobacteria* perform photosynthesis using chlorophyll *a* in much the same way that plants do, and many have elaborate internal photosynthetic membrane systems resembling the thylakoids of plant chloroplasts.

8. Cyanobacteria are nutritionally independent; they can be photosynthetic, rely on fermentation, or act as nitrogen fixers. They may exist colonially or live free as single cells. Filamentous forms show three cell types: vegetative cells, spores, and *heterocysts,* which serve as break points during reproduction and also as deposition sites for the oxygen-sensitive nitrogenase needed in nitrogen fixation.

9. *Gliding bacteria* (e.g., members of the genus *Beggiatoa*) are rod-shaped or filamentous, and locomote via a gliding motion whose mechanism remains largely unknown. Some soil-dwelling species are known to undergo daily vertical migrations between appropriate areas of oxygen and sulfur concentrations. Individual gliding bacteria aggregate to form fruiting bodies where thick-walled, desiccation-resistant spores or cysts are produced.

10. *Spirochetes* are coiled, cylindrical cells that possess unique structures called *axial filaments*. These filaments are composed of flagella positioned between the cell wall and an outer envelope and are responsible for the bacterium's locomotion. Many species are parasitic (e.g., *Treponema pallidum*, which causes syphilis in humans), while others are free-living in mud or water.

11. **Gram-negative rods** are a metabolically diverse group related primarily by the nature of their cell walls. Included is *Escherichia coli*, an important resident of the human digestive system. Also counted among this group are the causative agents of human plague, dysentery, cholera, and salmonella food poisoning. The tumors characteristic of *crown gall* in plants result from infection by a gram-negative bacterium whose plasmids are commonly used in recombinant DNA studies.

12. The *chlamydias* are extremely small bacteria whose complex reproductive cycle can occur only within the cells of other organisms. Many are agents of serious human diseases such as Rocky Mountain spotted fever, typhus, certain sexually transmitted diseases, and some forms of pneumonia.

13. The *gram-positive bacteria* include some species capable of forming *endospores*, a nonreproductive resting structure that replaces the parent cell and can endure harsh conditions while remaining dormant for very long periods of time. The endospore germinates to become a metabolically active bacterium whenever favorable conditions return.

14. Gram-positive rods include *Clostridium botulinum*, the agent that produces the highly potent toxin responsible for botulism poisoning in humans. Gram-positive cocci are abundant and common both outside and inside the human body. The genus *Staphylococcus* contains species causing respiratory, intestinal, and skin diseases, as well as species that produce the toxins responsible for toxic shock syndrome.

15. *Actinomycetes* possess a highly branched mat of filaments similar to that found in fungi, with which they were once classified. Some reproduce by forming spores, others by fission. The genus *Streptomyces* produces natural chemical defenses, including streptomycin and other antibiotics, during spore formation. This action may inhibit the growth of competing bacteria and fungi.

16. Members of the *mycoplasmas* lack cell walls entirely. They are the smallest cellular organisms known, and have less than half the DNA typical of other prokaryotes. This may represent the minimum level of genetic material necessary to maintain cellular existence. Virtually all mycoplasma species are parasites of animal or plant cells. However, since they lack cell walls, traditional antibiotics are ineffective at fighting their attacks.

17. Members of the domain Archaea, kingdom Archaebacteria, are recognized as related taxonomic groups based on three criteria. All exhibit a definitive lack of peptidoglycan in their cell walls, they have lipids of distinctive composition, and they possess great similarity in the base sequences of their ribosomal RNA. Their taxonomic difference from Bacteria and Eukarya is underscored by the fact that over half their genome is unlike any genes found in the other two domains.

18. The membrane lipid composition of Archaea is notable for its difference from all other prokaryotes and eukaryotes. Instead of *ester linkages* that connect glycerol to long-chain, unbranched fatty acids in the membrane lipids of Bacteria and Eukarya, the membrane lipids of Archaea contain branched long-chain hydrocarbons connected to glycerol by *ether linkages*.

19. Archaebacteria can be subdivided by habitat and lifestyle into four phyla: the hyperthermophiles (heat- and acid-loving), the methanogens (methane-producers), the extreme halophiles (salt-lovers), and a single-genus phylum containing only *Thermoplasma*.

20. *Hyperthermophiles* live in places where few other organisms can survive. Members of the genus *Sulfolobus* are common in hot sulfur springs, where they endure temperatures as high as 75°C. Although the pH in such habitats is typically 2 to 3, *Sulfolobus* can tolerate pH as low as 0.9.

21. *Methanogens* are obligate anaerobes that use methane production as a key step in metabolism. They are responsible for producing all of the biologically-generated methane gas released into our atmosphere (~2 billion tons annually). Habitats for methanogens range from hot volcanic vents to the digestive tracts of grazing mammals.

22. *Extreme halophiles* live in aquatic and terrestrial habitats nearly saturated with salt, where other organisms would quickly perish due to unfavorable osmotic conditions. These prokaryotes also live in the most alkaline habitats known to support life (pH = 11.5). Many species contain pink carotenoid pigments, which make them visible when present in large concentrations. Under low oxygen conditions they can make use of the photopigment retinal, combined with the protein *bacteriorhodopsin*, to trap light energy and form ATP without the aid of chlorophyll.

23. *Thermoplasmas* belong to a single genus of Archaebacteria that has a unique combination of characteristics. They lack cell walls, are both thermophilic and acidophilic, metabolize aerobically, and are found only in coal deposits. Along with the bacterial mycoplasmas, they have the smallest genome of any free-living organism, about 1100 kilobase pairs.

Activities (for answers and explanations, see page 221)

- For each of the prokaryote types listed below, give the domain and kingdom to which they belong, and cite one specific taxonomic characteristic of their group.

  a. Mycoplasmas *Bacteria Eubacteria, lack cell wall, smallest cell known, parasites.*

  b. Gram-positive rods *" " photosynthetic*

  c. Cyanobacteria

  d. Methanogens

  e. Gram-negative rods

  f. Hyperthermophiles

- Briefly describe what is meant by each of the following terms when used in reference to bacteria:

  a. Axial filament

  b. Antibiotics

  c. Fruiting body

  d. Heterocyst

Questions (for answers and explanations, see page 221)

1. Bacterial species capable of producing toxic substances or disease in other organisms belong to which of the following groups?
a. Gram-positive cocci
b. Chlamydias
c. Mycoplasmas
d. Spirochetes
e. All of the above

2. All of the following prokaryotes belong to the Eubacteria *except* the
a. methanogens.
b. spirochetes.
c. chlamydias.
d. cyanobacteria.
e. gliding bacteria.

3. Members of the bacterial group mycoplasma are uniquely characterized by
a. being among the largest of all known bacteria.
b. having more DNA than 80 percent of all other bacteria.
c. being the only bacterial group without parasitic forms.
d. entirely lacking cell walls.
e. none of the above

4. Which of the following statements correctly applies to the methanogens?
a. They do not typically cause disease.
b. They are prokaryotes.
c. They possess cell walls.
d. They are representatives of the Archaebacteria.
e. all of the above

5. *Escherichia coli*, a common bacterial resident in the human gut, is a member of which kingdom and subgrouping of prokaryotes?
a. Archaebacteria; extreme halophiles
b. Archaebacteria; hyperthermophiles
c. Eubacteria; gram-positive cocci
d. Eubacteria; gram-negative rods
e. Eubacteria; spirochetes

6. Bacteria of the genus *Streptomyces* are particularly notable because they
a. are important members of the phylum cyanobacteria.
b. form colonial aggregations and produce fruiting bodies.
c. locomote by means of flagella that contain axial filaments.
d. are the causative agents of the venereal disease gonorrhea.
e. produce antibiotics as chemical defenses.

Integrative Questions (for answers and explanations, see page 222)

1. In what ways is it accurate to say that the prokaryotes are "primitive" organisms, and in what ways is this a misleading statement?

2. In terms of overall size, the largest virus is _____ the smallest bacterium.

3. A bacteriophage virus infects a gram-positive rod bacterium. How many kingdoms of organisms are involved in this interaction?

4. Below is a list of potential habitats in which bacteria may be found. For each habitat, name a likely bacterial group that could reside there.

   a. Hot spring — _hyperthermophiles_

   b. Intestines of a cow — _methanogens_

   c. Within other cells — _many forms._

   d. Alkaline lake — _extreme halophile._

26
Protists and the Dawn of the Eukarya

CHAPTER LEARNING OBJECTIVES—after studying this chapter you should be able to:

❑ State the basic biological characteristics that separate the kingdom Protista from the Archaea, Bacteria, and all other Eukarya kingdoms.

❑ Identify the main characteristics of the protozoans, "funguslike" protists, and algae, and state the suspected evolutionary relationships of these groups to the other eukaryotic kingdoms.

❑ Describe the diversity of structure, locomotion, reproduction, and modes of nutrition in protists.

❑ Discuss the evolutionary events associated with the origin of eukaryotic cells from prokaryotes, and explain how endosymbiosis can account for the evolution of certain eukaryotic organelles.

❑ Explain the importance of contractile and food vacuoles in the lives of protists.

❑ Discuss the distinguishing characteristics of the following protozoans: flagellates, euglenoids, amoebas, actinopods, radiolarians, heliozoans, foraminiferans, apicomplexans, and ciliates.

❑ Describe the life cycle of the human malarial parasite *Plasmodium.*

❑ Identify the structures and functions of a typical ciliate body, and explain the method of locomotion in these protists.

❑ Discuss the distinguishing characteristics of the following "funguslike" protists: acellular slime molds, cellular slime molds, and water molds and their relatives.

❑ Differentiate between a plasmodium and a pseudoplasmodium, and apply these terms to the concept of a coenocytic body plan in the "funguslike" protists.

❑ Describe the process of cytoplasmic streaming observed in certain slime molds.

❑ Understand the functions of the following in the life cycles of acellular and cellular slime molds: sclerotium, sporangiophore, sporangium, swarm cells, and myxamoebas.

❑ Discuss the distinguishing characteristics of the following algal groups: dinoflagellates, diatoms, brown algae, red algae, and green algae.

❑ Relate the photosynthetic activity of algae to the total amount of photosynthesis that occurs on Earth.

❑ Identify the kinds of algal protists whose bodies have contributed to various forms of sedimentary rock formations.

❑ Explain the asexual and sexual reproduction of diatoms.

❑ Describe the purpose of the algal structure known as a holdfast, and state how this and other structures have allowed the brown algae to inhabit high-energy habitats.

❑ Discuss variation in photosynthetic pigment ratios in the red algae.

❑ Explain the concept of alternation of generations, and differentiate between isomorphic and heteromorphic forms of these life cycles.

❑ Define and correctly apply the following terms with respect to the reproduction of algae: "gamete," "spore," "zygote," "gametophyte," "sporophyte," and "sporocyte."

❑ Describe the difference between isogamous and anisogamous reproduction in algae.

❑ Differentiate between haplontic and diplontic heteromorphic alternation of generations in the green algae.

❑ Discuss the evolutionary trends observed in the transition from simple photosynthetic protists to the more complex forms of algae.

❑ Understand the general features of the proposed protistan phylogeny presented at the end of this chapter.

PROTISTA AND THE OTHER EUKARYOTIC KINGDOMS • THE ORIGIN OF THE EUKARYOTIC CELL • GENERAL BIOLOGY OF THE PROTISTS (pages 553–558)

Key Concepts

1. The kingdom *Protista* is a grab bag of organisms, defined mainly by exclusion. The general definition for protists is all eukaryotes not included among the plants, fungi, or animals.

2. Protists are mainly unicellular or colonial, although some are multicellular and achieve large sizes. Most are aquatic and include both freshwater and marine species as well as blood parasites. Others live in moist organic matter or damp soil.

3. All the other eukaryotic kingdoms are believed to have evolved from protists. The taxonomic line between each protist group and its related eukaryotic kingdom is often indistinct.

4. There are three important differences between fungi and the protists. Most fungi do not possess flagella, nor do they have visibly different male and female gametes, whereas protists do; and fungal cell walls contain chitin, which is absent from the cell walls of protists.

5. The main difference between algal protists and the plants is reproductive; plant embryos develop inside protective tissues formed by the parent plant, algae do not.

6. Animals are multicellular ingestive heterotrophs possessing an embryonic blastula stage and containing collagen and actin in their extracellular matrix. Protozoans are best characterized as unicellular ingestive heterotrophs that lack the other hallmarks of animals.

7. The evolutionary transition from prokaryotes to eukaryotes was likely a stepwise progession towards increasing cellular complexity, probably starting with the loss of the prokaryotic cell wall. This made possible the infolding of the flexible cell membrane, thus allowing for a substantial increase in cell surface area and the development of a cytoskeleton.

8. As the cell membrane folded inwards, membranous internal organelles developed including the nuclear envelope and digestive vesicles such as lysosomes. Ribosomes then attached to membranes forming the endoplasmic reticulum to yield a simple phagocyte. This hypothesis is supported by the existence today of simple protists called archezoans, which have a nucleus and cytoskeleton but no other membranous organelles.

9. *Endosymbiosis*, the condition in which one organism lives within the cells of another, played a key role in the further advancement of eukaryotes. As the world became more aerobic due to the photosynthetic action of cyanobacteria, the ingestion and incorporation of once free-living prokaryotes into evolving eukaryotic cells made possible organelles such as peroxisomes, mitochondria, and chloroplasts, all of which enhanced the metabolic capacity of eukaryotes.

10. Protists exhibit considerable metabolic diversity. Some are autotrophs, others absorptive heterotrophs, and still others ingestive heterotrophs. Some can switch between autotrophic and heterotrophic modes.

11. Means of locomotion are also diverse among protists, ranging from completely nonmotile to amoeboid movement and sophisticated ciliary or flagellar motion.

12. Freshwater protists have a more negative osmotic potential than their hypotonic environment, and some must therefore constantly pump out water gained through osmosis. They use specialized membranous vesicles called *contractile vacuoles* to cope with excess water. Experiments show that the rate at which the contractile vacuole operates changes according to the osmotic conditions in which the cell is placed.

13. *Food vacuoles* are another type of vesicle found in protists. They form around engulfed food particles; digestion occurs within the vesicle. Smaller vesicles then develop that increase the surface area for absorption of digested food into the cytoplasm.

14. The nature of the protist cell surface varies among taxonomic groups. Some have only a highly flexible plasma membrane, while others have additional coverings such as cell walls and various forms of secreted or accreted shells. Many protists also possess specialized organelles that make them sensitive to environmental stimuli such as light, chemicals, temperature, and touch.

15. Endosymbiosis is quite common among modern protists. Such relationships often, but not always, lead to mutualistic benefits of metabolism and protection for the protist and its partner. In protistan endosymbiosis, the partner is frequently (but not always) a bacterium or another protist.

16. Reproduction takes many forms among the protists. Asexual reproduction can employ binary fission, multiple fission, budding, or spore formation. Sexual reproduction may result in either a haploid or diploid multicellular generation, or both. Some protists practice a form of genetic recombination not directly related to reproduction.

17. Because of the extreme diversity of protists and the difficulties inherent in establishing and learning a meaningful biological taxonomy for this kingdom, your textbook groups protists by virtue of metabolic similarities, habitat type, locomotion, reproduction, and life cycles rather than by strict phyletic associations.

Activities (for answers and explanations, see page 222)

• In the space provided below, construct a simple diagram that depicts the probable evolutionary relationships between the kingdoms Archaebacteria, Eubacteria, Protista, Fungi, Plantae, and Animalia.

- Discuss the importance of the contractile vacuole to freshwater protists, and predict the change in rate of vacuole contraction that would occur after a protist is moved from an isosmotic solution to a hyperosmotic solution.

Questions (for answers and explanations, see page 220)

1. Which one of the following characteristics is *not* present in at least some members of the kingdom Protista?
   a. Prokaryotic cells
   b. Absorptive heterotrophy
   c. Endosymbiosis
   d. Multicellularity
   e. Reproduction by fission or budding

2. The main difference(s) between protozoans and animals is that
   a. animals are prokaryotes and protozoans are eukaryotes.
   b. animals are photoautotrophs and protozoans are chemoautotrophs.
   c. animals are multicellular ingestive heterotrophs and protozoans are unicellular ingestive heterotrophs.
   d. animals never exhibit mutualistic interactions, whereas many protozoans do.
   e. all of the above

3. Which of the following is *true* of endosymbiotic protists?
   a. The endosymbiotic relationship only forms with plants.
   b. The partner in the relationship is typically a bacterium or another protist.
   c. Very few protists engage in such relationships.
   d. Endosymbiotic protists usually reside on the exterior of their partner.
   e. The endosymbiotic relationship only lasts during the partner's breeding season.

4. Suppose you observe a freshwater protist resting in a sample of pond water for five minutes and count the number of times its contractile vacuole expels water from the cell. You then put this same protist into a solution of unknown solute concentration for the same amount of time and count the number of contractions, which is now twice what it was before. Which of the following *best* describes the nature of the second solution?
   a. Pond water
   b. Salt water
   c. Distilled water
   d. A 1:1 mixture of pond water and salt water

5. Which of the following modes of reproduction can be found in at least some protists?
   a. Binary fission
   b. Sexual reproduction
   c. Budding
   d. Spore formation
   e. All of the above

6. During the evolution of eukaryotes from prokaryotes, which of the following did *not* occur?
   a. Infolding of the flexible cell membrane
   b. Loss of the cell wall
   c. Endosymbiosis with once free-living prokaryotes
   d. Formation of new membranous internal organelles
   e. A switch from aerobic to anaerobic metabolism

PROTOZOAN DIVERSITY (pages 558–565)

Key Concepts

1. Protozoans are unicellular protists most of which ingest their food by endocytosis.

2. Thousands of protist species (spanning several phyla) possess at least one flagellum, and hence have the common name flagellates. They reproduce asexually by means of mitosis and cytokinesis.

3. Some flagellates are free-living predators that eat other protists, while others are specialized as internal parasites of animals such as insects and humans.

4. The anatomical structure of the flagellate group Choanoflagellida resembles a characteristic cell type found in sponges. The Choanoflagellida are therefore thought to be close evolutionary relatives of this evolutionarily ancient animal group.

5. Some flagellates are important human pathogens. A few are carried by insect vectors, including members of the genus *Trypanosoma*, which cause a lethal sleeping sickness in Africa. Others (e.g., *Trichomonas vaginalis*) are transmitted by sexual contact and cause venereal disease.

6. The *euglenoids* are flagellates, some of which are capable of photosynthesis. A typical representative is the genus *Euglena*. These common freshwater protists have a complex body plan and two flagella, and reproduce asexually by mitosis and cytokinesis. They are nutritionally flexible, being able to survive on photosynthetic activity when in the light and heterotrophy when in the dark.

7. The *amoebas* and their relatives are protozoans noted for the ability to form flexible extensions of their body mass, called pseudopods, which serve both in locomotion and feeding.

8. Amoebas are animallike and exist as predators, parasites, or scavengers. None are photosynthetic. Amoebas are not primitive organisms; their apparent simplicity is secondarily derived. They assimilate food by phagocytosis after surrounding it with pseudopods. Many are aquatic and highly adapted for life as bottom-dwellers, where a rich food supply is present.

9. The *actinopods* are all aquatic and have very thin pseudopods reinforced with microtubules. These structures greatly increase cell surface area, serve as feeding organs, help keep the organism afloat, and provide locomotion.

10. *Radiolarians* are a group of drifting marine actinopods. Their silica-based shells are often quite large and elaborate. Over hundreds of millions of years these shells have accumulated to form significant sedimentary deposits in tropical oceans. *Heliozoans* are freshwater actinopods; they differ mainly from the radiolarians by lacking an endoskeleton.

11. The *foraminiferans* consist of drifting marine protozoans with porous calcium carbonate shells. Threadlike nets of sticky pseudopods emerge from these pores and are used to capture planktonic food items.

12. Foraminiferans shed their old shells after mitotic reproduction. Massive numbers of discarded shells have contributed over millions of years to limestone deposits on the ocean floor and to sandy beaches. Because each geological period has had distinctive foraminiferan species, they are often used in dating sedimentary rock formations.

13. *Apicomplexan* protozoans are so named because they produce spores whose apical end contains many organelles. They are exclusively parasitic and often have complex life cycles. All have an amoeboid body form and lack a contractile vacuole. Osmotic cell expansion is limited by rigid cell walls.

14. Human malarial parasites of the genus *Plasmodium* are apicomplexans. Their life cycles involve multiple hosts (humans and female *Anopheles* mosquitoes), through which they cycle during different developmental stages.

15. *Plasmodium* parasites enter humans through the bite of an infected female mosquito. The parasites take up residence in the liver and lymph system, and then change form before reentering the bloodstream to attack red blood cells. When the human is bitten by an uninfected mosquito, the parasites are transferred from the human back to the insect and the life cycle is completed. This cycle is best interrupted by eliminating breeding sites for the mosquito host.

16. The *ciliates* are named for the presence of hairlike cilia on their surface. They represent a diverse and ecologically important group of protozoans.

17. Ciliates are all heterotrophic and have highly specialized body forms. Two types of nuclei are found in the ciliate cell: a large *macronucleus* whose DNA controls cell function, and up to 80 *micronuclei* that play an essential role in genetic recombination.

18. *Paramecium* is a representative ciliate genus. They possess a *pellicle*, which is an elaborate cell covering made from the outer plasma membrane and many associated alveoli, through which individual cilia protrude. The pellicle also contains needlelike defensive organelles called trichocysts, which can be rapidly expelled.

19. Although some of the cilia in paramecia are sensory in function, most aid in locomotion. The beating motion of these cilia can be directed to provide precise locomotory control, allowing the organism to move forward or backward and thus alter its direction in response to changing environmental conditions.

20. Reproduction in paramecia is usually by simple cell division. However, they also engage in an elaborate, nonreproductive sexual behavior called *conjugation*, where two cells reciprocally exchange genetic material derived from their micronuclei. After conjugation, the genetically recharged individuals separate and later undergo asexual cell division.

21. Some ciliates (along with other complex protists) exhibit such extreme specialization that they almost appear to function like multicellular animals. Specialized ciliate organelles include *cirri* (fused cilia) which permit highly coordinated walking movements, almost like animal appendages. *Myonemes* act like muscles within the cytoplasm to cause rapid movement of the entire cell or specific cell structures.

22. Even more complex cytoplasmic structures exist in some highly specialized ciliates that inhabit the digestive tracts of hoofed mammals. These structures resemble a skeleton, mouth, esophagus, anus, and gut, and demonstrate the degree of complexity possible within a single cell.

Activities (for answers and explanations, see page 222)

- Correctly associate the letters of each of the ciliate structures and functions listed below with the unlabeled lines on the following diagram of *Paramecium*.

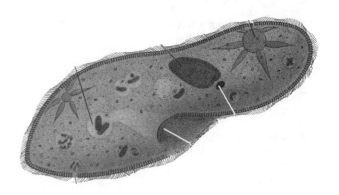

Structures	Functions
a. Food vacuole	*f.* Regulates water balance of cell
b. Pellicle	*g.* Nutrient digestion
c. Macronucleus	*h.* Cell covering to which cilia attach
d. Contractile vacuole	*i.* Defensive structures
e. Trichocysts	*j.* Contains DNA that controls cell functions

- Besides flagella, what locomotory-related structures are found in protozoans?

Questions (for answers and explanations, see page 222)

1. Which one of the following statements about protozoans is *false*?
   a. Most protozoans are capable of ingesting food by
      endocytosis.
   b. Apicomplexans are the only protozoan group without
      parasitic representatives.
   c. Foraminiferans and radiolarians are shelled proto-
      zoans.
   d. Some ciliates have organelles that permit coordinated
      "walking" movements.
   e. Although they appear structurally simple, amoebas
      are not primitive organisms.

2. All of the following are flagellated protists *except*
   a. *Euglena.*
   b. *Trypanosoma.*
   c. *Trichomonas.*
   d. Choanoflagellida.
   e. *Plasmodium.*

3. Foraminiferans, heliozoans, and radiolarians all share
   which of the following characteristics?
   a. They are photosynthetic.
   b. They use pseudopods in locomotion or feeding.
   c. They have calcium carbonate shells.
   d. They are parasites.
   e. They are flagellates.

4. Trichocysts are defensive structures found in the protozoan group known as the
   a. flagellates.
   b. amoebas.
   c. radiolarians.
   d. ciliates.
   e. all of the above

5. Microscopic examination of a ciliate could conceivably
   reveal all but one of the following structures. Select the
   exception.
   a. Pseudopod
   b. Cirrus
   c. Trichocyst
   d. Pellicle
   e. Myoneme

6. Although not closely related, the apicomplexans share a
   similar body form with the
   a. euglenoids.
   b. ciliates.
   c. flagellates.
   d. amoebas.
   e. none of the above

PROTISTS WERE ONCE CONSIDERED TO BE FUNGI
(pages 565–567)

Key Concepts

1. Unlike most true fungi, all "funguslike" protists possess
   flagella during some stage of their life cycle. They also
   lack the dikaryotic stage typical of some true fungi and
   have no chitin in their cell walls.

2. The phylum Myxomycota includes the *acellular slime
   molds*, while the phyla Dictyostelida and Acrasida are
   home to the *cellular slime molds*. At one time these superficially similar groups were lumped into a single phylum. In all three phyla, motile independent cells undergo striking organizational changes and aggregate to
   form fruiting bodies, which then produce spores. For
   ease of movement, slime molds prefer moist, cool habitats like forests.

3. The vegetative state of an acellular slime mold is called
   the *plasmodium*, a wall-less mass of cytoplasmic strands
   containing numerous diploid nuclei. This exemplifies
   the *coenocyte* body plan, in which many nuclei are surrounded by a single plasma membrane.

4. Acellular slime molds move by means of *cytoplasmic
   streaming*. In this type of locomotion one region of the
   outer cytoplasm becomes more fluid, causing the plasmodium to bulge in that area. Contractile proteins and
   microfilaments help control these streaming movements. Overall growth of the plasmodium can be almost
   indefinite under favorable conditions of food, moisture,
   and pH.

5. If conditions become unsuitable for continued growth,
   the plasmodium may either form a resistant structure
   called a *sclerotium* (which can revert back to a plasmodium upon the return of favorable conditions), or it may
   transform into a branched, spore-producing fruiting
   body called a *sporangiophore*.

6. Sporangiophores emerge from piled masses of the plasmodium. Diploid nuclei within each fruiting body
   divide meiotically and collect within *sporangia* located at
   the tip of the sporangiophore. The haploid nuclei then
   become surrounded by cell walls to form resistant
   spores.

7. These haploid spores are eventually shed and germinate, at which time they become flagellated *swarm cells*,
   which function as gametes. These can either assume a
   resting state (cysts) or fuse to create a diploid zygote
   that divides mitotically without wall formation to
   become a new, coenocytic plasmodium.

8. Unlike the acellular slime molds, the vegetative state of
   cellular slime molds is amoeboid. These *myxamoebas*,
   each containing a single haploid nucleus, engulf food
   particles and can exist indefinitely under favorable conditions. They reproduce asexually by mitosis and fission.

9. When conditions become unfavorable, Dictyostelid myxamoebas aggregate to form fruiting bodies in response to
   cAMP signals. The independent cells gather into a structure called a *pseudoplasmodium*, from which the fruiting
   bodies emerge, but each myxamoeba retains its individual identity. A pseudoplasmodium is therefore not coenocytic.

10. Haploid spores form within thick-walled cells on these
    fruiting bodies. The spores then germinate and produce
    new myxamoebas to complete this asexual reproductive
    cycle. Sexual reproduction also occurs in cellular slime
    molds. This process involves the direct fusion of two
    myxamoebas into a structure that later germinates to
    form new myxamoebas.

11. The phylum Oomycota is commonly known as the *water molds*. As their name suggests, water molds are aquatic. They are saprobic and are often found feeding on dead animals. Water molds are coenocytic, have flagellated reproductive cells, are diploid throughout most of the life cycle, and have cellulose cell walls.

12. Terrestrial species of Oomycota include the *downy mildews*. One terrestrial species, *Phytophthora infestans*, causes late potato blight, the disease that was responsible for the Irish potato famine of the mid-1840s.

Activity *(for answer and explanation, see page 222)*

• What do biologists mean when they refer to a "funguslike" protist as being coenocytic? How does the vegetative state of an acellular slime mold (the plasmodium) exemplify this concept? Why is the pseudoplasmodium of a cellular slime mold not considered coenocytic?

Questions *(for answers and explanations, see page 222)*

1. A major difference between the *vegetative* states of acellular and cellular slime molds is that
   a. acellular slime molds have haploid nuclei while cellular slime molds are diploid.
   b. acellular slime molds produce fruiting bodies and cellular slime molds do not.
   c. acellular slime molds use cAMP as a chemical cue for aggregation while cellular slime molds do not.
   d. acellular slime molds form a large, coenocytic mass of cytoplasm, whereas cellular slime molds usually exist as individual myxamoebas.

2. Which of the following is found in fungi but not in "funguslike" protists?
   a. Nuclei
   b. Plasma membranes
   c. Chitinous cell walls
   d. Membranous internal organelles
   e. All of the above

3. Which of the following are *not* members of the protist phylum Oomycota?
   a. Downy mildews
   b. Cellular slime molds
   c. Water molds
   d. *Phytophthora infestans*

4. All but one of the following processes or events occurs in at least one representative of the "funguslike" protists. Select the *exception*.
   a. Engulfing particulate foods
   b. Aggregation of cells in response to cAMP signals
   c. Sporangiophore development from the plasmodium stage

   d. Movement by means of cytoplasmic streaming
   e. Sexual reproduction by means of conjugation

5. Which of the following characteristics or structures of the "funguslike" protists is *incorrectly* associated with the group in which it occurs?
   a. Myxamoebas..................................cellular slime molds
   b. Sclerotium.....................................acellular slime molds
   c. Multiple diploid nuclei.................cellular slime molds
   d. Pseudoplasmodium.......................cellular slime molds
   e. Coenocytic.....................................water molds

6. In which of the following would you expect to find swarm cells but *not* myxamoebas?
   a. Acellular slime molds
   b. Cellular slime molds
   c. Water molds
   d. Downy mildews
   e. None of the above

THE ALGAE • PHYLOGENY OF THE PROTISTS
(pages 567–577)

Key Concepts

1. All algae are eukaryotic and photosynthetic. As a group, they are responsible for over half the total photosynthesis that occurs on Earth.

2. Unlike plants, no alga produces a multicellular embryo in a protective structure.

3. Extreme diversity is the rule for algal growth forms, which can be unicellular and relatively simple, or multicellular and very complex.

4. Algal life cycles are also highly varied. However, all phyla except for the red algae exhibit flagellated motile cells during at least one stage in the life cycle.

5. Biochemical characteristics are often used when classifying algae. The specific kinds of polysaccharide compounds used for storage, particular pigment combinations, and the substances found in cell walls are all important in this regard.

6. The phylum Pyrrophyta is dominated by the *dinoflagellates*, unicellular algae whose unique mixture of accessory and photosynthetic pigments give them a golden-brown color. They are of enormous importance in oceanic photosynthesis, and are common endosymbionts with corals.

7. Dinoflagellates have a peculiar anatomy, with two prominent grooves that run along the cell surface at right angles to one another. Each groove contains a flagellum that helps propel the alga through the water.

8. Enormous population explosions of dinoflagellates (e.g. *Gonyaulax*) cause "red tides," with a resultant accumulation of potent nerve toxins that may kill fish and other vertebrates, or accumulate in shellfish. Many species are also bioluminescent.

9. The phylum Chrysophyta contains the *diatoms* and their relatives. Most are single-celled and contain carotenoid pigments that give them a yellow or brownish color. Their principal storage products are the carbohydrate chrysolaminarin and several oils.

10. Most diatoms have radially symmetric silicon shells that fit together like the two halves of a box. Each species' shell is so distinctive that it is the only taxonomic character used to identify diatoms.

11. When reproducing asexually, each half of the diatom shell serves as the top half of the new organism. To avoid the ever-decreasing cell size that would result if this process continued indefinitely, sexual reproduction takes place. Gametes form, shed their cell walls, and fuse to make a zygote, which then grows substantially before a new cell wall is produced.

12. Like other marine protists with mineralized shells, diatoms have contributed to massive oceanic sedimentary rock formations. Diatomaceous earth, obtained from such rocks, has industrial uses for insulation, filtration, metal polishing, and as an insecticide.

13. *Brown algae* make up the phylum Phaeophyta. They earn their name from the combination of a yellow-brown carotenoid pigment called fucoxanthin and the green of chlorophyll. They are always multicellular and have branched filaments or leaflike growths called *thalli*.

14. Phaeophyta includes relatively small organisms such as *Ectocarpus* as well as the giant kelps of the genus *Macrocystis*, the largest of all algal species. Some are free-floating on the ocean's surface (e.g. *Sargassum*), while others are glued to the substrate by means of *holdfasts*. These holdfasts, coupled with specialized body forms, allow some brown algae to survive in high-energy habitats such as coastal intertidal zones.

15. Considerable tissue differentiation is observed in the larger brown algae. In addition to stalks and blades that are analogous to the stems and leaves of plants, food-conducting filaments resembling those of plants are also present in some species of brown algae.

16. The cell walls of brown algae contain large amounts of a sticky sugar polymer called *alginic acid.* In nature it holds cells together and provides the glue used by holdfasts; commercially it is used as an emulsifier in such products as ice cream and cosmetics.

17. *Alternation of generations* occurs in the life cycle of many algae and yields both multicellular diploid (2*n*) and multicellular haploid (*n*) stages. Mature diploid individuals produce spores, while mature haploid individuals produce gametes. If the stages are morphologically and genetically different from one another, the system is called *heteromorphic alternation of generations*. When the stages do not differ morphologically (even though they are genetically different), the system is known as *isomorphic alternation of generations.*

18. In multicellular plants, fungi, and protists, gametes are generally not produced directly by meiosis, as they are in animals. Instead, the multicellular haploid *gametophyte* generation makes gametes by mitosis and cytokinesis.

19. Fusion of the gametes produces a diploid *zygote*, which then undergoes mitotic divisions to produce the diploid *sporophyte* generation. Sporophytes produce haploid spores through meiosis, which occurs in specialized cells called *sporocytes*.

20. Both isomorphic and heteromorphic alternation of generations are seen in the brown algae. In the isomorphic forms, sporophyte and gametophyte are morphologically indistinguishable; an example is the brown alga *Ectocarpus*. In the heteromorphic species (e.g. *Laminaria*), distinct anatomical differences are evident between stages in the life cycle. In *Fucus* these differences are taken to the extreme, with the multicellular haploid phase being eliminated altogether.

21. Members of the phylum Rhodophyta, the *red algae*, are mostly multicellular and contain substantial amounts of accessory pigments phycoerythrin and phycocyanin, in addition to other accessory pigments and chlorophyll.

22. The quantity of phycoerythrin in a red alga may vary according to environmental light conditions. While the amount of chlorophyll is the same in both surface and deeper-water species, the quantity of red pigment is reduced if the alga grows near the surface. In deeper water, phycoerythrin accumulates in the alga. The "redness" of the alga thus depends on the ratio of chlorophyll to phycoerythrin.

23. The use of floridean starch as a photosynthetic storage product, and the lack of motile flagellated cells in any stage of the life cycle, are characteristics unique to the red algae. Agar, the laboratory substance used in cell and tissue cultures, is derived from a mucilaginous polysaccharide made by certain red algae.

24. Some red algae can enhance the formation of coral reefs by secreting calcium carbonate in and around their cell walls. Marine red algae have also been found to parasitize other red algae by injecting their nuclei into the host cell cytoplasm. The host's metabolism is then diverted to express the parasite's genes.

25. The Chlorophyta, or *green algae*, constitute the remaining phylum of plantlike protists. They are the only protists with the full complement of photosynthetic pigments characteristic of true plants, with which they form a monophyletic lineage. Chlorophylls *a* and *b* dominate, along with carotenoids and xanthophylls. Starch is the principle photosynthetic storage product.

26. Virtually every imaginable shape and body construction is seen within the green algae, from simple unicellular to colonial to advanced multicellular. Both uninucleate and multinucleate forms are known. Life cycles are also diverse and sometimes complex.

27. In sea lettuce (*Ulva lactuca*), an isomorphic life cycle is observed. The male and female gametes produced by the haploid gametophyte are the same size (i.e., they are *isogamous*).

28. Other green algae have heteromorphic life cycles. One variation on this theme is the *haplontic* life cycle (e.g. *Ulothrix*), in which the only diploid stage is the zygote itself (i.e., there is no multicellular diploid sporophyte). Zygotes act as sporocytes and directly undergo meiosis to produce spores.

29. Still other green algae are *diplontic* (like animals) and show no multicellular haploid stage. In these algae, meiosis produces gametes directly.

30. Important evolutionary patterns in the algae include a trend towards multicellularity along with a trend from isogamy towards anisogamy. Oogamy is an extreme example of the latter shown by the giant green alga *Oedogonium*. Here the female gamete is large and totally immobile, while the male gametes are small, free-swimming, and flagellated.

31. While admittedly complex, the phylogeny of the protists is becoming clearer as promising new molecular taxonomy techniques are developed. One possible result of this analysis may be that the concept of one protistan kingdom will eventually be replaced by an emphasis on the individual lineages present within this diverse and interesting assemblage of organisms.

Activities (for answers and explanations, see page 223)

- The algae encompass an enormous range of sizes, structural and metabolic complexities, and ecological lifestyles. List the characteristics that unify this otherwise diverse assemblage of organisms.

- Explain why, from a structural point of view, sexual reproduction is periodically necessary in diatoms.

- Review Figure 26.29 in your textbook (life cycle of *Ulva*) and then explain why this life cycle is considered to be isomorphic alternation of generations involving isogamy.

Questions (for answers and explanations, see page 223)

1. Select all of the following algal phyla that show *both* isomorphic and heteromorphic alternation of generations.
   a. Chlorophyta
   b. Rhodophyta
   c. Phaeophyta
   d. Chrysophyta
   e. Pyrrophyta

2. Which statement about algae is *true?*
   a. Brown algae regulate the manufacture of alginic acid depending on variations in wavelength of light.
   b. The quantity of phycoerythrin found in red algae can vary depending on light conditions.
   c. The quantity of chlorophyll *a* found in brown algae can vary depending on light conditions.
   d. Red algae regulate the deposition rate of floridean starch depending on light conditions.
   e. Diatoms make new silica shells as a function of light conditions.

3. Match each algal characteristic in the second column with an appropriate type of algae from the first column.

   ___a___ Red algae   *a.* Large amounts of phycoerythrin present

   ___f___ Green algae   *b.* Made of two half-shells fitted tightly together

   ___d,e___ Brown algae *c.* Common endosymbionts with corals

   ___b___ Diatoms   *d.* Cell walls contain alginic acid

   ___c___ Dinoflagellates *e.* Often glued to substrate via holdfasts

                          f. Chlorophyll *a* the dominant pigment

4. Red tides often produce deadly toxins that result in major fish kills. This phenomenon results from enormous population increases by species from which of the following?
   a. Green algae
   b. Red algae
   c. Brown algae
   d. Diatoms
   e. Dinoflagellates

5. Which of the following molecules, structures, or characteristics can be found in *all* representatives of the Chlorophyta?
   a. Chlorophyll
   b. Anisogamy
   c. Haplontic life cycle
   d. Diplontic life cycle
   e. Oogamy

6. Holdfasts and alginic acid are structures or substances characteristic of which algal phylum?
   a. Chlorophyta
   b. Rhodophyta
   c. Phaeophyta
   d. Chrysophyta
   e. Pyrrophyta

*Integrative Questions (for answers and explanations,
    see page 223)*

1. Your biology professor rushes into the classroom one
    day and announces that she has discovered a previously
    unknown organism, which she claims is a member of
    the kingdom Protista. Which of the following sets of
    characteristics would unequivocally associate this new
    creature with the other protists?
    a. Contains chlorophyll *a*, multicellular, photosynthetic,
    flagellated
    b. Diploid, multicellular, eukaryotic, switches between
    heterotrophy and autotrophy
    c. Eukaryotic, shows traits that exclude it from animal,
    plant, or fungal kingdoms
    d. Prokaryotic, possesses cilia, heterotrophic ingestive
    feeding
    e. all of the above

2. Consider the list of biological characteristics shown
    below. During a lifetime of studying all kinds of pro-
    tists, Dr. Amoeba says that the only one of these charac-
    teristics that he has *never* observed in protists is
    a. mitosis.
    b. nitrogen fixation.
    c. endosymbiosis.
    d. blood parasitism.
    e. autotrophy.

3. A protist exhibiting heteromorphic alternation of gener-
    ations with a haplontic life cycle would *not* have
    a. a multicellular haploid stage.
    b. a multicellular diploid stage.
    c. any gametes.
    d. any spores
    e. all of the above

4. A water mold is discovered growing on the surface of a
    freshwater green alga. Which two phyla of protists are
    involved in this situation?
    a. Myxomycota and Rhodophyta
    b. Foraminifera and Chrysophyta
    c. Apicomplexa and Pyrrophyta
    d. Actinopoda and Phaeophyta
    e. Oomycota and Chlorophyta

5. Which of the following statements about the kingdom
    Protista is *false*?
    a. Some protists locomote by means of ciliary or flagel-
    lar action.
    b. The sporophyte stage in some protists produces
    spores by mitosis← meiosis.
    c. The sporophyte stage in some protists is formed by
    mitotic division of the zygote.
    d. Some protists switch between autotrophic and het-
    erotrophic modes of metabolism.
    e. Both isomorphic and heteromorphic alternation of
    generations is known among the protists.

27

Plants: Pioneers of the Terrestrial Environment

CHAPTER LEARNING OBJECTIVES—after studying this chapter you should be able to:

❑ List the basic taxonomic characteristics of the kingdom Plantae, and explain how plants differ from other organisms.

❑ Apply the concept of alternation of generations to plant life cycles and modes of reproduction.

❑ Correctly use the terms "gametophyte," "gamete," "antheridium," "archegonium," "sporophyte," "sporangia," "spore," "meiois," "mitosis," and "fertilization" when discussing plant life cycles.

❑ Distinguish between the nontracheophytes and the tracheophytes, and describe the anatomical differences between them.

❑ Explain current ideas regarding the evolutionary origins of the plant kingdom and the sequence in which nontracheophytes and tracheophytes appeared on Earth.

❑ Understand the basic structure of plant vascular systems, and discuss the specific roles of xylem and phloem.

❑ Describe the basic characteristics of the nontracheophyte plants, and name the main phyla within this group.

❑ Explain the various stages in the moss life cycle and be able to use the terms "protonema," "rhizoid," "capsule," and "calyptra."

❑ Discuss the importance of tracheids in the evolution and function of tracheophyte plants.

❑ Distinguish between the nonseed and seed-bearing tracheophyte plants, and name the major phyla within each group.

❑ Describe the evolution of true roots and leaves from the simple ancestor *Rhynia* to the modern tracheophyte plants.

❑ Explain the difference between homospory and heterospory, and discuss the roles of megasporangia and microsporangia in these two processes.

❑ Discuss the fern life cycle, and explain how it differs from the moss life cycle.

❑ Differentiate between the gymnosperms and the angiosperms, and describe the importance of pollen in the reproduction of these plants.

❑ Name the four phyla of gymnosperms, and describe in detail the life cycle of a typical conifer.

❑ Discuss the origin and development of gymnosperm seeds using the terms "integument," "ovule," and "micropyle."

❑ Cite the main anatomical and reproductive characteristics of the angiosperms, and discuss the importance of the flower in angiosperm reproduction.

❑ Describe the parts of a flower, and distinguish between perfect and imperfect flowers.

❑ Understand the difference between a monoecious and a dioecious angiosperm.

❑ Explain the importance of pollination and fruit development in angiosperm reproduction.

❑ Describe the basic anatomical differences between monocot and dicot angiosperms, and cite examples of each.

❑ List the main trends in plant evolution from the nontracheophytes through the angiosperms.

THE PLANT KINGDOM (pages 582–585)

Key Concepts

1. The ancestors of land plants left the water nearly 500 million years ago and underwent rapid evolutionary change. Modern plants became established during the last 100 million years.

2. Plants are defined as multicellular, photosynthetic eukaryotes whose embryos develop within protective parental tissues. Their cell walls are made of cellulose, they have chloroplasts containing the pigments chlorophyll *a* and *b* as well as specific carotenoids, their storage carbohydrate is starch, and their life cycles have many commonalities.

3. Alternation of generations is a conspicuous feature of plant life cycles. A haploid spore develops by mitosis into the haploid multicellular plant, or *gametophyte*. This stage of the life cycle produces haploid gametes by mitosis.

4. Plant gametes fuse in the process of *syngamy* (fertilization) to form a diploid zygote, which then undergoes mitosis to form the multicellular diploid plant, or *sporophyte*. The sporophyte completes the plant life cycle by undergoing meiosis within specialized structures called *sporangia* to produce haploid spores.

5. The gametophyte and sporophyte stages of plants are always morphologically distinct. That is, plants exhibit heteromorphic alternation of generations.

6. There are twelve extant phyla of plants. The *nontracheophyte* plants include three phyla that lack, or have only rudimentary, internal tissues for water and nutrient transport. The remaining nine phyla comprise the *tracheophyte* plants, which all have well-developed transport tissues.

7. The earliest plant ancestors successfully made the transition from a purely aquatic existence to life on land. Becoming terrestrial required that plants overcome problems of structural support, desiccation, water and mineral transport, and reproduction.

8. True plants likely arose from the protist phylum Chlorophyta (green algae). Evidence includes similar pigment composition and thylakoid membrane anatomy, starch as the storage carbohydrate, and cell walls containing cellulose. Some green algae also have plant-like oogamous reproduction featuring a large stationary egg. Recent rRNA sequencing work corroborates this anatomical evidence.

9. Two groups of modern green algae, the stoneworts and members of the genus *Coleochaete*, are currently the best candidates for the closest plant relatives. Whatever the true ancestry, it is likely that ancestral algae formed a green mat along the margins of ponds or marshes, setting the stage for the emergence of plants on land.

10. The first nontracheophyte plants evolved from the green alga ancestor. One group of early nontracheophytes then evolved into the modern nontracheophytes, and another into the tracheophytes.

11. Nontracheophytes (the mosses, liverworts, and hornworts) have little or no water-transporting tissue; they therefore do not attain a large body size and often occur in dense masses. Water moves through this mass by capillary action, or else is splashed over the plants by environmental action. Minerals are distributed by simple diffusion.

12. Ferns, conifers, and flowering plants are tracheophytes. They have well-developed *vascular systems* with specialized conducting tissues: *xylem* (used for water and mineral transport) and *phloem* (used for transporting photosynthetic products).

13. Ironically, tracheophytes actually appear first in the fossil record, predating the nontracheophytes by at least 60 million years. This does not mean, however, that tracheophytes necessarily evolved before nontracheophytes. The discrepancy in fossil records likely results from the fact that tracheophytes are more likely to form fossils because of their anatomy and chemical composition.

14. Plants are distinguished from the algae largely on the basis of adaptations for life on land. These include protective coverings to prevent drying, specialized structures such as roots and rhizoids to gather water, support mechanisms that utilize turgor pressure, woody stems or thickened cell walls, and protection for the developing sporophyte embryo within parentally-derived tissues.

Questions (for answers and explanations, see page 223)

1. In plants, the process of mitosis leads to the formation of
   a. gametes.
   b. the sporophyte plant.
   c. the gametophyte plant.
   d. spores.
   e. More than one answer is correct.

2. A plant specimen is brought to you for examination. You notice that it has eukaryotic cells whose chloroplasts contain both chlorophyll *a* and *b*. Careful analysis of its cell nuclei indicate that all cells are haploid. Based on this information, the most likely group to which this specimen belongs is the
   a. mosses.
   b. ferns.
   c. green algae.
   d. conifers.
   e. Cannot determine from information given

3. A moss possesses all but one of the following plant characteristics. Select the *exception*.
   a. Spores
   b. Xylem and phloem
   c. Gametophyte tissue
   d. Cellulose in cell walls
   e. Chlorophyll *a* in chloroplasts

4. While searching through a collection of 400-million-year-old fossils you find one that is clearly a tracheophyte plant. However, no nontracheophyte fossils are present from this same time period. Which of the following is the best explanation for this observation?
   a. Nontracheophytes had not yet evolved 400 million years ago.
   b. Nontracheophytes did not coexist with tracheophytes.
   c. Tracheophytes fossilized better than nontracheophytes.
   d. Nontracheophytes were present at this time, but just not widely distributed.
   e. None of the above

5. The green alga *Ulva* has all but one of the following characteristics in common with a flowering plant. Select the *exception*.
   a. Similar photosynthetic pigments
   b. Cellulose cell walls
   c. Starch as a storage carbohydrate
   d. Heteromorphic alternation of generations
   e. Genetically different sporophytes and gametophytes

6. Which of the following plant groups lacks well-developed vascular tissue?
   a. Ferns
   b. Mosses
   c. Conifers
   d. Flowering plants
   e. None of the above

Activities (for answers and explanations, see page 223)

- List the similarities in structure and chemical composition that suggest an evolutionary link between the Chlorophyta and the true plants.

- On the following generalized diagram of a plant life cycle, identify the major morphological stages in the open boxes, and name the cell division process indicated by each arrow.

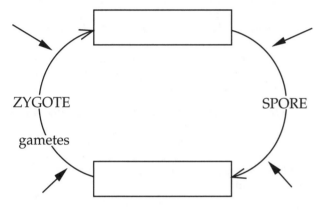

ZYGOTE   SPORE

gametes

NONTRACHEOPHYTES: LIVERWORTS, HORNWORTS, AND MOSSES   (pages 585–587)

Key Concepts

1. Nontracheophyte plants are widely distributed and highly successful. They are typically found growing as dense mats in moist habitats. Some grow in the soil, some on rocks or buildings, and some even grow on other plants. They differ from tracheophytes in that the conspicuous green plant that we recognize is the gametophyte generation, rather than the sporophyte.

2. Nontracheophytes are never large, mainly because they lack internal transport systems for water and minerals. Their size is also limited by reproductive needs: sperm cells must swim through rainwater or dew to fertilize the stationary egg.

3. Like other plants, nontracheophytes possess a waxy cuticle that helps retard water loss. The developing embryo is also protected within layers of maternal tissue. Specific morphology of the nontracheophytes is dictated by habitat features, particularly water availability. The gametophyte stage is especially variable, while the sporophyte stage is less so and thus more diagnostic of each species.

4. Nontracheophytes have heteromorphic generations that are nutritionally distinct. The gametophyte stage is photosynthetic and therefore nutritionally independent. The sporophyte usually does not photosynthesize and is always dependent upon the gametophyte for its development and nutrition.

5. During reproduction, haploid spores divide by mitosis to give rise to the multicellular gametophyte, within which specialized sex organs develop. The *archegonium* produces a single female gamete (egg), while each *antheridium* produces many flagellated male gametes (sperm). Since swimming sperm are involved, water is essential for fertilization to occur.

6. Once a diploid zygote is formed, mitotic divisions yield a multicellular sporophyte embryo that remains protected within the parental gametophyte tissue. The embryo eventually elongates and becomes a mature sporophyte, producing a *capsule* at its tip. Inside the capsule haploid spores develop by meiosis. The spores are released under appropriate conditions, germinate, and then mitotically divide to produce another gametophyte, completing the life cycle.

7. The nontracheophytes are subdivided into three phyla: *Hepatophyta* (liverworts), *Anthocerophyta* (hornworts), and *Bryophyta* (mosses).

8. Liverworts often have flattened, lobed gametophytes, although this morphology is not universal. Sex organs are produced on top of these flat plates, and rhizoids below. The sporophyte stalk rises well above the gametophyte to elevate the capsule for spore dispersal. Specialized coil structures called *elaters* help to release the mature spores.

9. *Marchantia* is a well-known genus of liverworts that reproduces vegetatively by forming *gemmae*, loose clumps of cells capable of developing into a whole new gametophyte plant. *Marchantia* can also reproduce asexually by simple fragmentation of the gametophyte.

10. The hornworts (so called because their sporophytes resemble horns) are structurally similar to liverworts. However, they differ from the liverworts and mosses in two important ways: their archegonia are embedded in gametophyte tissue, and their sporophytes are capable of almost indefinite growth, making them the tallest (up to 20 cm) known nontracheophytes. Hornworts sometimes harbor cyanobacteria that fix atmospheric nitrogen for use by the plant.

11. Mosses are the most diverse nontracheophytes. There are more species of mosses than liverworts and hornworts combined, and they can be found in almost every terrestrial habitat.

12. Moss spores germinate and mitotically develop into a filamentous structure called a *protonema*. The protonema then differentiates to produce nonphotosynthetic filaments called *rhizoids* that attach to the substrate. Differentiation of the protonema also produces the branched photosynthetic filaments ("leafy" aerial portions of the gametophyte) that we recognize as the typical moss.

13. Eggs and sperm develop in the archegonia and antheridia, respectively, located at the tips of the gametophyte. The flagellated sperm require water to swim to the egg. After sperm and egg unite, the resulting zygote mitotically develops into the elongated sporophyte,

which unlike the hornworts, grows from its apical end (as do the tracheophytes). A spore capsule eventually develops at the tip of the sporophyte within which meiosis occurs to produce spores. Inside the pointed cap of the capsule, are a set of humidity-sensitive teeth that facilitate spore dispersal during conditions favorable for germination.

14. A few species of mosses depart from the standard method of Bryophyte spore dispersal. For example, members of the genus *Sphagnum*, which account for much of the plant biomass found in tundra and arctic environments, have pressure-sensitive spore chambers that explosively release their spores.

15. A specialized cell type exists in many mosses that may represent an evolutionary precursor of some tracheophyte vascular tissue. When cells called hydroids die in a moss, they leave a hollow channel through which water can flow to other portions of the plant body, thus providing a limited method of water transport.

Activities (for answers and explanations, see page 223)

- Match the following nontracheophyte characteristics with the appropriate stage of the life cycle (i.e., sporophyte, gametophyte, both, or neither) in which they occur.

  _____ Sperm and eggs

  _____ Phloem

  _____ Chloroplasts

  _____ Waxy cuticle

  _____ Capsule

  _____ Archegonium

  _____ Diploid tissue

- Refer to Figure 27.8a in your textbook, which depicts a typical moss life cycle. Relate the relative percentage of this life cycle occupied by haploid and diploid tissues to the relative interdependence of the gametophyte and sporophyte generations in nontracheophytes.

Questions (for answers and explanations, see page 223)

1. Which of the following factors does *not* either limit the size to which nontracheophytes can grow, or the type of habitat where they are found?
   a. Swimming sperm
   b. Lack of vascular tissue
   c. Possession of a diploid sporophyte stage
   d. Little specialized supporting tissue
   e. None of the above

2. Which of the following characteristics do liverworts and hornworts have in common?
   a. Well-developed vascular systems
   b. Possession of a waxy cuticle
   c. Nutritionally independent sporophyte stages
   d. Diploid spores
   e. Gametes that form through meiosis

3. In which of the following nontracheophyte structures would you expect growth to occur at the base?
   a. Gametophyte of a moss
   b. Spore capsule of a liverwort
   c. Sporophyte of a moss
   d. Sporophyte of a hornwort
   e. Archegonium of a liverwort

4. Antheridia, archegonia, and haploid tissue cells all share which of the following?
   a. They do not occur at any stage in the life cycle of nontracheophyte plants.
   b. Each is directly involved in the production of liverwort spores.
   c. Hornworts have them, but mosses do not.
   d. All are found in the sporophyte stage of a hornwort.
   e. All are found in the gametophyte stage of a moss.

5. The easiest way to tell the difference between a hornwort and a moss is by examining
   a. the structure of the sporophyte.
   b. the structure of the gametophyte.
   c. the presence or absence of vascular tissue.
   d. Both *a* and *b*
   e. Both *b* and *c*

6. Indicate all of the following nontracheophytes in which you would expect to see a protonema at some stage in the life cycle.
   a. *Marchantia*
   b. A fossil liverwort
   c. All hornworts
   d. All mosses
   e. All liverworts

INTRODUCING THE TRACHEOPHYTES • THE SURVIVING NONSEED TRACHEOPHYTES (pages 587–595)

Key Concepts

1. Tracheophyte plants are a much larger and more diverse group than their nontracheophyte relatives. Nevertheless, all tracheophytes are linked by the presence of a common anatomical feature, the *tracheid*, which is a water-conducting element in xylem.

2. Tracheids confer two significant advantages on terrestrial plants: a mechanism for internal transport and a means of rigid support against gravity. These advantages helped set the stage for the permanent invasion of land by plants.

3. Tracheophyte phyla are divided into two groups: those that produce seeds and those that do not. The life cycles of nonseed tracheophytes are quite uniform, with the sporophyte constituting the large and obvious stage of the life cycle. The gametophyte stage is small but independent at maturity. The spores of nonseed tracheophyte plants represent a resting stage in the life cycle.

4. Nonseed tracheophytes still require an aqueous environment for reproduction because their flagellated sperm cells must swim to the egg.

5. Ancient tracheophytes first appear in the fossil record about 400 mya. Three surviving groups of nonseed tracheophytes, the lycopods (club mosses), horsetails, and ferns are present as fossils in the Devonian period between 345 and 400 mya. Seed plants first appear in the fossil record about 360 mya.

6. As terrestrial plant size increased, various trees (both nonseed and seed-bearing) became a dominant part of the landscape between 300 and 350 mya. This trend started with the large tree ferns, which subsequently gave way to the gymnosperms (conifers and their relatives). Angiosperm (flowering plant) fossils first appeared about 120 mya and then underwent an explosive evolutionary radiation to become the dominant plant life on Earth.

7. The now extinct members of the phylum Rhyniophyta were the first tracheophytes. They were clearly tracheophytes because they possessed tracheids, but they had neither true leaves nor roots.

8. Rhyniophytes were anchored below the soil surface by horizontal parts of the stem called *rhizomes*. Their stems had dichotomous aerial branching, and at the end of each branch the plant bore distinct sporangia, inside which clusters of four tightly packed spores developed. Fossil analysis of the genus *Rhynia* indicates that these structures were part of the diploid sporophyte plant.

9. The nonseed tracheophytes that evolved from the Rhyniophytes show major changes in anatomy, including true leaves and roots, and two distinct spore types.

10. According to Lignier's hypothesis, roots evolved from the dichotomous branching of early tracheophyte stems, some of which bent downward and penetrated the soil. Adaptation of these underground branches to the unique subterranean environment eventually altered them into true roots. Thus, the root and shoot systems of tracheophytes are homologous structures that evolved from the same organ.

11. True *leaves* are defined as flattened photosynthetic structures that attach laterally to the plant stem and possess vascular tissue. Two leaf types are found in tracheophyte plants: *simple* and *complex*.

12. Today, simple leaves are found only in the club mosses and horsetails. They are usually small, scalelike outgrowths of the stem with rarely more than a single strand of vascular tissue. Their emergence from the vascular tissue of the stem causes little change in the conformation of the stem's vascular structure.

13. Complex leaves are found only in the ferns and seed plants. Flattened stems with dichotomous branching, inside which substantial vascular and photosynthetic tissue developed, are thought to be the evolutionary precursors of complex leaves. Unlike simple leaves, major modifications in the stem vascular structure occur at the points where complex leaves emerge from the stem.

14. Nontracheophytes, as well as the ancient nonseed tracheophytes, are *homosporous* (having spores of only one morphological type). All other plant groups either are exclusively *heterosporous* (having spores with two distinct morphologies), or include a mix of species that employ one or the other spore-production mechanism. Heterospory evolved multiple times in several different plant lineages, underscoring its apparent selective advantage.

15. Heterosporous plants produce *megaspores* and *microspores*. Megaspores are generated in small numbers and develop into the large female gametophyte (*megagametophyte*) that produces eggs. Microspores are made in large numbers and develop into the smaller male *microgametophyte* that produces sperm.

16. Although extremely abundant in the Carboniferous period, today only a relatively few surviving species make up the phyla Lycophyta (club mosses) and Sphenophyta (horsetails or scouring rushes). They share a number of features, including having true roots and simple leaves. Both have homosporous and heterosporous species that exhibit heteromorphic alternation of generations. The sporophyte generation is the dominant stage in the life cycle.

17. Differences between the club mosses and horsetails include patterns of leaf arrangement, mechanisms of stem elongation (horsetails exhibit basal growth), and the location and design of sporangia. In addition, club moss sporangia assume a conelike appearance called a *strobilus*, whereas the sporangia of horsetails are arranged in whorls.

18. The previously problematic phylum Psilophyta (whisk ferns) is the taxonomic home for two genera of tracheophytes that lack roots and bear spores. *Psilotum* has only minute scales instead of leaves, and *Tmesipteris* has well-developed vascular tissue in its flattened photosynthetic surfaces. Both were once thought to be living relics of the Rhyniophyta, but on the basis of recent DNA sequencing they are now considered highly specialized, and relatively recent, offshoots from the more modern plants.

19. Ferns and seed plants typically have large leaves with more than one branching vascular strand. They also have stems and true roots. Their sporophytes are larger than, and independent of, the gametophyte. In ferns, the gametophyte is also independent of the sporophyte, whereas in seed plants it is not.

20. The phylum Pterophyta (ferns) contains about 12,000 species of large-leaved, nonseed tracheophytes that still require water for fertilization. The complex leaf of ferns, which is known as a frond, unfurls during development from a coiled "fiddlehead."

21. Fern sporangia are clustered on the underside of the frond in structures known as *sori*. Meiosis occurs in each sorus to produce haploid spores, which may disperse considerable distances. Most ferns are homosporous. Their spores germinate to form small, independent gametophytes.

22. Antheridia and archegonia develop on the fern gameto-phyte. Male gametes must swim to find the female gamete (often on another gametophyte) and fertilize it. The resulting zygote then develops into a multicellular sporophyte embryo within the gametophyte tissue. Shortly after beginning its development, the sporophyte sprouts roots and becomes nutritionally independent of the gametophyte stage.

23. The dominant stage in the fern life cycle is the sporo-phyte, which can become quite large and persist for many years. The flattened gametophyte is usually small and rather delicate. Some ferns produce a fleshy, tuber-ous gametophyte that must form an obligatory mutual-istic association with a fungus for the developing sporo-phyte to obtain its nutrition.

Activities (for answers and explanations, see page 224)

* Briefly describe how the Rhyniophyta, which have no true roots, could have given rise to the other nonseed tracheophytes, which do possess true roots.

* Compare and contrast the structure and function of sim-ple leaves with that of complex leaves, and indicate the plant groups where each leaf type is found.

Questions (for answers and explanations, see page 224)

Refer to the following diagram of a plant life cycle when answering Questions 1 and 2.

1. The type of reproduction depicted in this diagram is found in
   a. club mosses and hornworts only.
   b. ferns only.
   c. some ferns and all seed-bearing tracheophytes.
   d. seed-bearing tracheophytes only.
   e. nonseed tracheophytes only.

2. The specific events responsible for the switch between hap-loid and diploid stages in this life cycle are better known as
   a. meiosis and fertilization.
   b. mitosis and fertilization.
   c. meiosis and mitosis.
   d. meiosis and germination.
   e. fertilization and germination.

3. Match the characteristics on the left with the plant group(s) that possesses them on the right.
   a. Tracheids _____ Gymnosperms
   b. Spores _____ Psilopsids
   c. True roots _____ Club mosses
   d. Complex leaves _____ Horsetails
   e. Sori _____ Ferns

4. Megaspores are
   a. found in all plants.
   b. produced by the sporophyte generation of het-erosporous plants.
   c. produced by the gametophyte generation of het-erosporous plants.
   d. diploid structures that give rise to the megagameto-phyte.
   e. haploid structures that give rise to the sporophyte.

5. In all nonseed tracheophytes, meiosis takes place in
   a. sporangia.
   b. archegonia.
   c. antheridia.
   d. rhizomes.
   e. microspores.

6. Although they have significant anatomical advance-ments over the nontracheophytes, ferns are still limited to relatively moist habitats because
   a. their tracheids are rather poorly developed.
   b. their xylem tissue is seasonal.
   c. their reproduction requires water for fertilization.
   d. spore dispersal depends on heavy rains.
   e. frond tissue dries out easily.

THE SEEDS OF SUCCESS • THE GYMNOSPERMS: NAKED SEEDS • THE ANGIOSPERMS: FLOWERING PLANTS (pages 595–606)

Key Concepts

1. The seed plants include the *gymnosperms* (conifers and relatives) and the *angiosperms* (flowering plants). The angiosperms are the most recently evolved of the tra-cheophyte plants.

2. In both groups, the evolutionary trend toward a reduced gametophyte stage has continued. The haploid

gametophyte develops while still attached to the diploid sporophyte. Except for the most primitive seed plants, total reproductive independence from liquid water has been achieved by the elimination of swimming sperm.

3. Heterospory is the rule in seed plants. Megasporangia and microsporangia produce female and male spores, respectively, and are found on modified leaves grouped into strobili. Examples are cones and flowers.

4. The spores of seed plants are produced by meiosis within the sporangia, but the spores are not shed. Gametophyte development thus begins within the sporangium. Female gametophytes are retained within the megasporangium through maturity, while microgametophytes, commonly known as *pollen grains,* are released and distributed by wind or animal vectors.

5. Upon finding the appropriate portion of a sporophyte plant, the pollen grain forms a *pollen tube* that grows toward the female gametophyte. Pollen nuclei then travel down this tube to cause fertilization.

6. The diploid zygote that results from fertilization divides mitotically to form a sporophyte embryo while still inside the female gametophyte tissue. At some point this collection of sporophyte and gametophyte tissue goes dormant, resulting in the structure known as a *seed.*

7. Seeds often possess tissue from three plant generations. The protective seed coat is diploid sporophyte tissue from the original parent generation. Inside the seed coat is haploid gametophyte tissue from the second generation. Finally, the newly formed embryo is a third generation of diploid sporophyte tissue.

8. Unlike the embyros of nonseed plants, which develop directly into sporophytes, the seeds of gymnosperms and angiosperms provide a tough, protective coating in which the embryo can "rest" for many months or years before developing into a mature sporophyte plant.

9. The ability of seeds to rest for long periods of time, the presence of stored food reserves that can be used to initiate development of the new sporophyte, and specialized structures to facilitate dispersal have combined to make the seed plants an incredibly successful group.

10. The gymnosperms yield only to the angiosperms in their dominance of the land. Approximately 750 species of gymnosperms are grouped into four living phyla. *Cycadophyta* (cycads) are palmlike, *Ginkgophyta* (ginkgos) today exist as a single species of a once numerous group, *Gnetophyta* is a collection of diverse plants related by a unique type of vessel element cell, and *Coniferophyta* (the conifers), which is by far the most successful group of gymnosperms.

11. Gymnosperms exhibit secondary growth and include some of the tallest trees known, many of which are commercially valuable. Their impressive growth is despite the fact that tracheid cells are the only components in their xylem, which limits the extent of water conduction and structural support.

12. Clues about gymnosperm evolutionary origins are provided by transitional fossils from the Devonian period. Although these plants had distinct rhyniophyte and fernlike characteristics, the presence of pitted tracheids in their woody tissue clearly allies them with the gymnosperms. Gymnosperms remained the dominant forest plants worldwide until about 80 mya, and in some areas are still the primary flora.

13. The Coniferophyta is the largest of the gymnosperm phyla. All conifer species are heterosporous, with spore formation occurring in specialized strobili borne as distinct male and female cones.

14. The life cycle of pine is representative of conifers generally. Pollen grains (male gametophytes) are dispersed in vast numbers by wind, with successful pollen eventually landing on the appropriate female sporophyte tissue. A pollen tube grows to provide the final route of travel for the sperm nucleus to the egg.

15. A special layer of sporophyte tissue called *integument* houses the megasporangium. This integument ultimately becomes the seed coat. An *ovule* consists of the megasporangium, the integument, and the sporophyte tissue connecting them to the maternal plant. The *micropyle* is a small opening at the apex of the ovule that provides access for the pollen tube.

16. Since most conifer seeds lack a protective covering of fleshy fruit (they are "naked" seeds), their only protection comes from the tight packing of scales within the female cone. Some scales are so tightly packed that only severe environmental disturbances such as fire cause them to open. The conifer seeds that do have fleshy fruits (e.g., juniper and yew) are dispersed mainly by animals who seek the fruit as food.

17. The phylum Angiospermae (~275,000 species) represents the culmination of a trend toward increased size and independence of the sporophyte generation in plants. In addition to the tracheids found in other tracheophytes, their xylem also possesses specialized water-conducting cells called *vessel elements,* as well as *fibers,* which improve support.

18. *Double fertilization* is a reproductive characteristic that was formerly thought unique to the angiosperms, but that is now known to occur in the Gnetophyta as well. In this process the female gametophyte is penetrated by two male gamete nuclei (both from the same pollen grain). One sperm nucleus fertilizes the egg to form a diploid zygote that will become the next sporophyte generation. The other sperm nucleus fuses with a haploid nucleus (or nuclei) from the female gametophyte.

19. In Gnetophytes, double fertilization results in diploid ($2n$) tissue, but in angiosperms the resulting cell is triploid ($3n$) because two separate haploid gametophyte nuclei are involved in the fusion with the second sperm nucleus. Subsequent mitotic division of these cells, whether diploid or triploid, yields *endosperm* tissue that provides nutrients for the developing embryo. Thus, the most definitive characteristic of angiosperm reproduction is now considered to be the development of extensive triploid endosperm.

20. The angiosperms, or flowering plants, derive their name because they all produce specialized reproductive structures called flowers, inside which seeds form. Flowers

are a kind of strobilus derived from highly modified leaves. Flower parts bearing microsporangia are called *stamens* (male structures), and those bearing megasporangia are called *carpels* (female structures).

21. Each stamen contains two *anthers* that house the many sporangia where pollen is produced. Anthers are attached to the rest of the flower by filaments.

22. One or more fused carpels are referred to as a *pistil*, at the base of which resides the *ovary*. An ovary contains one or several ovules, where the female gametophyte is located. The stalk of a pistil is called the *style*, and its sticky terminal surface where the pollen attaches is known as the *stigma*.

23. Flowers have nonreproductive parts that are also derived from modified leaves. The upper leaves of a flower are called *petals* (collectively = *corolla*), while the lower ones are called *sepals* (collectively = *calyx*). Both play a role in attracting animal pollinators. Sometimes petals and sepals are indistinguishable, in which case they are called *tepals*, and at other times they are absent altogether. The stamens, carpels, corolla, and calyx are all attached to the main plant body by a central *receptacle*.

24. *Perfect* flowers contain both male and female reproductive parts as well as the characteristic nonreproductive structures. *Imperfect* flowers lack either carpels or stamens. *Monoecious* species have both male and female flowers on the same plant. *Dioecious* species have male and female flowers on separate plants.

25. Flowers often occur individually, but they may also be clustered together to form species-typical *inflorescences*.

26. Important evolutionary trends in flower structure include a reduction in the number and type of organs, specialization of petals and sepals, and a tendency toward bilateral symmetry accompanied by fusion of parts.

27. Selective pressures related to successful pollination have constrained and directed flower morphology. Long pistils and filaments are adaptations to improve contact with wind-blown pollen or to help deposit pollen on animal bodies for transport to another flower. Competition among pollen tubes "racing" to grow down the style may also have selected for longer female structures, thus facilitating the floral equivalent of mate selection in animals.

28. Many flowering plants have coevolved with their animal pollinators, which include insects (especially bees, beetles, moths, and butterflies), birds, and bats. These relationships can be highly species-specific. Animals visit flowers for food (pollen and nectar), and in so doing efficiently distribute pollen from one flower to another. Flower shape, size, and color have all been influenced by the evolutionary selective pressures exerted by the plant's need to attract pollinators.

29. A fruit develops from the ovary of a flowering plant to surround its seeds. *Simple fruits* (e.g., cherries) are those that develop from a single ovary, *aggregate fruits* (raspberries), from multiple carpels in a single flower, *multiple fruits* (pineapples), from inflorescences, and *accessory fruits* (apples), from parts other than the ovary.

30. Like gymnosperms, angiosperms exhibit heterospory and extreme reduction of the female gametophyte. Male gametophytes are the pollen grains, and the ovules are located within fused carpels.

31. Angiosperm zygotes develop an embryonic axis with one or two *cotyledons*, or seed leaves, which may absorb endosperm nutrients or be photosynthetically active after seed germination. Cotyledon number is used to subdivide the angiosperms into classes. The Monocotyledones have a single cotyledon, and the Dicotyledones have paired cotyledons. Monocots and dicots also differ with respect to the number of flower parts, vascular tissue arrangement, patterns of leaf veins, and secondary growth characteristics. Monocots include the grasses, lilies, and palms. Dicots are far more diverse, and include most herbs, weeds, vines, shrubs, and trees.

32. Angiosperms, along with two groups of gymnosperms (Gnetophytes and the long-extinct cycadeoids), evolved from a single ancestral species that produced no other groups. Ancient members of the magnolia and Chloranthaceae families are currently regarded as the most likely candidates for the first angiosperms.

33. A controversy continues over whether the first angiosperms were trees or herbs. Regardless, the predominant ground cover on Earth today is clearly herbaceous angiosperms. These hardy and opportunistic plants are usually the first colonizers of available ground, grow rapidly, and some have the capacity to withstand environmental extremes by storing photosynthate in specialized underground rhizomes or tubers.

Questions (for answers and explanations, see page 224)

1. Which of the following statements about seed-bearing tracheophytes is *false*?
   a. Heterospory is common in both gymnosperms and angiosperms.
   b. The triploid endosperm is found only in angiosperm seeds.
   c. Double fertilization occurs in most flowering plants.
   d. Gymnosperm sporophytes are nutritionally dependent on the gametophyte.
   e. The conifers are the largest and most diverse group of gymnosperms.

2. Raspberry is an example of
   a. a plant with accessory fruit.
   b. a plant with aggregate fruit.
   c. a monoecious gymnosperm.
   d. a nonseed tracheophyte.
   e. a tracheophyte that lacks vessel elements.

3. Which of the following events happens just prior to the fertilization of an egg cell in gymnosperms?
   a. Pollen is released from male cones.
   b. The female gametophyte withers and dies.
   c. The pollen tube forms.
   d. A sperm nucleus travels down the pollen tube.
   e. Sporophylls on the female cone open.

4. Consider a honeybee carrying pollen from one flower to another. The only pollen grains that have a reasonable

chance to fertilize the eggs contained in the ovules of the second flower are those that make contact with the
a. stigma.
b. anthers.
c. sepals.
d. receptacle.
e. filaments.

5. Which of the following would you *not* expect to find in the seed of a palm tree?
a. Haploid tissue
b. Integument
c. Two cotyledons
d. Triploid endosperm
e. An embryo

6. A certain flower has stamens, a corolla, and a calyx. It would therefore be classified as
a. a perfect flower.
b. an imperfect flower.
c. monoecious.
d. dioecious.
e. an inflorescence.

7. Ginkgos, cycads, and pines are all members of the
a. angiosperms.
b. gymnosperms.
c. Gnetophyta.
d. Angiospermae.
e. Coniferophyta.

Activities (for answers and explanations, see page 224)

- On the following diagram of a generalized flower, label the structures indicated by blank lines.

- In the space provided below, draw a simple diagram of a gymnosperm seed and label the three generations of tissues it contains.

Integrative Questions (for answers and explanations, see page 224)

1. Which of the following plants lack flagellated sperm cells?
a. Pine trees
b. Roses
c. Mosses
d. Ferns
e. More than one of the above

2. Compared with mosses, the anatomical structure in the angiosperm life cycle that has been most reduced in size is the
a. sporophyte.
b. gametophyte.
c. root system.
d. shoot system.
e. sporangium.

3. Seeds were a significant evolutionary development in plants because they
a. provided food for animal pollinators.
b. increased dependence of the gametophyte on the sporophyte.
c. improved the chances that a sporophyte embryo would survive and grow.
d. provided a developmental link between megasporangia and microsporangia.
e. none of the above

4. Which of the following statements most precisely describes a collection of organisms that contains a hornwort, a pine, and an apple tree?
a. They are all nontracheophytes.
b. They are all tracheophytes.
c. They are all gymnosperms.
d. They are all angiosperms.
e. They are all plants.

5. Which of the following is *not* an overall trend seen in the evolution of plants?
a. Increasing size and dominance of the sporophyte generation
b. Decreased dependence on liquid water for reproduction
c. A general trend toward increasing body size
d. A shift from homospory to heterospory
e. Decreasing complexity of vascular tissue

28
Fungi: A Kingdom of Recyclers

CHAPTER LEARNING OBJECTIVES—after studying this chapter you should be able to:

❑ Cite the main taxonomic characteristics used to distinguish the kingdom Fungi from all other eukaryotic organisms.

❑ Correctly assign representative fungi to one of four phyla (Chytridiomycota, Zygomycota, Ascomycota, Basidiomycota) or to a polyphyletic group called the imperfect fungi, and provide the basic structural and functional characteristics of these taxonomic groupings.

❑ Discuss the term "absorptive heterotrophy" as it applies to saprobic fungi, and appreciate the ecological importance of fungi as decomposers in the environment.

❑ Explain how fungi may play a role in the process of environmental bioremediation.

❑ Define and provide examples of facultative and obligate parasitism in the fungi.

❑ Provide several examples of mutualistic relationships in fungi, describe the organisms with which the fungi associate, and cite the benefits of these relationships to each organism.

❑ Understand the basic anatomy of a multicellular fungus, and correctly use the terms "hyphae," "mycelia," "septa," and "rhizoids."

❑ Describe four different mechanisms of asexual reproduction in fungi.

❑ Discuss the various stages in the life cycle of sexually reproducing fungi, and clearly distinguish between haploid and diploid conditions, mating types, dikaryotic cells, heterokaryotic cells, fruiting bodies, asci, and the various types of spores produced.

❑ Understand how conidia differ from other kinds of spores produced by fungi.

❑ Discuss the proposed evolutionary relationships between the chytrid fungi, the funguslike protists, and the animal kingdom.

❑ Differentiate between euascomycete and hemiascomycete fungi, and explain the role of ascospores, asci, and ascocarps in their life cycles.

❑ Discuss how yeast is an atypical fungus with regards to its anatomy, metabolic activity, and reproduction.

❑ Describe the basic anatomy of the basidiomycete fruiting body and explain the process of spore formation within this structure.

❑ Cite examples of how imperfect fungi are relevant to the lives of humans, and describe some of their unusual lifestyles.

❑ Discuss the ecological importance of fungal associations with plants, and describe the role of mycorrhizae in these relationships.

❑ Explain how a lichen differs structurally from a true fungus, and describe the ways that lichens reproduce.

❑ Discuss the reasons lichens are able to survive in habitats too harsh for other fungi and plants, and explain how lichens contribute to soil formation.

GENERAL BIOLOGY OF THE FUNGI (pages 610–615)

Key Concepts

1. The kingdom Fungi encompasses all eukaryotic, heterotrophic organisms with absorptive nutrition. Most are multicellular.

2. Fungi play a crucial ecological role as decomposers of organic matter, helping return nutrients to various biogeochemical cycles. Because they are such extraordinary chemical recyclers, fungi (along with some bacteria) are now being tested for possible use in the process of environmental *bioremediation*. The hope is that some human-generated toxic substances might be broken down by fungi into harmless metabolic byproducts.

3. Fungi inhabit all environments and attack virtually all other eukaryotes, including other fungi. Their life cycles are diverse and often complex.

4. The lifestyles of fungi fall into one of three categories: *saprobes* (live on dead matter), *parasites* (attack living matter), or *mutualists* (live in a mutually beneficial symbiotic association with other organisms).

5. All fungal species form spores, and the majority of species lack flagellated cells in any stage of the life cycle. Asexual reproduction is common. Sexual reproduction

is performed by various means, ranging from the fusion of filaments from different mating types to the union of actual gametes.

6. The fungal cell wall contains the polysaccharide *chitin*, which is also found in exoskeletons of arthropods (e.g., insects).

7. There are four recognized phyla of fungi: the *Chytridiomycota* (chytrids), the *Zygomycota* (conjugating fungi), the *Ascomycota* (sac fungi), and the *Basidiomycota* (club fungi). Another "catch-all" group, which does not form obvious sexual structures, is referred to as the *imperfect fungi*. However, as molecular taxonomy techniques improve, more of these imperfect fungi are being reclassified into one of the four other phyla.

8. Individual fungal cells form filaments called *hyphae*, which are partially divided by incomplete partitions of the filaments called *septa*. Hyphae grow rapidly to form a larger body, the *mycelium*, whose total growth may exceed one kilometer per day. Saprobic fungi use specially-modified hyphae known as *rhizoids* to anchor themselves to their food. Parasitic fungi may also have modified hyphae used to acquire nutrients from their host.

9. During sexual reproduction, many fungi generate elaborate fruiting bodies that produce and disseminate spores. A common example is mushrooms.

10. The fungal mycelium has an enormous surface-to-volume ratio, making it easy to absorb food and insuring that all cells within the mycelium are near the food source.

11. Fungi are very tolerant of extreme environmental conditions, including strongly hyperosmotic surroundings and low (subfreezing) or high (in excess of 50ºC) temperatures.

12. Saprobic fungi, along with bacteria, are the major decomposers of the biosphere, a function made possible by their absorptive nutrition. They are the principal decomposers of cellulose and lignin from plants, which most bacteria cannot metabolize. Their preferred carbon sources are sugars, and they obtain nitrogen from proteins or the breakdown products of protein. Essential B vitamins and minerals are also obtained from their food.

13. Parasitic fungi may be either *facultative* or *obligate*. Facultative parasites grow on their host organisms, but can also grow by themselves provided proper nutrients. Obligate parasites can grow only on their specific hosts, which are usually plants.

14. Some parasitic fungi have predatory adaptations, such as sticky traps or *constricting rings*, which they use to capture prey. Their prey are often tiny nematode worms. Once prey is captured, hyphae penetrate the prey to obtain nutrients.

15. *Lichens* are mutualistic associations of a fungus with either a unicellular alga or a cyanobacterium. The fungus provides protection and support to the autotroph, while the autotroph provides photosynthetic products as food to the fungus.

16. Other fungi form mutualistic associations with the roots of plants via structures called *mycorrhizae*. These fungi cause the plant's root hairs to swell, increasing the surface area for absorption of nutrients into the roots. They also help the plant retain water. By assisting the plant with its nutrient acquisition, the fungus also ensures food for itself. Seed germination in some plants (e.g., orchids) requires the presence of mycorrhizal fungi.

17. Insects are sometimes involved in mutualistic interactions with fungi. Leaf-cutting ants use plant leaves or flowers as a substrate for fungal growth in their colony. These fungal "farms" are meticulously tended by the ants, who use the fungus as food. Scale insects use the coverings of a parasitic fungus (genus *Septobasidium*) for protection from drying, as well as to deter predators. In the process, the insects provide food for this nonlethal fungal parasite.

18. Asexual reproduction in fungi is common and can occur in four ways. Haploid spores may be produced in specialized sporangia. Naked spores, or *conidia*, may also be formed outside of sporangia at the tips of individual hyphae. Unicellular fungi carry out simple cell division, which may produce two equivalent-sized cells or a tiny cell called a bud. Finally, some fungi reproduce by simple breakage of the mycelium, which then forms a new fungus.

19. Sexual reproduction in some fungi begins when hyphae of opposite *mating types* come into contact. Their hyphal cell walls are digested, and eventually the two haploid nuclei come to reside within a single hyphal cell, a condition that is both *dikaryotic* ($n + n$) and *heterokaryotic* (because the two nuclei are different). The heterokaryotic cell then typically develops into a heterokaryotic mycelium.

20. Within the heterokaryotic mycelium two haploid nuclei fuse to form a diploid zygote that then undergoes meiosis to produce haploid spores. The spores are modified, depending on the type of fungus, and germinate to produce new hyphae.

21. Three important features of sexual reproduction in most fungi are 1) the absence of motile gamete cells and thus, 2) a liberation from liquid water for fertilization, and 3) no true diploid tissue at any point in the life cycle (i.e., all cells except the zygote are either haploid or dikaryotic). Indeed, dikaryosis is perhaps the most significant genetic peculiarity of the fungi.

22. *Puccinia graminis*, a fungus that causes black stem rust in wheat, demonstrates three important points about the life cycles of fungal parasites: they are complex, they show extensive dikaryosis, and they often utilize multiple hosts.

23. *P. graminis* produces numerous spore types (*uredospores, teliospores, basidiospores,* and *aeciospores*) during its life cycle, all of which are well adapted for the peculiar ecological requirements of each stage. These spores include haploid, diploid, and dikaryotic types. In addition to wheat, the barberry plant also serves as a host for the fungus. While on barberry, specialized flask-shaped structures called *aecial primordia* are produced to attract insects for purposes of transfering reproductive structures known as *spermatia* from one flask to another. Figure 28.5 in your textbook provides a detailed life cycle for *P. graminis* showing the different spore types and their utilization of multiple host plants.

Questions (for answers and explanations, see page 225)

1. Which of the following would you *not* expect to find somewhere in the life cycle of a typical fungus?
   a. Haploid cells
   b. Diploid cells
   c. Spores
   d. Chloroplasts
   e. Dikaryotic cells

2. Fungal cell filaments called _____ combine to form the _____ .
   a. rhizoids, mycelium
   b. hyphae, mycelium
   c. basidiospores, hyphae
   d. rhizoids, fruiting body
   e. mycorrhizae, lichen

3. All fungi are absorptive heterotrophs. Which of the following is an adaptation shown by many fungi that greatly aids this mode of nutrient procurement?
   a. Dikaryosis
   b. Conjugation
   c. Complex life cycle
   d. Chitinous cell wall
   e. Large surface-to-volume ratio

4. Assume that two normal hyphal cells of different fungal mating types unite. After a period of time the cell walls between these cells will dissolve, producing a
   a. dikaryotic cell.
   b. fruiting body.
   c. zygote.
   d. dikaryotic cell which is also heterokaryotic.
   e. mycelium.

5. Which of the following statements about the fungal structures known as conidia is *false*?
   a. Conidia appear at the tips of hyphal cells.
   b. Conidia are formed within sporangia.
   c. Conidia are haploid.
   d. Formation of conidia is a type of asexual reproduction in fungi.
   e. All of the above

6. Which of the following statements about sexual reproduction in fungi is *false*?
   a. Motile gametes are present in all fungal species.
   b. Water is not required for fertilization to occur in fungi.
   c. There is no true diploid tissue in the life cycle of most sexually reproducing fungi.
   d. Sexual reproduction often begins with contact between opposite mating types.
   e. None of the above

Activities (for answers and explanations, see page 225)

- Complete the following diagram, which shows the generalized life cycle of a fungus, by adding the following terms in their appropriate locations: *spores, dikaryotic hyphae, meiosis, mitosis, fertilization.* Also, circle and label the portions of the life cycle corresponding to the haploid (*n*), diploid (2*n*), and the dikaryotic (*n* + *n*) stages.

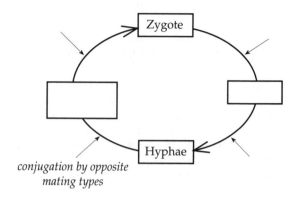

conjugation by opposite mating types

- Explain the functional differences between the following fungal structures: hyphae, mycelia, rhizoids, and mycorrhizae.

DIVERSITY IN THE KINGDOM FUNGI • FUNGAL ASSOCIATIONS (pages 615–624)

Key Concepts

1. The chytrids (phylum *Chytridiomycota*) are the most ancient fungi, possessing some protistlike features. They are all aquatic and have either parasitic or saprobic lifestyles. Many are unicellular; others develop into massive chains of filamentous branching cells (mycelia). Chytrids reproduce both asexually and sexually.

2. The chytrid genus *Allomyces* exhibits isomorphic alternation of generations. The diploid stage produces *zoospores*, which are diploid flagellated cells that arise by mitosis. Haploid zoospores are also produced by meiosis.

3. Each type of zoospore mitotically gives rise to an adult organism of that same ploidy. Mature haploid chytrids produce gametes in specialized *gametangia* located at the tips of the mycelial filaments. Chemical attractants produced by female gametes attract the swimming male gamete. Thick-walled resting sporangia are produced that can survive harsh conditions. The nuclei within these sporangia eventually undergo meiosis to produce new zoospores.

4. Chytrids possess the only flagellated cells found in the kingdom Fungi. This likely makes them the first group to have diverged from what is presumed to have been a flagellated funguslike protist ancestor. That ancestor likely also gave rise to the protist phylum Choanoflagellida and to the animals.

5. The phylum *Zygomycota* (conjugating fungi) numbers some 600 species. They possess hyphae that lack

cross walls, have no motile cells, and produce no fleshy fruiting body. The only diploid cell in the entire life cycle is the zygote. One important group of zygomycetes forms mycorrhizal associations with the roots of plants.

6. The black bread mold *Rhizopus stolonifer* is a zygomycete that can reproduce vegetatively or engage in sexual reproduction through conjugation. Asexual reproduction is characterized by the production of stalked sporangia within which large numbers of spores are produced.

7. Gaseous pheromones direct the conjugation process in *Rhizopus*. Hyphae of different mating types combine and produce gametangia that fuse to form zygotes. The zygotes then develop into resistant *zygospores*. New sporangia containing the spores produced by meiosis sprout from the zygospores. Spores germinate to yield the new haploid generation of hyphae.

8. About 30,000 species comprise the phylum *Ascomycota* (sac fungi). All produce a unique saclike structure called an *ascus* (plural *asci*), inside which meiosis occurs to produce *ascospores*. These spores then germinate to re-form the mycelial stage. Unlike zygomycetes, ascomycete hyphae have cross walls, but these walls contain perforations through which nuclei and other organelles can pass.

9. The ascomycete species that form a fruiting body called an *ascocarp* are known as the euascomycetes. Species that do not produce ascocarps are called hemiascomycetes.

10. Hemiascomycetes are generally very small, with many species being unicellular. The yeasts (e.g. genus *Saccharomyces*) are among the best known, having major commercial importance in the baking and brewing industries because of their ability to generate ethanol and carbon dioxide during alcohol fermentation. They are also used extensively in studies of molecular biology. Yeasts are atypical fungi in that they usually reproduce asexually, either by fission or *budding*. Conjugation takes place only occasionally, and there is no dikaryon stage.

11. The genus *Neurospora* (pink mold) is an example of an euascomycete fungus. Euascomycetes are multicellular with heterokaryotic hyphae. Sexual reproduction involves the passing of nuclei from one mycelium to another. The donor nuclei then divide independently of the host nuclei for some time. Nuclei only undergo fusion at the actual time of ascus formation.

12. Euascomycetes can be serious parasites of higher plants (e.g., chestnut blight and Dutch elm disease), as well as gourmet delicacies (morels and truffles). This group includes the cup fungi, whose large fruiting bodies are formed by tightly woven filaments of the dikaryotic mycelium.

13. Asexual reproduction in euascomycetes is by means of conidia, which develop by the millions on the tips of specialized hyphae and give molds their characteristic coloration. Sexual reproduction involves a dikaryon stage formed when male nuclei enter female hyphae of a compatible mating type. Once an ascus forms, nuclear fusion takes place and meiosis proceeds, resulting in the formation of ascospores.

14. The 25,000 species of club fungi comprise the phylum *Basidiomycota*. They have prominent fruiting structures (e.g., mushrooms) where spore formation occurs. Some are commercially important as foods, others are serious plant pathogens (e.g., brackets, smuts, and rusts).

15. Basidiomycete hyphae have walls with small pores. Dikaryotic hyphae undergo nuclear fusion and meiosis in a fruiting structure called a *basidium*, where haploid *basidiospores* are produced by mitosis. These spores are scattered and germinate to reestablish the haploid mycelium.

16. The fleshy fruiting structure of some basidomycetes is topped by a cap, or pileus, with gills on its underside. Basidia develop between these gills and discharge their spores for dispersal, most often by wind or water. Gill pattern and spore coloration are taxonomic characters used in the identification of basidiomycetes.

17. The *imperfect fungi* constitute a leftover grouping of some 25,000 fungus species that do not have obvious sexual stages. As such, they cannot be placed in the other four fungal phyla, which are distinguished from one another primarily using reproductive criteria. Whenever sexual structures are identified in an imperfect fungus, it is promptly reclassified into the appropriate phylum.

18. Some chemical by-products of imperfect fungi are of major importance to humans. Members of the genus *Penicillium* produce bacterial antibiotics; humans have usurped this property for our own medical advantage. Other species are responsible for the characteristic flavors of certain cheeses, soy sauce, and the alcoholic beverage known as saki. Still others produce carcinogens called aflatoxins when they grow on certain species of nuts.

19. Symbiotic associations between certain species of fungi and plants are common and ecologically important. One such association occurs in the form of *mycorrhizae*, which are fungal infections of plant roots. Ectomycorrhizae occur when the fungus wraps around the exterior of the root, and endomycorrhizae are formed when the fungus penetrates the root but leaves no visible sign on the root surface. Both serve to enlarge the root surface area, which leads to better water and nutrient absorption by the plant. Mycorrhizae also act like a sponge to help hold water near the plant roots.

20. Mycorrhizae are beneficial to both the fungus and the plant. The fungus derives food in the form of sugars and amino acids, while the plant receives improved absorption, useful growth hormones, and protection from attack by microorganisms. Indeed, this association may well have been one of the primary events that made possible the successful colonization of land by living things.

21. The approximately 13,500 "species" of lichens are not true organisms in the strict sense of the term. Instead, they are mutualistic associations between two different organisms: a fungus and either a unicellular alga or a cyanobacterium.

22. The combination of tough protection and structural support provided by the fungus, and the photosynthetic nutrition provided by the alga or bacterium, allows lichens to survive in extremely harsh environments. However, because they cannot eliminate the toxins associated with pollution, lichens do not typically occur near industrialized cities and are therefore good biological indicators of air pollution.

23. The fungal component of lichens is most often an ascomycete, but other classes of fungi are also known to form lichens. Some lichens, such as the reindeer "moss" *Cladonia subtenuis*, are important food items for large mammals. Lichens come in a variety of colors and forms, including crustlike (*crustose*), leafy (*foliose*), and shrubby (*fruticose*).

24. Lichens reproduce by simple fragmentation of the vegetative body (*thallus*), or by specialized structures known as *soredia*. The soredia consist of a few photosynthetic cells surrounded by fungal hyphae and are dispersed by wind or water to new locations where they develop into another lichen.

25. The fungal component of a lichen may produce its own characteristic fruiting bodies, but when spores are released from these structures they will lead only to the production of a new fungus, since no algal or bacterial cells are present.

26. In cross section, a lichen consists of a tightly matted upper layer of fungal hyphae, below which the algae or bacteria reside. Below this photosynthetic layer is another loosely arranged hyphal layer, and finally the rhizoids that attach the lichen to its substrate. This arrangement allows for good water retention and a modicum of structural support.

27. Lichens are often the first colonists to appear on bare rock. By acidifying their environment during growth, lichens help break down rock and initiate or hasten the formation of soils. Because of their incredible resistance to cold and dryness, lichen species can outnumber plant species as much as 100-fold in harsh environments like Antarctica.

Questions (for answers and explanations, see page 225)

1. Select all of the following fungal groups that contian members likely to appear on your dinner plate as an *intended* part of the meal!
   a. Basidiomycetes
   b. Zygomycetes
   c. Ascomycetes
   d. Imperfect fungi
   e. None of the above

2. Soy sauce, Roquefort cheese, and penicillin all have a common link to which of the following fungal groups?
   a. Basidiomycetes
   b. Zygomycetes
   c. Chytrids
   d. Imperfect fungi
   e. None of the above

3. A specimen of fungus is brought to you for identification. Upon close examination, you discover that its hyphae are completely septate and it has gills on the underside of the pileus. To which fungal group does it most likely belong?
   a. Basidiomycetes
   b. Zygomycetes
   c. Ascomycetes
   d. Chytrids
   e. None of the above

4. A process known as karyogamy, where haploid nuclei fuse, occurs in certain groups of fungi during sexual reproduction. Which of the following statements about this process is *true*?
   a. It occurs only in the hemiascomycetes, such as yeast.
   b. The rate of growth of the fungal mycelium is maximum at this time.
   c. It takes place just prior to ascus formation in euascomycetes.
   d. It denotes the point where hyphal cell walls dissolve and the remaining cell becomes heterokaryotic.
   e. The characteristic flavor of some cheeses and alcoholic beverages is the direct result of this process.

5. If you subjected the soredium of a lichen to microscopic and chemical analysis, which of the following would you *never* expect to find?
   a. Chlorophyll *a* in the algal component
   b. Chitin in the cell walls of the fungal component
   c. Eukaryotic cells
   d. A fungal fruiting structure actively producing spores
   e. DNA

6. Although there are over 13,500 known "species" of lichens, unlike other members of the kingdom Fungi, none of them are true organisms. This is because lichens
   a. are actually a mutualistic association between two different organisms.
   b. are really viruses, which are not considered organisms.
   c. have no means of self-reproduction.
   d. cannot live in harsh conditions.
   e. do not contain nucleic acids.

7. Which of the following statements about mycorrhizae is *true*?
   a. They are a type of lichen that can withstand very cold temperatures.
   b. They are ecologically important associations between fungi and plant roots.
   c. They form in the gill regions of basidiomycete fungi.
   d. They develop from mitotic division of zygospores.
   e. They are parasitic on dead aquatic animals.

8. Common molds such as *Neurospora* and the fungi known as truffles share which of the following traits?
   a. They are both ascomycetes.
   b. They both produce ascospores.
   c. Their spores will germinate to form a mycelium.
   d. Meiosis occurs inside an ascus to yield spores.
   e. All of the above

Activities (for answers and explanations, see page 225)

- For a variety of reasons, yeasts are considered somewhat peculiar fungi. List a few characteristics of the genus *Saccharomyces*.

- Refer to Figure 28.14c in your textbook, which shows a photograph of a bracket fungus. Explain the purpose of this bracket structure, and identify where in the basidiomycete life cycle the bracket occurs.

- Folklore has it that one can determine directions in the woods by seeing on which side of tree trunks the lichens grow. Suggest a possible environmental reason why this observation has biological validity.

Integrative Questions (for answers and explanations, see page 225)

1. Which of the following is true of *all* fungi?
   a. They are multicellular
   b. They are heterotrophic and eukaryotic
   c. They obtain nutrients by absorption
   d. Both *a* and *b*
   e. Both *b* and *c*

2. You find a small, flattened organism that looks like a leafy plant growing on an otherwise bare rock surface. Microscopic analysis shows numerous threadlike, eukaryotic cells surrounding what are clearly bacterial cells. The best description for this organism is a(n)
   a. mushroom.
   b. alga.
   c. foliose lichen.
   d. truffle.
   e. club fungus.

3. A saprobic fungus could utilize which of the following objects as a food source?
   a. Spoiled cheese
   b. A dead fly
   c. Overripe fruit
   d. Animal feces
   e. All of the above

4. The fungal part of a lichen and members of the genus *Neurospora* share which of the following characteristics?
   a. Haploid hyphae
   b. Autotrophic nutrition
   c. Asexual reproduction via budding
   d. Obligate parasitism
   e. None of the above

29

Protostomate Animals: The Evolution of Bilateral Symmetry and Body Cavities

CHAPTER LEARNING OBJECTIVES—after studying this chapter you should be able to:

❑ Describe the general characteristics that define the animal kingdom and explain our current understanding of how animals evolved from protists.

❑ Explain how division of labor among colonial protists is thought to have led to the evolution of animals.

❑ Differentiate between protostomate and deuterostomate animals.

❑ Explain how characteristics of symmetry, embryology, morphology, and physiology are used to classify animals, and in what ways these factors have influenced the evolution of animal body plans.

❑ Differentiate between the types of body symmetry in animals.

❑ Discuss the meaning of the terms "diploblastic" and "triploblastic," and know what types of animals fall into each category.

❑ Describe the different types of embryological development in animals using the terms "radial cleavage," "spiral cleavage," "determinate cleavage," and "indeterminate cleavage," and know the fate of the embryonic blastopore in protostomes and deuterostomes.

❑ Define the term "coelom," and describe the different types of animal body plans that are based on this structure.

❑ Discuss the evolutionary sequence of feeding adaptations in animals.

❑ List the basic characteristics of the sponges, and explain why these animals are considered to be similar to ancestral protists.

❑ List the basic characteristics of the cnidarians and ctenophores, and explain in what ways these animals are more complex than sponges.

❑ Discuss alternation of generations as seen in the cnidarians, differentiate between polyp and medusa body forms, and explain the mechanisms of reproduction in each.

❑ Recognize representatives of the cnidarian classes, and list the basic characteristics of each.

❑ Describe the significance of bilateral symmetry in the evolution of complex animal body forms.

❑ Discuss the evolutionary significance of complete digestive systems and hydrostatic skeletons in roundworms and rotifers.

❑ Describe the body plans of flatworms and how flatworm life cycles have evolved.

❑ Identify common representatives of the annelid worms, and discuss how their body plans allow more complex movements than are possible with an unsegmented body.

❑ Explain the evolutionary significance of exoskeletons, their advantages and disadvantages, and in what animal phyla they occur.

❑ Provide reasons why arthropods, especially the insects, are such a diverse and successful assemblage of animals.

❑ Discuss the defining characteristics of representative arthropod phyla and classes.

❑ Discuss the phenomenon of metamorphosis, and describe the different kinds of metamorphosis found in insects.

❑ Describe the ways in which the molluscan body plan differs from that of arthropods.

❑ List the distinguishing characteristics of representative mollusk groups.

❑ Explain how cephalopod mollusks differ from other molluscan groups.

❑ Discuss the main trends in protostome evolution and how these trends compare with those found in deuterostomes.

DESCENDANTS OF A COMMON ANCESTOR • FOOD PROCUREMENT: THE KEY TO THE ANIMAL WAY OF LIFE • CLUES TO EVOLUTIONARY RELATIONSHIPS AMONG ANIMALS (pages 628–631)

Key Concepts

1. The animal kingdom consists entirely of multicellular, eukaryotic heterotrophs that ingest their food. The kingdom has its evolutionary roots in a single lineage of colonial protists. Supporting evidence comes from commonalities among extant animals in the structure of their rRNA and extracellular matrix molecules.

2. Protistan ancestors of the animals may have resembled modern day colonies of *Volvox*. Division of labor and the specialization of cell types for movement, nutrition, and reproduction that likely occurred in ancient colonial protists probably account for the evolutionary origins of many basic animal characteristics.

3. Multicellularity and the differentiation of cell types in animals involves the integration of at least three complex processes. First, genes must be regulated, the differentiation status of cells must then be transmitted during division, and differentiated cells must finally be arranged in predictable spatial patterns.

4. Animals are more diverse than other organisms. Their high degree of complexity requires the acquisition of large quantities of food through energetically expensive activities. Animal diversity is thus primarily linked to specializations for acquiring food, which have in turn led to adaptations in body form, locomotion, sensory systems, and behavior.

5. The relationship between an animal's overall *body plan* and the functioning of its component parts provides clues to the ways that animals can live in and use their environments.

6. Several important components of body plans are considered in animal classification. These include patterns of embryological development, comparative morphology and physiology, and information from the fossil record.

7. *Symmetry* is one aspect of an animal's body plan. Symmetrical animals can be divided into similar halves along at least one plane.

8. *Asymmetric* animals, represented by the sponges, have no plane of symmetry. *Spherical symmetry* is seen mainly in protists, but is also present in some animals. *Radial symmetry* refers to bodies that yield similar halves when cut in any plane along the main body axis. If the body is modified so that only two planes will produce similar halves, the animal has *biradial symmetry*.

9. A *bilaterally symmetrical* animal can be divided into mirror images only by a single plane that passes through the midline of the body. Bilateral symmetry is characteristic of animals that move through their environment mainly in one direction.

10. Anatomical terms used to describe bilaterally symmetric body plans include *anterior* (front), *posterior* (back), *dorsal* (top or side without a mouth), and *ventral* (bottom or side with a mouth). The development of sensory structures and central nervous system tissues in the anterior end of the animal (*cephalization*) is closely allied with bilateral symmetry.

11. The early developmental stages of animal embryos appear to have evolved quite slowly (they are *evolutionarily conservative*) and are therefore useful in determining evolutionary relationships among animals.

12. The number of distinct cell layers that develop in animal embryos is diagnostic. *Diploblastic* animals have only two layers (ectoderm and endoderm), while *triploblastic* animals also have a third, or middle layer of mesoderm.

13. Except for the simplest animals, all other animals belong to two main lineages: the *protostomes* and the *deuterostomes*. Because of their peculiar development, sponges are classified separately from the protostomes and deuterostomes and are neither diplo- nor triploblastic. Cnidarians, ctenophores and some protostomes are diploblastic. All deuterostomes, and most protostomes, are triploblastic.

14. Other developmental patterns further separate the protostome and deuterostome lineages. Protostomes exhibit *determinate* cleavage of the fertilized egg—early divisions irreversibly limit the fate of embryonic cells. If separated, individual protostome cells cannot produce an entire embryo. Protostomes also exhibit spiral cleavage in the early embryo.

15. Cleavage in deuterostome embryos has a radial pattern and is *indeterminate*. When separated, embryonic deuterostome cells can develop into complete embryos.

16. A final embryological difference between the two main animal lineages is the fate of the embryonic blastopore. In protostomes the mouth arises from or near the blastopore, whereas in deuterostomes the blastopore becomes the anus.

17. The presence or absence of a body cavity is another important feature of animal body plans. *Acoelomate* animals lack a body cavity altogether; their digestive systems are completely surrounded by mesenchyme tissue. *Pseudocoelomate* animals have a body cavity with muscles on only one side (a *pseudocoel*) that is derived from the embryonic blastocoel.

18. *Coelomate* animals have a body cavity (a *coelom*) surrounded by muscles that forms directly from embryonic mesoderm. The internal organs of coelomates hang within the body cavity inside pouches of mesodermal tissue called the *peritoneum*.

Questions (for answers and explanations, see page 225)

1. All but one of the following are factors that help determine animal body plans. Select the *exception*.
   a. Patterns of embryological development
   b. Type of body symmetry
   c. Whether or not the animal is heterotrophic
   d. Type of body cavity
   e. Specializations in feeding mechanism

2. An animal is known to be a deuterostome. Which of the following does *not* apply to this animal?
   a. It is diploblastic.
   b. If a coelom is present, it formed from the embryonic gut.
   c. Its embryos exhibit indeterminate cleavage.
   d. Its early embryonic cleavage pattern was radial.
   e. Three distinct tissue layers were present while it was an embryo.

3. The primary reason animals are more diverse than plants is the considerable variation in animal
   a. body plans.
   b. embryology.
   c. symmetry.
   d. multicellularity.
   e. methods of food acquisition.

4. Cephalization is a characteristic mainly associated with which of the following types of body symmetry in animals?
   a. Asymmetric
   b. Radial
   c. Biradial
   d. Bilateral
   e. Spherical

5. If an animal is a protostome, then it also
   a. has spiral embryonic cleavage.
   b. has determinate embryonic cleavage.
   c. has a mouth that develops from the blastopore.
   d. is likely to be triploblastic.
   e. All of the above

6. An animal is divided along its main body axis to produce similar halves. Which of the following types of symmetry could apply?
   a. Spherical
   b. Radial
   c. Radial or biradial
   d. Bilateral
   e. Biradial or bilateral

7. Which of the following is *not* a group of animals thought to have arisen from the ancestral colonial protists during the early stages of animal evolution?
   a. *Volvox*
   b. Ctenophores
   c. Cnidarians
   d. Protostomes
   e. Sponges

Activities (for answers and explanations, see page 226)

- Explain the basic differences among the acoelomate, pseudocoelomate, and coelomate body plans found in animals.

- Complete the following table comparing characteristics of the protostomes and deuterostomes.

Characteristics	Protostome	Deuterostome
Cell type		
Mode of nutrition		
# of embryonic layers		
Type of cleavage		
Cleavage pattern		
Fate of blastopore		

EARLY ANIMAL DIETS: PROKARYOTES AND PROTISTS • SIMPLE AGGREGATIONS: SPONGES AND PLACOZOANS (pages 631–633)

Key Concepts

1. The primary food sources for early animals were *phytoplankton* (small floating algae) and the extensive algal mats that covered the shallow sea bottoms. Morphological adaptations for feeding on algae favored the evolution of larger, more mobile animals. As they evolved and became more abundant, some of these larger animals also began to feed on small animals.

2. The presence of large grazing animals opened evolutionary avenues for carnivores. This selective pressure in turn led to the development of protective shells, burrowing, and better locomotion among prey species.

3. One of the earliest animal groups to evolve from the colonial protists was the phylum *Porifera* (sponges). Because they are so much like colonial protists, the lineage leading from the ancestral protists to the modern sponges must have separated very early from the one that gave rise to the other animals.

4. Sponges, which are aggregations of many different types of specialized cells, lead a *sessile* lifestyle (attached to the substrate). The 10,000 species of sponges are exclusively aquatic, with the vast majority being marine.

5. Flagellated collar cells called *choanocytes* are unique to the sponges. They move water and food particles into the animal via numerous small pores on the body wall, and then out again through a larger opening at the top.

6. Wandering amoebalike cells located between the epidermis and the choanocyte layer provide a primitive level of communication between sponge cells. Structural support in sponges is provided by mineralized or proteinaceous spicules that form a complex skeletal pattern within the animal's body.

7. Despite their cellular specialization, sponges lack true tissue-level organization or any regular form of body symmetry.

8. The most important selective force shaping sponge morphology is pattern of water movement. Shallow waters with strong wave action favor flattened forms, while larger, more upright forms are found in calm, deeper waters.

9. Reproduction in sponges can be either sexual or asexual. A single individual produces both sperm and eggs, which are released into the water column. Asexual reproduction is by budding or other mechanisms that yield fragments capable of forming a new sponge.

10. The phylum *Placozoa* represents another distinct evolutionary branching from the lineage that gave rise to the rest of the animal kingdom. It consists of only two species of very small, asymmetric animals with no body cavity, no distinct tissues, and no organs. Their bodies contain at most a few thousand cells and only four cell types.

11. Placozoans have a flattened body plan consisting of two layers of flagellated cells that cover a fluid-filled cavity. They feed on bacteria and protists either by external digestion or by endocytosis. Placozoans are widely distributed in the world's tropical oceans.

Activities (for answers and explanations, see page 226)

- Refer to Figure 29.5 in your textbook and identify the function(s) of each of the following structural components of a sponge:

Amoebocytes:

Choanocytes:

Incurrent pores:

Osculum:

Spicules:

- Recall the embryological and developmental characteristics of animal lineages discussed in the first sections of this chapter, and then explain why sponges and placozoans do not meet the criteria set forth that define diploblastic, triploblastic, protostome, and deuterostome animals.

Questions (for answers and explanations, see page 226)

1. Which of the following is *not* a good explanation for why grazers, rather than carnivores, were the first animals to evolve?
   a. Animal biomass is generally not as good a form of nutrition as is plant biomass.
   b. The most abundant food at the start of animal evolutionary history was large populations of algae.
   c. Early animals were likely too small to have been very effective predators.

d. Anatomical specializations for chasing down and catching prey had not yet evolved.
e. All of the above

2. The increase in the diversity of simple grazing animals was followed by the rapid evolution of larger, more mobile predators. Which of the following is *not* a useful defense mechanism against predation?
   a. Improved locomotion
   b. Protective shells
   c. Burrowing in mud or sand
   d. Increased rate of new species formation

3. Sponges have, in general, a very simple body plan. Which of the following statements about sponge structure or function is *false*?
   a. Choanocytes are flagellated cells that play a role in food acquisition.
   b. Large species are usually found in areas of heavy wave action, where more food can be obtained.
   c. Individual sponges are both male and female.
   d. As a whole, the phylum Porifera shows no regular pattern of body symmetry.
   e. Water enters a sponge through incurrent pores, and exits via one or more oscula.

4. Which of the following is *not* a trait shared by both sponges and placozoans?
   a. Asymmetric body plan
   b. No true tissues and organs
   c. Flagellated cells
   d. Mineralized structural support
   e. Limited numbers of specialized cell types

5. Which of the following statements about sponges and placozoans is *false*?
   a. Neither has true tissue-level organization.
   b. Both are considered to have asymmetric body plans.
   c. Each is given its own phylum-level status in the animal kingdom.
   d. Both have many different types of cells in their bodies.
   e. Both are groups that branched off from the main animal lineages early in evolution.

THE EVOLUTION OF CELL LAYERS: THE CNIDARIANS (pages 633–638)

Key Concepts

1. The phylum *Cnidaria* consists of animals with two distinct cell layers (i.e., they are diploblastic), no body cavity, and radial symmetry. They are an ancient group whose lineage split off from the mainline of animal evolution sometime in the Precambrian.

2. The 10,000 species of cnidarians (sea anemones, jellyfish, corals, and their relatives) are almost all marine. They are named for the possession of a unique cell type called the *cnidocyte*, which contains a dischargeable stinging structure known as a *nematocyst*. This stinging mechanism enables cnidarians to subdue large prey items and provides defense against predators.

3. The cnidarian body plan is more complex than that of sponges or placozoans. Distinct tissue layers are present, including a modified epithelium, which forms tentacles and a nerve net to integrate body activities. A single opening serves as both mouth and anus for the *gastrovascular cavity*, a blind-ended sac that aids in digestion, gas exchange, and circulation.

4. Many cnidarians exhibit alternate body forms. The life cycle often begins with an asexual *polyp* stage, a sessile cylindrical stalk with mouth and tentacles directed upward. The other body form, or *medusa*, has a downward-directed mouth and tentacles and is a free-living, sexual stage. Much of the cnidarian body consists of a gelatinous material called *mesoglea*, which forms a middle body layer almost devoid of cells and therefore has very low metabolic needs.

5. The ciliated cnidarian larva that results from the union of sperm and egg is called a *planula*. After a brief period spent swimming freely in the water column, it settles to the bottom and transforms into a polyp.

6. *Hydrozoa* is a class of cnidarians in which the polyp dominates the life cycle. This group contains the only freshwater members of the phylum. Some species are solitary; others are colonial with a high degree of polyp specialization and a communal gastrovascular cavity. Hydrozoans such as the siphonophores (e.g., Portugese man-of-war) combine medusae and polyps to form a complex, free-floating animal.

7. The medusa stage dominates the life cycle of the exclusively marine class *Scyphozoa* (jellyfish). These animals are active swimmers with long tentacles specialized for defense and prey capture. Individual animals are either male or female. Their planular larvae form through sexual reproduction and then undergo a brief polyp stage before once again becoming free-swimming medusae.

8. The 6,000 species of the class *Anthozoa* include the sea anemones and corals. They are all marine. In this group the medusa stage is completely absent; both asexual and sexual reproduction are performed exclusively by the polyp.

9. Anemones are typically solitary and sedentary, although a few can crawl or swim, while corals are usually colonial and exclusively sessile. Corals produce an external skeleton of calcium carbonate that supports the colony. The buildup of these skeletons over geologic time has played an important role in the formation of large oceanic reefs.

10. Many corals live in clear tropical oceans. They survive in these relatively nutrient-poor waters by forming symbiotic relationships with photosynthetic dinoflagellates. These tiny protists significantly contribute to the nutritional base of the coral colony in return for the defense provided by the animal's stinging cells.

11. The cnidarians' long evolutionary history and success are due to their combination of simple body plans, low metabolic rates because the largely inert mesogleal layer dominates their body, and their ability to capture large prey. They can thus survive in environments with lower encounter rates of prey than animals with greater

metabolic needs. Associations with photosynthetic symbionts further enhance the lives of some colonial cnidarians.

Activities (for answers and explanations, see page 226)

• Compare and contrast the anatomy and foraging mechanisms of cnidarians with those of sponges.

• For each cnidarian class listed below, indicate with a checkmark whether the medusa or the polyp is the dominant stage in the life cycle.

Class	Medusa	Polyp
Hydrozoa		
Anthozoa		
Scyphozoa		

• The diagram below shows the polyp stage from a generalized cnidarian life cycle. Label the following structures: ectoderm, endoderm, mesoglea, gastrovascular cavity, mouth, and tentacles.

Questions (for answers and explanations, see page 226)

1. Compared with the sponges, which of the following does *not* represent a difference in body plan found in the cnidarians?
   a. Presence of a distinct body cavity
   b. Radial symmetry
   c. Possession of nematocysts
   d. Alternation of body forms
   e. Gastrovascular cavity

2. Corals are common representatives of the cnidarian class _____ .
   a. Scyphozoa
   b. Anthozoa
   c. Hydrozoa
   d. Porifera
   e. Placozoa

3. Which of the following traits is *not* shared by both sea anemones and jellyfish?
   a. Nematocysts located on tentacles
   b. Possession of a gastrovascular cavity
   c. Dominant stage in life cycle is a medusa
   d. Engage in sexual reproduction
   e. Are carnivorous

4. Which of the following is a reason the cnidarians are an evolutionarily successful group?
   a. Simple body plans
   b. Ability to capture large prey
   c. Low metabolic needs
   d. Symbiotic associations with photosynthetic organisms
   e. All of the above

5. Which of the following characteristics is unique to the phylum Cnidaria?
   a. Sexual reproduction
   b. Symbiotic associations with other organisms
   c. Sedentary body forms
   d. Nematocysts
   e. Nervous tissue

THE EVOLUTION OF BILATERAL SYMMETRY • PROTOSTOMES AND DEUTEROSTOMES: AN EARLY LINEAGE SPLIT • PROTOSTOMES: A DIVERSITY OF BODY PLANS • WORMLIKE BODIES: MOVEMENT THROUGH SEDIMENTS (pages 638–643)

Key Concepts

1. Most animals exhibit bilateral symmetry, which probably first arose in simple, flattened, crawling organisms not unlike the modern placozoans.

2. The diploblastic phylum *Ctenophora* (comb jellies) contains about 100 species of bilaterally symmetric marine carnivores. They represent an early split from the lineage that gave rise to all other bilateral animals.

3. Ctenophores are superficially similar to the cnidarians, but have important structural differences. Ctenophore guts have one-way flow by virtue of two anal pores that open opposite the mouth. Body movement is created by the action of eight rows of fused ciliary plates (ctenes), instead of musclar action as in the cnidarians.

4. The comb jellies lack stinging cells. They capture prey using sticky filaments attached to their tentacles. Ctenophores are primarily plankton feeders and do not pursue large food items as do cnidarians. Sexual reproduction is simple: a miniature ctenophore develops directly from a fertilized egg and slowly grows to adult size.

5. The next major split in animal evolutionary lineages produced the two dominant, monophyletic animal groups present today; the *protostomes* and *deuterostomes*. All are bilaterally symmetric and triploblastic. All protostomes are united by shared, derived traits of central nervous system construction and larval feeding mechanisms that are different from those that unite the deuterostomes.

6. Most of the animals alive today are protostomes. Diversification of protostome body plans has been driven largely by advantages gained from rapid movement in both predators and prey. One such movement-related feature is the *hydrostatic skeleton*, which works by creating forces through the action of opposing muscle groups on the constrained, incompressible fluids inside a body.

7. Rotifers (phylum *Rotifera*) are tiny protostomate animals not much larger than ciliate protists. The 1,800 species possess well-developed internal organs including a complete digestive tract. Their pseudocoel is greatly reduced and does not function as a hydroskeleton. All rotifers are aquatic (mostly freshwater) and very active. Their locomotion is made possible by beating cilia.

8. Food is pulled into a rotifer's mouth on water currents established by the beating of cilia in the *corona*, a feeding organ located atop the animal's head. Organic particles are swept into the *mastax*, where grinding teeth help process the food.

9. Members of the phylum *Acanthocephala* (spiny-headed worms) are obligate parasites, primarily of freshwater fishes. They have no gut, absorbing food directly through the body wall. The life cycle of many acanthocephalans involves multiple hosts, the behavior of which is modified by the worm to increase the chances that its life cycle will be successfully completed.

10. There are 16 wormlike protostome phyla, all of which are bilaterally symmetrical, legless, soft-bodied, and much longer than they are wide. Such a body form enables a worm to move easily through loose marine sediments.

11. Roundworms (phylum *Nematoda*) have a thick cuticle overlying their epidermis, beneath which are located well-developed longitudinal muscles. Circular muscles are lacking, resulting in somewhat inefficient thrashing movements. Roundworms are among the most abundant of all animals, with over 20,000 species known and perhaps as many as a million more yet to be identified. Thousands or millions of individuals can be found inhabiting very small volumes of organic matter.

12. The diets of roundworms are varied. Some are predators on protists and very small animals; others are significant parasites of plants and animals, including humans. The roundworm *Trichinella spiralis* causes the sometimes fatal disease trichinosis. In the case of human infection, pigs usually serve as the intermediate host for the parasite.

13. Horsehair worms (phylum *Nematomorpha*) are internal parasites of both terrestrial and aquatic insects and crabs. Adults have a highly reduced and perhaps nonfunctional gut, and thus these worms may feed only in their larval stages.

14. Loriciferans (phylum *Loricifera*) are representative of several groups of worms that inhabit loose ocean sediments. Loriciferans were discovered only in 1983. They are small animals whose biology remains poorly understood.

15. Two phyla of protostomes, the flatworms and ribbon worms, have traditionally been considered ancestral to nearly all other lineages. Your textbook adopts a new interpretation based on recent evidence suggesting that the supposedly ancestral body forms of these animals are actually derived traits.

16. Members of the phylum *Platyhelminthes*, or flatworms, lack a body cavity but have more complex internal organs than the cnidarians or ctenophores. Their flattened body plan permits efficient respiratory gas exchange because all cells are near the body surface. The digestive system of flatworms contains a mouth that opens into a blind-ended sac, which is often highly branched to increase the surface area available for nutrient absorption.

17. The class Turbellaria probably most closely resembles the ancestral flatworms. *Dugesia*, the common freshwater planarian, is representative of the group. Chemosensory organs are located on the anterior end of the animal, along with two light-sensitive eye spots. The presence of longitudinal nerve cords and a modest brain represent additional improvements for sensory control and locomotion in the flatworm body plan.

18. Some flatworms are carnivorous, others are scavengers. However, the bulk of the 25,000 species of flatworms are parasites. Tapeworms (class *Cestoda*) and certain flukes (class *Trematoda*) are common internal parasites of animals, especially vertebrates. Other flukes (class *Monogenea*) are important external parasites.

19. Complex life cycles with multiple hosts often characterize parasitic flatworms. These features are thought to help defeat the host's defense mechanisms and to increase the flatworm's likelihood of completing its life cycle.

20. The 900 species of ribbon worms (phylum *Nemertea*) are mostly marine and resemble flatworms in many respects. A notable exception is that nemerteans possess a complete digestive tract with a mouth at one end and an anus at the other. This one-way flow of food permits specialization of the digestive tract, leading to increased digestive efficiency.

21. The fluid-filled body cavity, or *rhyncocoel*, of ribbon worms contains a *proboscis* armed with a sharp stylet. The proboscis can be ejected with considerable force during prey capture, allowing paralytic toxins to be injected. It also assists in burrowing.

Activities *(for answers and explanations, see page 227)*

- The following schematic diagram shows a longitudinal section through an unidentified bilaterally symmetric animal. Explain why, based solely on the information available in this diagram, you can conclude that this mystery animal is *not* a member of the phylum Platyhelminthes.

Anterior   Posterior

- Assuming that the above schematic animal is a roundworm, draw the orientation of the muscles that operate its hydroskeleton, and briefly explain the kind of body motions their contractions would produce.

Questions *(for answers and explanations, see page 227)*

1. Which of the following statements is *true* of all flatworms?
   a. Flatworms are biradially symmetric.
   b. Flatworms have a complete digestive system.
   c. Flatworms tend to have large, thickened bodies.
   d. Flatworms are triploblastic.
   e. All of the above

2. Although relatively simple, *Dugesia* shows significant differences in body plan compared with the cnidarians and ctenophores. Which of the following is *not* one of those differences?
   a. Longitudinal nerve cords
   b. Complete digestive tract
   c. Anterior chemosensory organ
   d. Light-sensitive eye spots
   e. Highly branched gastrovascular cavity

3. Ribbon worms are members of the phylum
   a. Platyhelminthes.
   b. Nemertea.
   c. Nematoda.
   d. Rotifera.
   e. Cestoda.

4. Hydrostatic skeletons are found in all but one of the following animal groups. Select the *exception*.
   a. Roundworms
   b. Nemerteans
   c. Rotifers
   d. Nematodes
   e. Ribbon worms

5. The rhyncocoel is a body plan feature found in which of the following groups?
   a. Trematoda only
   b. Rotifera and Nematoda
   c. Rotifera only
   d. Nemertea and Cestoda
   e. Nemertea only

6. Consider the diagram of a rotifer shown below:

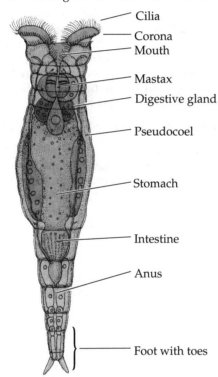

Cilia
Corona
Mouth
Mastax
Digestive gland
Pseudocoel
Stomach
Intestine
Anus
Foot with toes

Which aspects of this animal's body plan (apparent from this diagram) are also found in the phylum Nematoda?
   a. Complete digestive tract
   b. Bilateral symmetry
   c. Pseudocoelom body cavity
   d. Complex organ structure
   e. All of the above

COELOMATE PROTOSTOMES • SEGMENTED BODIES: IMPROVED LOCOMOTION (pages 643–650)

Key Concepts

1. The complete encircling of the coelomic cavity by both longitudinal and circular muscles, combined with body segmentation for added precision of movement, gave animals a new and highly effective means of controlling body shape. The segmented coelomate body evolved several times among protostome and deuterostome phyla.

2. There are also lineages of unsegmented coelomate worms. The phyla *Sipuncula* (peanut worms) and *Echiura* (echiurian worms) contain exclusively marine species who burrow in sand or mud, or make their homes in coral reef skeletons. They have either tentacles or a proboscis on their anterior ends that serve as feeding organs.

3. Pogonophorans (phylum *Pogonophora)* are deepwater, bottom-dwelling marine animals with tentacles that can be drawn down well into the coelom. There are about 145 species, all of which have been discovered only recently. The largest live near oceanic hydrothermal vents and use their large coelomic cavities to harbor colonies of specialized bacteria that fix carbon using energy obtained from hydrogen sulfide gas.

4. The 15,000 members of the phylum *Annelida* (annelid worms) are found in freshwater, marine, and terrestrial habitats. They exemplify the advantages gained by repeated segmentation of the coelom. Complex body shapes can be attained, the coordination of which is enhanced by separate nerve centers in each segment.

5. Annelids lack a rigid external covering, and their flexible body wall is key to the locomotory process. This thin wall is also essential for gas exchange, which limits annelids to moist habitats.

6. The majority of annelids belong to the class *Polychaeta.* The class is almost entirely marine. Its members have well-developed anterior eyes and tentacles. Thin outgrowths known as *parapodia* emerge from each segment, serving respiratory as well as locomotory functions. Stiff bristles called setae also assist in locomotion.

7. Polychaetes usually have distinct male and female sexes. The larval stage, known as a *trochophore*, adds body segments as it grows, eventually metamorphosing into an adult worm.

8. The class *Oligochaeta* is composed mainly of terrestrial or freshwater annelids. The common earthworm is an example. Oligochaetes are virtually naked by comparison with polychaetes, lacking all external appendages except for a few setae. Most are scavengers.

9. All oligochaetes are *hermaphroditic* (an individual has both male and female sex organs). When two individuals copulate, sperm is exchanged between each member of the pair. Eggs are laid externally in a cocoon, which later hatches to release miniature worms.

10. Members of the class *Hirudinea* (leeches) show many similarities in body form and reproduction with the oligochaetes, from which they probably evolved. However, leech body segments are not divided into distinct compartments as in the annelids. Their movement is thus quite different. Leeches move with the aid of suckers located at each end of the body.

11. Most leeches are external parasites of other animals. They have toothed jaws that can penetrate the host's outer surface. An anticoagulant is then injected into the wound to maintain blood flow. This property of leeches was once widely used in early forms of medicine, and is still employed today under specific circumstances to reduce fluid pressure in damaged tissues.

12. Most of the 100,000 species of modern mollusks (phylum *Mollusca*) possess heavily calcified shells into which they can retract their body to avoid predators. These shells may be large and rather ornate. Early mollusks were wormlike with only calcified spicules on their dorsal surface for protection.

13. Although highly modified within specific groups, the basic molluscan body plan consists of a large muscular *foot* for locomotion, a covering of specialized tissue called the *mantle* that protects the internal organs and secretes the shell, and a *visceral mass* of internal organs proper.

14. An important anatomical character of mollusks is the *radula*, a rasping feeding structure located in the anterior end of the animal. In some forms it has become modified for drilling or as a piercing dart attached to poison glands.

15. Mollusks have a complete digestive tract and a simple nervous system. Their coelom is much reduced. The fluid-filled sinuses of the open circulatory system are important elements in the animal's hydraulic skeleton.

16. The monoplacophorans (class *Monoplacophora*) have a single, caplike shell instead of a series of interconnected plates. Once thought extinct, they are believed to be the ancestors of all other mollusks. They have multiple gills, well-developed muscles, and a repeating excretory system.

17. Chitons (class *Polyplacophora*) have conspicuous plates that connect along the back to form the shell. They are marine grazers that spend most of their lives firmly attached to algae-covered rocks.

18. One lineage of early mollusks developed a hinged shell that covered the entire body. The class *Bivalvia* (bivalves) includes the clams, oysters, scallops, and mussels. Many have large, heavy shells that have so reduced locomotion that they are quite sedentary. Food is filtered from the water by the gills, which serve both respiratory and feeding functions.

19. A second lineage of early mollusks developed into the gastropods (class *Gastropoda*). These animals use their large muscular foot to move about slowly. Gastropods undergo a characteristic larval torsion during development, which ends up rotating the visceral mass 180 degrees relative to the foot. The result is that both the anus and mouth come to lie above the head, producing a U-shaped digestive tract and nervous system.

20. Species richness is high among gastropods. Familiar crawling forms include snails, slugs, whelks, limpets, abalones, and nudibranchs. Some are of substantial commercial value. Other, less familiar forms such as the sea butterflies and heteropods are active swimmers. The only terrestrial mollusks are gastropods.

21. Cephalopod mollusks (class *Cephalopoda*) are the most complex members of the phylum, and include such species as the squids and octopuses. They evolved from yet a third lineage of ancestral mollusks.

22. The cephalopod foot is modified into a head with tentacles, a siphon for creating propulsive water jets, and a well-developed brain. Cephalopods have large gills inside the mantle cavity and a beak for capturing and subduing prey.

23. Cephalopods have sophisticated mechanisms for regulating the air/water mixture in their shell cavity, making them the first animals to move vertically within the water column. They also have an exit tube that can forcibly direct water out from the shell, allowing them to swim rapidly and be active predators.

24. Sophisticated locomotory abilities and a pelagic (open ocean) lifestyle have favored the evolution of elaborate sensory organs among cephalopods. Their eyes are in many ways comparable to those of humans and other vertebrates.

25. Members of the genus *Nautilus* are living representatives of the earliest cephalopods. They possess an external shell and have distinct limitations for controlling their buoyancy. Evolution altered the body plans of subsequent cephalopod groups, substantially reducing the size of the shell and substituting other mechanisms of buoyancy control, such as fast swimming.

Activities (for answers and explanations, see page 227)

- Next to the arthropods, mollusks are the most diverse group of invertebrate animals. List three reasons that the mollusks have been so evolutionarily successful.

  a.

  b.

  c.

- List several ways in which the body plans of oligochaete and polychaete worms are similar and several ways in which they differ.

Questions (for answers and explanations, see page 227)

1. Match each of the characteristics listed below with the molluscan group(s) that show those characteristics.

   a. Advanced eyes   _____   Monoplacophorans

   b. Larval torsion   _____   Cephalopods

   c. Pelagic hunting   _____   Gastropods

   d. Calcified shells   _____   Polyplacophorans

   e. Multiple gills   _____   Bivalves

2. When dissecting a squid, which of the following would you *not* expect to find?
   a. A beak
   b. Numerous interconnected plates forming a shell
   c. Well-developed gills inside the mantle cavity
   d. Large grasping tentacles
   e. A well-developed brain and visual system

3. The group of mollusks that is believed to have evolved earliest is the class
   a. Bivalvia.
   b. Polyplacophora.
   c. Gastropoda.
   d. Cephalopoda.
   e. Monoplacophora.

4. A meal containing snails, clams, and octopus represents which of the following classes of mollusks?
   a. Bivalvia only
   b. Polyplacophora only
   c. Gastropoda, Bivalvia, and Cephalopoda
   d. Cephalopoda and Gastropoda
   e. Monoplacophora only

5. Which of the following structures is *absent* from a typical gastropod mollusk (e.g., a garden snail)?
   a. Protective shell
   b. Head
   c. Radula
   d. Mantle
   e. None of the above

6. The cephalopods are complex mollusks. Which of the following is *not* a characteristic of this group?
   a. Grasping tentacles
   b. Sophisticated vision
   c. Sessile lifestyle
   d. Propulsive water siphon
   e. Large brain

7. Suppose you have a box before you that contains a representative of all annelid worm species. If you blindly reach into this box and pull out one individual, odds are that it will be a member of the class
   a. Oligochaeta.
   b. Polychaeta.
   c. Pogonophora.
   d. Hirudinea.
   e. Sipuncula.

8. The combination of a true coelom and repeating body segmentation allows the annelids to do which of the following that anatomically "simpler" worms cannot?
   a. Move through loose marine sediments
   b. Attain complex body shapes and thus locomote more precisely
   c. Feed as external parasites on other animals
   d. Be hermaphroditic
   e. Inject paralytic poisons into their prey

ARTHROPODS: SEGMENTED EXTERNAL SKELETONS • THEMES IN PROTOSTOME EVOLUTION (pages 650–658)

Key Concepts

1. The development of *exoskeletons* was a major evolutionary change in animal body form. Originating from a thickening of the outer body covering through incorporation of the strong, flexible polysaccharide *chitin*, exoskeletons later became hardened to assume supportive, protective, and locomotory functions.

2. The large group known collectively as the *arthropods* are the most diverse animals with exoskeletons. Their hardened external coverings are jointed, especially the appendages. This design, in conjunction with specialized internal musculature, permits rapid movement. Other jointed appendages acquired sensory, feeding, or reproductive functions.

3. A hardened exoskeleton makes hydrostatic movements virtually impossible. Arthropod coeloms have thus acquired a new function: they now act as body cavities in which blood bathes the animal's internal organs.

4. Exoskeletons are not without limitations. Because they are impermeable to gases, specialized respiratory systems such as gills or trachea had to evolve in arthropods. Moreover, since they are of fixed size, exoskeletons constrain the animal's growth and must be molted periodically to accommodate increases in body size. This temporarily renders the animal unprotected and hinders its movement until a new skeleton forms.

5. Arthropods are a highly diverse lineage, due largely to the great adaptability of their jointed exoskeleton. The locomotory, supportive, protective, and antidesiccation benefits of this body plan allowed arthropods to successfully colonize terrestrial environments. The arthropods account for the largest percentage of all animal species, numbering well over one million.

6. Four phyla of arthropods are known, three of which have living representatives today. Trilobites (phylum *Trilobita*) are an extinct group of arthropods that flourished until the end of the Paleozoic; why they disappeared remains a mystery. Although heavily armored, they had a simple repetitive body plan reminiscent of the annelids.

7. The living arthropods are classified into three phyla: *Chelicerata*, *Crustacea*, and *Uniramia*.

8. The classes *Pycnogonida* (sea spiders), *Merostomata* (horseshoe crabs), and *Arachnida* (spiders, scorpions, mites, and relatives) are all members of the phylum Chelicerata. The bodies of all 63,000 known species are

divided into two regions, an anterior portion with mouth parts and a posterior region with four pairs of walking legs. Sea spiders are a small group of exclusively marine chelicerates. Horseshoe crabs are locally common along certain seashores, and have remained essentially unchanged for many millions of years.

9. Arachnids are highly successful terrestrial relatives of horseshoe crabs. Most have simple life cycles where miniature adult forms hatch directly from fertilized eggs. Others give birth to live young. Some are parasitic, but many are predators (e.g., spiders) and use their well-developed vision or elaborate silken webs to capture prey.

10. Crustaceans, with over 40,000 species, are the dominant marine arthropods. They include such familiar forms as sowbugs, crabs, shrimp, and lobsters. The body plan of most consists of a head, thorax, and abdomen, with appendages attached to each. A protective fold of the exoskeleton, the *carapace*, covers the head and other body segments. Separate sexes engage in copulation, and in some species the eggs hatch into a free-living larval form called a *nauplius*.

11. Barnacles (class *Cirripedia*) are specialized crustaceans in which the adult is housed inside a secreted calcium shell attached to the substrate. The animal stands on its head within this shell, and feeds by extending its modified appendages into the water currents to gather food particles.

12. Members of the phylum *Uniramia* have bodies that are divided into two or three regions. The posterior region of these animals typically has many more segments and appendages than does the anterior. Many unirames are terrestrial and possess specialized air-breathing systems for delivering oxygen to their internal tissues.

13. The Uniramia contains two dominant subphyla, the *Myriapoda* and the *Insecta*. The most notable members of the myriapods are the millipedes and centipedes. These 13,000 species have long, flexible bodies with well-formed heads, segments, and appendages. Millipedes are scavengers, while centipedes are active predators.

14. The 75 species of onychophorans (subphylum *Onychophora*) are also unirames. Their bodies are not segmented, and they share anatomical traits with both the annelids (internal organ structure, hydrostatic skeleton and flexible chitinous body covering) as well as the arthropods (circulatory system and simple appendages).

15. Another group of unirames includes the 550 living species of water bears (subphylum *Tardigrada*). Their body plan is somewhat similar to onychophorans, but since they are much smaller they lack circulatory and respiratory systems. Water bears live in marine sediments and wet soils. Neither the water bears nor the onychophorans are ancestral to the other arthropods, although they may resemble early arthropod body forms.

16. The bodies of insects (subphylum Insecta) are divided into head, thorax, and abdominal segments. A single pair of antennae is attached to the head and three pairs of legs are connected to the thorax. Specialized tubes called *tracheae* allow gas exchange by bringing air directly to the body tissues.

17. Insects are incredibly abundant and diverse. The 1.5 million described species probably represent only a fraction of those living today. They are found in virtually all habitats, although very few live in the oceans. Insects make their living through predation, scavenging, and parasitism.

18. Insects that hatch as miniature adults and then grow by simply increasing their size are said to have *simple development*. In other species, hatchlings are radically different from the adult form and undergo a developmental change known as *metamorphosis*. In metamorphosis, subsequent molts produce new immature stages called *instars*. If changes between these instars are gradual, the metamorphosis is *incomplete*, but if the changes are dramatic, then the metamorphosis is said to be *complete*.

19. Associated with complete metamorphosis is extreme specialization of each stage in the life cycle. Adults are often best adapted for reproduction and dispersal, whereas larvae are adapted for growth and feeding. A specialized inactive phase called the *pupa* represents a transitional form between the larva and the adult.

20. Entomologists recognize some 28 orders of insects, all of which can be classified as either wingless or winged. The winged insects can be further subdivided into those that do not fold their wings over the back, and those that do. The ability to fold the wings means that they are out of the way while the insect is not in flight. This adaptation has allowed the folded-wing groups to invade many more habitats than the fixed-wing groups. The folded-wing groups have thus become extremely abundant and diverse.

21. Insect diversity is in part explained by the ancient origins of this class, which permitted them to colonize relatively unexploited terrestrial habitats where they underwent extensive evolutionary adaptation in body form in the absence of significant competition from other animal phyla.

22. The fluid-filled cavities of protostomes were excellent precursors for hydrostatic skeletons that, with appropriate changes in musculature, enabled sophisticated movements. Subsequent segmentation of the body further enhanced mobility, ultimately leading to swimming forms.

23. Hard, external protective coverings originally evolved in response to predation pressure. Once in place, these coverings were incorporated as key elements of new locomotory systems.

24. Since many protostome lineages originated at times when only small organisms or particulate food was available, numerous sessile forms that actively extract food from moving water arose and are still represented today. Despite the relative ease of feeding, sessile body forms require specializations for finding mates, for the dispersal of gametes and larvae, and to deal with the intense competition for space. The emergence of coloniality was one evolutionary solution to these problems.

Activities (for answers and explanations, see page 227)

- On the following drawing of a crab larva, or nauplius, identify as many characteristics as you can that designate this animal as a member of the arthropod phylum Crustacea.

- Protostome evolution was intimately linked to the aquatic environment, and only much later did colonization of the land occur. List three lines of evidence that support this view of a watery origin for the protostomes.

a.

b.

c.

Questions (for answers and explanations, see page 227)

1. Lobsters, millipedes, and butterflies all share which of the following traits?
   a. Parapodia
   b. Setae
   c. Gas exchange across the skin
   d. Jointed appendages
   e. Complete metamorphosis

2. All but one of the following animals has an exoskeleton. Select the *exception*.
   a. Millipede
   b. Barnacle
   c. Honeybee
   d. Copepod
   e. Earthworm

3. One day while vacationing at the beach, your friend brings you an unidentified animal that was found in a local tidepool. You note that it has a true coelom, numerous appendages, and only male sex organs. It also lacks any evidence of an exoskeleton. To which of the following groups does this animal most likely belong?
   a. Polychaeta
   b. Pogonophora
   c. Oligochaeta
   d. Crustacea
   e. Insecta

4. When grasshoppers hatch from eggs, they look like miniature adults. Over time they pass through several molting stages until they eventually reach full maturity. This pattern of insect development is known as
   a. complex development.
   b. incomplete metamorphosis.
   c. complete metamorphosis.
   d. pupal development.
   e. simple development.

5. Onychophorans and tardigrades are similar to earthworms in that they have
   a. jointed appendages.
   b. a hardened exoskeleton.
   c. a flexible body covering.
   d. a well-developed respiratory system.
   e. all of the above

6. The bulk of arthropods belong to the phylum Uniramia, and of these, the majority of species belong to the
   a. Myriapoda.
   b. Insecta.
   c. Cirripedia.
   d. Tardigrada.
   e. Trilobita.

Integrative Questions (for answers and explanations, see page 228)

1. Which of the following statements is true of all bilaterally symmetric animals?
   a. They are ingestive heterotrophs.
   b. They have some form of digestive system.
   c. They are multicellular.
   d. They are triploblastic.
   e. All of the above

2. Ants, snails, and earthworms all share which of the following characteristics?
   a. They are protostomes.
   b. They are acoelomate animals.
   c. They are biradially symmetric.
   d. They have excellent vision.
   e. None of the above

3. The body plans of rotifers and squid are alike in that they both
   a. have spherically symmetrical bodies.
   b. possess complete digestive tracts.
   c. are pseudocoelomate animals.
   d. have radial cleavage of the embryo.
   e. all of the above

4. A lobster eats a clam. An octopus then eats the lobster. Which phyla of animals are involved in this feeding sequence?
   a. Crustacea only
   b. Mollusca only
   c. Mollusca and Uniramia
   d. Crustacea and Mollusca
   e. Uniramia and Crustacea

5. The extinct group of animals known as the trilobites are most closely related to which of the following?
   a. Crabs
   b. Earthworms
   c. Snails
   d. Sponges
   e. Sea anemones

6. Which of the following animals does *not* exhibit cephalization?
   a. Cockroach
   b. Nautilus
   c. Jellyfish
   d. Earthworm
   e. Lobster

30

Deuterostomate Animals: Evolution of Larger Brains and Complex Behaviors

THREE BODY PARTS: AN ANCIENT BODY PLAN • ECHINODERMS: COMPLEX RADIAL SYMMETRY (pages 663–669)

Key Concepts

1. Deuterostomes are characterized by having indeterminate cleavage, mesoderm that forms as an outpocketing of the embryonic gut, and an anus that forms directly from the blastopore. All are triploblastic and possess well-developed body cavities.

2. Although deuterostome species diversity is low compared with protostomes, the deuterostomes are of special interest to us because we are of that lineage.

3. The *lophophorate* animals are sessile, filter-feeding, mainly marine deuterostomes that possess a specialized feeding and respiratory structure called a *lophophore*. It typically consists of one or two rows of ciliated hollow tentacles, which are located near the mouth and are under muscular control.

4. The lophophorate body plan consists of an anterior *prosome*, middle *mesosome*, and posterior *metasome*. Each of these body compartments has its own coelomic cavity, known respectively as the *protocoel*, *mesocoel*, and *metacoel*. The gut is U-shaped, with the anus emerging near the mouth, but outside of the tentacles.

5. The phylum *Phoronida* is a small group (15 species) of bottom-dwelling, marine worms that live in secreted chitinous tubes. Particulate food is filtered through the external lophophore and transferred to a stomach located at the animal's posterior end.

6. The brachiopods (phylum *Brachiopoda*) superficially resemble bivalve mollusks, but they are actually sedentary lophophorates whose feeding structure is housed inside a hinged shell. Once very numerous, today only about 350 species survive in marine habitats.

7. The moss animals (phylum *Bryozoa*) are colonial lophophorates. They secrete "houses" from which individuals emerge to feed. Food is passed internally between individual animals by connecting tissues. The colony forms through the asexual reproduction of a single founding member. Sexual reproduction yields free-ranging larvae that ultimately settle to form new colonies.

8. The 10 living species of pterobranchs (phylum *Pterobranchia*) are little changed from the ancestral lophophorates. They are small and sedentary. Their body plan is clearly lophophorate in its basic features, but pterobranchs also have a proboscis surrounded by a collar of tentacles that serves both respiratory and feeding functions. Some species are colonial and others solitary.

9. Although the basic lophophorate body plan is mainly associated with sedentary or only slightly mobile lifestyles, other tripartite deuterostomes such as the arrow worms (phylum *Chaetognatha*) have become quite mobile. The 100 species of arrow worms have a streamlined, bilaterally symmetric body with two pairs of lateral fins and a taillike caudal fin.

10. Arrow worms are small marine carnivores that swim in short rapid bursts. This allows them to feed among masses of plankton and larval fish in the open ocean, where they are often the dominant predators. Some arrow worms engage in daily vertical migrations of several hundred meters to avoid their own predators.

11. The 7,000 living species of echinoderms (phylum *Echinodermata*) have combined a radially symmetric body plan with an active lifestyle. A significant change among echinoderms was the calcification of an internal skeleton that provides protection from predators.

12. Nearly all living echinoderms have bilaterally symmetric, ciliated larvae that feed in the plankton before settling to the substratum and metamorphosing into radial adult forms.

13. The *water vascular system* is a diagnostic trait of echinoderms. Seawater fills a system of modified coelomic cavity channels, which are under muscular control. This system coordinates the action of hundreds of *tube feet*, which cause movement. A connection to the outside environment is provided by way of a pore called the sieve plate.

14. *Pelmatozoans* are a subphylum of the echinoderms that were once very abundant, but today are represented only by the sea lilies and feather stars (class *Crinoidea*).

15. Sea lilies (80 species) attach themselves to the substratum and stand upright by means of a flexible stalk of calcareous discs from which extend many jointed arms. Feeding is accomplished by the muscular contraction of the arms and tube feet, which funnel food into the animal's mouth. The tube feet are also used for gas exchange and excretion.

16. The 600 species of feather stars are much like sea lilies except that they have flexible appendages. These appendages grasp the substratum during feeding or resting and then let go to allow walking or swimming movements.

17. *Eleutherozoans* represent the other subphylum of modern echinoderms. The 2,000 species of brittle stars (class *Ophiuroidea*) are like crinoids in outward appearance, but usually have many fewer arms (often only five). They feed on organic material from the surface of marine sediments, and have only a single opening to the digestive system that acts as both mouth and anus, a characteristic not shared with other eleutherozoans.

18. Sea stars (class *Asteroidea*) are probably the most familiar echinoderms. Although they somewhat resemble the brittle stars, the arms of sea stars are much less flexible. They are important predators in many marine communities.

19. The tube feet of sea stars use both adhesive substances and hydraulic suction to facilitate locomotion and feeding. Suction is generated by the interplay of internal musculature and the water vascular system. The combined force of a sea star's many hundreds of tube feet working in unison can exhaust the powerful muscles of the bivalves, gastropods, fish, and worms on which sea stars prey.

20. Some sea stars have an additional adaptation for feeding on bivalves: they evert their stomach into an opening in the prey's shell. The sea star then secretes digestive enzymes into the soft tissues of the bivalve, ingests the food, and retracts its stomach before moving on.

21. All other echinoderms lack arms. Sea daisies (class *Concentricycloidea*) are a little-known group that apparently feeds on bacteria found on rotting wood in deep ocean waters.

22. Sea urchins (class *Echinoidea*) are round-bodied, herbivorous echinoderms that possess a fused internal skeleton. The skeleton is covered with defensive spines that may secrete toxic substances.

23. Sea cucumbers (class *Holothuroidea*) have a more elongated, soft body. They lack spines and possess large feathery feeding tentacles that protrude from the mouth. The tentacles are covered with a sticky mucus to which food adheres.

Activities (for answers and explanations, see page 228)

- For each of the deuterostome groups listed below, indicate with a "yes" or "no" whether that group possesses a lophophore.

  _____   Pterobranchs

  _____   Bryozoans

  _____   Chaetognathans

  _____   Phoronids

  _____   Brachiopods

  _____   Echinoderms

- What is the major functional difference between the feeding mechanism of a sea star and that of a brachiopod?

- Which group tends to be the more mobile, lophophorates or echinoderms? What anatomical feature is responsible for this marked difference in locomotory ability?

Questions (for answers and explanations, see page 228)

1. The general evolutionary trend among the deuterostomes can best be stated as
   a. a decrease in the number of body segments from three to one.
   b. an increase in the size and importance of the lophophore.
   c. a reduction in the size and importance of the lophophore combined with a more mobile lifestyle.
   d. a strong trend toward a more sedentary lifestyle.
   e. a loss of mobility due to the enlargement of the lophophore.

2. Which one of the following is _not_ a valid characteristic of the moss animals?
   a. Indeterminate cleavage of the early embryo
   b. Tripartite body plan with presence of a well-developed lophophore
   c. Segmented coelomic cavity
   d. Asexual formation of larvae, which disperse to form new colonies

   e. Common tissue connections between colonial individuals

3. The phoronids differ from the arrow worms mainly in that
   a. the phoronids are completely sessile, whereas arrow worms are active swimmers.
   b. arrow worms lack the obvious lophophore present in the phoronids.
   c. phoronids are filter feeders and arrow worms are predators.
   d. phoronids secrete chitinous tube shells and arrow worms do not.
   e. all of the above

4. The brachiopods bear a superficial resemblance to which of the following protostome groups?
   a. Annelid worms
   b. Round worms
   c. Bivalve mollusks
   d. Cephalopod mollusks
   e. Crustaceans

5. During the dissection of a sea lily, you would expect to find all of the following _except_
   a. a water vascular system.
   b. diploid cells.
   c. some form of coelomic cavity.
   d. a lophophore.
   e. tube feet.

6. Anatomically, the echinoderms that are most similar to feather stars are the
   a. sea lilies.
   b. sea stars.
   c. sea urchins.
   d. sea cucumbers.
   e. brittle stars.

7. All but one of the following echinoderm classes belong to the subphylum Eleutherozoa. Select the _exception_.
   a. Asteroidea
   b. Concentricycloidea
   c. Crinoidea
   d. Ophiuroidea
   e. Holothuroidea

8. A friend returns from the beach one day and shows you the skeleton of what he claims is a sea star. Upon close examination you inform him that what he really has is the internal skeleton of a sea urchin. Which of the following observations allows you to make this discrimination?
   a. There are numerous tube feet still attached to the skeleton.
   b. The skeleton shows evidence of once having had many large arms attached.
   c. There is a long flexible stalk made of calcareous discs.
   d. There is a large hole at one end through which feeding tentacles once protruded.
   e. The skeleton is rounded, completely enclosed, and has five rows of holes for the tube feet.

9. The fossil record of echinderms is substantial. Which of the following are reasons for this large number of echinoderm fossils?
   a. Their internal skeletons fossilize well.
   b. They are abundant in the world's oceans.
   c. They have been successful deuterostomes for over 400 million years.
   d. They are bottom-dwellers and live in areas that promote fossilization.
   e. All of the above

MODIFIED LOPHOPHORES: NEW WAYS OF FEEDING • THE ORIGIN OF THE VERTEBRATES: SUCKING MUD (pages 669–674)

Key Concepts

1. Other deuterostomes modified the lophophore and its coelomic cavity to create a new tripartite body plan consisting of a proboscis, collar, and trunk. The acorn worms (phylum *Hemichordata*) include 70 species whose lophophore has been replaced by a proboscis modified for digging. A sticky mucus coating on the proboscis facilitates prey capture. Behind the mouth is a pharynx that opens to the outside through *pharyngeal slits* surrounded by respiratory membranes.

2. The chordates (phylum *Chordata*) are a sister lineage to the lophophorate deuterostomes that modified the respiratory function of pharyngeal slits to make them into feeding devices.

3. All chordates have bilateral symmetry, gill slits at some developmental stage, a ventral heart, a tail that extends beyond the anus, and a dorsal hollow nervous system and supporting rod called a *notochord*. In some chordates the notochord is lost during metamorphosis or is replaced by an internal skeletal structure.

4. The tunicates (subphylum *Urochordata*) represent a group of 2,500 sessile marine species that are probably most similar to the ancestral chordates. As adults they bear little resemblance to other chordates since their pharynx is modified to form a large feeding basket. However, as free-swimming larvae they possess the typical chordate characteristics of pharyngeal gill slits, a hollow nerve cord, and a notochord.

5. Sea squirts (class *Ascidiacea*) are the most common tunicates. Two large siphons direct water in and out of the large pharyngeal basket, where respiration and feeding take place. Some species are solitary, others are colonial.

6. The larvaceans (class *Appendicularia*) are a small group of tunicates that spend their entire lives as free-roaming individuals in the plankton.

7. Lancelets (subphylum *Cephalochordata*) are small, fish-like, sediment-dwelling deuterostomes that have pharyngeal gill baskets used to filter food from the mud. They retain a notochord throughout life.

8. Many of the early chordates extracted prey from muddy ocean bottoms. Filtering food from mud is difficult because so much nonfood must be ingested. However, pharyngeal baskets with their many openings and large surface area are ideal for collecting small food particles from the mud and then eliminating the indigestible sediments.

9. About 500 mya, several additional chordate changes in body plan improved mud feeding. The *vertebral column*, a jointed dorsal supporting structure, replaced the more primitive notochord in the animals called vertebrates (subphylum *Vertebrata*). The development of external armor also allowed vertebrates to feed on the surface mud with less concern for predators.

10. The typical chordate body plan has a centralized vertebral column with two pairs of appendages attached. Chordates possess a large coelom in which the internal organs, including well-developed circulatory and respiratory systems, are grouped.

11. It is thought that the first vertebrates may have evolved in estuaries. Modern vertebrate fishes have the capacity to maintain nearly constant internal salt concentrations as environmental salinities change, a common occurrence in estuaries. Since predators on early vertebrates likely could not handle large salinity changes, life in estuaries provided relative protection from predation for the first vertebrates.

12. The first vertebrates were bottom-dwelling, mud-sucking animals. They ultimately gave rise to the jawless fishes (class *Agnatha*). Early agnathans such as the ostracoderms were small and heavily armored, and probably made their living feeding on carrion. A once abundant group, the agnathans are today represented by only two groups, the lampreys and hagfishes. They have a tough scaly skin instead of the ostracoderm armor and feed on carrion or by parasitizing other fish.

13. Jaws evolved from the skeletal gill arches of ancestral agnathans. Jaws represent a major change in the vertebrate body plan because they permit the capture and handling of living prey. The development of teeth further enhanced the usefulness of jaws by allowing hard body parts, as well as soft tissues, to be consumed. Chewing also aids digestion and improves the nutritional value of food.

14. Placoderms (class *Placodermi*) were the most important group of early jawed fishes. Although now extinct, these heavily armored, often large, and highly maneuverable animals were important predators of the Devonian seas.

15. The sharks, skates, rays, and their relatives comprise the cartilaginous fishes (class *Chondrichthyes*), a largely marine group characterized by a skeleton made entirely of pliable *cartilage*. Their skin is tough and leathery, and is covered with hardened projections that increase its durability.

16. The loss of heavy armor, as seen in the placoderms, coincided with an increase in rapid swimming ability in the Chondrichthyes. Also contributing to increased speed and maneuverability was the evolution of paired pelvic and pectoral fins, which improved balance and steering, along with a large caudal fin to provide propulsion.

17. Although most of today's sharks, skates, and rays are predatory, the largest living species are actually filter feeders that specialize in collecting enormous quantities of plankton.

18. The other major lineage of fishes evolved in fresh water. The bony fishes (class *Osteichthyes*) have skeletons made of calcified bone rather than cartilage. There is a single gill chamber covered by a hardened flap, the movement of which increases water flow over the gill surfaces.

19. Early bony fishes had lunglike sacs that supplemented gill respiration. Subsequent evolution altered these sacs to become *swim bladders*, which now serve as adjustable gas-filled organs for buoyancy.

20. Changes that improve swimming speed and maneuverability of bony fishes include loss of the heavy armor characteristic of ancient fishes. Their external surface is now covered by smooth, flat, thin scales, which provide adequate protection with little weight.

21. Remarkable diversity characterizes the 20,000 species of bony fish. Although a majority still inhabit fresh water, many have reinvaded the ocean, and virtually every feeding style including predation, grazing, and scavenging is represented.

22. The bony fish are behaviorally complex. Many are solitary, but open water species often form large schools for protection from predators and to improve food finding. Numerous species build nests, court mates, and actively defend their young.

23. Because fish eggs are heavier than water, they typically tend to sink. Many fish species thus return to shallow waters for breeding purposes, where they attach their eggs to the substrate. In species where the eggs are small enough, they may simply be discharged into surface waters where the larvae rapidly develop before the eggs can sink.

Activities (for answers and explanations, see page 228)

- Describe the major elements of the skeleton of a modern bony fish. What one skeletal factor differs most notably between these animals and the agnathan fishes?

- Discuss the main advantages of hinged jaws and why this evolutionary change stimulated an extensive radiation of the ancient agnathan fishes into more modern jawed forms.

Questions (for answers and explanations, see page 228)

1. Which one of the following is *not* a valid characteristic of the acorn worms?
   a. Determinate cleavage of the early embryo
   b. Tripartite body plan
   c. Segmented coelomic cavity
   d. Mucus-coated proboscis used for feeding
   e. Absence of a lophophore

2. The primary anatomical characteristic that preadapted pharyngeal slits for use as feeding structures was their
   a. large surface area.
   b. location anterior to the mouth.
   c. ability to provide support for the internal skeleton.
   d. bilateral symmetry.
   e. none of the above

3. Lancelets are like larvaceans in that
   a. the adults of both are primarily sedentary.
   b. neither retains a notochord throughout life.
   c. both are vertebrates.
   d. both have jaws.
   e. each evolved from some ancestral tunicate.

4. Which of the following statements correctly describes the placoderm fishes?
   a. Placoderm fishes arose directly from the tunicates.
   b. Placoderm fishes are members of the class Chondrichthyes.
   c. Placoderm fishes had large, formidable teeth.
   d. Placoderm fishes were heavily armored and jawless.
   e. A few placoderm species still live in modern oceans.

5. Which of the following groups of chordates evolved prior to the appearance of cartilaginous fishes?
   a. Bony fishes and sea squirts
   b. Tunicates, lancelets, and agnathans
   c. Tunicates, lancelets, and bony fishes
   d. Agnathans and bony fishes
   e. None of the above

6. The most probable origin of fish jaws was from the
   a. paired pelvic fins of bony fishes.
   b. pharyngeal slits of tunicates.
   c. notochord of larvaceans.
   d. gill arches of agnathans.
   e. supportive cartilage of the caudal fin in sharks.

7. The lungs of early bony fishes evolved into swim bladders. This change had the primary effect of
   a. improving feeding efficiency.
   b. supplementing gill respiration.
   c. aiding reproduction.
   d. providing buoyancy.
   e. making it easier to capture large prey.

8. Which of the following is *not* a key element in the basic vertebrate body plan?
   a. Internal skeleton
   b. Pharyngeal basket
   c. Dorsal spinal cord
   d. Segmented muscles
   e. Organs suspended in the coelom

COLONIZING THE LAND • SURVIVING DINOSAUR LIN-EAGES • THE TERTIARY: MAMMALS DIVERSIFY
(pages 674–684)

Key Concepts

1. The lobe-finned fishes (subclass *Crossopterygii*) were the first land vertebrates. The structure of their limbs, as well as their possession of lunglike sacs developed from swim bladders, preadapted them for the transition to land. Long believed to be extinct, a living crossopterygian (*Latimeria chalumnae*) was discovered in the 1930s. Because it lives at great depths, the swim bladder of this large predator contains fats rather than gases.

2. Amphibians (class *Amphibia*) evolved from some ancient crossopterygian-like ancestor during the Devonian. Early amphibians were probably quite clumsy on land, but rapidly evolved limb modifications that improved their locomotion.

3. Three orders of amphibians, comprising about 4,500 species, still exist today: caecilians (order *Gymnophiona*), frogs and toads (order *Anura*), and salamanders (order *Urodela*). They are widely distributed in tropical and temperate regions.

4. Amphibian reproduction and respiration remain tightly linked to the aquatic environment. The larvae of virtually all species are exclusively aquatic, eventually metamorphosing into adults that spend some portion of their life on land. The land-based portion of the life cycle is still limited to relatively moist habitats because lung breathing is supplemented with cutaneous respiration, which requires a moist body surface.

5. Although the reasons are unknown, amphibian populations are currently declining on a worldwide basis. The cause of this decline is an area of active research.

6. The transition from an amphibious existence to a fully terrestrial lifestyle was made possible by two events: the development of tough, water-impermeable skin, and the advent of a shelled egg through which gases, but not water, can freely pass. The reptiles (class *Reptilia*) were the first vertebrates to develop both traits.

7. The reptilian amniote egg houses an embryo surrounded by three membranes: the amnion, the chorion, and the allantois. This fluid-filled package is encased inside a leathery, water-retarding shell. Stored food in the form of yolk provides the developing embryo with nutrients, and little parental care is provided once the eggs are laid.

8. The heavily keratinized skin of reptiles is relatively water-impermeable, so gas exchange takes place exclusively via the lungs. Changes in respiratory and circulatory musculature, as well as additional heart chambers, make gas exchange more efficient than in amphibians. Reptiles can thus sustain more activity than amphibians.

9. The roughly 6,000 species of modern reptiles belong to three subclasses. *Chelonia* includes the turtles and tortoises. They have a heavy shell into which the head and limbs can be withdrawn for protection. Most are slow-moving vegetarians that are at least partially aquatic, but a few are exclusively terrestrial.

10. The subclass *Sphenodontida* contains only two lizardlike species known as the tuataras. The subclass Squamata contains all true lizards and the snakes. Most lizards have four legs that allow better terrestrial motion than amphibian limbs do. Lizard feeding modes include insectivory, herbivory, and carnivory. Snakes are completely legless, and have therefore evolved entirely new methods for locomotion. They are mainly carnivorous.

11. Another amniote lineage, the thecodonts (class *Archosauria*), diverged from the other reptiles during Mesozoic times and gave rise to the subclass *Crocodylia*, a collection of semi-aquatic predators that includes the crocodiles, alligators, gavials, and caimans.

12. A second group of thecodonts produced the dinosaur subclasses *Ornithischia* and *Saurischia*. These diverse and sometimes enormous animals ruled the earth for many millions of years inhabiting land, sea, and air. The dinosaurs evolved a more vertical positioning of the legs and special ventilatory muscles that enabled the lungs to operate simultaneously with locomotion. These important traits have been maintained in both birds and mammals.

13. One branch of the dinosaur lineage lives on today in the birds (subclass *Aves*). Birds are characterized by having feathers: highly modified reptilian scales that serve as insulators and increase the available surface area for flight. The best known avian fossil is *Archaeopteryx*, which had well-developed feathers but still retained obvious reptilian characteristics in its skull, limbs, breastbone, and tail.

14. Many anatomical features of modern birds are adaptations for weight reduction associated with flight. Radical changes in bone structure, breast musculature, metabolism, brain structure, respiration, and circulation evolved among birds. Many of these changes helped to make flight possible or more efficient.

15. The avian amniote egg also is freed from a requirement for external water. It has a rigid calcified shell and the same internal membranes found in reptilian eggs. Most bird eggs are laid in nests, which the parents tend. Some species hatch *altricial* (helpless) young, while others produce *precocial* (independent) offspring. Low reproductive rates combined with high levels of offspring survival are characteristic of birds.

16. As a group, birds have a highly varied diet. Herbivores and insectivores are common, and among the carnivorous species both active predators and scavengers exist. Sizes range from tiny hummingbirds to several living and extinct forms as large as or larger than humans.

17. The mammals (class *Mammalia*) emerged from yet another reptilian lineage during the Mesozoic. A major evolutionary adjustment involved skeletal simplification through loss or fusion of bones, particularly those of the jaw. Limbs came to be positioned under the body more than in reptiles, allowing improved locomotion.

18. Other key mammalian features include hair or blubber for insulation, mammary glands, sweat glands, and a four-chambered heart. The number of teeth is smaller in mammals than in reptiles, but mammalian dentition is more diverse.

19. Milk secreted by the female's mammary glands nourishes the young, a trait unique to mammals. Fertilization is internal, and most mammals undergo at least some development within the female's uterus prior to birth.

20. There are about 4,000 living species of mammals. Collectively, they show the largest size range of any animal group, with representatives as small as shrews and as large as the blue whale, the largest animal ever to live on Earth.

21. The mammalian subclass *Prototheria* is represented by a single order *Monotremata*, which contains the duck-billed platypus and spiny anteaters. These mammals lay an external, reptilelike egg. After hatching, the young nurse from their mother as do other mammals.

22. The subclass *Theria* contains two other mammal groups, the *marsupials* and the *eutherians*. Marsupials (e.g., koalas and kangaroos) give birth to their young very early in development. The young then crawl into the mother's ventral pouch, or marsupium, to complete their development. Marsupials are well adapted for prolonged droughts and variable habitats. Although once widely distributed, they are now restricted mainly to Australia and South America.

23. Eutherians represent by far the largest number of mammal species and are arranged into 16 orders. Unlike marsupials they undergo extensive development within the female's uterus prior to birth.

24. The largest eutherians are marine mammals (whales), but the most abundant are small terrestrial species such as rodents, bats, and insectivores. These latter groups alone account for more than 50 percent of all eutherian species.

Activities (for answers and explanations, see page 228)

- Use the following abbreviations to indicate which vertebrate groups possess(ed) the listed characteristics.
  (A = amphibians, B = birds, D = dinosaurs, L = lobe-finned fishes, M = mammals, R = modern reptiles)

  _____ Lungs

  _____ Aquatic larvae

  _____ Wings

  _____ Feathers

  _____ Uterine development

  _____ Terrestrial egg

  _____ Water-impermeable skin

- *Archaeopteryx*, although generally accepted as the earliest fossil bird, still exhibits many reptilian features. List four of these features and tell how each has been modified in modern birds.

a.

b.

c.

d.

Questions (for answers and explanations, see page 229)

1. The transition from aquatic to terrestrial lifestyles required many adaptations in the vertebrate lineage. Which of the following is *not* one of those adaptations?
   a. Switch from gill respiration to air-breathing lungs
   b. Improvements in water resistance of skin
   c. Development of feathers for insulation
   d. Alteration in mode of locomotion
   e. Adjustments to reproductive physiology

2. The eggs of reptiles and birds have many anatomical similarities. Which of the following is *not* a common feature of both?
   a. Developing embryo resides in an air-filled chamber
   b. Tough exterior shell for protection of embryo
   c. Ability to pass gases between egg and environment
   d. Complex system of internal membranes
   e. Abundant supply of yolk to nourish developing embryo

3. Dinosaurs dominated the Earth for nearly 150 million years before becoming extinct. Which of the following vertebrate groups is arguably a living representative of the dinosaur lineage?
   a. Amphibians
   b. Birds
   c. Lobe-finned fishes
   d. Snakes
   e. Mammals

4. All mammals nourish their young with milk manufactured by the mother, but not all are born after an extended period of development inside the female's uterus. Which of the following mammals emerges from the uterus in the *least* developed condition?
   a. Mouse
   b. Dolphin
   c. Human
   d. Kangaroo
   e. Bat

5. Modern amphibians include all of the following groups except for the
   a. caecilians.
   b. frogs.
   c. salamanders.
   d. toads.
   e. caimans.

6. Refer to the following diagram of an amniote egg:

Of the structures depicted in this diagram, which is responsible for providing the developing embryo with food?
   a. Amnion
   b. Allantois
   c. Yolk sac
   d. Shell
   e. Chorion

7. Consider the evolutionary sequence of change in vertebrate limb structure shown by Figure 30.29 in your textbook. Of what significance is the trend toward limbs being located more directly under the body?
   a. Improved thermoregulation
   b. More efficient feeding
   c. Better locomotion
   d. Increased egg laying ability
   e. None of the above

8. Amphibians remain tightly linked to moist habitats and bodies of water because of their
   a. food requirements.
   b. respiratory requirements.
   c. reproductive requirements.
   d. both a and b
   e. both b and c

HUMANS EVOLVE: EARTH'S FACE IS CHANGED • DEUTEROSTOMES AND PROTOSTOMES: SHARED EVOLUTIONARY THEMES (pages 684–692)

Key Concepts

1. The *primates* are the order of mammals to which humans belong. Primates have undergone significant radiation and evolutionary change since their appearance in the late Cretaceous.

2. Primate characteristics stem largely from adaptations for an arboreal lifestyle. Flexible limbs, dexterous hands with opposable thumbs, nails instead of claws, forward-facing eyes for improved depth perception, and small litters all facilitated life in the trees.

3. Living *prosimians* are similar to the earliest primates. They are represented today by lemurs, tarsiers, and their relatives. *Anthropoid* primates, which include monkeys, apes, and humans, diverged from prosimians some 55 mya. New World and Old World lineages emerged within the anthropoid lineage, and it is from the Old World group that apes emerged about 35 mya.

4. The first apes were largely arboreal, but later came down to the ground as climates cooled and forested habitats began to fragment 15 to 20 mya. Some of these apes underwent morphological changes. For instance, the extinct genus *Ramapithecus* had reduced muzzle size and shortened canines, features that suggest the beginnings of a lineage leading to humans.

5. There are four living genera of apes: gorillas *(Gorilla)*, chimpanzees *(Pan)*, orangutan *(Pongo)*, and gibbons *(Hylobates)*. Their dentition is quite different from that of humans and their relatives. In apes chewing occurs in an up and down fashion, whereas in humanlike primates chewing is from side to side. The teeth of *Ramapithecus* show clear evidence of side to side chewing.

6. Humans and their relatives (family *Hominidae*) began diverging from an apelike ancestral primate (possibly *Ramapithecus*) approximately 5 mya. The oldest known hominid fossil is "Lucy," a member of the species *Australopithecus afarensis*, which dates back 3.5 million years. Her species is thought to be the progenitor of all subsequent hominid groups.

7. Other australopithecine species also existed, apparently occurring simultaneously with *A. afarensis* but in different parts of the African continent. These include a robust form, and a more slender form called *A. africanus*. Both disappeared suddenly about 1.5 mya.

8. The australopithecines evolved *bipedalism*, in which only the hindlimbs are used for locomotion. Ground mobility was improved by modification of the pelvis and hindlimb musculature. Forelimbs became free to carry large objects, and better vision was facilitated by being able to see over the surrounding tall grasses.

9. Coincident with the evolution of bipedalism were changes in hand structure, which allowed for precise grasping motions. Foot modifications also improved walking and running. Both changes increased the ability of hominids to exploit their ground habitat.

10. Members of the genus *Homo* evolved from the australopithecines about 2.5 mya. *Homo* was a contemporary of *A. africanus* for at least half a million years.

11. *Homo habilis* was the first member of the genus to appear. Its body and brain size had nearly doubled over that of the australopithecines, and it was the first hominid to use tools for obtaining food.

12. From a basic diet of plant matter, *H. habilis* expanded its feeding habits to include animals. An increased division of labor between the sexes, with males hunting cooperatively and females harvesting plant foods while simultaneously tending infants, laid the foundation for more modern hominid societies.

13. The next hominid species, *Homo erectus*, appeared 1.6 mya and spread rapidly throughout Africa and Asia. *H.*

erectus was as large as modern humans, cooked its food (which included large mammals), and made specialized stone tools.

14. Our own species, *Homo sapiens*, replaced *H. erectus* in tropical Africa 500,000 years ago. Remnant populations of *H. erectus* survived for another quarter million years before disappearing completely. *H. sapiens* continued earlier hominid evolutionary trends of increasing brain size and social complexity.

15. With increased sociality came improved communication, leading to better coordination of group hunting and exploitation of variable or patchy resources. More effective communication skills may have also enhanced social status and thus reproductive success.

16. A new trait that appeared with *H. sapiens* was religion and a concept of life after death. Grave sites were provisioned with food, clothing, and tools, presumably intended for use in the afterworld.

17. Several varieties of *H. sapiens* have existed. Neanderthal people lived between 75,000 and 30,000 years ago. They were short, stocky, and created excellent hunting weapons.

18. The more recent Cro-Magnon people had bodies much like our own. They appeared between 50,000 to 100,000 years ago and briefly overlapped with Neanderthals, whom they may have exterminated. About 20,000 years ago their range expanded rapidly across Asia and into North America.

19. Cro-Magnons used sophisticated tools and were responsible for elaborate cave paintings depicting animal hunts. Their efficient cooperative hunting probably caused the extinction of numerous large mammals of that time.

20. Increasing brain size and the development of formal language ultimately led to *culture*, which then became an integral component of human evolution. Culture allowed traditions and knowledge to be passed rapidly from one generation to the next through teaching and observation, bypassing the slower process of evolutionary change through genetic mechanisms.

21. Changes in the pattern of human tool use reflect the role of culture in moving us from our cooperative hunting origins to our current lifestyles. *Agriculture* benefited greatly from cultural transmission, originating independently in the Middle East and Asia some 10,000 years ago, and then later independently in Central and South America. Steady increases in human population size are directly related to the more reliable and abundant food resources that resulted from pastoralism (herding animals) and agriculture.

22. Deuterostome evolution parallels that of protostomes in several ways: bottom feeding in marine sediments dominated the early evolution of each group; compartmentalization of the coelomic cavity and improvements in feeding structures occurred; and improvements in locomotion and buoyancy opened up new habitats for exploitation.

23. Both groups invaded the terrestrial world, but with different results. Exoskeletons provided protection and support for protostomes, but limited growth. The internal skeletons of deuterostomes also provided support against gravity, but did not hinder growth. The consequence is that vertebrates have become the dominant, large animals on land.

24. Some lineages of terrestrial vertebrates have returned to an aquatic lifestyle. Many of these aquatic reinvasions produced specialized filter-feeding forms.

25. Terrestrial herbivores must be relatively large to handle and process tough-bodied land plants in sufficient quantities to sustain growth and metabolism. The appearance of large herbivores in turn selected for large carnivores. This evolutionary cycle was at least temporarily halted by the appearance of humans and their use of tools for cooperative hunting and defense.

Activities (for answers and explanations, see page 229)

- Rank the following hominid species in terms of their appearance in the fossil record from most ancient (**1**) to most recent (**6**).

  _____ *Homo erectus*

  _____ *Homo sapiens* - Neanderthal

  _____ *Homo sapiens* - Cro-Magnon

  _____ *Australopithecus africanus*

  _____ *Australopithecus afarensis*

  _____ *Homo habilis*

- Which characteristics that predisposed the early primates for life in the trees are still present in modern humans?

Questions (for answers and explanations, see page 229)

1. The species to which the hominid fossil called "Lucy" belongs possessed all but one of the following traits. Select the *exception*.
   a. Bipedal locomotion
   b. Well-developed language and culture
   c. Hands capable of precision grasping
   d. Feet modified for walking and running
   e. Ability to carry objects with forelimbs

2. The first hominid species to regularly eat animal foods collected through cooperative hunting was
   a. *A. afarensis.*
   b. *A. africanus.*
   c. *A. robustus.*
   d. *H. habilis.*
   e. *H. sapiens.*

3. The development of language and culture in hominids was beneficial because
    a. it provided each species with a foundation for more efficient social interactions.
    b. it helped increase the status of individuals within the group.
    c. hunting and farming methods could be passed easily from one generation to the next.
    d. it complemented the division of labor between the sexes.
    e. All of the above

4. The now-extinct hominid most like ourselves is
    a. *H. habilis.*
    b. *H. erectus.*
    c. Cro-Magnon.
    d. Neanderthal.
    e. *A. robustus.*

5. Which of the following statements about human evolution is *false*?
    a. Bipedalism was a hominid adaptation for life on the ground.
    b. Increases in the size of hominid brains preceded the appearance of language and culture.
    c. The spread of pastoralism is a relatively recent human event that is largely the result of culture.
    d. The extinction of the Cro-Magnon people was caused by the emergence of the Neanderthals.
    e. Humans are not the direct descendants of modern-day chimpanzees.

6. All but one of the following is a trend in deuterostome evolution. Select the *exception*.
    a. Improvements in locomotion
    b. Development of a rigid exoskeleton
    c. Compartmentalization of the coelomic cavity
    d. Improvements of feeding structures
    e. Invasion of the terrestrial world

Integrative Questions (for answers and explanations, see page 229)

1. Match the relevant method of feeding with each deuterostome group listed.

    a. Filters with lophophore   _____ Chondrichthyes

    b. Bites with toothed jaws   _____ Agnatha

    c. Sucks mud or blood   _____ Asteroidea

    d. Digging proboscis   _____ Brachiopoda

    e. Extrudes stomach into prey _____ Hemichordata

2. Which of the following statements about deuterostome evolution is *true*?
    a. Lophophores are common in the phylum Echinodermata.
    b. Amphibians are the immediate evolutionary ancestors of the fishes.
    c. The evolution of jaws is but one of several important changes in feeding specializations.
    d. Larvaceans and lancelets gave rise to the echinoderms.
    e. A major advance in deuterostome evolution was the loss of the coelomic cavity.

3. A bony fish eats a frog. The fish is eaten by an eagle. The eagle is eaten by a tiger. How many different classes of vertebrates are involved in this feeding sequence?
    a. One
    b. Two
    c. Three
    d. Four
    e. Five

4. Which of the following characteristics is present at some stage in the life of modern humans?
    a. Bilateral symmetry
    b. Pharyngeal slits
    c. Internal skeleton
    d. Notocord
    e. All of the above

5. Choose the one set of characteristics that applies to *both* humans and sea stars.
    a. Eukaryotic, ingestive heterotroph, tube feet, radial symmetry, internal skeleton
    b. Chordate, radial symmetry, ingestive heterotroph, triploblastic
    c. Bilaterally symmetrical, lophophorate, vertebrate, produces terrestrial egg
    d. Deuterostomate, indeterminate cleavage of embryo, true coelomic cavity
    e. Water vascular system, internal skeleton, jaws, triploblastic

Answers and Explanations

Chapters 25–30

CHAPTER 25 – BACTERIA AND ARCHAEA: THE PROKARYOTIC DOMAIN

Note: The page numbers listed with the section titles refer to the Study Guide; numbers in parentheses for each question refer to the relevant key concepts.

WHY THREE DOMAINS • GENERAL BIOLOGY OF THE PROKARYOTES • PROKARYOTES IN THEIR ENVIRONMENT (pages 168–170)

- (5, 8–10, 13, 17, 26)
  1. Invasiveness is a measure of how easily bacteria can multiply within a host, while toxigenicity is a measure of the level of injurious chemicals the bacterium produces.
  2. Facultative anaerobes are those prokaryotes that can switch between aerobic and anaerobic metabolism, depending on environmental conditions.
  3. Chemoautotrophic prokaryotes use energy obtained from the oxidation of inorganic molecules to fix CO_2.
  4. Gram staining is a technique used to identify different categories of bacteria based on the chemical nature of their cell walls, which differentially bind to the staining molecules.
  5. These are the possible shapes of bacterial cells. "Coccus" = spherical, "bacillus" = rod-shaped, and "spiral" = curvilinear.

1. Choice *c*. Koch's postulates test for the pathogenic nature of microorganisms. In bacteria, both gram-positive and gram-negative bacteria may be pathogenic. (8–10, 25)
2. Choice *a*. The purple color indicates a positive reaction to the Gram stain. The rod shape indicates a bacillus type of bacterium. (5, 8–10)
3. Choice *d*. All three items represent important differences between eukaryotic and prokaryotic flagella. (7)
4. Choice *a*. The cyanobacteria use chlorophyll *a* as their primary photosynthetic pigment. (15)
5. (15–18)
   b: photoautotrophy
   d: photoheterotrophy
   c: chemoautotrophy
   a: chemoheterotrophy
6. Choice *c*. Although prokaryotes lack membranous internal organelles, many nevertheless have extensive internal membrane systems to carry out metabolic functions. (6)

7. Choice *c*. (20)
8. Choice *c*. An example of this combination of invasiveness and toxigenicity would be the bacterium that causes diptheria. (26)

PROKARYOTIC PHYLOGENY AND DIVERSITY • THE BACTERIA • THE ARCHAEA (pages 170–172)

- (4, 7, 8, 11, 13, 16, 17, 20, 21)
  a. Mycoplasmas: Bacteria, Eubacteria, lack cell walls, smallest cells known, parasites
  b. Gram-positive rods: Bacteria, Eubacteria, some make endospores, some toxic
  c. Cyanobacteria: Bacteria, Eubacteria, photosynthetic, fermentation, nitrogen fixers
  d. Methanogens: Archaea, Archaebacteria, obligate anaerobes, methane production
  e. Gram-negative rods: Bacteria, Eubacteria, metabolically diverse
  f. Hyperthermophiles: Archaea, Archaebacteria, tolerate low pH, high temperatures
- (8–10, 15)
  a. A unique protein structure composed of flagella that assists the movement of spirochetes.
  b. Chemicals manufactured by some Eubacteria that kill other cells, such as bacteria or fungi. Humans have harnessed this method of chemical defense for use in medical treatment against bacterial infections.
  c. Fruiting bodies form from aggregations of gliding bacteria and produce thick-walled spores or cysts, which are highly resistant to drying out.
  d. Heterocysts are found in filamentous forms of cyanobacteria. They serve as break points during reproduction and as important storage sites for oxygen-sensitive enzymes needed in nitrogen fixation.

1. Choice *e*. All of these groups have at least some toxic or disease-causing members. (10, 12, 14, 16)
2. Choice *a*. The methanogens are members of the Archaebacteria. (21)
3. Choice *d*. The absence of any kind of cell wall is unique to this group of bacteria. (16)
4. Choice *e*. All of the statements listed apply to methanogens. (16)
5. Choice *d*. (11)
6. Choice *e*. The antibiotic streptomycin was first isolated from these bacteria. (15)

Integrative Questions

1. Prokaryotes are primitive in the sense that they are structurally like the first types of cells to arise on Earth. They lack the sophisticated membrane-bounded organelles found in eukaryotic cells, which somewhat limits their cellular capacities. Referring to prokaryotes as primitive is also misleading because these organisms exhibit a wide array of metabolic adaptations and are capable of living in a number of rather harsh environments. The prokaryotes are extremely abundant and evolutionarily very successful organisms.

2. Even the largest viruses are VERY much smaller than the smallest cells!

3. *One.* Since viruses are acellular, they do not have kingdom status. The gram-positive bacterium is a member of the kingdom Eubacteria.

4. a. Hot springs: *hyperthermophiles*
   b. Intestines of a cow: *methanogens, gram-negative rods*
   c. Within other cells: *many forms, excluding the archae-bacteria*
   d. Alkaline lake: *extreme halophiles*

CHAPTER 26 – PROTISTS AND THE DAWN OF THE EUKARYA

Note: The page numbers listed with the section titles refer to the Study Guide; numbers in parentheses for each question refer to the relevant key concepts.

PROTISTA AND THE OTHER EUKARYOTIC KINGDOMS • THE ORIGIN OF THE EUKARYOTIC CELL • GENERAL BIOLOGY OF THE PROTISTS (pages 175–176)

● (3)

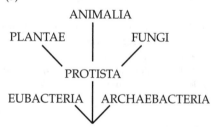

● The contractile vacuole is used to regulate water balance in some protists. In fresh water, the environment is hypoosmotic to the cell and water thus enters the protist's body. The contractile vacuole then pumps this excess water out. In an isosmotic solution, there should be no net accumulation of water, so the vacuole should contract only rarely. When moved to a hyperosmotic environment the cell will lose water through osmosis, so no pumping activity should be seen. (12)

1. Choice *a.* All protists are eukaryotic. (1, 2)
2. Choice *c.* Multicellularity is the main difference between animals and protozoans. (6)
3. Choice *b.* (15)
4. Choice *c.* Increased pumping of the contractile vacuole indicates that the cell is in a solution that is even more hypoosmotic than pond water; in this case, distilled water is the best answer. (12)
5. Choice *e.* (16)
6. Choice *e.* All protists perform aerobic metabolism. (7–10)

● (16–18)

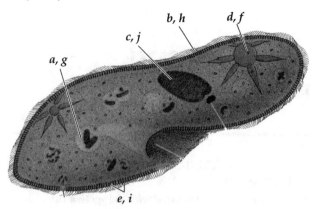

● Cilia are used for locomotion in the ciliates, as are cirri and their associated myonemes. Amoebas use pseudopods. (17–19, 21)

1. Choice *b.* The Apicomplexa include only parasitic forms of protists. (13)
2. Choice *e.* *Plasmodium* is an apicomplexan. (2–5, 13)
3. Choice *b.* Pseudopods are a prominent feature of actinopod and foraminiferan anatomy. (9–12)
4. Choice *d.* Trichocysts are the defensive structures of ciliates such as *Paramecium.* (18)
5. Choice *a.* Pseudopods are found in amoebas and actinopod protists, not ciliates. (7, 9, 18)
6. Choice *d.* Both phyla have an amoeboid body plan. (7, 13)

PROTISTS WERE ONCE CONSIDERED TO BE FUNGI (pages 178–179)

● A protist is coenocytic when it takes the form of a wall-less mass of cytoplasm that contains numerous diploid nuclei. This aptly describes the plasmodium, or vegetative body form, of an acellular slime mold. The pseudoplasmodium of a cellular slime mold is not coenocytic because, although individual myxamoebas do aggregate to form a large mass, or fruiting body, they retain their individual cellular identities by virtue of the plasma membranes that remain between individual cells. (2–3, 8–9)

1. Choice *d.* (3, 8–9)
2. Choice *c.* "Funguslike" protists have flagella, while true fungi do not. Both possess membranous internal organelles, but only fungi have chitin in their cell walls. (1)
3. Choice *b.* Cellular slime molds belong to the phylum Dictyostelida. (2, 11–12)
4. Choice *e.* Conjugation is common among certain animallike protists, bacteria, and fungi, but not the funguslike protists. (1–12)
5. Choice *c.* Cellular slime molds retain their individual cell identities and therefore have only a single haploid nucleus. (9)
6. Choice *a.* (7–8)

THE ALGAE • PHYLOGENY OF THE PROTISTS
(pages 179–182)

- They are all eukaryotic and photosynthetic and have a zygotic stage that is independent of parental protection. (1–5, 17–18)

- When diatoms reproduce asexually, each half of the shell is used as a template for a new complementary half. If the smaller half were used indefinitely, the cell would eventually become too small to be functional. Periodic sexual reproduction avoids this problem by forming a zygote, which grows substantially in size before resuming the process of asexual reproduction. (9–11)

- The life cycle is isomorphic because the mature gametophyte and sporophyte stages are morphologically indistinguishable. It is isogamous because the male and female gametes look alike. (17, 27)

1. Choices *a* and *c*. Although most chlorophytes are heteromorphic, *Ulva* is a genus with isomorphic forms. (17, 27–28)

2. Choice *b*. The amount of phycoerythrin varies at different water depths (i.e., different light conditions). (21–22)

3. (6, 9–10, 16, 21, 24)
   a: Red algae
   f: Green algae
   d, e: Brown algae
   b: Diatoms
   c: Dinoflagellates

4. Choice *e*. Dinoflagellates, a kind of pyrrophyte algae, are the cause of red tides. (8)

5. Choice *a*. Chlorophyll *a*, while not exclusive to the green algae, is the only characteristic listed that is found in all members of Chlorophyta. (25)

6. Choice *c*. (14–16)

Integrative Questions

1. Choice *c*. As nebulous as it sounds, this is the only description given that fits all protists! The other choices either contain characteristics not found in protists (e.g., prokaryotic), or characteristics found in only some protists (e.g., flagellated).

2. Choice *b*. Only certain members of the kingdom Eubacteria are capable of nitrogen fixation.

3. Choice *b*. In haplontic life cycles, the only diploid stage is the zygote.

4. Choice *e*. The green alga is Chlorophyta and the water mold is Oomycota.

5. Choice *b*. Sporophytes do produce spores, but by the process of meiosis, not mitosis.

CHAPTER 27 – PLANTS: PIONEERS OF THE TERRESTRIAL ENVIRONMENT

Note: The page numbers listed with the section titles refer to the Study Guide; numbers in parentheses for each question refer to the relevant key concepts.

THE PLANT KINGDOM (pages 183–185)

1. Choice *e*. Mitosis occurs in several phases of the plant life cycle (see the diagram in the Activities section). Of the choices mentioned, the only place it does not occur is during the formation of spores, which is achieved through meiosis. (3–4)

2. Choice *e*. The information given is insufficient to determine a specific plant group. All plants are eukaryotic, have both chlorophyll *a* and *b*, and have both single-celled as well as multicellular haploid forms. (2–4)

3. Choice *b*. In the tracheophyte plants xylem tissue conducts water and minerals, while phloem carries the products of photosynthesis. Since mosses are nontracheophytes, they lack these tissues. (6, 10–11)

4. Choice *c*. Nontracheophytes did evolve before tracheophytes. However, because tracheophytes fossilize more easily due to their structure, in older rock formations they are more likely to be preserved than are nontracheophytes. (13)

5. Choice *d*. Unlike plants, *Ulva* has isomorphic alternation of generations. (2–5, 8)

6. Choice *b*. Mosses are nontracheophytes. (6, 11)

- Similar photosynthetic and accessory pigments; starch as the storage carbohydrate; cellulose cell walls; oogamous reproduction. (8)

- (3–4)

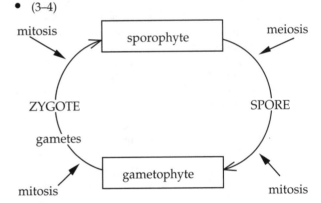

NONTRACHEOPHYTES: LIVERWORTS, HORNWORTS, AND MOSSES (pages 185–186)

- *Gametophyte:* sperm and eggs (5)
  Neither: phloem (2)
  Gametophyte: chloroplasts (4)
  Both: waxy cuticle (3)
  Sporophyte: capsule (6)
  Gametophyte: archegonium (5)
  Sporophyte: diploid tissue (6)

- More than 75% of the life cycle is occupied by the haploid gametophyte, which forms the "leafy" part typically recognized as a moss. This is to emphasize that the sporophyte of nontracheophytes is smaller, shorter-lived, and nutritionally dependent on the gametophyte. (1, 5–6)

1. Choice *c*. Water-related reproductive factors and those concerning physical support are what limit the absolute size of nonvascular plants; having a diploid sporophyte stage is not one of these factors. (1–2)

2. Choice *b*. All plants, including the hornworts and liverworts, have a water-resistant waxy cuticle. (3)

3. Choice *d*. Unlike the sporophytes of mosses and liverworts, and unlike gametophytes, the hornwort sporophyte grows at its base. (13)

4. Choice *e*. Antheridia and archegonia are the gamete-producing structures that develop on the haploid gametophyte stage of mosses and other nontracheophytes. (5–6, 8–13)

5. Choice *d*. Both sporophytes and gametophytes are distinctly different in these two groups of nonvascular plants. Neither possesses vascular tissue. (10, 12)

6. Choice *d*. A protonema is the filamentous structure that forms from the mitotic divisions of a germinating moss spore. (12)

INTRODUCING THE TRACHEOPHYTES • THE SURVIVING NONSEED TRACHEOPHYTES (pages 186–188)

• The branching of Rhyniophyte stems was dichotomous. Some branches may have penetrated the soil, and any adaptations for increased surface area and uptake of water or support of the aerial portions of the plant would have been greatly favored by evolution. In time, these adaptations would have accumulated to produce the features shown by the true roots of higher plants. (10)

• Simple leaves and complex leaves are flattened, vascularized photosynthetic structures attached to the plant stem. Simple leaves are small and scalelike, and hardly distort the stem vascular tissue at their point of separation. They are found only in the club mosses and horsetails. Complex leaves are present only in the ferns and seed plants. Their large photosynthetic surfaces are dichotomously branched and highly vascularized. The vascular tissue radically alters the stem vascular system at its point of departure. (11–13)

1. Choice *c*. This is alternation of generations with heterospory, which is found in some ferns and all seed-bearing tracheophytes. (14–15)

2. Choice *a*. Meiosis is how the plant goes from diploid to haploid, and fertilization is how it goes from haploid to diploid. (3–4, 14–15)

3. (1–3, 9, 13, 19, 21).
Gymnosperms: *a, b, c, d*
Psilopsids: *a, b*
Club mosses: *a, b, c*
Horsetails: *a, b, c*
Ferns: *a, b, c, d, e*

4. Choice *b*. Megaspores are the large spores produced by heterosporous plants that subsequently develop into the haploid female gametophyte (megagametophyte). They are produced through meiosis by the diploid sporophyte plant. (14–15)

5. Choice *a*. Meiosis is the process that produces spores in plants. Spores are made in specialized structures called sporangia. (14, 19–21)

6. Choice *c*. Ferns have a vascular system, so water transport is not the issue here. The fact that ferns have swimming sperm cells is the reproductive factor that still limits them to moist habitats. (22)

THE SEEDS OF SUCCESS • THE GYMNOSPERMS: NAKED SEEDS • THE ANGIOSPERMS: FLOWERING PLANTS (pages 188–191)

1. Choice *d*. The sporophyte plant is the dominant stage in gymnosperm life cycles. It is the gametophyte that is dependent on the sporophyte. (2, 4)

2. Choice *b*. Raspberry plants have fruits formed from the multiple carpels of a single flower. (29)

3. Choice *d*. The pollen (sperm) nucleus must travel down the pollen tube before entering the female gametophyte tissue. (4–6)

4. Choice *a*. The receptive female part of a flower is the sticky stigma. Pollen grains that alight here can form a pollen tube and compete for access to the ovules contained in the ovary. (22, 27)

5. Choice *c*. Palms belong to the Monocotyledones subdivision of the angiosperms. They possess everything in the list except for a second cotyledon. (31)

6. Choice *b*. Since it lacks carpels, this flower is imperfect. (20–24)

7. Choice *b*. (10)

• (20–24)

• (7)

Seed coat (parental, 2n)
Embryo (3rd gen., 2n)
Inside seed coat (2nd gen., n)

Integrative Questions

1. Choice *e*. Both the angiosperms and gymnosperms lack swimming sperm cells, and thus have been freed of the requirement for liquid water to reproduce. Pine trees and roses are examples.

2. Choice *b*. There has been a strong tendency in plant evolution to reduce the size of the gametophyte generation, a pattern that reaches its culmination in the angiosperms.

3. Choice *c*. Seeds are a way of protecting and providing initial nourishment to the sporophyte embryo.

4. Choice *e*. Hornworts are nontracheophytes, pines are gymnosperms, and apple trees are angiosperms. The best answer is therefore that they are all plants.

5. Choice *e*. The first four answers are all valid trends in plant evolution. The last choice is just the opposite of the actual trend observed in plant vascular tissue.

CHAPTER 28 – FUNGI: A KINGDOM OF RECYCLERS

Note: The page numbers listed with the section titles refer to the Study Guide; numbers in parentheses for each question refer to the relevant key concepts.

GENERAL BIOLOGY OF THE FUNGI (pages 192–194)

1. Choice *d*. No fungi are photosynthetic. (6)

2. Choice *b*. (8)

3. Choice *e*. A large surface-to-volume ratio maximizes contact between fungal cells and their food source. (10)

4. Choice *d*. Since the original cells are of different mating types, and because the new cell has two distinct nuclei, the new cell is both heterokaryotic and dikaryotic. (19)

5. Choice *b*. Conidia are also known as naked spores because they form at the tips of hyphal cells and not in specialized sporangia. (18)

6. Choice *a*. Fungi lack motile gametes. (5, 21)

● (19–20)

fertilization — Zygote — *meiosis*

Dikaryotic Hyphae — Spores

n + n — Hyphae — *n*

conjugation by opposite mating types — *mitosis*

2n — n

● Hyphae consist of haploid fungal cells that result from the germination and growth of spores. As they grow and multiply, the hyphae form a mass of threadlike filaments called a mycelium, which may have a cottony texture or form a more solid structure like a fruiting body. Rhizoids are extensions of the mycelium that anchor some fungi to the substrate, while mycorrhizae are mutualistic associations between fungal hyphae and the roots of plants. (8, 16)

DIVERSITY IN THE KINGDOM FUNGI ● FUNGAL ASSOCIATIONS (pages 194–197)

1. Choices *a, c, d*. The fungal groups that might appear on your dinner plate include the edible mushrooms in the class Basidiomycetes, fungi involved in producing certain cheeses or soy sauce in the imperfect fungi, and truffles and morels in the class Ascomycetes. (8, 12, 14, 17, 18)

2. Choice *d*. All three are caused by, or come from, imperfect fungi. (17, 18)

3. Choice *a*. The pileus, or cap, is a common characteristic of the club fungi (Basidiomycetes). The gills are where

the spores form, and basidiomycete hyphal cells are septate. (14–16)

4. Choice *c*. As the name suggests, karyogamy is the fusion of two nuclei. It occurs prior to ascus formation in euascomycetes, as well as in conjunction with zygote formation in other groups of fungi. (8, 11)

5. Choice *d*. Lichens reproduce vegetatively by breaking off a piece of their existing structure (the thallus), or by producing a special dispersal structure called the soredium. Fungal fruiting bodies are not observed in the reproduction of lichens. (24, 25)

6. Choice *a*. Since lichens are an association between a fungus and some photosynthetic alga or bacterium, they are therefore excluded from true organismal status. (21)

7. Choice *b*. (19)

8. Choice *e*. All of the traits listed are common to the class Ascomycetes, of which truffles and *Neurospora* are members. (8, 11, 12)

● The yeasts are generally very small and unicellular. They reproduce primarily by asexual means (either by fission or budding), and there is never a dikaryotic stage. They also engage in alcohol fermentation. (10)

● The bracket is the fruiting body, or basidium, of this fungus. It forms as a specialized outgrowth of the mycelium. Within certain hyphae of the basidium, dikaryotic cells undergo nuclear fusion to form zygotes. The zygotes then undergo meiosis to form basidiospores, which are shed from the underside of the bracket. (14–16)

● Since lichens have a photosynthetic component to their mutualistic relationship (the alga or bacterium), it makes sense that they should grow best on the side of trees that get the most sun. In the Northern Hemisphere this is the southern side, in the Southern Hemisphere the northern side. However, lichens are also sensitive to moisture and other environmental conditions, so these rules are not ecological absolutes. (21–27)

Integrative Questions

1. Choice *e*. Some fungi, such as the yeasts, are unicellular.

2. Choice *c*. The description provided best describes a foliose (leafy) species of lichen.

3. Choice *e*. Saprobes feed on all sorts of dead organic matter.

4. Choice *a*. Most fungi, including the two groups mentioned here, have haploid hyphal cells.

CHAPTER 29 – PROTOSTOMATE ANIMALS: THE EVOLUTION OF BILATERAL SYMMETRY AND BODY CAVITIES

Note: The page numbers listed with the section titles refer to the Study Guide; numbers in parentheses for each question refer to the relevant key concepts.

DESCENDANTS OF A COMMON ANCESTOR ● FOOD PROCUREMENT: THE KEY TO THE ANIMAL WAY OF LIFE ● CLUES TO EVOLUTIONARY RELATIONSHIPS AMONG ANIMALS (pages 199–200)

1. Choice *c*. All animals are heterotrophic, so this criterion does not really contribute to body plan, while all of the others do. (1, 4–7)

2. Choice *a*. All deuterostomes are triploblastic. (12, 13)

3. Choice *e*. Diversity in the methods for how food is acquired is thought to have influenced much of animal evolution. (4)

4. Choice *d*. Bilateral animals tend to move through the environment. Cephalization is important in the control and coordination of this locomotion. (9, 10)

5. Choice *e*. (12–15)

6. Choice *c*. Similar body halves could be obtained with either radial or biradial symmetry. Spherical symmetry has no main body axis along which to cut, and bilateral symmetry produces mirror-image halves. (7–9)

7. Choice *a*. *Volvox* is a colonial protist, not an animal. All the others are animals and therefore have their evolutionary origins in the ancestral colonial protists. (1, 2)

• (17, 18) The body cavity, or coelom, is a diagnostic feature of animal bodies. Acoelomate animals have no cavity at all. Pseudocoelomate animals have a partial fluid-filled cavity in which internal organs reside. Coelomate animals have a complete body cavity, with internal organs slung in sacs of mesentery tissue.

• (1, 12–16)

Characteristics	Protostome	Deuterostome
Cell type	-- both eukaryotic --	
Mode of nutrition	-- both ingestive heterotrophs --	
# of embryonic layers	two or three	three
Type of cleavage	determinate	indeterminate
Cleavage pattern	spiral	radial
Fate of blastopore	mouth	anus

EARLY ANIMAL DIETS: PROKARYOTES AND PROTISTS • SIMPLE AGGREGATIONS: SPONGES AND PLACOZOANS (pages 200–201)

• (5–9)

Amoebocytes are wandering cells that provide simple communication within the sponge body.

Choanocytes are flagellated collar cells that move water and food into the body cavity of a sponge for feeding.

Incurrent pores are the numerous small openings in a sponge's body through which water enters.

Osculum is the larger opening in a sponge's body through which water, wastes, and gametes exit.

Spicules are mineral- or protein-based structural elements that help support a sponge's body.

• Since sponges and placozoans lack true tissues, they are neither diplo- nor triploblastic. Therefore, all the other embryological criteria that qualify animals as either protostomes or deuterostomes (cleavage pattern, cell fate, blastopore development) also do not apply. (9–11)

1. Choice *a*. In fact, the opposite is often true. The main point here is that algae and bacteria were about the only food available to early animals, and so grazers were the first to evolve. Predators on these grazers came later. (1, 2)

2. Choice *d*. While the rate of new species formation may lead to diversity, it does not protect the individual animal from being attacked or eaten. (1, 2)

3. Choice *b*. Because they are not structurally robust, heavy wave action would destroy large, upright sponges. Thus we find smaller, flattened forms in near-shore or intertidal areas subjected to waves. (8)

4. Choice *d*. Placozoans do not have support structures like the spicules of sponges. (6, 10, 11)

5. Choice *d*. While sponges do indeed have many different cell types, placozoans have very few. (4, 10)

THE EVOLUTION OF CELL LAYERS: THE CNIDARIANS (pages 201–203)

• Cnidarians have tentacles complete with stinging cells to subdue and capture relatively large prey. Sponges are filter feeders that extract particulate matter from water currents moving through their porous bodies. (2, 11)

• (4, 6–8)

Class	Medusa	Polyp
Hydrozoa		✔
Anthozoa		✔
Scyphozoa	✔	

• (3–4)

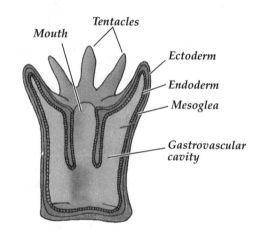

1. Choice *a*. Cnidarians have made several body plan improvements, including radial symmetry, nematocysts, alternating body forms, and a gastrovascular cavity. They do not, however, possess a true coelom, or body cavity. (1–4)

2. Choice *e*. Corals and sea anemones are members of the cnidarian class Anthozoa. (8)

3. Choice *c*. While the dominant stage in the life cycle of jellyfish is the medusa, anemones have lost this stage altogether. (7–8)

4. Choice *e*. (11)

5. Choice *d*. Nematocysts, the stinging cells found in cnidocytes, are the only item in this list that the cnidarians possess exclusively. (2)

THE EVOLUTION OF BILATERAL SYMMETRY • PROTOSTOMES AND DEUTEROSTOMES: AN EARLY LINEAGE SPLIT • PROTOSTOMES: A DIVERSITY OF BODY PLANS • WORMLIKE BODIES: MOVEMENT THROUGH SEDIMENTS (pages 203–205)

- The main clue to why this animal cannot be a flatworm is the presence of a continuous digestive tract running from the anterior end of the animal to its posterior end. The phylum Platyhelminthes lacks a complete digestive tract. (16)

- Roundworms have only longitudinal muscles, which run the length of the animal's body. They contract in alternating waves down each side, resulting in a rather ungainly, but nevertheless effective type of side to side writhing motion. (11)

1. Choice *d*. Choices *a* to *c* are all false; exactly the opposite is true in each case. Flatworms are, however, triploblastic animals. (5, 6, 16)
2. Choice *b*. Although having changed from the cnidarian body plan, flatworms such as *Dugesia* have yet to develop a complete digestive system. (1–4, 16, 17)
3. Choice *b*. (20)
4. Choice *c*. Hydrostatic skeletons are fluid-filled sacs that, when combined with appropriate musculature, can produce motion. Rotifers, however, rely on the beating of cilia for most of their body movements. (7)
5. Choice *e*. (20, 21)
6. Choice *e*. All of these characteristics are found in both the rotifers and roundworms. (7, 11)

COELOMATE PROTOSTOMES • SEGMENTED BODIES: IMPROVED LOCOMOTION (pages 205–207)

- (12–15)
  a. Heavily calcified shells offering protection against all but the toughest predators.
  b. Body plans that are adapted for many different lifestyles.
  c. Complete digestive tract, improved circulatory system, and better sensory organs.

- Polychaetes are mostly marine, segmented, have parapodia and well-developed sensory systems. Their sexes are distinct and they have a trochophore larval stage that undergoes metamorphosis to become an adult. Oligochaetes are aquatic and terrestrial, lack parapodia, and have limited sensory abilities. They are hermaphrodites, and their young hatch as fully developed, miniature worms. Both groups have hydrostatic skeletons with circular and longitudinal muscles. (1, 6–9)

1. (12–25)
  Monoplacophorans: *d, e*
  Cephalopods: *a, c*
  Gastropods: *b, d*
  Polyplacophorans: *d*
  Bivalves: *d*

2. Choice *b*. Interconnected plates that form a shell are characteristic of the polyplacophorans. All the other traits are found in cephalopods such as the squid. (17, 21–25)
3. Choice *e*. (16)
4. Choice *c*. Snails are gastropods, clams are bivalves, and octopuses are cephalopods. (18–21)
5. Choice *e*. A garden snail possesses all of these structures. (13, 19)
6. Choice *c*. Cephalopods are predatory, active swimmers. (22–24)
7. Choice *b*. The majority of annelids are polychaetes, so if the box does contain a fair representation of all annelid species, odds are the one selected will be a polychaete. (6)
8. Choice *b*. Annelids have successfully combined true coelomic cavities and repeating segmentation to more precisely control body movements compared with the anatomically simpler worms. (4, 5)

ARTHROPODS: SEGMENTED EXTERNAL SKELETONS • THEMES IN PROTOSTOME EVOLUTION (pages 207–210)

- The fact that we are told this is a crab larva indicates that it is also a crustacean arthropod. From the diagram itself we see evidence of a jointed exoskeleton; notice that there are distinct segments to the body, in particular, an obvious head and abdomen (there is also a thorax, but it is less evident). Moreover, there are paired appendages that attach to specific body segments. Other obvious features are the well-developed sensory organs (eyes, antennae) located on the anterior end of the animal. (1–3, 10)

- (22-24)
  a. The presence of fluid-filled cavities, which make for effective hydroskeletons.
  b. Hard external protective coverings for defense against large, rapidly swimming aquatic predators.
  c. Numerous sessile (and still mainly aquatic) representatives.

1. Choice *d*. Lobsters, millipedes, and butterflies are all arthropods, and therefore have exoskeletons with jointed appendages. (1, 2)
2. Choice *e*. Earthworms have a very thin cuticular layer that functions in gas exchange and is also part of their hydroskeleton. (1–4, and information from previous section also)
3. Choice *a*. All of the animals considered in this section have true coeloms, so that piece of information is not useful. The observation that this animal lacks an exoskeleton eliminates all of the crustaceans and insects from consideration. The presence of appendages and the fact that the specimen came from a shallow tidepool eliminates the pogonophores as well as the oligochaetes. Only the polychaetes have single-sexed individuals that also meet the other criteria. (1–24, and information from previous section also)
4. Choice *e*. (18)
5. Choice *c*. (14, 15)
6. Choice *b*. (17)

Integrative Questions

1. Choice *e*. All of these statements correctly apply to every bilaterally symmetric animal.

2. Choice *a*. None are acoelomates, all are bilaterally symmetric, and neither snails nor earthworms have very good vision. However, all three are indeed protostome animals.

3. Choice *b*. Neither are spherically symmetrical, only rotifers are pseudocoelomates, and since neither is a deuterostome, they can't have radial cleavage of the embryo. However, both do possess complete digestive tracts.

4. Choice *d*. Lobsters are crustaceans, while clams and octopuses are mollusks.

5. Choice *a*. Of those listed, crabs are the most closely related to trilobites because they are both arthropods.

6. Choice *c*. Cephalization is the accumulation of neural tissue in the anterior end of the animal. The only animal listed that lacks some obvious form of this characteristic is the jellyfish, a cnidarian.

CHAPTER 30 – DEUTEROSTOMATE ANIMALS: EVOLUTION OF LARGER BRAINS AND COMPLEX BEHAVIORS

Note: The page numbers listed with the section titles refer to the Study Guide; numbers in parentheses for each question refer to the relevant key concepts.

THREE BODY PARTS: AN ANCIENT BODY PLAN • ECHINODERMS: COMPLEX RADIAL SYMMETRY (pages 211–214)

- *no*: Pterobranchs (8)
  yes: Bryozoans (7)
  no: Chaetognathans (9)
  yes: Phoronids (5)
  yes: Brachiopods (6)
  no: Echinoderms (11–23)

- Brachiopods are filter feeders; they use the cilia on their internalized lophophore to strain food from the water. Sea stars are active predators and scavengers that use the action of their tube feet to overpower prey such as bivalve mollusks; they may also evert their stomachs to digest this food externally. (6, 11)

- The echinoderms are much more mobile than the mostly sedentary lophophorates. The echinoderm water vascular system and specialized internal musculature coordinates the action of the tube feet to achieve motion. (3, 13)

1. Choice *c*. Most simple deuterostomes are filter-feeding lophophorates. As other feeding mechanisms have evolved, animals have tended to become larger and more mobile. (3–13)

2. Choice *d*. While new bryozoan colonies do in fact form from a single larva that settles out from the plankton, the larva itself is generated by *sexual* reproduction. (7)

3. Choice *e*. (5, 9–10)

4. Choice *c*. The hinged shells of the two groups appear somewhat similar. (6)

5. Choice *d*. Lophophores are characteristic of several other phyla, but not the echinoderms. (3, 15)

6. Choice *a*. The sea lilies and feather stars together make up the class Crinoidea. (14–16)

7. Choice *c*. Crinoids belong to the Pelmatozoans. All other echinoderms are Eleutherozoans. (14, 17)

8. Choice *e*. Sea stars do not have a rounded, fused internal skeleton. They also have distinct arms, which are lacking in the sea urchins. (18–22)

9. Choice *e*. (11–23)

MODIFIED LOPHOPHORES: NEW WAYS OF FEEDING • THE ORIGIN OF THE VERTEBRATES: SUCKING MUD (pages 214–215)

- The skeleton of a modern bony fish would have a well-formed skull with hinged jaws, a vertebral column, two sets of paired pectoral and pelvic fins, a caudal (tail) fin, and perhaps a dorsal fin. It is the presence of hinged jaws that makes this skeleton so obviously different from that of agnathan (jawless) fishes. (9–20)

- Hinged jaws are advantageous because they allow the bearer to catch and manipulate larger, more mobile prey. They also might be used as weapons for self-defense. Hinged jaws were much more adaptable than the sucking aparatus of the agnathans, and so the early jawed fishes underwent rapid evolutionary radiation into previously unexploited feeding habitats. (12, 13)

1. Choice *a*. Determinate cleavage is characteristic of protostomes, not deuterostomes. (1)

2. Choice *a*. Filter feeding requires a large surface area in order to gain enough food to sustain larger animals. (8)

3. Choice *e*. Larvaceans are a type of tunicate, and the lancelets (like all other chordates) evolved from tunicate ancestors. (4, 6–7)

4. Choice *c*. Placoderms were among the first large predatory fishes with hinged jaws; formidable teeth were an important part of their feeding apparatus. (14)

5. Choice *b*. Cartilaginous fishes include the sharks and their relatives. Bony fishes either evolved simultaneously with these fishes, or perhaps slightly after. Lancelets, tunicates, and agnathans all appeared before the sharks. (4, 7, 15)

6. Choice *d*. The hyoid gill arches of agnathan fishes are believed to be the anatomical precursors of jaws in vertebrates. (13)

7. Choice *d*. Some of the other answers are perhaps justifiable as indirect effects, but improved buoyancy is the primary role of the swim bladder. (19)

8. Choice *b*. Pharyngeal baskets are present in the prevertebrate animals. (7–10)

COLONIZING THE LAND • SURVIVING DINOSAUR LINEAGES • THE TERTIARY: MAMMALS DIVERSIFY (pages 216–218)

- (Refer to all key concepts from this section)
  Lungs: *A, B, D, L, M, R*
  Aquatic larvae: *A, L*
  Wings: *B, D, M*
  Feathers: *B*
  Uterine development: *M*
  Terrestrial egg: *B, D, M, R*
  Water-impermeable skin: *B, D, M, R*

- (13)
  a. Teeth have been completely lost.
  b. Bones of the tail have been lost.
  c. Breast bone (sternum) has been enlarged to accommodate flight muscles.
  d. Digits of hand have been reduced and fused to form base for attachment of flight feathers.

1. Choice *c*. Feathers are certainly useful for insulation but had nothing to do with the transition to life on land, since birds evolved from terrestrial vertebrates. (13)
2. Choice *a*. The embryo resides inside a *fluid*-filled chamber created by the amnion. (7, 15)
3. Choice *b*. Many vertebrate systematists now believe that the avian lineage is a direct descendant of the dinosaurs. (12, 13)
4. Choice *d*. Kangaroos are marsupials whose young are born after only a minimum of uterine development. They then crawl into the mother's ventral pouch to complete their growth. (22)
5. Choice *e*. Caimans are a type of semiaquatic reptile. (3, 11)
6. Choice *c*. (7)
7. Choice *c*. The limb position of mammals raises the body off of the ground for better locomotion than is possible in the reptiles. (17)
8. Choice *e*. Amphibian eggs and larvae must develop in water, and adult amphibians must stay moist because they lack desiccation-resistant skin. (4)

HUMANS EVOLVE: EARTH'S FACE IS CHANGED • DEUTEROSTOMES AND PROTOSTOMES: SHARED EVOLUTIONARY THEMES (pages 218–220)

- (6–18)
  4: *Homo erectus*
  5: *Homo sapiens* – Neanderthal
  6: *Homo sapiens* – Cro-Magnon
  2: *Australopithecus africanus*
  1: *Australopithecus afarensis*
  3: *Homo habilis*

- (2)
  Flexible limbs, opposable thumbs, depth perception, flattened nails, and small numbers of offspring in humans are all holdovers from the early primate arboreal lifestyle.

1. Choice *b*. Lucy was a very early hominid who lived long before formal language and culture appeared. (6, 14–15, 20)

2. Choice *d*. One of the important expansions in lifestyle of *Homo habilis* was the incorporation of animal protein in the diet, which provided better nutrition and spurred cooperative hunting. (11, 12)
3. Choice *e*. (15, 20)
4. Choice *c*. Both Cro-Magnon and Neanderthal were of our own species, *Homo sapiens*, but Cro-Magnon was anatomically more like us and definitely survived into the more recent past. (16–18)
5. Choice *d*. Actually, just the reverse is true, although exactly how big an influence Cro-Magnon had on the disappearance of Neanderthal is not entirely clear. (17, 18)
6. Choice *b*. Exoskeletons are a characteristic of certain protostome phyla. (22–24)

Integrative Questions

1. *b*: Chondrichthyes
   c: Agnatha
   e: Asteroidea
   a: Brachiopoda
   d: Hemichordata
2. Choice *c*. Although jaws were a very important event in vertebrate evolution, significant feeding modifications also occurred in other deuterostome lineages (e.g., loss of lophophores in hemichordates, mobile mud-sucking in agnathans, etc.).
3. Choice *d*. Fish, amphibian, bird, mammal.
4. Choice *e*. They are all chordate characteristics, and humans are chordates.
5. Choice *d*. This is the only set of characteristics listed that accurately fits *both* humans and sea stars.

31

The Flowering Plant Body

CHAPTER LEARNING OBJECTIVES—after studying this chapter you should be able to:

❏ Describe the characteristic features of the two classes of angiosperms, the Dicotyledones and the Monocotyledones.

❏ List the three principle plant organs and describe their functions.

❏ Describe the characteristic features of stems, including nodes, internodes, apical buds, lateral buds, and leaf primordia.

❏ Describe the general structure of roots and the specific adaptations seen in taproot and fibrous root systems.

❏ Characterize the structure of a typical leaf, and differentiate between simple and compound leaves.

❏ Name and characterize the general functions of the three tissue systems in vascular plants.

❏ Name and discuss the various modifications of roots, stems, and leaves.

❏ Explain the relationships between primary cell walls, secondary cell walls, and the middle lamellae.

❏ Describe the features and functions of pit pairs in plants.

❏ Describe the characteristic features and functions of the following plant cell types, and be able to recognize these cell types: parenchyma, sclerenchyma, collenchyma, tracheids, vessel elements, sieve tube elements, and companion cells.

❏ Name and characterize the common simple and compound tissue types seen in the vascular plants.

❏ Explain what a meristem is, characterize where the major meristems are located in the plant body, and differentiate between primary and secondary growth.

❏ Describe the formation of the procambium, the ground meristem, and the protoderm, and explain the activities of these meristems in producing the mature tissues of the plant root and stem.

❏ Characterize and be able to identify the major structural features of the root system, and discuss typical differences seen in monocot and dicot roots.

❏ Characterize and be able to identify the major structural features of the stem, and discuss typical differences seen in monocot and dicot shoots.

❏ Characterize and be able to identify the major structural features associated with secondary growth in woody plants.

❏ Characterize and be able to identify the major structural features associated with leaves.

❏ Explain the significance of the differences in the leaf anatomy of C_3 and C_4 plants.

❏ Differentiate between the two types of reaction wood, tension wood and compression wood, and discuss the occurrence and functions of each.

FLOWERING PLANTS: DEFINITION AND EXAMPLES • ORGANS OF THE PLANT BODY • LEVELS OF ORGANIZATION IN THE PLANT BODY (pages 698–705)

Key Concepts

1. The *flowering plants,* or *angiosperms* (phylum *Anthophyta*), have vascular tissue, including xylem composed of vessel elements and fibers, sexual reproduction involving double fertilization, and seeds with endosperm. *Monocots* (Monocotyledones) and *dicots* (Dicotyledones) are the two classes of the Angiospermae.

2. The monocots (grasses, lilies, orchids, and palms) are the narrow-leaved flowering plants with one *cotyledon* (embryonic leaf).

3. Monocots usually have leaves with parallel venation, vascular bundles scattered throughout the stem, and floral parts in multiples of three.

4. *Dicots* have two cotyledons, and include the remaining broad-leaved flowering plants.

5. Dicots usually have leaves with netlike venation, vascular bundles arranged in a ring within the stem, and floral parts in fours or fives.

6. Plant parts such as branches or leaves act as semi-independent units. As a result of this modular design, control systems in plants are decentralized.

7. Plant growth is restricted to specific regions of active cell division called *meristems*. Meristems at the tips of shoots and roots result in elongation of the plant body. Unlike the

230

determinant growth seen in animals, the activity of meristems results in indeterminant growth in plants.

8. The vascular plant body consists of three organs, the *root system*, *leaves*, and *stem*. Leaves and stem combined are called the *shoot system*.

9. The main functions of these organs are as follows: leaves: are the sites of photosynthesis; roots provide anchorage and water/nutrient absorption; stems display leaves and transport materials between the roots and leaves.

10. *Nodes* are the attachment sites of leaves to stems; stem regions between nodes are called *internodes*.

11. *Taproot* and *fibrous* are the two main root system types found in flowering plants. The taproot system, common in dicots, has a single main root that often serves as a food storage organ. Fibrous root systems, seen in monocots and some dicots, create very large surface areas for absorption of water and minerals.

12. The end of each root is covered by a *root cap* that protects it from abrasion. Behind the root cap is the *root apical meristem* where dividing cells cause the root to grow in length. Some new cells also replace lost root cap cells.

13. *Adventitious roots* arise from stems or leaves and may help to prop up the plant (corn, for example) or to establish a detached plant part in the soil.

14. *Lateral buds* are found at nodes on stems and can develop into branches.

15. Stems are highly modified in many species. Potato *tubers* are portions of stems; strawberry *runners* are horizontal stems that can form new plants asexually.

16. The *shoot apical meristem* within the terminal or *apical bud* at the tip of each stem causes the stem to grow in length.

17. *Leaf primordia* within each bud develop into mature leaves.

18. In addition to photosynthesis, leaves also carry out other important metabolic activities and, in succulent plants, store water. Tendrils in peas and squash and spines in cacti are modified leaves.

19. A typical leaf consists of a flattened *blade* and a *petiole*, or stalk, that attaches the blade to the stem.

20. *Simple leaves* have a single blade; *compound leaves* have multiple blades that can be pinnately or palmately arranged. The venation of a leaf blade can be parallel or netlike in arrangement.

21. A *tissue* is an organized collection of functionally and structurally similar cells. Some tissues consist of only one cell type; other tissues are composed of several cell types.

22. The vascular plant body consists of three tissue systems, the *vascular tissue system*, the *dermal tissue system*, and the *ground tissue system*.

23. The vascular tissue system consists of *xylem* and *phloem*. Its function is to conduct materials throughout the plant body. The dermal tissue system protects the body surface. The ground tissue system has a diversity of roles.

Questions (for answers and explanations, see page 263)

1. Match the plant organs on the left with the functions they perform on the right.

_____	Root system	*a.* Water absorption
_____	Leaves	*b.* Photosynthesis
_____	Stem	*c.* Nutrient absorption
		d. Transport
		e. Anchorage

2. A plant living in an environment with low soil moisture and abrasive soil particles would likely have
   a. a taproot system with well-developed root caps.
   b. a taproot system with poorly developed root caps.
   c. a fibrous root system with well-developed root caps.
   d. a fibrous root system with poorly developed root caps.
   e. either a taproot or a fibrous root system, but with no root apical meristems.

3. Which of the following is *not* typical of the plant group to which rice and palms belong?
   a. More than one embryonic leaf
   b. Leaves with parallel veins
   c. Fibrous root systems
   d. Scattered vascular bundles
   e. Floral parts in multiples of three

4. Characterize each of the following plant structures as modified roots, stems, or leaves.
   a. Potato tubers   _____
   b. Runners in strawberry plants   _____
   c. Cactus spines   _____
   d. Tendrils in peas   _____
   e. Prop roots in corn   _____

Activities (for answers and explanations, see page 263)

- On the following diagram of a simple plant, identify a node, an internode, a lateral bud, an apical bud, a leaf petiole, and a leaf blade, and circle the location of a root apical meristem.

- Contrast control systems and growth in plants and animals and explain evident differences in terms of the structural organization of plants.

PLANT CELLS • PLANT TISSUES AND TISSUE SYSTEMS
(pages 705–709)

Key Concepts

1. The polysaccharides (primarily cellulose) secreted by the cell prior to cell expansion become the *primary cell wall*. Additional layers of polysaccharides (including other materials like lignin and suberin) laid down by some cells after elongation become the *secondary cell wall*.

2. The *middle lamella* cements plant cell walls together.

3. Cell walls lie outside the plasma membrane of the cell. Openings in the wall are called *pits*.

4. Pits usually occur in pairs (called *pit pairs*) with a middle lamella, a thinned primary wall, and no secondary wall between the two cells. Pit pairs are traversed by *plasmodesmata*, facilitating the movement of materials between neighboring cells.

5. *Parenchyma cells* are the most numerous cells in the plant body and are alive when their primary functions are performed. Parenchyma cells are not elongated, have a large central vacuole, and lack secondary cell walls.

6. A major function performed by parenchyma cells is storage of lipids as droplets within the cytoplasm or of starch in specialized plastids called *leucoplasts*. Parenchyma cells containing chloroplasts occur in the leaves and are the primary sites for photosynthesis.

7. In some regions, called *meristems*, parenchyma retains an ability to divide mitotically.

8. *Sclerenchyma* cells are dead at functional maturity. They provide structural support by means of their thick secondary cell walls. Sclerenchyma cells are of two types: elongated *fibers*, usually occurring in bundles, and variously shaped *sclereids*, particularly common in bark.

9. *Collenchyma* cells are elongated living cells with thickened primary cell walls and no secondary walls. The flexibility of petioles and stems is due in part to tissues rich in collenchyma cells.

10. *Tracheids* are spindle-shaped xylem cells found mostly in gymnosperms. Dead at maturity, tracheids, interconnected by numerous pits, form the hollow conduits through which water and minerals travel.

11. *Vessel elements* are a type of xylem cell found mostly in angiosperms. Vessel elements are dead at functional maturity, and larger than tracheids. Vessel elements lose most of their end walls and form hollow tubes called *vessels* for water transport.

12. Unlike xylem, phloem cells are alive at functional maturity. The *sieve tube element* is the main phloem cell type. *Sieve tubes* are stacks of sieve tube elements arranged end-to-end and interconnected through their perforated end walls, called *sieve plates*.

13. As sieve tube elements mature, the central vacuole membranes break down, nuclei and some other organelles disappear, and the sieve tube elements become specialized for transport of organic nutrients. In some flowering plants small, nucleated *companion cells* may regulate the activities of adjacent sieve tube elements.

14. Simple tissues consist of a single cell type. Simple tissues include parenchyma, sclerenchyma, and collenchyma.

15. Xylem and phloem are compound tissues. In addition to tracheids and vessel elements, xylem may include parenchyma cells and sclerenchyma cells, and may perform the functions of support, transport, and storage.

16. Phloem includes sieve tube elements, companion cells, fibers, sclereids, and parenchyma cells.

17. The vascular tissue system has two main functions: the transport of food from sites of production (*sources*) in the leaves to sites of utilization or storage (*sinks*), and movement of water and minerals upward from the roots to the stem and leaves.

18. The dermal tissue system includes the cell layers covering the plant. *Epidermis* is a layer one or more cells thick covering the entire plant body. Epidermis is covered by a waxy *cuticle* in the stems and leaves, but not the roots. *Periderm* covers the roots and stems of older woody plants.

19. The ground tissue system includes mainly parenchyma cells, with some collenchyma and sclerenchyma. Its functions include storage, support, photosynthesis, and other activities.

Questions (for answers and explanations, see page 263)

1. What would be the correct order of structures you would encounter in moving from inside a parenchyma cell into a living, immature sclerenchyma cell?
   a. Primary cell wall, secondary cell wall, middle lamella, secondary cell wall, primary cell wall
   b. Primary cell wall, middle lamella, secondary cell wall, primary cell wall
   c. Primary cell wall, middle lamella, primary cell wall, secondary cell wall
   d. Secondary cell wall, primary cell wall, middle lamella, primary cell wall, secondary cell wall
   e. Primary cell wall, middle lamella, primary cell wall

2. To which of the following cell types would this description apply: cell alive at maturity, elongated shape, and a thick primary cell wall, but lacking a secondary cell wall?
   a. Collenchyma cell
   b. Sclerenchyma cell
   c. Parenchyma cell
   d. Tracheid
   e. Vessel element

3. Which of the following characteristics do tracheids, sclerenchyma, and vessel elements have in common?
   a. Are dead at functional maturity
   b. Lack primary cell walls
   c. Lack secondary cell walls
   d. Mainly transport water
   e. Mainly transport organic nutrients

4. Which of the following characteristics do parenchyma, collenchyma, and sieve tube elements have in common?
   a. Contain leucoplasts
   b. Lack primary cell walls
   c. Lack nuclei
   d. Transport water
   e. Are alive at functional maturity

5. Choose *all* of the following that are primary functions of the dermal tissue system.
   a. Protection
   b. Transport
   c. Food storage
   d. Reduction of water loss
   e. Absorption of water and minerals

Activities (for answers and explanations, see page 263)

- In the following figure showing several plant cells, label an example of a primary cell wall, a secondary cell wall, and the middle lamella.

- In the following figure showing cells of the phloem, label a companion cell, a sieve tube element, and a sieve plate.

GROWTH: MERISTEMS AND THEIR PRODUCTS • LEAF ANATOMY TO SUPPORT PHOTOSYNTHESIS • SUPPORT IN A TERRESTRIAL ENVIRONMENT   (pages 709–716)

Key Concepts

1. *Meristems* are regions where cells divide to form new tissues. *Primary growth*, due to the activity of the root and shoot apical meristems, produces the primary tissues of the plant.

2. A *vascular cambium*, a cylinder of tissue located between the xylem and phloem, produces new xylem cells to the inside and new phloem cells to the outside and results in growth in circumference — *girth* — of roots and stems.

3. A *cork cambium* located in the bark of woody plants produces cells impregnated with the waxy substance *suberin* to augment the dermal tissue system as girth increases.

4. *Secondary growth*, due to the activity of lateral cambia, such as the vascular and cork cambia, produces the secondary tissues of the plant.

5. Unlike animals, plants show *indeterminate growth*, because they continue to grow throughout their lives. Based on the growth strategies of plants, we can categorize plants as *annuals*, *biennials*, or *perennials*.

6. The root apical meristem gives rise to three primary cylindrical meristems, the inner *procambium*, the *ground meristem*, and the outermost *protoderm*. The procambium produces the vascular tissues, the ground tissues arise from the ground meristem, and the protoderm develops into the dermal tissue system.

7. The apical and primary meristems constitute the root's *zone of cell division*. Just above this zone is the *zone of elongation*, where cells are increasing in length. The older cells are found in the *zone of differentiation*, where they become structurally and functionally specialized.

8. The epidermis of the root develops from the protoderm. Epidermal cells are flattened and may possess *root hairs* that increase surface area and aid in absorption of water and minerals.

9. The *root cortex* contains parenchyma cells that frequently serve as food storage sites and may form an association with fungal species to produce a *mycorrhiza*. Mycorrhizae improve absorption of water and minerals by the root.

10. The innermost cell layer of the cortex is the *endodermis*, a single layer of cells each ringed by a waterproof seal containing the waxy substance *suberin* that prevents the inward passage of water. To enter the *vascular cylinder*, water must pass through the plasma membrane of an endodermal cell.

11. The vascular cylinder, or *stele*, consists of three tissues: the *pericycle*, the *xylem*, and the *phloem*.

12. Undifferentiated cells making up the pericycle give rise to lateral roots and provide additional cells to increase the girth of the root.

13. Xylem occupies the center of the root in dicots; in monocots a layer of parenchyma cells called *pith* lies internal to the xylem.

14. Phloem is found in cylinders between the arms of xylem.

15. Like the root apical meristem, the shoot apical meristem also gives rise to a procambium, ground meristem, and protoderm.

16. Growth of dicot stems occurs in a zone of elongation below the shoot apical meristem. Grasses and some other monocots have *basal meristems*, an adaptation that helps the plant deal with grazing by herbivores.

17. Leaves develop from *leaf primordia* laid down by the shoot apical meristem.

18. Stem vascular tissue occurs in bundles, generally forming a cylinder in dicots, and uniformly distributed throughout the stem in monocots.

19. In dicots, a storage tissue called *pith* is located internal to the vascular tissue. Another cylinder of storage tissue called *cortex* is located external to the vascular tissue. Pith, cortex, and tissue between the vascular bundles make up the ground tissue system of the stem.

20. In stems, the vascular cambium is a single layer of undifferentiated cells located between primary xylem and phloem, but also extending between the vascular bundles to form a continuous cylinder of meristematic tissue. The vascular cambium produces secondary xylem *(wood)* to the inside, and secondary phloem to the outside.

21. *Vascular rays*, formed by cells derived from the vascular cambium, are rows of radially arranged parenchyma cells that connect the storage tissues of the stem (cortex and pith) with the vascular tissues.

22. As the stem increases in diameter, some cells within the vascular cambium divide tangentially to give rise to additional cambium cells, thus increasing the cylinder's circumference.

23. Trees from temperate regions have wood with *annual rings* that result from differences in xylem cell size due to seasonal variation in the availability of water.

24. The xylem tracheids and vessel elements produced in the spring are larger and have thinner walls than the xylem cells produced in the summer months.

25. *Heartwood* results from deposition of resins in the older, more central xylem. The younger, more peripheral xylem called *sapwood* is mainly responsible for water and mineral transport in the stem.

26. Secondary growth in both the root and stem displaces the cortex and epidermis, both of which are eventually lost.

27. A *cork cambium* produces layers of cork cells containing suberin to replace the lost epidermis and provide waterproofing.

28. Cells produced to the inside by the cork cambium are called the *phelloderm*. Cork cells, cork cambium, and phelloderm together constitute the *periderm*.

29. *Lenticels* are openings in the epidermis/periderm of the stem that allow gas exchange.

30. Chloroplast-containing parenchyma cells within the leaf form two layers of photosynthetic cells, an upper layer of elongated cells called the *palisade mesophyll* and a lower layer of irregularly shaped, loosely packed cells called the *spongy mesophyll*.

31. Vascular bundles form an extensive network of veins that bring the plant's transport system to within a short distance of each cell.

32. A nonphotosynthetic epidermal layer, covered with a waxy cuticle, prevents water loss from the leaf.

33. Openings called *stomata* (sing. *stoma*), are located in the epidermis. They permit the needed exchange of CO_2 and O_2 between the leaf interior and the external atmosphere. Specialized pairs of *guard cells* change shape to regulate the size of the stomatal opening.

34. C_4 plants have a modified leaf anatomy in which photosynthetic cells are arranged concentrically around each vein. The outer mesophyll layer and the inner *bundle sheath* layer have different types of chloroplasts that carry out different biochemical reactions.

35. Terrestrial plants depend on two types of support systems: turgor within cells due to a relatively high pressure potential, or cells containing walls reinforced with lignin, such as secondary xylem (wood).

36. *Reaction wood* is specialized secondary xylem that maintains trees in an upright position. It occurs in two different forms, *compression wood* and *tension wood*.

37. Compression wood, found in gymnosperms, occurs in compressive areas (like the lower sides of limbs). It contains thicker and shorter tracheids, with more lignin and less cellulose, and expands as it develops to push the branch upward.

38. Tension wood, found in angiosperms, occurs in expansive areas (like the upper side of limbs). It contains fibers with thicker walls, containing less lignin and more cellulose, and contracts as it develops, pulling the branch upward.

Questions *(for answers and explanations, see page 264)*

1. Choose *all* of the following structures that do *not* result from the activity of the procambium.
   a. Guard cell
   b. Sieve tube element
   c. Cortex cell
   d. Vessel element
   e. Collenchyma

2. Which of the following does *not* show a correct derivation of cellular types?
   a. Root apical meristem → procambium → tracheid
   b. Root apical meristem → ground meristem → parenchyma
   c. Root apical meristem → ground meristem → companion cell
   d. Root apical meristem → protoderm → periderm
   e. Root apical meristem → procambium → sclerenchyma

3. Which one of the following structures found in roots performs a function analogous to that of the lateral buds of stems?
   a. Epidermis
   b. Cortex
   c. Endodermis
   d. Pericycle
   e. Vascular tissue

4. Choose all of the following that are *not* complex tissues.
   a. Epidermis
   b. Phloem
   c. Endodermis
   d. Xylem
   e. Pericycle

5. Which of the following regions are most responsible for water transport?
   a. Sapwood
   b. Wood formed in the spring
   c. Wood formed in the summer
   d. Heartwood
   e. Bark

6. Based on your knowledge of the location of xylem and phloem within the vascular cylinder of the stem, which of the following would best describe the locations of these tissues in vascular bundles of veins within leaves?
   a. Phloem should be above xylem in the same vein
   b. Xylem should be above phloem in the same vein
   c. Veins only contain xylem
   d. Veins only contain phloem
   e. Xylem and phloem are in separate veins

7. Lenticels are functionally similar to which of the following structures?
   a. Guard cells
   b. Root hairs
   c. Stomata
   d. Vascular rays
   e. Periderm

8. In proceeding from the vascular cambium toward the periphery of a mature woody stem, which of the following could be the correct sequence of tissue types encountered?
   a. Primary phloem → secondary phloem → cortex → phelloderm → cork cambium → cork
   b. Secondary phloem → primary phloem → phelloderm → cork cambium → cork
   c. Secondary phloem → primary phloem → cortex → phelloderm → cork cambium → cork
   d. Secondary phloem → primary phloem → pith → phelloderm → cork cambium → cork
   e. Primary phloem → secondary phloem → cortex → epidermis

9. Plants with the C_4 photosynthetic pathway (C_4 plants) differ in leaf anatomy from plants using the C_3 photosynthetic pathway in the following way:
   a. C_4 plants have specialized guard cells.
   b. C_4 plants have two concentric layers of photosynthetic cells around each vein.
   c. C_4 plants only have stomata on the lower leaf surface.
   d. C_4 plants only have a palisade mesophyll layer.
   e. C_4 plants have a much thicker waxy cuticle covering the epidermis of the leaf.

10. Which of the following statements about compression wood is *not* true?
    a. Compression wood has thicker and shorter tracheids than tension wood.
    b. Compression wood is found on the underside of limbs.
    c. Compression wood tracheids have more lignin and less cellulose.
    d. Compression wood is found in gymnosperms.
    e. Compression wood contracts as it develops.

11. Categorize the life cycle of each of the following plants as either annual, perennial, or biennial.

    _____ Carrot

    _____ Red maple

    _____ Rice

    _____ Coconut palm

Activities *(for answers and explanations, see page 264)*

- On this photomicrograph of a portion of a typical dicot root, label the location of the cortex, endodermis, pericycle, xylem, and phloem.

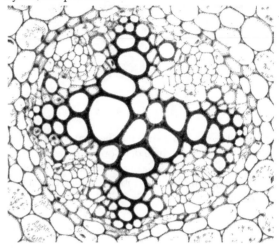

- On the following figure of a cross-sectioned leaf, label the location of (A) a palisade mesophyll cell, (B) a spongy mesophyll cell, (C) a guard cell, (D) a stoma, (E) an epidermal cell, (F) the cuticle, (G) a vein, and (H) a bundle sheath cell.

Is this leaf from a C_3 or a C_4 plant? _____

- On the following photomicrograph of a vascular bundle of a monocotyledonous stem, label the location of (A) a parenchyma cell, (B) a vessel element, (C) a companion cell, (D) a sieve tube cell, and (E) a sclerenchyma cell.

- On the following figure showing the tissues and regions of the root tip, label the three developmental zones, the epidermis, a root hair, the stele, the cortex, the protoderm, the ground meristem, the procambium, the apical meristem, and the root cap.

32

Transport in Plants

UPTAKE AND TRANSPORT OF WATER AND MINERALS • TRANSPORT OF WATER AND MINERALS IN THE XYLEM (pages 720–726)

Key Concepts

1. The *osmotic potential* of any solution is negative or zero and depends on the solute concentration of the solution.

2. Systems (solutions or cells) with the same osmotic potential (solute concentration) are *isosmotic*. A system with a more negative osmotic potential is *hyperosmotic* to a system with a less negative osmotic potential, which is *hypoosmotic* to it.

3. Water moves from a solution with a lesser solute concentration (less negative osmotic potential) to a solution with a greater solute concentration (more negative osmotic potential) if no other factors are operating.

4. As water enters a plant cell, a positive pressure potential, called *turgor pressure,* develops because of the expansion of the cytoplasmic portion of the cell against the rigid cell wall.

5. Equilibrium is reached when the tendency of water to enter the plant cell due to osmotic potential is balanced by the repulsion of water from the cell due to the pressure potential.

6. The overall tendency of a solution to take up water is called its *water potential* (ψ). Water potential is the sum of the osmotic potential (ψ_0), usually negative, and the pressure potential (P), usually positive.

$$\psi = \psi_0 + P$$

7. The water potential, osmotic potential, and pressure potential of pure water under atmospheric pressure are all equal to zero. Water always moves toward the region of most negative water potential.

8. Facilitated diffusion by *carrier proteins* moves some minerals across cell membranes and down their concentration gradient. Carrier proteins are also used in active transport to move minerals against a concentration gradient.

9. In plants, active transport involves *proton pumps* that move hydrogen ions (H^+) from the inside to the outside of the cell.

10. Proton pump activity creates an electrical potential across a typical plant cell membrane of about −120 millivolts (mV). Cations, such as K^+, move through membrane channels into the cell due to the net negative charge of the inside of the membrane. Other ions, such as Cl^-, are cotransported into the cell with H^+ by a symport system.

11. In *patch clamping*, mild suction is used to attach a small section of membrane to the end of a micropipette. Ion channels in the isolated membrane can be studied by measuring changes in electric potential across the membrane.

12. Water and minerals can enter the plant from the soil by diffusing through the meshwork of cell walls and intercellular spaces called the *apoplast*.

13. The meshwork of living cells and their interconnections *(plasmodesmata)* is called the *symplast*. Entrance to the symplast is controlled by selectively permeable cell membranes.

14. As water enters the plant from the soil, it follows a water potential gradient. Water has a tendency to flow from the cortex to the stele because the water potential of the cortex is less negative than the water potential of the stele.

15. Because the suberin-containing *Casparian strips* of the endodermis seal off the apoplast of the cortex from the apoplast of the vascular cylinder, water and its dissolved minerals must enter the symplast by passing through the cytoplasm of endodermal cells.

16. Endodermal membrane proteins regulate the kinds and amounts of minerals entering the stele.

17. Xylem cells (tracheids and vessel elements) are dead at maturity and, thus, a part of the apoplast.

18. Minerals are actively pumped out of living parenchyma and pericycle cells within the stele and enter the xylem. Parenchyma cells with numerous mitochondria and specializations to increase the surface area of the plasma membrane are called *transfer cells*.

19. As transfer cells pump minerals into the xylem conduits, the water potential within the xylem becomes more negative and water passively reenters the apoplast by osmosis.

20. *Eduard Strasburger* conducted research demonstrating that the rise of sap in the stem can occur without the presence of roots. The rise is not caused by the pumping of living cells within the stem but rather seems to involve the leaves.

21. *Capillary action* by itself can cause water to rise only to about 40 cm within the xylem conduits of the plant.

22. *Root pressure* results because of the passive movement of water into the roots due to a water potential gradient.

23. *Guttation*, the loss of liquid water through openings in the leaves, occurs under conditions when water is in abundance and is evidence for the existence of root pressure. Root pressure also causes sap to ooze from cut stumps.

24. *Transpiration* is evaporative water loss from leaves.

25. As water evaporates from the surface of a mesophyll cell and is lost from the leaf, the water potential of the mesophyll cell becomes more negative. As water diffuses into the mesophyll cell from an adjacent xylem conduit, a *tension* or "pull" results on the water column within the xylem.

26. Tension develops within water columns because of the great *cohesion* of water molecules created by hydrogen bonding between them (refer to Chapter 2).

27. As water moves throughout the plant, from the soil into the root cortex, into the stele, up the xylem conduits, into mesophyll cells, and out into the dry air surrounding the leaf, it is following a water potential gradient, always going from regions of less negative water potential to regions of more negative water potential.

28. Minerals are moved passively with water in the xylem.

29. *Per Scholander* used a pressure bomb to show that the xylem sap was under tension in all cases where sap was ascending in a plant. The sap tension measurements were great enough to account for the movement of sap in the tallest plants.

Questions (for answers and explanations, see page 264)

1. In a plant root, a cell type in which water *cannot* move via the apoplast is the
   a. epidermis.
   b. endodermis.
   c. pericycle.
   d. vessel element.
   e. parenchyma.

2. In a plant root, a cell type in which water *cannot* move via the symplast is the
   a. epidermis.
   b. endodermis.
   c. pericycle.
   d. vessel element.
   e. parenchyma.

3. If the pressure potential is +0.16 megapascals (mPa) and the osmotic potential is −0.24 megapascals, then the water potential would be
   a. +0.4 mPa.
   b. +0.08 mPa.
   c. −0.08 mPa.
   d. +0.16 mPa.
   e. −0.24 mPa.

4. If you were to order the water potential of the following root cells/regions from least to most negative, which cell/region would be third?
   a. Xylem conduit
   b. Soil next to root
   c. Cortex apoplast
   d. Stele apoplast
   e. Transfer cell

5. If solutions A and B are separated by a membrane permeable only to water, and if the following are the osmotic and pressure potentials of the two solutions (in megapascals), in which direction will there be a net movement of water?

Solution	Osmotic Potential	Pressure Potential
A	−0.1	0.0
B	−0.2	+0.1

   a. From A to B
   b. From B to A
   c. No net movement
   d. First from A to B, then from B to A
   e. First from B to A, then from A to B

6. Based on your response to Question 5, use the terms "isosmotic," "hyperosmotic," or "hypoosmotic" to describe solutions A and B.

7. What is the *minimum* number of cell membranes that a water molecule must move through in getting from the soil into a xylem vessel element?
   a. 0
   b. 1
   c. 2
   d. 6
   e. Many more than 6

8. Root pressure is caused by
   a. the active transport of water into the xylem by living cells next to the xylem.
   b. the passive movement of water into the xylem because of the less negative water potential there.
   c. the passive movement of minerals into the xylem from living cells next to the xylem.
   d. the active transport of minerals into the xylem by living cells next to the xylem and the passive movement of water into the xylem.
   e. the passive movement of water into the xylem because of the greater pressure potential within living cells next to the xylem.

9. Examine the following diagram of a leaf cross-section. Which letter (a to e) indicates the area where the water potential would be *least* negative?

10. The movement of water up the stems of tall plants is *least* dependent on which of the following factors?
    a. Guttation
    b. Transpiration
    c. Cohesiveness of water molecules
    d. Tension within columns of water molecules
    e. Evaporative water loss from leaves

Activity *(for answer and explanation, see page 265)*

- In the following figure showing a portion of a cross-section through a root, label regions corresponding to the epidermis, cortex, endodermis, pericycle, and xylem. Also, draw the paths that a water molecule could take from a root hair to the xylem by remaining entirely within the symplast or by traveling through both the symplast and apoplast.

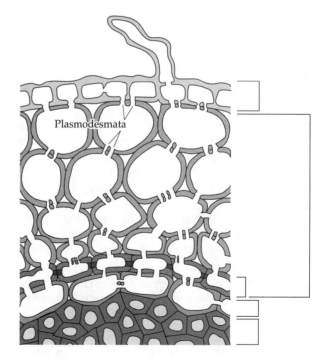

Plasmodesmata

TRANSPIRATION AND THE STOMATA • TRANSLOCATION OF SUBSTANCES IN THE PHLOEM (pages 726–731)

Key Concepts

1. The waxy *cuticle* covering the leaf epidermis minimizes transpiration but also prevents exchange of carbon dioxide and oxygen between leaf tissues and the atmosphere.

2. *Stomata* (singular: *stoma*) are openings in the leaf created by a pair of specialized epidermal cells called *guard cells*. Unlike other epidermal cells, guard cells have chloroplasts. To conserve water, stomata are closed at night or when water is limited.

3. Potassium ions move into guard cells and make their water potential more negative. Water then passively diffuses into the guard cells, which swell, become turgid, and change shape so as to create the stomatal opening. As potassium ions move out of the guard cells, water passively follows, the cells become less turgid, and the stoma is closed or reduced in size.

4. Shape change of guard cells is regulated by special radially oriented cellulose microfibrils so that stomatal opening occurs when the cells take up water and stomatal closing occurs when they lose water.

5. The carbon dioxide level within the intercellular spaces of the leaf influences potassium ion movement into guard cells. Low carbon dioxide levels lead to stomatal opening; high carbon dioxide levels lead to stomatal closing.

6. *Abscisic acid* is released by certain cells within the leaf in response to water loss and an increasingly negative water potential. Abscisic acid may cause guard cells to release potassium ions, thus closing the stomata and preventing further water loss.

7. Light may also influence the opening and closing of stomata due to light activation of proton pumps recently discovered in guard cells. The activity of the pumps would influence the movement of K^+ and Cl^- ions into guard cells.

8. Unlike most plants, the *succulent plants* in the family Crassulaceae have their stomata open at night and closed during the day.

9. At night when stomata of crassulaceans are open, carbon dioxide combines with *PEP* (phosphoenolpyruvic acid) to become chemically fixed in the form of certain organic acids (e.g., malic acid and aspartic acid). This causes the plant tissues to become acidic at night.

10. During the day, when stomata of crassulaceans are closed, the organic acids are broken down to release the carbon dioxide and make it available for photosynthesis. The set of chemical reactions for fixing and releasing carbon dioxide is referred to as *crassulacean acid metabolism (CAM)*.

11. Because CAM plants can keep their stomata closed during the day they lose less water through transpiration. This adapts them for life in hot, dry habitats.

12. Unlike transport in the xylem, phloem transport involves living cells, can occur in both directions simultaneously, and is an energy-requiring process.

13. Sap-feeding aphids have been useful for obtaining samples of phloem sap. After they insert their stylet into a sieve tube, the aphid can be detached from its stylet, which continues to exude sap.

14. The *pressure flow model* is the most widely accepted explanation for transport of phloem sap. It depends on loading of sugars into the phloem sieve tubes in *source regions* (usually the leaves) and unloading of sugar from the phloem in *sink regions* (usually the roots). Phloem loading and unloading is an ATP-requiring process.

15. Sugars and other solutes produced in leaf mesophyll cells leave the symplast and enter the apoplast near a vein.

16. Specific organic molecules are selectively transported from the apoplast into the symplast of phloem cells, frequently companion cells.

17. In some species, sucrose is transported from mesophyll to phloem within the apoplast and is loaded from there into sieve tubes.

18. Loading of phloem with organic materials makes the osmotic potential and the water potential of these cells more negative. Water moves osmotically into the phloem sieve tubes and increases their pressure potential.

19. Secondary active transport involving a sucrose-proton symport loads sucrose into the phloem. Sucrose is co-transported with protons from apoplast to symplast, as the protons move down a proton gradient established by primary active transport.

20. In sink regions, solutes are actively unloaded from the sieve tubes and this makes the water potential of these cells more negative. Water passively moves from the sieve tubes into the surrounding tissue and reduces the pressure potential in the sieve tubes.

21. The gradient of pressure potential from source regions to sink regions causes the bulk flow of water and dissolved solutes.

22. In some plants, like sugar maples, excess photosynthate stored in the trunk and twigs over the winter is moved into the xylem the following spring.

Questions (for answers and explanations, see page 265)

1. As a leaf loses water, a series of events typically occurs leading to the closing of stomata. Which of the following shows the correct sequence of these events?
   a. Potassium ions enter guard cells, water passively enters guard cells, abscisic acid is released by leaf cells, pressure potential in guard cells decreases
   b. Potassium ions enter guard cells, water passively leaves guard cells, pressure potential in guard cells decreases
   c. Water passively leaves guard cells, pressure potential in guard cells decreases, abscisic acid is released by leaf cells, potassium ions leave guard cells
   d. Potassium ions leave guard cells, water passively leaves guard cells, pressure potential in guard cells decreases, abscisic acid is released by leaf cells
   e. Abscisic acid is released by leaf cells, potassium ions leave guard cells, water passively leaves guard cells, pressure potential in guard cells decreases

2. Which of the following explains why the leaf cells of plants within the family Crassulaceae have a lower pH at night than during the day?
   a. These plants keep their stomata open at night but closed during the day.
   b. Carbon dioxide in these plants is chemically fixed in organic acids during the night, but these acids are metabolized during the day.
   c. These plants have a buildup of abscisic acid at night.
   d. Because their stomata are closed during the day, excess carbon dioxide within the leaf forms carbonic acid.
   e. The premise statement is incorrect; the leaves of CAM plants have a lower pH during the day than at night.

3. Which of the following characteristics applies to both xylem and phloem transport?
   a. Follows a water potential gradient
   b. Involves only living cells
   c. Can occur in both directions
   d. Involves active transport of solute with passive movement of water
   e. Is a passive, non-energy-requiring process

4. Which of the following cell types would be the *third* cell type a sugar molecule is likely to encounter on its route from its site of production in a chloroplast to its site of storage in the root?
   a. Mesophyll cell
   b. Companion cell
   c. Sieve tube element
   d. Parenchyma
   e. Endodermis

5. According to the pressure flow model, select *all* of the following choices where active transport is involved in movement of sugar and other organic solutes.
   a. From mesophyll cell to apoplast within leaves
   b. From apoplast to companion cell within leaves
   c. From companion cell to sieve tube element within leaves
   d. Between sieve tube elements within stems
   e. From sieve tube element to surrounding root cells

Activities (for answers and explanations, see page 265)

• In the following figure showing two stomatal cycles, label the cycle typical of a CAM plant and the cycle typical of most other plants. Circle on the appropriate cycle the times when you would expect the cytoplasm of leaf tissue to have its highest pH.

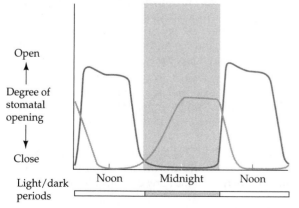

• In the figure below showing movement of sucrose through the symplast and apoplast of the leaf, label a phloem parenchyma cell, a companion cell, a mesophyll cell, a bundle sheath cell, and a sieve-tube element. Indicate on the drawing where you would expect to find cell membranes with an abundance of sucrose–proton symports.

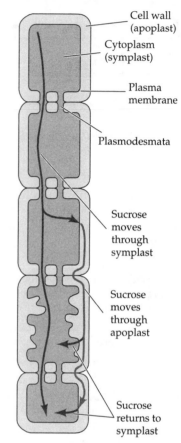

• In the figure below showing a sucrose-proton symport in the cell membrane of a sieve tube cell, label the side of the membrane that is the symplast and the side that is the apoplast.

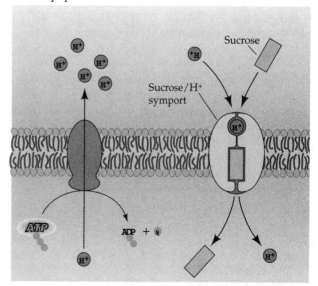

33
Plant Responses to Environmental Challenges

CHAPTER LEARNING OBJECTIVES—after studying this chapter you should be able to:

❏ Contrast how animals and plants react to infection.

❏ Explain what phytoalexins are and their role in the hypersensitive reaction in plants.

❏ Discuss the role of pathogenesis-related proteins in plant–pathogen interactions and the involvement of salicylic acid in their production.

❏ Describe the highly specific defense mechanism involving gene-for-gene resistance and how resistance and avirulence genes interact to confer disease resistance.

❏ Explain how grazing by herbivores affects the photosynthetic production of plants.

❏ Describe some common adaptations of plants to grazing by herbivores.

❏ Define what secondary products are and list the major types of secondary products produced by plants.

❏ Explain how the secondary product canavanine protects plants from herbivores.

❏ Describe three major ways that plants protect themselves from their own toxic secondary products.

❏ Define what a xerophyte is, and compare the rate of growth and photosynthetic efficiency of xerophytes relative to other plants.

❏ Describe important adaptations seen in the leaves of xerophytes.

❏ Describe important adaptations seen in the roots of xerophytes.

❏ Discuss the problems faced by plants living in waterlogged soils and some of their adaptations such as pneumatophores.

❏ Describe important adaptations, such as aerenchyma, that are seen in submerged plants.

❏ Define what a halophyte is, and discuss important adaptations seen in halophytes.

❏ Explain why halophytes and xerophytes have many similar adaptations and describe these adaptations.

❏ Explain what the major processes are that result in soils laden with heavy metals.

❏ Characterize the responses that plant species have shown to life in soils contaminated with heavy metals.

❏ Characterize important adaptations that plants show to serpentine soils.

PLANT–PATHOGEN INTERACTIONS • PLANTS AND HERBIVORES: BENEFITS AND LOSSES (pages 735–741)

Key Concepts

1. Plants and their pathogens have coevolved. Pathogens have mechanisms to attack plants; plants have evolved defense mechanisms. Signaling between plants and pathogens is beginning to be understood at the chemical level.

2. The protective outer covering of plants may include cork or an epidermis covered with suberin or waxes. The pathogen must penetrate these barriers.

3. Whereas animals generally attempt to repair infected tissue, plants prevent the spread of the pathogen by isolating the diseased area with nonliving tissue. This may involve deposition of polysaccharides against the cell wall to seal plasmodesmata and prevent spread of the pathogen to adjacent tissue. Additional lignin can then be laid down upon the built-up polysaccharide to better isolate the diseased area.

4. *Phytoalexins* are a class of chemicals providing plants with some nonspecific protection against fungi and some bacteria.

5. *Pathogenesis-related (PR) proteins* include enzymes that attack the cell wall of pathogens or alert other plant cells to the presence of the pathogen. Some of the breakdown products of the pathogen that are produced by the PR proteins may induce additional defensive responses from the plant under attack.

6. A *hypersensitive reaction* in plants occurs when cells near the site of infection release phytoalexins and then die. The remaining "dead spot," or *necrotic lesion*, isolates the microbial infection, while the rest of the plant remains infection-free.

7. Export of *salicylic acid* from an infected plant part may serve as a signal for the production of PR proteins. PR proteins may limit the spread of the infection.

8. Highly specific defense mechanisms in some plants involve *gene-for-gene resistance*. This system involves a large number of *R (resistance) genes* in plants that are related to numerous *Avr (avirulence) genes* present in specific pathogens. Dominant *R* genes favor resistance. Since *Avr* genes produce a product that allows the plant to detect the presence of the pathogen, dominant *Avr* genes reduce the effectiveness of the pathogen.

9. If a plant has a dominant allele for an *R* gene that corresponds to an *Avr* gene represented in a pathogen by a dominant allele, it will be resistant to that pathogen strain, even if no dominant allele pairings exist for other *R, Avr* genes. Most gene-for-gene interactions elicit the hypersensitive response.

10. *Grazing* tends to increase the photosynthetic production of certain plant species. In a grazed plant, minerals absorbed by the root system need not be divided among as many leaves, changes in the source-to-sink ratio favor transport of photosynthetic products away from the leaves, and removal of the older, outer leaves makes more light available to the inner, younger leaves.

11. Many plants have developed adaptations to grazing by herbivores. Removal of apical buds by grazing usually leads to the development of lateral buds and results in a bushier plant. Grazed plants may have a longer growing season than ungrazed plants and produce more fruits.

12 *Secondary products* are special chemicals produced by plants that influence the behavior of other organisms. Some secondary products serve as attractants of animal pollinators, repellents of herbivore grazers or fungi, or growth inhibitors of competing plant species.

13. Most secondary plant products are of low molecular weight and help plants compensate for their inability to move.

14. Alkaloids affect herbivore nervous systems.

15. Some nitrogen- and sulfur-containing secondary plant products are carcinogenic or cause nerve damage and pain in herbivores.

16. Phenolics are distasteful to herbivores, interfere with the functioning of the nervous system, or act as fungicides.

17. Quinones reduce competition by inhibiting growth of other plants.

18. Terpenes act as fungicides and insecticides, but some also attract pollinators.

19. Steroids mimic animal hormones, thereby preventing normal development of insect herbivores.

20. Flavonoids attract pollinators and seed dispersers.

21. Canavanine is an amino acid chemically very similar to arginine. Some plants use it as a nitrogen storage product. Herbivores that consume it are poisoned when the canavanine is incorporated into their proteins instead of arginine.

22. Toxic secondary plant products are usually isolated in vacuoles within cells if the products are hydrophilic, or

within special tubes called *laticifers* if the products are hydrophobic.

23. Some plants store precursors of toxic secondary plant products in one compartment and the enzymes that will activate them in a different compartment. Mixing of enzyme and substrate occurs only when herbivores damage tissue.

24. Some plants have modified enzymes or receptor molecules that do not recognize the toxic secondary plant product. For example, many canavanine-producing plants have tRNA-charging enzymes that can distinguish between canavanine and arginine and only bind arginine.

Questions (for answers and explanations, see page 265)

1. Complete the following table by indicating whether dominant alleles will be <u>present</u> or <u>absent</u> given the disease-resistance conditions specified.

| | | Pathogen avirulence gene | |
		Dominant allele	Dominant allele
Plant host cell resistance gene	Dominant allele	DISEASE	RESISTANCE
	Dominant allele	DISEASE	DISEASE

Which combination of gene pairs is likely to elicit the hypersensitive response in a plant? Explain your answer.

2. Which of the following is *not* a plant adaptation to grazing by herbivores?
   a. Production of spines and thorns
   b. Production of canavanine
   c. Resistance to salt toxicity
   d. Stem growth from basal meristems
   e. A bushy growth form.

3. Which of the following is *not* a typical use of secondary plant products?
   a. Reduce competition from neighboring plants
   b. Promote pollination
   c. Attract animals
   d. Energy storage within the plant
   e. Repel animals
   f. Inhibit fungal growth
   g. Affect the normal development of insects

4. Order the following events that are hypothesized to occur in the hypersensitive reaction following microbial infection in plants.

_____ Production of a necrotic lesion

_____ Synthesis of pathogenesis-related proteins

_____ Release of salicylic acid

_____ Secretion of phytoalexins

_____ Death of infected tissue

5. Which of the following is *not* a typical mechanism used by plants to protect themselves from their own toxic secondary products?

   a. Storage of hydrophobic secondary products in laticifers
   b. Production of the secondary product in an inactive form
   c. Activation of the secondary product only in damaged tissue
   d. Modified enzymes or receptors that are able to recognize the secondary product
   e. Isolation of the secondary product in a special compartment

6. Discuss some of the ways in which grazing may actually increase photosynthetic production in plants.

7. Discuss how the prevalent production of secondary products by plants may be an adaptation to their sedentary lifestyle.

8. Explain how the secondary product canavanine protects plants from herbivores.

DRY ENVIRONMENTS • WHERE WATER IS PLENTIFUL AND OXYGEN IS SCARCE • SALINE ENVIRONMENTS • HABITATS LADEN WITH HEAVY METALS • SERPENTINE SOILS   (pages 741–746)

Key Concepts

1. *Xerophytes* are plants with structural or biochemical adaptations for life in dry environments. Some desert plants do not show any xerophytic adaptations and complete their life cycles during periods of the year when water is available.

2. Xerophytic leaf adaptations include a thickened waxy cuticle, stomata in sunken, hair-bordered pits, fleshy leaves in which water is stored, or loss of leaves during dry periods. Xerophytic adaptations reduce water loss but also restrict photosynthesis, so xerophytes tend to grow slowly.

3. Some xerophytes also show root adaptations, including those with a very long taproot that is able to reach underground water supplies, species in which the root system dies back during dry periods, and species, such as the cacti, with very shallow but extensive fibrous root systems.

4. In general, xerophytes fix more grams of carbon by photosynthesis per gram of water lost through transpiration than nonxerophytes do.

5. The osmotic potential in root cells of water-stressed plants tends to be very negative due to the accumulation of the amino acid *proline* within these cells. The increased water potential gradient allows these plants to extract more water from the soil.

6. Floating plants tend to have their stomata restricted to the upper epidermis of the leaf, unlike the more normal situation where stomata are generally found on the lower surface.

7. The leaves of many submerged or floating plants have mesophyll layers containing large air spaces. Such layers are called *aerenchyma* and provide buoyancy as well as oxygen storage.

8. Plant species living in waterlogged soils may have slow-growing root systems that do not penetrate deeply. Alcoholic fermentation may supply some ATP in these root cells when insufficient oxygen is available for aerobic respiration.

9. *Pneumatophores* are extensions of the root systems of swamp-dwelling plants that emerge from the water and help to absorb oxygen for the submerged root system.

10. *Halophytes* are plants that have evolved adaptations for growing in saline (salty) environments, such as land bordering the sea or agricultural areas where *salinization* has occurred.

11. The large negative water potential of saline soils makes it more difficult for plant root systems to absorb water. Also, increased sodium ion (and sometimes chloride ion) concentration is toxic to many plants.

12. Many halophytes have evolved a resistance to salt toxicity so they can maintain relatively high levels of sodium ions in their cells. The accumulation of sodium in the cells of halophytes makes their own water potential more negative and facilitates water absorption from the soil. Some halophytes have salt glands in their leaves that secrete excess salt.

13. Some halophytes, like xerophytes, accumulate proline within their cells. Other adaptations seen in both halophytes and xerophytes are succulence, high root-to-shoot ratios, sunken stomata, reduced leaf area, and thick cuticles. Many succulent plants use crassulacean acid metabolism (CAM).

14. *Heavy metals* such as copper, nickel, lead, cadmium, chromium, and zinc usually make soil toxic to most plants.

15. Heavy metal contamination can be due to natural geological processes; acid rain, which releases toxic aluminum ions in soil and water; or the deposition of tailings from mining operations.

16. In some plant species, individuals become genetically adapted to tolerate contamination by a specific metal, usually the most common one. Even though they take up the contaminant and accumulate it in their tissues, they are tolerant to it.

17. Plant species that evolve tolerance to heavy metals are able to thrive in those habitats because of reduced competition with other, nontolerant species.

18. *Serpentine soils* are deficient in calcium, have higher levels of magnesium than more fertile soil, and may also contain relatively high levels of heavy metals. Some plant species are adapted to living on serpentine soils by being highly efficient at absorbing calcium and excluding or tolerating magnesium.

Questions (for answers and explanations, see page 266)

1. Which of the following is *not* an adaptation seen in a typical xerophyte?
   a. Leaves with aerenchyma
   b. Stomata in pits
   c. Leaves that can be curled up
   d. Ability to accumulate proline in root cells
   e. Thickened cuticle

2. Mangrove trees grow submerged in salt water. Which of the following adaptations would *not* be useful given their environment?
   a. Salt glands
   b. Ability to develop large negative water potential within root cells
   c. Stomata on upper leaf surface
   d. Pneumatophores
   e. Thickened cuticle

3. Xerophytes have many features in common with halophytes. Which one of the following is *not* a feature that these two plant types have in common?
   a. Accumulation of proline
   b. Succulence
   c. Thick cuticle
   d. Tolerance of high sodium ion concentration
   e. Large root-to-shoot ratio

4. Which of the following statements about serpentine soils or the plants that are adapted to live on them is *not* true?
   a. Serpentine soils are deficient in magnesium.
   b. Plant growth on serpentine soils is sparse.
   c. Plants adapted to grow on serpentine soils are efficient at absorbing calcium.
   d. Heavy metals may be abundant in serpentine soils.
   e. Some plants will grow on serpentine soils if the mineral content is adjusted.

5. Match the term on the left with the *most* appropriate description on the right.

   a. Proline   _____   Seen in mangrove roots

   b. Aerenchyma   _____   Can result from mining activities

   c. Pneumatophore   _____   Can increase aluminum content of soils

   d. Tailings   _____   Many cacti show this

   e. Succulence   _____   Seen in floating plants

   f. Acid rain   _____   Affects the osmotic potential of the cell

6. Describe what salinization is, its principal causes, and the problems that it creates for plants.

7. List three examples of heavy metals:

   _____

   _____

   _____

34

Plant Nutrition

LIFESTYLES AND THE ACQUISITION OF NUTRIENTS • MINERAL NUTRIENTS ESSENTIAL TO PLANTS • SOILS AND PLANTS (pages 750–756)

Key Concepts

1. *Nutrients* are chemical raw materials obtained by organisms from their environment.

2. *Deficiency symptoms* result when an organism lacks enough of a needed nutrient. For example, nitrogen deficiency in plants causes a yellowing of leaves called *chlorosis* because nitrogen is essential for chlorophyll synthesis.

3. Carbon and oxygen obtained from atmospheric carbon dioxide enter living systems through photosynthesis by *autotrophs* (mostly plants, but also some protists and bacteria). Hydrogen is supplied by water, and also fixed into biological form by photosynthesis.

4. Nitrogen from the inert dinitrogen gas N_2 is fixed into organic form by specialized soil bacteria.

5. *Mineral nutrients*, including sulfur, phosphorus, magnesium, and iron, are obtained by plants in ionic form from the soil.

6. *Autotrophs* make their own organic food from inorganic nutrients. A *heterotroph* is any organism that requires preformed organic molecules to live. *Herbivores* eat plants and *carnivores* eat herbivores or other carnivores.

7. Most autotrophs are photosynthetic and use light energy to produce organic compounds. Some autotrophs are *chemosynthetic* and use the energy stored in reduced

inorganic molecules, like H_2S, to synthesize organic compounds. All chemosynthesizers are bacteria.

8. Plants are *sessile* (stationary) organisms that must grow in order to move and exploit new resources.

9. Plants are more selective than heterotrophs in their procurement of nutrients, eliminating much of the need for waste removal.

10. An *essential element* is an element required directly for normal growth and reproduction, and for which there is no substitute. Lack of an essential element results in characteristic *deficiency symptoms*.

11. In addition to hydrogen, oxygen, carbon, and nitrogen, essential elements include magnesium (chlorophyll constituent; cofactor for many enzymes), iron (constituent of oxidation–reduction molecules), phosphorus (ATP constituent; involved in enzyme regulation), potassium and chlorine (maintenance of electric neutrality and proper osmotic balance) and calcium (processing of hormonal and environmental cues; affects membranes and the cytoskeleton; constituent of the middle lamella).

12. *Macronutrients* are essential elements needed in concentrations of at least 1 mg/g dry matter; *micronutrients* are needed in concentrations of less than 100 µg/g dry matter.

13. Plant physiologists identify essential elements by transplanting seedlings into solutions that lack a single ingredient and looking for deficiency symptoms.

14. Soil includes both living (roots, bacteria, and fungi) and nonliving components and can be divided into horizontal layers called *horizons*, which vary with depth.

15. The *A horizon* is the uppermost soil zone containing most of the living and dead organisms. The A horizon is depleted of minerals by leaching.

16. The *B horizon* is the next deepest zone, where minerals leached from the A horizon accumulate.

17. The *C horizon* is the the deepest zone, containing the original parent material from which the soil develops.

18. Most plant roots reside in the A horizon. Some deep-rooted plants extend into the B horizon, but rarely into the C horizon.

19. *Mechanical* and *chemical weathering* of rock produces soil. Hydrolysis is a key first step in chemical weathering of soil minerals.

20. *Laterization*, common in tropical soils, results in infertile soils because silica and soluble nutrients are leached, while insoluble aluminum and iron compounds remain in the A horizon.

21. In semiarid areas, a hard layer of calcium carbonate may form in the B horizon because water is removed rapidly from the soil by evaporation or plant roots, leaving mineral nutrients behind. Because of the high nutrient content of these soils, irrigation can support very successful agriculture.

22. The physical and chemical properties of soils depend on the amounts and types of clay particles they contain.

23. *Ion exchange* refers to the the exchange of protons, either released by roots or derived from the ionization of carbonic acid ($H_2CO_3 \rightarrow HCO_3^- + H^+$), with K^+, Mg^{2+}, and Ca^{2+} ions bound to the negatively charged clay particles, making these minerals available to plants.

24. Inorganic fertilizers are usually described by their N–P–K values, where, for example, a 5–10–5 fertilizer contains 5% nitrogen, 10% phosphorus, and 5% potassium.

25. Application of calcium-containing material, usually calcium carbonate, or liming can reverse acidification of soil caused by decomposition of organic material. Liming releases protons from soil particles by ion exchange and the excess protons can then be leached from the soil. Liming also increases availability of Ca^{2+}.

26. *Humus* is the carbon-rich, dark material resulting from the death and decay of organisms, especially plants. *Mull* is a fertile, alkaline humus; *mor* is an infertile, acidic humus.

27. As soils age, clay content tends to increase, soluble minerals decrease, and biomass and humus reach a peak and then decrease.

Questions (for answers and explanations, see page 266)

1. Match the appropriate letters for the roles in plant metabolism listed on the right to the plant nutrients listed on the left.

_____ Magnesium	a. Oxidation–reduction molecules
_____ Iron	b. Chlorophyll constituent
_____ Phosphorus	c. Important enzyme cofactor
_____ Potassium	d. Helps maintain osmotic balance
_____ Nitrogen	e. Constituent of all amino acids
	f. Needed for chlorophyll synthesis
	g. ATP constituent

2. In the beginning of the growing season, deep plowing is a common agricultural practice in many parts of the world. Which of the following statements best explains the need for this practice?
a. Deep plowing mixes mull and mor soils.
b. Deep plowing reduces ion exchange in soils.
c. Deep plowing promotes laterization.
d. Deep plowing mixes the A and B horizons.
e. Deep plowing brings the C horizon to the surface.

3. Which of the following statements about the cycling of nutrients is *not* true?
a. Nitrogen is fixed into organic form by bacteria.
b. Some chemosynthetic autotrophs use H_2S to synthesize organic compounds.
c. Heterotrophs must obtain their organic compounds directly from autotrophs.
d. Carbon, hydrogen, and oxygen first become part of living material via an autotroph.
e. Both photosynthetic and chemosynthetic autotrophs can fix carbon.

4. Which of the following statements correctly characterizes changes in $CaCO_3$, clay, and biomass during the long-term aging of soil?
   a. $CaCO_3$ gradually decreases, clay gradually increases, and biomass peaks and then declines.
   b. $CaCO_3$ gradually decreases, clay and biomass gradually increase.
   c. Clay gradually increases, $CaCO_3$ and biomass peak and then decline.
   d. All three gradually decrease.
   e. All three gradually increase.

5. Which of the following is *not* involved in the phenomenon called ion exchange?
   a. Negatively charged clay particles
   b. pH
   c. Cations like K^+, Mg^{2+}, and Ca^{2+}
   d. H_2S
   e. Ionization of carbonic acid

6. Match the appropriate letters for characteristics listed on the right with the soil horizons listed on the left.

   _____ A horizon   a. Leaching occurs here

   _____ B horizon   b. Most plant roots are here

   _____ C horizon   c. Most plant nutrients are here

                              d. Soil formation occurs here

                              e. Most humus is here

                              f. Fewest roots are here

7. In the following figure showing aging of soils, label the curves representing changes in biomass, humus, $CaCO_3$, and clay.

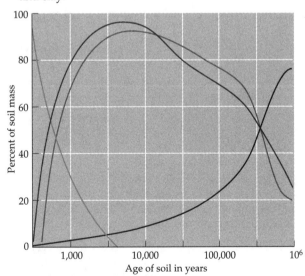

8. Discuss the features of plant nutrient acquisition that relate to their status as sessile organisms.

9. In terms of their potential for supporting agriculture, compare the characteristics of tropical soils subject to laterization with soils more typical of semiarid regions.

NITROGEN FIXATION • SULFUR METABOLISM • HETEROTROPHIC AND CARNIVOROUS SEED PLANTS (Pages 756–762)

Key Concepts

1. Dinitrogen gas (N_2) is highly unreactive and only a few species of bacteria, called the *nitrogen fixers*, are able to convert N_2 to ammonia (NH_3).

2. All organisms depend on the nitrogen fixers for the world's supply of chemically available nitrogen. Some nitrogen is fixed through industrial means, and some is fixed in the atmosphere by lightning, volcanic eruptions, and forest fires.

3. *Rhizobium* and several other species of soil bacteria can fix nitrogen when they form *root nodules* in association with certain plants, such as the legumes.

4. Species of cyanobacteria are important nitrogen fixers in the oceans (along with some other photosynthetic bacteria), in fresh water, and in association with fungi (lichens), or with ferns, cycads, or bryophytes.

5. The filamentous bacteria called *actinomycetes* fix nitrogen in association with the roots of many shrub species, such as alder.

6. Shrubs with nodules containing nitrogen-fixing bacteria are important pioneering species in many plant communities.

7. *Nitrogenase* is the enzyme responsible for catalyzing the nitrogen-fixing reaction. Nitrogenase contains molybdenum and iron atoms.

8. While dinitrogen is bound to nitrogenase, three successive pairs of hydrogens are added to it. With each pair added, one covalent bond between the two nitrogen atoms breaks until two ammonia molecules are formed and released from the enzyme.

9. Nitrogen fixation requires much ATP and strong reducing agents to transfer the hydrogens to dinitrogen.

10. Nitrogenase is sensitive to oxygen and is only catalytically active in an anaerobic environment.

11. *Mutualistic symbiosis* occurs when two different species associate closely and interact in some way so that both species benefit from the interaction. Nodule formation and nitrogen fixation depend on mutualistic symbiosis between *Rhizobium* and an appropriate plant species, usually a legume.

12. *Rhizobium*–legume nodule formation begins when root hairs stimulated by *Rhizobium* cells invaginate to form an *infection thread*. At the same time, the bacteria produce substances that cause formation of a special region in the cortex called a *primary nodule meristem*.

13. The infection thread with enclosed bacteria grows into the root cortex where it enters a cell in the primary nodule meristem. The thread now bursts and releases the bacteria into the cell's cytoplasm.

14. Bacteria increase in size and develop an elaborately folded internal membrane system to become *bacteroids*.

15. The cortex cell produces a type of hemoglobin, called leghemoglobin, creating a low oxygen area around the bacteroids.

16. Most industrial nitrogen fixation now occurs by an energy-intensive method called the *Haber process*, in which dinitrogen and hydrogen gases are combined to form ammonia. Genetic engineering of plants may someday change this situation.

17. However, providing adequate ATP and reducing power and creating an oxygen-free environment within plant cells will be much more difficult than simply inserting the nitrogenase genes into a higher plant.

18. Plants can use the ammonium ion (NH_4^+) resulting from ammonia produced by the nitrogen fixers, but often prefer nitrogen in the form of nitrate (NO_3^-).

19. *Nitrification* is the oxidation of NH_4^+ to nitrate by certain chemosynthetic soil bacteria called the *nitrifying bacteria*.

20. Other nitrifying bacteria include *Nitrosomonas* and *Nitrosococcus*, which convert NH_4^+ into nitrite, NO_2^-, and *Nitrobacter*, which oxidizes nitrite to nitrate (NO_3^-).

21. Plants do not incorporate the nitrogen in nitrate directly into amino acids; instead they reduce nitrate to ammonia, a process called *nitrate reduction*.

22. The steps in which nitrite is reduced to ammonia take place within chloroplasts and result in the synthesis of various amino acids.

23. All organically fixed nitrogen needed by animals must be obtained from plants.

24. *Denitrification* carried out by aerobic bacteria (genera *Bacillus* and *Pseudomonas*) living anaerobically converts fixed nitrogen, usually in the form of nitrate (NO_3^-), to dinitrogen gas in the atmosphere.

25. Nitrogen fixation, nitrification, nitrate reduction, denitrification, and other processes constitute the *nitrogen cycle*.

26. Sulfur is a constituent of the amino acids cysteine and methionine, as well as several other important molecules, such as coenzyme A.

27. Plants take up sulfur in the form of sulfate ions (SO_4^{2-}). Sulfate is reduced and incorporated into cysteine.

28. Some *heterotrophic plants* are parasites, such as the mistletoes, dodders, and Indian pipe.

29. Some heterotrophic plants are carnivores: they obtain additional nitrogen and energy by capturing and digesting insects. Examples include the Venus's-flytrap, sundews, and pitcher plants.

Questions (for answers and explanations, see page 267)

1. Which of the following groups contain *no* species that are able to fix nitrogen?
   a. Cyanobacteria in the ocean and fresh water
   b. Soil bacteria including *Rhizobium*
   c. Cyanobacteria in lichens
   d. Aerobic bacteria in the genera *Bacillus* and *Pseudomonas*
   e. Filamentous bacteria called actinomycetes

2. Nitrogenase
   a. is insensitive to oxygen.
   b. contains magnesium.
   c. releases two NH_3 molecules as products.
   d. requires an aerobic environment.
   e. catalyzes an oxidation reaction that yields ATP.

3. Which of the following statements about *Rhizobium*–legume nodule formation is *not* true?
   a. *Rhizobium* can only fix nitrogen after it becomes a bacteroid within a root cortex cell.
   b. *Rhizobium* induces invagination of root hairs.
   c. Within an infection thread, *Rhizobium* is still extracellular to the plant.
   d. The infection thread can fuse with any root cell of an appropriate legume species.
   e. Bacteroids are surrounded by plant-produced hemoglobin.

4. The least challenging problem facing biotechnologists trying to develop a recombinant DNA alternative to the Haber process is
   a. inserting the nitrogenase gene into a plant cell.
   b. creating an anaerobic environment within the plant cell.
   c. providing adequate reducing power for the reaction.
   d. providing an adequate ATP supply for the reaction.

5. Which of the following statements about nitrification is *not* true?
   a. *Nitrobacter* oxidizes nitrite to nitrate.
   b. *Nitrosomonas* and *Nitrosococcus* convert ammonium ions to nitrite.
   c. Nitrification reactions are energy-producing (exergonic) reactions.
   d. The nitrifying bacteria are chemosynthetic autotrophs.
   e. Heterotrophic plants are more directly dependent on the nitrifying bacteria for usable nitrogen than autotrophic plants.

Fill-in the blanks to complete the sentences in the next three questions.

6. The nitrogen-containing molecule that is most typically incorporated into amino acids by plants is _____.
   The organelle in the plant where this process occurs is the _____.

7. In the metabolism of nitrogen and sulfur in plants, nitrate is to sulfate, as ammonia is to _____.

8. Some plants have a heterotrophic nutritional mode.
   Mistletoe is a _____ plant.
   Venus flytrap is a _____ plant.

9. In the next set of figures showing nodule formation, label the following structures: root hair, cortex cell, *Rhizobium* bacteria, infection thread, bacteroids, and nodule.

(a)

(b)

(c)

(d)

In plant material shown in which of these figures would you expect leghemoglobin to be most abundant? _____

10. Discuss the problems inherent in current attempts to use biotechnology to "equip" plants with the the ability to fix their own nitrogen.

11. In the figure below, label the boxes to show the location of each of the following steps in the nitrogen cycle: nitrification, nitrate reduction, denitrification, and nitrogen fixation.

Which of the compounds shown in this cycle are most reduced? _____

Integrative Question (for answers and explanations, see page 267)

Match the appropriate letters for the generalized chemical reactions listed below with the organisms that can accomplish them.

Denitrifying bacteria   _____

Nitrifying bacteria   _____

Nitrogen fixers   _____

Plants   _____

Heterotrophs   _____

a. Dinitrogen to ammonia
b. Ammonium to nitrite
c. Nitrite to nitrate
d. Nitrate to nitrite
e. Nitrite to ammonium
f. Ammonia to amino acids
g. Sulfate reduction
h. Ammonia to dinitrogen

35

Regulation of Plant Development

FOUR MAJOR REGULATORS OF PLANT DEVELOPMENT • FROM SEED TO DEATH: AN OVERVIEW OF PLANT DEVELOPMENT • SEED DORMANCY AND GERMINATION • GIBBERELLINS: A LARGE FAMILY OF HORMONES
(pages 765–773)

Key Concepts

1. Plant development—the progressive changes in the plant through time—is regulated by the environment, hormones, the pigment phytochrome, and the plant's genome.

2. Environmental cues, such as changes in light or temperature, are mediated by plant hormones and phytochrome.

3. *Plant hormones* are produced in specific parts of the plant and are transported to different parts of the plant where they have their effects. Plant hormones usually exert multiple influences on plant development. Major plant hormones include auxin, gibberellins, cytokinins, abscisic acid, and ethylene.

4. Phytochrome is acted on directly by light and, unlike hormones, is not transported beyond the cell where it resides, but has its action locally.

5. Development is ultimately determined by the plant's genome. The genome encodes phytochrome and the enzymes that catalyze reactions producing the plant's hormones and mediating the plant's response to its hormones.

6. Many events in the life cycle of flowering plants are subject to developmental regulation, including seed dormancy, germination, utilization of endosperm, cell division and elongation, flowering, pollen tube development, fruit development, fruit ripening, plant dormancy, and senescence.

7. Plant development results from the effects of the environment, hormones, and phytochrome on three cellular processes: division, expansion, and differentiation.

8. In some species, seed *germination* can only occur after a period of *dormancy*. Three major mechanisms maintain dormancy: exclusion of water and oxygen from the embryo, mechanically restraining the embryo, and chemical inhibition of development.

9. Modification of the seed coat, called *scarification*, may break dormancy by allowing water to pass through the impermeable seed coat, or by removing the restraining effect of the coat on embryo growth.

10. Agents of scarification include abrasion by the environment, digestive enzymes, and fire.

11. Chemical *inhibitors* may also maintain dormancy. The concentration of inhibitors is reduced by leaching or, as with the common inhibitor *abscisic acid*, by competition with growth promoters.

12. Some seeds need to be dried or may require a period of cold temperatures before they will germinate. Refrigerating seeds to hasten germination is called *stratification*.

13. Some seeds will not germinate until a specific amount of time passes (called *afterripening*); others simply require water, presence or absence of light, or some other environmental factor.

14. Light-requiring seeds usually germinate near the soil surface. They tend to be small and have little stored food.

15. Dark-requiring seeds must be buried in order to germinate. They tend to be large and contain much stored food.

16. Dormancy seems to be advantageous in three different ways: it helps plants survive adverse conditions, increases chances of finding suitable new habitats, and aids in dispersal.

17. Seeds initiate germination by absorbing water in a process called *imbibition*.

18. After imbibition, metabolic changes occurring in the seed include increases in the rates of respiration, RNA synthesis, and protein synthesis. After the radicle emerges from the seed, DNA synthesis begins.

19. Lipids (fats and oils) are the main energy storage molecules in the majority of seeds; starch is also a common energy storage molecule in seeds.

20. Amino acids are usually stored in seeds in polymerized protein form. When the embryo needs amino acids, it hydrolyzes protein.

21. In some species, developing embryos secrete *gibberellins*, which diffuse through the endosperm to the *aleurone layer*, located beneath the seed coat.

22. Under stimulation by gibberellin, proteins in bodies called *aleurone grains* break down into constituent amino acids. These amino acids are used to synthesize digestive enzymes.

23. Digestive enzymes are released from the aleurone layer into the endosperm, where they catalyze the hydrolysis of energy storage polymers. Building block molecules can now be absorbed by the embryo.

24. *Gibberellins* are a family of chemically similar plant hormones produced by higher plants and some fungi, such as the "foolish seedling" disease fungus, *Gibberella fujikuroi*.

25. In 1925 *Eiichi Kurosawa* showed that cell-free extracts of culture media in which *Gibberella fujikuroi* had been growing caused stem elongation in rice seedlings.

26. The gibberellin known as *gibberellin A_1* controls stem elongation in many species; many other gibberellins are intermediates in the synthesis of gibberellin A_1. *Gibberellin A_3* is a commercially available gibberellin.

27. Dwarf plant strains have a recessive mutant allele that prevents them from making their own gibberellin A_1. They grow normally in response to external gibberellin application.

28. In some species, gibberellin secreted by seeds promotes development of the fruit containing the seeds.

29. During their second year, many biennial plants secrete gibberellins that induce rapid shoot elongation and development of flowers, a process called *bolting*.

30. Other effects of gibberellins in different species include fruit development from unfertilized flowers, seed germination, and budding in spring.

31. Gibberellin secreted by the embryo leads to metabolism of the seed endosperm in some species (as considered in Chapter 33).

32. Stem elongation in most species results from interaction between gibberellin and *auxin*.

Questions (for answers and explanations, see page 267)

1. Four players in regulation of plant development are the environment, hormones, phytochrome, and the plant's genome. Discuss their interaction.

2. Which of the following is *not* an agent of scarification?
   a. Digestive enzymes
   b. Fire
   c. Afterripening
   d. Abrasion by soil particles
   e. Stream bed abrasion

3. In the spaces provided below, arrange the letters for the following events in their correct sequence as they would occur during normal seed germination.

   a. Emergence of radicle   b. Breaking of dormancy
   c. DNA synthesis   d. Imbibition
   e. RNA synthesis   f. Protein synthesis

4. In the spaces provided below, arrange the letters for the following events in their correct sequence as they would occur during the phase of seed germination leading up to the metabolism of endosperm nutrients in a species with an aleurone layer.

___ → ___ → ___ → ___ → ___ → ___

a. Aleurone grains release amino acids

b. Gibberellin secretion by embryo

c. Building-block molecules absorbed by embryo

d. Synthesis of digestive enzymes

e. Hydrolysis of endosperm molecules

f. Diffusion of gibberellins to aleurone

5. Which of the following is *not* an effect of gibberellins on plant development in at least some species?
a. Stem elongation
b. Fruit development
c. Induction of bolting
d. Phototropism
e. Metabolism of endosperm in seeds

6. Which of the following were *not* observed in studies done on certain dwarf strains of plants?
a. Applications of extracts of normal strains promoted growth of dwarf strains.
b. Dwarf strains grew normally if additional fertilizer was applied.
c. Application of gibberellin A_1 promoted growth of dwarf strains.
d. Gibberellin caused little additional growth of normal strains.
e. Dwarf strains are homozygous recessive for genes involved in gibberellin synthesis.

7. The following figure shows a seed in the process of germinating shortly after it has imbibed water. In this figure, label the aleurone layer, the endosperm, and the embryo.

What do the small arrows shown in the figure represent?

What will be the next step in this process, due to the movement shown by these arrows?

AUXIN: A HORMONE WITH MANY EFFECTS • MASTER REACTIONS TO HORMONES • CYTOKININS: HORMONES ACTIVE FROM SEED TO SENESCENCE • ETHYLENE: A GASEOUS HORMONE THAT PROMOTES SENESCENCE (pages 773–782)

Key Concepts

1. Auxin, or *indoleacetic acid*, mediates a variety of plant responses as diverse as *phototropism* and *apical dominance*.

2. Phototropism is a light-directed growth response in plants, either toward light, as in shoots, or away, as in roots.

3. Charles Darwin and his son studied phototropism using the leaf sheaths of grass seedlings, called *coleoptiles*, and determined that the coleoptile tip senses light direction.

4. In 1926, *Frits Went* showed that a chemical diffuses from the tip of the coleoptile and that its distribution in the coleoptile below the tip results in phototropism.

5. The tip produces auxin and the cells in the growth region below the tip respond to the chemical by elongating.

6. Illumination from the side causes auxin produced in the tip to move laterally toward the shaded side. The unequal auxin distribution reaching the growth region causes more elongation on the shaded side, making the tip bend toward the light.

7. *Gravitropism*, a plant growth response directed by gravity, results because auxin produced in the shoot tip moves toward the lower side of a tilted shoot. Greater elongation on the lower side causes the tip to curve up and away from gravity.

8. The effects of auxin on vegetative plant development include the induction of root formation in cut stems, the delay of petiole *abscission* by auxin produced in the leaf, and the maintenance of dormancy in lateral buds by auxin produced in the apical bud, a process called *apical dominance*.

9. The chemical *2,4-D* is a *selective herbicide* that kills dicots at concentrations that do not affect monocots.

10. In strawberries, the fruits, called *achenes*, produce auxin, which induces growth of the receptacle. Auxin treatment can bring on fruit development without fertilization, a process called *parthenocarpy*.

11. Plant cell walls are composed of cellulose *microfibrils* embedded in a matrix of other polysaccharides. Each microfibril consists of about 250 cellulose molecules.

12. In growing plant cells, uptake of water and expansion of the vacuole increase hydrostatic pressure within the cell and stretch the cell wall. Walls must be loosened before they can be stretched.

13. Reversible bending of the coleoptile is called *elasticity*; irreversible bending is *plasticity*. Auxin increases plasticity of coleoptiles.

14. Auxin loosens cell walls, perhaps by increasing the hydrogen ion concentration within the cell wall. Decreased pH may activate an enzyme in the cell wall that breaks linkages of microfibrils to one another or with components of the polysaccharide matrix.

15. Recent evidence suggests that auxin and perhaps other plant hormones may not act directly in causing their effects, but first may need to bind to specific receptor proteins.

16. Work with tissue cultures and intact plants clearly shows that auxin can cause undifferentiated cells like pith to develop into specific cell types.

17. The ratio of auxin and *cytokinins* determines the nature of organ formation in many plants; a high proportion of auxin leads to root formation, a high proportion of cytokinins leads to bud and shoot formation.

18. Whereas auxin stimulates cell elongation, cytokinins stimulate cell division; an appropriate combination of the two produces rapid growth of tissues.

19. Cytokinins are produced in the roots and moved to other parts of the plant.

20. *Kinetin*, a synthetic cytokinin extracted from aged herring sperm DNA, increases the rate of cell division in cultures of carrot root cells.

21. *Zeatin* and *isopentenyl adenine* are two cytokinins naturally found in plants.

22. The following are some of the effects of cytokinins: germination of light-requiring seeds kept in the dark; inhibition of stem elongation; stimulation of lateral bud growth; expansion of leaves; delayed senescence of leaves; alteration of the distribution of nutrients within the plant.

23. The gas *ethylene* promotes senescence in plants, including abscission of leaves and ripening of fruit.

24. Leaf senescence and abscission are helpful in areas with cold winters where photosynthesis would be minimal and leaves could be damaged by freezing.

25. During leaf senescence, resources in the leaves (like amino acids) are reclaimed by the plant.

26. During reproduction in many species, adult plants die because all available resources have been allocated to seed production.

27. In response to ethylene, chlorophyll degrades and cell walls break down in ripening fruit.

28. All parts of the plant produce ethylene, especially ripening fruit.

29. Carbon dioxide has effects antagonistic to those of ethylene; ethylene promotes fruit ripening, carbon dioxide delays it.

30. The apical hook of many dicot seedlings kept in the dark (i.e., underground) is maintained by production of ethylene in the hook area.

31. Under stimulation from ethylene, stems are inhibited from elongating and lose their normal gravitropism.

Questions (for answers and explanations, see page 268)

1. Experiments done by Charles Darwin and his son on plant phototropism showed that
   a. auxin is produced in the tip of the coleoptile.
   b. the tip of the coleoptile is the light receptor of the plant.
   c. within coleoptiles, auxin moves laterally away from the source of the light.
   d. more cell elongation takes place on the shaded side of the plant.
   e. phototropism would occur even if the coleoptile tip was covered.

2. A coleoptile is suspended in a horizontal position and illuminated as shown below. Which of the following arrows shows the expected growth response of the coleoptile under these conditions?

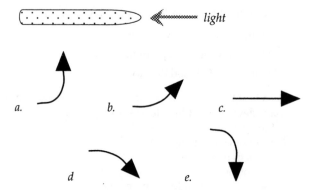

3. A coleoptile tip is severed from its base and an agar block is placed between the tip and base. After a period of time with illumination from the right side, the block is removed and cut into two pieces, *x* and *y*. Pieces *x* and *y* are positioned on the cut bases of two other *dark-grown* coleoptiles. From the following diagrams, select the letter showing the expected response of the coleoptile bases.

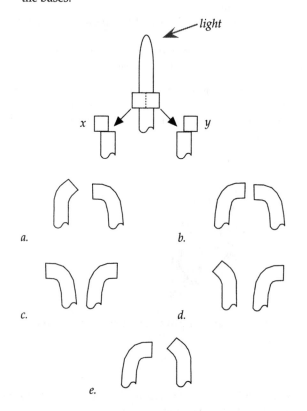

4. You mix auxin of different concentrations with agar and make three types of blocks (*x, y, z*), but you get the blocks mixed up. To determine the relative concentrations of auxin in the blocks, you place the blocks on the cut bases of coleoptiles as shown below (note carefully the positioning of the blocks). Auxin concentration in the blocks, from highest to lowest, is:

a. *x, y, z*
b. *x, z, y*
c. *y, x, z*
d. *y, z, x*
e. *z, y, x*

5. Number the following steps associated with auxin-induced cell elongation in the sequence in which they are hypothesized to occur.

_____ New microfibrils are deposited on the inner cell wall surface.

_____ An enzyme is activated within the cell wall.

_____ pH decreases within cell walls.

_____ Hydrostatic pressure in vacuole stretches the cell wall.

_____ Auxin binds to receptor protein.

_____ Linkages between microfibrils are broken.

6. Which of the following events would likely occur if a plant tissue culture is treated with a solution containing a relatively high concentration of cytokinin and a relatively low concentration of auxin?
a. Bud and shoot formation; rapid cell expansion; decreased cell division
b. Bud and shoot formation; slow cell expansion; increased cell division
c. Bud and shoot formation; rapid cell expansion; increased cell division
d. Root formation; rapid cell expansion; decreased cell division
e. Root formation; slow cell expansion; increased cell division

7. Which of the following is *not* a normal plant response to cytokinins?
a. Seed germination
b. Lateral bud growth
c. Abscission of leaves
d. Leaf expansion
e. Redistribution of nutrients

8. You wish to delay the ripening of stored fruit. Which of the following would *not* help accomplish this objective?

a. Increasing the ventilation within the storage bins
b. Removal of ripe fruit from the bins
c. Decreasing the atmospheric pressure where the fruit is stored
d. Controlling the photoperiod where the fruit is stored
e. Decreasing the atmospheric ethylene concentration where the fruit is stored

9. Explain the observation that in the past, trees near gas street lights would lose their leaves sooner than trees farther from the lamps.

ABSCISIC ACID: THE STRESS HORMONE • LIGHT AND PHYTOCHROME • PLANT RESEARCH TODAY: GENES AND GENE PRODUCTS   (pages 782–785)

Key Concepts

1. *Abscisic acid* is a plant growth substance that is in high concentration in winter-dormant buds and also in water-stressed leaves.

2. In preparation for winter, plants undergo many changes leading to winter dormancy.

3. Formation of dormant buds with wax-covered bud scales protects the growth regions of stems.

4. Other features of winter dormancy are cessation of stem growth, leaf abscission, and an increase in the solute concentration of transport systems, lowering the freezing point.

5. Night length is monitored by plants and used to initiate the process of winter dormancy when nights become longer.

6. Studies done on plant species whose seeds require exposure to light in order to germinate show that blue light and red light induce germination, but green light does not.

7. Far-red light reverses the effect of previous red light exposure and prevents germination.

8. Seeds given successive exposures to red and far-red light respond to the last treatment received: if red, they germinate; if far-red, they remain dormant.

9. *Phytochrome*, a bluish pigment, is a protein that exists in two interconvertible forms, P_r and P_{fr}.

10. P_r absorbs best at a wavelength of 660 nm (red light) and is converted into P_{fr}; P_{fr} absorbs best at a wavelength of 730 nm (far-red light) and is converted into P_r.

11. Flowering plant seedlings kept in the dark do not synthesize chlorophyll and are called *etiolated*.

12. Etiolated shoots elongate more rapidly than shoots of seedlings kept in the light. This response helps bring the shoot to the soil surface rapidly.

13. The hook on the shoot of an etiolated dicot seedling helps protect the apical bud from abrasion by pulling instead of pushing the apical bud through the soil.

14. Phytochrome in an etiolated seedling is in the P_r form. Light exposure converts phytochrome P_r to P_{fr} and causes stem elongation to slow, the hook to straighten, chlorophyll synthesis to begin, and leaves to expand.

15. Work on *Arabidopsis* has led to the identification of five genes encoding different phytochromes in that species. Interaction between these different forms may "fine-tune" a plant's growth response to different light conditions.

16. Phytochromes in solution in the cytosol appear to activate membrane-bound *G proteins* and produce two effects: conversion of guanosine triphosphate (GTP) into *cGMP* (cyclic guanosine monophosphate), and opening of channels to calcium ions (Ca^{2+}). Ca^{2+} then binds to *calmodulin*, and calcium-calmodulin and cGMP activate specific genes.

17. The small flowering plant *Arabidopsis thaliana* is being studied extensively to identify developmental mutants. It produces very many small seeds, has a short generation time, and a genome about the size of *Drosophila's*.

Questions (for answers and explanations, see page 268)

1. Which of the following is *not* characteristic of changes associated with winter dormancy in many species of temperate plants?
   a. Development of wax-covered bud scales
   b. High concentration of abscisic acid in buds
   c. Decrease in solute concentration of sap
   d. Leaf abscission
   e. Decrease in the rate of stem growth

2. Two seedlings, A and B, are etiolated because they have been kept in the dark since germination. Seedling A is placed in light of wavelength 660 nm (red light); seedling B is placed in light of wavelength 730 nm (far-red light). Which of the following comparative statements about the seedlings is *true*?
   a. After light exposure, seedling A has more P_r than B.
   b. After light exposure, seedling B has more P_{fr} than A.
   c. Before light exposure, the phytochrome of both seedlings is in the P_r form.
   d. After light exposure, seedling B will begin chlorophyll synthesis.
   e. After light exposure, seedling A will develop a "hook" at its apex.

3. Which of the following does *not* normally occur when an etiolated seedling is exposed to light?
   a. Chlorophyll synthesis begins.
   b. The rate of shoot elongation increases.
   c. The hypocotyl hook straightens (in species with a hypocotyl hook).
   d. Phytochrome P_r is converted to P_{fr}.
   e. Leaf expansion begins.

4. From the following list, select the first action that phytochrome seems to have in causing a developmental effect in a plant.

   a. Open calcium channels in the cell membrane
   b. Activate G proteins
   c. Convert GTP to cGMP
   d. Activate specific genes
   e. Form calcium-calmodulin

5. Dormant lettuce seeds were exposed to the following succession of brief periods of red light (R) and far-red light (FR) light: FR-R-FR-R-FR-R-FR. Which of the following is most likely to occur?
   a. Most of the seeds will germinate.
   b. Most of the seeds will not germinate.
   c. About one-half will germinate.
   d. The seeds will germinate but development will cease unless the epicotyl is exposed to red light.
   e. The seeds will germinate but development will cease unless the epicotyl is exposed to far-red light.

6. What are some of the features of *Arabidopsis thaliana* that are making it the plant biologist's *E. coli*?

Integrative Questions (for answers and explanations, see page 268)

1. Match the major site of production on the right with the plant growth substance produced at that site on the left.

   Abscisic acid _____   a. Embryo

   Auxin   _____   b. Buds

   Cytokinin   _____   c. Leaves

   Ethylene   _____   d. Fruit

   Gibberellin _____   e. Root tips

                           f. Shoot tips

2. Match the main effect on the right with the plant hormones on the left.

   Abscisic acid _____   a. Hydrolysis of endosperm

   Auxin   _____   b. Promotes winter dormancy

   Cytokinin   _____   c. Stimulates cell division

   Ethylene   _____   d. Stimulates cell elongation

   Gibberellin _____   e. Promotes fruit ripening

                           f. Causes petiole abscission

36
Reproduction in Flowering Plants

❏ Compare and contrast the advantages and disadvantages of sexual and asexual reproduction in plants.

❏ Identify and describe the following parts and regions of a complete flower: carpels, stamens, petals, sepals, pistil, ovary, style, stigma, filament, anthers, corolla, calyx, and receptacle.

❏ Identify and describe the structure of the ovule.

❏ Identify and describe the structure of the male and female gametophytes in flowering plants.

❏ Describe the process of pollination and some of the methods that plants use for transport of pollen to female flower parts.

❏ Describe the events and consequences of double fertilization in flowering plants, including formation of the sporophyte embryo and endosperm.

❏ Identify and describe the structures and stages associated with early development in a typical dicot.

❏ Identify and describe the structures associated with a dicot seed.

❏ Describe some of the variation associated with fruit structure and dispersal in flowering plants, and give an example of each.

❏ Differentiate between indeterminate and determinate growth relative to the activities of vegetative apical meristems, floral meristems, and inflorescence meristems.

❏ Explain how photoperiodism is involved in controlling the onset of flowering in many plants.

❏ Define the terms "day-neutral plant," "short-day plant," and "long-day plant," and interpret the results of data from flowering studies to classify plants relative to these terms.

❏ Explain how it was determined that photoperiodic plants measure night length rather than day length.

❏ Describe the involvement of phytochrome in the photoperiodic timing mechanism.

❏ Describe the characteristic features of circadian rhythms.

❏ Characterize the evidence that argues for the existence of a flowering hormone.

❏ Define the term "vernalization" and give an example of it.

❏ Explain what is meant by the term "vegetative reproduction" and give some common examples of it.

❏ Define the term "apomixis" and give an example of it.

❏ Describe important agriculture practices involving asexual reproduction, such as the production of slips and grafts, and the prospects for the development of artificial seeds.

MANY WAYS TO REPRODUCE • SEXUAL REPRODUCTION (pages 789–794)

Key Concepts

1. Meiosis and recombination of gametes associated with sexual reproduction leads to increased genetic diversity. Increased genetic diversity may help to better adapt a population to new environments.

2. *Carpels, stamens, petals,* and *sepals* are the parts (modified leaves) of a complete flower.

3. Female flower parts are called *pistils.* Each pistil consists of one or more carpels and can be divided into an *ovary* (containing one or more *ovules*), an elongated *style,* and a pollen-receptive *stigma.*

4. Ovules consist of a *megasporangium,* a *stalk* suspending the ovule within the ovary, and protective layers called *integuments.* The opening through the integuments into the megasporangium is called the *micropyle.*

5. Male flower parts are called *stamens* and consist of a *filament* and two *anthers,* each with four *microsporangia.*

6. Whorls of petals, often colored, are called the *corolla.* Whorls of sepals, usually green, are called the *calyx.*

7. The stem tip bearing the flower parts is called the *receptacle.*

8. Within the megasporangium, the diploid megasporocyte undergoes meiosis, producing four haploid cells, three of which die. The remaining cell divides mitotically three times to produce eight nuclei.

9. The female gametophyte, called an *embryo sac*, consists of eight nuclei organized into seven cells. The *egg* and two *synergids* are located at the micropyle end of the embryo sac. Two *polar nuclei* are located within a large central cell. Three *antipodal cells* are located at the other end of the sac.

10. Within anthers, *microsporocytes* divide meiotically, each producing four haploid *microspores*. Each microspore divides once mitotically to form a pollen grain. Thus, the male gametophyte consists of one cell with two nuclei.

11. One nucleus within the pollen grain, called the *generative nucleus*, divides to produce two *sperm nuclei*; the remaining nucleus is the *tube nucleus*.

12. Transfer of pollen to the stigma of different flowers can involve transport by wind (grasses), water (some aquatic angiosperms), or animal pollinators.

13. *Self-pollinating* species transfer pollen directly from anther to stigma within the same flower.

14. The pollen tube grows from the surface of the stigma down through the style and into the female gametophyte, usually following a chemical gradient. The tube nucleus remains close to the growing end of the pollen tube.

15. The two sperm nuclei are released into the embryo sac. One sperm nucleus fuses with the egg to form the diploid zygote; the other sperm nucleus fuses with the two polar nuclei to form a triploid nucleus. This process, called *double fertilization*, is only found in angiosperms.

16. The zygote divides mitotically to form the new *sporophyte embryo*; the triploid nucleus divides to form the nutritive tissue called *endosperm*. The remaining cells in the embryo sac degenerate.

17. After fertilization, the endosperm accumulates starch, lipids, and proteins, the integuments develop into a seed coat, and the carpel becomes the wall of the fruit surrounding the seed.

18. Using a combination of new techniques, *in vitro* fertilization can now be accomplished in flowering plants. Its use should enable the study of many interesting problems in plant development.

19. The first division of the zygote produces one cell that will form a filamentlike supporting structure called the *suspensor* and another cell that becomes the *embryo*.

20. The single cell layer covering the embryo becomes the *protoderm*; the protoderm develops into the *epidermis*.

21. In dicots, two embryonic leaves called *cotyledons* develop and give the embryo a heart-shaped appearance.

22. During the *torpedo stage*, the embryo elongates and internal tissue differentiation begins.

23. The elongated region below the attachment of the cotyledons is called the *hypocotyl*.

24. The *shoot apex* is located between the cotyledons. The *root apex* is on the opposite end of the hypocotyl. Apical meristems on the shoot and root apex result in *primary growth* of the plant.

25. During seed development in some species (like peas and peanuts), the cotyledons absorb most of the endosperm; in other species (like castor beans), the endosperm remains intact.

26. Minimally, *fruit* includes the ovary wall and its seeds, but also may consist of other parts of the flower.

27. Some fruit is structurally modified to facilitate dispersal by wind (dandelions), animals (burdocks), or water (coconuts).

28. Nutritive material associated with the fleshy fruit may be used to attract animal species, which then act as agents of dispersal for the seeds.

Activities (for answers and explanations, see page 269)

- Label all cells and nuclei of the embryo sac shown below with the correct numbers of the following terms: 1) synergids, 2) egg cell, 3) polar nuclei, and 4) antipodal cells.

- Label the drawing of a developing plant embryo shown below with the correct letters for the following parts: a) suspensor, b) root apex, c) seed coat, d) hypocotyl, e) shoot apex, f) cotyledon, g) endosperm.

Questions (for answers and explanations, see page 269)

1. Pistils are formed from one or more
   a. ovaries.
   b. ovules.
   c. carpels.
   d. anthers.
   e. megagametophytes.

2. A flower without a calyx would lack
   a. carpels.
   b. stamens.
   c. petals.
   d. sepals.
   e. a receptacle.

3. A species has flowers with 12 petals, six sepals, six stamens, three carpels, 18 ovules, and one ovary. How many seeds can this species produce per flower?
   a. 1
   b. 3
   c. 6
   d. 12
   e. 18

4. Crested wheat grass has a diploid number ($2n$) of 14. Which of the following is the correct sequence of numbers for the chromosomes you would find in an egg cell, an endosperm cell, and a megagametocyte from this species?
   a. 7, 14, 28
   b. 7, 21, 14
   c. 7, 28, 14
   d. 14, 28, 21
   e. 7, 21, 7

5. Which of the following is *not* a mechanism used for pollination of cross-pollinating species?
   a. Wind
   b. Insects
   c. Direct transfer from anther to sigma
   d. Water
   e. Birds

6. The male gametophyte in flowering plants consists of a maximum of ____ cell(s) with ____ nuclei. The number of chromosomes in the nuclei is (n or $2n$) ____ .

7. Compare and contrast the relative ease with which *in vitro* fertilization has been accomplished in animals and plants.

8. Comment on the statement that in many flowering plant species, paternal genes disperse a greater distance from the parental plant than do maternal genes.

TRANSITION TO THE FLOWERING STATE • PHOTOPERIODIC CONTROL OF FLOWERING (pages 794–800)

1. During vegetative growth the plant shows indeterminate growth—the apical meristems continue to produce leaves, lateral meristems, and *internodes* (the stem region between the nodes where leaves and buds form).

2. Under certain conditions, an apical meristem becomes an *inflorescence meristem*. Inflorescence meristems produce leafy structures called *bracts*, internodes, and new lateral meristems that are either additional inflorescence meristems or *floral meristems*.

3. Floral meristems produce whorls of sepals, petals, stamens, and carpels. Floral meristems and some inflorescence meristems show determinate growth.

4. *Organ identity genes*, analogous to animal's homeotic genes, work together to determine the types of floral whorls (sepals, petals, stamens, or carpels) produced by floral meristems. Mutations of these genes result in plants with incorrectly constructed flowers.

5. *Photoperiodism* is the regulation of a biological process by the daily light–dark period.

6. Many plants time the occurrence of their flowering through photoperiodism and will only flower when the day length is greater or less than some *critical day length*.

7. *Short-day plants (SDP)* flower when the day length is less than the critical day length. They usually flower in the late summer or early spring.

8. *Long-day plants (LDP)* flower when the day length is greater than the critical day length. They usually flower in the midsummer.

9. Some plants, called *short-long-day plants*, only flower when long days follow short days. They usually flower before midsummer.

10. Some plants, called *long-short-day plants*, only flower when short days follow long days. They usually flower in the fall.

11. Plants whose flowering is unaffected by day length are called *day-neutral* plants.

12. Experiments with *cocklebur*, an SDP, showed that photoperiodic plants actually measure the length of darkness, not light, in determining the photoperiod.

13. In these experiments, either the day or night period was varied while the other period was kept constant. Cocklebur only flowers if the night length is more than 9 hours. A single long night will cause flowering in this species.

14. Other experiments showed that brief exposure to light during the dark period caused an LDP to flower and prevented an SDP from flowering, but a brief period of darkness during the light period had no effect.

15. That red light is the most effective wavelength in night-interruption experiments, and that far-red light nullifies a previous exposure to red light, suggests the involvement of phytochrome in night length determination.

16. *Circadian rhythms* are cycles of biological events that occur approximately every 24 hours. *Sleep movements* are examples of such rhythms in plants.

17. The period of a circadian rhythm (time from one cycle to the next) is insensitive to temperature, although the amplitude (difference between high and low measurements) is reduced at lower temperatures.

18. Circadian rhythms can persist in constant light or darkness, but they can also be *entrained* by light–dark periods other than 24 hours.

19. Circadian rhythms can be permanently phase-shifted by exposure to light during the dark period.

20. The molecular and cellular basis of the *biological clock* remains unknown, as does how light can reset the clock.

21. Night length determination occurs in the leaves. A single leaf can cause flowering if exposed to an appropriate night length.

22. Grafting experiments have suggested that leaves produce a hypothetical chemical messenger called *florigen*, which moves to other parts of the plant and induces flowering there. Grafting of a single induced leaf can cause flowering in the recipient plant in some species.

Questions (for answers and explanations, see page 269)

1. In the following figure showing shoots resulting from three types of meristems (*a, b, c*), label the following structures: vegetative meristem, inflorescence meristem, floral meristem, leaf, internode, bract, carpel, petal, sepal, and stamen.

(a)

(b)

(c)

Which shoot system (*a, b,* or *c*) would be abnormal in a plant with a mutation in an organ identity gene? _____

Explain which shoot systems can show indeterminate growth, determinate growth, or both.

2. The next figure shows the results of exposing four plants to different light–dark periods to discover if photoperiod affects flowering in this species. Day and night length are shown as the white–dark portions of each 24-hour bar. Plant 3 was exposed to a 15-hour night interrupted by a brief flash of white light shown as a small white rectangle.

Which of the following statements is true? This species is a

a. short-day species with critical day length of 12 hours.
b. short-day species with a critical day length between 12 and 15 hours.
c. short-day species with a critical day length between 9 and 12 hours.
d. long-day species with a critical day length between 12 and 15 hours.
e. long-day species with a critical day length between 9 and 12 hours.

3. A long-day species has a critical day length of 14 hours. Select letters of all of the following treatments likely to induce flowering in this species. Note: Day and night length are shown as the white–dark portions of each 24-hour bar; *fr* indicates a flash of far-red light, *r* indicates a flash of red light.

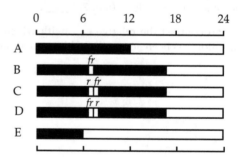

4. If a species is a long-short-day plant, it is likely to flower in

a. early spring.
b. before midsummer.
c. midsummer.
d. late summer.
e. fall.

5. The following figure shows the results from a series of experiments on henbane. Day and night length are shown as the white–dark portions of each 24-hour bar. Plant 3 was exposed to a short flash of white light during its dark period; the lights were briefly turned off during the light period for plant 4.

Based on these results, which of the following conclusions is *not* valid?

a. Henbane is a long-day plant (LDP).
b. The critical period for henbane is greater than 12 hours.
c. To determine when to flower, henbane measures day length.
d. Henbane probably flowers in midsummer.
e. Henbane is not day-neutral.

6. Which of the following conclusions is *not* verified by studies done on the regulation of flowering?

a. Phytochrome may be involved in the determination of night length by plants.
b. Night length is determined within the leaves.
c. Plants exposed to the correct photoperiod produce a flowering hormone, florigen.
d. If a flowering long-day plant (LDP) is grafted to a short-day plant (SDP) kept under long days, the SDP will flower.
e. If a flowering short-day plant (SDP) is grafted to a long-day plant (LDP) kept under short days, the LDP will flower.

7. Discuss the evidence for a flowering hormone in plants.

VERNALIZATION AND FLOWERING • ASEXUAL REPRODUCTION (pages 800–802)

Key Concepts

1. *Vernalization* represents the events that occur in some plants during periods of cold temperature that eventually lead to normal flowering.

2. A self-fertilizing individual heterozygous at one locus can produce offspring with all three genotypes, but it cannot produce individuals with new alleles.

3. A *clone* is a group of individuals with identical genotypes, usually all produced by the same parent.

4. Asexually reproducing individuals produce a clone of offspring. Asexual reproduction is a good strategy for a well-adapted species in a stable environment.

5. *Vegetative reproduction* involves propagation by the vegetative parts of a plant: the stems, leaves, or roots.

6. Horizontal parts of the stem that have become modified for vegetative reproduction are *stolons* (strawberries), *tubers* (potatoes), and *rhizomes* (bamboo).

7. Vertical parts of the stem that have become modified for vegetative reproduction are *bulbs* (lilies and onions) and *corms* (crocuses, gladioli, etc.).

8. Leaves can also produce new plantlets (as in the genus *Kalanchoe*, for example), as can roots, as in *root suckers* (grasses, aspens, poplars).

9. *Apomixis* is asexual production of seeds. By skipping both meiosis and fertilization, the ovule develops into a seed; the ovary wall develops into a fruit. Pollination is sometimes needed for endosperm production or to synchronize the timing of development.

10. *Slips* are cut pieces of stem that can be placed into soil and allowed to develop roots and grow.

11. In *grafting*, a piece of plant, called the *scion*, is attached to a root-bearing host plant, called the *stock*. The cambium of the scion eventually becomes joined to the cambium of the stock via fused wound tissue.

12. Grafting is very important in agriculture because it allows the combination of a shoot system that produces desirable fruit with a hardy root system.

13. Artificial seeds may be important in the future as a way to deliver plant embryos resulting from recombinant DNA and tissue culture techniques.

Questions (for answers and explanations, see page 270)

1. A self-pollinating plant is heterozygous for a single gene. What percentage of the offspring of this plant will have genotypes different from that of this plant?
a. 0%
b. 25%
c. 50%
d. 75%
e. 100%

2. Match the plants on the right to the plant parts on the left that they use in vegetative reproduction.

_____ Stolons	a. Potatoes
_____ Rhizomes	b. Onions, lilies
_____ Bulbs	c. Crocuses
_____ Tubers	d. Strawberries
_____ Corms	e. Bamboo
_____ Root suckers	f. Aspens, poplars

3. Which of the following statements about apomictic reproduction (apomixis) is *not* true?
   a. Apomixis is a form of asexual reproduction.
   b. Apomixis is likely to occur in a species that is well-adapted to a stable environment.
   c. Apomixis produces offspring that are genetically identical to the parents.
   d. Pollination is never required in apomixis.
   e. Apomixis results in a clone of offspring.

4. Which of the following is *not* characteristic of the application of grafting?
   a. The cambium of the scion and stock must join.
   b. The stock usually has a hardy root system.
   c. Both scion and stock must produce wound tissue.
   d. The scion usually has desirable fruit.
   e. The slip is usually treated with root-inducing chemicals.

5. Discuss the construction of an artificial seed as presented in the textbook.

Answers and Explanations

Chapters 31–36

CHAPTER 31 – THE FLOWERING PLANT BODY

Note: The page numbers listed with the section titles refer to the Study Guide; numbers in parentheses for each question refer to the relevant key concepts.

FLOWERING PLANTS: DEFINITION AND EXAMPLES • ORGANS OF THE PLANT BODY • LEVELS OF ORGANIZATION IN THE PLANT BODY (pages 230–231)

1. (8, 9)
   Root system: *a, c, e*
   Leaves: *b*
   Stem: *d*

2. Choice *c*. A fibrous root system provides the plant with a large absorptive surface area and well-developed root caps would protect the root tip meristems from abrasion. (11, 12)

3. Choice *a*. Rice and palms are monocots. As the name implies, monocots have a single embryonic leaf or cotyledon. They also usually have leaves with parallel venation, vascular bundles scattered throughout the stem, and floral parts in multiples of three. (1–5)

4. (13, 15, 18)
   Potato tubers: *stem*
   Runners in strawberry plants: *stem*
   Cactus spines: *leaves*
   Tendrils in peas: *leaves*
   Prop roots in corn: *roots*

• (10, 12, 14–17, 20)

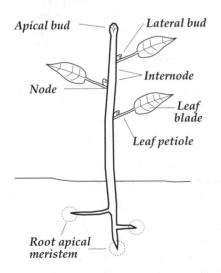

PLANT CELLS • PLANT TISSUES AND TISSUE SYSTEMS (pages 232–233)

1. Choice *c*. Parenchyma cells have only primary cell walls; sclerenchyma cells have both, with the secondary wall laid down to the inside of the primary cell wall. (1, 2, 5, 8)

2. Choice *a*. (5–11)

3. Choice *a*. (8, 10, 11)

4. Choice *e*. (5, 6, 9, 13)

5. Choices *a, d, e*. Epidermis protects the plant body, retards evaporative water loss in the shoot system, and absorbs water and minerals in the root system. (18)

• (1, 2)

• (12, 13)

• (6, 7)
   Because plant parts such as branches and leaves act as somewhat autonomous modules, the control mechanisms in plants are more decentralized than in animals. Also, plant growth is restricted to specific meristematic areas; growth in animals is not as regionalized.

GROWTH: MERISTEMS AND THEIR PRODUCTS • LEAF ANATOMY TO SUPPORT PHOTOSYNTHESIS • SUPPORT IN A TERRESTRIAL ENVIRONMENT (pages 233–236)

1. Choices *a, c, e*. Guard cells as part of the epidermis are derived from the protoderm; cortex and collenchyma cells come from the ground meristem. (6–8, 15)

2. Choice *c*. Companion cells are vascular tissue and derive from the procambium. (6–8, 15)

3. Choice *d*. Both lateral buds and pericycle are regions that give rise to appendages of the plant body. (11, 12)

4. Choices *a, c,* and *e*. Complex tissues are composed of more than one cell type. This includes xylem and phloem. (8–14)

5. Choices *a* and *b*. Water transport does not occur in bark and very little occurs in heartwood. More water transport can occur in the sapwood formed in the spring because its tracheids and vessel elements are larger than those of sapwood formed in the summer. (20–25)

6. Choice *b*. Within stems, xylem is located to the inside of phloem, so within the vascular bundle of a vein, xylem will be above phloem. (20, 31)

7. Choice *c*. Like stomata on leaves, lenticels are openings on the surface of stems. (29, 33)

8. Choice *b*. Secondary phloem is interior to primary phloem, and usually the cortex has been lost and replaced by tissues derived from the cork cambium, with phelloderm to the inside and cork to the outside. (18–28)

9. Choice *b*. C_4 plants have an outer layer of mesophyll cells and an inner bundle sheath layer around each vein. (34; also refer to Chapter 8)

10. Choice *e*. Compression wood expands as it develops and helps hold the branch in position. (36–38)

11. (5)
    Carrot: *biennial*
    Red maple: *perennial*
    Rice: *annual*
    Coconut palm: *perennial*

- See Figure 31.18*c* on page 711 in the textbook. (6–14)
- (34)

This is a C_4 plant—it has the outer layer of mesophyll and inner bundle sheath layer around each vein seen in C_4 plants.

- See Figure 31.20 on page 712 in the textbook. (18–20)
- (6–11)

CHAPTER 32 – TRANSPORT IN PLANTS
Note: The page numbers listed with the section titles refer to the Study Guide; numbers in parentheses for each question refer to the relevant key concepts.

UPTAKE AND TRANSPORT OF WATER AND MINERALS • TRANSPORT OF WATER AND MINERALS IN THE XYLEM (pages 237–239)

1. Choice *b*. Endodermal cells are surrounded by a water-tight endodermal "gasket" that isolates the apoplast of the stele from the apoplast of the root cortex. (12–19)

2. Choice *d*. Since vessel elements are dead at maturity, they are not part of the root's symplast. (12–19)

3. Choice *c*. The water potential is the sum of the osmotic potential (usually negative) and the pressure potential (usually positive), so WP = –0.24 + 0.16 = –0.08 mPa. (3–7)

4. Choice *e*. There is a water potential gradient from outside to inside the root. The correct ranking is (from less to more negative): soil next to root, cortex apoplast, transfer cell, stele apoplast, xylem conduit. (3, 12–19)

5. Choice *c*. Since the water potential is the same on both sides of the membrane,
   $WP_A = –0.1 + 0.0 = WP_B = –0.2 + 0.1 = –0.1$ mPa,
   the net movement of water will be zero. (3–8)

6. Solution A is hypoosmotic to solution B; solution B is hyperosmotic to solution A. (1–3)

7. Choice *c*. Water could travel to the endodermis layer via the apoplast, pass through the outer endodermis cell membrane, and enter the stele apoplast by diffusing through the inner endodermis membrane. (12, 13)

8. Choice *d*. Minerals are actively pumped into the xylem conduits; water follows osmotically. (18–22)

9. Choice *e*. Water potential will be least negative in the xylem conduits and most negative outside the leaf. (23–27)

10. Choice *a*. Guttation is caused by root pressure. It is only important to the movement of water in quite small plants. (20–27)

- (12–19)

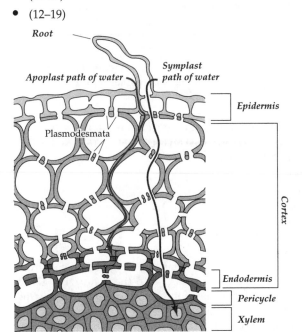

Root

Apoplast path of water

Symplast path of water

Plasmodesmata

Epidermis

Cortex

Endodermis

Pericycle

Xylem

TRANSPIRATION AND THE STOMATA • TRANSLOCATION OF SUBSTANCES IN THE PHLOEM (pages 240–241)

1. Choice *e*. As leaf cells lose water, they release abscisic acid, which causes potassium ions to leave guard cells, decreasing their pressure potential and closing stomata. (2–7)

2. Choice *b*. As carbon dioxide is fixed into organic acids at night the pH decreases; metabolism of these acids during the day (with the carbon dioxide going into the citric acid cycle) causes the pH to increase. (8–11)

3. Choice *a*. Choices *b*, *c*, and *d* are true for phloem transport only; choice *e* is true for xylem transport only. (12)

4. Choice *c*. Sugar molecules diffuse through the symplast of mesophyll cells, are pumped into the apoplast near a vein, reenter the symplast of a companion cell, and then move into a sieve tube element. (14–19)

5. Choices *b* and *e*. Active transport is used to selectively move sugars and other organic compounds from the leaf apoplast into the phloem symplast. Unloading of sugars from the phloem sieve tube elements in sink areas such as the roots also occurs by active transport. (14–19)

- (8–11)

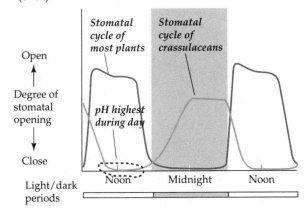

Stomatal cycle of most plants

Stomatal cycle of crassulaceans

Open

Degree of stomatal opening

pH highest during day

Close

Light/dark periods

Noon   Midnight   Noon

- (14–19)

Sucrose-proton symports in cell membrane

Cell wall (apoplast)

Plasma membrane

Cytoplasm (symplast)

Plasmodesmata

Sucrose moves through symplast

Sucrose moves through apoplast

Sucrose returns to symplast

Mesophyll cell

Bundle sheath cell

Phloem parenchyma cell

Companion cell

Sieve tube element

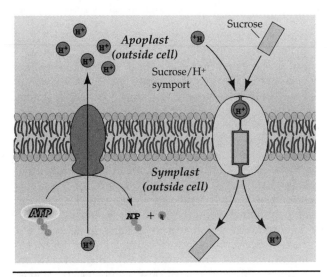

Sucrose

Apoplast (outside cell)

Sucrose/H⁺ symport

Symplast (outside cell)

CHAPTER 33 – PLANT RESPONSES TO ENVIRONMENTAL CHALLENGES

Note: The page numbers listed with the section titles refer to the Study Guide; numbers in parentheses for each question refer to the relevant key concepts.

PLANT–PATHOGEN INTERACTIONS • PLANTS AND HERBIVORES: BENEFITS AND LOSSES (pages 242–244)

1. (8, 9)

		Pathogen avirulence gene	
		Dominant allele *absent*	Dominant allele *present*
Plant host cell resistance gene	Dominant allele *present*	DISEASE	RESISTANCE
	Dominant allele *absent*	DISEASE	DISEASE

For a particular pair of *R* (resistance) and avirulence genes, both the plant and the pathogen must have dominant alleles if the plant is to show resistance. Most gene-for-gene resistance involves the hypersensitive response.

2. Choice *c*. Resistance to salt toxicity is an adaptation seen in halophytes and is not related to grazing by herbivores. (10, 11)

3. Choice *d*. Secondary plant products are chemicals produced by the plant but used to influence another organism. They are not used by the plant directly, as in energy storage. (12–24)

4. (4–7)
   3: Production of a necrotic lesion
   5: Synthesis of pathogenesis-related proteins
   4: Release of salicylic acid
   1: Secretion of phytoalexins
   2: Death of infected tissue

5. Choice *d*. Plants with this adaptation usually have modified enzymes and receptor molecules that do not recognize the toxic secondary product. That is, they respond to the normal product and can distinguish it from the toxic secondary product, as in canavanine-producing plants. (22–24)

6. After a plant has been grazed, minerals absorbed by the roots (especially nitrogen) need not be shared by as many leaves, so the remaining leaves are better supplied. Grazing often removes the older, slower growing, outer leaves so that more light is available to the younger, faster growing, inner leaves. Also, many grazed plants respond by producing multiple shoots and becoming bushier. (10, 11)

7. The major functions of secondary products in plants are to attract, resist, and inhibit other organisms. Since plants cannot move, they cannot flee from herbivores or disease-causing organisms and are often dependent on animals for pollination and seed dispersal services. They need these secondary products to influence the behavior of these animals with whom they are coevolving. (12, 13)

8. Canavanine is very similar to arginine. In many herbivores, the enzyme responsible for charging the tRNA molecule specific for arginine cannot distinguish between arginine and canavanine. Consequently, canavanine is occasionally substituted for arginine during protein synthesis. This produces proteins with modified tertiary structures and reduced biological activities. (21, 24)

DRY ENVIRONMENTS • WHERE WATER IS PLENTIFUL AND OXYGEN IS SCARCE • SALINE ENVIRONMENTS • HABITATS LADEN WITH HEAVY METALS • SERPENTINE SOILS   (pages 244–245)

1. Choice *a*. Mesophyll with large air spaces, called aerenchyma, is characteristic of submerged plants with floating leaves, not xerophytes. (1–7)

2. Choice *c*. Any halophytic adaptation would be useful; stomata in the upper epidermis is characteristic of submerged plants with floating leaves. (6–13)

3. Choice *d*. Tolerance to high sodium ion concentration is useful to halophytes, but not xerophytes. All other adaptations would serve both plant types. (10–13)

4. Choice *a*. Serpentine soils are deficient in calcium and have too much magnesium. (18)

5. (5, 9, 13, 15)
   Seen in mangrove root: *c. Pneumatophore*
   Can result from mining activities: *d. Tailings*
   Can increase aluminum content of soils: *f. Acid rain*
   Many cactus show this: *e. Succulence*
   Seen in floating plants: *b. Aerenchyma*
   Affects the osmotic potential of the cell: *a. Proline*

6. Salinization is any process that increases the salt content of soil. Agricultural salinization results from irrigation in hot, dry regions with high evaporation rates. In these regions, salt buildup produces soils with large negative osmotic potentials that create osmotic problems for plants. Sodium and chloride are also toxic to plants in high concentration. (10–13)

7. Heavy metals include such elements as copper, nickel, lead, cadmium, chromium, and zinc. (14)

CHAPTER 34 – PLANT NUTRITION

Note: The page numbers listed with the section titles refer to the Study Guide; numbers in parentheses for each question refer to the relevant key concepts.

LIFESTYLES AND THE ACQUISITION OF NUTRIENTS • MINERAL NUTRIENTS ESSENTIAL TO PLANTS • SOILS AND PLANTS   (pages 246–248)

1. (11)
   Magnesium: *b, c*
   Iron: *a, f*
   Phosphorus: *g*
   Potassium: *c, d*
   Nitrogen: *a, b, e, g*

2. Choice *d*. There is a constant leaching of nutrients from the A horizon downward into the B horizon. Deep plowing moves some of these nutrients back up into the A horizon. (14–17, 19, 22, 24)

3. Choice *c*. Some heterotrophs, like the carnivores, can obtain organic compounds by consuming other heterotrophs. (3, 4, 6, 7)

4. Choice *a*. During the long-term aging of soil, $CaCO_3$ gradually decreases, clay gradually increases, and biomass peaks and then declines. (27)

5. Choice *d*. Hydrogen sulfide has little to do with ion exchange. (23)

6. (14–18, 26)
   A horizon: *a, b, e*
   B horizon: *c*
   C horizon: *d, f*

7. (27)

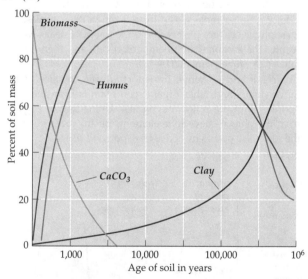

8. Most sessile organisms are aquatic and depend on the movement of water to bring them nutrients. Plants avoid depleting their supply of nutrients by growth of roots, leaves, and stems, and by dispersal of their seeds. (8)

9. Laterization involves the removal of mineral nutrients by leaching so that only oxidized, insoluble iron and aluminum compounds remain. Consequently, these soils are very poor in nutrients and are not suitable for agriculture. In contrast, soils in semiarid regions have retained their soluble nutrients because of the limited rainfall and high rates of evaporation. These soils are suitable for agriculture provided water is available for irrigation. (20, 21)

NITROGEN FIXATION • SULFUR METABOLISM • HETEROTROPHIC AND CARNIVOROUS SEED PLANTS (pages 248–250)

1. Choice *d*. These organisms are denitrifying bacteria. (1–8, 19)

2. Choice *c*. All other statements are false. (7–10)

3. Choice *d*. Infective threads fuse preferentially with root *cortex* cells. (11–15)

4. Choice *a*. The technology for inserting genes into plasmids and transforming plant cells is well established; changing the genetics of higher plants to create a suitable environment for nitrogenase (no oxygen, lots of ATP and reducing agents) will be far more difficult. (16, 17)

5. Choice *e*. Heterotrophic plants are *less* dependent on nitrogen obtained from nitrification since they receive some nitrite and nitrate through their parasitic or carnivorous nutritional modes. (18–20, 28, 29)

6. The nitrogen-containing molecule that is most typically incorporated into amino acids by plants is *ammonia*. The organelle in the plant where this process occurs is the *chloroplast*. (21, 22)

'7. In the metabolism of nitrogen and sulfur in plants, nitrate is to sulfate, as ammonia is to *cysteine*. (20, 21, 27)

8. Mistletoe is a *parasitic* plant.
   Venus flytrap is a *carnivorous* plant.(28, 29)

9. (12–15)

Nodules shown in figure *d*.

10. Insertion of the genes to produce nitrogenase into crop plants will be much less difficult than providing the necessary anaerobic conditions for nitrogenase to function properly. Strong reducing agents and an energy source are also required. (16, 17)

11. (18–25)

NH_3 and NH_4^-

Integrative Question

Denitrifying bacteria:	*h*
Nitrifying bacteria:	*b, c*
Nitrogen fixers:	*a*
Plants:	*d, e, f, g*
Heterotrophs:	*none*

CHAPTER 35 – REGULATION OF PLANT DEVELOPMENT

Note: The page numbers listed with the section titles refer to the Study Guide; numbers in parentheses for each question refer to the relevant key concepts.

FOUR MAJOR REGULATORS OF PLANT DEVELOPMENT • FROM SEED TO DEATH: AN OVERVIEW OF PLANT DEVELOPMENT • SEED DORMANCY AND GERMINATION • GIBBERELLINS: A LARGE FAMILY OF HORMONES (pages 251–253)

1. Environmental cues such as temperature or light changes cause changes in the plant's production of specific hormones; in the case of phytochrome, this may cause the conversion of one form of phytochrome to another. Hormones and phytochrome affect rates of cell division, elongation, and differentiation. The plant genome codes for phytochrome and for the enzymes that moderate the plant's developmental responses to environmental cues. (1–5)

2. Choice *c*. Scarification is the chemical or mechanical modification of the seed coat. Afterripening refers to the amount of time that must pass before seeds can germinate in some species, regardless of whether scarification is required. (9, 10)

3. *b*. Breaking of dormancy → *d*. Imbibition → *e*. RNA synthesis → *f*. Protein synthesis → *a*. Emergence of radicle → *c*. DNA synthesis (8, 17, 18)

4. *b*. Gibberellin secretion by embryo → *f*. Diffusion of gibberellins to aleurone layer → *a*. Aleurone layer grains

release amino acids → *d*. Synthesis of digestive enzymes → *e*. Hydrolysis of endosperm molecules → *c*. Building-block molecules absorbed by embryo (19–23)

5. Choice *d*. Gibberellins are involved in all listed phenomena except phototropism. (21, 24–32)

6. Choice *b*. Dwarf strains were genetically unable to produce gibberellin, so additional fertilizer would have no effect. (24–27)

7. (21–23)

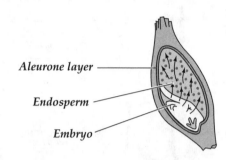

Aleurone layer
Endosperm
Embryo

The small arrows represent gibberellins released from the embryo diffusing through the endosperm to the aleurone layer.

Gibberellins cause protein-containing bodies called aleurone grains to be hydrolyzed into amino acids. These amino acids are then assembled into digestive enzymes in the aleurone layer. The digestive enzymes are secreted into the endosperm and the reserve polymers located there are broken down.

AUXIN: A HORMONE WITH MANY EFFECTS • MASTER REACTIONS TO HORMONES • CYTOKININS: HORMONES ACTIVE FROM SEED TO SENESCENCE • ETHYLENE: A GASEOUS HORMONE THAT PROMOTES SENESCENCE (pages 253–255)

1. Choice *b*. Statements *a*, *c*, and *d* are true, but were not demonstrated by the Darwin's' studies. Statement *e* is false. (1–3)

2. Choice *b*. Phototropism and gravitropism will both be at work in determining the response of the coleoptile. As the coleoptile grows upward in response to gravity, its lower surface will become more illuminated than the upper surface, so it will assume an intermediate position relative to gravity and light gradients. (5–7)

3. Choice *e*. After illumination from the side, more auxin will be on the left side of the tip than the right side. Therefore, more auxin will diffuse into block *x* than block *y*. With the blocks positioned as shown, the left base should bend to the right more than the right base should bend to the left. (5–7)

4. Choice *b*. Clearly, the auxin concentration of block *z* is greater than *y*, since the base with *z* bends more than the base with *y*. The rightmost base shows that more bending occurs under *x* than *z*, so the correct order is *x*, *z*, *y*. (5–7)

5. *1:* Auxin binds to receptor protein.
   2: pH decreases within cell walls.
   3: An enzyme is activated within the cell wall.
   4: Linkages between microfibrils are broken.

5: Hydrostatic pressure in vacuole stretches the cell wall.
6: New microfibrils are deposited on the inner cell wall surface. (11–15)

6. Choice *b*. Cytokinin promotes bud and shoot formation and increased cell division; auxin promotes root formation and increased cell elongation. (15–18)

7. Choice *c*. Cytokinin *delays* senescence (and abscission) of leaves. (17–22)

8. Choice *d*. Anything to reduce atmospheric ethylene concentration would help. Photoperiod does not normally effect ripening of fruit. (23–26)

9. Trees near old gas lamps were exposed to some ethylene from the lamps. Ethylene promotes leaf senescence and loss (abscission). (23–25)

ABSCISIC ACID: THE STRESS HORMONE • LIGHT AND PHYTOCHROME • PLANT RESEARCH TODAY: GENES AND GENE PRODUCTS (pages 255–256)

1. Choice *c*. Solute concentration of the sap is *increased* to prevent freezing during the winter. (1–4)

2. Choice *c*. Seedlings kept in the dark have most of their phytochrome in the P_r form. Exposure to light of wavelength 660 nm converts P_r to P_{fr} and light of 730 nm converts P_{fr} to P_r, so seedling A will have mostly P_{fr} and seedling B mostly P_r. P_{fr} initiates the reversal of etiolation, so *seedling A should begin chlorophyll synthesis and a straightening of the "hook"* (*both* dark-grown seedlings will have a hook initially). (9–14)

3. Choice *b*. The rate of shoot elongation is greatest when the seedling is in the dark, and *decreases* upon exposure to light. (9–14)

4. Choice *b*. First G proteins are activated by the phytochrome. These in turn open calcium channels and convert GTP to cGMP. Ca^{2+} binds calmodulin, and the calcium-calmodulin and cGMP turn on specific genes. (16)

5. Choice *b*. Far-red light prevents seed germination. Plants respond to the last light treatment they received (far-red in this case). (5–8)

6. *Arabidopsis thaliana* is small and produces many seeds that can be grown in petri dishes. It has a short generation time, a small genome (about the size of *Drosophila's*), and many developmental mutants have thus far been identified. (17)

Integrative Questions

1. Abscisic acid: *b*
   Auxin: *c, f*
   Cytokinin: *e*
   Ethylene: *d*
   Gibberellin: *a*

2. Abscisic acid: *b*
   Auxin: *d, f*
   Cytokinin: *c*
   Ethylene: *e*
   Gibberellin: *a*

CHAPTER 36 – REPRODUCTION IN FLOWERING PLANTS

Note: The page numbers listed with the section titles refer to the Study Guide; numbers in parentheses for each question refer to the relevant key concepts.

MANY WAYS TO REPRODUCE • SEXUAL REPRODUCTION (pages 257–259)

- (9–11, 15, 16)

- (15–17, 19–24)

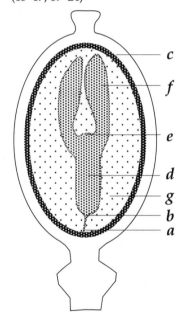

1. Choice *c*. A pistil is a single fused carpel or a group of them. Carpels can include one or more ovules. (2–7)
2. Choice *d*. The calyx consists of the sepals. (2–7)
3. Choice *e*. Each ovule contains one megagametophyte and can develop into one seed. (2–9)
4. Choice *b*. An egg cell is haploid (n), endosperm is triploid ($3n$), and a megagametocyte, being part of the sporophyte generation, is diploid ($2n$). (8, 9, 15, 16)
5. Choice *c*. Direct transfer of pollen from anther to stigma would most likely result in self-pollination, not cross-pollination. (12, 13)
6. The male gametophyte in flowering plants consists of a maximum of *one* cell with *three* (one tube and two sperm) nuclei. The number of chromosomes in the nuclei is *n*. (10, 11)
7. *In vitro* fertilization is relatively easy to accomplish in animals because their gametes are released from the organism and easy to collect and combine. Plant gametes are integral parts of other structures (pollen tubes and the embryo sacs) and are difficult to separate and may depend on interaction with these structures to develop normally. (9, 11, 14, 18)
8. Paternal genes are transported in the pollen grain, but also in the seed. Maternal genes are transported only in the seed. In general, pollen is dispersed greater distances than are seeds. (11–13, 27)

TRANSITION TO THE FLOWERING STATE • PHOTOPERIODIC CONTROL OF FLOWERING (pages 259–261)

1. (1–4)

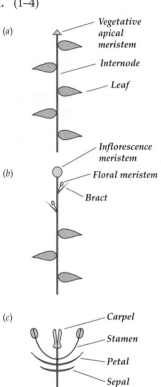

Shoot c, the shoot (flower) produced by a floral meristem would be abnormal in a plant mutant for an organ identity gene.

Vegetative meristems show indeterminate growth; they can continue indefinitely to give rise to leaves, internodes, and additional vegetative meristems. Floral meristems produce only flowers and no additional meristems, so they show only determinate growth. Inflorescence meristems can show either indeterminate or determinate growth, depending on whether they produce additional inflorescence meristems or only floral meristems.

2. Choice *b*. Since the day length has to be less than a minimum value, this is a short-day plant. Since it flowers with days of 12 but not 15 hours, the critical day length is greater than or equal to 12 but less than 15 hours. (5–8)
3. Choices **D** and **E**. A long-day species with a critical day length of 14 hours requires more than 14 hours of light (less than 10 hours of dark) to flower. E provides those conditions, but so does D, where a brief red flash creates two short nights. The far-red flash in B has no effect and negates the red flash in C. (5–8, 12–15)
4. Choice *e*. Long-short-day plants flower in the fall when short days follow long days. (9, 10)

5. Choice *c*. Interrupting a long day with a brief period of dark, as in plant 4, did not prevent flowering, but interrupting a long night with a flash of light did, as in plant 3. These observations suggest that night length, not day length, is being measured. (12–15)

6. Choice *c*. The existence of florigen is only a hypothesis. (24, 30, 31)

7. If a photoperiodically induced leaf is immediately removed from the plant, it will not flower. If the photoperiodically induced leaf is grafted to another plant, the second plant will flower. In studies of plants grafted together, induction of flowering in one plant causes the other plants to flower as well. All of these studies suggest that a hormone is produced in photoperiodically induced leaves, and then transported to the rest of the plant. (22)

VERNALIZATION AND FLOWERING • ASEXUAL REPRODUCTION (pages 261–262)

1. Choice *c*. Since the plant is self-pollinating, the genetic cross is a heterozygote mated with a heterozygote. This will produce about 50% homozygous individuals. (2)

2. Stolons: *d. Strawberries*
   Rhizomes: *e. Bamboo*
   Bulbs: *b. Onions, lilies*
   Tubers: *a. Potatoes*
   Corms: *c. Crocuses*
   Root suckers: *f. Aspens, poplars* (6–8)

3. Choice *d*. Pollination is required in some species to initiate the development of endosperm or to time developmental events. (9)

4. Choice *e*. Slips are cut stems that are used to form new plants, without grafting, by inducing root formation directly, typically using auxin. (10–12)

5. Artificial seeds would contain a multicellular somatic embryo, derived asexually in tissue culture, embedded in a water-soluble gel, and encapsulated by a protective covering. The gel and covering would dissolve after planting; materials such as nutrients, fungicides, and pesticides could be added to the gel. (13)

37

Physiology, Homeostasis, and Temperature Regulation

CHAPTER LEARNING OBJECTIVES—after studying this chapter you should be able to:

❑ Define "homeostasis" and explain its importance to animal physiology.

❑ Differentiate between the concepts of control and regulation in homeostasis.

❑ Name the four animal tissue types, and give examples of each.

❑ List the primary organ systems in the vertebrate body, and describe their functions.

❑ Explain how set points, feedback control, and error signals influence the regulation of homeostasis.

❑ Discuss the difference between negative and positive feedback control mechanisms, and compare them to the process of feedforward control.

❑ Use the analogy of a mechanical thermostat to explain how biological thermostats work in animals.

❑ Explain why animals need to control their body temperature.

❑ Describe the concept of Q_{10} and relate this concept to animal thermoregulation.

❑ Differentiate between the terms "metabolic rate," "acclimatization," and "metabolic compensation" as they apply to animal thermoregulation.

❑ Define and correctly apply the terms "homeotherm," "poikilotherm," "heterotherm," "ectotherm," and "endotherm."

❑ Interpret graphs depicting the results of laboratory or field experiments on the metabolic response of various animals to changes in environmental temperature.

❑ Describe the difference between physiological and behavioral mechanisms for thermoregulation, and provide specific examples of each.

❑ Explain the structure and thermoregulatory function of countercurrent heat exchangers in animals.

❑ For a representative endotherm, define the terms "basal metabolic rate," "thermoneutral zone," "upper critical temperature," and "lower critical temperature."

❑ Discuss the various physiological and morphological adaptations shown by animals for life in hot or cold climates.

❑ Explain the central role of the hypothalamus in controlling animal thermoregulation.

❑ Describe the role of hypothalamic thermal set points in normal thermoregulatory behavior.

❑ Discuss the cause and function of the fever response, and explain why under certain circumstances fevers can be good for organisms.

❑ Define "hypothermia" and explain how torpor and hibernation are used by some animals as thermoregulatory adaptations in cold climates.

❑ Explain the role of circannual rhythms in regulating animal hibernation.

THE NEEDS OF THE ONE VERSUS THE NEEDS OF THE MANY • ORGANS AND ORGAN SYSTEMS (pages 808–814)

Key Concepts

1. Physiology is the study of how animals function. Single-celled organisms and simple multicellular animals function by direct exchange with their external environment. This requirement distinctly limits the size, distribution, and overall complexity of an organism.

2. Complex multicellular animals can maintain a stable *internal environment* optimal for physiological function, regardless of external conditions. Internal stability allows for cellular specialization, which further enhances stability. The maintenance of a stable internal environment is called *homeostasis*.

3. Homeostasis is under precise regulatory control through a complex network of tissues, organs, and organ systems that gather information about relevant environmental stimuli and then coordinate appropriate

271

physiological and behavioral responses. The endocrine and nervous systems are of particular importance in this regard.

4. Cells associate with one another according to common functions. Masses of specialized cells with a common purpose are called tissues, of which four general types are recognized. Most organs contain examples of all four tissue types.

5. *Epithelial tissues* form dense sheets of tightly compacted cells that cover the outer body surface and the lining of hollow internal organs. Epithelial cells are used in movement, secretion, absorption, transport, and protection. They can be *simple* (single layer) or *stratified* (many layers) and are often rapidly replaced due to wear and tear.

6. *Connective tissues* provide support and reinforcement for other tissue types. They typically consist of loose arrays of cells embedded in an extracellular matrix. Protein fibers such as *collagen* or *elastin* provide matrix flexibility in some connective tissues (e.g., tendons and ligaments). Other connective tissues include mineralized *bone* (body support), *adipose tissue* (fat storage), and *blood* (gas transport).

7. *Muscle tissue* is made of cells that contract; this enables the movement of organs, limbs, and other body parts. Muscle tissue is subdivided into *skeletal muscle* (large body movements), *smooth muscle* (movement of internal organs except the heart), and *cardiac muscle* (heart contractions).

8. *Nervous tissue* consists of two basic cell types. *Neurons* are highly diverse cells that process and integrate electrochemical information. Neurons are specialized for many different information-handling tasks, including detection, relay, analysis, and reaction. *Glial cells* provide supporting functions for neurons.

9. Your textbook uses the human body to exemplify basic animal organ systems. The *nervous system* includes all of the nervous tissue in a body. In humans, the brain and spinal cord make up the *central nervous system*. The *peripheral nervous system* rapidly conveys information among sensory organs, the central nervous system, and effectors (muscles or glands).

10. The *endocrine system* transfers information more slowly than nervous tissue using specialized chemical signals. Molecules called hormones are manufactured by ductless glands and are then transported via the blood to target cells elsewhere in the body. Because of their common roles in communication, many important connections exist between the nervous and endocrine systems.

11. The *skin system* (including hair and nails) is the largest organ of the body. It acts as a first line of defense against disease-causing organisms and serves as a barrier to water loss. A rich complement of nervous tissue also gives skin a sensory function.

12. The *skeletal system* supports and protects soft body parts. It interacts with the voluntary muscle system to produce large body movements. Involuntary muscle movements are not associated directly with skeletal elements; they generate the movements of the gut and other internal organs, as well as the beating of the heart.

13. Organs of the *reproductive system* produce both the sex cells and the sex hormones. Additional organs for storage and delivery of sex cells exist in many species, as do specialized internal and external organs used for offspring development (e.g., breasts, uterus, etc.).

14. The *digestive system*, or gut, is a continuous tube divided into specialized segments. Each segment performs different nutrient-processing functions such as secretion, digestion, movement, absorption, and elimination of solid wastes.

15. The *respiratory system*, or *gas exchange system*, provides a way for oxygen to enter the body and for carbon dioxide to exit. Complex air passages and large exchange surfaces are common features of respiratory systems.

16. The *circulatory system* connects all cells and organ systems in the body via a continuous system of vessels. Blood, the fluid transport medium, consists of both cellular and plasma components. Nutrients, oxygen, hormones, heat, and wastes are pumped through this system by the action of a muscular heart.

17. The *lymphatic system* represents a second set of circulatory organs, although it lacks a muscular pump and does not form a complete circuit of vessels. Its job is to mobilize defense mechanisms during immune responses and return lost circulatory fluids to the blood.

18. The *excretory system* removes nitrogenous wastes from the body, and also helps maintain proper water and salt balance. The kidney is the primary excretory organ of vertebrates.

Questions (for answers and explanations, see page 382)

1. Homeostasis involves the coordinated responses of many physiological processes. Which of the following contributes to the proper maintenance of homeostasis in animal bodies?
   a. pH buffering of the blood
   b. Control of blood flow to the skin
   c. Elimination of nitrogenous wastes
   d. Production of hormones
   e. All of the above

2. Your eyes are _____ elements of the _____ system.
   a. regulatory; skeletal
   b. sensory; nervous
   c. control; excretory
   d. regulatory; respiratory
   e. sensory; lymphatic

3. Endocrine organs release hormones that influence many of the physiological systems involved in homeostasis. Endocrine organs are mostly composed of which tissue type?
   a. Epithelium
   b. Connective
   c. Muscle
   d. Nervous
   e. All of the above

4. Which of the following is a characteristic unique to connective tissue?

a. Highly modified cells that show the special property of contractility
b. Diverse anatomy with distinct specializations for information transfer
c. Lines the inner surfaces of the intestines and lungs in mammals
d. Loose array of cells embedded in some kind of extra-cellular matrix
e. None of the above

5. The two human organ systems that communicate via electrical signals and transfer essential materials between all other organ systems are the
   a. respiratory and digestive systems.
   b. lymphatic and circulatory systems.
   c. endocrine and lymphatic systems.
   d. muscular and endocrine systems.
   e. nervous and circulatory systems.

6. Which one of the following animal organs or organ systems is *not* directly involved in the functions of information control, protection, support, or movement?
   a. Skeletal
   b. Excretory
   c. Muscle
   d. Skin
   e. Nervous

Activities (for answers and explanations, see page 382)

- Your skin, like most other organs in the human body, contains representatives of all four vertebrate tissue types. Suggest specific functions for each of the following tissues found in human skin:

  Epithelium:

  Connective:

  Muscle:

  Nervous:

- For each of the organ systems listed on the left, choose its primary function from the list on the right.

  _____ Excretory   a. Electrical communication

  _____ Circulatory   b. Remove nitrogenous waste

  _____ Skeletal   c. Process and absorb food

  _____ Muscle   d. Chemical communication

  _____ Nervous   e. Produce and deliver gametes

  _____ Endocrine   f. Protection and support

  _____ Digestive   g. Contractile movements

  _____ Reproductive   h. Deliver nutrients to body cells

CONTROL, REGULATION, AND HOMEOSTASIS • TEMPERATURE AND LIFE • MAINTAINING OPTIMAL BODY TEMPERATURE (pages 814–822)

Key Concepts

1. Maintaining homeostasis requires both control and regulation. Control allows for a change in system status, while regulation maintains a system within certain limits. A crucial factor in the regulatory process is the need for information concerning the current status of both external and internal environments.

2. A *set point* is the desired condition of a system. *Feedback* mechanisms provide information about current conditions relative to the set point. The difference between the set point and the current condition of a system is the *error signal*, the nature of which indicates what corrective actions are needed to restore homeostasis.

3. *Controlled systems* include those molecules, cells, tissues, organs, and organ systems that change their characteristics in response to sensory input and subsequent commands from the *regulatory systems*. Feedback mechanisms are essential for regulation since they are how systems detect, process, integrate, and act upon relevant information.

4. *Negative feedback* (the most common form of regulation) occurs when the end result of a process feeds back to reduce or reverse the process, leading to a stable condition approximating the set point. The regulatory principles involved in negative feedback are exemplified by the mechanical model of a household thermostat.

5. Thermostats have an adjustable set point that is compared to current room temperature through information conveyed by a sensor. Depending on the type of error signal produced, either the furnace or the air conditioner is turned on. When room temperature returns to the set point, negative feedback shuts down the system.

6. *Positive feedback* does not regulate a system within stable limits. Rather, it produces an amplification of the initial perturbation that is reinforced until the process achieves its end. Sexual arousal is one example.

7. *Feedforward* information is used to change the set point of a regulatory mechanism by anticipating physiological changes and causing responses that counteract those changes.

8. Maintenance of proper body temperature is of great physiological importance to animals. They must avoid having body temperatures drop below $0°C$ (where ice crystals form) or rise above $45°C$ (where most proteins denature). Even within this 45-degree window, small changes in temperature can significantly influence physiological function.

9. The temperature dependence of any reaction system is represented by its Q_{10} value. Q_{10} values result from dividing the rate of a chemical process at temperature T ($°C$) by the rate of the same process at temperature $T - 10°C$.

10. Most physiological processes have Q_{10} values between 2 and 3, meaning that for every $10°C$ rise in temperature, the rate of the process doubles or triples. A Q_{10} value of 1 indicates that a process is not temperature-sensitive.

11. Q_{10} values can differ for the individual reactions in complex metabolic pathways. Thus, minor changes in body temperature can lead to severe disruptions of homeostasis. This places a premium on animals being able to compensate for even the slightest change in body temperature.

12. *Metabolic rate* is the sum total of energy turnover in an animal's cells. Q_{10} analysis predicts that metabolic rates should change with seasonal fluctuations in environmental temperature. In reality, metabolic rates often do not change with season, a phenomenon referred to as *acclimatization*. Similar metabolic rates across seasons are achieved through *metabolic compensation*, in which cellular biochemistry is adjusted by the use of different enzymes with appropriate temperature optima.

13. Physiologists do not use the terms "cold-blooded" or "warm-blooded" when discussing animal thermoregulation. Instead, animals that regulate their body temperature at constant levels are called *homeotherms*. Animals whose body temperature changes with fluctuations in environmental temperature are called *poikilotherms*. *Heterotherms* are animals that regulate their body temperature at certain times, but not at others.

14. Animals are also classified by thermoregulatory mechanisms. *Ectotherms* depend on external sources of heat to maintain an elevated body temperature. *Endotherms* produce and retain sufficient metabolic heat to regulate their body temperature without external heating.

15. Under strict laboratory conditions, an ectotherm's body temperature and metabolic rate will rise and fall in direct proportion to increases or decreases in environmental temperature. By contrast, an endotherm of similar mass will exhibit a constant body temperature over the same range of environmental temperatures because it is able to adjust its metabolic rate to compensate for heat loss.

16. Under real-world conditions, ectotherms can maintain relatively constant body temperatures by using behavioral thermoregulatory mechanisms such as shuttling between sunlight and shade (desert lizard example). Endotherms also engage in various types of behavioral thermoregulation, including sophisticated activities such as nest construction and huddling in birds, or the wearing of clothes by humans.

17. Body temperature in both endotherms and ectotherms is strongly influenced by regulation of blood flow to the skin. For example, marine iguanas retain heat while foraging in cold water by constricting peripheral blood flow. They obtain additional heat by shunting blood to the skin while basking in the sun. Humans can likewise lose or retain metabolic heat by regulating peripheral blood flow.

18. Metabolic heat production is not unique to endotherms. Some insects rely on muscle contraction to generate heat for flight on cold nights. Honeybees engage in clustering behavior to metabolically heat the sensitive brood portions of their hive in winter. Some snakes can increase the incubation temperature for their eggs through isometric muscle contractions.

19. Sustained muscle activity produces large amounts of heat. If retained within the body core, this heat helps maintain body temperature in excess of ambient temperature.

20. *Countercurrent heat exchangers* are found in some animals, including several species of fast-swimming tunas and sharks that live in cold waters. Vessels carrying warm blood from the swimming muscles run parallel to vessels carrying cold blood into the peripheral muscles. Heat is transferred from the outgoing vessels to incoming ones, thereby concentrating heat in the body core. An elevated core body temperature in these fish allows the swimming muscles to provide much higher levels of sustained power output than ambient water temperature would otherwise permit.

Activities (for answers and explanations, see page 382)

- Refer to Figure 37.10 in your textbook, which shows a schematic diagram of a home heating and cooling system. Explain what would constitute an error signal for this system, and describe how the thermostat would respond to this signal to adjust the internal temperature of the house appropriately.

- Discuss the difference between the concepts of acclimatization and metabolic compensation as they relate to temperature regulation in animals.

Questions (for answers and explanations, see page 382)

1. Resetting the thermoregulatory set points in an animal about to undergo a long-term, predictable change in metabolism (e.g., hibernation) is an example of
   a. negative feedback.
   b. positive feedback.
   c. countercurrent exchange.
   d. feedforward regulation.
   e. more than one of the above

2. During childbirth, the pressure exerted on the mother's cervix by the emerging infant leads to increased contraction of the uterus. This interaction could be explained by
   a. negative feedback.
   b. feedforward information.
   c. positive feedback.

d. controlled homeothermy.

e. metabolic compensation.

3. In a laboratory experiment, metabolic rate as a function of environmental temperature is measured for a mouse and a lizard of comparable size. The results are shown in the graph below. Which of the following statements about these results is *false*?

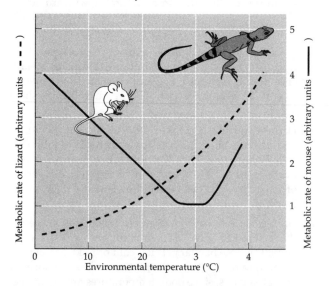

a. Below 25°C, the lizard's metabolic rate is directly proportional to environmental temperature.

b. Below 25°C, the mouse's metabolic rate is inversely proportional to environmental temperature.

c. The overall pattern of change in metabolic rate for the mouse is characteristic of ectotherms.

d. The overall pattern of change in metabolic rate for the lizard is characteristic of ectotherms.

e. The mouse and the lizard have different physiological mechanisms for thermoregulation.

4. The reaction rate for an important metabolic process in human digestion is measured as 200 units at a temperature of 30°C. If the Q_{10} for this reaction is 1.5, what would the rate be at 20°C?

a. 300

b. 200

c. 150

d. 133

e. Cannot determine from information given

5. Lizards cannot metabolically produce or retain enough heat to maintain continuous elevated body temperatures, but they are able to hold their body temperatures somewhat constant by appropriately positioning themselves in the environment. Select the terms that *best* apply to these thermoregulatory capabilities in lizards.

a. Endothermic, physiological heterotherm

b. Endothermic, behavioral poikilotherm

c. Endothermic, physiological homeotherm

d. Ectothermic, behavioral poikilotherm

e. Ectothermic, behavioral homeotherm

6. In an environment with an ambient temperature less than an animal's core body temperature, which of the following would be an *inappropriate* physiological or behavioral response if the animal needed to eliminate excess body heat?

a. Wallowing in a pool of water

b. Preventing blood flow to peripheral vessels

c. Reducing the metabolic activity of voluntary skeletal muscles

d. Seeking shelter in a shaded area

e. All of the above

7. In fast-swimming, cold-water fish such as sharks and tunas, which of the following contribute to the generation and maintenance of body core temperatures in excess of ambient water temperatures?

a. High rates of activity in their powerful swimming muscles

b. An ability to rapidly acclimatize to the surrounding water

c. Specializations in the size and location of their blood vessels

d. Metabolic rates that are insensitive to temperature change

e. Both *a* and *c*

THERMOREGULATION IN ENDOTHERMS • THE VERTEBRATE THERMOSTAT (pages 822–829)

Key Concepts

1. Endotherms generate more metabolic heat than do ectotherms. This is because endotherms expend more energy counteracting ion "leakage" across their cell membranes. Endotherms routinely maintain large temperature differentials between their bodies and the environment.

2. Endotherms adjust their metabolic rate to maintain a constant body temperature. The *thermoneutral zone* is a range of ambient temperatures within which an endotherm's metabolic rate is low and independent of temperature. The *basal metabolic rate*, which represents the rate of metabolism needed to carry out essential physiological functions when the animal is at rest, does not change within the thermoneutral zone.

3. Lower and upper critical temperatures bound the thermoneutral zone of endotherms. Metabolism increases as environmental temperature drops below the lower critical limit until a maximum metabolism is reached. Continued decreases in ambient temperature result in net heat loss and a reduction in body temperature.

4. Metabolic rate increases above the upper critical temperature are due to increased energy expenditures associated with active mechanisms of heat loss such as panting and sweating.

5. Both mammals and birds produce metabolic heat through shivering. Heat forms during the chemical breakdown of ATP as opposing muscle groups are simultaneously contracted.

6. Many mammals also engage in nonshivering heat production during which *brown fat* is metabolized. In brown fat cells, the protein *thermogenin* uncouples oxidative phosphorylation to release heat without the production of ATP. This process is common in newborn

infants and adult mammals that hibernate or are acclimatized to cold.

7. Special morphological adaptations for thermoregulation characterize animals that live in cold climates. Most are adaptations for reducing heat loss, either by improving various aspects of insulation or reducing surface-to-volume ratios through rounder bodies and shorter appendages.

8. Endothermic animals can also regulate heat loss by controlling peripheral blood flow. Wolves conserve heat in winter by limiting blood flow to their feet. In this way just enough heat is transferred to keep the foot tissue from freezing.

9. Excess heat production can be a problem for cold-adapted or large-bodied endotherms. One way to lose excess heat is to expose patches of bare skin to air, ice, or cool water. Evaporative cooling is another mechanism for heat loss. Because liquid water requires a substantial heat input to become gas, sweating or panting removes body heat. However, sweating and panting also lead to water loss and require a significant energy output, which explains the rise in metabolic rate at temperatures above the upper critical limit.

10. Thermoregulatory mechanisms are ultimately under the control of a small region of the brain called the *hypothalamus*. In addition to its role as a biological thermostat, the vertebrate hypothalamus also regulates many other homeostatic systems.

11. The hypothalamus generates thermal set points against which its own temperature is compared. Temperature information about other areas of the body is obtained through peripheral nervous connections.

12. Different thermoregulatory processes often have different hypothalamic set points. For example, the temperature at which sweating occurs differs from the ones that stimulate constriction of peripheral blood vessels or shivering.

13. Unlike a household thermostat, which responds in a simple on-off fashion, the hypothalamus can respond proportionally to the difference between set points and hypothalamic temperature. The thermoregulatory responses of animals are thus proportional to how much the hypothalamic temperature differs from a specific set point.

14. Information about peripheral skin temperature is used to anticipate possible changes in body temperature, an example of feedforward regulation.

15. Fevers represent upward shifts in the thermoregulatory set point that cause an increase in body temperature. They result from exposure to *pyrogens*, substances from viruses or bacteria that stimulate a change in the set point. Although dramatic increases in body temperature are potentially life-threatening, slight fevers probably aid the body's efforts to ward off infection.

16. Fever symptoms are produced when pyrogens stimulate cells of the immune system to release *interleukins*, chemicals that raise the hypothalamic set point for metabolic heat production. Interleukins in turn cause the release of *prostaglandins*, which are responsible for many symptoms of the illness. Aspirin is a potent inhibitor of prostaglandin release. Taking aspirin for a fever thus reduces symptoms of the illness, but does not eliminate the actual cause.

17. Temperature settings can also be lowered on the biological thermostat. *Hypothermia* (below normal body temperature) results either from natural downward adjustments in set points or from physiological insult. Extreme hypothermia can be lethal to most endotherms.

18. Because metabolism is temperature-dependent, hypothermia conserves metabolic energy. Many endothermic species engage in regulated hypothermia to survive prolonged periods of cold and food shortages.

19. Hummingbirds, pocket mice, and other small endotherms show an adaptive form of hypothermia called *daily torpor*. They lower their body temperature as much as 10 to 20°C below normal during daily periods of inactivity in order to avoid exhausting their energy reserves.

20. *Hibernation* is a seasonal form of regulated hypothermia that can last for a few days or many weeks. It is used by animals whose habitat gets too cold or food-depleted for them to remain active. Hibernation results in enormous energy savings, since body temperatures are maintained at levels close to freezing with metabolic rates one-fiftieth of normal basal metabolism. Only one bird hibernator is known; all others are mammals.

21. Hibernation is controlled by inherent *circannual rhythms*, which divide the animal's year into the active season (where hibernation is not physiologically possible) and the hibernation season. Hibernating animals periodically revive from their state of reduced metabolism, at which time they may ingest stored foods to replenish fat reserves.

22. Mammalian hibernation likely evolved as a natural extension of reductions in hypothalamic set points that occur in all animals during sleep.

Activities (for answers and explanations, see page 382)

- On the axes provided below, draw a graph of metabolic rate versus environmental temperature for an endothermic animal and indicate each of the following:

  ✔ upper and lower critical temperatures

  ✔ thermoneutral zone

  ✔ basal metabolic rate

- The graph you drew should show an increase in metabolic rate both above and below the thermoneutral zone. It is easy to understand why an endothermic animal's metabolic rate should rise when environmental temperature falls below the lower critical temperature (i.e., increased metabolism generates more body heat). What physiological phenomenon explains the apparently counterintuitive observation that metabolic rate also rises when environmental temperature exceeds the upper critical limit?

Questions *(for answers and explanations, see page 383)*

1. One cold winter day your little brother urges you to come see a small animal shivering outside on the front porch. Before arriving, you suspect that the animal you are about to see is
   a. a bird or mammal approaching the upper end of its thermoneutral zone.
   b. a lizard that has exhausted its supply of brown fat.
   c. a bird or mammal whose temperature is below the lower critical limit.
   d. a bird that is operating at its basal metabolism.
   e. a mammal with a severely damaged hypothalamus.

2. The upper critical temperature for an endothermic animal is
   a. the point at which it achieves its maximum rate of heat production.
   b. the maximum temperature at which it can maintain a constant metabolic rate.
   c. equivalent to its body temperature during hibernation.
   d. the temperature at which the animal enters into torpor.
   e. the point at which nonshivering heat production commences.

3. Evaporative cooling is an effective way to increase heat loss, but carries with it the important physiological drawback(s) of
   a. increasing water uptake, which can lead to bloating.
   b. accelerating the rate of brown fat deposition.
   c. dramatically lowering the hypothalamic thermal set point.
   d. stimulating the release of pyrogens that lead to fever.
   e. increased use of ATP and substantial loss of water.

4. Which of the following statements about hypothalamic function is *false*?
   a. Circulating blood temperature is monitored by the hypothalamus.
   b. Hypothalamic thermal set points never change.
   c. The hypothalamus is a part of the central nervous system.

d. Hibernation and torpor involve downward shifts in hypothalamic thermal set points.
   e. Different thermoregulatory responses have different hypothalamic set points.

5. The difference between the concepts of daily torpor and hibernation is that
   a. daily torpor is a rhythmic phenomenon and hibernation is not.
   b. daily torpor involves an upward shift in thermal set points and hibernation a downward shift.
   c. daily torpor occurs only in mammals and hibernation only in birds.
   d. daily torpor lasts for shorter periods than does hibernation.
   e. All of the above

6. Substances called _____ are released when an animal is exposed to certain bacteria or viruses, resulting in an upward shift in the hypothalamic thermal set point that leads to _____ .
   a. pyrogens; fever
   b. pyrogens; reduced metabolism
   c. brown fat; fever
   d. thermogenins; basal metabolism
   e. ATP; increased metabolism

7. Which of the following is *not* a morphological adaptation for life in the cold?
   a. Decreased surface-to-volume ratio of body
   b. Shortened appendages
   c. Thicker insulation
   d. Metabolism of brown fat
   e. Waterproof feathers and fur

Integrative Questions *(for answers and explanations, see page 383)*

1. In which of the following animals might you expect to see some type of behavioral thermoregulation?
   a. Humans
   b. Elephants
   c. Marine iguanas
   d. Birds
   e. All of the above

2. Some animals rely heavily on the metabolism of brown fat to assist in their thermoregulation. Brown fat is a type of _____ tissue.
   a. epithelial
   b. connective
   c. nervous
   d. muscle
   e. None of the above

3. A small rodent is at rest on a day when the environmental temperature is 30°C. Its body temperature is 37°C, and it is expending no energy for digestion, reproduction, or thermoregulation. Metabolically speaking, this individual is most likely
   a. in its thermoneutral zone.
   b. exhibiting a fever.
   c. at its upper critical limit of metabolism.
   d. hibernating.
   e. having a metabolic rate above basal.

38

Animal Hormones

CHAPTER LEARNING OBJECTIVES—after studying this chapter you should be able to:

❏ Define the term "hormone," and explain the relationship between hormones, target cells, and receptors.

❏ Explain the difference between autocrine and paracrine local hormones, and contrast their action with that of circulating hormones.

❏ Discuss the action of the paracrine hormone histamine, and relate its function to the phenomenon of anaphylactic shock.

❏ Describe the action of the paracrine hormones known as growth factors.

❏ Provide functional reasons that the same hormone can cause different responses in different types of cells, and evolutionary reasons that the same hormone molecules are used for different purposes in divergent animal groups.

❏ Explain the difference between endocrine and exocrine glands.

❏ Cite the experimental evidence that explains the role of ecdysone and juvenile hormone in the control of insect growth and development.

❏ Describe the hormonal differences in growth and development between insects that show simple development and those that exhibit complete metamorphosis.

❏ List the major endocrine glands of humans, their primary hormone products, and the main function of each hormone.

❏ Discriminate between the anterior and posterior lobes of the pituitary, and describe their anatomical connections with the hypothalamus.

❏ Explain the role of the portal blood vessels in mediating the interaction of the hypothalamus and pituitary.

❏ Differentiate between the roles of hypothalamic releasing hormones and release-inhibiting hormones.

❏ Identify abnormalities in human growth and metabolism resulting from the action of hormones.

❏ Explain the negative feedback control mechanism regulating the action of thyroxine, and discuss the causes of hypo- and hyperthyroidism.

❏ Describe the hormonal feedback control mechanisms for glucose metabolism and blood calcium availability in humans.

❏ Identify the major sex steroids in both males and females, state where they are produced in the body, and discuss their effects on growth and development.

❏ Describe the sequence of hormonal events that triggers the onset of puberty and sexual development.

❏ Explain the hormonal and genetic conditions that produce hermaphrodites and pseudohermaphrodites in humans.

❏ Discuss important biochemical differences between water-soluble and lipid-soluble hormones.

❏ Compare and contrast the cellular modes of action of water-soluble and lipid-soluble hormones.

❏ Describe the general features of hormone signal transduction pathways that utilize second messenger systems, with special emphasis on the roles of receptor proteins, G proteins, cAMP, protein kinases, adenylate cyclase, and Ca^{2+}.

❏ Explain the phenomenon of signal amplification and its relationship to the cascade of chemical reactions triggered by hormones.

❏ Discuss why lipid-soluble hormones act more slowly and produce longer-lasting cellular responses than do water-soluble hormones.

HORMONES AND THEIR ACTIONS • HORMONAL CONTROL OF MOLTING AND DEVELOPMENT IN INSECTS
(pages 832–837)

Key Concepts

1. *Hormones* are chemical signals produced and released by cells in one part of the body, but which have their effects on *target cells* located elsewhere. Only those tar-

get cells that possess appropriate chemical *receptors* can respond to hormones. The response may be developmental, physiological, or behavioral.

2. Hormone signals can be grouped by the distance over which the signal operates. Local hormones are released in small quantities and stimulate effects in the immediate vicinity of their release point. They rarely enter the blood supply because of rapid uptake or enzymatic degradation by nearby target cells.

3. *Autocrine* hormones are local hormones with receptors on the hormone-releasing cell itself. A common function of such hormones is to shut down their own production as part of negative feedback systems.

4. *Paracrine* hormones are local chemical messages that act in the vicinity of their release. *Histamine* is a paracrine hormone that promotes the inflammation associated with immune responses. Histamine makes blood vessels more permeable, allowing protective blood proteins and white blood cells to enter damaged tissues.

5. Local response to histamine is protective, but side effects may occur if the response becomes widespread. Many simple allergy symptoms (e.g., hay fever) are due to histamine activity. In some cases, however, histamine response is excessive. This can cause the circulatory and respiratory systems to cease functioning, which leads to the life-threatening condition called *anaphylactic shock*. Injections of another hormone, *epinephrine,* reverse the effects of histamine and can counteract anaphylactic shock.

6. *Growth factors* are another large class of paracrine hormones that help to repair tissue damage or provide chemical cues that direct growth and development. About 50 types are known, some of which aid in the growth, development, or repair of nervous, epidermal, and vascular tissue.

7. While the various types of local hormones act in the immediate vicinity of their release, circulating hormones are transported by the blood substantial distances from their site of production. The same circulating hormone can have very different effects on different target cells depending on the type of receptors present. For example, epinephrine causes heart muscle to beat faster, but stimulates metabolic changes in liver tissue.

8. Chemical signalling is an ancient, efficient, and evolutionarily conservative feature of animal physiology. Successful hormone molecules and their associated "second messenger" substances have changed little through the course of evolution, although their functions have changed substantially. The hormones thyroxine and prolactin are examples.

9. Many local and circulating hormones are released from discrete clusters of secretory cells called glands. *Exocrine glands* produce nonhormone substances and release them outside of the body via a duct system. *Endocrine glands* lack ducts and release their hormone products directly into the extracellular fluid of the body.

10. The episodic molting of insect exoskeletons (growth), and the transition from juvenile instars to adult forms (development) are controlled by three hormones, as demonstrated by Wigglesworth's experiments with the bug *Rhodnius*. His experiments showed that insect brain cells produce a *brain hormone*, which is stored in paired head structures called *corpora cardiaca*. After the animal has been stimulated by a blood meal, the corpora cardiaca release their store of brain hormone.

11. Brain hormone triggers the *prothoracic gland* to produce and release a second hormone called *ecdysone*, which is directly responsible for initiating the physiological events associated with molting. Wigglesworth demonstrated the chemical basis of this interaction by noting the effects of connecting decapitated instars of various ages and feeding status (see textbook Figure 38.3).

12. Wigglesworth also showed that other brain structures called *corpora allata* are responsible for determining the transition from juvenile to adult forms. These structures produce *juvenile hormone*. When present, juvenile hormone leads to the production of another juvenile instar.

13. Normally juvenile hormone production ceases between the fourth and fifth instars in *Rhodnius*, causing the transition from juvenile to adult forms. However, if early instars have their corpora allata removed, they will immediately molt into adults. By contrast, if older instars are attached to younger ones still possessing active corpora allata, the older animals molt into juvenile instars and not the adult form.

14. Development in insects such as *Rhodnius* is referred to as *incomplete metamorphosis* because there is a gradual progression from hatchling to adult. Juvenile hormone has a more complex interaction with development in insects that undergo *complete metamorphosis* (e.g. moths and butterflies) where distinct larval, pupal, and adult forms exist.

15. Textbook Figure 38.4 shows the hormonal regulation of complete metamorphosis in the silkworm moth *Hyalophora cecropia.* Larval endocrine cells produce high concentrations of juvenile hormone, and thus the episodic release of ecdysone results in successive molts that produce more larval forms. When larvae reach a certain age, juvenile hormone concentrations fall and the pupal molt follows. Since pupae produce no juvenile hormone, the product of the next molt is the adult form.

16. Juvenile hormone provides a natural mechanism for controlling insect pest species. Exposing larval forms to artificially high concentrations of synthetic juvenile hormone prevents them from ever maturing into adults, effectively stopping reproduction.

Activities (for answers and explanations, see page 383)

- Based on the results of Wigglesworth's experiments, indicate on the following diagram of *Rhodnius* the relative anatomical positions of the neuroendocrine cells that produce brain hormone, the corpora allata, and the prothoracic gland.

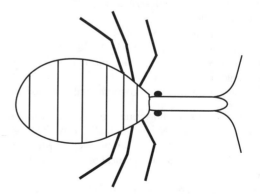

- Match the following terms with the appropriate description

_____ Histamine	a.	Produces secretions for use inside the body
_____ Endocrine gland	b.	Contains receptors specific for certain hormones
_____ Target cell	c.	Young insects remain subadults in its presence
_____ Ecdysone	d.	A local paracrine hormone mediating inflammatory responses
_____ Juvenile hormone	e.	Triggers physiological responses leading to molting

- Consider the story about hormonal regulation of "macho" and "wimpy" cichlid fish found on page 832 of your textbook. Given this description of hormonal effects on the physiology and behavior of males, explain where in the body of a cichlid you would expect to find receptors for the hormones involved in changing a "wimpy" male into a "macho" male.

Questions (for answers and explanations, see page 383)

1. The hormone vasopressin is produced by a gland located at the base of the human brain. Its primary effect is to influence water reabsorption by the kidneys. Which of the following accurately describes vasopressin and its action?
   a. It is a local hormone whose concentration is rarely detectable in the blood.
   b. It circulates through the blood until encountering receptors in the brain.
   c. It is a local hormone that acts in much the same way as histamine does on blood vessels.
   d. It is a circulating, endocrine hormone whose receptors are located on target cells in the kidneys.
   e. It is an exocrine hormone produced by target cells in the brain.

2. Histamine is correctly described as a
   a. circulating hormone.
   b. paracrine local hormone.
   c. receptor protein found on target cells in blood vessels.
   d. hormone responsible for the molting of insect exoskeletons.
   e. chemical released by the corpora allata of *Rhodnius*.

3. The same circulating hormone can have different effects on adjacent target cells in an animal's body because
   a. different cells can have different receptor proteins that initiate specific biochemical responses to the hormone.
   b. the hormone is modified by the blood and can therefore influence many cell types.

   c. the extracellular matrix is modified by the hormone, which influences the response of target cells.
   d. histamine is released in response to the hormone, which changes the permeability of blood vessels.
   e. None of the above

4. If the corpora cardiaca of a fourth instar of *Rhodnius* are surgically removed, it will
   a. develop into an adult.
   b. molt but remain a juvenile.
   c. not molt and remain a fourth instar.
   d. molt and return to a third instar.
   e. none of the above

5. If a fourth instar of *Rhodnius* is partially decapitated one week after receiving a blood meal, it will develop into a fifth instar and not an adult. This occurs because
   a. the bug makes no ecdysone.
   b. the source of brain hormone has been removed.
   c. the corpora allata are no longer present.
   d. juvenile hormone concentrations remain high.
   e. none of the above

6. Synthetic forms of juvenile hormone can be used as an effective weapon against insect pests because
   a. juvenile hormone poisons the adult insects.
   b. adults will not mate when juvenile hormone is present.
   c. larvae refuse to feed in the presence of juvenile hormone and die before molting into adults.
   d. juvenile hormone prevents the molting of successive larval instars.
   e. juvenile hormone prevents insect larvae from molting into reproducing adults.

VERTEBRATE ENDOCRINE SYSTEMS (pages 838–849)

Key Concepts

1. Vertebrate hormones and their glandular sources are numerous and varied. Your textbook focuses mainly on human hormones.

2. Over a dozen endocrine glands are found in humans, many of which produce more than one hormone. Several nonglandular tissues, such as the lining of the intestinal tract or certain cells of the nervous system, also secrete hormones.

3. The *pituitary* is sometimes called the "master gland" of the vertebrate endocrine system, producing or storing many different hormones. In humans it is a blueberry-sized organ located at the base of the skull near the back of the roof of the mouth.

4. The pituitary is divided into two lobes. The *posterior pituitary* derives from an outpocketing of the developing brain. It retains neural connections via the pituitary stalk with nerve cell bodies in the hypothalamus that produce *neurohormones*. The *anterior pituitary* is formed from the embryonic mouth cavity, is made primarily of epithelial tissue, and has endocrine functions.

5. *Vasopressin (antidiuretic hormone* or *ADH)* is a neurohormone released by the posterior pituitary. It is a small peptide manufactured by nerve cells in the hypothala-

mus and released from the posterior pituitary in response to low blood pressure or high salt content in the blood. Its primary action is to facilitate water reabsorption by the kidneys.

6. *Oxytocin,* the other neurohormone of the posterior pituitary, is also a small peptide. It plays a crucial role in human reproduction by stimulating uterine contractions during childbirth and by promoting the release of milk from the mother's mammary glands.

7. The anterior pituitary produces two classes of protein-based hormones. *Tropic hormones* control the activity of other endocrine glands in the body; they include thyrotropin, adrenocorticotropin, luteinizing hormone, and follicle-stimulating hormone.

8. The second class of anterior pituitary hormones exerts influence on nonendocrine tissues. Chief among these is *growth hormone*, which indirectly stimulates growth by causing the liver to produce growth-regulating factors called *somatomedins* that influence the development of bone and cartilage.

9. Overproduction of growth hormone in children leads to gigantism, while underproduction causes dwarfism. Dwarfism is now routinely treated with human growth hormone produced by genetically engineered bacteria.

10. *Prolactin* is another anterior pituitary hormone. In humans it has effects on the production and secretion of milk in females and the endocrine function of testes in males. In other mammals it helps maintain pregnancy in females.

11. *Melanocyte-stimulating hormone* is yet another anterior pituitary product. Its function is obscure in humans, but in fishes, amphibians, and reptiles that are able to change color, it stimulates the redistribution of melanin in pigmented skin cells.

12. *Endorphins* and *enkephalins* are also hormones produced by the anterior pituitary; they serve as the body's natural painkillers. They are derived by precise enzymatic cleavage of a large parental protein called pro-opiomelanocortin, which also gives rise to other anterior pituitary hormones.

13. The anterior pituitary really acts as an intermediary in hormonal control, rather than as a master gland. Neurohormonal releasing hormones such as *thyrotropin-* and *gonadotropin-releasing hormones* are first manufactured by the hypothalamus and then secreted into the *portal blood vessels* that connect the hypothalamus with the anterior pituitary. They then stimulate the release of pituitary hormones into the general circulatory system. There are also hypothalamic release-inhibiting hormones that can halt pituitary hormone release.

14. The *thyroid gland* is a bilobed structure surrounding the trachea. It produces *thyroxine* which regulates cell metabolism in mammals. Thyroxine also influences growth and development by enhancing amino acid uptake and protein synthesis. Each molecule of thyroxine contains four molecules of iodine. The thyroid also makes T_3, which is the same molecule with only three iodine molecules. Thyroxine can be enzymatically converted into T_3. T_3 is the more active hormone in body cells.

15. The release of thyroxine is controlled by both the pituitary and the hypothalamus. Hypothalamic cells release thyrotropin-releasing hormone, which stimulates the pituitary to release the circulating hormone *thyrotropin*. Thyrotropin causes target cells in the thyroid gland to release thyroxine. Thyroxine feeds back negatively onto the pituitary, making it less sensitive to thyrotropin-releasing hormone and slowing the release of thyrotropin.

16. Malfunction of the thyroid can lead to a condition called *goiter* in which the gland increases to enormous proportions. This happens either because of *hyperthyroidism* (excess thyroxine production) or *hypothyroidism* (depressed thyroxine production). Both conditions involve abnormalities in the negative feedback control of thyroid activity.

17. Hyperthyroid goiter occurs when the anterior pituitary does not respond properly to the negative influence of excess thyroxine, and the thyroid gland becomes overactive. Hypothyroid goiter occurs when circulating thyroxine levels are too low, usually because iodide (an important component of the thyroxine molecule) is lacking in the diet. The gland grows larger through overstimulation because there is too little functional thyroxine to inhibit thyrotropin production by the pituitary.

18. The thyroid gland also produces *calcitonin*, which helps regulate blood calcium levels in mammals. Calcitonin stimulates *osteoblast* cells, which lay down new bone, and decreases the activity of *osteoclast* cells, which breaks down existing bone. Calcitonin thus acts to reduce the level of circulating calcium in the blood.

19. The *parathyroid glands* are located on the surface of the thyroid. They secrete *parathormone*, which functions as an antagonist to calcitonin by stimulating osteoclasts to dissolve bone and release more calcium. Parathormone also promotes absorption of calcium by the kidneys and digestive tract.

20. The *pancreas* plays an important role in glucose metabolism through the action of two antagonistic hormones, *insulin* and *glucagon*. Insulin, produced by cells in the pancreatic *islets of Langerhans*, promotes both the use of blood glucose by cells and the storage of excess glucose as glycogen or fat. Glucagon, made by other cells in the islets, acts to mobilize stored glycogen for conversion to glucose when blood glucose levels fall.

21. The disease *diabetes mellitus* results from a failure of the pancreas to manufacture sufficient quantities of functional insulin, or from a lack of response in target cells. This causes abnormally high blood glucose levels and deprives cells of the fuel they need for metabolism. The condition can often be treated with insulin replacement therapy.

22. *Somatostatin* is a pancreatic hormone that inhibits the release of insulin and glucagon and slows digestive activity within the gut. This functionally extends the time over which nutrients can be absorbed and used by the body. Somatostatin also serves as a hypothalamic neurohormone that inhibits the release of growth hormone from the anterior pituitary.

23. An adrenal gland rests atop each kidney, and each is divided into two regions. The *adrenal medulla* forms the core of this gland. Its primary hormone product is *epinephrine,* which is associated with the fight-or-flight reactions to stress. Epinephrine promotes increased heart and breathing rates, elevates blood pressure, and redirects blood flow to skeletal muscles. The adrenal medulla also manufactures *norepinephrine,* which counteracts the physiological effects of epinephrine.

24. The *adrenal cortex* makes up the outer layer of each adrenal gland. It produces many steroid hormones that fall into three classes: *glucocorticoids, mineralocorticoids,* and *sex steroids.* Steroid hormones are synthesized from cholesterol.

25. Glucocorticoids, such as *cortisol,* influence blood sugar levels and general metabolism. Cortisol contributes to short-term management of stress by reducing glucose demand in nonessential cells, increasing blood pressure, promoting fat mobilization, and blocking immune system reactions.

26. Cortisol release is regulated by the pituitary hormone *adrenocorticotropin,* which is in turn regulated by its own hypothalamic releasing hormone. Chronic stress, which causes prolonged elevated cortisol levels, does permanent damage to the brain regions involved in the negative feedback regulation of cortisol release.

27. Mineralocorticoids, such as *aldosterone,* help regulate salt and ion balance, thus ensuring the proper osmotic conditions necessary for homeostasis.

28. Sex steroids are also manufactured by the adrenal cortex, but their roles in promoting sexual development and reproductive activity are relatively minor compared to the gonadal sex hormones.

29. *Androgens* are the male sex steroids, with *testosterone* being the dominant one. *Estrogens* and *progesterone* constitute the female steroids, with *estradiol* representing the main form of estrogen.

30. The presence or absence of sex hormones determines whether a developing fetus will become male or female. After birth, sex hormones regulate sexual maturation and the development of secondary sexual characteristics.

31. In humans, an XY combination of sex chromosomes causes the undifferentiated gonads of a seven-week-old fetus to begin producing androgens, which leads to a male reproductive system. If the fetus is genetically XX, no androgens are made and a female system forms. Thus, in humans the default developmental pattern is female in the absence of androgens.

32. In other vertebrates such as birds, where the presence of estrogens initiates female development, the default sex in the absence of appropriate female steroids is male.

33. Abnormalities in sexual development can occur due to errors in hormonal control. *Hermaphrodites* are individuals who possess both male and female gonads, while *pseudohermaphrodites* are those with the external genitalia of one sex and the gonads of another.

34. Sex hormones have their most dramatic developmental effects at the time of puberty. Hypothalamic-releasing hormones cause the pituitary to secrete the gonadotropins *luteinizing hormone* and *follicle-stimulating hormone.* In males these hormones cause the testes to produce androgens, and in females they stimulate the ovaries to begin secreting estrogens and progesterone.

35. These hormonal events not only lead to gonadal maturation, but also produce dramatic changes in secondary sexual characteristics. In males, the voice deepens, skeletal muscle mass increases, the penis enlarges, and facial, pubic, and body hair develops. In females, the breasts, vagina, and uterus all enlarge, the pelvis broadens, there is increased subcutaneous fat deposition, pubic hair develops, and the menstrual cycle commences.

36. There has been a dramatic increase in the use of synthetic androgens *(anabolic steroids)* by young men and women seeking to increase muscle mass for athletic or aesthetic purposes. The presence of these hormones interferes with normal sexual development and also poses serious health risks in the forms of increased rates of cancer, heart disease, kidney damage, sterility, and mental disorders.

37. Many nonendocrine tissues and organs in the body also produce hormones. Examples include hormones made by the digestive tract that help the gut process food, and *atrial natriuretic hormone,* which is released by the heart and influences kidney function to lower blood volume and blood pressure.

Questions (for answers and explanations, see page 383)

1. Which of the following sets of vertebrate hormones are all produced by the anterior pituitary?
   a. Somatostatin, vasopressin, insulin
   b. Oxytocin, prolactin, adrenocorticotropin
   c. Prolactin, growth hormone, enkephalins
   d. Thyroxine, calcitonin, thyrotropin
   e. Estrogen, glucagon, progesterone

2. A drop in blood temperature would be detected by the _____ , which would chemically stimulate the _____ to release the hormone _____, leading eventually to a rise in metabolic rate.
   a. hypothalamus; anterior pituitary; thyrotropin
   b. posterior pituitary; adrenal glands; adrenaline
   c. anterior pituitary; hypothalamus; thyroxine
   d. hypothalamus; pancreas; insulin
   e. anterior pituitary; posterior pituitary; prolactin

3. If you wanted to incapacitate a male dog's ability to regulate water reabsorption in its kidneys, and at the same time minimize the impairment of other physiological functions, which of the following structures would you need to selectively destroy?
   a. Parathyroid glands
   b. Anterior pituitary
   c. Hypothalamus
   d. Posterior pituitary
   e. Adrenal glands

4. A blood sample taken from a normal woman 30 minutes after eating a large meal would likely reveal elevated quantities of
   a. glucagon.
   b. cortisol.
   c. insulin.
   d. parathormone.
   e. aldosterone.

5. A rat is kept in a small cage with many other rats and subjected to mild electric shocks every half-hour for several days. Which of the following hormones would you expect to find in abnormally high concentrations in this animal?
   a. Cortisol
   b. Testosterone
   c. Vasopressin
   d. Calcitonin
   e. Progesterone

6. A woman conceives a child who is genetically XY and whose body cells are insensitive to female sex hormones. Which of the following best describes this child on its eighteenth birthday?
   a. It will be a hermaphrodite.
   b. It will be a pseudohermaphrodite.
   c. It will be a normal female.
   d. It will be a normal male.
   e. It will be look like a normal male, but be sterile.

7. At the onset of puberty in both males and females, large amounts of _____ are released by the _____.
   a. testosterone and estrogen; anterior pituitary
   b. aldosterone and cortisol; adrenal glands
   c. gonadotropin-releasing hormones; hypothalamus
   d. somatostatin and insulin; pancreas
   e. oxytocin and vasopressin; posterior pituitary

8. All but one of the following hormones are related by a common chemical characteristic. Select the *exception*.
   a. Epinephrine
   b. Estrogen
   c. Progesterone
   d. Testosterone
   e. Cortisol

Activities (for answers and explanations, see page 384)

- In the following space, diagram the negative feedback control system that regulates the release of thyroxine in humans. Be sure your drawing includes all of the relevant hormones, releasing factors, neural control centers, and endocrine glands involved in this process.

- Based on your understanding of the anatomical relationship between the hypothalamus and the pituitary gland, suggest a mechanism for how a baby's cry can rapidly stimulate a nursing mother to release milk from her breasts.

MECHANISMS OF HORMONE ACTION (pages 849–854)

Key Concepts

1. For hormones to cause a response, they must alter the internal biochemistry of their target cells. This is accomplished in two fundamentally different ways, depending on whether the chemical messenger is a *water-soluble hormone* (peptides and proteins) or a *lipid-soluble hormone* (steroids and thyroxine).

2. Water-soluble hormones do not readily pass through the membranes of target cells, and their receptors are on the cell surface. Each receptor has a binding domain projecting outside the cell membrane, a trans-membrane domain anchoring it to the membrane, and a catalytic domain extending into the cytoplasm.

3. When the hormone attaches to the binding domain of its receptor, the catalytic domain typically activates a *second messenger* molecule such as cyclic adenosine monophosphate, or *cAMP*. cAMP then activates (or inactivates) important protein kinases within the cell that ultimately trigger the cell's response to the hormone. This sequence of events is known as a *signal transduction pathway*.

4. When the hormone binds to its receptor, the receptor undergoes a conformational change that causes a second membrane protein to become active. This activated *G protein* binds *guanosine triphosphate (GTP)* inside the cell. One subunit of the G protein then dissociates and activates the enzyme *adenylate cyclase*, which converts ATP into cAMP. Active G proteins are inactivated when GTP is hydrolyzed to form GDP, ending the hormone's effect on its target cell.

5. G proteins are diverse. Some activate adenylate cyclase, which in turn triggers many other cell reactions, while others inactivate adenylate cyclase and produce an inhibitory effect. G proteins represent important control points in cell metabolism. For example, the presence of excitatory and inhibitory G protein–coupled hormone receptors in the same cell allows for control through the action of antagonistic hormones.

6. The cAMP second messenger stimulates (or inhibits) the action of intracellular enzymes called *protein kinases*, which are responsible for catalyzing the transfer of phosphates from ATP to specific proteins. cAMP also interacts with phosphoprotein phosphatases, another class of enzymes that remove the phosphates from phosphorylated proteins.

7. The influence of epinephrine on liver cells demonstrates the effects of second messenger systems. Epinephrine binds to its receptor on the cell surface and triggers production of cAMP, which then activates *protein kinase A (PKA)*. The presence of active PKA permits the phosphorylation of two important enzymes, glycogen synthase (which converts glucose into glycogen) and glycogen phosphorylase (which converts glycogen into glucose).

8. Phosphorylating glycogen synthase inactivates the enzyme, thus halting glycogen production. However, the addition of phosphate to glycogen phosphorylase activates this enzyme, thereby increasing the conversion of glycogen into glucose. The cAMP second messenger thus stimulates both the shutdown of glycogen production and the acceleration of glycogen breakdown when the body's need for glucose is high.

9. Extreme amplification of hormone signals is possible with second messengers. At each step in the biochemical cascade of reactions, there is a 10-fold to 100-fold increase in the number of molecules activated. Thus, for each molecule of epinephrine that binds to the cell membrane, billions of glucose molecules may be made available for use by the cell.

10. As the hormone concentration is reduced near the cell surface, the chain of events that produced cAMP is disrupted, leading to a decline in the intracellular concentration of cAMP through enzymatic degradation. In the case of epinephrine and liver cells, this removes the inhibition on glycogen formation, which in turn lowers the concentration of free glucose. Moreover, as cAMP levels fall, its inhibitory effect on phosphoprotein phosphatase is removed, leading to the dephosphorylation of the enzymes controlling glycogen metabolism.

11. Second messenger molecules other than cAMP are also used in response to hormones such as norepinephrine, vasopressin, and GnRH. *Inositol trisphosphate* (IP₃) and *diacylglycerol* (DAG) are second messengers derived from a membrane phospholipid *(phosphatidylinositol)* through the action of specific G proteins. These two messengers have different modes of action that are synergistic. DAG activates protein kinase C, while IP₃ causes release of intracellular Ca^{2+} ions, which in turn further activate PKC.

12. Different signal transduction pathways and their attendant second messenger systems can act in complementary (or antagonistic) ways to regulate cell metabolism. The biochemical response of liver cells to epinephrine and norepinephrine is an example of complementary action using both the cAMP and phosphatidylinositol signal transduction pathways.

13. Ca^{2+} ions play a direct role in many other cell activities, including control of membrane ion channels, secretory processes, and nerve and muscle cell function. They also interact with specific binding proteins such as *calmodulin* to mediate other biochemical responses in cells, including activation of certain protein kinases.

14. Cell surface receptors can also have effects on the nucleus and gene expression. Some cAMP-stimulated PKA travels to the nucleus where it activates *cAMP response element–binding* protein (*CREB*). CREB then binds with *cAMP response element (CRE)* to stimulate gene transcription.

15. *Gluconeogenesis*, where glucose is made from amino acids, is stimulated in liver cells by the action of epinephrine. The process requires many enzymes not normally active in the cell. Epinephrine triggers CREB activation, which in combination with PKC translocated to the nucleus, causes gene transcription of the needed gluconeogenic enzymes.

16. Many types of growth factors can also stimulate gene expression by interacting with cell surface receptors known as *receptor tyrosine kinases*. The outer surface of these molecules serves as the hormone receptor, while the inner surface acts as an enzyme to activate intracellular signalling molecules. The activated second messengers then have their effects on gene expression.

17. The lipid-soluble hormones, which include thyroxine and the steroid hormones produced by the gonads and adrenal cortex, do not bind to surface receptors on the plasma membrane. Instead, they pass through the membrane and enter directly into the cell cytoplasm.

18. Lipid-soluble hormones bind with specific receptor proteins inside the cytoplasm or nucleus. The proteins then undergo conformational changes, allowing the protein–hormone complex to associate with a chromosome and influence DNA transcription.

19. Because they penetrate the cell and directly influence gene expression, the actions of lipid-soluble hormones tend to be slower and longer-lasting than those of water-soluble hormones, which bind to their receptors on the cell surface and initiate short-lived changes in membrane ion permeability or enzyme activity.

Questions (for answers and explanations, see page 384)

1. Which of the following statements about hormones and second messenger systems is *false*?
   a. cAMP is one of the most common second messenger molecules.
   b. Activated G proteins help to activate the enzyme adenylate cyclase.
   c. Second messenger systems are not involved with most lipid-soluble hormones.
   d. When calmodulin binds Ca^{2+}, it is activated and in turn can activate certain protein kinases.
   e. Thousands of hormone molecules are required to activate a few intracellular enzymes.

2. The cascade of regulatory steps associated with second messenger systems results in
   a. the conversion of cAMP to ATP, which is then used by the cell for energy.
   b. conformational changes in numerous protein kinases within the cell.
   c. a remarkable amplification of the effect of a single hormone molecule.
   d. Both *a* and *b*
   e. Both *b* and *c*

3. Which of the following does *not* play a significant role in the signal transduction pathway for glucose metabolism stimulated by the hormone epinephrine?

a. Epinephrine surface receptors
b. cAMP
c. Protein kinase A
d. Receptor tyrosine kinase
e. GTP

4. Lipid-soluble hormones differ from water-soluble hormones in all but one of the following ways. Select the *exception*.
   a. Mechanism of cellular action
   b. Chemical structure
   c. Location of receptor binding site on the target cell
   d. Basic role as a communication signal
   e. Ability to penetrate the cell membrane

5. Which of the following is *not* considered a type of second messenger molecule in the biochemical responses of cells to water-soluble hormones?
   a. DAG
   b. cAMP
   c. IP$_3$
   d. cGMP
   e. ATP

6. Target cells in animals tend to respond faster and for a shorter time to the presence of water-soluble hormones than they do to lipid-soluble hormones. This is mainly due to differences in the
   a. type of target cells involved.
   b. amount of hormone released.
   c. cellular mechanism of action of the two hormone types.
   d. availability of calcium ions in the cell.
   e. None of the above

Activities (for answers and explanations, see page 384)

• Using the list of molecules given below, indicate the proper sequence of events that occurs when a water-soluble hormone stimulates the production of a cAMP-based second messenger signal transduction pathway inside a cell. *Note:* several of these events may occur in parallel.

   _____ Inactive G protein

   _____ Active adenylate cyclase

   _____ Hormone

   _____ cAMP

   _____ ATP

   _____ Active G protein

   _____ Receptor molecule

   _____ Protein kinase

   _____ GTP

   _____ Phosphoprotein phosphatase

   _____ Inactive adenylate cyclase

• A certain cell in mice is known to respond to two different hormones, one water-soluble and the other lipid-soluble. Suppose that equal concentrations of these two hormones are given to a mouse at the same time.
   a. Which hormone will exert its cellular effects first? Explain your answer.

   b. Which hormone is more likely to have its primary effect on gene expression in this cell? Explain your answer.

Integrative Questions (for answers and explanations, see page 384)

1. Consider the following vertebrate hormones: oxytocin, adrenocorticotropin, prolactin, insulin, and somatostatin. Which of the following do all five have in common?
   a. They are all released by the posterior pituitary.
   b. They are all released by the anterior pituitary.
   c. They are all neurohormonal releasing factors.
   d. They are all made of amino acids.
   e. They are all sex hormones.

2. A sample of radioactively labeled prolactin is injected into an adult female mouse, which is then analyzed for the location of this hormone in her tissues. This marked sample would most likely be found
   a. attached to the chromosomal proteins of kidney cells.
   b. inside the nucleus of hypothalamic cells.
   c. bound to protein receptors on the plasma membrane of mammary gland cells.
   d. attached to molecules of cAMP within the pancreas.
   e. none of the above

3. Pregnant women require the proper functioning of many important hormones to maintain their own health and that of the fetus. In humans, the hormone _____ is largely responsible for sustaining pregnancy, while _____ helps protect the calcium content of the mother's bones.
   a. progesterone; calcitonin
   b. testosterone; parathormone
   c. aldosterone; thyroxine
   d. estrogen; cortisol
   e. oxytocin; insulin

4. If you wanted to interfere with the cellular mode of action of the hormone thyroxine, which of the following would be most effective?
   a. Block action of G proteins
   b. Inhibit hormone receptors in the cytoplasm
   c. Inactivate calmodulin
   d. Inactivate adenylate cyclase
   e. Remove all GTP

39
Animal Reproduction

CHAPTER LEARNING OBJECTIVES—after studying this chapter you should be able to:

❑ Differentiate between asexual and sexual reproduction in animals, and explain the advantages and disadvantages of each strategy.

❑ Distinguish among the asexual forms of reproduction known as budding, regeneration, and parthenogenesis.

❑ Explain the difference between the germ cell line, gametes, and other cells of an animal body.

❑ Describe the various cellular stages in the processes of spermatogenesis and oogenesis, and know whether each stage is diploid or haploid.

❑ Differentiate among the following terms: "gonads," "testes," "ovaries," "sperm," "eggs," "ova," "zygotes," and "fertilization."

❑ Discuss the difference between dioecious and monoecious animals, and explain how these terms apply to simultaneous and sequential hermaphrodites.

❑ Describe the advantages and disadvantages of internal versus external fertilization, and discuss the various mating strategies that animals use to achieve both types of fertilization.

❑ Distinguish among oviparity, viviparity, and ovoviviparity.

❑ Describe the anatomy and function of all the human male and female reproductive structures discussed in this chapter.

❑ Differentiate between emission and ejaculation in males, and explain the hormonal control of male sexual function.

❑ Explain the hormonal control of female sexual function, and relate this to events that occur in the ovaries and uterus during the menstrual cycle.

❑ Characterize the differences between a menstrual cycle and an estrous cycle.

❑ Describe the four phases of human sexual response and the physiological characteristics associated with each phase for both males and females.

❑ List the various forms of contraception available to humans and discuss their efficacy.

❑ Give examples of how technology can be used to help solve problems of human infertility.

❑ Describe the major events that occur in both mother and baby during each trimester of a human pregnancy.

❑ Discuss the role of the placenta in supporting the development of a human fetus.

❑ Explain the sequence of events that occurs during human birth, and describe how oxytocin helps control this process.

ASEXUAL REPRODUCTION • SEXUAL REPRODUCTION (pages 857–865)

Key Concepts

1. Sperm and eggs are part of an almost continuous line of germ cells whose existence is interrupted only by periodic episodes of meiosis. In this sense, animals are simply devices for the reproduction of their germ cell lines.

2. Sexual reproduction is the norm for animals. Gametes are produced that contain a precise subset of the parental genome. When two gametes combine, a new individual develops whose genetic makeup is unlike that of either parent. This genetic diversity is then operated upon by natural selection during periods of environmental change.

3. Many animals also reproduce asexually. Asexual reproduction generates new individuals directly from pre-existing ones, with offspring genetically identical to their parents.

4. Efficiency is a hallmark of asexual reproduction since little energy is required, mates need not be found, and all members of a population are potentially able to have offspring. The only limiting factor is the animal's ability to convert available resources into new offspring.

5. Sponges and cnidarians use a form of asexual reproduction called *budding*, where new individuals arise as external mitotic outgrowths of older ones.

6. Animal cells capable of asexually producing a complete, new individual are said to be *totipotent*. Totipotency is rare in animals. However, some clusters of mature cells can make a whole new organism through *regeneration*. For example, sea stars and some annelid worms can lose a portion of their body and then regenerate the lost part, or have the lost part regenerate a new body.

7. A third kind of asexual reproduction is *parthenogenesis*, the development of offspring from unfertilized eggs. Many arthropods and some species of fish, amphibians, and reptiles utilize this process.

8. Most parthenogenetic species also reproduce sexually. Which mechanism they use often depends on environmental conditions. In aphids, parthenogenesis occurs when the habitat is favorable and stable. During times of environmental instability, aphids employ sexual reproduction to promote genetic variability among their offspring. In bees and wasps parthenogenesis is part of sex determination, with unfertilized eggs becoming males and fertilized eggs females.

9. Some parthenogenetic species still require mating. The mating process itself serves only to stimulate egg development, and no fertilization actually takes place. A few parthenogenetic species have no males in their populations; the females mate with males of closely-related species.

10. Sex generates enormous genetic variability among offspring due to the random assortment of alleles into *gametes* that occurs during meiosis and *gametogenesis*. These genetically unique gametes combine during *fertilization* to form a *zygote*, the genotype of which is also unique. By selecting for offspring whose genetic make-up best adapts them for a particular environment, natural selection operates on the genotypic diversity produced by sexual reproduction.

11. The tiny haploid gametes of males are called *sperm*. They are motile cells powered by a beating flagellum. *Eggs*, or *ova*, are produced by females. They are non-motile cells with relatively large stores of cytoplasm, organelles, and energy reserves.

12. Sperm and eggs are produced by the *gonads*, or primary sex organs. Male gonads are called *testes*, female gonads are *ovaries*. Accessory sex organs are usually present in the form of various ducts, canals, glands, and structures for the delivery and receipt of gametes. The primary and accessory sex organs combined represent the animal's reproductive system.

13. The diploid germ cell line proliferates mitotically from early embryonic tissues, and only comes to reside within the gonads later in development. *Oogonia* are the germ cells of females and *spermatogonia* those of males.

14. The process of *spermatogenesis* yields new sperm cells. Within the seminiferous tubules of mammalian testes, diploid *primary spermatocytes* undergo the first division of meiosis to produce haploid secondary spermatocytes. The second meiotic division results in four haploid spermatids. Complete spermatogenesis takes about ten weeks in humans, during which spermatocytes and spermatids are nourished by specialized cells lining the seminiferous tubules.

15. *Oogenesis* is the process of gamete formation in females. *Primary oocytes* develop from a subset of oogonia and enter meiosis I, where they arrest during prophase. Each oocyte then grows in size, acquiring all the nutrients and cytoplasmic structures needed for development should fertilization occur.

16. Once a month throughout the reproductive lifespan of human females, at least one primary oocyte resumes meiosis. After meiosis I, one daughter cell retains a majority of the cytoplasmic material and develops into a large *secondary oocyte*; the other becomes the *first polar body*. Meiosis II also divides the secondary oocyte unequally, producing a large, haploid *ootid* as well as a *second polar body*. The polar bodies then degenerate, leaving a well-provisioned ovum. In humans and many other species, the second meiotic division does not occur until after fertilization.

17. *Dioecious* animals are those with distinct males and females. *Monoecious*, or *hermaphroditic*, species are those where the same individuals possess both male and female reproductive structures. Hermaphrodites can also result from developmental abnormalities in characteristically dioecious species.

18. *Simultaneous hermaphrodites* are both male and female at the same time. They usually cannot fertilize themselves, and thus require another individual for mating. However, hermaphroditic species that have a low probability of encountering potential mates, such as tapeworms, are able to self-fertilize.

19. *Sequential hermaphrodites* are first one sex, then later in life change to the other sex. This strategy helps prevent inbreeding in situations where siblings live together, or where a dominant male controls all the matings within a group of females, as with some tropical fish.

20. Fertilization can either be external or internal. *External fertilization* only occurs in aquatic habitats. Adaptations for external fertilization include the production of large numbers of gametes, mechanisms to synchronize mating attempts, and highly coordinated sexual behaviors such as the long migrations of salmon.

21. *Internal fertilization* is essential for terrestrial animals in order to avoid drying out of the sperm, but it is also used by many aquatic species. A major advantage of internal fertilization is protection of the offspring during development.

22. Special accessory sex organs facilitate *copulation*; the process by which sperm move from the male's reproductive tract into the female's. Most males of internally fertilizing species have a tubular *penis* that is inserted into the female's *vagina* (specialized sperm deposition canal) or *cloaca* (combined urinary, digestive, and reproductive chamber) prior to sperm release.

23. A few animals that use internal fertilization forego direct copulation. In salamanders and scorpions, males deposit a protected *spermatophore* (sperm packet) somewhere in the environment. The female then collects and inserts this packet into her reproductive tract. Male squids and spiders also use spermatophores, but have specially modified tentacles or legs that they use to insert the sperm packet into a female.

24. Even though they deliver sperm in spermatophores, most male insects also possess *genitalia* (external sex organs), which include a tubular penis that fits precisely into the female's reproductive opening. This adaptation may exist because of the prolonged copulatory times

required to achieve successful fertilization, or to prevent cross-species fertilization attempts. Insect penises are also often equipped with devices that scoop out the spermatophores left by other males, thus increasing the chances that the new male will be the one who fertilizes the eggs.

25. Developing embryos require oxygen and other essential nutrients, metabolic waste removal, and a suitable physical environment. Meeting these requirements necessitates different strategies depending on whether an animal is aquatic or terrestrial.

26. Modern fish are the closest living relatives to the ancestral aquatic vertebrates, and most still use external fertilzation exclusively. Sharks and rays have evolved internal fertilization and carry out sperm transfer while being held together by organs called *claspers*. In amphibians, fertilization is still aquatic and external, but is made more efficient by a behavior called *amplexus*, where a male holds on tightly to a female while she releases her eggs and by simultaneously releasing sperm increases the probability of fertilization.

27. Reptiles were the first vertebrates to solve the problem of terrestrial reproduction. They utilize a shelled egg (as do birds), which is permeable to gases but not to water. Since sperm must reach this egg before the shell forms, internal fertilization is required. Some reptiles accomplish this by use of hemipenes, while birds and mammals utilize erectile penises. Most mammals have relinquished the shelled egg in favor of internal development of the embryo within the female's body.

28. *Oviparous* animals lay external eggs that develop independently of the female parent. Terrestrial oviparous animals such as insects, reptiles, birds, and a few mammals coat their eggs with tough external shells for protection from desiccation and predators. They may also provide parental care in the form of nests or guarding behavior. After hatching, the offspring may receive additional care until they achieve independence.

29. *Viviparous* animals retain the embryo within the mother's body for some portion of its development. Most viviparous animals are mammals. Marsupial mammals give birth to very underdeveloped offspring that then complete their development in an external pouch located on the mother's abdomen.

30. *Eutherian mammals* complete their embryonic growth inside the mother's uterus and are born in a relatively advanced state. A distinguishing feature in the uterine development of these animals is the presence of a complex anatomical connection between the fetal and maternal blood supplies.

31. Some species of fish, amphibians, and reptiles are *ovoviviparous*. They produce shelled eggs, but instead of laying them externally, the eggs are retained inside the female's body until the young hatch.

Activities (for answers and explanations, see page 384)

- For each of the stages in spermatogenesis shown next, indicate whether the nucleus of the cell is diploid (2*n*) or haploid (*n*).

_____Primary spermatocytes

_____Spermatids

_____Spermatogonia

_____Secondary spermatocytes

_____Functional sperm

- How is it possible for a sea star to reproduce both regeneratively and sexually? Which of these two processes results in "offspring" that are genetically identical to the parent?

Questions (for answers and explanations, see page 385)

1. Asexual reproduction is a good strategy to use in stable environments because
   a. gametogenesis is most efficient under these conditions.
   b. offspring are genetically preadapted for the stable environment because they are identical to their parents.
   c. asexual parthenogenesis produces a large amount of genetic diversity.
   d. animal cells tend to be more totipotent under stable conditions.
   e. all of the above

2. If you compared the genetic makeup of an animal produced by parthenogenesis with that of its father, which of the following results would you expect?
   a. About a 75 percent degree of genetic similarity
   b. 100 percent genetic similarity, except in cases of highly variable environments
   c. No genetic similarity at all, except for that due to mutations
   d. Genetic similarity would vary depending on the species.
   e. The comparison is inappropriate, since parthenogenetic animals have no fathers.

3. An important difference between a sperm and an egg is
   a. their relative sizes.
   b. the amount of cytoplasm they contain.
   c. whether or not they are motile.
   d. which sex produces them.
   e. All of the above

4. If you were examining the gonads of a squirrel and found functional sperm cells, what would this *unambiguously* tell you about the animal?
   a. Its age
   b. Its sex
   c. How many offspring it has already produced
   d. Whether or not it is a hermaphrodite
   e. None of the above

5. Select the set of terms that best describes a male squid.
   a. Monoecious, asexual, parthenogenetic
   b. Totipotent, regenerative, sequential hermaphrodite
   c. Dioecious, makes spermatogonia, utilizes a spermatophore
   d. Internal fertilization, simultaneous hermaphrodite, makes oogonia
   e. Produces diploid gametes, external fertilization, copulates

6. Which of the following statements about animals that utilize external fertilization is *false*?
   a. They are divided about equally between terrestrial and aquatic species.
   b. Many produce large numbers of gametes to ensure successful reproduction.
   c. The behaviors associated with mating are often highly synchronized.
   d. The probability of any one egg being fertilized and developing into an adult can be low.
   e. None of the above

7. Which of the following statements about animal reproduction is *false*?
   a. Species that reproduce sexually cannot also reproduce asexually.
   b. Viviparity, but not ovoviviparity, is common in mammals.
   c. Male insects can remove spermatophores deposited in a female by other males.
   d. Oogenesis and spermatogenesis both occur in simultaneous hermaphrodites.
   e. The cells that initiate budding in sponges and cnidarians are totipotent.

8. Which of the following would likely be present in the body or lifestyle of a terrestrial male animal that reproduces sexually?
   a. Internal fertilization
   b. An intromittent organ as part of the genitalia
   c. Synchrony of reproductive physiology with the female
   d. Gametogenesis
   e. All of the above

REPRODUCTIVE SYSTEMS IN HUMANS AND OTHER MAMMALS (pages 865–877)

Key Concepts

1. In most mammals, including humans, the male's testes are located in an external pouch of skin called the *scrotum*. Involuntary muscle contractions adjust scrotum size to maintain proper thermal conditions for spermatogenesis.

2. Each testis contains a large volume of tightly coiled *seminiferous tubules* where spermatogenesis takes place. Surrounding the tubules are clusters of cells where male sex hormones are produced. *Sertoli cells* nurture the developing spermatids near the lumen of the tubules, and clusters of *Leydig cells* produce male sex hormones. Fully formed sperm move into a storage organ called the *epididymis*, where they mature and become motile.

3. Newly formed spermatids undergo substantial modification before becoming functional sperm. They acquire a flagellum for movement, a midpiece with mitochondria to generate the ATP needed for swimming, and a head region containing the genetic material. At the tip of the sperm head is the *acrosome*, which carries enzymes needed to gain access to the egg cell.

4. The *vas deferens* is a tube connecting the epididymis with the *urethra*. The urethra runs the length of the penis and also serves as the exit for urine from the bladder. The penis is largely made of spongy connective tissue. When engorged with blood it becomes erect to facilitate insertion into the female's vagina. Some mammalian species (not humans) also have a bone in the penis.

5. The *glans penis* is an area of thin, highly sensitive skin at the tip of the penis. In humans it is covered by the foreskin, which is often removed at birth by circumcision. The glans penis is responsive to sexual stimulation.

6. The climax of the male sex act occurs in two phases. During *emission*, sperm mixes with fluids from the *seminal vesicles, prostate gland*, and *bulbourethral glands* to form *semen*, which collects in the urethra at the base of the penis.

7. Seminal vesicle fluid makes up two-thirds of the semen volume. It contains mucus, proteins, and fructose to fuel the sperms' mitochondria, as well as *prostaglandins*, which stimulate contractions of the female's reproductive tract. Prostate fluid comprises most of the remaining semen volume. It is alkaline to neutralize the acidity of the female's reproductive tract, and also contains clotting enzymes that make the semen a gelatinous mass. The very small amount of alkaline fluid released from the bulbourethral glands neutralizes acidity in the male's urethra and lubricates the tip of the penis.

8. *Ejaculation* follows emission. It occurs when wavelike muscle contractions at the base of the rigid penis move upward, forcing semen through the urethra and out of the body.

9. Male sexual function is controlled by several hormones. *Testosterone* from the Leydig cells maintains spermatogenesis and triggers development of male secondary sexual characteristics. *Gonadotropin-releasing hormone (GnRH)* from the hypothalamus, along with *luteinizing hormone (LH)* and *follicle-stimulating hormone (FSH)* from the anterior pituitary, are involved with testosterone in a negative feedback system between the hypothalamus, anterior pituitary, and Sertoli cells to regulate spermatogenesis (see textbook Figure 39.11). Another hormone (*inhibin*) is produced by the Sertoli cells and functions only to inhibit the production of GnRH and FSH.

10. The female gonads (*ovaries*) are paired structures that release eggs (*ova*) directly into the lower body cavity. The egg is then collected into one of the paired *oviducts* (*fallopian tubes*) whose ciliated lining propels the egg into the *uterus*, or womb. This thick-walled, muscular cavity is where the embryo develops. An opening at the bottom of the uterus called the *cervix* leads to the *vagina*, which serves as the birth canal as well as a receptacle for the penis and sperm during copulation.

11. The opening of the vagina is surrounded by two sets of folded skin, the delicate *labia minora* and the thicker *labia majora*. The anterior ends of the labia minora enclose a small bulb of erectile tissue called the *clitoris,* which is the sensory homolog of the glans penis. The clitoris plays an important role in female sexual arousal.

12. Sperm deposited in the vagina swim through the cervix, uterus, and most of the oviduct, where they fertilize the egg. Early embryonic development of the zygote produces a blastocyst, which moves down the oviduct and implants itself in the internal lining of the uterus, or *endometrium.* This initiates development of the *placenta,* which exchanges nutrients and wastes between the mother and her developing baby.

13. Human ovaries contain nearly a million primary oocytes at birth. This number is reduced to about 200,000 by the onset of puberty. During a woman's reproductive lifespan, only about 450 oocytes actually mature into functional eggs. *Menopause,* which occurs near age 50, marks the end of female fertility, as the maturation of oocytes ceases.

14. A developing egg and the layer of ovarian cells surrounding it form a functional unit called the *follicle.* Up to a dozen follicles develop in a woman's ovaries each month during her reproductive lifespan, but usually only one ruptures to release its egg in the process called *ovulation.* After ovulation, the remaining follicular cells form the *corpus luteum,* which remains inside the ovary and functions as an endocrine gland producing estrogen and progesterone.

15. Ovulation is only one of several events in the human female's menstrual cycle, so named for the conspicuous sloughing off of the endometrial lining known as *menstruation.* The menstrual cycle really consists of two coordinated cycles: one in the ovary, which prepares an egg for release, the other in the uterus, which prepares it for pregnancy.

16. The reproductive cycles of female mammals vary by species. The human cycle averages 28 days from one menstruation event to the next. Unlike humans, most female mammals reabsorb their endometrial lining rather than sloughing it off through menstruation.

17. *Estrous cycles* occur in mammals that do not menstruate. A striking behavior in these species is the increased sexual receptivity of females around the time of ovulation, referred to as estrus, or "heat." In humans, sexual receptivity is not physiologically correlated with stages of the reproductive cycle or season (i.e., there is no estrus).

18. Several hormones released by the hypothalamus and anterior pituitary play critical roles in coordinating the timing of the human female's menstrual cycle. The hypothalamus increases its output of gonadotropin-releasing hormone at the time of puberty, which leads to increased outputs of follicle-stimulating hormone (FSH) and luteinizing hormone (LH) by the anterior pituitary. Ovarian development follows, and the rapid rise in estrogen and progesterone secretion that results initiates the cycles of follicular growth and uterine development that continue monthly until the onset of menopause.

19. The ovarian and uterine cycles commence at the onset of menstruation. Just prior to this event, the anterior pituitary secretes increased levels of FSH and LH in response to rising GnRH levels, which causes the development of new follicles within the ovaries and the production of estrogen. Eventually one of these follicles matures and the others wither away.

20. The mature follicle begins producing larger amounts of estrogen, which causes the endometrium to grow. Increased estrogen levels negatively feed back on the anterior pituitary to reduce gonadotropin release. About day 12 of the cycle this negative feedback is reversed and a surge of LH is released by the pituitary, triggering ovulation.

21. The corpus luteum, which develops from the remaining follicle cells in the ovary, continues to secrete estrogen and rapidly increases its production of progesterone to help maintain the endometrial lining. Both hormones have negative effects on pituitary gonadotropin release, effectively shutting down the production of any new follicles.

22. If the egg is not fertilized, the corpus luteum degenerates by about day 26 of the cycle, dramatically reducing estrogen and progesterone concentrations. This initiates the breakdown of the endometrial lining, leading to the onset of menstruation. It also removes the inhibition on pituitary gonadotropins, allowing new follicles to start developing.

23. If fertilization does occur, the developing blastocyst begins secreting a new hormone called *human chorionic gonadotropin.* hCG maintains the corpus luteum in an active state of steroid production so that the newly implanted embryo can survive until a functional placenta has formed. The placenta then takes over the production of progesterone, which maintains the uterine lining and prevents the pituitary from secreting additional gonadotropins, thus shutting down the ovarian cycle for the duration of pregnancy.

24. Human sexual responses are similar in men and women, and consist of four distinct phases. The *excitement* phase is marked by elevated heart and breathing rates, increased blood pressure, and enhanced muscle tension. In women the clitoris swells and becomes more sensitive, and the vaginal walls secrete lubricating fluids. In men the penis fills with blood and becomes erect.

25. The *plateau* phase in women shows further increases in heart and breathing rates. The clitoris now retracts, and the sensitivity once focused there spreads across the external genitals. Men also show increased breathing and heart rates, the diameter of the glans increases, testes swell as the scrotum tightens, and small amounts of lubricating fluids are released from the penis.

26. The *orgasm* phase in women begins with several long contractions (2 to 4 seconds each) in the outer third of the vagina, followed by a rapid series of shorter ones. Unlike men, women may experience multiple orgasms in rapid succession, which can last for several minutes. A man's orgasm is stimulated by frictional pressure on the shaft and glans of the penis. This stimulates massive contractions of muscles in the genital area that expel the semen.

27. During the *resolution* phase, blood flow and breathing rates return to normal in both sexes. This takes from 5 to 10 minutes if a woman has had an orgasm, but may take 30 minutes or more if she has not. Resolution in men is more immediate, marked by rapid shrinking in penis size within minutes after ejaculation. Unlike women, men experience a *refractory period* of 20 minutes or more during which they are physiologically incapable of achieving a full erection or ejaculation regardless of the intensity of sexual stimulation.

28. The simplest and most natural forms of human birth control are *abstinence* (avoiding sex altogether), *coitus interruptus* (where the penis is withdrawn before ejaculation occurs) and the *rhythm method* (avoiding sex around the time of ovulation). Abstinence is very effective, but failure rates are high for both the rhythm method and coitus interruptus.

29. Most other forms of human birth control interfere with fertilization and have much higher success rates than natural methods. Forms of *sterilization* such as vasectomy or tubal ligation are highly effective, but are usually irreversible. Slightly less effective but reversible ways to prevent the union of sperm and egg include mechanical barriers such as *condoms, diaphragms,* and *cervical caps* or *vaginal sponges*. These devices are typically used in conjunction with *chemical spermicides* (e.g., contraceptive foams and jellies).

30. Another highly effective method of contraception is the *birth control pill*, which uses hormonal manipulations of the ovarian cycle to prevent maturation and release of eggs.

31. Methods for preventing the actual implantation of a fertilized egg are also used as birth control. The *intrauterine device (IUD)* is a small plastic structure inserted into the uterus that interferes with implantation.

32. *Abortion* is yet another way of preventing birth. It can occur spontaneously (a "miscarriage") as the result of developmental abnormalities in the fetus or problems with implantation. Abortions can also be induced medically through a variety of methods.

33. Because of the cyclic nature of female fertility, it is much easier to control egg production than it is to reliably interfere with the continuous manufacture of sperm by males. Human contraceptive techniques have therefore focused on women more than men. Moreover, since it requires only a single functional sperm to achieve fertilization, it is much easier to reversibly halt female fertility than it is to render a male temporarily sterile.

34. Reproductive problems of many kinds may prevent humans from having children naturally. Low sperm production or poor sperm motility are common problems in men, while blockage of the oviducts or *endometriosis* (proliferation of the endometrial lining outside of the uterus) are sometimes encountered in women. Parents may also be the carriers of recessive genetic diseases that could emerge in their children.

35. Increasing numbers of human couples are seeking help in the form of *assisted reproductive technologies (ARTs)* when they cannot produce children on their own. ARTs include methods such as *in vitro fertilization*, where sperm and egg are united outside of the mother and the early embryo is then implanted back into the uterus. Alternatively, *gamete intrafallopian transfer (GIFT)* can be used to bypass blockages of the upper oviduct by injecting gametes directly into the lower oviduct where fertilization can occur naturally.

36. Many parasitic bacteria, yeasts, protozoans, and viruses have evolved to take advantage of the close contact that occurs between partners during reproduction and to obtain new hosts. The pathologies produced by such parasites are collectively known as *sexually-transmitted diseases (STDs)*.

Questions (for answers and explanations, see page 385)

1. Mature mammalian sperm are stored in the _____ prior to their release during ejaculation.
   a. seminiferous tubules
   b. scrotum
   c. vas deferens
   d. epididymis
   e. testes

2. The function of the prostate gland in human males is to
   a. initiate the muscular contractions leading to emission.
   b. initiate the muscular contractions leading to ejaculation.
   c. produce a fructose solution used to fuel the mitochondria in the sperm's midpiece.
   d. swell with blood and cause the penis to become erect.
   e. secrete alkaline fluids that neutralize the acidity of the female's reproductive tract.

3. Which of the following best represents the normal path of a sperm cell as it makes its way from the point of entry into the female's reproductive tract to the place where fertilization typically occurs?
   a. Cervix, vagina, ovary, oviduct
   b. Vagina, cervix, uterus, oviduct
   c. Uterus, oviduct, vagina, cervix
   d. Ovary, oviduct, cervix, vagina
   e. Vagina, uterus, cervix, oviduct

4. For approximately how long during the human female's menstrual cycle are progesterone concentrations high enough to maintain the uterus in a proper condition for pregnancy?
   a. All of the cycle
   b. None of the cycle
   c. Less than one-half of the cycle
   d. More than one-half of the cycle
   e. Two days either side of midcycle

5. If you needed to take a blood sample during the time in a woman's menstrual cycle when the concentrations of her gonadotropic hormones would be at their *lowest* levels, which of the following days on average would be the best choices for sampling?
   a. 1 to 5
   b. 5 to 10
   c. 10 to 15
   d. 15 to 20
   e. 20 to 25

6. A couple have been trying to conceive a child for over a year without success. Recently, the man's doctor informed him that his seminal vesicles are defective. This means that

   a. the man makes functional sperm, but they are not delivered because he is unable to ejaculate.

   b. the man's sperm lack functional flagella.

   c. the fluid needed to lubricate the man's urethra for easier sperm travel is defective.

   d. the man's semen lacks the fructose fuel needed for the sperm to swim.

   e. the acidity in the woman's reproductive tract kills the sperm before they can reach the egg.

7. Which of the following human contraceptive methods prevents the implantation of a fertilized egg?

   a. IUD

   b. Diaphragm

   c. Birth control pill

   d. Condom

   e. Contraceptive jelly

8. Which of the following statements about human sexual response is *false*?

   a. Increased breathing rate is common during the excitement phase of both men and women.

   b. Like women, men often experience multiple orgasms in rapid succession.

   c. The plateau phase in women is marked by retraction of the clitoris.

   d. A man's penis engorges with blood and becomes rigid during the excitement phase.

   e. The resolution phase is typically shorter in women who have achieved orgasm.

9. All but one of the following is a biological or medical reality for human reproduction. Which one is *not*?

   a. Negative feedback control of the hypothalamus by sex hormones

   b. Contraception through complete chemical blockage of sperm production

   c. Surgical transfer of gametes from one area of the reproductive tract to another

   d. Transmission of disease-causing agents during sexual contact

   e. Susceptibility of reproductive organs to cancer

Activities (for answers and explanations, see page 385)

• The following graph depicts the average blood concentrations of three circulating hormones collected from 100 healthy adult women. Each hormone plays a role in regulating some aspect of the menstrual cycle. Identify the most likely hormone that each curve represents, and explain the criteria used in making your choices.

• Where applicable, indicate the structure in the reproductive systems of both male and female humans that corresponds with each of the following.

   1. Site of gametogenesis:

      Male _____   Female _____

   2. Primary source of sex steroids:

      Male _____   Female _____

   3. Main organ of sexual sensory stimulation:

      Male _____   Female _____

   4. Storage organ for mature gametes:

      Male _____   Female _____

   5. Organ where most fetal development occurs:

      Male _____   Female _____

FROM CONCEPTION TO BIRTH IN HUMANS (pages 877–880)

Key Concepts

1. *Gestation*, or pregnancy, in mammals represents the time from fertilization to birth. Its length generally correlates positively with body size. Human gestation is 266 days, or about nine months, and is divided into three 90-day trimesters.

2. The first trimester begins with fertilization and the formation of the zygote. This single cell then undergoes a series of rapid cell divisions to become an *embryo*.

3. The early embryonic stage known as the *blastocyst* (about six days old) is what implants in the uterine lining. The blastocyst quickly produces hCG to maintain the corpus luteum and its supply of progesterone and estrogen needed to maintain the pregnancy.

4. Organ development follows implantation and is the main event of the first trimester. At this time the limbs and all major organ systems of the embryo start to develop.

5. A critical organ that forms early in the first trimester is the *placenta*, which arises from an interaction between endometrial and embryonic tissues.

6. The embryo is connected to the placenta by the *umbilical cord*, which contains blood vessels that run between the embryo and the regions of the placenta where nutrients and wastes are exchanged between mother and developing child. The placenta also begins producing progesterone to supplement the corpus luteum.

7. By the eighth week of pregnancy the embryo resembles a miniature human and is called a *fetus*. Because of the high rate of cell division at this stage, the fetus is highly susceptible to drug use or nutritional deficiencies in the mother's diet. At this time a multitude of hormonal changes occur in the mother's body that cause major physiological shifts, producing the classic "symptoms" of pregnancy such as morning sickness, mood swings, and unusual food cravings.

8. The second trimester is a period of rapid fetal growth, along with the first signs of fetal movement. As estrogen and progesterone secretions increase in the placenta, chorionic gonadotropin levels drop and the corpus luteum degenerates, taking with it many of the unpleasant symptoms of pregnancy.

9. Fetal growth continues throughout the third trimester. Final maturation of the fetus's internal organs occurs in preparation for birth. Important among these are the complete functioning of its digestive system, glycogen storage in the liver, urine production by the kidneys, and regular sleep and wake cycles in the brain.

10. Slow, rhythmic *Braxton–Hicks contractions* (false labor) occur throughout pregnancy. Actual childbirth, or *parturition*, begins with the onset of major contractions in the uterine muscles, otherwise known as true labor or stage one of delivery. The mucous plug that has been covering the cervix is shed, and the fluid-filled sack that surrounds the fetus ruptures.

11. Labor is influenced by numerous physiological factors, including mechanical stimulation of the uterine musculature, hormonal shifts leading to an abundance of estrogen, and the release of oxytocin from the posterior pituitary. Positive feedback loops (see textbook Figure 39.23) stimulate and intensify labor contractions, which often last up to 15 hours in first-time pregnancies.

12. Delivery is the second stage of labor. It commences when the cervix is fully dilated and the baby's head passes through the vaginal opening. As soon as it clears the birth canal, the infant begins breathing and its umbilical cord is clamped and surgically severed. Expulsion of the placenta and remaining portions of the umbilical cord follows in the process known as afterbirth.

Activity (for answer and explanation, see page 386)

* Arrange the following terms or events related to human pregnancy in their proper chronological sequence from first to last:

parturition; initial placental development; afterbirth; embryo becomes a fetus; blastocyst implantation; embryo formation; fertilization; rapid fetal growth; mature fetal organ systems become functional.

Questions (for answers and explanations, see page 386)

1. The _____ is the embryonic stage in humans that implants in the uterus to stimulate development of a(n) _____ .
   a. corpus luteum; placenta
   b. fetus; embryo
   c. fetus; umbilical cord
   d. blastocyst; fetus
   e. blastocyst; placenta

2. Which of the following statements about human pregnancy and fetal development is *false*?
   a. The blastocyst is a stage in fetal development that appears early during the third trimester.
   b. A vital connection between the fetal and maternal blood supplies occurs at the placenta.
   c. Most of the growth in fetal size occurs during the second and third trimesters.
   d. All major organ systems have formed by the end of the fetus's first three months of life.
   e. The corpus luteum degenerates sometime during the second trimester.

3. The role of the corpus luteum and the hormones it produces *ends*
   a. at the time of embryonic implantation in the uterus.
   b. with the formation of the placenta.
   c. during the second trimester of pregnancy.
   d. just prior to parturition.
   e. after parturition is complete.

4. The majority of human organ formation occurs
   a. entirely during the last trimester of pregnancy.
   b. after implantation and during the first trimester of pregnancy.
   c. shortly before the onset of true labor.
   d. due to the stimulation generated during Braxton–Hicks contractions.
   e. under the influence of oxytocin and prolactin.

5. An important factor contributing to the onset and maintenance of labor contractions is
   a. increased contractility of the uterine muscles from increased oxytocin levels.
   b. mechanical pressure on the cervix generated by the emerging baby's head.
   c. stretching of the uterus during the final stages of fetal growth.
   d. increased contractility of the uterine muscles due to elevated estrogen levels.
   e. all of the above

Integrative Questions (for answers and explanations,
see page 386)

1. Which of the following animals qualifies as a sexually
   reproducing, oviparous species?
   a. Human
   b. Chicken
   c. Kangaroo
   d. Sea star
   e. None of the above

2. If you examine the reproductive system of a normal
   human female before she undergoes puberty, which of
   the following would you *not* expect to find?
   a. Corpus luteum
   b. Ovaries
   c. Primary oocytes
   d. Oviducts
   e. Estrogen

3. Which of the following does *not* apply to a sexually
   mature and reproductively normal human male?
   a. Has cells that undergo gametogenesis
   b. Lacks oviducts and a clitoris
   c. Can be permanently sterilized by vasectomy
   d. Contains no cells capable of totipotency
   e. Produces offspring genetically indentical to himself

4. Keeping a normal, sexually mature woman in her mid-
   twenties on continuous high doses of progesterone will
   a. cause her to ovulate repeatedly within each menstrual
   cycle.
   b. stimulate the production of mature follicles.
   c. prevent her from producing mature follicles and ovu-
   lating.
   d. cause the endometrial lining of her uterus to slough
   off.
   e. stimulate milk production in her mammary glands.

40

Animal Development

CHAPTER LEARNING OBJECTIVES—after studying this chapter you should be able to:

❏ Discuss the roles of determination and differentiation in animal development.

❏ Describe the cellular events associated with fertilization and the formation of a zygote in various types of animals.

❏ Explain the mechanisms that permit species-specific fertilization and prevent multiple fertilizations of the same egg.

❏ Distinguish between a fast block and slow block to polyspermy.

❏ Compare the relative contributions to the zygote of genetic and cytoplasmic materials that are made by sperm and egg.

❏ Discuss how fertilization causes rearrangements of the egg cytoplasm and what consequences these changes have for early embryonic development.

❏ Describe the effects of egg yolk concentration on early cleavage patterns in animals.

❏ Describe the overall pattern of cleavage, including changes in cell number and cell size, that occurs during early development in different types of animals.

❏ Explain how the anterior–posterior and the bilateral axes are determined in the early amphibian embryo.

❏ Describe the developmental sequence from the fertilized egg through the gastrula stage for the sea urchin, frog, chicken, and human using the following terms: "blastula," "blastomere," "blastocoel," "blastopore," "blastocyst," "blastodisc," "trophoblast," "archenteron," "gastrulation," "dorsal lip," "primitive streak," "gray crescent," and "yolk plug."

❏ Explain the concept of fate maps, and relate this idea to differences between mosaic development and regulated development in animals.

❏ Describe the developmental events that leads to the production of monozygotic twins in humans.

❏ Show the derivation of the three primary germ layers in vertebrates, and list the structures they form in the adult.

❏ Explain what kinds of animals are included in the protostomes and the deuterostomes, and describe a key developmental difference between these two groups.

❏ Describe important features seen during gastrulation in frogs, birds, reptiles, and mammals.

❏ Explain how the studies of Spemann and Mangold on the dorsal lip of the frog blastopore and transplantation of embryonic tissues contributed to our understanding of embryonic induction.

❏ Describe the process of neurulation as seen in vertebrates, including formation of the notochord, neural tube, and somites.

❏ Identify the extraembryonic membranes present in the amniotic egg and discuss the function of each.

❏ Discuss how extraembryonic membranes contribute to the formation of the mammalian placenta.

❏ Explain how extraembryonic membranes permit the detection of genetic diseases in humans, and distinguish between the medical techniques known as amniocentesis and chorionic villus sampling.

❏ Explain what is meant by growth and discuss the two types of growth curves seen in animals.

FERTILIZATION: CHANGES IN SPERM AND EGG ● CLEAVAGE: REPACKAGING THE CYTOPLASM   (pages 883–890)

Key Concepts

1. *Development* is the progressive change in the structure and function of an individual as it passes through the stages of its life cycle.

2. The embryo stage begins with the *zygote*, which results from the union of two gametes during *fertilization*, and

continues until the organism becomes independent of the nutrients provided by its mother.

3. During development, cell number increases, cytoplasmic resources are selectively partitioned, and the relative position of cells changes within the embryo, leading to a cascade of gene expression, growth, and other important developmental events.

4. Fertilization involves a complex series of events that produce a diploid zygote and initiate embryonic development. The sequence begins with the juxtaposition of sperm and egg.

5. Animals that utilize external fertilization, such as the sea urchin, have several mechanisms to insure fertilization of the egg by the proper sperm. Species-specific chemotaxic signals are released by the egg and attract the sperm to it. Once in contact with the jelly coat surrounding the egg, acrosomal enzymes are activated in the sperm, allowing it to digest its way through the jelly until the sperm head encounters the proteinaceous *vitelline envelope* surrounding the egg.

6. An *acrosomal process* grows from the sea urchin sperm and makes contact with species-specific receptors on the vitelline envelope utilizing the binding protein *bindin*. A fertilization cone then forms to facilitate entry of the sperm into the egg.

7. Animals that use internal fertilization, such as humans and other mammals, insure species-specificity mainly through complex mating behaviors. Egg-sperm recognition mechanisms are therefore not highly developed.

8. Like all other cells, the mammalian egg is bounded by a plasma membrane. Immediately surrounding this membrane is a glycoprotein layer called the *zona pellucida*, which is covered by follicle cells within a jellylike matrix collectively known as the *cumulus*.

9. When human sperm initially enter the female's reproductive tract they are incapable of penetrating the protective layers surrounding an egg. Sperm activation occurs inside the uterus through a process called *capacitation*. Capacitated sperm utilize newly-activated acrosomal enzymes to help disperse the cells of the cumulus, improving the chances of sperm penetration.

10. Once inside the cumulus, linkage reactions occur between enzymes on the sperm head and substrate molecules in the zona pellucida. Additional enzymes from the acrosome digest their way through this layer, and the sperm eventually fuse with the egg plasma membrane. Binding proteins called fertilins now cause fusion of the sperm and egg membranes to permit entry of the sperm in the egg cell. *Blocks to polyspermy* that prevent more than one sperm nucleus from entering the egg are initiated at this point.

11. The *fast block* to polyspermy happens within a second of egg penetration. The rapid uptake of sodium ions triggered by sperm penetration changes the electrical potential across the egg's plasma membrane, thereby preventing the entry of additional sperm.

12. A *slow block* to polyspermy also occurs over a period of 20 to 30 seconds. In sea urchins, cortical vesicles residing just below the plasma membrane are stimulated to release their enzymes when sequestered calcium ions flood the egg cytoplasm following sperm penetration. This causes the bond between the plasma membrane and the vitelline envelope to break. Water flows by osmosis into this newly created space, raising the vitelline envelope and forming the *fertilization membrane*, which further blocks sperm entry.

13. Postfertilization release of calcium within the egg also triggers increased metabolic activity. Oxygen consumption and protein synthesis both increase as cytoplasmic pH changes. The egg nucleus now completes its second meiotic division before fusing with the sperm nucleus.

14. In frogs and many other animals, nutrient-laden *yolk* is concentrated in one end of the egg, called the *vegetal hemisphere*. The other end of the egg, which contains the nucleus but relatively little yolk, is called the *animal hemisphere*.

15. Upon fertilization, which always occurs in the animal hemisphere, frog eggs undergo an asymmetric rearrangement of cytoplasmic material that sets the stage for bilateral symmetry in the developing embryo. The *gray crescent* forms as a diffusely pigmented band opposite the site of sperm entry, establishing the anterior–posterior axis of the body.

16. The embryo now undergoes a series of mitotic cell divisions called *cleavage*. Early cleavage is synchronized and the total number of daughter cells, or *blastomeres*, doubles with each division cycle. Cleavage ends with the formation of the *blastula*, a hollow ball of cells with a fluid-filled central cavity called the *blastocoel*.

17. In most animals, cleavage is rapid and results in an increase in cell number, but no growth in the volume of the embryo. Consequently, cells become smaller and smaller and the ratio of nuclear to cytoplasmic volume increases.

18. *Complete* cleavage happens when the cleavage furrow completely bisects the egg mass. It occurs in eggs that contain little yolk (sea urchin) or modest amounts of yolk (frog). The resulting blastomeres are all about the same size at the 8-cell stage in sea urchins. In frogs, the first two cleavages form four equal-sized cells. The third cleavage is transverse and closer to the animal pole, forming blastomeres of unequal size at the 8-cell stage.

19. In large eggs with much yolk, like those of fish, reptiles, and birds, yolk-free cytoplasm is restricted to a very small part at one end of the egg. Cleavage in these animals is therefore *incomplete* and the embryo develops as a *blastodisc* on top of the yolk mass.

20. A special type of incomplete cleavage occurs in insects such as fruit flies. *Superficial* cleavage results when early mitosis does not include cytokinesis. The many nuclei formed by this process migrate to the surface of the embryo, where they eventually develop plasma membranes and form a single layer of cells surrounding a yolky core.

21. Complete cleavage can be further subdivided based on the position of mitotic spindles in successive cell generations. If spindles form at right angles, cleavage is said to be *radial* (sea urchin and frog). If the spindles are not at right angles, then cleavage is *spiral* (mollusks). In *rotational* cleavage (mammals), spindles form at right angles but in a different sequence than in sea urchins.

22. Mammalian development differs in important ways from that of other animals. Cell division in mammals is slower, and genes (mostly maternal) are expressed during cleavage. At the 8-cell stage, cells maximize their surface contact, form tight junctions with each other, and become a compact mass of cells. By the 32-cell stage, an inner cell mass is surrounded by an exterior sac of cells called the *trophoblast*, which eventually becomes part of the placenta.

23. Until the blastocoel forms, the zona pellucida continues to surround the mammalian embryo to prevent premature implantation in the oviduct (an ectopic pregnancy). Once in the uterus the embryo emerges from the zona pellucida, at which time the trophoblast cells initiate uterine implantation and the formation of a placenta.

24. During cleavage large numbers of new cells are formed around a central cavity. However, none of these new blastomeres has differentiated, nor has much of their genome been expressed (except in mammals). During the next major stage of development the blastomeres, until now kept from interacting by the blastocoel, will move and form new associations that will lead to cellular differentiation and important inductive interactions. These cell movements are so regular that individual blastomeres can be followed and their developmental changes charted to produce *fate maps*.

25. In animals with *mosaic development*, such as roundworms and clams, each blastomere in the early embryo is necessary to form a portion of the adult body. Other animals, such as vertebrates and sea urchins, show *regulative development*, in which the loss of certain early embryonic cells is compensated for by the remaining cells, and development forms a normal adult.

26. *Monozygotic twins* are formed when the cells of a young embryo in a species with regulated development are physically separated into two groups. Each group then gives rise to a complete embryo that is genetically identical to its twin. In humans and other mammals, whether the twins will have separate placentas or share a common placenta depends on whether the separation occurs before or after trophoblast formation.

Questions (for answers and explanations, see page 386)

1. Match the structures on the left with the appropriate descriptors on the right.

_____	Blastocoel	*a.*	A hollow sphere of cells
_____	Fertilization	*b.*	Glycoprotein layer of mammalian egg
_____	Acrosome	*c.*	Yolk-rich half of frog egg
_____	Zona pellucida	*d.*	Contains enzymes for egg penetration
_____	Capacitation	*e.*	Cells from early cleavage
_____	Blastula	*f.*	Activates sperm for possible fertilization
_____	Radial cleavage	*g.*	Deuterostome cell division
_____	Blastomeres	*h.*	Internal cavity of blastula
_____	Trophoblast	*i.*	Union of sperm and egg
_____	Vegetal hemisphere	*j.*	Outer cells of early mammalian embryo

2. Select the correct curve describing changes in the ratio of nuclear volume to cytoplasmic volume during the cleavage stage of development and the correct statement about changes in blastomere size during that stage.

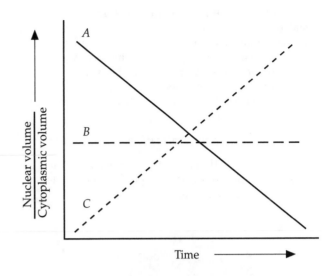

a. Curve *A*; blastomeres decrease in size during this stage
b. Curve *B*; blastomeres decrease in size during this stage
c. Curve *C*; blastomeres decrease in size during this stage
d. Curve *A*; blastomeres increase in size during this stage
e. Curve *C*; blastomeres increase in size during this stage

3. In which of the following animal eggs would you expect to find a zona pellucida covered by a cumulus?
a. Human
b. Frog
c. Sea urchin
d. Chicken
e. All of the above

4. In sea urchins, which of the following events associated with fertilzation occurs just *prior* to the actual penetration of the egg by a sperm?
a. Capacitation
b. Growth of the acrosomal process
c. Release of species-specific chemotaxic cues
d. Formation of the fertilization cone
e. Formation of the fertilization membrane

5. If you could selectively inhibit the release of calcium ions within a sea urchin egg, which of the following processes would be altered?
a. Fast block to polyspermy
b. Slow block to polyspermy
c. Formation of the acrosomal process
d. Postfertilization metabolic activity
e. More than one of the above

6. The formation of monozygotic twins in humans is possible because mammals, like other vertebrates, exhibit
a. mosaic development.
b. regulated development.
c. complete cleavage.
d. superficial cleavage.
e. capacitation.

7. Which of the following statements about the mammalian blastocyst is *not* true?
   a. The trophoblast gives rise to the embryo proper.
   b. Maternal genes are expressed during cleavage.
   c. The blastocyst implants in the mother's uterus.
   d. Early mammalian development is slow.
   e. Tight junctions are common between the outer layer of cells in the blastocyst.

Activities (for answers and explanations, see page 386)

• In the following diagram of a frog blastula, label the blastoderm, blastocoel, and the poles of the vegetal and animal hemispheres.

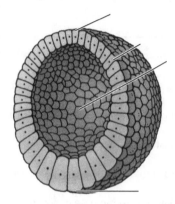

• Diagram the animal embryos formed by complete and incomplete cleavage for the 2-, 4-, and 8-cell stages of development. Explain why these embryos look so different from one another at the 8-cell stage.

GASTRULATION: PRODUCING THE BODY PLAN • NEURULATION: INITIATING THE NERVOUS SYSTEM   (pages 890–897)

Key Concepts

1. *Gastrulation* results in the formation of a three-layered embryo called a *gastrula*. During gastrulation, the rate of cell division is less than during cleavage and the total volume of the embryo changes little, but massive cell movements may occur by processes such as *invagination*.

2. During gastrulation, large numbers of surface cells move into the interior of the embryo. This movement produces an outer germ layer called the *ectoderm*, an inner germ layer lining the gut called the *endoderm*, and a middle layer between the endoderm and ectoderm called the *mesoderm*.

3. Ectoderm gives rise to the epidermis of the skin and epidermal derivatives like hair, feathers, and scales, the lin-

ing of the oral and nasal cavities, sweat, oil, and milk glands, and the nervous system.

4. Endoderm produces the lining of the digestive tract, the respiratory passages, and most of the major internal organs.

5. Mesoderm forms the skeleton, muscles, circulatory system, heart, blood, gonads, kidneys, and the coverings of internal organs.

6. In the sea urchin, mesoderm arises from mesenchyme cells. *Primary mesenchyme* forms when cells from the vegetal pole of the last blastula break free, bulge into the blastocoel as their protein layers hydrate, and then actually migrate into the blastocoel. Invagination of surface cells from the vegetal pole starts at a spot called the *blastopore*, and continues until the primitive gut or *archenteron* has formed.

7. Cells at the tip of the archenteron then break free and enter the blastocoel to form *secondary mesenchyme*. Active shape changes of the cytoskeleton in secondary mesenchyme helps move the archenteron forward within the embryo until it touches the ectoderm, where the mouth will form. The blastopore eventually becomes the anus.

8. Sea urchin gastrulation illustrates the general pattern seen in *deuterostome* animals (echinoderms and vertebrates), where the anus forms first from the blastopore and the mouth appears secondarily. In the *protostomes* (annelid worms, arthropods, and mollusks), the blastopore gives rise to the mouth first and the anus forms secondarily.

9. In frogs, invagination begins at a point just below the gray crescent called the *dorsal lip of the blastopore*. Sheets of surface cells move toward the blastopore in a process called *epiboly*, then stream inward to form the archenteron (which eventually displaces the blastocoel). During this movement the dorsal lip expands to become a complete circle. Additional cells pour over the dorsal lip to become new mesoderm.

10. By the end of gastrulation the blastopore has narrowed. The yolk plug, which is composed of vegetal hemisphere cells located ventral to the blastopore and displaced by the expanding archenteron, has become internalized as part of the endoderm. All three germ layers have now been determined for the specific tissue types they will eventually become. The gastrula also exhibits both dorsal–ventral and anterior–posterior organization.

11. *Hans Spemann* found that the gray crescent region of frog eggs contains cytoplasmic factors essential for normal gastrulation. If the zygote is made to divide such that only one of the new cells receives gray crescent material, only that cell will develop normally.

12. Spemann and *Hilde Mangold* conducted delicate embryonic transplants to test the hypothesis that the fate of cells is determined by gastrulation. In early gastrulas, they found that transplanted pieces always took on the characteristics of the area to which they were moved, regardless of their "normal" fate. Transplants from late gastrulas, however, became structures predicted by their fate maps, regardless of where they were trans-

planted. Thus, the process of gastrulation does determine the fate of cells.

13. Additional experiments by Mangold and Spemann using the dorsal lip of the blastopore showed that the fate of tissues is frequently determined by interaction with other tissues in the embryo through a process called *embryonic induction*, and that the dorsal lip was the *primary embryonic organizer*.

14. Developmental biologists know that embryonic organization involves a cascade of inductive tissue interactions, but the actual molecular mechanisms of determination and induction are essentially unknown and remain an active area of research.

15. In birds and reptiles, the large nutrient supply constrains blastula development to a small disc of cells on the surface of the yolk. In chickens, the upper layer of cells in the blastula is called the *epiblast*, from which all embryonic tissues will develop.

16. The posterior epiblast thickens at the start of gastrulation, eventually forming the *primitive streak* as cells migrate to the midline of the embryo. A depression called the *primitive groove* develops along the length of the streak and serves as the blastopore. Through it cells migrate into the blastocoel to become endoderm and mesoderm.

17. No archenteron forms in bird embryos, but endoderm and mesoderm still migrate to form the gut and other structures. The anterior end of the primitive groove thickens to form *Hensen's node*, which like the dorsal lip in frogs determines the fate of cells that pass over it.

18. The common evolutionary connection to reptiles that is shared by birds and mammals is reflected in their similar embryological development. The mammalian trophoblast is equivalent to the avian epiblast. Cells from the inner mass of a mammal embryo separate to yield the *hypoblast*, which forms a layer surrounding the blastocoel. Hypoblast cells contribute to the formation of extraembryonic membranes. The remaining inner mass of cells becomes the epiblast, the upper layer of which forms the amnion that will surround the developing embryo. Gastrulation of the epiblast then proceeds as in birds.

19. Formation of the vertebrate nervous system occurs during *neurulation*, the developmental stage that follows gastrulation. An internal tubular structure forms from a sheet of initially external ectoderm. Whereas the location of the blastopore determines the anterior–posterior axis, neurulation forms the bilateral axis.

20. A cartilagenous *notochord* develops from mesoderm called *chordomesoderm*, which lies closest to the embryonic midline. The notochord gives structural support to the forming embryo.

21. A *neural plate* of thickened ectoderm forms on the dorsal surface of the gastrula above the notochord. A groove then develops and the edges of the plate begin to fold. The plate rolls up to produce the cylindrical *neural tube*, which becomes covered by the neural folds. The spinal cord is derived from the neural tube, with the most anterior portion becoming the brain. Improper neural tube development in humans can lead to such conditions as spinal bifida and anencephaly.

22. Mesoderm also forms lateral blocks of cells along the neural tube called *somites*. Some somite cells develop into muscles, while others become the vertebrae that make up the backbone.

Questions (for answers and explanations, see page 386)

1. Whereas the location of the _____ determines the anterior–posterior axis of the embryo, the _____ determines the embryo's bilateral symmetry.
   a. blastopore; neural plate
   b. hypoblast; blastopore
   c. blastopore; primary mesenchyme
   d. vegetal hemisphere; notochord
   e. primitive streak; epiblast

2. Which one of the following structures is *not* formed as a result of gastrulation in frogs?
   a. Yolk plug
   b. Archenteron
   c. Anus
   d. Gray crescent
   e. Endoderm

3. For each of the following organisms, select from the choices listed on the right the structure or location where invagination begins during formation of the gastrula.

_____ Sea urchin	a.	Primitive streak
_____ Frog	b.	Blastopore
_____ Bird	c.	Archenteron
	d.	Yolk plug
	e.	Dorsal lip of blastopore

4. Which of the following statements about development in a cow is *false*?
   a. It follows the deuterostome plan.
   b. A hypoblast, epiblast, and amnion are present in the embryo.
   c. There is no notochord.
   d. Neurulation initiates development of the spinal cord and brain.
   e. Somites give rise to muscle and vertebrae.

5. The heart, ovaries, and skeleton of a human female have which of the following in common?
   a. They all form from embryonic endoderm.
   b. They all form from embryonic ectoderm.
   c. They all form from embryonic mesoderm.
   d. They all form from the archenteron.
   e. They all form from the neural tube.

6. You stain cells at the *ventral* lip of a frog blastopore with an intense dye. These cells will most likely become part of the
   a. gut tube.
   b. nervous system.
   c. notochord.
   d. ectoderm.
   e. yolk plug.

7. Why did Hans Spemann call the dorsal lip of the blastopore the embryonic organizer?

    a. It is the point where gastrulation begins.

    b. It becomes part of the notochord.

    c. It becomes part of the nervous system.

    d. It leads to the the establishment of the embryonic axes.

    e. It induces a secondary site of gastrulation if transplanted to another embryo.

Activities (for answers and explanations, see page 387)

- In the following diagram of a sea urchin gastrula, label the ectoderm, endoderm, primary mesenchyme, secondary mesenchyme, blastopore, blastocoel, and archenteron.

- Label the listed structures in the following longitudinal diagram of a late frog gastrula: blastopore, area where brain will form, yolk plug, notochord, neural plate, archenteron, ectoderm, mesoderm, and endoderm. Also, indicate the (A) anterior, (P) posterior, (D) dorsal, and (V) ventral ends of the embryo.

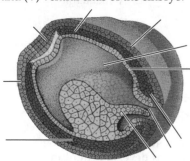

EXTRAEMBRYONIC MEMBRANES • DEVELOPMENT CONTINUES THROUGHOUT LIFE (pages 897–900)

Key Concepts

1. Four *extraembryonic membranes* surround the developing embryos of reptiles, birds, and mammals. These membranes help regulate the exchange of nutrients and waste between the embryo and its environment.

2. Extraembryonic membranes are key components of the *amniotic egg*, a major adaptation for terrestrial life in the vertebrate lineage. The development of the amniotic egg was one of the most significant events in the evolutionary transition from amphibians to reptiles because it freed terrestrial vertebrates from the need for aquatic reproduction.

3. In bird eggs, the *yolk sac* is an extraembryonic membrane that forms around the large nutrient yolk mass. The yolk sac develops from endodermal cells incorporated into the hypoblast. Although the sac is continuous with the gut, yolk does not pass directly into the gut. Instead, blood vessels lining the outer surface of the sac absorb nutrients and deliver them to the embryo.

4. Another extraembryonic membrane, the *amnion*, surrounds the entire embryo and encases it in a fluid-filled chamber (the *amniotic cavity*) that prevents desiccation, physically cushions the developing animal, and buffers it from environmental changes. The amnion develops from a union of the ectoderm and mesoderm.

5. A continuous extraembryonic membrane called the *chorion* forms where the expanding mesoderm and ectoderm meet below the yolk. The chorion serves as the inner lining of the egg shell and regulates gas exchange with the external environment.

6. The *allantois* is a fourth extraembryonic membrane. Arising from the endoderm, it serves as a sac for the deposition of metabolic wastes produced as the embryo grows.

7. Although viviparous, mammals also possess extraembryonic membranes. The first one to develop is the trophoblast, which surrounds the mass of embryonic cells. After emerging from the zona pellucida, the trophoblast interacts directly with the endometrial lining of the uterus to facilitate implantation of the embryo and initial development of the placenta.

8. The trophoblast sends out many small projections, called villi, to increase the surface area for contact with the mother's blood vessels. Since there is no yolk sac in mammals, this membrane becomes incorporated into the trophoblast to yield the chorion. It is the interaction between the chorion and the uterine lining that ultimately produces the placenta.

9. Coincident with formation of the chorion, the epiblast splits off cells that give rise to the amnion. Eventually an amniotic cavity completely encloses the embryo. An allantois also forms in mammals, but its importance is species-specific and depends on the efficiency of waste exchange across the placenta. In humans, allantoic tissues contribute primarily to the formation of the umbilical cord.

10. During human development, embryonic cells slough off and accumulate in the amnionic fluid. *Amniocentesis* is a medical procedure used to sample this fluid and culture embryonic cells to check for the presence of metabolic and genetic disorders.

11. Since amniocentesis cannot produce reliable results prior to the 16th week of pregnancy (too late for some abortions to be performed), an alternative method of sampling younger embryonic cells called *chorionic villus sampling* has been developed. CVS samples cells directly from the chorion, and the results can be available as early as the eighth week of pregnancy.

12. Growth, an irreversible increase in size, characterizes the development of most animals and continues well past the embryonic stage, usually throughout life.

13. Many mammals, including humans, show an S-shaped growth curve. There are growth spurts at infancy and puberty, interspersed with periods of little growth during childhood and adulthood.

14. Other animals, notably many species of arthropods, grow throughout life. However, since growth is limited by an exoskeleton, the growth curve appears as a step-function punctuated by periods of molting.

15. Overall growth is accomplished without the need for structural changes in organs because developmental influences among neighboring cells continue throughout life. When such control mechanisms are lost, a common end result is cancer, where unrestricted cell growth occurs in inappropriate areas.

16. Determination and differentiation of cells also continues throughout life. Examples from humans include the production of new blood cell types by undifferentiated bone marrow cells, and the body's response to injury, where cells in the area of the wound can dedifferentiate and then regenerate new cell types according to local needs.

Questions (for answers and explanations, see page 387)

1. The extraembryonic membranes known as the chorion and allantois are found in all but one of the following animals. Select the *exception*.
   a. Trout
   b. Bear
   c. Chicken
   d. Alligator
   e. Human

2. During its development, the human embryo is contained within a fluid-filled chamber bounded by the membranous
   a. yolk sac.
   b. endometrium.
   c. amnion.
   d. chorion.
   e. allantois.

3. Which of the following extraembryonic structures in a chicken egg is responsible for the physiological function of gas exchange with the external environment?
   a. Trophoblast
   b. Yolk sac
   c. Amnion
   d. Chorion
   e. Allantois

4. As the maternal nutrients that cross the placenta are to a developing mammal embryo, so the contents of the _____ is to a developing reptile or bird embryo.
   a. egg shell
   b. yolk sac
   c. amnion
   d. chorion
   e. allantois

5. Which of the following statements about animal growth is *true*?
   a. It begins early in development and typically continues throughout life.
   b. It can be constant or show periods of slowed and then accelerated activity.
   c. It is usually accomplished without the need for structural change in organs.
   d. When control of growth is lost, pathologies such as cancer can result.
   e. All of the above

Activities (for answers and explanations, see page 387)

- Label the following diagram of a generalized amniotic egg. Why was the development of this structure so important to the evolution of terrestrial vertebrates?

- Draw an idealized growth curve for a fast growing animal such as a mouse.

- Draw an idealized growth curve for an insect species with four episodes of molting.

Integrative Questions (for answers and explanations,
*   see page 387)*

1. Which of the following is *not* considered a part of animal development?
   a. Fertilization
   b. Cleavage
   c. Gastrulation
   d. Neurulation
   e. Amniocentesis

2. Number the following processes to show their position in the normal sequence of animal development.

   _____ Neurulation

   _____ Cleavage

   _____ Fertilization

   _____ Invagination at the blastopore

   _____ Blastula formation

   _____ Mesoderm formation

3. Match the organisms on the right with all of the appropriate developmental descriptors on the left.

   _____ Trophoblast embryo   a. Chick

   _____ Spherical blastocoel   b. Sea urchin

   _____ Anus forms first   c. Frog

   _____ Vitelline envelope   d. Human

   _____ Gray crescent   e. Fruit fly

   _____ Primitive streak

   _____ Continuous growth curve

41

Neurons and Nervous Systems

NERVOUS SYSTEMS: CELLS AND FUNCTIONS (pages 903–907)

Key Concepts

1. The function of animal *nervous systems* is to detect and process environmental information, and then convert this information to signals that control muscle response. Nervous systems are notable for their ability to perform many different tasks simultaneously.

2. *Neurons* are the functional cellular units of nervous systems. They perform the basic functions of receiving, encoding, and transmitting information.

3. Information transfer in nervous systems is both electrical and chemical. Electrical signals are used to transfer information within (and occasionally between) nerve cells, while chemical signals are commonly used to communicate between cells.

4. *Sensors* (e.g., eyes and ears) convert stimuli from within or outside the animal into messages that the nervous system can use to cause behavioral or physiological responses by way of *effectors* (e.g., muscles and glands).

5. The structure and complexity of nervous systems varies among species, and generally correlates positively with mobility and the need to process larger amounts of information. As a rule, larger and more complex nervous systems permit more sophisticated functional control.

6. The less structurally complex animals, such as cnidarians, have "simple" nervous systems composed of nerve nets. More complex nervous systems (various worm phyla and arthropods) have centralized neural tracts

with clusters of neurons called *ganglia*, which often serve the function of primitive brains. Such moderately complex nervous systems are capable of producing remarkable control and behavior. Indeed, much of our knowledge about basic nerve cell function comes from studies of these animals.

7. In humans and other vertebrates, the enormous collection of neurons called the *brain* and *spinal cord* make up the *central nervous system*. The CNS connects with the *peripheral nervous system* (PNS), which transfers neural information to and from the central system.

8. For vertebrates of comparable body size, the total number of neurons, development of certain brain areas, and overall structural complexity of the brain positively correlate an animal's behavioral repertoire and ability to learn and remember.

9. Neurons have excitable plasma membranes that generate electrical signals called *nerve impulses*. Neurons typically consist of four regions: a cell body, dendrites, an axon, and axon terminals.

10. The *cell body* of a neuron contains the nucleus and most of the cell's organelles. A number of long extensions, or processes, emerge from the cell body.

11. *Dendrites* are the most abundant neuronal processes. They are usually fairly short and collect incoming signals from other neurons.

12. *Axons* are typically elongated processes that connect the cell body with its target cell(s). There is usually only one axon per neuron, and its length may range from less than a millimeter to many meters, depending on function and species. Axons conduct their electrical signals at speeds of many meters per second, making the nervous system a rapid communication system compared to the chemically mediated hormonal system.

13. *Axon terminals* are swellings found at the end of tiny processes that branch off the ends of each axon. They are storage sites for molecules called *neurotransmitters*, which are used to chemically transfer information from the neuron to its target cell.

14. Neurochemical communication occurs at the *synapse*, a 25 nm intercellular gap (cleft) between the axon terminal and an adjacent cell membrane. Neurotransmitter is released from the *presynaptic membrane*, diffuses across the cleft, and binds to *postsynaptic membrane* receptors on the target cell. By summing individual synaptic events from many inputs, neurons can rapidly integrate large quantities of information.

15. *Glial cells* are the other cell type in nervous tissue. They differ from neurons by lacking the ability to generate and conduct electrical impulses, although some do communicate electrically through specialized contacts called *gap junctions*. Glial cells serve a number of accessory functions for neurons, including physical support and spatial orientation during development, electrical insulation, nourishment, and consumption of foreign particles or cell debris.

16. The insulative role of certain glial cells is particularly important in vertebrates. Both *Schwann cells* in the PNS and *oligodendrocytes* in the CNS produce *myelin*, a fatty substance that wraps around axons to speed the conduction of nerve impulses.

17. *Astrocytes* are specialized star-shaped glial cells that create the *blood–brain barrier*. The smallest blood vessels supplying the brain are covered by astrocytes, which provide protection against leakage of toxic substances from the blood that might cause damage to sensitive brain cells.

18. Since the number of brain cells is fixed at birth and damaged cells cannot be regenerated, the blood–brain barrier helps reduce the loss of these important cells. The barrier is not infallible, however, because lipid-soluble substances such as anesthetics and alcohol can penetrate the blood–brain barrier and produce toxic effects.

19. Even the simplest animal nervous systems have dozens of neurons and hundreds of synapses. In the complex nervous systems of vertebrates, there are billions of neurons and many times that number of synapses. These multitudes of cells and connections are grouped into well-defined neural circuits that control specific functions.

Questions (for answers and explanations, see page 387)

1. The portion of a neuron that stores and releases chemicals used to communicate with other cells is called the
   a. dendrite.
   b. cell body.
   c. axon.
   d. axon terminal.
   e. synapse.

2. If you randomly make a cross section through a neuron somewhere along the length of its structure, your cut will most likely pass through the
   a. dendrite.
   b. cell body.
   c. axon.
   d. axon terminal.
   e. synapse.

3. An animal is highly mobile and performs complex behaviors. Which of the following would you likely *not* expect to find as part of its nervous system?
   a. Neurons
   b. Glial cells
   c. Ganglia
   d. Nerve net
   e. Brain

4. Imagine that in front of you are the brains of two vertebrates of approximately equal body size. You are told that one is from a fish, and the other from a mammal. What is a reliable measure that you can use to correctly identify which brain belongs to which animal?
   a. Overall brain size
   b. Development of certain brain areas
   c. Total number of brain cells
   d. Complexity of neural connections
   e. All of the above

5. Most toxic substances are prevented from entering the brain by the blood–brain barrier. The cells responsible for creating this barrier are called _____ and belong to a type of neural tissue called _____ .
    a. dendrites; glial cells.
    b. oligodendrocytes; myelin.
    c. astrocytes; glial cells.
    d. axons; Schwann cells.
    e. ganglia; synapses.

6. A particular disease of the nervous system specifically involves chemical synapses in the area where neurotransmitter is stored and released. In other words, this disease affects the
    a. axon terminals of the presynaptic cell.
    b. axon terminals of the postsynaptic cell.
    c. cell bodies of both the pre- and postsynaptic cells.
    d. dendrites of the postsynaptic cell.
    e. axon of the presynaptic cell.

Activities (for answers and explanations, see page 388)

- On the neuron diagram that follows, label the four basic anatomical regions and briefly describe the function of each.

- Refer to Figure 41.1 in your textbook. Explain how the differences in nervous system structure between a sea anemone and a squid can produce the differences in behavioral complexity and overall body control indicated for these species.

NEURONS: GENERATING AND CONDUCTING NERVE IMPULSES (pages 907–916)

Key Concepts

1. A neuron's *membrane potential* is the electric charge that exists across its plasma membrane. When the cell is inactive, this charge is known as the *resting potential*. At rest most neurons are about 60 millivolts (mV) more negative on the inside of the cell.

2. *Action potentials* represent a rapid reversal of the resting potential in response to an electrical, chemical, or physical stimulus. During this event the inside of the cell becomes temporarily positive with respect to the outside. Action potentials last for only 1 to 2 milliseconds and sweep down the length of a neuron, providing the basis for electrical communication.

3. *Voltage* (potential) is the tendency for electrical charges to move between two points, such as the poles of a battery or opposite sides of a membrane. Electric *current* is caused by the flow of electrons between areas of opposite charge. In neurons, current is caused by the movement of charged ions such as Na^+, K^+, Cl^-, and Ca^{2+} across the membrane.

4. The hydrophobic properties of phospholipid bilayers prevent charged ions from freely passing through membranes. Ions must therefore pass through the neuron membrane via specialized protein *pumps* or *ion channels*.

5. Pumps move charged ions across membranes against the ion's concentration gradient. This is an energy-requiring process. In neurons, the sodium–potassium pump is the primary pumping mechanism.

6. The sodium–potassium pump expels three Na^+ ions from the cell for every two K^+ ions that enter. This exchange works to maintain high K^+ concentrations inside the cell and high Na^+ concentrations outside. Although the differential ion concentrations produced by the pump influence membrane potential, they do not create the resting potential.

7. The resting potential actually results from the interaction of charged ions with specific ion channels. These channels selectively permit charged ions to pass in either direction across the membrane, but always with net movement down the concentration gradient for a given ion.

8. Many ion channels can change their shape, which either inhibits or permits ion flow; they are said to be gated. *Voltage-gated* channels alter their shape in response to changes in membrane potential. *Chemically-gated* channels change shape when they bind appropriate signaling molecules.

9. Open potassium channels are more abundant in neuron membranes than are open sodium channels, and K^+ ions therefore diffuse out of the cell more rapidly than Na^+ diffuses in. This produces an excess negative charge on the inside of the cell, due mostly to the charges on chloride ions and protein molecules that remain. When the voltage difference becomes large enough (about –60 mV for most neurons), further K^+ diffusion stops and the resting potential is established.

10. The resting potential (a voltage difference across the membrane) means that the neuron is *polarized*. If the inside of a neuron becomes more negative than resting, the membrane is said to be *hyperpolarized*. If the inside becomes less negative (i.e., more positive), the membrane has *depolarized*. Whether hyperpolarization or depolarization takes place depends on the action of specific gated ion channels.

11. Hyperpolarization typically results from the opening of gated potassium channels. More K⁺ ions diffuse out of the cell (down their concentration gradient), thus increasing the negative charge on the inside of the membrane. Gated chloride channels can also causing hyperpolarization, allowing Cl⁻ ions to flow into the cell. Depolarization usually occurs when gated sodium channels open, allowing Na⁺ ions to enter the cell where their concentration is lower. This reduces the magnitude of the negative charge on the inside of the membrane, and may even reverse it to a positive charge.

12. The coordination of gated ion channels producing depolarization and then hyperpolarization forms the basis of action potentials. Because membranes are inherently poor conductors of electricity, the action potential must be continuously regenerated in successive membrane patches to successfully pass down an axon.

13. Action potentials last only 1 to 2 msec. They are commonly measured by inserting a microelectrode into a neuron and comparing its readings with one placed outside the cell. When referenced to the external electrode, an action potential first involves a rapid depolarization from the resting potential of –60 mV to a peak reversal state of +50 mV. This is followed by a slower hyperpolarization that eventually restores the original resting potential.

14. If measured at successive locations along the length of the axon, the action potential follows exactly the same series of events at each location. Thus, action potentials are an all-or-nothing, self-regenerating phenomenon.

15. The depolarization phase of an action potential is due to the rapid opening of gated sodium channels. When a threshold voltage more positive than the resting potential is reached, all the sodium gates in a patch of membrane open. Because the action of the sodium–potassium pump has produced a high concentration of Na⁺ ions on the outside of the cell, sodium rushes into the cell. In less than a millisecond, the membrane potential is reversed to the peak positive voltage of the action potential.

16. At the peak of the action potential spike, gated sodium channels close and remain closed for at least a few milliseconds, long enough for the normal leakage of K⁺ ions out of the cell to restore the resting potential. In many neurons, voltage-gated potassium channels also open, allowing K⁺ ions to rush out of the cell down their concentration gradient. This flow hyperpolarizes (i.e., repolarizes) the membrane. As the membrane nears its resting potential, increasing numbers of the potassium channels close until the resting potential is restored.

17. A *refractory period*, during which no new action potentials can be produced, follows immediately after an action potential. This effect is explained by assuming the presence of two types of voltage-gated sodium channels: activation and inactivation gates. Activation gates open and inactivation gates close at the onset of an action potential, but because the activation gates respond faster, the membrane depolarizes. Eventually the activation gates switch back to their closed condition, halting the influx of Na⁺ ions. Meanwhile, the inac-

tivation gates remain closed for several milliseconds before spontaneously reopening, thus rendering the membrane temporarily incapable of generating a new action potential.

18. The concentration difference of Na⁺ ions across the neuron membrane represents the "battery" that powers action potentials. Because few ions must move for an action potential to occur (relative to the available ion pools inside and outside the cell), the battery is not easily run down. The sodium–potassium pump can easily restore ionic balance even in the face of a hundred or more action potentials per second.

19. Action potentials propagate along axons because the current flow produced by ion movements stimulates the depolarization of adjacent membrane patches. The current associated with an action potential always brings adjacent patches to the threshold voltage, and so the action potential is self-regenerating. Action potentials normally propagate in only one direction along an axon since the refractory period experienced by the previously-stimulated patch of membrane prevents depolarization in the reverse direction.

20. Axon size is one factor that determines the conduction velocity of action potentials. For invertebrates such as squids or insects, the diameter of the axon is positively correlated with the speed of action potentials. Neurons involved in quick-response behaviors such as escape therefore tend to have large-diameter axons.

21. There are anatomical limits on the absolute diameter that axons can achieve. In large animals such as vertebrates, increased conduction velocity is achieved not by increasing axon diameter, but by surrounding axons with *myelin*. This electrically insulating material, manufactured by Schwann cells, gives the shiny appearance to white matter in the central nervous system.

22. The myelin cover along an axon is not continuous. It is absent at regular intervals, exposing the bare axon membrane at sites known as *nodes of Ranvier*. Action potentials are generated only at these nodes, where ions can actually move across the neuron membrane. Current flows within the axon between successive nodes to stimulate a new action potential, giving the appearance that the regenerating action potential is jumping from one node to the next. Such *saltatory conduction* can increase action potential conduction velocities to a maximum of 120 m/s.

23. In humans and other vertebrates, high-speed myelinated axons are part of the sensory systems that detect sharp pain. Sensory axons that transmit information about temperature or aching pain tend to be slow-speed and unmyelinated.

24. An important technique developed in the 1980s for studying neuron function is *patch clamping*. Patch clamps make it possible to study ion movements through individual ion channels, providing insights into the molecular basis of membrane electrical properties that can not be achieved with traditional microelectrode recordings.

Questions (for answers and explanations, see page 388)

1. The rapid depolarization of a neuron during the first half of an action potential is due to the
   a. exit of K^+ ions from the cell through gated potassium channels.
   b. rapid reversal of ion concentration caused by the action of the sodium–potassium pump.
   c. movement of both Na^+ and K^+ ions through appropriate open channels.
   d. entry of Na^+ ions into the cell through gated sodium channels.
   e. none of the above

2. One microelectrode is inserted into a healthy neuron and another is placed outside the cell. If the neuron is then hyperpolarized by a chemical stimulus, what will happen to the membrane potential recorded by the electrodes?
   a. It will become more positive.
   b. It will become more negative.
   c. It will first become more positive, then return to normal.
   d. It will not change.
   e. Cannot determine from the information given

3. If a chemical is applied to a neuron that inactivates all voltage-gated potassium channels, and then immediately thereafter the resting membrane potential is depolarized beyond threshold, which of the following will occur?
   a. A normal action potential will be generated.
   b. The neuron will hyperpolarize.
   c. Open potassium channels will compensate for the activity of the voltage-gated channels.
   d. The cell will depolarize, but may not repolarize to the resting membrane potential.
   e. K^+ ions will diffuse into the cell in large numbers.

4. The refractory period of a neuron results from
   a. the shutting down of the sodium–potassium pump.
   b. activation of voltage-gated potassium channels.
   c. closing of inactivated voltage-gated sodium channels.
   d. activation of membrane receptors on the neuron surface.
   e. the inability of charged ions to pass through the myelin covering the axon.

5. Which of the following statements about the behavior of action potentials or neurons is *false*?
   a. In ants, small-diameter axons have higher conduction velocities than large-diameter axons.
   b. In humans, myelinated axons have higher conduction velocities than unmyelinated axons.
   c. Current flow inside the axon is an important mechanism in saltatory conduction.
   d. In the myelinated axons of fish, action potentials are only generated at the nodes of Ranvier.
   e. Refractory periods occur in both myelinated and unmyelinated neurons.

6. The unmyelinated axon of a vertebrate neuron
   a. could possibly be associated with a slow pain sensor.
   b. conducts action potentials more slowly than a myelinated axon.

c. has gated ion channels all along its length.
d. lacks nodes of Ranvier.
e. All of the above

Activities (for answers and explanations, see page 388)

- On the diagram of an action potential shown below, properly label the axes of the graph and indicate the parts of the trace where the gated sodium channels and gated potassium channels are mostly open. Also indicate the resting membrane potential and the apparent threshold voltage for this neuron.

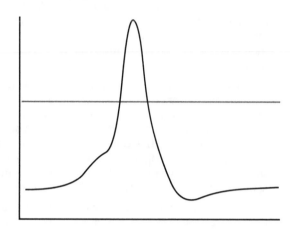

- The diagram below depicts a portion of membrane from a neuron. Explain the role of the various ions, channels, and pumps shown in creating the resting membrane potential of a neuron.

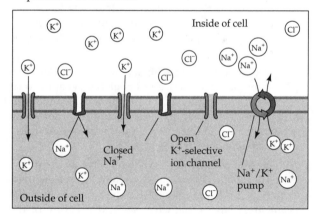

NEURONS, SYNAPSES, AND COMMUNICATION
• NEURONS IN CIRCUITS (PAGES 916–924)

Key Concepts

1. Neurons interact with neurons and other target cells at *synapses*. A one-way flow of information occurs at chemical synapses from *presynaptic membrane* to *postsynaptic membrane*.

2. *Neurotransmitters* are the signaling molecules used at chemical synapses. They are stored within vesicles located in the axon terminal of the presynaptic cell, and released into the *synaptic cleft* upon arrival of an action potential. Neurotransmitters bind to receptors on the postsynaptic membrane, where they cause specific chemically-gated ion channels to open or close and thus alter membrane polarization.

3. *Acetylcholine* is a common vertebrate neurotransmitter used at modified synapses known as *motor end plates*, which form at the junctions of motor neuron axons and skeletal muscle (a *neuromuscular junction*). Motor end plates have highly convoluted surfaces to maximize the area for acetylcholine receptor sites. Relatively few action potentials can thus stimulate the contraction of a muscle fiber.

4. The release of neurotransmitter is mediated by voltage-gated calcium channels, which occur at no other location in neurons except the synapse. Action potentials cause calcium channels to open, flooding the axon terminal with Ca^{2+} ions. This causes the synaptic vesicles to fuse with the presynaptic membrane and release their contents into the synaptic cleft. The neurotransmitter then moves to the postsynaptic cell by simple diffusion.

5. Postsynaptic cell membranes, such as those found at motor end plates, dendrites, and most regions of nerve cell bodies, have few voltage-gated sodium channels. They are thus unable to produce action potentials. Instead, the chemically-gated channels at these sites perturb the postsynaptic cell's resting potential just enough that the stimulation spreads across a portion of the postsynaptic cell. If the stimulus is sufficiently large, this perturbation reaches an area containing more voltage-gated sodium channels where an action potential can be generated.

6. Each action potential arriving at a neuromuscular junction stimulates the release of about 100 synaptic vesicles, each containing some 10,000 neurotransmitter molecules. This quantity is sufficient to generate an action potential in the postsynaptic membrane and produce a muscle fiber twitch.

7. Although some synapses are always excitatory (e.g., the motor end plate), others cause an inhibition of the postsynaptic cell. *Excitatory postsynaptic potentials* (EPSPs) occur at synapses that cause a depolarization of the postsynaptic membrane, making it easier for an action potential to occur. *Inhibitory postsynaptic potentials* (IPSPs) are generated at synapses that produce hyperpolarizations of the postsynaptic cell, making action potentials less likely to occur.

8. Whether a synapse is excitatory or inhibitory depends on the chemical nature of the postsynaptic receptors and not on the neurotransmitter itself. Thus, the same neurotransmitter can be both inhibitory (stimulates gated potassium channels) and excitatory (stimulates gated sodium channels), depending on its location within the nervous system.

9. The decision-making process of an individual neuron is accomplished by summing all of its excitatory and inhibitory inputs, and using this information to vary the rate at which it produces action potentials. This summation and integration of incoming information occurs in both space and time. It takes place at the *axon hillock*, where the axon emerges from the cell body.

10. High concentrations of voltage-gated sodium channels occur at the axon hillock, making it the first place on the neuron cell body where the spreading EPSPs and IPSPs encounter membrane that is capable of generating action potentials. Because of the decrease in intensity of postsynaptic potentials that occurs over distance, synapses that make contact with the postsynaptic cell nearer the axon hillock have more influence on whether the cell fires an action potential.

11. Many chemical synapses are fast-acting, responding within a few milliseconds of neurotransmitter contact. Others act more slowly, taking hundreds of milliseconds to respond. This occurs because the binding of neurotransmitter to receptor causes the activation of second messenger systems involving cAMP. Chemically mediated slow synapses are involved in the opening of ion channels, the control of membrane pumps, enzyme activation, and the induction of gene expression.

12. Some synapses form between axon terminals of one neuron and the axon of another. These synapses influence the amount of neurotransmitter released from the presynaptic cell. Depending on their effect, such influences are either termed *presynaptic excitation* or *presynaptic inhibition*.

13. Electrical synapses, or gap junctions, also occur between some neurons. Here the neurons have an intercellular gap of only 2 to 3 nanometers and are physically linked by connexons. Electrical currents or chemical ions actually pass through the connexons, permitting extremely rapid transmission in either direction.

14. Although very fast, gap junctions are not common in vertebrate nervous systems because they do not allow for summation, they limit the number of synaptic contacts that can be made with any given neuron, and they cannot be inhibitory. All of these features run counter to the high degree of integration and precise control characteristic of complex nervous systems.

15. More than 25 vertebrate neurotransmitter substances are known, but none are better understood than acetylcholine, which functions at all the voluntary (skeletal) neuromuscular synapses in the peripheral nervous system of vertebrates. Both acetylcholine and epinephrine control the peripheral nervous system's contact with smooth (involuntary) muscle, and each plays a role in the central nervous system as well. However, neither is the dominant neurotransmitter molecule of the brain.

16. The primary neurotransmitters of the vertebrate brain are simple amino acid substances such as *glutamic acid* and *aspartic acid*, which are excitatory, and *glycine* and *gamma-aminobutyric acid* (GABA), which are inhibitory. Amino acid derivatives such as *dopamine, norepinephrine,* and *serotonin* are also important to brain function, along with a number of peptide molecules. Recent evidence also indicates that the gases carbon monoxide and nitric oxide are chemical messengers between some neurons.

17. Neuropharmacology is the study and development of drugs that affect the nervous system. The action of acetylcholine is a good example of how neurotransmitters and related drugs can have powerful and divergent effects depending on the type of receptors present in their target tissues.

18. Within the central nervous system *nicotinic* acetylcholine receptors are excitatory, while *muscarinic* receptors are inhibitory. Chemical analogs of acetylcholine (e.g., nicotine from cigarettes or compounds found in certain toxic mushrooms) can have dramatic influences on nervous system activity when they bind to these receptors causing physiological and behavioral responses.

19. *Curare* is another natural substance that influences acetylcholine activity. Curare binds to and inactivates acetylcholine receptors in skeletal muscle, rendering the muscle incapable of contraction. It is used on hunting arrows by some native peoples in South America, and medicinally to treat severe muscle spasms. Another medicinal compound called *atropine* blocks the muscarinic receptors in heart and other muscle tissue. It can be used to increase heart rate, dilate the pupil of the eye, or decrease gut spasms.

20. An important part of synaptic control is the need to inactivate neurotransmitters and to clear the synapse of old transmitters before new ones arrive. Three mechanisms are used to inactivate neurotransmitters.

21. The first involves the actual chemical destruction of the neurotransmitter while still in the synaptic cleft. For example, acetylcholine is inactivated by the enzyme *acetylcholinesterase*. The absence of acetylcholinesterase leads to uncontrolled muscle spasms, and perhaps death, due to continuous stimulation of the neuromuscular junction. Many potent nerve toxins and insecticides have acetylcholinesterase inhibitors as their active ingredient.

22. A second method is to simply allow the neurotransmitter to diffuse away from the synaptic cleft. A third approach is to actively reabsorb the neurotransmitter into the presynaptic cell, where it is metabolically degraded or repackaged into new synaptic vesicles.

23. Networks, or neuron circuits, can interact in complex ways to process and integrate information. *Monosynaptic reflexes* are the simplest forms of neural circuits. They have a single synapse that separates the incoming signal from the outgoing information. The involuntary knee-jerk response in humans is an example.

24. When the doctor's hammer taps the tendon covering the knee, stretch receptors made of *muscle spindles* stimulate action potentials in sensory neurons. Their axons enter the spinal cord to synapse on individual motor neurons. The motor neurons send action potentials through their axons. These action potentials exit the spinal cord and travel to the extensor muscle. The action potentials cause the extensor muscle to contract, extending the leg.

25. Although this reflex circuit is involuntary, signals reaching the brain also provide conscious information that the knee has been struck. This occurs through spinal *interneurons* whose cell bodies (located in the *gray matter*

of the spinal cord) are stimulated by synaptic input from branches of the entering sensory axons. The axons of interneurons travel up the spinal cord through the myelinated *white matter* to appropriate brain centers. Decisions are made about this input and then relayed back to the motor neurons. This system provides voluntary control of leg position.

26. *Polysynaptic reflexes* are also common in involuntary responses. In the knee-jerk example, at the same time that excitatory signals are sent to the extensor muscle, inhibitory signals are sent to the leg flexor by way of additional sensory input onto other motor neurons. This prevents the flexor from contracting, a motion that would oppose the action of the extensor.

27. More complex reflexes, such as escape responses, result from the layering of additional polysynaptic circuits. The initial control of these behaviors is, however, still orchestrated primarily within the spinal cord, and usually does not require conscious intervention through higher brain functions.

28. Only the most basic of neural circuits have been studied thoroughly. The neural mechanisms of complex activities such as reasoning, perception, and memory remain largely unknown. Nevertheless, patterns of neural activity associated with certain nervous system functions have been documented.

29. The neural basis of simple forms of learning has been studied in the sea hare *Aplysia*. This mollusk has a monosynaptic gill siphon withdrawal reflex that protects the delicate gill tissue from damage. *Habituation* of the withdrawal reflex to repeated touching of the siphon is achieved by decreasing the amount of neurotransmitter released at the axon terminals of presynaptic cells, thus reducing the level of response.

30. The *Aplysia* siphon withdrawal reflex can also undergo *sensitization* if mild electric shocks are applied to the animal's tail. In this case an interneuron releases serotonin onto the presynaptic cells, stimulating a second messenger system that causes calcium channels to remain open longer for any one action potential. This means that more neurotransmitter is released and a larger postsynaptic response occurs.

Activities (for answers and explanations, see page 388)

- Match the following chemicals with their appropriate functions in the vertebrate nervous system.

_____Curare	*a.* Inhibitory neurotransmitter in the brain
_____Acetylcholine	*b.* Stimulates fusion of synaptic vesicles with presynaptic membrane
_____Glycine	*c.* Blocks inhibitory effects of muscarinic receptors in heart muscle
_____Calcium	*d.* Inactivates acetylcholine receptors in skeletal muscle
_____Atropine	*e.* Has both excitatory and inhibitory influences

- The active ingredients in many nerve gases belong to a class of chemicals called anticholinesterases. Suggest a possible synaptic mechanism to explain how these chemicals can do such dreadful damage to an animal's nervous system.

- The following diagram represents a monosynaptic reflex loop. Label the sensory neuron, motor neuron, and effector in this simple neural circuit. Also indicate which part of the circuit is the sensory side, and which the motor side.

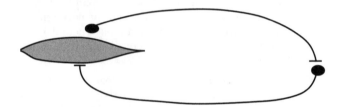

Questions (for answers and explanations, see page 388)

1. Certain neurotransmitters affect the behavior of post-synaptic neurons by
   a. altering their membrane potentials through the action of gated ion channels.
   b. stimulating the release of calcium into the postsynaptic cell.
   c. initiating the bonding of membrane proteins that form connexons.
   d. causing the active uptake of neurotransmitter molecules across the postsynaptic membrane.
   e. none of the above

2. Which of the following statements about the process of summation in a neuron is *false*?
   a. Slight perturbations of the membrane potential spread across the postsynaptic cell body.
   b. Axons that terminate closer to the axon hillock have more influence on the summation process.
   c. Summation consists essentially of comparing the total of excitatory and inhibitory postsynaptic inputs.
   d. The concentration of voltage gated sodium channels is highest in the dendrites of the postsynaptic cell.
   e. All of the above

3. All but one of the following chemicals are neurotransmitters that function in the human brain. Select the *exception*.
   a. Dopamine
   b. Glycine

c. Atropine
d. Glutamic acid
e. Serotonin

Refer to the following diagram when answering Questions 4 and 5.

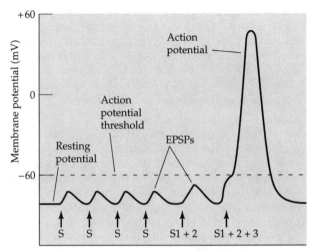

4. The neural phenomenon depicted by this diagram is best described as
   a. spatial summation.
   b. temporal summation.
   c. presynaptic excitation.
   d. presynaptic inhibition.
   e. synaptic transmission.

5. The electrical events labeled as EPSPs are the result of
   a. hyperpolarizations of the postsynaptic membrane.
   b. depolarizations of the postsynaptic membrane.
   c. hyperpolarizations of the presynaptic membrane.
   d. depolarizations of the presynaptic membrane.
   e. the action of anticholinesterases.

6. If you could selectively neutralize the active transport mechanisms at the axon terminals of a neuron, which of the following processes would you most significantly impact?
   a. Fusion of synaptic vesicles with the presynaptic membrane
   b. Uptake of released neurotransmitter by the presynaptic cell
   c. Operation of voltage-gated potassium channels
   d. Release of calcium within the postsynaptic cell
   e. Diffusion rate of neurotransmitter across the synaptic cleft

7. The one-way flow of information between presynaptic and postsynaptic cells is a characteristic of most kinds of neuron interactions. A notable exception to this rule is the
   a. neuromuscular junction.
   b. integration of postsynaptic potentials by the axon hillock.
   c. nicotinic receptor system of vertebrate neurons.
   d. motor end plate.
   e. gap junction.

8. Which of the following statements about the neural basis of the knee-jerk reflex in humans is *false*?
   a. The leg extensor is stimulated to contract by the motor neuron in this reflex loop.
   b. The only synapses between neurons involved in this loop occur in the spinal cord.
   c. The basic response is a polysynaptic circuit containing several interneurons.
   d. Conscious action can modify the basic reflex because of the action of interneurons.
   e. It is very simple by comparison to other neural circuits such as those involved in learning or memory.

9. In *Aplysia*, a difference between habituation and sensitization of the gill siphon withdrawal reflex is the
   a. number of neurons involved in the controlling circuit.
   b. amount of neurotransmitter released.
   c. number of synapses involved.
   d. behavioral response of the animal.
   e. all of the above

Integrative Questions (for answers and explanations, see page 389)

1. For a randomly selected brain cell taken from a human, which of the following would most likely be *true*?
   a. It would have a cell body.
   b. It would be sensitive to one or more neurotransmitters.
   c. It would contain synaptic vesicles in its axon terminals.
   d. It would have sodium-potassium pumps in its plasma membrane.
   e. All of the above

2. Put the following events in their correct sequence:
   a. Opening of calcium ion channels in the presynaptic membrane
   b. Inactivation of neurotransmitter in the synaptic cleft
   c. Arrival of an action potential at the axon terminal
   d. Release of neurotransmitter from the presynaptic cell
   e. Fusion of synaptic vesicles with the presynaptic membrane

3. If you electrically stimulate a resting neuron in the middle of its axon and cause it to fire an action potential, which of the following will happen?
   a. The action potential will propagate toward the axon terminal.
   b. The action potential will propagate toward the cell body.
   c. The action potential will propagate toward both the axon terminal and the cell body.
   d. The action potential will not propagate at all.
   e. The neuron will hyperpolarize.

4. The resting membrane potential for neuron A is –70 mV, while the resting potential for neuron B is –50 mV. The threshold voltage for the production of an action potential is –35 mV for both neurons. Which of the following statements is *false*?
   a. Neuron A must depolarize by 35 mV to reach the threshold voltage.
   b. Neuron B must hyperpolarize by 15 mV to reach the threshold voltage.
   c. The inside of both neurons is negatively charged with respect to the outside.
   d. A single EPSP received by neuron A would cause it to depolarize slightly.
   e. A single IPSP received by neuron B would cause it to hyperpolarize slightly.

42

Sensory Systems

CHAPTER LEARNING OBJECTIVES—after studying this chapter you should be able to:

❑ Appreciate why different animal species have unique sensory worlds.

❑ Define sensory transduction and explain the possible mechanisms that can lead to receptor, generator, and action potentials in various types of sensors.

❑ Know the three basic categories of biological sensors.

❑ Describe how sensation is related to the locations of CNS neurons that receive nerve impulses from a sensor.

❑ Discuss the concept of sensory adaptation, and explain why different types of sensors show different levels of adaptation.

❑ Describe the function of chemosensors, and explain the molecular mechanism for the transduction of chemical stimuli.

❑ Differentiate between olfaction and gustation, and discuss the role of pheromones in arthropod chemosensation.

❑ Explain the neural mechanisms for human smell and taste.

❑ Describe the function of mechanosensors, and some examples of the ways that mechanoreception is used in sensory systems.

❑ Discuss the types of mechanoreceptors present in human skin.

❑ Describe several sensory systems that employ hair cells for mechanoreception, and explain the role of hair cells in each of these sensors.

❑ Describe the parts of the mammalian ear and explain their functions.

❑ Discuss the structure and function of the mammalian cochlea, and explain how sound pitch is transduced by this organ.

❑ Differentiate between conduction deafness and nerve deafness in humans, and provide possible causes for each.

❑ Describe the function of photosensors, and provide some examples of the ways that photoreception is used by animals.

❑ Explain the molecular mechanism of photodetection by rhodopsin and the second messenger system that leads to receptor potentials in photoreceptors.

❑ Discuss the structure and function of the arthropod compound eye, and compare it with that of the vertebrate eye.

❑ Describe the process of visual accommodation in the vertebrate eye.

❑ Differentiate between the fovea and blind spot of the vertebrate eye and discuss the function of each.

❑ Explain the difference between a rod and a cone, discuss their functions, and describe their relative importance in nocturnal and diurnal animals.

❑ Name the five layers of neurons in the vertebrate eye, and describe the anatomical and functional relationships among these layers.

❑ Relate the function of photoreceptors and bipolar, ganglion, horizontal, and amacrine cells to the general process by which visual patterns are detected in the vertebrate eye.

SENSORS, SENSORY ORGANS, AND TRANSDUCTION
• CHEMOSENSORS: RESPONDING TO SPECIFIC MOLECULES (pages 927–934)

Key Concepts

1. The sensory world is different for each animal species. How the world "appears" to a given animal depends on the specific environmental stimuli that its sensory cells detect.

2. Sensors *transduce* (convert) various kinds of stimuli into action potentials, the universal language of nervous systems. Sensors are keyed to specific forms of energy such as chemicals, mechanical forces, or light, which can come from the external or internal environment.

3. All sensory information is coded as action potentials. The perceived sensation associated with these action potentials depends on where specifically in the central nervous system the sensory information is targeted. The phantom limb effect illustrates this point.

4. Not all sensory information communicated to the CNS reaches our conscious attention. Many internal body conditions are continuously monitored by the senses as part of homeostasis without our ever being consciously aware of them.

5. Specialized sensory organs that gather and focus stimuli may be associated with sensory cells. These organs facilitate the transduction process and help to collect, amplify, dampen, or filter the stimulus. Although the details

of transduction vary with different sensory cells, a general pattern is common to all types of sensors.

6. A stimulus activates receptor proteins on the sensor cell membrane and changes the membrane's ion permeability, either as a direct result of the receptor protein's opening or closing an ion channel, or because the protein sets into motion a series of intracellular events that indirectly affect the channels. These ion movements produce a *receptor potential*, the magnitude of which is directly proportional to stimulus intensity.

7. Receptor potentials are converted into action potentials in two ways. The first is for the sensor itself to generate an action potential. Receptor potentials in these sensors are called *generator potentials*, because they spread to a region of cell membrane containing voltage-gated sodium channels where an action potential is generated. Crayfish stretch sensors provide an example.

8. The second way is for the sensor to release neurotransmitter onto an adjacent neuron, which causes the production of an action potential in the second cell. Photosensors are good examples of this transduction mechanism. In both mechanisms, the magnitude of receptor or generator potentials determines the frequency of action potentials fired by the sensor or its associated neuron.

9. *Adaptation* occurs when sensors reduce their response to continuous stimulation. By adapting to background conditions, the sensor can respond to stimulus changes that represent new information. Different categories of sensors have different degrees of adaptation. Sensors that monitor critical parameters such as pain and balance do not adapt at all, or adapt more slowly than those that monitor less critical stimuli such as light and sound.

10. *Chemosensors* are responsible for detecting chemical stimuli that produce the senses of taste and smell, as well as monitoring internal conditions such as blood carbon dioxide levels. Chemosensation is the oldest and most ubiquitous of senses and can have powerful effects on behavior and physiology.

11. Arthropods use chemical signals called *pheromones* (external hormones) to attract mates. When a female silkworm moth secretes the pheromone bombykol, males detect the stimulus using sensors on their antennae and fly toward its source. The male's response is so acute that one molecule of bombykol can generate several action potentials in a single chemosensor. Arthropods may also have chemosensory hairs located on their bodies, especially the feet. These hairs contain sensor cells that detect food and other important chemical features of the environment.

12. *Olfaction* is the sense of smell. Vertebrate olfactory receptors are located on sensor cells embedded in epithelium that lines the nasal cavity. The dendrites (olfactory hairs) of these cells project from the lining and are covered by a thin layer of mucus into which odorant chemicals diffuse to make contact with the hairs. Nasal congestion causes a loss of smell due to the proliferation of the mucus coating, making it difficult for chemicals to contact the hairs.

13. A second messenger-signalling system mediates the transduction of olfactory stimuli. Odor molecules bind to specific receptors on the sensor hairs. This activates G proteins, which in turn activate enzymes such as adenylate cyclase, causing an increase in the concentration of the second messenger cAMP. Gated sodium channels open in response to cAMP, depolarizing the cell and generating an action potential.

14. Odor intensity (concentration of odorant molecules) is coded for by the number of action potentials generated by sensor cells. The subjective impression of different kinds of smells comes from the activation of various combinations of receptors and sensors. Recent evidence indicates that a large family of genes codes for the wide variety of olfactory receptor types present in vertebrates.

15. *Taste buds* are the chemosensors responsible for *gustation*, the sense of taste. They are found primarily in the mouths of terrestrial vertebrates, although they also occur on the skin of aquatic animals that forage in murky waters where vision is of little use.

16. Clusters of taste sensors are embedded in the tongues of humans and other mammals. Many are associated with the raised papillae that give the tongue its bumpy texture. Tiny hairlike projections called microvilli are connected to the taste sensors and emerge through a pore in the top of each taste bud.

17. Unlike olfactory sensors, taste sensors are not neurons. The cell bodies of the sensory neurons that actually relay taste information to the central nervous system are located at the base of each cluster of taste sensor cells.

18. As chemicals make contact with receptors on the microvilli of taste sensor cells, receptor potentials are generated. The sensor cells release neurotransmitter onto the dendrites of the sensory neurons below each taste bud. These neurons then generate the action potentials that travel to the central nervous system.

19. The epithelial surface of the tongue is subjected to much physical abuse. Taste buds and their sensor cells are therefore replaced every few days, but the sensory neurons remain to form synapses with new sensor cells.

20. In general, specific regions of the human tongue are responsible for recognizing the four basic categories of taste (bitter, sour, salty, and sweet). In reality, these areas overlap to a large extent, and many varieties of molecules can be detected by our taste buds. Subtle discriminations within each basic flavor category are possible due to the interplay of gustatory and olfactory senses, which is why your sense of taste suffers during a cold.

21. Snakes use their tongues to smell. The flicking tongue gathers airborne molecules and delivers them directly onto olfactory receptors located in the roof of the mouth. In this way the animal makes repeated, rapid samplings of the air, which would not otherwise be possible due to its slow rate of respiration.

Activities (for answers and explanations, see page 389)

- Describe the difference between a receptor potential and a generator potential in a sensory cell. What are the main functions of these potentials?

- When a person receives a sharp blow to the head, he or she sometimes reports "seeing stars" and "hearing bells" when in fact there are no such physical stimuli occurring at the time. Explain how the mechanisms of sensory transduction and sensation can account for this phenomenon.

Questions (for answers and explanations, see page 389)

1. According to the definition in your textbook, which of the following is sensory transduction?
   a. Sex pheromone from a female moth produces action potentials in the antennal nerve of a male.
   b. Receptor potentials in taste sensors stimulate action potentials in associated sensory neurons.
   c. An active G protein in olfactory neurons causes production of cAMP.
   d. Sugars in a food source bind to receptors on the chemosensory hairs of a fly's foot.
   e. None of the above

2. An electrode is inserted into a chemosensory nerve leading away from a taste bud in the mouth of a dog. A mild acid solution is then continuously flushed over the sensors associated with this nerve. Initially the nerve responds to this stimulation but over time ceases to carry action potentials. This observation would best be explained by
   a. transduction.
   b. adaptation of the sensor cell.
   c. depletion of neurotransmitter in the sensory nerve.
   d. second messenger influences that increase cell membrane potentials.
   e. death of the sensor cell.

3. Which of the following is *not* a neuron?
   a. Human olfactory sensor
   b. Moth antennal chemosensor
   c. Human taste sensor
   d. Grasshopper foot chemosensor
   e. Snake olfactory receptor

4. If the concentration of cAMP suddenly increases in a human olfactory sensor cell, which of the following will happen?
   a. The membrane potential will become more negative.
   b. The membrane potential will become more positive.
   c. The membrane potential will remain unchanged.
   d. The sensor will show immediate signs of adaptation.
   e. The person will perceive a complex mixture of smells.

5. It's cold season and you have developed a serious case of nasal congestion. The reason that food loses much of its flavor under these conditions is that
   a. adaptation of chemosensors is more rapid.
   b. the transduction process in your taste buds has been inhibited.
   c. you cannot make sufficient quantities of cAMP.
   d. olfactory sensitivity is decreased by epithelial swelling and increased mucus buildup.
   e. mucus from the olfactory epithelium coats your taste buds.

6. Which of the following statements about human gustation is *false*?
   a. Taste sensors are relatively short-lived cells due to the high degree of abrasion they encounter.
   b. A single sensory neuron is associated with each taste bud in the vertebrate tongue.
   c. Lemon juice is tasted most clearly on the sides of the tongue.
   d. Receptors on the microvilli of taste sensors bind to food molecules.
   e. Humans perceive only three categories of tastes: sweet, sour, and bitter.

7. The chemosensory molecules used for mate attraction in arthropods are otherwise known as
   a. hormones.
   b. pheromones.
   c. protein receptors.
   d. G proteins.
   e. second messengers.

8. Which of the following sensory structures is capable of transduction?
   a. Snake olfactory sensor
   b. Dog chemosensor
   c. Insect chemosensory hair
   d. Human taste bud
   e. All of the above

MECHANOSENSORS: STIMULI THAT DISTORT MEMBRANES (pages 934–940)

Key Concepts

1. *Mechanosensors* respond to mechanical forces that distort cell membranes. Some mechanosensors are associated with the senses of touch, pain, pressure, or position. Others act as stretch receptors in blood vessels, muscles, or joints. Highly modified mechanosensors called hair cells are involved in balance and hearing.

2. When mechanical forces distort the sensor membrane, changes in ion permeability occur that alter the cell's membrane potential. The number of action potentials fired in response to this deformation is directly proportional to the stimulus intensity.

3. Several types of mechanoreceptors are found in human skin. Near the surface are *Meissner's corpuscles*, sensitive mechanoreceptors that detect light touch and adapt rapidly to a stimulus. *Expanded-tip tactile sensors* adapt more slowly than Meissner's corpuscles; they provide steady-state information about touch. Different body areas have different densities of these receptors, explaining why the fingertips are more touch-sensitive than the skin on a person's back.

4. Other mechanosensors are found deep within the skin. Physical displacement of epidermal hairs is detected by nerve endings that wrap their dendrites around the bases of hair follicles. Rapid skin vibrations are detected by sensory neurons called *Pacinian corpuscles*, which are encased in concentric layers of connective tissue. They adapt rapidly to stimuli because this layering efficiently redistributes pressure that deforms the nerve endings and therefore stimulates.

5. *Stretch sensors* respond to the forces associated with the stretching of vessel walls, bone joints, and muscles. *Muscle spindles* are examples of stretch sensors that contribute to the maintenance of body position through their connection with neural reflex loops. *Golgi tendon organs* are stretch sensors found in the tendons and ligaments that monitor the forces exerted by a contracting muscle.

6. *Hair cells* are a nonneuronal class of mechanosensors present in many animal sensory systems. All hair cell varieties possess *stereocilia* that project from the cell surface. When bent in one direction they hyperpolarize the cell membrane. If bent the other way they depolarize it. Depolarization causes the release of neurotransmitter onto neighboring sensory neurons, which fire action potentials.

7. The lateral line system of fishes uses hair cells to detect pressure changes associated with water movement. Lateral line hair cells are located in a canal just beneath the skin surface. Their stereocilia protrude into the canal and are encased in a gelatinous structure called the cupula. Water movement through the canal causes cupulae to move and stereocilia to bend. This induces electrical activity in associated sensory neurons.

8. *Statocysts* are the position-sensing organs of many invertebrates. Hair cells line the closed chambers of these organs, inside which are dense structures called *statoliths*. As the animal changes position, gravity causes the statoliths to move and the stereocilia to bend. Statolith position, as measured by the response of hair cells, indicates the animal's body position in space.

9. Vertebrate equilibrium is monitored by two other types of hair cell organs. The inner ear has three fluid-filled *semicircular canals*, each with a terminal swelling called an ampulla. Each ampulla contains many hair cells encased in a cupula. As the head moves, so does the fluid in the canals. This causes movement of the cupula and bending of the hair cells. Action potentials are then fired by associated sensory neurons.

10. The vertebrate *vestibular apparatus* contains two statocyst-like chambers used to detect head position. The floors of these chambers are covered with hair cells, on top of which lie many calcium carbonate *otoliths*. Gravity pulls on the otoliths to bend the stereocilia, generating action potentials that inform the nervous system about head position.

11. The sense of hearing *(audition)* also uses hair cells as transducers. Airborne pressure waves are converted into action potentials within the ear. These are then interpreted by the brain as sound.

12. In humans and other vertebrates, several accessory structures are used to collect, direct, and amplify sound waves. *Ear pinnae* are the external structures that collect sound, funneling it into the auditory canal where the pressure waves cause the *tympanic membrane* to vibrate. The tympanic membrane separates the auditory canal from the air-filled middle ear. Air pressure in the middle ear is equalized through a connection with the oral cavity called the *eustachian tube*, which keeps the tympanic membrane from inappropriately bulging in or out.

13. Connected to the tympanic membrane in the middle ear is a series of three small bones called *ear ossicles*. Membrane vibrations are first transferred to the *malleus* (hammer), which is physically attached to the tympanic membrane, then to a fulcrum bone called the *incus* (anvil), and finally to the *stapes* (stirrup). This system of bony levers acts to amplify sound 20-fold.

14. The stapes is attached to the *oval window*, a membrane that connects the middle ear with the fluid-filled inner ear. Vibration of the stapes causes the oval window to vibrate. Because fluids are relatively incompressible, this creates movement within the inner ear. Thus, airborne pressure waves are converted into the oscillatory movements of inner-ear fluids.

15. The coiled inner ear is known as the *cochlea*. It is within the cochlea that the transduction of pressure waves into action potentials takes place.

16. The cochlea consists of three parallel chambers separated by several membranes. Two of these chambers are interconnected at the distal end of the cochlea, making them one large chamber divided lengthwise by the *basilar membrane*. Atop the basilar membrane rests the *organ of Corti*, the structure that contains the hair cells. *Reissner's membrane* forms a cavity that serves a nutritional function for the organ of Corti but Reissner's membrane is not involved in sound transduction.

17. Sounds of different frequencies, or pitches, cause the oval window to produce unique fluid oscillations inside the cochlea, which are absorbed by displacement of the *round window*. These oscillations cause specific regions of the basilar membrane to flex and stimulate the organ of Corti. The hair cells in these areas are attached to another, fixed membrane called the *tectorial membrane*. This produces a shearing force that bends the stereocilia. The resulting deformation of hair cells leads to action potentials that travel to the brain.

18. The vibration of the basilar membrane is "tuned" to specific frequencies of sound. Low-frequency sounds cause the distal end of the membrane to vibrate, while high-frequency sounds stimulate vibrations in the proximal end. Only those hair cells located near regions of the basilar membrane stimulated by particular frequencies will respond. The action potentials produced by different combinations of hair cells are perceived as different sounds in the brain.

19. The loss of auditory ability, or deafness, can result from two general causes. *Conduction deafness* is a mechanical problem caused by the failure of the tympanic membrane or ear ossicles to properly transfer sound vibrations from the middle ear to the inner ear. It is usually the result of repeated ear infections that damage the middle ear tissues, or age-related stiffening of the movable ear ossicles.

20. *Nerve deafness* is caused by physical damage to the delicate transduction mechanism within the cochlea, usually as a result of prolonged exposure to excessively loud sounds such as those from jet engines or highly amplified music. It can occur at any age and is irreversible. Nerve deafness may also result from damage to auditory pathways in the nervous system.

Questions (for answers and explanations, see page 389)

1. All but one of the following mechanosensor systems employs hair cells as its transducer. Select the *exception*.
   a. Organ of Corti
   b. Statocyst
   c. Vestibular apparatus
   d. Meissner's corpuscle
   e. Lateral line

2. In which of the following would you find the malleus?
   a. Knee joint of a human
   b. Middle ear of a dog
   c. Surface skin layer of a fish
   d. Foot of a fly
   e. Skeletal muscle of a bird

3. Mechanosensors have which of the following features *in common* with other types of sensory transducers?
   a. Mechanical deformation of the membrane is the transducing stimulus.
   b. Hair cells are utilized to convert a physical stimulus into action potentials.
   c. The stimulus leads to a change in ion permeability of the sensor cell membrane.
   d. They are integral sensory components of monosynaptic reflex loops.
   e. None of the above

4. Statocysts are to the vestibular system as expanded-tip tactile receptors are to

   a. Pacinian corpuscles.
   b. stretch receptors.
   c. the organ of Corti.
   d. the tympanic membrane.
   e. hair cells.

5. Which of the following statements about mechanosensors is *false*?
   a. Hair cells are found in the semicircular canals of the human ear.
   b. Cupulae are present both in the vestibular apparatus of mammals and the lateral line system of fishes.
   c. Different body areas have different proportions of the various skin touch receptors.
   d. Hearing and balance in humans are both mediated by the same kind of mechanosensor.
   e. All mechanosensors are neurons that fire action potentials in response to a physical force.

6. A man acquires a viral infection of the inner ear that permanently damages only the proximal end of the basilar membrane. Which of the following conditions would result?
   a. He would suffer conduction deafness.
   b. He would permanently lose some of his high-frequency hearing.
   c. He would permanently lose some of his low-frequency hearing.
   d. He would have difficulty maintaining proper balance.
   e. He would become painfully sensitive to even very low-intensity sounds.

7. A constant light touch on the skin of your finger is at first sensed by your brain, but then rapidly fades away. This effect is the result of adaptation by _____ .
   a. Reissner's membrane.
   b. Pacinian corpuscles.
   c. Meissner's corpuscles.
   d. expanded-tip tactile sensors.
   e. the tectorial membrane.

Activities (for answers and explanations, see page 390)

• Match each of the following mechanosensors with the most appropriate description.

____Organ of Corti   *a.* Pressure sensor in fishes

____Pacinian corpuscle   *b.* Monitors the forces muscles exert on bones

____Golgi tendon organ   *c.* Senses balance in humans

____Statocyst   *d.* Structure of the vertebrate cochlea

____Vestibular apparatus   *e.* Skin vibration sensor

____Lateral line   *f.* Position sensor in lobster

- The diagram below shows some of the structures associated with the mammalian inner ear. Indicate which of these structures help detect head position, acceleration, and deceleration, and explain specifically how this is accomplished.

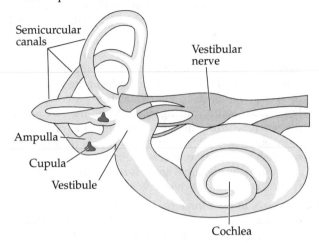

- The heavy line below represents the basilar membrane of a human ear encased within the cochlea. (Note that it has been stretched out flat, similar to Figure 42.14 in your textbook). The positions of other relevant inner-ear structures are also shown.

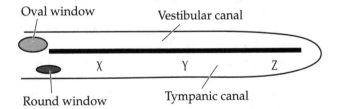

Predict the relative pitch of the three sounds that a person would hear if the basilar membrane were to vibrate first at point X, then Y, and finally Z.

Location	Perceived pitch of sound
X	_____
Y	_____
Z	_____

PHOTOSENSORS AND VISUAL SYSTEMS: RESPONDING TO LIGHT • SENSORY WORLDS BEYOND OUR EXPERIENCE (pages 940–949)

Key Concepts

1. Photosensitivity is the ability to detect light. Although the sophistication of visual systems ranges from those capable only of simple light–dark discriminations to those that code color and pattern, all animals detect light using a single photoreactive molecule called *rhodopsin*.

2. Rhodopsin consists of a nonabsorbing protein called *opsin* coupled with a light-absorbing group called *11-cis-retinal*. The whole molecule is situated in the plasma membrane of a photosensor cell.

3. When rhodopsin absorbs a photon of light, the 11-*cis*-retinal undergoes a conformational change to become the isomer all-*trans*-retinal. This strains the bond between opsin and retinal, causing the molecule to split apart in a process known as bleaching. The bleached molecule is no longer photosensitive. After bleaching, retinal spontaneously reconverts to the 11-*cis* isomer. This allows it to combine with the isolated opsin and regenerate a functional rhodopsin molecule.

4. The conformational change of rhodopsin involves a series of chemical intermediates, including photoexcited rhodopsin, which causes a cellular response by triggering a cascade of reactions that change ion flows and membrane potential in the photosensor.

5. Each vertebrate photosensor has thousands of rhodopsin molecules embedded in the membranes of its outer segments. The membrane potential of an inactive photosensor is considerably less negative than most neurons because the membrane is more permeable to Na^+ ions. Sodium channels are kept open in the resting state by the presence of a high concentration of cGMP molecules.

6. Photosensor cells do not produce action potentials. Instead, they hyperpolarize when stimulated by light. This occurs because the photoactivation of rhodopsin causes a cascade of chemical events that leads to the closing of sodium channels.

7. Photoexcited rhodopsin activates a G protein called *transducin*, which in turn activates a phosphodiesterase enzyme. This enzyme converts cGMP to 5'-GMP, removing the molecule needed to keep the sodium channels open. With Na^+ ion flow into the cell interrupted, the membrane hyperpolarizes.

8. Although this mechanism seems a roundabout way to change membrane potential, its power lies in the enormous signal amplification possible through the repeated action of each transducin and phosphodiesterase molecule. Thus, a single photon of light can result in the closing of over a million sodium channels.

9. Simple invertebrates such as flatworms have photosensitive cells organized into primitive eye cups. The system does not provide image vision, but does allow the animal to discriminate between areas of light and dark.

10. Arthropods have more complex visual systems capable of detecting patterns, images, and movement. These animals have *compound eyes* that consist of many individual optical units known as *ommatidia*. Depending on the species, the number of ommatidia in each eye varies from only a few to over 10,000.

11. Each ommatidium has a lens system that focuses light on photosensitive *retinula cells*. Microvilli on each retinula cell contain rhodopsin molecules that capture photons. In some insects, the microvilli overlap to form a central structure called the *rhabdom*. The receptor potentials generated by the combined retinular cells within a single ommatidium are then sent to the brain. Because each ommatidium is fixed within the compound eye and views only a limited portion of space, the animal probably sees a fragmented image of the world.

12. The best image-forming eyes are found in mollusks and vertebrates. Their eyes operate like a camera, and are strikingly similar in both structure and function considering the independent evolution of each group.

13. Vertebrate eyes are fluid-filled spheres bounded by a sheet of tough, opaque connective tissue called the *sclera*. An area in the front of the sclera is transparent, forming the *cornea,* which allows light to enter the eye.

14. Immediately inside the cornea is the *iris*, a pigmented tissue that is under nervous control. Changes in the size of the iris regulate the diameter of its central aperture, called the *pupil*. Pupillary size determines the amount of light that passes through the eye to reach the photosensors.

15. A crystalline *lens* lies behind the iris, helping to focus light rays on the photosensors. An important function of the lens is *accommodation*, which is how the eye adjusts its focus for objects at different distances. Strong *ciliary muscles* attached to the lens either move it back and forth within the eye (fishes, amphibians, and reptiles) or cause it to change shape (birds and mammals). This adjusts the optical properties of the lens and refocuses the light rays. In humans, age contributes to a reduction in lens elasticity, making accommodation difficult and producing the need for corrective lenses.

16. The *retina* consists largely of neural tissue that is connected directly to the brain. It contains the photosensor cells, the density of which varies across the retinal surface. A *fovea* is the place on the retina where photosensors are in particularly high concentration, thus providing good visual acuity. The number, location, and shape of the fovea varies with species.

17. The neurons within the retinal tissue overlie the photosensor cells. No photosensors occur at the spot on the retina where the axons from these neurons join to exit the eye. This area is known as the *blind spot*. You do not normally detect the visual "hole" created by your blind spot because the brain fills in this space based on information from photosensors in the surrounding area.

18. Two major types of photosensors are found in vertebrate retinas, each named because of its shape. *Rod cells* are the more abundant, numbering nearly 100 million in humans. They are very sensitive to low levels of light, but make no contribution to color vision. *Cone cells* are much less numerous (about 3 million in humans). They are only activated by relatively high light levels and produce color vision.

19. The retinal distribution and number of rods and cones varies with the animal's lifestyle. Diurnal animals have cone-dominated foveas, with rod cells pushed mostly to the peripheral areas of the retina. This provides excellent visual acuity during daylight hours, but not at night. The entire retina of some diurnal animals contains only cone cells. Nocturnal animals tend to have rod-dominated foveas, and so their visual acuity is best at night. Some nocturnal animals have only rods in their retinas, and thus lack color vision.

20. There are at least three types of vertebrate cones. Each detects a different color because of the slightly different wavelength absorption properties of their opsin molecules. Retinal actually absorbs the photon, but it is the opsin that determines where in the color spectrum retinal absorbs best. Human cones absorb maximally in the blue, green, and red wavelengths.

21. Five layers of neurons (four in addition to the rods and cones) are found in the vertebrate retina. These neurons overlie the photosensors, so light must pass through them before encountering the rods and cones. Behind the rods and cones of diurnal animals is a dark layer of epithelium that captures stray light and prevents backscattering, which would blur the image.

22. Many nocturnal animals, for which efficient light gathering is a must, have a highly reflective layer behind the photosensors called the *tapetum*. This structure returns uncaptured photons to the rods for possible detection. While it enhances visual sensitivity, acuity is decreased. Light reflected from the tapetum back through the eye is what produces the "eye shine" of animals illuminated by a car's headlights at night.

23. Photosensors form the outermost layer of retinal neurons. The next inward layer consists of *bipolar cells,* which make direct synaptic contact with more than one rod or cone. As the membrane potential of the photosensors changes, neurotransmitter is released onto the bipolar cells. Like rods and cones, the bipolar cells do not fire action potentials.

24. Several bipolar cells make synaptic contact with a single ganglion cell, whose axon exits the eye and goes directly to the brain. When a bipolar cell is stimulated, it releases neurotransmitter onto the ganglion cell, which may then fire an action potential. The rate at which action potentials are produced by ganglion cells varies directly with the pattern and quantity of light striking the photosensors it monitors.

25. Sandwiched between the bipolar and ganglion cell layers are *amacrine* and *horizontal* cells, which communicate laterally across the retina to increase sharpness, contrast, and overall light sensitivity. Horizontal cells link neighboring

photosensor–bipolar cell pairs, while amacrine cells connect adjacent bipolar–ganglion cell combinations.

26. The information processing achieved by these complex retinal connections helps reduce the total amount of neural information sent to the brain. Each ganglion cell receives input from hundreds of adjacent photosensors and tens of bipolar cells, which together form the *receptive field* of that ganglion cell.

27. Ganglion receptive fields are mapped by measuring the cell's response to a spot of light passing over the section of retina it monitors. These fields are circular and consist of two concentric rings called the *center* and the *surround*. The cell's response to a stimulus varies depending on whether the spot hits the center or the surround.

28. Some ganglion cells are excited by light in the center of their receptive fields but not the surrounds; these cells have *on-center* receptive fields. If a stimulus in the surround but not the center is excitatory, the receptive field is called *off-center*. For large spots of light that cover portions of both the center and surround, the response of the center dominates. This pattern allows the ganglion cells to inform the brain about light–dark contrasts.

29. The interconnection of bipolar and ganglion cells made possible by the horizontal cells provides the underlying circuitry for ganglion receptive fields. Since any one photosensor or bipolar cell can be connected to several ganglion cells, the receptive fields of adjacent ganglion cells often overlap.

30. Forms of electromagnetic radiation beyond the human visual spectrum can be sensed by other animals. Ultraviolet radiation contains shorter wavelengths than our visual system can respond to, but many insects see quite well in this region of the spectrum. Flowering plants take advantage of this sensitivity and use ultraviolet markers to attract insect pollinators.

31. Wavelengths longer than humans can detect are also sensed by some animals. Infrared radiation, which we sense as heat, is detected by pit vipers using special organs located near their eyes. This information can be used to hunt prey based on the radiation they emit, even when there is not enough light to see.

32. Echolocation is a mechanism used by bats, dolphins, and a few other vertebrates to create acoustic images of the world. These animals emit high-intensity (and often high-frequency) sounds and then listen for the echoes that return as those sounds reflect off of objects in the environment.

33. *Electrosensors* are modified mechanosensors found in certain fish that live in murky waters. Some are associated with the lateral line system and others are separate organs. They produce weak electric fields that are used to detect and capture prey, as a means of defense, or to communicate with other individuals.

34. Magnetic sensitivity is also a viable sensory system in some animals, although the exact mechanism for magnetoreception remains unclear. Its primary function appears to be the gathering of directional information from the Earth's magnetic field for use in orientation.

Questions (for answers and explanations, see page 390)

Questions (for answers and explanations, see page 390)

1. In which of the following animals would you expect to find the protein opsin?
   a. Human
   b. Bird
   c. Salamander
   d. Butterfly
   e. All of the above

2. Photopigments from the eye of a cow are extracted and labeled with a substance that turns bright blue in the presence of 11-*cis*-retinal. A sample of these labeled pigments is then exposed to bright white light for 20 minutes. Which of the following statements about the results of this treatment is *false*?
   a. The retinal side group will have dissociated from the opsin protein.
   b. The sample will be bright blue.
   c. The sample will show little or no blue color.
   d. Little light-absorbing rhodopsin will be present.
   e. Most of the retinal will be in the all-*trans* isomeric form.

3. Which of the following cells in the human visual system send information *directly* to the brain?
   a. Ganglion cells
   b. Bipolar cells
   c. Amacrine cells
   d. Horizontal cells
   e. Rods and cones

4. A photon of light striking a human rod cell causes that cell to hyperpolarize because
   a. the rate of conversion of 5'-GMP to cGMP is increased.
   b. the flow of Na^+ ions into the cell rises dramatically.
   c. transducin causes inactivation of the enzyme phosphodiesterase.
   d. cGMP levels fall, causing sodium channels to close.
   e. none of the above

5. Which of the following is *not* an element of the vertebrate eye?
   a. Lens
   b. Horizontal cell
   c. Retinula cell
   d. Iris
   e. Sclera

6. Through which of the following cell layers must a photon of light pass before striking a cone cell in the eye of a hawk?
   a. Amacrine
   b. Bipolar
   c. Horizontal
   d. Ganglion
   e. All of the above

7. Changing the focus of a vertebrate eye by making adjustments to the shape or position of the lens is called
   a. bleaching.
   b. isomerization.
   c. hyperpolarization.
   d. accommodation.
   e. transduction.

8. Which of the following statements about animal vision is *false*?
   a. The tapetum aids in the rapid regeneration of bleached rhodopsin molecules.
   b. There are several types of cones in vertebrate retinas, each of which detects different wavelengths.
   c. An ommatidium is only one visual element within the compound eye of an arthropod.
   d. Photosensors in simple invertebrates detect light–dark contrasts, but do not produce image vision.
   e. Although retinal actually absorbs the photon, the wavelength it absorbs best is determined by opsin.

Activities (for answers and explanations, see page 390)

- Label each of the following structures on the diagram of a human eye shown below: sclera, cornea, iris, pupil, lens, retina, fovea, blind spot.

- Below is the receptive field of a retinal ganglion cell.

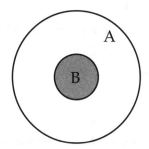

Previous studies have revealed that this is an off-center receptive field. Suppose you stimulate the portion of the retina associated with this receptive field using a small spot of light, and then record the action potentials the ganglion cell produces. Predict the cell's response to each of the following experimental conditions:

1. The light is flashed briefly at point A.

2. The light is flashed briefly at point B.

3. The light is moved slowly from point A to point B.

Integrative Questions (for answers and explanations, see page 390)

1. Which of the following sensory systems would you *not* expect to find in a pit viper?
   a. Olfaction
   b. Infrared detection
   c. Echolocation
   d. Gustation
   e. Vision

2. Hearing, touch, balance, and temperature sensitivity are all related by a common transduction mechanism known as
   a. photoreception.
   b. mechanoreception.
   c. chemoreception.
   d. electroreception.
   e. magnetoreception.

3. After a day of vigorous sailing you come ashore and find that your walk is somewhat unsteady. When you sit down to rest, the world seems to spin a little even though you are perfectly still. The sensory basis underlying this uneasy feeling is related to overstimulation of
   a. the visual system.
   b. the vestibular system.
   c. the olfactory system.
   d. the auditory system.
   e. none of the above

4. You discover a new species of animal which appears to be a very primitive type of invertebrate. Which of the following kinds of sensory systems would you almost certainly expect it to possess?
   a. Chemical
   b. Visual
   c. Auditory
   d. Magnetosensory
   e. All of the above

43

The Mammalian Nervous System: Structure and Higher Functions

CHAPTER LEARNING OBJECTIVES—after studying this chapter you should be able to:

❏ Distinguish between a neuron and a nerve, and relate their functions to overall nervous system activity.

❏ Differentiate between the central and peripheral nervous systems, and properly apply the concepts of afferent and efferent information flow to the peripheral system.

❏ Relate ideas of voluntary and involutary (autonomic) control to both afferent and efferent nervous system activity.

❏ Distinguish among the hindbrain, midbrain, and forebrain areas of the vertebrate brain, and discuss the primary neural functions of each.

❏ Compare and contrast the roles of the diencephalon and telencephalon in various vertebrate groups.

❏ Describe the general anatomy, the complexity, and the importance of the cerebral cortex in birds and mammals.

❏ Describe the basic anatomy of the spinal cord and diagram how various neural pathways enter and exit the cord.

❏ Discuss the role of interneurons in spinal cord function.

❏ Identify the anatomical locations of the reticular and limbic systems in vertebrates, and explain their functions.

❏ Label the frontal, temporal, parietal, and occipital lobes of the cerebral cortex, and discuss the functions of each.

❏ Distinguish between the functions of the primary motor cortex and the somatosensory cortex, and explain how regions of the body "map" onto these brain areas.

❏ Describe the role and significance of association cortex in the human brain.

❏ Compare and contrast evolutionary trends in cortical development among vertebrate groups.

❏ Describe the structure and function of the sympathetic and parasympathetic branches of the autonomic nervous system.

❏ Give specific examples of how the sympathetic and parasympathetic branches of the autonomic nervous system interact to regulate body functions.

❏ Discuss the organization of visual receptive fields in the occipital cortex, and explain the neural basis for binocular vision.

❏ Describe neural changes associated with the different stages of sleep, and distinguish between the physiological and mental events associated with non-REM and REM sleep in humans.

❏ Name the various forms of learning in humans and other vertebrates, and explain the experimental evidence that provides some insight into the neural mechanisms underlying some of these types of learning.

❏ Identify the various forms of memory in humans and relate these to the role of the hippocampus.

❏ Describe the process of lateralization in the human brain, and explain how studies of individuals with a severed corpus callosum shed light on this process.

❏ Discuss the neuroanatomical systems involved in human language.

THE NERVOUS SYSTEM: STRUCTURE, FUNCTION, AND INFORMATION FLOW • FUNCTIONAL SUBSYSTEMS OF THE NERVOUS SYSTEM
(pages 952–960)

Key Concepts

1. The cells that make up the nervous systems of complex animals are grouped into well-defined regions that control specific body functions. The brain is a region of particular importance.

2. For its size, the human brain is the most complex structure known. Within it resides many of our individual characteristics and our personalities. The brain senses

bodily needs, makes decisions about appropriate actions, and then initiates and coordinates responses. It also monitors and remembers information.

3. Neurons are the functional units of the brain. The human brain contains about 10 billion neurons, enabling us to handle vast amounts of information with great flexibility.

4. An axon is the conducting process of a single neuron; it carries very specific neural information. By contrast, a *nerve* is a bundle of axons from many different neurons. Nerves simultaneously convey many different kinds of information both to and from locations in the body.

5. The human nervous system is grouped into two main divisions. The brain and spinal cord make up the *central nervous system* (CNS). Information is brought to and from the CNS by an extensive *peripheral nervous system* (PNS), which connects to the CNS through the spinal and cranial nerves.

6. The PNS is functionally divided into two subsystems. The *efferent* side carries information from the central nervous system to effector organs in the periphery. The *afferent* side relays information from sensors located in the periphery to the central nervous system.

7. The afferent and efferent subsystems can be further divided based on whether or not we are consciously aware of the information being transmitted. Information from some peripheral sensors (e.g., eyes and ears) reaches our consciousness, while we remain unaware of information from our internal organs and metabolic processes. The efferent system likewise has a *voluntary* component that consciously controls skeletal muscle, as well as an *involuntary* or *autonomic* component that directs glands and smooth muscle.

8. In addition to neural input from the PNS, chemical information is also relayed to the CNS through the action of hormones. The neurohormones of the hypothalamus and pituitary are important examples of these chemical influences on nervous system function.

9. The first evidence of a developing nervous system in vertebrates is the embryonic neural tube. The anterior end of this tube subdivides into three regions: the *hindbrain*, *midbrain*, and *forebrain*. The remaining part of the neural tube forms the spinal cord, from which the spinal nerves emerge.

10. The hindbrain gives rise to two structures, the *medulla* and the *pons*, that maintain important basic physiological functions such as breathing, circulation, swallowing, and vomiting. They also act as intermediaries for information passing between the spinal cord and higher brain areas. The *cerebellum* is a third hindbrain structure that coordinates or refines the motor control of many basic behaviors.

11. The midbrain develops above the hindbrain and becomes a processing center for auditory and visual information. All information traveling between the spinal cord and the forebrain must also pass through the midbrain.

12. The forebrain develops atop the midbrain. The central portion of the forebrain is called the *diencephalon*. It contains an important sensory relay station called the *thalamus* and the main regulatory center for physiological function, the *hypothalamus*.

13. The diencephalon is surrounded by the *telencephalon*, which consists of two *cerebral hemispheres* (collectively, the *cerebrum*). It is the largest of all the brain structures in birds and mammals, and is responsible for controlling sensory perception, learning, memory, and conscious behavior. Telencephalon size increases steadily with the evolutionary progression from simple vertebrates through birds and mammals.

14. A key point to remember is that the organization of the neural axis is linear in vertebrates. Communication between the spinal cord and the forebrain must therefore first pass through both hindbrain and midbrain structures before it reaches the telencephalon.

15. Primitive autonomic functions are localized in the *brain stem* (which includes the medulla, pons, midbrain, and diencephalon) or even lower in the spinal cord. More complex control functions reside in the telencephalon. As a result, animals higher on the phylogenetic scale (e.g., mammals) rely heavily on telencephalic functions for their behavior, while those on the lower end (e.g., fishes) function with little or no telencephalic control.

16. The nervous system can be modeled as a parallel processor that contains many functional subsystems, with any one anatomical structure making a contribution to multiple subsystems.

17. The spinal cord is a conduit for information between the periphery and the brain, integrating much of this information along the way. In cross section, the spinal cord shows a butterfly-shaped region of gray matter, which contains neuron cell bodies, surrounded by white matter made of axons that run up and down the cord.

18. The gray matter is divided into *dorsal* and *ventral horns*, which are connected to the spinal nerves by way of the dorsal and ventral roots, respectively. Spinal nerves carry both incoming and outgoing information, with afferent signals entering through the dorsal roots and efferent signals exiting via the ventral roots.

19. *Interneurons* reside entirely in the gray matter and connect afferent axons with the cell bodies of efferent neurons. They also relay information up and down the spinal cord between the brain and neuronal cell bodies whose axons exit through spinal nerves.

20. Spinal interneurons are largely responsible for the information processing and integration that goes on in the spinal cord. They are involved in spinal circuits that coordinate reflexive actions and help to generate repetitive motor patterns such as walking and swimming.

21. An anatomically or neurochemically distinct group of brain neurons is called a *nucleus*. The *reticular system* is a brain stem structure running through the core of the medulla, pons, and midbrain. It contains many discrete brain nuclei that help control balance, coordination, sensitivity to pain, and state of arousal.

22. Reticular system nuclei exhibit high levels of activity during wakefulness and low activity during sleep or coma. Because of their involvement in controlling the arousal state, this collection of brain nuclei is also known as the reticular activating system.

23. Brain stem damage often has catastrophic effects, but the extent of the effects depends on the location of the injury relative to the reticular system. If injury is at midbrain or higher levels, so that alerting information from the reticular system cannot reach the forebrain, the result is coma. If injury is to the neural axis below the reticular system, normal patterns of wakefulness and sleep continue even though loss of sensation and motor function do occur.

24. The forebrains of fishes, amphibians, and reptiles are dominated by the diencephalon, with only a small amount of telencephalon present. In birds and mammals, the telencephalon is highly elaborated to form the *neocortex*, which completely covers the diencephalon.

25. The primitive portion of the avian and mammalian forebrain is known as the *limbic system*. It controls basic physiological drives, instincts, and emotions such as pain, pleasure, and rage; it is also believed to play a role in learning. In humans, the *hippocampus* is the portion of the limbic system that mediates the transfer of short-term memory to long-term memory.

26. The cerebral hemispheres dominate the brains of higher vertebrates and in humans cover all other brain regions except the cerebellum. The thin *cerebral cortex* forms a sheet of gray matter within which reside most of the neuron cell bodies in the nervous system. Ridges (*gyri*) and valleys (*sulci*) result from the convoluted folding of the cerebral cortex, helping bring the total surface area to about 1 m^2.

27. Lobes of the cerebral cortex are present in each hemisphere and are dedicated to specific control functions that can be fairly easily documented and mapped. However, most of the cerebral surface, including large areas of the lobes, appears to serve as *association cortex*, where higher-order processing that is harder to define and localize takes place.

28. The *temporal lobe* is located on the side of the head, near the temple. It processes auditory information (especially spoken language) and has association areas involved with the recognition, identification, and naming of objects (e.g., faces). *Agnosias*, caused by damage to the temporal lobe, are disorders where one is aware of a stimulus but cannot identify it.

29. The *occipital lobe* is located in the back of the brain; it receives and processes visual information. Association areas within this lobe organize the visual world and facilitate translation of visual experience into language.

30. The *parietal lobe* is found atop the brain. It contains a region called the *primary somatosensory cortex* that receives information about touch and pressure from the whole body surface via the thalamus. The amount of somatosensory cortical space devoted to a specific body region is positively correlated with the degree of sensory discrimination in that region.

31. Association areas of the parietal lobe attend to complex stimuli, but this attention is not symmetrical between the hemispheres. *Contralateral neglect syndrome* results from damage to the right parietal lobe and causes afflicted individuals to neglect stimuli from the left visual field and left side of the body. As evidence of the cortical asymmetry, complementary damage to the parietal lobe in the left hemisphere does not produce neglect of the right side.

32. The *frontal lobe* is located near the front of the brain and is separated from the parietal lobe by the deep *central sulcus*. A portion of the frontal lobe sends motor commands to the voluntary muscles. Each body region is mapped onto an area of this lobe called the *primary motor cortex*. Like the somatosensory cortex, the motor cortex accurately represents the relative importance of motor control in various body regions. Association areas within the frontal lobe are generally involved with planning.

33. Cortical size increases not only in the evolutionary progression from fish to more complex vertebrates, but cortical size varies considerably within the mammals. For example, a difference in cortical size can be observed between rodents and primates. The increase in cortical size parallels advancements in behavioral repertoire between groups.

34. The most dramatic increase in cortical size and function has occurred during the last several million years of human evolution, and strongly correlates with advances in the complexity of our behavior. While humans do not have the largest brains in absolute size, we do have the largest ratio of brain size to body size found in vertebrates. We also have the most highly developed cerebral cortices, including the largest relative amount of association cortex.

Questions (for answers and explanations, see page 391)

1. The secretion of hormones from an endocrine gland is most likely under the control of which of the following components of your nervous system?
   a. Autonomic
   b. Voluntary
   c. Dendritic
   d. Limbic
   e. Reticular

2. The structure leading from the mammalian inner ear to the midbrain that includes the axons of auditory neurons from the cochlea would best be called
   a. a brain nucleus.
   b. the auditory nerve.
   c. the thalamus.
   d. the hindbrain.
   e. the parietal lobe.

3. Imagine that you have become microscopic in size and are walking over the surface of the cerebral cortex in a human brain. The peaks and valleys you cross are
   a. parts of the brain stem.
   b. the hippocampus.
   c. foldings of the cortex, which increase surface area.
   d. bundles of interneurons in the reticular formation.
   e. structural elements of the limbic system.

4. A person who is aware of an auditory stimulus but cannot identify it may have incurred damage to the _____ of the brain.
   a. frontal lobe
   b. parietal lobe
   c. occipital lobe
   d. temporal lobe
   e. reticular system

5. A nerve impulse traveling up the spinal cord to the brain would first pass through the _____ before reaching the telencephalon.
   a. pons
   b. medulla
   c. midbrain
   d. diencephalon
   e. All of the above

6. You stub your toe and immediately perceive the sensation of pain, which is a function of neural activity in your brain. This sensation would not be possible without the action of
   a. afferent neurons.
   b. efferent neurons.
   c. interneurons.
   d. afferent neurons and interneurons.
   e. efferent neurons and interneurons.

7. Damage to the back of the brain would most likely impair which of the following functions of the human cerebral cortex?
   a. Hearing
   b. Vision
   c. Speech
   d. Hand movements
   e. Pressure sensation in the feet

8. Which of the following animals would likely have the *smallest* ratio of telencephalon size to body size?
   a. Fish
   b. Amphibian
   c. Reptile
   d. Bird
   e. Mammal

9. Which of the following is *not* part of the central nervous system?
   a. Brain stem
   b. Spinal gray matter
   c. Neuronal cell body of a sensory afferent
   d. Spinal interneuron
   e. Cerebellum

10. In which of the following animals would serious damage to the telencephalon have the most impact on conscious behavior?
    a. Shark
    b. Lizard
    c. Frog
    d. Turtle
    e. Rat

Activities (for answers and explanations, see page 391)

- This textbook chapter began with the sad story of Phineas Gage and his terrible accident. Briefly explain the change in this man's behavior in terms of the damage he sustained to his higher brain structures and functions.

- On the mid saggital view of an adult human brain shown below, label the indicated structures and briefly describe their functions.

- Consider Figure 43.3 in your textbook. At approximately what point in the embryonic neural development of humans does the relative importance of the forebrain change from the pattern seen in fish and reptiles? Explain your answer.

INFORMATION PROCESSING BY NEURONAL CIRCUITS • UNDERSTANDING HIGHER BRAIN FUNCTIONS IN CELLULAR TERMS (pages 960–970)

Key Concepts

1. The autonomic nervous system, which controls gland and involuntary muscle activity, includes the *sympathetic* and *parasympathetic* subdivisions. These subdivisions work opposite one another to cause changes in effector activity.

2. One common example is the "fight-or-flight" response. Here the sympathetic pathways stimulate increases in blood pressure, heart rate, and cardiac output in preparation for defense or escape. The parasympathetic pathways cause the same physiological parameters to decrease, returning them to normal levels after the threat has passed.

3. It is not always the sympathetic pathway that increases activity, nor the parasympathetic that decreases it. For example, digestion is speeded up by the parasympathetic and slowed by the sympathetic. Nevertheless, these pathways still work in opposition to one another.

4. The neural circuits of parasympathetic and sympathetic pathways consist of two cell types. The first cell in the circuit is located in either the brain stem or the spinal cord and is called a *preganglionic neuron*. Its axon uses acetylcholine as a neurotransmitter.

5. Preganglionic cells synapse onto *postganglionic neurons,* which reside in a cluster (ganglion) of functionally related neurons located somewhere outside the spinal cord. The axons of postganglionic cells terminate on effector organs. If they are part of the sympathetic division they use norepinephrine as their synaptic neurotransmitter; if part of the parasympathetic, they use acetylcholine.

6. In effectors that receive input from both parasympathetic and sympathetic pathways, the target cells respond in opposite ways to acetylcholine and norepinephrine. One neurotransmitter causes depolarization and an increase in activity, and the other leads to hyperpolarization and a decrease in activity. The control of cardiac pacemakers and digestive muscles are examples.

7. Anatomical differences also exist between the two autonomic divisions. Preganglionic neurons of the sympathetic system are located in the thoracic and lumbar regions of the spine, and those of the parasympathetic in the brain stem and lower spinal segments. Postganglionic sympathetic neurons are mostly situated in two chains adjacent to the spinal cord, while those of the parasympathetic are found close to the target organ.

8. Visual information travels from the retina of the eye through the optic nerve to a relay station in the thalamus. From there it travels to the occipital cortex, where the brain combines information from many ganglion cells to form cortical receptive fields.

9. The work of Hubel and Wiesel revealed that there are several types of receptive fields in the visual cortex. The fields of *simple cells* respond best to a bar of light with fixed orientation that stimulates one particular location on the retina. *Complex cells* also respond to bars of fixed orientation, but they do so at many locations across the retina. Some complex cells also prefer bars that move in a particular direction.

10. The ganglionic receptive fields that supply information to simple cortical cells are arranged close together on the retina, while those of complex cortical cells are more widely scattered. It appears that the brain creates its image of the visual world by analyzing the patterns of light edges that stimulate various regions on the retina.

11. *Binocular vision* is a product of the brain's integrating visual information from both eyes, which are situated on the head so as to view the same object from slightly different angles.

12. The *optic chiasm* is the point just below the hypothalamus where the optic nerves join and then separate again. The axons of cells that view the right visual field in both eyes go to the left cerebral hemisphere, and those that view the left visual field go to the right hemisphere.

13. Both simple and complex cells in the primary visual cortex are organized into alternating columns that gather information from the left eye, then the right, and so on. Those cells on the border between columns receive input from both eyes, and are thus *binocular cells.* They measure distance by determining the disparity between where an object falls on the retinas of the two eyes.

14. Specific visual images are likely the product of simultaneous interactions among many retinal and cortical cells. This process is enhanced by neural input from memory and other sensory channels, which may in part justify the large amount of cerebral cortex devoted to association areas.

15. Few details about the underlying circuitry of complex nervous system functions are known. The study of higher brain functions such as learning, memory, consciousness, language, and dreaming are particularly difficult problems, although some progress has been made in understanding the mechanisms of these neural phenomena.

16. Sleep occurs in all birds and mammals, and probably other vertebrates as well. Yet the neural reasons for why animals sleep and how sleep cycles are controlled are not clear. The *electroencephalogram* (EEG), which monitors regional brain activity, and the *electromyogram* (EMG), which records the electrical activity associated with muscle movements, provide tools for studying these questions.

17. EEG and EMG traces from sleeping mammals reveal two broad categories of sleep. *Slow-wave sleep* is characterized by low-frequency, high-amplitude EEG waves. In *rapid-eye-movement* (REM) sleep, slow-wave EEG patterns are absent and the ocular muscles cause rapid eye twitches.

18. In humans, slow-wave sleep is called *non-REM* and is divided into four stages of progressively deeper sleep. A person enters sleep in Stage 1 non-REM and then progresses to stage 4. Stages 3 and 4 are the deep, restorative, slow-wave form of sleep.

19. Each non-REM cycle is followed by a REM sleep period during which humans experience vivid dreams and nightmares. At this time descending neural input from the brain inhibits the movement of voluntary muscles, presumably to prevent the acting out of dreams.

20. In humans, there are usually four or five cycles of non-REM/REM sleep each night, with non-REM accounting for about 80 percent of sleep time and REM sleep the other 20 percent. While the function of REM sleep is not known, it is probably a rather basic, cellular-level phenomenon given that it involves about the same percentage of total sleep time in all mammals.

21. Important neural differences exist between non-REM and REM sleep. During non-REM sleep, neurons in the thalamus and cerebral cortex decrease their activity and responsiveness (they become hyperpolarized). At the same time, these cells begin to fire periodic bursts of action potentials that synchronize to produce the slow-wave EEG pattern.

22. Hyperpolarization of the brain stem nuclei that bring on sleep has recently been linked to the release of adenosine, whose concentration in the extracellular brain fluid increases as ATP concentration decreases. Thus, high adenosine levels may help hyperpolarize cells that produce deep, restorative sleep, which in turn leads to the opportunity for renewal of ATP energy in the brain.

23. Coincident with the transition from non-REM to REM sleep, inactive brain stem nuclei resume activity, slow-wave EEG patterns disappear, and an activity pattern resembling that of an awake brain emerges. The cerebral cortex can once again process information, which allows dreaming to occur, but normal sensory and motor pathways remain inhibited.

24. *Learning*, the modification of behavior by experience, and *memory*, the ability to retain learned information in the nervous system, are present to some degree in all animals. They are, however, most highly developed in humans, making possible such things as language, culture, art, and science.

25. *Associative learning* occurs when unrelated stimuli are linked to the same response. A simple example is the *conditioned reflex*, in which a novel stimulus is associated with another stimulus that normally triggers the response. Eventually the novel stimulus will by itself stimulate the response, as with Pavlov's experiments on the neural control of salivation in dogs.

26. Conditioned reflexes can now be used to study the underlying neural circuitry mediating learning, as shown by Thompson's work on cerebellar control of the eye blink reflex in rabbits.

27. Learning that leads to lasting memory likely involves long-lasting synaptic changes. In the mammalian hippocampus, high-frequency neural stimulation facilitates *long-term potentiation* leading to increased synaptic sensitivity. This results from increased calcium ion entry into the postsynaptic cell. The opposite effect, called *long-term depression*, appears after continuous, low-level stimulation of synapses that causes decreased calcium entry. Such understanding may one day lead to treatments for learning and memory loss in age-related conditions such as *Alzheimer's disease*.

28. Like other complex neural phenomena, human memory is not well understood. It is clear that stimulation of certain brain regions evokes vivid memories, although memories do not reside in any one brain location. It seems that the neural connections responsible for stimulating memory have a multiplicity of pathways within the brain.

29. There are several categories of memory. *Immediate memory* recalls events that are happening now; it is nearly perfect but lasts only a few seconds. *Short-term memory* retains less information, but can last for many minutes.

30. Repetitive use or reinforcement facilitates the transfer of short-term memory into *long-term memory*, which can last for years. The hippocampus is an integral part of the neural circuitry that allows the formation of long-term memory in humans. Without it the individual cannot remember learning tasks or facts for more than 10 to 15 minutes, although the motor skills associated with these actions are retained.

31. *Declarative memory* is the recollection of people, places, events, and things, while *procedural memory* is the recollection of how to perform a motor task. Studies of humans who have lost portions of their hippocampus show that these two forms of learning and memory are independent and involve different neural mechanisms.

32. The ability of humans to acquire and use language resides in the cerebral hemispheres. More precisely, in 97 percent of people these skills reside in the left hemisphere, a neural phenomenon known as *lateralization*.

33. The human cerebral hemispheres are connected by a mass of white matter called the *corpus callosum*, which allows the transfer of neural information from one hemisphere to the other. Surgical (e.g., to control epilepsy) or accidental severing of the corpus callosum produces "split-brain" individuals, studied by Sperry and colleagues.

34. The case of a split-brain patient whose corpus callosum is cut to control severe epilepsy shows that mechanical knowledge contained in the right hemisphere cannot be expressed in language, which is controlled by the left hemisphere, but the mechanical act itself may still be carried out since this motor function is governed by the right hemisphere. In some instances language cannot even be used as a mechanism to communicate with the right hemisphere in these patients.

35. Further experiments with patients suffering from *aphasias* (deficits in the ability to understand or use words) indicate that normal language skills depend on the proper flow of information between different language areas in the left cerebral cortex. *Broca's area* is a part of the motor cortex that is essential for speech. *Wernicke's area* is part of the temporal lobe related to sensory aspects of language perception and production. The *angular gyrus* is essential for integrating spoken and written language. Damage to any of these areas can result in aphasia.

36. If damage occurs to the left cerebral hemisphere while a person is still a child and the process of language lateralization has not been completed, the right hemisphere will take over the motor and sensory functions related to speaking and writing with little or no detectable loss of language skills.

37. The question of how neural circuits can lead to extremely complex brain phenomena such as consciousness remains beyond current technology, and perhaps even the conceptual framework of modern neuroscience. What is apparent is that such complexity does not simply result from the simultaneous operation of all possible neurons and circuits in the nervous system.

Activities (for answers and explanations, see page 391)

- Explain the neural reason why it is probable that a sleepwalker is *not* in REM sleep. What unique facial behavior characterizes the portion of the human sleep cycle during which dreams occur?

- Refer to textbook Figure 43.16 and explain how Thompson's work on the eye blink reflex in rabbits permits the conclusion that this neural pathway involves a cerebellar brain nucleus.

- Explain why surgically cutting the corpus callosum in the human brain would relieve symptoms of severe epilepsy without destroying basic brain function.

Questions (for answers and explanations, see page 392)

1. In the fight-or-flight response, input from the _____ pathway stimulates an increase in heart rate and blood pressure, whereas input from the _____ pathway returns these parameters to normal levels.
   a. sympathetic; parasympathetic
   b. parasympathetic; sympathetic
   c. autonomic; central
   d. central; autonomic
   e. postganglionic; preganglionic

2. What do the kinds of neural stimulation that lead to penile erection, bladder contraction, and pupillary constriction all have in common?
   a. They are controlled by the forebrain.
   b. They are controlled by the sympathetic branch of the autonomic nervous system.
   c. They are controlled by the parasympathetic branch of the autonomic nervous system.
   d. The nerve impulses involved pass through the optic chiasm.
   e. The nerve impulses involved pass through Wernicke's area.

3. If you wanted to find a complex cell in a cat's visual system, you would need to look in
   a. Broca's area.
   b. the spinal cord.
   c. the reticular activating system.
   d. the occipital cortex.
   e. the somatosensory cortex.

4. In which of the following would the issue of disparity be a factor?
   a. Sleep and dreams
   b. Binocular vision
   c. Language comprehension
   d. Declarative memory
   e. Long-term potentiation

5. In which of the following structures would you expect to find the axon terminals of a postganglionic cell from the sympathetic division of the autonomic nervous system?
   a. Lungs
   b. Heart
   c. Pupil of the eye
   d. Stomach
   e. All of the above

6. A friend wakes you from your sleep, and you immediately have the sensation of having just experienced a vivid dream. If you had been making EMG and EEG recordings of muscle and neural activity at the time you were awakened, which of the following would be *true*?
   a. Your hands and feet might have been twitching slightly.
   b. Your eyes would have been twitching.
   c. Most of your voluntary body muscles would have been inactive.
   d. Your cerebral cortex would have been active as though you were awake.
   e. All of the above

7. Memories associated with the skills that allow a person to sight-read and perform a piece of music they have never seen before reside in
   a. Wernicke's area.
   b. Broca's area.
   c. the corpus callosum.
   d. the right cerebral hemisphere only.
   e. many areas of the brain.

8. Observations of patients with aphasias indicate that
   a. only Broca's area is essential for normal language skills.
   b. language skills depend on proper neural flow between the temporal lobes and motor cortex.
   c. the right hemisphere dominates the production and use of language in humans.
   d. the corpus callosum is not necessary for interhemispheric transfer of information in the brain.
   e. most people can control language functions from either cerebral hemisphere.

9. Which of the following areas of the brain are especially important in controlling aspects of the stages of sleep in mammals?
   a. Thalamus and cerebral cortex
   b. Limbic system and spinal cord
   c. Broca's area and Wernicke's area
   d. Occipital cortex and angular gyrus
   e. Hippocampus and corpus callosum

Integrative Questions (for answers and explanations, see page 392)

1. You see a piece of chocolate and reach down to pick it up. As you begin chewing the candy, glands in your mouth start secreting saliva. Which of the following elements of your nervous system is involved in this sequence of events?
   a. Central
   b. Peripheral
   c. Autonomic
   d. Parasympathetic
   e. All of the above

2. In which of the following cortical lobes can association areas and long-term memory be found?
   a. Temporal
   b. Parietal
   c. Frontal
   d. Occipital
   e. All of the above

3. Which of the following statements about nervous systems and higher brain function is *false*?
   a. Memory is diffusely spread throughout the cerebral cortex.
   b. REM sleep is associated with dreaming in humans.
   c. The forebrain is highly developed in fish and amphibians.
   d. Basic reflexes like breathing and vomiting are coordinated by the brainstem.
   e. The neural basis of consciousness remains unclear.

4. The neural process of speaking a written word begins in _____ of the human brain.
   a. the primary visual cortex
   b. Wernicke's area
   c. Broca's area
   d. the motor cortex
   e. the corpus callosum

44

Effectors

❑ Define the term "effector" and give examples of the kinds of things that effectors do in animals.

❑ Compare and contrast the molecular structure of microfilaments with that of microtubules, and discuss the roles of each in controlling animal effectors.

❑ Describe the structure and function of animal cilia and flagella.

❑ Explain the molecular basis for axoneme function.

❑ Discuss the roles of microfilaments, cytoplasm, plasmagel, and plasmasol in controlling the amoeboid movements of cells.

❑ Describe the structural and functional differences between the three types of vertebrate muscle tissue.

❑ Differentiate between muscles, muscle fibers, and myofibrils.

❑ Explain the structure and function of a sarcomere.

❑ Discuss the roles of actin, myosin, troponin, and tropomyosin in the sliding filament theory of muscle contraction.

❑ Describe the molecular events that occur when a sarcomere contracts, and discuss the importance of ATP in this process.

❑ Explain the role of a motor unit in muscle contraction.

❑ Describe the events that lead to a muscle twitch, and explain how twitches are related to the phenomena of tetanus and tonus.

❑ Discuss the anatomical and physiological differences between fast-twitch and slow-twitch fibers.

❑ Explain how the special anatomy of cardiac muscle leads to its rhythmic contractions.

❑ Compare the structural and functional properties of hydroskeletons, exoskeletons, and endoskeletons.

❑ Discuss the advantages and disadvantages associated with hydroskeletons, exoskeletons, and endoskeletons.

❑ Distinguish between the bones of the axial and appendicular skeleton in humans.

❑ Describe the roles of osteoblasts, osteoclasts, and osteocytes in building, repairing, and remodeling bone.

❑ Identify the various types of bones and joints in a human skeleton, and explain the structural and functional importance of each.

❑ Explain the concept of antagonistic muscle action and distinguish between a flexor and an extensor muscle.

❑ Discuss the roles of ligaments and tendons in the action of endoskeletons.

❑ Describe how endoskeletal action can be analyzed using the lever concept and the general principles of skeletal structure that emerge from this analysis.

❑ List several of the nonmuscle effectors found in animals and give their functions.

CILIA, FLAGELLA, AND CELL MOVEMENT
(pages 973–977)

Key Concepts

1. *Effectors* allow animals to respond to sensory stimuli. Both internal and external responses are possible, and they can take the form of locomotion, chemical secretion, or energy emission.

2. Cell movements range from the actions of cilia and flagella to the contractions of muscle. *Microtubules* and *microfilaments* are the cellular structures responsible for creating movement. They consist of long protein molecules that slide past one another. Microtubules make cilia and flagella move, while microfilaments generate larger-scale movements of muscle.

3. *Cilia* are tiny, hairlike appendages used to move entire cells, push mucus or fluids past the linings of epithelial tissue, or filter food from the environment.

4. Cilia push against a fluid like a swimmer's arms push against water during the breaststroke. The *power stroke* occurs when the outstretched cilium moves backward through the fluid. The cilium folds during the *recovery stroke*, creating minimal drag as it is pulled forward.

5. If cilia are fixed in place, as is the case for cells lining the respiratory tract and portions of the female reproductive system, the fluids that overlie them are moved in the same direction as the power stroke.

6. In eukaryotes, *flagella* are essentially long cilia that occur singly or in small clusters. Flagella move in a whiplike

fashion that creates more power than can be produced by an individual cilium. Flagella can be used to move entire cells such as sperm.

7. Microtubules are responsible for the movement of cilia and flagella. They contain long chains of the globular polypeptide *tubulin* and are always found in a characteristic ring of nine pairs. Sometimes one or two extra microtubules are found in the center of this ring, connected to it by radial spokes. The entire complex is called an *axoneme*.

8. Axoneme movement occurs when adjacent microtubules slide past one another. Chemical crossbridges between microtubules are made of the protein *dynein*, a mechanoenzyme that generates force and undergoes a conformational change when it hydrolyzes ATP in the presence of Ca^{2+} ions.

9. Since microtubules are fixed to the cell at their bases, the sliding movement produces a bend in the cilium or flagellum rather than an elongation. This motile mechanism is an inherent property of the axoneme and requires no external input.

10. Microtubules also cause changes in cell shape through mechanical effects on the cytoskeleton. By changing the length of tubulin chains, intracellular events such as spindle formation during mitosis and directed cell growth in neurons can be coordinated.

11. Microfilaments are made largely of the protein *actin*. They are used to support cilia and to direct cell movements. Acting in combination with another microfilament made of the protein *myosin*, actin generates the contractile forces used to cleave dividing cells, facilitate phagocytosis and pinocytosis, and cause muscle contraction.

12. *Amoeboid movement* is another microfilament-controlled activity. It is characteristic of certain protists and the mobile phagocytes within animal bodies. Actin and myosin microfilaments contract in precise patterns just beneath the plasma membrane of these cells to regulate the distribution of thick *plasmagel* and thin *plasmasol*. The plasmasol is pushed out to form a pseudopod, which eventually stops growing as the cytoplasm turns to plasmagel.

13. Controlled amoeboid motion is based on two factors. The first is the cytoplasm's ability to alternate between plasmasol and plasmagel forms. The second is the coordinated contraction of microfilaments that directs cytoplasmic streaming to form a pseudopod.

Activities (for answers and explanations, see page 392)

• In mammals, portions of the respiratory tract are lined with cilia. Suggest a physiological function for these cilia and briefly explain the molecular basis for how that function is accomplished.

• Match each term with its best description.

_____ Axoneme   *a.* Mechanoenzyme bridge between microtubules in cilia

_____ Recovery stroke   *b.* Central structure of a flagellum

_____ Myosin   *c.* Cell shape change caused by microfilaments

_____ Dynein   *d.* Bending of a cilium to reduce drag forces in water

_____ Pinocytosis   *e.* Microfilament protein often associated with actin

Questions (for answers and explanations, see page 392)

1. ATP is needed for the movements of cilia and flagella because
    a. the microtubules are made of ATP.
    b. ATP is a precursor of tubulin, from which the microtubules are then made.
    c. its hydrolysis provides the energy needed to change the shape of dynein.
    d. ATP is used to connect the base of the cilium or flagellum with the cell.
    e. none of the above

2. Detailed analysis of a cilium from the human respiratory tract would reveal all but one of the following. Select the *exception*.
    a. Tubulin
    b. Plasmagel
    c. Microtubules
    d. Axoneme
    e. Dynein

3. Which of the following statements is *false*?
    a. Effectors are involved only in the production of cell movement.
    b. Microtubules play more of a role in flagellum movement than do microfilaments.
    c. Cilia are often found associated with epithelial tissue in vertebrates.
    d. Flagella are essentially large cilia found singly or in small clusters.
    e. Axonemes are present both in cilia and flagella.

4. Microfilaments play a major role in all but one of the following cell activities. Select the *exception*.
    a. Phagocytosis
    b. Muscle contraction
    c. Amoeboid movement
    d. Flagellum movement
    e. Cell cleavage

5. Which of the following always occurs when an axoneme moves?
    a. Microfilaments form crosslinkages with microtubules.
    b. Tubulin dissociates under the regulation of mechanoenzymes.
    c. A flagellum completes its recovery stroke.
    d. Phagocytosis or pinocytosis occurs.
    e. Adjacent microtubules slide past one another.

6. A scientist gently treats a flagellar axoneme with digestive proteins that separate the cross-links and spokes but leave the microtubules and dynein arms intact. If she then exposes the isolated microtubules to ATP and calcium ions the result should be
   a. the complete breakdown of dynein but not tubulin.
   b. the complete breakdown of tubulin but not dynein.
   c. the breakdown of both tubulin and dynein.
   d. bending of the whole structure.
   e. elongation of the whole structure.

7. If the connection that anchors the microtubules of a cilium to a cell is severed, then the
   a. cilium will continue to function normally.
   b. cilium will beat like a flagellum.
   c. cilium will no longer bend.
   d. microtubules will no longer slide past one another.
   e. power stroke will occur but not the recovery stroke.

MUSCLE CONTRACTION: ACTIN AND MYOSIN INTERACTIONS (pages 977–984)

Key Concepts

1. Muscle cells are specialized for contraction. Their abundance and diversity make them the most important effectors in animals.

2. Muscle cells have high densities of the microfilaments actin and myosin. All three vertebrate muscle types share a common mechanism of contraction that depends on the sliding action of these microfilaments.

3. *Smooth muscle* is under autonomic control and provides the contractile activity of internal organs. It helps regulate food digestion, blood flow, and excretion.

4. Smooth muscle cells are simple compared to other types of muscle. They are long and spindle-shaped, contain a single nucleus, and possess a loose network of microfilaments.

5. Smooth muscle occurs in sheets with gap junctions between adjacent cells. Action potentials can therefore travel rapidly through the entire sheet. Smooth muscle cells also respond to stretching by depolarizing their membranes, making it easier for an action potential to occur, which leads to contraction. Smooth muscle has the capacity to contract in proportion to the magnitude of a stretching stimulus.

6. *Cardiac muscle* is found only in the heart. Cardiac muscle appears striped due to the regular arrangement of actin and myosin bundles. Cardiac muscle cells (also known as muscle *fibers*) form a meshwork of cells in which the individual fibers are not easily separated by the high pressures developed within the heart.

7. Cardiac muscle fibers are uninucleate cells joined together by *intercalated disks*, which provide good adhesion. Gap junctions are also present, allowing direct electrical communication between cells for rapid transmission of action potentials.

8. Although influenced by the autonomic nervous system, the beating of the heart does not depend on direct neural influence. Certain cardiac muscle fibers are *myogenic*; they initiate contraction on their own and act as a pacemaker for the heart. A completely isolated vertebrate heart can beat for hours due to this pacemaking property, making heart transplant surgery possible .

9. *Skeletal* (or *striated*) *muscle* controls all voluntary movements. Skeletal muscle cells are large and contain many nuclei. Bundles of these muscle fibers constitute a whole muscle (e.g., biceps).

10. Just as a whole muscle is made of many muscle fibers, a fiber is made of many *myofibrils*. Each myofibril is in turn a bundle of microfilaments arranged in a highly regular fashion.

11. The myofibril consists of an array of thick myosin filaments surrounded by several thin filaments of actin. The two filament types overlap in places, creating a banded pattern (the striations) along the myofibril. Repeating units of banding, called *sarcomeres*, are evident.

12. The sarcomere is the basic unit of muscle contraction. The widths of certain characteristic bands making up each sarcomere shorten as the microfilaments slide past one another during muscle contraction.

13. *Z lines* on either end of the sarcomere represent the outer boundary anchor points for the actin filaments. A band of myosin filaments (the *A band*) is centered in the middle of the sarcomere.

14. As the actin and myosin move past each other, the entire sarcomere shortens. The light spaces where actin and myosin filaments do not overlap (*H band* and *I band*) shorten as the A band approaches the Z lines. The *M band*, a dark stripe within the H band, contains proteins that hold the myosin filaments in a hexagonal arrangement.

15. The fine structure of actin and myosin suggests a mechanism for how sarcomeres contract. Myosin molecules are made of two long polypeptides coiled together, each with a globular head at one end that contains a binding site capable of forming a crossbridge with actin. A myosin filament contains many myosin molecules. Actin filaments consist of two chains of monomers wound together in a helix. Two strands of *tropomyosin*, a third protein, are coiled around the actin chains and block access to the myosin binding sites.

16. The *sliding-filament theory* explains the molecular events that occur when actin and myosin interact and a sarcomere shortens. The head regions on each myosin subunit bind to actin molecules on an adjacent microfilament, forming a chemical crossbridge between the two chains. When this bond forms, the myosin heads change orientation and the two filaments move past each other by about 5 to 10 nm.

17. ATP binds to myosin at these crossbridge sites, causing the myosin to release the actin. Energy gained from the subsequent hydrolysis of ATP reorients the myosin head so that it can form another bond with actin. ATP is thus required to break the bonds between microfilaments, not to make them. This explains the phenomenon of *rigor mortis*, where the body stiffens following death because ATP is no longer being produced to break the actin and myosin bonds that cause muscle contraction.

18. A *motor unit* includes all of the muscle fibers innervated by a single motor neuron. The timing and coordination of actin–myosin interactions within a motor unit begins at the nerve–muscle synapse. Neurotransmitter released at the neuromuscular junction stimulates changes in the membrane potential of muscle fibers in the motor unit. This leads to depolarization and the firing of an action potential that spreads rapidly across the entire cell surface.

19. The plasma membrane of muscle fibers plunges deep into the cytoplasm to form a system of *T-tubules* that conducts action potentials throughout the muscle cell. The T-tubules communicate with a membranous network called the *sarcoplasmic reticulum* that surrounds every myofibril. The sarcoplasmic reticulum serves as a storage site for Ca^{2+} ions.

20. Action potentials travel quickly through the T-tubules and cause the sarcoplasmic reticulum to release Ca^{2+} ions. This stimulates yet another chemical reaction that directly mediates the interaction of actin and myosin.

21. At rest, tropomyosin covers the myosin binding sites on each actin molecule within the filament. Another protein, called *troponin,* is found attached at regular intervals along the tropomyosin strand. Calcium entering the cytoplasm of the muscle cell upon depolarization binds to troponin and causes it to change shape. This causes the attached tropomyosin to move away from the actin binding sites it was covering, preparing the actin for interaction with myosin.

22. When the muscle cell repolarizes, the calcium ion channels close. Molecular pumps then recollect Ca^{2+} ions into the sarcoplasmic reticulum. This returns troponin to its original conformation, allowing tropomyosin to once again cover the myosin binding sites. The muscle fiber can no longer sustain contraction and returns to its resting condition (i.e., it relaxes).

23. Smooth muscle lacks troponin and tropomyosin, but its contraction is still mediated by Ca^{2+}. Action potentials, hormones, and stretching stimulate entry of Ca^{2+} into the cell, where it combines with the protein *calmodulin,* which acts as a second messenger. Calmodulin activates the enzyme myosin kinase, which phosphorylates myosin heads so that they can bind with actin to cause contraction. Actin–myosin interactions are halted when Ca^{2+} is removed from the cell, calmodulin becomes unbound, and the second messenger signaling system is halted.

24. *Twitches* are the minimal units of contraction caused by a single action potential within the muscle fiber.

25. If the action potentials that generate twitches are separated by enough time, each twitch appears as an all-or-none event (i.e., a complete contraction and relaxation cycle). However, if the action potentials arrive in rapid succession, the cell does not have enough time to complete its relaxation before the next contraction phase begins. The muscle tension created increases in a graded response proportional to the level of neural stimulation.

26. If the stimulation rate is very high, the muscle cell never has time to relax and remains in a completely contracted state called *tetanus*. Tetanus can be maintained only as long as the supply of ATP persists. If the energy supply fails, the muscle fatigues and tension declines.

27. The graded response of whole muscle depends on the number of motor units activated. Low levels of activation are continuously maintained in many muscles, a condition known as *tonus*. Muscle tone is possible without fatigue because precise neural control means that only a small number of fibers are involved at any one moment.

28. Muscle fibers are categorized according to the speed of their twitch. *Fast-twitch fibers* are found in white muscle, which lacks oxygen-binding myoglobin and has few mitochondria or blood vessels. The white breast muscle of a chicken can rapidly produce high maximum tension, but it fatigues quickly and thus cannot sustain this tension. White muscle is useful during short bursts of powerful anaerobic muscle activity, which explains why chickens are not strong flyers. Fast-twitch fibers are also abundant in human sprinters and weight-lifters.

29. *Slow-twitch fibers* are found in dark muscle, which has high concentrations of myoglobin, mitochondria, and blood vessels. They are slow to reach maximum tension, but capable of sustaining that tension for long periods of time without fatigue. They are also well supplied with fat and glycogen reserves for ATP production. Dark muscle is the endurance muscle associated with the chicken's legs. It is used in sustained aerobic activity, such as when the bird runs or when a human swims long distances.

30. Although there can be some adjustment through physical training, the proportion of fast- and slow-twitch fibers in human skeletal muscle is determined mostly by a person's genetic background.

Questions (for answers and explanations, see page 392)

1. Smooth muscle controls the action of all but one of the following effectors. Select the *exception*.
   a. Human thumb
   b. Rat intestine
   c. Chicken blood vessel
   d. Frog urinary bladder
   e. Cow stomach

2. Bundles of _____ make up a _____ , which in turn is bundled to create a _____ .
   a. muscle cells; myofibril; whole muscle
   b. microtubules; muscle cell; myofibril
   c. microtubules; whole muscle; microfilament
   d. microfilaments; myofibril; whole muscle
   e. None of the above

3. Microfilaments can be found in
   a. skeletal muscle.
   b. cardiac muscle.
   c. smooth muscle.
   d. all three vertebrate muscle types.
   e. every muscle type except cardiac.

4. When a sarcomere contracts,
   a. the H band and I band both widen.
   b. the A band stays the same, but the Z lines get closer together.
   c. the Z lines move farther apart.
   d. the A band widens.
   e. the H band shortens and the I band widens.

5. Which of the following statements about the sliding-filament theory is *false*?
   a. Ca^{2+} ions play a crucial role in facilitating the binding of actin and myosin.
   b. The separation of myosin from actin requires ATP.
   c. Troponin is a double-stranded protein that covers the myosin binding sites on actin microfilaments.
   d. The headpiece of a myosin molecule binds to special receptor sites on the actin filament.
   e. Rigor mortis results from a breakdown in the mechanism that releases bound microfilaments.

6. Before stimulating the release of calcium ions that causes muscle contraction in skeletal muscle, an action potential must pass through the
   a. sarcoplasmic reticulum.
   b. T-tubules and sarcoplasmic reticulum.
   c. plasma membrane and T-tubules.
   d. plasma membrane, T-tubules, and sarcoplasmic reticulum.
   e. none of the above

7. The release of Ca^{2+} ions is essential for initiating the sliding action of skeletal muscle microfilaments because
   a. calcium is needed to complete the structural organization of the myofibril.
   b. the enzyme that produces ATP is calcium-dependent.
   c. it is the calcium that depolarizes the muscle cell.
   d. tropomyosin cannot bind to actin in the absence of Ca^{2+} ions.
   e. calcium binds to troponin, causing it to change shape and expose myosin binding sites on the actin.

8. When a person accidentally touches a live electric wire and passes current through his body, many muscles become locked in a state of continuous contraction known as
   a. tonus.
   b. tetanus.
   c. twitching.
   d. depolarization.
   e. myogenesis.

9. The leg muscles of champion marathon runners have a large proportion of _____ fibers, as well as _____.
   a. slow-twitch; little or no glycogen and fat reserves
   b. fast-twitch; lots of glycogen and fat reserves
   c. slow-twitch; very few mitochondria
   d. fast-twitch; abundant mitochondria
   e. slow-twitch; abundant mitochondria

10. The intercalated discs of _____ muscle _____ .
   a. smooth; make possible the movements of intestines and glands
   b. smooth; provide strong mechanical adhesion and rapid electrical communication
   c. skeletal; are the basis for all voluntary muscle action
   d. skeletal; make possible both fast twitches and slow twitches.
   e. cardiac; provide strong mechanical adhesion and rapid electrical communication

Activities (for answers and explanations, see page 393)

- In the following diagram of a relaxed sarcomere, label the actin and myosin filaments, as well as all the major bands and lines visible.

- Now sketch this same sarcomere as it would appear in a contracted state. What essentially has happened to cause this change in the shape of the sarcomere?

SKELETAL SYSTEMS PROVIDE RIGID SUPPORTS • OTHER EFFECTORS IN THE ANIMAL KINGDOM (pages 984–992)

Key Concepts

1. For muscles to generate useful motion they must have supporting structures against which to direct their contractile forces. Such support is provided by *skeletal systems*, of which there are three basic types.

2. *Hydrostatic skeletons* result from the containment of incompressible fluids inside a muscular body cavity. When the muscles contract, the fluid moves and changes the body shape. Opposing sets of circular and longitudinal muscles produce alternating elongation and shortening of the body, allowing controlled movements such as those of a sea anemone.

3. Earthworms have a compartmentalized hydrostatic skeleton made of many repeating segments. Locomotion is achieved by passing alternating waves of circular and longitudinal muscle contraction down the length of this segmented body.

4. Hydroskeletons can also be used to create jet propulsion. By contracting muscles surrounding an open-ended cavity, squids and octopuses force water out of their bodies under pressure. The force generated by this moving water pushes the animal in the opposite direction.

5. *Exoskeletons* are hardened outer body surfaces to which internal muscles attach. Muscle contraction causes the exoskeleton to bend at hinged areas called *joints*.

6. The shell of a mollusk is the simplest kind of exoskeleton. It is made of one or only a few heavily mineralized plates. These plates, which can be quite massive, cover most of the animal's body and grow with it, often producing growth rings. Shells offer good protection for the soft hydroskeletal components and delicate internal body organs.

7. The complex exoskeletons of arthropods are made of a layered *cuticle* that forms tough plates covering all body areas and appendages. Hardening material reinforces the plates everywhere except the joints, which remain flexible.

8. The thin outer *epicuticle* of an exoskeleton is covered with a waxy coating to prevent desiccation. The thicker, tougher *endocuticle* contains the polysaccharide *chitin* for strength and may be further reinforced with calcium salts. Muscles attached to the inner surfaces of the exoskeleton contract to create movement around the hinged joints.

9. Although it provides excellent protection against drying and physical damage, once in place the arthropod exoskeleton cannot increase its size. To grow, an animal must *molt* its exoskeleton and secrete a larger one. Immediately after molting the animal passes through a brief period of vulnerability until its newly formed exoskeleton hardens.

10. The vertebrate *endoskeleton* is an internal framework of hinged bones that attach to the skeletal muscles. Platelike, rodlike, and tubelike bones articulate at joints to permit a wide range of motion.

11. The human skeleton contains over 200 bones and is subdivided into the trunk region, or *axial skeleton*, and the outer region, or *appendicular skeleton*. The appendicular skeleton attaches to the axial skeleton at two locations, the pectoral and pelvic girdles. Endoskeletons are not as protective as exoskeletons, but this disadvantage is more than offset by the endoskeleton's ability to grow continuously throughout life.

12. *Cartilage* and *bone* are the connective tissues that make up the endoskeleton. They are good structural tissues because of their tough extracellular matrices, which are reinforced with collagen fibers.

13. Cartilage is a strong yet flexible material that confers stiffness, resiliency, and padding to endoskeletons. It is used to shape certain features such as the nose, larynx, and ears, or to join bone surfaces together in regions that experience high impact, strain, and repeated movement. Cartilage is the principal component of the embryonic skeleton in most vertebrates, but is replaced with bone as development progresses. The cartilaginous

fishes (sharks and rays) are an exception, retaining a cartilage-based skeleton throughout life.

14. The extracellular matrix of bone is heavily mineralized with calcium phosphate, making it hard and rigid. Endoskeletons also represent the body's major reserve of calcium ions, which are used in muscle and nerve function as well as many other physiological activities. Soluble calcium in body fluids and skeletal calcium are in a dynamic equilibrium that is under hormonal control.

15. Bone-producing cells called *osteoblasts* lay down new matrix on bone surfaces. Eventually they become encased in bone, but remain connected to other living cells via canals that pass through the bone matrix. At this stage the cells are called *osteocytes*, and they help to regulate the dynamic equilibrium of calcium balance in the extracellular fluid.

16. *Osteoclasts* are the cells that erode or reabsorb bone. By tunneling into the bone, osteoclasts create room for osteoblasts to remodel old bone with new matrix. How this activity is controlled is not understood, but it may be related to the pattern of stress forces experienced by the bone.

17. *Membranous bone* develops as plates of connective tissues directly under the skin (e.g., skull), whereas *cartilage bone* begins as a cartilage mold of the skeletal structure and then eventually ossifies into mineralized bone (e.g., long bones of limbs).

18. Bone may be *compact* (solid and hard) or *cancellous* (spongy). The location and function of the bone usually determines whether it will be compact or cancellous, and bones with mixtures of both structures are common.

19. Cancellous bone is lightweight yet rigid. It is well suited for the large heads of long bones that must provide surface area for articulation and muscle attachment without adding weight. The internal meshwork of cancellous bone also provides high resistance to the compression forces encountered at skeletal joints.

20. Compact bone makes up the outer shaft of the hollow long bones. Maximum strength against bending, coupled with the minimum weight that results from having a hollow center, provides excellent biomechanical advantage to these tubelike bones.

21. Compact bone is also called *Haversian bone* because of the concentric rings of bone matrix that form around central blood vessels. Osteocytes within each *Haversian system* are connected to one another, but isolated from other systems. Because of these independent areas of fused bone matrix, Haversian bone is highly resistant to fracturing.

22. Since muscles can only exert force while contracting, endoskeletal joints must be controlled by antagonistic pairs of muscles. A *flexor* muscle bends the joint, where two bones come together, and the *extensor* straightens it.

23. Bones are held together at joints by *ligaments*, and muscles are attached to bone by *tendons*. Both are flexible, strong straps of connective tissue. Ligaments may also hold tendons in place to redirect the force being exerted by a muscle.

24. Several types of joints are found in the human endoskeleton. *Hinged joints* such as the knee or elbow flex well but do not rotate. *Ball-and-socket joints* like the shoulder and hip permit movements in almost any direction. *Pivotal joints* such as the elbow are specialized for rotation. Still other types of joints that permit some rotation, but not as much as ball-and-socket, are found in the bones of the hand.

25. Important mechanical features of endoskeletons are understood by viewing them as *levers*, which are simple machines that distribute forces over a distance. All levers have a *fulcrum* (pivot), a *load arm*, and a *power arm*.

26. Levers with a short load arm and a long power arm produce strength, which is ideal for activities such as digging and chewing. If the load arm is long and the power arm short, strength gives way to speed. The limb bones of running and flying animals tend to be of this second form. Comparing the mechanical structure of the jaw and the lower leg clearly demonstrates these concepts.

27. Effectors not related to movement are also used under special circumstances. Some are specialized for food capture, others for defense, and still others for communication or sensory functions.

28. *Nematocysts* are a type of effector used by cnidarians to capture prey or repel enemies. These cells contain coiled harpoons, often tipped with toxin, that are forcibly ejected from the animal's body.

29. Changes in body color are accomplished by pigment-containing effector cells called *chromatophores*. The response of these cells is under neural and hormonal control. Most chromatophore cell types are fixed in place, but others are capable of amoeboid movement. Still others change shape via the action of attached muscles.

30. Glands also qualify as effectors because they release chemicals that produce responses. Chemicals such as pheromones are used in communication, others regulate physiological functions like digestion, and still others play a role in prey capture or defense.

31. The defensive chemicals produced by certain snakes, frogs, mollusks, and pufferfish are deadly toxins that function by inhibiting key aspects of nerve and muscle function. Other chemicals, such as the mercaptan emitted by skunks, are irritating substances with relatively mild and reversible effects.

32. Some effectors produce electric pulses that are used to sense the environment, communicate, capture prey, or repel predators. Most electrical probing of the environment is done using low levels of voltage and power. However, the modified muscle cells in the electric organs of certain eels can generate pulses of high-voltage, high-power electricity strong enough to stun prey.

Activities (for answers and explanations, see page 393)

- Select from the right column all those terms that apply to each item in the left column.

   _____ Osteoblast   *a.* Cell that lays down new bone

   _____ Nematocyst   *b.* Muscle that causes the leg to straighten

   _____ Extensor   *c.* Effector cell that causes color change

   _____ Exoskeleton   *d.* Found in crabs and insects

   _____ Chromatophore   *e.* Found only in cnidarians

- On the diagram of a human endoskeleton shown below, circle the names of the labeled bones belonging to the axial skeleton.

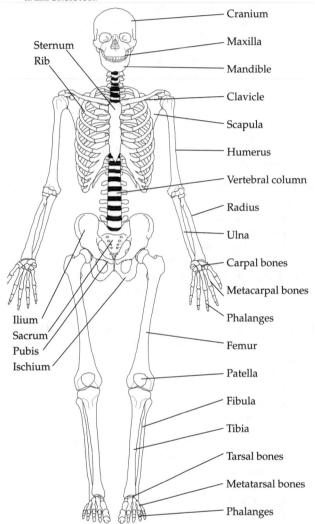

Cranium
Maxilla
Mandible
Clavicle
Scapula
Humerus
Vertebral column
Radius
Ulna
Carpal bones
Metacarpal bones
Phalanges
Femur
Patella
Fibula
Tibia
Tarsal bones
Metatarsal bones
Phalanges

Sternum
Rib
Ilium
Sacrum
Pubis
Ischium

- The following is a drawing of a greyhound dog, a breed prized for its impressive running speed. Even though you cannot see its bones, what obvious endoskeletal features related to speed are apparent from the drawing?

Questions (for answers and explanations, see page 393)

1. Which of the following statements about animal skeletons is *false*?
   a. Skeletal motions are mainly the result of smooth and cardiac muscle activity.
   b. Muscle forces are exerted against some sort of supporting structure.
   c. Articulated joints allow motion to occur in endoskeletons and exoskeletons.
   d. Only endoskeletons contain bone.
   e. Exoskeletons must be periodically molted for the animal to grow in size.

2. The extension and retraction of a sea anemone's body is mechanically analogous to _____ , and demonstrates the workings of a(n) _____ .
   a. tightening a screw; endoskeleton
   b. moving a hinge; exoskeleton
   c. using a pulley and rope; endoskeleton
   d. squeezing a water balloon; hydroskeleton
   e. none of the above

3. The primary difference between an endoskeleton and an exoskeleton is
   a. whether both circular and longitudinal muscles are present.
   b. the kind of bones each skeleton has.
   c. the presence or absence of joints.
   d. the specific locations of ligaments and tendons.
   e. whether or not the skeleton is on the outside of the body.

4. All but one of the following bones are found in the human appendicular skeleton. Select the *exception*.
   a. Finger bones
   b. Upper arm bone
   c. Jawbone
   d. Ankle bones
   e. Kneecap

5. _____ are responsible for regulating extracellular calcium balance by releasing or sequestering Ca^{2+} ions in the bone matrix.
   a. Osteoblasts
   b. Osteoclasts
   c. Osteocytes
   d. Osteoclasts and osteocytes
   e. Osteoblasts, osteoclasts, and osteocytes

6. A football player suffers a knee injury that damages the tissue holding his upper and lower leg bones together. The damaged tissue is probably a kind of
   a. muscle.
   b. tendon.
   c. ligament.
   d. cuticle.
   e. cartilage.

7. Which of the following statements about skeletons and muscles is *false*?
   a. Bones with long power arms and short load arms are found in the skeletons of powerful animals.
   b. The limb bones of fast-flying birds tend to have long load arms.

c. When a flexor muscle contracts, its opposing extensor muscle relaxes and is stretched.
   d. A common category of appendicular skeletal injury in humans is damage to the vertebral disks.
   e. Both extensor and flexor muscles are needed to operate regions of the skeleton adapted for power.

8. Which of the following is probably the most common type of animal effector organ not related to movement?
   a. Light organ
   b. Chemical organ
   c. Auditory organ
   d. Chromatophore
   e. Electric organ

9. All but one of the following animals possesses a type of hydroskeleton. Select the *exception*.
   a. Butterfly
   b. Squid
   c. Earthworm
   d. Sea anemone
   e. Snail

Integrative Questions (for answers and explanations, see page 393)

1. In a hydrostatic skeleton, antagonistic control of movement is achieved by
   a. the use of ligaments as straps over tendons to redirect forces from a skeletal muscle.
   b. flexible hinge joints where the skeleton has not hardened.
   c. repeated segmentation of the body.
   d. waves of action potentials sweeping through the smooth muscle of the body cavity.
   e. the opposing action of circular and longitudinal muscles.

2. Which of the following effectors is *not* found in humans?
   a. Smooth muscle
   b. Electric organ
   c. Glands
   d. Jointed endoskeleton
   e. Cilia

3. Cilia bend in a rowing motion, while flagella create whiplike movement. This difference is due mainly to
   a. the differences in length between the two types of effectors.
   b. differences in the arrangement of their microtubules.
   c. the kind of muscles that cause the motion to occur.
   d. differences in the rate at which each effector takes up Ca^{2+} ions.
   e. the presence or absence of calmodulin.

4. Which of the following would you *not* expect to find in both a skeletal muscle cell and a cardiac muscle cell?
   a. Actin filaments
   b. Myosin binding sites
   c. Potassium channels that close slowly after an action potential
   d. Large numbers of mitochondria
   e. Troponin

45

Gas Exchange in Animals

RESPIRATORY GAS EXCHANGE • ADAPTATIONS FOR GAS EXCHANGE (pages 995-1005)

Key Concepts

1. Metabolic production of ATP is accomplished through the oxidation of nutrient molecules by cells. This process requires oxygen (O_2) and produces the waste product carbon dioxide (CO_2).

2. Specialized cells such as muscle can incur an oxygen debt for short periods of time and still function properly. However, many cells, especially neurons, cease metabolic activity and die shortly after their oxygen or ATP supplies are eliminated.

3. Whole-animal gas exchange (not to be confused with the biochemical process of cellular respiration) is how gases are exchanged between animals and their environments.

4. Many animals use breathing to move gases in and out of their bodies, but the actual exchange of gases takes place across specialized internal respiratory membranes. Other animals accomplish gas exchange without breathing, absorbing and releasing gases across external respiratory membranes.

5. The rate of gas exchange necessary to support metabolism varies greatly with species and physiological conditions. For example, different species subjected to identical physiological conditions can have very different respiratory rates, as can the same species subjected to different conditions.

6. Regardless of respiratory anatomy, simple *diffusion* is the only mechanism for gas exchange in animals; active transport is never involved. Gas exchange is subject to a variety of important physical limitations.

7. Whether an animal respires in air or water is one physical limitation that influences the structure of gas-exchange mechanisms. It is easier to respire in air because air contains more oxygen by volume than does water. In addition, oxygen diffuses faster in air than in water. It also takes more energy to move water across respiratory exchange surfaces than it does to move an equal volume of air.

8. Physical limits on the rate of O_2 diffusion in water also affect the size, shape, and location within the body of animal cells. All respiring eukaryotic cells must be located within a millimeter or less of the nearest oxygen source, otherwise the slow diffusion of O_2 through an aqueous medium severely limits the rate of cellular respiration.

9. Many small invertebrates have solved this diffusion problem by having a thin body plan. Others have a centralized hollow body cavity through which water circulates. Larger, more complex animals such as vertebrates have specialized gas-exchange tissues with a very large surface area, combined with an internal circulatory system for transporting gases throughout the body.

10. Temperature is another factor limiting respiration in aquatic breathers. As temperature rises, the O_2 content of water drops. However, because of Q_{10} effects, the metabolic rate of aquatic ectotherms increases as water temperature rises, creating a need for more O_2. Water-breathing animals are thus forced to expend energy to extract greater amounts of O_2 from the water as temperature rises, or else decrease their activity.

11. Air-breathing animals face similar O_2 constraints with respect to altitude. *Barometric pressure* is the pressure caused by the weight of all atmospheric gases combined. Since the quantity of atmosphere overhead decreases with increasing altitude, both barometric pressure and O_2 concentration decrease with altitude.

12. The *partial pressure* of a gas indicates what proportion of total barometric pressure is due to that gas. As with barometric pressure, partial pressure decreases with altitude. Because gas exchange occurs by diffusion, which is driven by concentration or pressure differences across respiratory membranes, the reduction in the partial pressure of oxygen (P_{O_2}) that occurs at high altitude makes gas exchange more difficult.

13. Diffusion rates for O_2 and CO_2 are similar given eqivalent gas concentrations. However, CO_2 typically leaves the body more readily than O_2 enters. This occurs for air breathers because CO_2 concentrations are very low in the atmosphere, and for water breathers because CO_2 solubility in water is high.

14. *Fick's law of diffusion* describes the rate at which substances diffuse between two locations (see textbook page 999 for the equation). Since substances diffuse faster in air than in water, animals can increase the diffusion rate by breathing air whenever possible. The only other parameter that animals can influence directly to increase diffusion rate is the size of the respiratory surface area over which gas exchange occurs.

15. *External gills* are specialized elaborations of the body surface that greatly enlarge the membrane area available for gas exchange in water. They are efficient but delicate, and their external position exposes them to many hazards. *Internal gills* are protected by body coverings and water is pumped over them.

16. Air-breathing animals must prevent their respiratory membranes from drying out. The actual gas-exchange surfaces of vertebrate air breathers are elaborate, highly divided air sacs contained within expandable bags called *lungs*. The gas-exchange surfaces of air-breathing insects are finely branched tubes that penetrate the animal's entire body.

17. Increased respiratory membrane surface area is not the only way to improve gas exchange. Fick's law suggests that animals can also maximize the concentration gradient of gases across these membranes by 1) continuously exposing the external side to fresh respiratory medium and 2) maintaining the lowest possible O_2 and the highest possible CO_2 concentrations on the internal side of the membrane. Membranes can also be kept very thin to minimize diffusion distance.

18. An animal's *gas-exchange system* includes the respiratory membranes where diffusion occurs, the mechanisms used to *ventilate* the external surfaces of those membranes with respiratory medium, and the mechanisms used to *perfuse* the internal side of the respiratory surfaces with blood.

19. In the respiratory system of insects, air enters the body through openings called *spiracles*. It then travels through a system of air tubes called *tracheae* until it reaches finely branched tubes known as *air capillaries*, where diffusion actually takes place. Insect tracheal systems are so extensive that all body cells are well supplied with O_2.

20. The high rates of metabolism that the tracheal system permits allow insects to maintain very active lifestyles, but the limits on diffusion imposed by the small diameter of the tubes and dead-end airways keeps insects from becoming large animals.

21. Fish have paired internal gills that are supported by bony *gill arches* and covered with protective *opercular flaps*. Lining the arches are hundreds of interdigitated rows of *gill filaments* pointing in the direction of water flow. Each filament has many tiny, delicate surface folds called *lamellae*, which greatly increase the surface area for gas exchange.

22. Gas exchange in fishes is maximized by ventilating both sides of the respiratory membranes. The external surface is ventilated with water using a two-pump mechanism. The mouth cavity acts as a *positive-pressure pump* to push water over the gills. Slightly out of phase with the mouth, the opercular flaps act as a *negative-pressure pump* to pull water over the gills. The result is a virtually

continuous, unidirectional flow of water over the gill filaments.

23. On the internal side of the gill respiratory membranes, *afferent blood vessels* carry blood to the lamellae and *efferent blood vessels* take it away. Since blood perfusion is in the opposite direction of water movement, a *countercurrent flow* is established. This antiparallel flow creates more efficient gas exchange between the water and blood than would be possible with parallel flow.

24. *Countercurrent exchangers* are the physiological and anatomical mechanisms associated with antiparallel flow. They generally act to maximize the concentration differences between the blood and the environment, thereby maximizing diffusion. Countercurrent exchangers are used not only for gas exchange, but also for concentrating urine in the kidneys and for retaining body heat.

25. Most air-breathing vertebrates have lungs with branching airways that terminate in clusters of blind-ended microscopic sacs where gas exchange actually takes place. Air movement must therefore be *tidal* (moving in and out of the lungs along the same path) rather than unidirectional. Tidal respiration is relatively inefficient compared to one-way air flow with countercurrent exchange.

26. Birds are unique among vertebrate air breathers in having unidirectional air flow through their lungs. This is made possible by tiny airways known as *parabronchi*, from which emerge countless air capillaries that greatly increase the surface area for gas exchange.

27. Bird lungs expand and contract very little during breathing. To help move air through the lungs, birds have large *air sacs* at the front and back of the body. These air sacs act as bellows to create unidirectional flow. A single breath of air remains in the avian respiratory system for two complete inhalation–exhalation cycles.

28. Upon first inhalation, air passes through the tubular *trachea* (windpipe) and into smaller airways called *bronchi* before entering the posterior air sacs. During first exhalation, this air moves from the posterior air sacs into the lungs via the parabronchi. The second inhalation forces the air from the lungs into the anterior air sacs while new air refills the posterior sacs. During the second exhalation the old air leaves the anterior air sacs and exits the body.

29. Birds enhance gas exchange by having crosscurrent blood flow on the internal side of the respiratory membranes. With blood flow and air flow operating at right angles, more O_2 can be extracted from a given volume of air than is possible in the tidal breathing of mammals, reptiles, or amphibians. This increased respiratory efficiency permits the sustained high metabolic rates and energy-intensive behaviors (e.g., flight at high altitudes) characteristic of birds.

30. Vertebrate lungs have their evolutionary origins as blind-ended outpocketings of the digestive tract. In bony fishes these outpocketings serve as air bladders. In terrestrial vertebrates they have evolved into lungs.

31. All terrestrial vertebrates except birds utilize tidal breathing. The *tidal volume*, or amount of air moved per breath during a normal breathing cycle, averages about 500 ml in adult humans. The *inspiratory reserve volume* represents the air gathered in excess of the tidal volume during deep inhalation. More than the tidal volume can also be exhaled; this constitutes the *expiratory reserve volume*.

32. Some *residual volume* remains in the lungs even after extreme exhalation because the lungs do not collapse completely. *Total lung capacity* is therefore the sum of tidal volume, residual volume, and inspiratory and expiratory reserve volumes.

33. Tidal breathing constrains respiratory efficiency. Fresh air entry is limited to one-half of the respiratory cycle, and once inside the lungs, the fresh air mixes with stale air from the residual volume. Mixing of air greatly reduces P_{O_2} and therefore limits diffusion. *Dead space* refers to the lung volume that is not ventilated with fresh air, and consists of the residual volume and some or all of the expiratory reserve volume.

34. Tidal breathing also eliminates the possibility of countercurrent exchange. This means the P_{O_2} in blood leaving the lungs of tidal breathers can never exceed that of the exhaled air, whereas in fishes and birds the blood leaving the respiratory surfaces can have a P_{O_2} almost as high as that of the environment. Despite these limitations, tidal breathing mechanisms are adequate to serve the respiratory needs of mammals.

Activities (for answers and explanations, see page 393)

- Suggest a reason that P_{O_2} is lower on top of Mount Everest in the Himalayas than it is on top of Mount Kilimanjaro in Africa, and what effect this would have on an air-breathing animal.

- If you were to design an apparatus that would enable human divers to rebreathe the same air by allowing that volume of air to exchange gases with the surrounding water, what three essential design features would you need to build into your apparatus?

  a.

  b.

  c.

• Below is a diagram of a countercurrent exchanger from a fish gill. The values shown indicate percent saturation of O_2. Based on this information alone, indicate which tube represents the flow of blood and which represents the flow of water. Indicate the direction in which each of these fluids is flowing, and show the direction of net O_2 diffusion. Explain the reasoning behind your choices.

Questions (for answers and explanations, see page 394)

1. Which of the following statements about gas exchange and respiration is *false*?
   a. Whole-body respiration refers to gas exchange between the animal body and its environment.
   b. Temperature is an important limiting influence on gas exchange in air-breathing animals.
   c. The chemical breakdown of nutrients to form ATP is an oxygen-requiring metabolic process.
   d. The diffusion of gases in air is faster than their diffusion in water.
   e. The concentration of O_2 in air is greater than it is in a similar volume of water.

2. As water temperature increases,
   a. O_2 content of the water decreases.
   b. barometric pressure decreases.
   c. CO_2 content of the water increases sharply.
   d. the diffusion rate of O_2 decreases.
   e. all of the above

3. All but one of the following are valid reasons that breathing air is easier than breathing water. Select the *exception*.
   a. A given volume of air is easier to move through the respiratory system than the same volume of water.
   b. Air holds more oxygen per unit of volume than does water.
   c. O_2 diffuses faster in air than it does in water.
   d. Temperature increases affect the O_2 content of water more than they do that of air.
   e. Water breathers have more difficulty getting rid of CO_2 because it does not dissolve well in water.

4. An aquatic animal has no specialized respiratory surfaces, and must rely on simple diffusion alone to carry out gas exchange. Which of the following would *not* be characteristic of this animal?
   a. Its body would be very thin or possibly have a hollow, water-filled central cavity.
   b. Its metabolic rate would be rather low.

   c. It would be a fast-swimming, active predator.
   d. None of its cells would be more than one millimeter from the outside world.
   e. Water temperature would have a large influence on its rate of gas exchange.

5. External gills, tracheae, and lungs all share which of the following sets of characteristics?
   a. Part of gas exchange system; exchange both CO_2 and O_2; increase surface area for diffusion
   b. Used by water breathers; based on countercurrent exchange; use negative pressure breathing
   c. Exchange only O_2; are associated with a circulatory system; found in vertebrates
   d. Are associated with spiracles; used only by air breathers; relatively inefficient respiratory rates
   e. Found in insects; employ positive-pressure pumping; based on crosscurrent flow

6. Which of the following would you expect to find in an insect's respiratory system?
   a. Lamellae
   b. Gill arches
   c. Spiracles
   d. Countercurrent exchanger
   e. Parabronchi

7. A water-borne molecule of oxygen arriving at the gills of a fish would encounter which of the following structures *prior* to entering the fish's blood?
   a. Opercular flap, gill arch, gill filament, pleural cavity
   b. Gill filament, air capillary, parabronchus
   c. Tracheae, bronchioles, alveoli
   d. Gill filaments, lamellae, blood capillary wall
   e. None of the above

8. During the second inhalation in a sequence of breaths taken by a bird, air is
   a. exiting the body directly from the posterior air sacs.
   b. exiting the body directly from the lungs.
   c. entering the mouth and going straight to the lungs.
   d. filling both the anterior and posterior air sacs.
   e. entering the mouth and going straight to the anterior air sacs.

9. All but one of the following animals utilizes tidal breathing. Select the *exception*.
   a. Bear
   b. Salmon
   c. Frog
   d. Lizard
   e. Whale

10. Which of the following represents a *larger* volume of air than is normally found in the resting tidal volume of a human lung?
    a. Residual volume
    b. Inspiratory reserve volume
    c. Expiratory reserve volume
    d. Total lung capacity
    e. All of the above

MAMMALIAN LUNGS AND GAS EXCHANGE
(pages 1005–1008)

Key Concepts

1. Air enters the mammalian respiratory system through the oral or nasal cavities, which join in the pharynx. One opening from the pharynx leads to the esophagus, which passes food to the stomach. The other opening leads to the *trachea*, a cartilage-reinforced tube that directs air toward the lungs. At the head of the trachea is the *larynx*, which contains the vocal cords.

2. The trachea branches into two smaller airways called *bronchi*, one of which leads to each lung. The bronchi branch repeatedly and decrease in size, eventually becoming small-diameter *bronchioles*. Located at the end of the bronchioles are clusters of blind-sac respiratory surfaces called *alveoli*.

3. The airways connecting mammalian lungs with the outside world act as conduits for gases (i.e., they are dead spaces). It is only in the alveoli where actual diffusion of gases occurs between the lungs and the blood.

4. Although individually tiny, the over 300 million alveoli in human lungs collectively possess an enormous surface area (~70 m^2). Each alveolus is made of extremely thin cells and is surrounded by small, thin-walled blood vessels. This arrangement produces diffusion distances of less than 2 μm, which greatly enhances the rate of gas exchange.

5. *Surfactant* is produced by cells in the alveoli. Its function is to reduce the surface tension of liquids covering the alveolar walls. Surface tension results from the cohesive forces between molecules of a liquid; it makes fluid surfaces act like an elastic membrane. Surfactant interferes with these cohesive forces and thus reduces surface tension

6. Although surface tension contributes to the elastic recoil aiding exhalation, it can make inhalation difficult if it is not reduced by surfactant. The condition known as *respiratory distress syndrome* in premature infants is an example. These babies are born prior to complete lung development, and thus they do not produce the necessary surfactant to reduce surface tension in the alveoli.

7. Mucus is also produced by the epithelial cells lining mammalian airways. Beating cilia in these airways create a *mucus escalator* that moves the mucus, along with any trapped particulate matter, away from the delicate alveoli. This material eventually reaches the pharynx, where it is swallowed.

8. Inhaled pollutants such as cigarette smoke immobilize the cilia and cause the mucus escalator to cease operation. Hacking and coughing result as the body attempts to clear the airways of accumulated debris.

9. Mammalian lungs are suspended inside the *thoracic cavity* and are virtually surrounded by the ribs. The thoracic cavity is lined by the *pleural membranes*, which enclose each lung within its own *pleural cavity*. Below the lungs is located the dome-shaped muscular *diaphragm*, which separates the thoracic cavity from the abdominal cavity.

10. Inhalation is an active process in mammals. As the diaphragm contracts, it flattens and moves downward into the abdominal cavity. At the same time, the rib cage expands because of the contraction of *intercostal muscles* between the ribs. Both actions increase the volume of the pleural cavities, creating a negative pressure (suction) that pulls air into the lungs.

11. If the seal formed by the pleural cavity is penetrated, as can occur with a puncture wound to the thoracic region, air leaks into the cavity as it tries to expand. This prevents the lung from inflating properly, and may lead to lung collapse.

12. Exhalation in mammals is usually a passive process. It begins when the diaphragm and intercostals relax, thereby reducing the volume of the thoracic cavity and pushing air out of the lungs. During periods of strenuous activity when gas exchange demands are high, muscle groups located between the ribs can contract to forcibly expel air from the lungs.

Activities (for answers and explanations, see page 394)

- Draw a simple diagram that shows how negative pressure is generated during the inhalation phase of human breathing, and how this leads to expansion of the lungs with fresh air. Is this an active or a passive process? Explain your answer.

- Arrange the following structures of the human respiratory system in the sequence that a molecule of oxygen would encounter them during inhalation.

alveoli	*bronchioles*	*trachea*	*blood capillaries*
bronchi	*larynx*	*nasal cavity*	*pharynx*

Questions (for answers and explanations, see page 394)

1. Which of the following statements about the mammalian respiratory system is *false*?
   a. The structural integrity of the thoracic cavity is essential to generate negative pressure during inhalation.
   b. Surfactant reduces the surface tension in the alveoli.
   c. Inhalation is a passive process in mammalian lungs, while exhalation requires muscle contraction.
   d. It is generally not as efficient as the respiratory exchange system of birds.
   e. The diaphragm and the intercostal muscles help produce negative pressure for human breathing.

2. Which of the following structures is the actual gas exchange surface in the mammalian respiratory system?
   a. Bronchus
   b. Alveolus
   c. Bronchiole
   d. Trachea
   e. Nasal cavity

3. A key *difference* between the respiratory systems of mammals and birds is that
   a. mammals utilize diffusion and birds do not.
   b. birds utilize diffusion and mammals do not.
   c. mammals have a unidirectional flow of air and birds do not.
   d. birds have a unidirectional flow of air and mammals do not.
   e. mammals have parabronchi and birds do not.

4. The function of the mucus elevator in the mammalian respiratory system is to
   a. move surfactant from the bronchi to the alveoli.
   b. produce negative pressure during inhalation.
   c. stimulate contraction of rib musculature during forcible exhalation.
   d. trap and remove particulate matter that has entered the respiratory system.
   e. none of the above

5. The condition known as respiratory distress syndrome in premature human infants is caused by
   a. lack of surfactant in the alveoli.
   b. lack of surfactant in the bronchi.
   c. puncture of the thoracic cavity.
   d. failure of the diaphragm to contract.
   e. failure of the intercostal muscles to contract.

BLOOD TRANSPORT OF RESPIRATORY GASES
• REGULATING BREATHING (pages 1011–1013)

Key Concepts

1. The blood circulatory system perfuses the internal surface of respiratory membranes. By removing O_2 as fast as it diffuses across the membrane, the circulatory system helps keep the oxygen-concentration low on the internal surface, which encourages further diffusion. O_2 is then circulated throughout the body for use by cells.

2. The liquid fraction of blood, called *blood plasma*, carries only a small fraction of the O_2 picked up at the respiratory membranes. The oxygen-carrying capacity of blood is greatly enhanced by the oxygen-binding pigment molecule hemoglobin, large numbers of which are contained within specialized *red blood cells*.

3. Four polypeptide subunits make up a molecule of hemoglobin, each of which surrounds an iron-containing heme group. The heme group reversibly binds O_2 and holds it within the red blood cell, thereby maximizing the concentration difference of free O_2 between the two sides of the respiratory membrane.

4. The P_{O_2} of the local environment determines the ability of hemoglobin to pick up or release oxygen. An S-shaped oxygen-binding curve describes this relationship. It shows that when P_{O_2} is high, hemoglobin molecules carry their maximum load of O_2, and that at low P_{O_2} values hemoglobin releases its O_2 cargo.

5. This relationship exists because of shape changes that occur when hemoglobin subunits bind to oxygen. At low P_{O_2} levels, only one subunit of hemoglobin will bind to O_2. Once a single molecule of O_2 is bound, the hemoglobin changes shape, which makes it easier for the next O_2 to bind. Further binding requires only slight increases in P_{O_2}, and each addition of O_2 causes further shape changes in the hemoglobin molecule. This interaction between the binding of O_2 to one subunit of hemoglobin and the change in binding affinity of another subunit is known as *positive cooperativity*.

6. The cooperativity mechanism is adaptive because of the steep slope produced in the oxygen-binding curve. When fully loaded hemoglobin molecules reach normal body tissues, they usually give up about 25 percent of their O_2 load. However, it only takes a slight drop in the P_{O_2} of those tissues to stimulate a massive unloading of O_2 from the hemoglobin. Such drops in P_{O_2} often occur because of increased metabolic activity by cells, exactly the situation where more O_2 is required.

7. *Myoglobin* is a special oxygen-carrying pigment found in muscle. It has only one polypeptide chain, and its affinity for O_2 is much higher than hemoglobin's, so it will pick up and retain O_2 at much lower P_{O_2} values (i.e., it plots to the left of the hemoglobin oxygen-binding curve). Myoglobin provides an oxygen reserve when metabolic demands are high and blood flow may be restricted because of muscle activity. Dark muscle, which is used for extended bouts of flying, running, or diving, contains much myoglobin.

8. Hemoglobin function is regulated by structural properties of the polypeptides in the molecule, as well as by chemical characteristics of the cellular environment in which hemoglobin is found.

9. In adult humans, hemoglobin is made up of two pairs of different polypeptides. Two alpha chains and two beta chains are found in each molecule. They combine to produce the characteristic oxygen-binding properties of human hemoglobin described above.

10. Fetal hemoglobin differs from adult hemoglobin in that the beta chains are replaced by two gamma chains. This alteration allows fetal hemoglobin to pick up O_2 from the mother's blood even though both are at the same P_{O_2}. Thus, like myoglobin, the fetal hemoglobin oxygen-binding curve plots to the left of the adult curve.

11. Mammals adapted for life at high altitudes, such as llamas and vicuñas, also have left-shifted oxygen-binding curves. This allows their hemoglobin to saturate at the very low P_{O_2} levels characteristic of the high mountains where they live.

12. The *Bohr effect* describes the influence of pH on hemoglobin function. In tissues undergoing high metabolic activity, acidic metabolites are produced that lower pH. Under these conditions, the oxygen-binding curve shifts to the right, so hemoglobin will more easily give up its O_2 in areas of high cell metabolism.

13. *Diphosphoglyceric acid*, a normal cell metabolite, is found in especially high concentrations within human red blood cells. The concentration of diphosphoglyceric acid increases in response to sustained exercise and during acclimation to high altitudes. It reversibly binds to deoxygenated hemoglobin and causes it to change shape, making the hemoglobin more likely to release O_2. This also produces a shift to the right in the oxygen-binding curve.

14. Humans thus employ a fundamentally different strategy from llamas and vicuñas for altering hemoglobin function at high altitude. However, the end result (delivering more O_2 to the body cells) is the same.

15. The blood circulatory system also has responsibility for removing waste CO_2 from the tissues and transporting it to the gas-exchange surfaces. Although CO_2 readily dissolves in water, very little of it is actually transferred in the blood as dissolved CO_2.

16. Most CO_2 is transported in the blood in the form of *bicarbonate ions* (HCO_3^-). CO_2 reacts with water to form carbonic acid (H_2CO_3), which then dissociates into H^+ and HCO_3^-. This reaction proceeds rather slowly in the blood plasma, but is greatly accelerated inside red blood cells by the enzyme *carbonic anhydrase*.

17. As CO_2 enters the blood from metabolizing tissues, it diffuses into red blood cells and is quickly converted into bicarbonate ions, which then return to the plasma. This keeps the CO_2 concentration low in the blood, promoting further diffusion from tissues to blood. A small amount of CO_2 is also carried directly by deoxygenated hemoglobin in the form of *carboxyhemoglobin*.

18. Because of breathing, CO_2 concentration in the airspace of the lungs is kept low relative to its concentration in the blood. Thus, upon arriving at the alveoli, CO_2 in the blood readily diffuses into the alveolar air space. This stimulates the movement of HCO_3^- from the plasma into the red blood cells, where H^+ and HCO_3^- reassociate to form H_2CO_3, which then converts to CO_2 and H_2O through the action of carbonic anhydrase. The newly reformed CO_2 leaves the red blood cells and enters the aleveoli from the plasma.

19. Although breathing results from many complex, coordinated movements involving several different muscle groups, it is nevertheless an autonomic function. Ventilatory rhythms are automatically maintained by the central nervous system to provide the proper depth and frequency of breathing for current metabolic conditions.

20. Neurons in the medulla of the brain stem are primarily responsible for the control of breathing. These cells fire as the active phase of inhalation begins. At the peak of inhalation the neurons cease firing, and the passive recoil of respiratory musculature produces exhalation. In situations that demand high respiratory rates, expiratory neurons in the medulla also fire to drive active exhalation.

21. The *Hering–Breuer reflex* is an override mechanism that is activated during intense respiration to prevent ventilatory muscles from damaging the lung tissue. Stretch receptors in the lungs detect overextension and send action potentials to the brain along the vagus nerve, inhibiting the activity of respiratory neurons in the medulla.

22. Neurons located in higher brain centers also influence breathing rhythm. The control provided by these neural elements allows for proper breathing during speech, eating, and coughing.

23. Humans and most other mammals are not well-equipped to detect low levels of blood P_{O_2}. Instead, mammals are very sensitive to increases in the blood P_{CO_2}, because it is CO_2 that varies most in response to changing metabolic or environmental conditions.

24. Blood P_{CO_2} is sensed primarily by detectors located in the vicinity of the medullary neurons controlling ventilatory rhythm. What P_{O_2} sensitivity there is resides in nodes of tissue found on the aorta and carotid arteries of the heart. These vessels continuously receive large volumes of blood and are equipped with chemosensors to detect drops in P_{O_2}. If activated, the chemosensors send signals directly to the respiratory centers of the brain.

Questions (for answers and explanations, see page 394)

1. Which of the following statements about hemoglobin is *false*?
   a. Hemoglobin is what allows the blood to carry a large volume of O_2.
   b. Its capacity to hold O_2 is influenced by pH and chemical features of the environment.
   c. It contains a single polypeptide chain with very high affinity for O_2.
   d. In humans, fetal hemoglobin is structurally different from adult hemoglobin.
   e. It is packaged inside of red blood cells.

2. Which of the following would normally be present in the blood plasma of a mammal?
   a. Red blood cells
   b. Bicarbonate ions
   c. Water
   d. Oxygen
   e. All of the above

3. The oxygen-binding curve for myoglobin is steep and shifted to the left of the same curve for hemoglobin. This indicates that
   a. people with myoglobin are acclimated for life at high altitudes.
   b. myoglobin occurs in high concentration in the human fetus.
   c. myoglobin does not function well at high percent saturation of oxygen.
   d. myoglobin picks up and releases O_2 at lower P_{O_2} values than does hemoglobin.
   e. none of the above

4. A shift in the O_2 binding capacity of hemoglobin as a function of pH is known as
   a. oxygen binding
   b. the Hering–Breuer reflex.
   c. the Bohr effect.
   d. respiratory ventilation.
   e. none of the above

5. The primary respiratory effect of diphosphoglyceric acid is to
   a. shift the oxygen-binding curve for myoglobin to the right of normal.
   b. shift the oxygen-binding curve for hemoglobin to the left of normal.
   c. convert CO_2 into HCO_3^- so that more CO_2 can be removed from body tissues.
   d. alter the pH of the blood plasma during times of reduced metabolic activity.
   e. change the shape of hemoglobin and make it more likely to release O_2 at the P_{O_2} of the tissues.

6. The main reason CO_2 concentrations in the alveoli of mammalian lungs are usually rather low is that
   a. CO_2 is rapidly removed from the lungs through breathing.
   b. carbonic anhydrase converts the CO_2 into carbonic acid.
   c. CO_2 rapidly diffuses from the alveolar air into the blood.
   d. CO_2 spontaneously converts to bicarbonate ions once in the alveolar air.
   e. carboxyhemoglobin binds to CO_2 as soon as it enters the alveolar air.

7. All but one of the following are involved in the neural control of ventilation. Select the *exception*.
   a. Neurons in the medulla
   b. The vagus nerve
   c. Chemosensors on the surface of the medulla
   d. Contraction state of the diaphragm
   e. Stretch receptors in the lungs

8. A person receives a serious cut on the back of the neck that is deep enough to severely damage the brain stem in the lower medulla. The respiratory effect of this type of wound would likely be
   a. catastrophic, and it might result in a complete cessation of breathing.
   b. to reduce respiration to an irregular rhythm.
   c. to shift the oxygen-binding curve of hemoglobin to the left.
   d. to stimulate the production of additional diphosphoglyceric acid.
   e. almost undetectable.

Activities (for answers and explanations, see page 394)

- Match each of the following features of respiratory control with its best description.

  ___Carbonic anhydrase   a. Dark muscle pigment with high oxygen-binding affinity

  ___Myoglobin   b. Tendency for hemoglobin to release O_2 as P_{O_2} decreases

  ___Oxygen binding   c. Brain stem structure responsible for regulating ventilatory rhythm

  ___Carboxyhemoglobin   d. Deoxygenated blood pigment bound to CO_2

  ___Medulla   e. Important gas exchange enzyme found in red blood cells

- Examine the family of oxygen-binding curves shown below.

Which of these curves probably belongs to a normal adult human? _____
Which of these curves probably belongs to an adult human acclimated to high altitude? _____

Which of these curves probably belongs to a llama? _____

Explain the reasoning behind your choices.

Integrative Questions (for answers and explanations, see page 395)

1. An animal has external gills. Which of the following *must* also be true of this animal?
   a. It is a fish.
   b. It is aquatic.
   c. It is an invertebrate.
   d. It lives at high altitude.
   e. It is an insect.

2. All but one of the following respiratory structures or functions is present in birds. Select the *exception*.
   a. Lungs
   b. Large air sacs
   c. Tidal respiration
   d. Crosscurrent flow of air and blood
   e. Parabronchi

3. A jellyfish, an earthworm, and a lion all use which one of the following in their gas exchange systems?
   a. Countercurrent flow
   b. Diffusion
   c. Negative-pressure pump
   d. Positive cooperativity
   e. Carboxyhemoglobin

4. Which of the following is *smaller* than an alveolus from a human lung?
   a. Red blood cell
   b. Molecule of hemoglobin
   c. Molecule of oxygen
   d. Diameter of smallest blood capillary in lung
   e. All of the above

46

Circulatory Systems

CHAPTER LEARNING OBJECTIVES—after studying this chapter you should be able to:

❑ Compare the circulatory needs of simple aquatic animals with those of more complex and terrestrial animals.

❑ Describe the basic differences between gastrovascular cavities, open circulatory systems, and closed circulatory systems.

❑ Explain the structure and function of a representative invertebrate circulatory system.

❑ Describe the role of the heart, vessels, and blood in a closed circulatory system, and discuss the circulatory advantages closed systems have over open systems.

❑ Distinguish between the pulmonary and systemic circuits of a closed circulatory system.

❑ Compare the structure and function of arteries, veins, arterioles, venules, and capillaries.

❑ Describe the evolution of vertebrate hearts and circulatory systems from fish through mammals, and state the advantages that accrued with each advancement.

❑ Identify the major structures of the human heart, give their functions, and trace the flow of blood through the heart.

❑ Describe the various phases of the cardiac cycle, and explain the importance of systole and diastole in this process.

❑ Discuss the use of an electrocardiogram to measure heart function, and explain several abnormal heart conditions that can be identified using this technique.

❑ Describe the anatomy and neural connections that produce the myogenic properties of the human heart.

❑ Explain how heartbeat rate is controlled by the autonomic nervous system.

❑ Discuss the cause and prevention of stroke as well as several forms of heart disease.

❑ Describe the control mechanisms for regulating the exchange of materials between capillaries and body tissues.

❑ Explain how the architecture of capillaries influences their function and regulation.

❑ Compare the structure and function of the lymphatic system with that of the blood circulatory system.

❑ Explain how blood contained within the venous system is returned to the heart for recirculation.

❑ Name the cellular and acellular parts of blood and explain their functions.

❑ Describe the cellular and biochemical events that lead to blood clotting and vessel repair.

❑ Explain what is meant by "autoregulation" of the circulatory system.

❑ Discuss the hormonal and neural feedback control mechanisms that influence cardiovascular function in humans.

❑ Explain the physiological changes that occur during the diving response of marine mammals, and discuss why these responses are advantageous.

CIRCULATORY SYSTEMS: VESSELS, BLOOD, AND A PUMP (pages 1016–1021)

Key Concepts

1. Multicellular organisms depend heavily on a division of labor among their body cells. In all but the simplest animals, supplying the needs of distantly separated cells is accomplished by internal transport systems.

2. Most transport systems consist of a muscular pump called the heart, a network of interconnected vessels, and a transport fluid called blood. Collectively, these components are referred to as a *circulatory* or *cardiovascular system*.

3. Simple transport systems do not have vessels; some do not even have pumping mechanisms. If the animal is small, unassisted environmental fluid movements coupled with diffusion are sufficient to transport substances past all body cells.

4. Many cnidarians utilize nonvascular transport. For example, every cell in a hydra's body is in direct contact with the environment either through its external surface or through its *gastrovascular cavity*, a blind-ended sac that serves as both a digestive and a transport organ.

5. The gastrovascular cavities of some invertebrates are highly branched. For very thin animals, this puts every cell within diffusion distance of the transport system. Flatworms are examples of this body design.

6. Larger animals, whose bodies cannot be adequately supplied by simple diffusion from a gastrovascular cavity, must have some type of circulatory system that transports fluids to specific regions within the body.

7. Increased body size and the inability of simple diffusion to handle transport needs is a particularly acute problem for terrestrial animals because the external medium (air) can no longer serve a transport function without causing the animal to dry out. These animals need circulatory systems to maintain the proper composition of their intercellular fluids.

8. In some internal transport mechanisms, body movements cause the intercellular fluid to percolate through the tissues outside of the circulatory vessels. This is known as an *open circulatory system* and is characteristic of animal groups such as mollusks and arthropods.

9. The movement of fluids through an open circulatory system is usually aided by the action of a muscular pumping heart, which propels the fluid through short vessels that open directly into major regions of the body. After trickling through the tissues, the fluid collects in the body cavity and reenters the heart either through valved holes called *ostia* (arthropods), or with the aid of open vessels (mollusks).

10. The relative inefficiency of open circulatory systems limits the metabolic activity and body size of animals. However, groups such as insects have respiratory systems (tracheal tubes) that are separate from their circulatory systems. This anatomical separation permits insects to lead more active lives than they could with an open circulatory system that served both respiratory and circulatory functions.

11. *Closed circulatory systems* keep the transport fluid separate from the intercellular fluid in a network of closed vessels. One or more pumping hearts create pressure to move blood within the vascular network.

12. Closed circulatory systems have important design advantages that lead to an overall metabolic superiority over open circulatory systems. Closed systems provide more rapid transport, they direct blood flow to specific areas of the body, and they provide a boundary that constrains the movement of specialized cellular elements and large blood molecules.

13. Different levels of sophistication are seen in closed circulatory systems. Earthworms have a relatively simple closed system consisting of large dorsal and ventral vessels that direct the blood along the animal's body. These vessels are connected by five contractile vessels that act as pumping hearts. Within each body segment, the large vessels branch into smaller vessels, where exchange occurs.

14. Vertebrates have sophisticated closed circulatory systems. The multichambered vertebrate heart has valves that prevent the backflow of blood when the chambers contract. There has been an increase in the number of heart chambers during evolution so that modern fishes have two chambers, amphibians and most reptiles have three, and crocodilians, birds, and mammals have four.

15. Increased heart complexity allows for the separation of blood flow into two distinct pathways. All vertebrates except fishes have a *pulmonary circuit* that sends blood to the respiratory surfaces and then returns it to the heart. The *systemic circuit* then sends this oxygenated blood to the body cells. Waste-laden, deoxygenated blood is collected from the body tissues and returned to the heart before once again entering the pulmonary circuit for reoxygenation.

16. Vessels of the vascular system are classified in terms of their size and whether they direct blood to or from the heart. *Arteries* are large-diameter vessels that take blood away from the heart. *Veins* are the large vessels that return blood to the heart. Arteries and veins branch repeatedly to become smaller *arterioles* and *venules*, respectively.

17. *Capillaries* are the smallest diameter vessels; they connect the arterioles to the venules in both the pulmonary and systemic circuits. The exchange of materials between the intercellular fluid and the vascular system takes place only across the capillary walls.

18. The fish heart has two chambers, consisting of a heavily muscularized pump called the *ventricle* attached to a less muscular pump known as the *atrium*. Deoxygenated blood from the body returns to the heart through veins and collects in the atrium, which pumps it into the ventricle. The blood is then pumped from the ventricle to the gills, where it flows through extensive capillary beds to accomplish gas exchange. Once oxygenated, the blood travels through a large dorsal artery called the *aorta*, which distributes blood to other arteries and arterioles that branch throughout the body.

19. Most of the pressure generated by ventricular contraction in fishes is dissipated by the high resistance of the tiny gill capillaries. Blood traveling through a fish's body is therefore under relatively low pressure, putting distinct upper limits on the rate of metabolic exchange possible.

20. Lungfish provide transitional anatomical evidence for how other vertebrate groups evolutionarily solved the problem of low systemic blood pressure. Lungfish have two atria and an air bladder that is modified into a primitive lung. The lung forms as an outpocketing of the gut. Oxygenated blood leaving the gills enters the lung and is then directed back to the heart, forming a rudimentary pulmonary circuit. The two atria serve primarily to separate oxygenated blood bound for the body from deoxygenated blood heading to the gills; they also keep the systemic blood pressure higher than it is in other fish.

21. Adult amphibians have three-chambered hearts. Separate atria receive deoxygenated blood from the body and oxygenated blood from the lungs. The two types of blood are then partially mixed as they flow through the single ventricle, whose anatomy keeps the blood types mostly separated.

22. The main advantage of the amphibian system is that oxygenated blood returning to the heart from the lungs can be repressurized before traveling to the body tissues. This dramatically increases the rate at which nutrients and oxygen can be delivered to body cells, thereby increasing metabolic potential.

23. Most reptiles also have three-chambered hearts, but a septum partially divides their ventricle to reduce the mixing of pulmonary and systemic blood supplies. This further increases the pressure and efficiency of the entire circulatory system. The crocodilians have a complete partition in the ventricle, which yields a four-chambered heart.

24. Because their metabolic rates are typically lower than that of birds or mammals, reptiles and crocodilians do not have to breathe continuously. In addition, many often spend extended periods of time underwater without breathing. These animals have evolved special anatomical features of the heart and arteries that permit the switching of blood flow between pulmonary and systemic circuits.

25. Reptiles and crocodilians can alter blood flow because they have two separate aortas, one of which can receive blood from either side of the ventricle. When not breathing, blood can be shunted from the ventricle pressurizing the pulmonary circuit to the aorta feeding the systemic circuit. In crocodiles, which have two separate ventricles, the two aortas are connected by a channel that shunts blood from one aorta to another under appropriate pressure conditions.

26. Four-chambered hearts are characteristic of all birds and mammals. The complete separation of pulmonary and systemic circuits creates independent pumping mechanisms for the lungs and the body tissues, which greatly increase transport efficiency.

27. Other important advantages of four-chambered hearts include: providing the highest possible oxygen concentration for the systemic blood supply, maximizing gas exchange by sending to the lungs blood that has the lowest possible O_2 concentration and the highest possible CO_2 concentration, and permitting differential pressurization of the pulmonary and systemic circuits.

Questions (for answers and explanations, see page 395)

1. All but one of the following structures can be found in the transport system of a butterfly. Select the *exception*.
   a. Heart
   b. Artery
   c. Ostia
   d. Capillaries
   e. Transport fluid

2. Like other animals, hydras and chickens need to obtain oxygen from their environment and eliminate the waste products of cellular metabolism. Which of the following transport structures is found in *both* of these animals?
   a. Complex network of transport vessels that supply all regions of the body
   b. Gastrovascular cavity to facilitate transport to internal body tissues
   c. Powerful muscular pump to move transport fluids
   d. Atria, ventricles, and capillaries
   e. None of the above

3. For a variety of reasons, closed circulatory systems are considered to be more efficient than open circulatory systems. Which of the following is *not* one of those reasons?
   a. Closed systems rely exclusively on simple diffusion for transport, whereas open systems rely on pumping mechanisms.
   b. Transport within closed systems is more rapid than in open systems.
   c. Blood is directed to specific areas of need in closed systems, but not in open systems.
   d. Closed systems create a boundary between transport fluids and the body tissues.
   e. Closed systems operate under higher pressure than do open systems.

4. Through which of the following vessels would the largest volume of blood flow during a one-minute period in a particular animal?
   a. A capillary in the lung
   b. A venule leaving a capillary bed in the right foot
   c. An artery connected to the heart
   d. An arteriole feeding a capillary bed in the liver
   e. Cannot determine from the information given

5. An important difference between a fish heart and a lizard heart is that the
   a. fish heart is part of an open circulatory system, and the lizard heart is not.
   b. fish heart has two chambers, and the lizard heart has three.
   c. fish heart has many ostia, whereas the lizard heart has none.
   d. fish heart pumps blood to the lungs, but the lizard heart pumps blood to the gills.
   e. fish heart pumps oxygenated blood, and the lizard heart pumps deoxygenated blood.

6. In which of the following would you likely record the highest blood pressure?
   a. The ventricle supplying blood to the gills of a fish
   b. The anterior dorsal artery of an ant
   c. The pulmonary vein of a frog
   d. The ventricle supplying blood to the systemic circuit of a bird
   e. All of these locations would have the same pressure.

7. Between the time it leaves the heart and the next time it reenters the heart, blood traveling through the systemic circuit of a cow will pass through
   a. arteries and veins.
   b. arteries, veins, and capillaries.
   c. arteries, arterioles, and veins.
   d. arteries, arterioles, venules, and capillaries.
   e. arteries, arterioles, capillaries, venules, and veins.

8. Which of the following statements about the circulatory systems of birds and mammals is *false*?
   a. Blood pressure on the arterial side of the systemic system is relatively high.

b. Only oxygenated blood travels through the systemic circuit.

c. The pulmonary circuit carries blood to and from the lungs.

d. The pumping systems for the pulmonary and systemic circuits are independent.

e. Blood in the right atrium never mixes with blood in the left atrium.

9. During the period of time when a reptile is not breathing,

a. little blood is pumped from the heart to the lungs.

b. blood is shunted from the systemic to the pulmonary circuit.

c. the metabolic rate of the animal increases dramatically.

d. the resistance in its pulmonary circuit is lower than in its systemic circuit.

e. All of the above

Activities (for answers and explanations, see page 395)

- Match each of the following cardiovascular structures with its best description.

____Ventricle	*a.* Muscular pump that sends blood to the lungs or gills
____Ostia	*b.* Smallest diameter vessels in the pulmonary circuit
____Atrium	*c.* Vessel through which blood passes just prior to entering a capillary
____Venule	*d.* Collecting holes in the pumping structure of insect cardiovascular systems
____Capillaries	*e.* Receives blood returning to the heart from the lungs of a frog
____Arteriole	*f.* Blood passes through this vessel just prior to entering a vein

- Explain how the open circulatory system of a grasshopper functions, and how it differs from the closed circulatory system of an earthworm.

- The following is a schematic diagram of a mammalian circulatory system. Indicate with arrows the correct flow of blood through this system. Identify the atria and the ventricles of the heart, the pulmonary and systemic circuits, and show one example of an artery and a vein. Also indicate the location of a capillary bed.

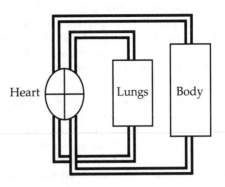

THE HUMAN HEART: TWO PUMPS IN ONE • THE VASCULAR SYSTEM: ARTERIES, CAPILLARIES, AND VEINS (pages 1021–1030)

Key Concepts

1. The human heart has four chambers: a right and left atrium, and a right and left ventricle. Each atrium is paired with a ventricle to form two independent pumping systems. Atria collect blood returning to the heart by veins, and ventricles pump blood away from the heart through arteries.

2. Because it is a pressurized system, valves are needed to prevent backflow of blood when the atria and ventricles relax. *Atrioventricular valves* prevent backflow between the ventricles and atria. The *pulmonary* and *aortic valves* prevent backflow between their respective arteries and the ventricles.

3. The right atrium receives systemic blood from large veins (the *superior* and *inferior venae cavae*), which drain the upper and lower body, respectively. This blood is pumped into the right ventricle, which sends it to the lungs via the *pulmonary artery*. Blood returns from the lungs through the *pulmonary vein* and is collected by the left atrium, which pumps it into the left ventricle. The blood is then pumped back through the systemic circuit via the aorta.

4. High resistance caused by the large number of arterioles and capillaries in the systemic circuit requires the left ventricle to pump harder than the right ventricle, which services the smaller quantity of capillary beds in the lungs. Higher pressures are therefore produced in the systemic circuit than in the pulmonary circuit.

5. The *cardiac cycle* involves alternating contractions of the atria and ventricles. The period of ventricular contraction is called *systole*, while ventricular relaxation is called *diastole*.

6. The beating sound made by a human heart is caused by the closing of heart valves. Atrioventricular valves close with a "lub" sound as the ventricles contract. When the ventricles relax, the aortic and pulmonary valves close with a "dub" sound. Defective valves produce abnormal sounds called murmurs.

7. Blood pressure is measured in large arteries located near the heart (usually in the upper left arm) using a sphygmomanometer and stethoscope and listening for

the pulsing sound of blood moving through the arteries. By constricting the arterial flow with the pressure cuff and then listening for the reappearance or disappearance of pulsing sounds, systolic and diastolic pressures can be determined. These are reported as a ratio of systolic to diastolic pressure, which for a young adult is typically around 120/80 mm Hg.

8. Some cardiac muscle cells are capable of spontaneous, or *myogenic,* contraction without direct stimulation by the nervous system. The resting membrane potential of these pacemaker cells slowly depolarizes until an action potential is initiated, which then rapidly spreads via gap junctions throughout the entire heart muscle to produce coordinated, synchronous contractions.

9. Heartbeat rate is controlled by the release of neurotransmitters from the autonomic nervous system. Acetylcholine released from parasympathetic nerves slows the depolarization rate of pacemaker cells and leads to a reduction in heartbeat rate. If overactive, this control system can produce a *vagal reaction,* where heartbeat rate slows to the point that fainting results.

10. Sympathetic nerves release norepinephrine onto pacemaker cells. This causes the depolarization rate and contraction strength of muscle cells to increase, leading to increased heartbeat rate and ventricular pressure. The release of epinephrine (adrenaline) from the adrenal glands is also stimulated by these changes, further contributing to the excitatory effect on the heart.

11. The *sinoatrial node* is the collection of pacemaker cells that initiates contraction of the atria. It is located at the junction between the superior vena cava and the right atrium.

12. Since there are no gap junctions between the atria and the ventricles, a second pacemaker region called the *atrioventricular node* located at the base of the atria regenerates action potentials in the ventricles. The *bundle of His* channels these action potentials from the AV node down the length of the ventricles. *Purkinje fibers* then disperse the signal throughout the ventricular muscle mass.

13. The slight time delay caused by the regeneration of action potentials in the AV node separates the contractions of the atria and ventricles. The atria thus contract first, filling the ventricles with blood. The ventricles then contract to expel blood from the heart.

14. An *electrocardiogram* (EKG or ECG) measures the collective electrical activity of cardiac muscle during contraction and relaxation. EKG waves are divided into regions that correspond to various stages in the depolarization and repolarization of muscle cells within the heart. By analyzing the pattern of these waves, abnormal heart conditions such as *tachycardia* (heart rate over 100 beats/min), *ventricular fibrillation* (uncoordinated contraction of ventricles), and *heart block* (failure of ventricular stimulation) can be detected. Textbook Figure 46.10 provides details.

15. The walls of arteries and arterioles contain elastic fibers that permit vessel expansion under the increased pressure of ventricular contraction. These fibers recoil when the ventricles relax, helping to smooth the flow of blood. Arteries and arterioles are known as *resistance vessels* because their diameters, and thus their resistances, can be regulated through the action of smooth muscles located in the vessel walls.

16. Cells exchange nutrients, oxygen, and wastes with the blood across capillary walls. Arterioles are connected with venules by capillary beds. These beds are so extensive that no body cell is more than one or two cell diameters away from a capillary.

17. Thin, permeable walls characterize capillaries. Because the total cross-sectional area of a capillary bed is large compared to the arterioles feeding it, flow rate is greatly reduced and pressure lowered in the capilllaries. This facilitates diffusion across the capillary wall.

18. Capillaries are permeable to small molecules and water, but not to large molecules such as blood proteins. Pressure tends to force water and small molecules out of the capillary in a process called *filtration*. The large molecules that remain inside the vessel create an osmotic potential that pulls water back into the capillary.

19. Since blood pressure is highest on the arterial side of a capillary, water loss is also greatest there. By contrast, osmotic forces dominate on the low-pressure venous side of the bed. The net movement of water within a capillary bed is determined by the interaction between these opposing forces.

20. During the inflammatory response stimulated by *histamine,* arterioles expand and blood flow increases to the injured area. This increases blood pressure in capillaries. At the same time the permeability of capillaries and venules increases, allowing water to move out of these vessels and into the surrounding tissues. The result is swelling, or *edema,* which can be countered by the use of *antihistamine* drugs.

21. Decreases in the osmotic potential of blood can also lead to capillary water loss. The disease kwashiorkor is characterized by extreme edema and results from severe protein deficiency. In response to this deficiency, blood proteins are degraded, removing the molecules responsible for creating the osmotic potential that draws water back into the blood.

22. Capillary architecture, concentration gradients, and the chemical characteristics of substances transported in the blood all determine whether small molecules will cross the capillary wall. The capillaries of some tissues, such as those associated with the blood–brain barrier, are particularly selective about what will pass. Capillaries found in the digestive tract and kidneys allow large molecules to pass through pores in the capillary membrane. Still others utilize endocytosis.

23. The *lymphatic system* returns intercellular fluid lost by seepage from capillaries back to the blood. *Lymph* (as this fluid is known once it enters the lymphatic system) is chemically similar to blood but has less protein and contains no red blood cells. Lymph is returned to the blood by a separate system of lymphatic vessels that begin as fine capillaries in the tissues and eventually collect into the *thoracic duct,* which empties into the superior vena

cava. The propelling force for lymph movement is the contraction of skeletal muscle. Backflow is limited by valves within the lymphatic vessels.

24. Lymph nodes are associated with the major lymphatic vessels of mammals and birds, and play an important role in immune defenses. In addition to being sites for lymphocyte production and phagocytosis, they also act as mechanical filters to trap foreign particles, a function that can result in their becoming collection sites for metastasized cancer cells.

25. Blood returns to the heart through the venous system, which is characteristically under low pressure. The low pressure allows blood to accumulate in the veins, which are referred to as *capacitance vessels* because of their capacity to store blood. In humans, up to 80 percent of total blood volume is in the veins at any one time.

26. Gravity helps move blood through veins located above the heart. However, venous blood below the heart must be pushed or pulled against gravity. If this does not occur, blood accumulates in the lower body, the brain becomes deprived of oxygen, and fainting results. Fainting is a self-correcting response because it changes body position and thus redistributes the blood.

27. Like the lymphatic system, the action of skeletal muscle is important for moving blood through the venous system, especially for veins located below the heart. Valves keep blood flow unidirectional as the milking action of the muscles gradually pushes the blood back to the heart.

28. During vigorous exercise, muscular pumping action on the veins is enhanced. This pushes an increased volume of blood back to the heart, causing the cardiac muscle to be stretched. The heart responds by contracting more forcefully, a relationship known as the *Frank–Starling law*.

29. A ventilatory pump helps return venous blood to the heart as the negative pressure created by contraction of the diaphragm and intercostal muscles pulls blood and lymph toward the chest. A few veins, especially large ones located near the heart, also contain smooth muscle that can assist in the return of venous blood.

30. Arterial disease is now responsible for one-half of all human deaths in developed countries. *Atherosclerosis* (hardening of the arteries) occurs over many years through the accumulation of fatty deposits called *plaque* at sites of endothelial damage within the arterial system. As the vessel diameter narrows, calcium deposits form and reduce the elasticity of the vessel wall.

31. A blood clot, or *thrombus*, forms if clotting agents in the blood come in contact with exposed plaque. The result can be complete blockage of the vessel. If the blockage develops in a *coronary artery* supplying blood to the heart, then the condition is called a *coronary thrombosis*, which often leads to a *coronary infarction* (heart attack).

32. A dislodged piece of a thrombus, called an *embolus*, can travel through the vascular system and block other vessels, producing an *embolism*. Small-diameter arteries of the brain are particularly susceptible. When these vessels are blocked, brain cells normally nourished by their blood supply will die; this is the physiological basis of a *stroke*.

33. Both genetic and lifestyle factors influence the likelihood of developing arterial disease. High-fat diets, smoking, lack of exercise, obesity, certain medical conditions, and high blood pressure (*hypertension*) are all risk factors.

Questions (for answers and explanations, see page 395)

1. The inferior and superior venae cavae of humans have which of the following in common?
   a. Carry venous blood
   b. Empty into the right atrium
   c. Are capacitance vessels
   d. Drain the systemic blood circuit
   e. All of the above

2. A man visits his doctor and learns that his blood pressure is 125/50 mm Hg. This indicates that his blood pressure is
   a. slightly higher than normal when the ventricles are relaxed.
   b. slightly higher than normal when the atria are contracted.
   c. lower than normal when the ventricles are contracted.
   d. lower than normal when the ventricles are relaxed.
   e. all of the above

3. During tachycardia,
   a. heartbeat rate exceeds 100 beats/min.
   b. the ventricles contract in an uncoordinated fashion.
   c. blood pressure drops precipitously.
   d. the ventricles fail to contract at all.
   e. coronary arteries are blocked.

4. An artificial electric stimulus is simultaneously provided to both the sinoatrial node and the atrioventricular node. This would cause
   a. the atria to beat first, then the ventricles.
   b. the ventricles to beat first, then the atria.
   c. all four chambers of the heart to contract simultaneously.
   d. the left side of the heart to beat first, then the right side.
   e. heart murmurs.

5. The coronary arteries of humans are susceptible to all but one of the following conditions. Select the *exception*.
   a. Atherosclerosis
   b. Stroke
   c. Plaque deposits
   d. Embolism
   e. Thrombosis

6. Which of the following statements about capillaries is *false*?
   a. They directly link the arteries of the systemic blood supply with the veins in the same circuit.
   b. Diffusion of gases and nutrients occurs between the blood and tissues across capillary walls.
   c. Capillary walls are highly permeable to small molecules, but not to large blood proteins.
   d. Histamines increase the permeability of capillary walls.
   e. Filtration is the process in which blood pressure forces water and small molecules out of capillaries.

7. Which of the following is *not* associated with the lymphatic system?
   a. Thoracic duct
   b. Specialized transport vessels
   c. Red blood cells
   d. Large proteins
   e. Elements of the immune system

8. Which of the following is *not* a mechanism that helps move venous blood back to the heart in humans?
   a. The pull of gravity
   b. Activity of parasympathetic nerves
   c. Pumping action of skeletal muscle
   d. Specialized valves that prevent backflow of blood
   e. Ventilatory pumping

9. A patient is brought into a hospital suffering from severe histamine-induced edema. This means that
   a. the patient has suffered a stroke.
   b. the patient has suffered a coronary infarction.
   c. too much blood has collected in the venous system.
   d. intercellular fluids have accumulated in the body tissues due to leaky capillaries and venules.
   e. none of the above

Activities (for answers and explanations, see page 396)

- On the following diagram, label the indicated structures of the human heart.

- Number the following cardiovascular elements in the sequence that they would be encountered while traveling through the bloodstream of the human body. Begin your journey at the left ventricle.

  ___ Left ventricle   ___ Right ventricle

  ___ Atrioventricular valve   ___ Aortic valve

  ___ Pulmonary valve   ___ Pulmonary arterioles

  ___ Pulmonary venule   ___ Aorta

  ___ Vena cavae   ___ Left atrium

  ___ Right atrium   ___ Systemic capillaries

  ___ Pulmonary capillaries   ___ Pulmonary artery

  ___ Pulmonary vein   ___ Systemic arterioles

  ___ Systemic venules

- On the following graph of human blood pressure (measured in the left ventricle), indicate which sections of the curve correspond to systole and which to diastole.

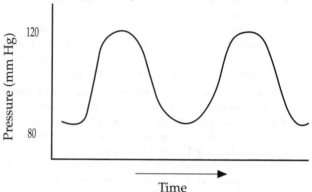

BLOOD: A FLUID TISSUE • CONTROL AND REGULATION OF CIRCULATION (pages 1030–1036)

Key Concepts

1. Blood is a type of connective tissue. It consists mainly of *plasma*, an aqueous substance of complex chemical composition, within which are suspended a variety of cellular elements.

2. When a sample of whole blood is centrifuged, the cellular fraction moves to the bottom of the tube and the plasma remains at the top. The *packed cell volume*, or *hematocrit*, is normally about 40 percent in humans (less in women than in men). It can vary widely with health, physical conditioning, and acclimation to high altitude. The cellular fraction of blood contains three important cell types.

3. *Erythrocytes*, or red blood cells, comprise the largest percentage of the hematocrit. Packed with hemoglobin, their function is to transport gases between the respiratory surfaces and the body cells.

4. Red blood cells are produced at a rate of over two million every second by *stem cells* located in the bone marrow. Production rate is controlled by the hormone *erythropoietin*, which is released by the kidney in response to low oxygen levels.

5. Just before entering a capillary from the stem cells, red blood cells lose many of their internal organelles as an adaptation for increased hemoglobin content. Mature cells are shaped like biconcave discs. This increases the surface area available for gas exchange, and makes the cell flexible so that it can more easily squeeze through capillaries.

6. *Leukocytes*, or white blood cells, are less abundant than red blood cells but come from the same lineage of stem cells (called *totipotent*). Leukocytes are important com-

ponents of the body's defense mechanisms, functioning in the search for foreign particles and the phagocytosis and destruction of those particles. They also play a role in antibody formation.

7. White blood cells are capable of amoeboid movement. They spend much of their time outside the vascular system following chemical cues to sites of infection or cell damage.

8. *Platelets* are produced by a second lineage of stem cells (called *pluripotent*) in the bone marrow. Their concentration in the blood is intermediate to that of red and white blood cells. Platelets are tiny cell fragments packed with the enzymes and chemicals needed to seal leaks in the vascular tissue.

9. Platelets are activated by contact with the collagen fibers that are exposed when blood vessels are damaged. Upon activation the platelet swells, becomes irregularly shaped, and sticks to the damaged site. Activated platelets working together create a plug at the damaged site that temporarily patches the leak.

10. Many chemical *clotting factors* are involved in the complex cascade of reactions that ultimately leads to the complete repair of damaged vascular tissue. The process begins with platelet activation, which stimulates the circulating inactive enzyme *prothrombin* to be converted into its active form *thrombin*.

11. Thrombin causes the blood protein *fibrinogen* to polymerize and form threads of *fibrin*, which settle at the injury site to form a framework upon which a stable clot can develop. This promotes the growth of scar tissue.

12. Plasma has a large concentration of dissolved ions, gases, proteins, and nutrient molecules. Important ions include Na^+ and Cl^-, and common nutrients are glucose, amino acids, lipids, and lactic acid. Plasma is very similar in composition to intercellular fluid, except that the plasma has a higher concentration of protein.

13. In addition to the proteins involved in clotting, the plasma also includes *albumin*, the protein responsible for maintaining proper osmotic potential in the capillaries. Antibodies, hormones, and various carrier molecules such as *transferrin*, which moves iron from the gut to its site of usage, are also present in the plasma.

14. *Autoregulatory mechanisms* in each capillary bed control the constriction and dilation of arterioles supplying the bed. Changes in oxygen content, waste production, pH, and blood flow in the capillaries influence the arterial pressure of blood leaving the heart.

15. Autoregulation is intrinsic to the tissue served by a capillary bed. However, blood flow through capillaries is also influenced by neural and hormonal input. These control mechanisms affect the contraction of smooth muscle in the arteries and the arterioles that supply the capillary bed.

16. Activities that increase cell metabolism in tissues also increase blood flow to those tissues. Precapillary sphincter muscles form cuffs around the arterioles feeding capillary beds. As O_2 concentration decreases, or CO_2 and metabolite concentrations increase, the smooth muscle sphincters relax to allow more blood into the bed.

17. Most arteries and arterioles receive direct neural input from the sympathetic division of the autonomic nervous system. In the vessels of skeletal muscle, sympathetic nerves release acetylcholine and cause the smooth muscle to relax, leading to dilation and increased blood flow. Elsewhere, sympathetic nerves release norepinephrine, which causes smooth muscle to contract and thus restrict blood flow.

18. Hormones such as epinephrine, angiotensin, and vasopressin also cause arterioles to constrict. In particular, they influence the arterioles feeding peripheral capillary beds, reducing blood flow to the extremities. During times of metabolic stress this response channels more blood to essential internal organs such as the brain, heart, and kidneys.

19. Cardiovascular control centers in the medulla are the source of autonomic nerve signals that regulate blood flow. These centers receive input about changes in blood pressure from stretch receptors located in the aorta and the carotid arteries, which supply blood to the brain. This information helps determine whether blood flow to the periphery will be restricted.

20. Chemosensors in the carotid arteries and aorta detect low O_2 levels and stimulate the cardiovascular control centers to increase blood pressure. Neural input from higher brain centers, which regulate emotions and anticipate physiological needs, can also affect cardiovascular control.

21. Heartbeat rate is slowed by neural inputs during the diving reflex of marine mammals. These animals can dramatically reduce heartbeat rate, which, when coupled with a restriction of peripheral blood flow, helps to conserve O_2 for the operation of heart and brain.

22. Overall cardiovascular control is thus maintained by two mechanisms: autoregulatory systems that provide for the needs of local tissues, and the control of autonomic nerves and hormones that respond to whole-body metabolic demands.

Questions (for answers and explanations, see page 396)

1. The rate of new red blood cell production in the _____ is controlled by the _____ .
   a. stem cells of bone marrow; blood protein fibrinogen
   b. blood plasma; nutrients glucose and lactic acid
   c. packed cell volume; hormone erythropoietin
   d. capillaries; protein transferrin
   e. stem cells of bone marrow; hormone erythropoietin

2. Which of the following statements about leukocytes is *false*?
   a. They are capable of amoeboid movement.
   b. They are tiny cell fragments that initiate the chain of events leading to clot formation.
   c. They often leave the vascular system following chemical cues toward damaged cells.
   d. They engulf foreign particles in the blood through the process of phagocytosis.
   e. There are many fewer of them in the packed cell volume than there are erythrocytes.

3. Before fibrin threads are produced to lay the groundwork for clot formation,
   a. fibrinogen causes platelets to become activated.
   b. red blood cells lose their hemoglobin and stimulate the activation of erythropoietin.
   c. platelets are activated and prothrombin is converted to thrombin.
   d. transferrin is converted into prothrombin, which activates the platelets.
   e. none of the above

4. The autoregulatory mechanisms that control the dilation and constriction of capillaries are influenced by the
   a. pH of the blood.
   b. O_2 content of the blood.
   c. production of metabolic wastes in neighboring tissues.
   d. rate of blood flow through the capillaries.
   e. all of the above

5. Which of the following statements about the control of circulation in humans is *false*?
   a. Sympathetic nerve input to skeletal muscle causes the blood vessels in the muscle to dilate.
   b. Adrenaline produces arteriole constriction, especially in peripheral capillary beds.
   c. Blood flow can be regulated by autonomic nerve signals emanating from cardiovascular control centers in the medulla of the brain.
   d. Hormones such as angiotensin and vasopressin cause venules to constrict.
   e. Carotid artery chemosensors detect low O_2 levels in the blood and promote increased blood pressure.

6. A whale dives for food while carrying a radio transmitter that allows you to measure a number of parameters related to its cardiovascular function. Which of the following is likely to be true about this animal during its dive?
   a. Heartbeat rate will decrease, and peripheral blood flow will decrease.
   b. Heartbeat rate will increase, and peripheral blood flow will decrease.
   c. Heartbeat rate will decrease, and the number of new red blood cells produced will increase.
   d. Heartbeat rate will increase, and activity of autoregulatory mechanisms in the brain will decrease.
   e. Heartbeat rate will decrease, and peripheral blood flow will increase.

7. You perform an experiment in which a drug is supplied to a normal human that causes selective dilation of *only* the precapillary sphincter muscles. Which of the following effects is most likely to occur?
   a. Blood flow through the capillary beds will decrease.
   b. The permeability of the capillary walls will decrease.
   c. A chemical cascade leading to clotting will be initiated.
   d. Blood flow through the capillary beds will increase.
   e. Heartbeat rate will slow dramatically and capillary blood pressure will rise.

8. The medullary cardiovascular control center receives neural input from
   a. higher brain centers involved with emotions and stress.
   b. chemosensors on the aorta and carotid arteries.
   c. stretch sensors in the arteries.
   d. chemosensors in the medulla.
   e. all of the above

Activities (for answers and explanations, see page 396)

- Match each of the following blood components with its best description.

  ____Plasma
  ____Erythrocyte
  ____Leukocyte
  ____Platelet
  ____Thrombin
  ____Albumin
  ____Hematocrit

  a. Protein that influences blood osmotic potential in capillaries
  b. Causes polymerization of molecules needed to stabilize a blood clot
  c. Contains large concentration of hemoglobin
  d. Aqueous component of blood containing many dissolved chemicals
  e. Cell fragments that initiate the events associated with blood clotting
  f. Important blood element used in the body's defense mechanisms
  g. Red and white blood cell volume after whole blood is centrifuged

- Diving is a strenuous activity, yet when marine mammals dive, their heart rate actually decreases and blood flow to most of their internal organs is reduced. Explain the cardiovascular reasons that this type of diving response is adaptive for marine mammals.

Integrative Questions (for answers and explanations, see page 396)

1. If you compare blood from the superior vena cava of a human with blood from the aorta of a dog, the main difference between the two samples of blood would be
   a. the types of cells in the cellular fraction.
   b. the presence or absence of platelets.
   c. the concentration of O_2.
   d. the presence or absence of plasma.
   e. the presence or absence of prothrombin.

2. Which of the following characteristics of animal circulatory systems can be found in a clam?
   a. Closed circulatory system
   b. Pumping heart
   c. Red blood cells
   d. Pulmonary valves
   e. None of the above

3. Which of the following statements about animal circulatory systems is *false*?
   a. Leukocytes typically contain large amounts of hemoglobin.
   b. The hormone erythropoietin controls red blood cell production.
   c. Circulation is not necessary in cnidarians because they can utilize direct diffusional exchange with their environment.
   d. In arthropods, circulatory fluids return to the heart through valved holes called ostia.
   e. The right and left atria in a human heart receive blood from the venous supply.

4. If too much blood collects in a person's venous supply because of prolonged inactivity, this may
   a. cause a decrease in the production of erythropoietin.
   b. cause the person to suffer a brain embolism and stroke.
   c. lead to the development of atherosclerosis or hypertension.
   d. deprive the brain of oxygen and cause the person to faint.
   e. block neural activity in the Purkinje fibers of the heart.

47

Animal Nutrition

CHAPTER LEARNING OBJECTIVES—after studying this chapter you should be able to:

❑ Explain the differences between the terms "autotroph," "heterotroph," "saprotroph," "detritivore," "predator," "carnivore," "herbivore," "omnivore," "filter feeder," and "fluid feeder."

❑ Distinguish between the processes of intracellular and extracellular digestion.

❑ Describe how the term "calorie" is used to understand nutrition, and distinguish it from a "kilocalorie" and a "Calorie."

❑ Discuss the relative importance of carbohydrates, fats, and proteins as energy sources for animals.

❑ Explain the difference between the terms "undernourished," "overnourished," and "malnourished," and cite examples of the problems caused by these conditions in humans.

❑ Discuss the metabolic reasons that certain amino acids and fats are essential in the diets of animals.

❑ Distinguish between macro- and micronutrients in animals, and explain their roles in nutrition.

❑ Describe the role of vitamins in animal nutrition, explain how they function, and give examples of some problems that result from vitamin deficiencies in humans.

❑ Provide examples of common feeding adaptations in herbivores and carnivores.

❑ Explain how the structure and function of teeth provide useful information about the diets of vertebrates.

❑ Describe the basic structure and function of a complete digestive tract.

❑ Discuss the role of hydrolytic enzymes in animal digestion, and explain how these enzymes are stored and activated.

❑ Describe the structure and function of the vertebrate gut, with special emphasis on the human digestive system.

❑ Explain the role of the liver and pancreas in aiding vertebrate digestion.

❑ Discuss the ways that carbohydrates, proteins, and fats are digested and absorbed in the vertebrate gut.

❑ Describe modifications to the basic vertebrate gut design seen in ruminant herbivores and explain the function of these modifications.

❑ Explain how an intrinsic neural reflex differs from a reflex mediated by the central nervous system, and give examples of each from the human digestive system.

❑ Discuss the roles of the hormones gastrin, secretin, cholecystokinin, glucagon, insulin, cortisol, and epinephrine in the control of digestion and nutrient availability in humans.

❑ Explain the digestive purpose of lipoproteins, and differentiate among the various kinds of lipoproteins found in humans.

NUTRIENT REQUIREMENTS (pages 1038–1045)

Key Concepts

1. *Heterotrophs* obtain energy and structural molecules from preformed organic nutrients. *Autotrophs* harness solar or inorganic chemical energy to manufacture their own organic nutrients. All animals are heterotrophs.

2. Energy in the chemical bonds of food is ultimately transferred to ATP, which is used in all aspects of cell metabolism. These energy conversions are never 100 percent efficient; some energy is lost in the form of heat.

3. A useful physiological unit of heat energy is the *calorie*, defined as the energy needed to raise the temperature of one gram of water one degree Celsius. A calorie is

equivalent to 4.184 joules. Most physiological processes produce or use large amounts of energy. It is therefore convenient to think in terms of a *kilocalorie* (= 1,000 calories). The dietician's *Calorie*, indicated with a capital "C," equals one kilocalorie.

4. An animal's metabolic rate dictates the amount of energy it must acquire through feeding and digestion. Energy is obtained through the metabolism of fats, carbohydrates, or proteins. Fats provide roughly twice as much energy per gram (9.5 kcal) as do carbohydrates (4.2 kcal) or proteins (4.1 kcal).

5. Nutrients are often metabolized immediately upon uptake by body cells. However, since most animals do not eat continuously, yet require energy at a steady rate to fuel cell

metabolism, nutrients must also be stored for later use when the animal is not actively feeding.

6. Liver and muscle cells store carbohydrates in the form of glycogen, but glycogen's structural properties do not allow it to be stored in large quantities. Fats, on the other hand, not only provide more energy than do carbohydrates but can also be stored more compactly. Fats thus represent a better long-term source of energy reserves. Protein is not used for long-term energy storage, although it is broken down to yield energy under conditions of metabolic stress.

7. An animal is *undernourished* if its food intake does not provide sufficient energy to meet its metabolic needs. This shortfall must be made up by self-consumption, which begins by using stored glycogen and fat, and if necessary proceeds to the breakdown of body proteins. Since proteins are integral components of enzymes, muscles, neurons, and blood, sustained undernourishment rapidly leads to impaired body function and eventually death.

8. Human undernourishment frequently occurs in underdeveloped countries, and now affects over one billion people world wide. Undernourishment also disables thousands of people in Western developed nations where the self-imposed starvation known as *anorexia nervosa* (a psychological aversion to fat) is on the increase.

9. *Overnourishment* occurs when an animal consistently acquires more energy than it needs for metabolism. The excess is stored as glycogen and fat, which may be of great importance to seasonal hibernators and migratory species that rely on these reserves to survive extended periods of food deprivation.

10. Human overnourishment can lead to serious health risks. High blood pressure, heart attack, and diabetes are a few of the problems associated with abnormal weight gain. Excess weight puts a heavy burden on the heart, which must pump harder to supply the extra blood needed to sustain more body mass.

11. Unlike plants, animals cannot directly synthesize required organic molecules. Carbon skeletons, such as the acetyl group ($COCH_3$), must first be obtained through the metabolism of fats, carbohydrates, and proteins. These skeletons are then rearranged to form the complex molecules needed by animals.

12. *Malnutrition* occurs when an animal cannot obtain sufficient quantities of a required carbon skeleton, even though sufficient calories are available in the diet. Skeletons such as the acetyl group are rarely in short supply. Others, such as the amino acids used in protein synthesis, are often limited in available food resources.

13. Certain amino acids are synthesized by animals through the transfer of an amino group to the acetyl skeleton. Most animals, however, cannot directly synthesize all needed amino acids. These must be obtained through the diet.

14. Required dietary amino acids are called *essential amino acids*. If one or more essential amino acids are absent from the diet, protein formation is impaired. Different species have different essential amino acid require-

ments. Herbivores generally require fewer essential amino acids than do carnivores.

15. Eight essential amino acids are required in the human diet: isoleucine, leucine, lysine, methionine, phenylalanine, threonine, tryptophan, and valine. All are made available to the body when meat, eggs, and milk are eaten regularly. However, one or more is usually lacking in any particular type of plant food. Vegetarians must therefore carefully select food combinations to insure a proper balance of essential amino acids.

16. The human body does not store large quantities of free amino acids, yet it is continuously synthesizing new proteins. Thus, all of the amino acids needed for protein synthesis must be made available in each meal.

17. Several factors prevent animals from using dietary protein to make their own protein directly: macromolecules do not readily cross cell membranes, protein structure is highly species-specific, and foreign proteins are attacked by the animal's immune system. These problems are avoided by digesting dietary proteins into their component amino acids and then reassembling them inside cells.

18. Virtually all the lipids needed for the construction of cell membranes and sex hormones can be assembled by combining acetyl groups with a dietary source of *essential fatty acids*, notably linoleic acid.

19. Certain mineral elements are also required by animals, and these needs are highly species-specific. Elements needed in large quantities (e.g., calcium, phosphorus, and sodium) are called *macronutrients*. Those needed only in trace amounts (e.g., iron, copper, and zinc) are termed *micronutrients*.

20. Some minerals are used directly as ions to stimulate the contraction of muscle or to generate action potentials in the nervous system. Others function as enzyme cofactors, are structural components of important electron carrier molecules and hemoglobin, or help to maintain proper osmotic balance in tissues.

21. A special group of essential nutrients are the *vitamins*. They are generally small, organic molecules that function mainly as coenzymes; all are required for normal physiological function. Unlike essential amino acids and fatty acids, vitamins are only needed in small amounts. They are divided into two broad categories based on solubility properties.

22. *Water-soluble vitamins* include the B vitamins: thiamin (a coenzyme necessary for cellular respiration), niacin (a component of the electron transfer molecules NAD and NADP), and riboflavin (which forms the oxidation–reduction site of FAD and FMN). Vitamin C (ascorbic acid) plays an important role in the formation of collagen.

23. *Fat-soluble vitamins* include vitamin A (retinol), which is essential for visual pigment formation, vitamin E, which may protect cellular membranes from oxidation, vitamin K, which plays a crucial role in blood clotting, and vitamin D (calciferol), which regulates calcium metabolism. The need for vitamin D may have played a role in the evolution of geographic variation in human skin color.

24. If water-soluble vitamins are acquired in excess of metabolic needs, the excess is simply excreted through the urine. By contrast, excess fat-soluble vitamins tend to accumulate in body tissues, and can eventually build to toxic levels.

Activities (for answers and explanations, see page 397)

- Match each of the following nutrients or nutritional conditions with its best description.

  ___Anorexia nervosa   a. Breakdown of nutrients into chemical building blocks

  ___Malnutrition   b. Dietary element required in relatively large amounts

  ___Vitamin C   c. Found in citrus fruits and also known as ascorbic acid

  ___Heterotrophy   d. Condition resulting from an insufficient intake of calories

  ___Digestion   e. Results from an insufficient intake of essential nutrients

  ___Undernourished   f. Psychological aversion to fat

  ___Macronutrient   g. Process of obtaining energy from preformed organic food

- Is it possible for an animal to be simultaneously overnourished and undernourished? Explain your answer.

- Is it possible for an animal to be overnourished and at the same time malnourished? Explain your answer.

Questions (for answers and explanations, see page 397)

1. For reasons mainly associated with storage and energetic efficiency, a migrating bird carries most of its energy reserves in the form of
   a. glycogen.
   b. fat.
   c. protein.
   d. macronutrients.
   e. micronutrients.

2. Which of the following qualifies as an essential fatty acid for animals?
   a. Acetyl group
   b. Ascorbic acid
   c. Isoleucine
   d. Linoleic acid
   e. Niacin

3. A chipmunk takes in 50 kcal of food per day, but this food is completely lacking in the B vitamin thiamin. It also has a total metabolic rate of 75 kcal per day. Which of the following appropriately describes the energetic and nutrient status of this animal?
   a. Slightly undernourished and also malnourished
   b. Greatly undernourished but not malnourished
   c. Slightly overnourished but not malnourished
   d. Greatly overnourished and also malnourished
   e. Perfectly balanced nutritional and energetic condition

4. Certain amino acids are essential in the diet of animals because
   a. they prevent overnourishment.
   b. they are cofactors and coenzymes that are required for normal physiological function.
   c. the animal cannot directly synthesize them through the transfer of an amino group to an appropriate carbon skeleton.
   d. animals need these substances in order to make the stored fats used during hibernation and migration.
   e. none of the above

5. Sodium, a substance important for the proper functioning of muscle and nerves, is nutritionally classified as
   a. a mineral micronutrient.
   b. an essential amino acid.
   c. an essential fatty acid.
   d. a water-soluble vitamin.
   e. a mineral macronutrient.

6. Which of the following statements about animal nutrition is *false*?
   a. Essential fatty acids from the diet are combined with acetyl groups to form lipids.
   b. Excess fat-soluble vitamins (such as A, D, and K) are largely excreted through the urine.
   c. A person on a 1,000 Calorie-per-day diet is actually taking in 1 million calories of energy.
   d. Glycogen is a common energy storage molecule found in animal liver and muscle cells.
   e. All animals are heterotrophs, but not all heterotrophs are animals.

7. A person has been starving for five days. Which of the following energy reserves would probably be present in the *lowest* concentration in this person's body?
   a. Carbohydrates
   b. Proteins
   c. Fats
   d. Both proteins and fats
   e. Both carbohydrates and proteins

8. All of the following are water-soluble vitamins *except* for
   a. retinol.
   b. niacin.

c. riboflavin.

d. ascorbic acid.

e. thiamin.

ADAPTATIONS FOR FEEDING • DIGESTION
(pages 1045–1049)

Key Concepts

1. Heterotrophs show diverse adaptations for gaining nutrients, and the ecological niche of an organism is largely determined by its morphological, behavioral, and biochemical adaptations for feeding. Microbes and fungi are *saprotrophs*, obtaining nutrients through absorption of dead organic matter. Some animals are *detritivores*, gathering energy by ingesting the remains of dead organisms. Animals that actively ingest other living organisms are called *predators*.

2. *Carnivores* prey on other animals, while *herbivores* prey on plants. *Omnivores* prey on both animals and plants. *Filter feeders* strain small organisms or particles from the surrounding medium, while *fluid feeders* draw blood or other body fluids from their prey.

3. Simple grazing is a common form of herbivory, but specialized feeding and locomotory structures such as an elephant's trunk or a hummingbird's wings are also used. Since vegetation is often difficult to digest and low in nutrient energy, herbivores have evolved many adaptations for efficiently collecting and processing large quantities of plant material.

4. Carnivores have remarkable powers of prey detection, and have coupled this sensory ability with stealth, speed, and power. Since their prey will often resist being consumed, many carnivores have specialized adaptations for immobilizing, killing, and ingesting prey through the use of sticky surfaces, poisons, and external digestive enzymes.

5. A hallmark of vertebrates is their specialized teeth, which aid in food acquisition and the initiation of digestion. Although vertebrate teeth are sufficiently distinct to allow identification of individual taxa, they nevertheless all share a common three-layered structural plan.

6. Each tooth has a crown, or biting surface, and a root, which anchors the tooth into the jaw. A hard calcium phosphate layer called *enamel* covers the crown. A layer of bony material known as the *dentine* underlies the enamel and extends into the root. Beneath the dentine is the *pulp cavity*, which contains nerves, blood vessels, and the cells that produce dentine.

7. The specific function of a tooth is reflected in its shape. Incisors are the sharp, flattened teeth used for cutting, chopping, and gnawing. Canines are daggerlike and are used for stabbing, ripping, and shredding. Cheek teeth (molars and premolars) are used for shearing, crushing, and grinding.

8. Most of the food ingested by heterotrophs is too large to be absorbed directly into body cells. Food must therefore undergo *digestion*, a process that chemically breaks down nutrients into their molecular subunits, before it can be absorbed.

9. A few animals, such as sponges, engulf food particles and carry out intracellular digestion. Most animals, however, first perform *extracellular digestion*. This occurs inside a specialized digestive compartment, or gut, within the body.

10. Even though they are contained within the gut, nutrients undergoing digestion are still technically outside of the body because they have not yet crossed cell membranes. Degradative enzymes are secreted into the gut and catalyze the breakdown of nutrients into their building-block molecules, which are then absorbed across cell walls.

11. The simplest digestive system is the gastrovascular cavity of cnidarians. This gut type consists of a blind-ended sac with a single opening that serves as both an entrance for food and an exit for wastes. Extracellular digestion in the gut is supplemented by intracellular digestion after food is absorbed through endocytosis.

12. Flatworm gastrovascular cavities are more complex than those of cnidarians. They are highly branched to increase the surface area for nutrient absorption. The cavity opening is attached to an extendable *pharynx* that is deployed during feeding.

13. Some internal parasites, such as tapeworms, have no digestive system at all. Instead, they live inside the gut of their host and directly absorb the predigested food that surrounds them.

14. All animals more advanced than the sponges, cnidarians, and flatworms have a tubular gut with two distinct openings. Undigested food enters through the mouth, and digestive wastes called *feces* exit through the anus.

15. Tubular guts show regional specializations for particular functions. Food enters at the anterior end, or *foregut*, through the mouth and *buccal cavity*. The initial breakdown of food begins through the action of teeth or mandibles, although some animals simply ingest large chunks of food that are then swallowed whole. Additional mechanical breakdown can occur in the gizzard (if present) where muscular contractions, sometimes aided by small stones, grind food into even smaller pieces.

16. Food then passes into a *stomach* or *crop* that functions mainly as a storage chamber. Depending on the species, some chemical digestion and absorption may also occur here.

17. Most digestion and absorption occurs in the *midgut* or *intestine*, which receives the well-mixed and partially digested food from the storage chambers. The intestinal wall is lined with cells that secrete digestive enzymes; it is also associated with accessory glands that produce other digestive chemicals.

18. The last stop for food in the digestive system is the *hindgut*, which reclaims most of the water and ions released during digestion. The hindgut also acts as a temporary storage site for feces, which are eventually passed out of the body through the anus. This occurs by contractions of the muscular *rectum* in a process called *defecation*.

19. The midgut and hindgut regions of many animal digestive systems harbor colonies of symbiotic bacteria or protists that cooperate in the digestion of food. These colonies are especially important to herbivores that feed exclusively on plant materials containing high concentrations of hard-to-digest cellulose.

20. Absorptive surfaces in the midgut of many animals contain elaborate structures to increase surface area. Earthworms have a simple longitudinal infolding of the intestinal wall called a typhlosole. Sharks have a spiral valve that forces food to take a longer path through the intestine. Other vertebrates have highly folded intestinal walls lined with *villi,* which are in turn covered with *microvilli* that greatly increase the available area for nutrient absorption.

21. Proteins, carbohydrates, and fats are all digested by specific *hydrolytic enzymes* that catalyze the breakdown of macromolecules into their constituent amino acids, monosaccharides, monoglycerides, and fatty acids.

22. Digestive enzyme names all end with the suffix "-ase," and each is named according to its function. Thus, a carbohydrase degrades carbohydrates, a protease hydrolyzes proteins, etc. Whether the enzyme cleaves the macromolecule on the outside or the inside is indicated by the prefixes "exo-" and "endo-," respectively.

23. Many digestive enzymes are produced in an inactive form called a *zymogen* and then stored within secretory cells of the gut lining or accessory organs. This protects the animal's body cells from self-digestion. The inactive enzyme is later released into the gut, where it is activated by changes in pH or the action of other enzymes.

24. The gut also avoids digesting itself by virtue of a lubricating, protective mucus layer that isolates the gut cells from the action of digestive enzymes. *Ulcers* develop in the gut wall whenever this protective layer fails. Instead of mucus, insects secrete a chitin tube around food and digestive enzymes to protect the gut from damage.

Questions (for answers and explanations, see page 397)

1. For which of the following organisms do the terms "heterotroph," "carnivore," and "filter feeder" all apply?
   a. Mushroom
   b. Blue whale
   c. Lion
   d. Bison
   e. Termite

2. Consider the digestive characteristics of a sheep. Which of the following sets of descriptions applies to this animal?
   a. Saprobe, engulfs food, performs intracellular digestion
   b. Autotroph, synthesizes organic nutrients, performs extracellular digestion
   c. Heterotroph, absorbs organic nutrients, carnivore
   d. Detritivore, ingests food, performs intracellular digestion
   e. Herbivore, ingests food, performs extracellular digestion

3. Many of the toothed whales (e.g., dolphins) have a large number of conical teeth, but no flattened molars like those found in humans. This suggests that the toothed whales
   a. carefully chew their food into a fine pulp before swallowing.
   b. rip or tear their food and probably swallow it whole.
   c. are herbivorous and feed mostly on easily digested algae.
   d. strain large volumes of water through their teeth and filter out the small fish.
   e. none of the above

4. Which of the following statements about the feeding and digestive adaptations of animals is *false?*
   a. Feeding morphology and the biochemistry of digestion largely determine an animal's ecological role.
   b. Tooth structure provides important clues about the lifestyle of living and extinct vertebrates.
   c. The simplest animal digestive systems are the gastrovascular cavities of cnidarians and flatworms.
   d. The digestive system is entirely absent from some species of internal parasites that live in the host's digestive tract.
   e. Herbivores possess more diverse feeding adaptations than carnivores because the prey of herbivores is so elusive and difficult to capture.

5. The _____ is primarily a storage chamber within the digestive system, while the _____ reabsorbs water, ions, and generates the feces.
   a. buccal cavity; midgut
   b. crop; midgut
   c. stomach; hindgut
   d. buccal cavity; hindgut
   e. hindgut; rectum

6. A drug is given to an animal that completely blocks the absorption of nutrients within the digestive system. Which of the following structures is most severely impacted by this treatment?
   a. Intestine
   b. Buccal cavity
   c. Crop
   d. Stomach
   e. Esophagus

7. Although their feeding strategies are rather divergent, the digestive systems of elephants and hummingbirds both contain all but one of the following materials. Select the *exception.*
   a. Zymogens
   b. Feces
   c. Proteases
   d. Dentine
   e. Nucleases

8. After a lunch that includes a cheeseburger with "the works" and fries, which of the following can be found actively participating in the digestive process within your gut?
   a. Peptidase
   b. Lipase
   c. Protease
   d. Carbohydrase
   e. All of the above

9. The typhlosole of an earthworm and the spiral valve of a shark are both adaptations for
   a. capturing and ingesting prey items.
   b. improving absorption of nutrients by the intestine.
   c. protecting the gut from its own digestive enzymes.
   d. filtering food items from the environment.
   e. preventing ulcers.

Activities (for answers and explanations, see page 397)

• In the following table, list the important structures associated with each section of the tubular gut, and indicate the basic function(s) of these structures.

Section	Structures	Functions
Foregut		
Midgut		
Hindgut		

• Examine Figure 47.7 on page 1047 of your textbook. Diagram b shows the lower jaw and teeth of a human and a cat. What structural differences in dentition exist between these two species, and what is the functional significance of these differences?

STRUCTURE AND FUNCTION OF THE VERTEBRATE GUT (pages 1049–1057)

Key Concepts

1. Although the digestive tracts of different vertebrates are highly specialized, all of the segments have the same general tissue organization. Beginning with the *lumen*, or central cavity of the gut, and working outward, four distinct tissue layers can be identified.

2. Immediately adjacent to the lumen is the *mucosa*. In certain portions of the gut the mucosa is highly folded to increase the surface area for absorption. Many mucosal cells also secrete lubricating mucus or digestive enzymes.

3. The *submucosa* lies just outside the mucosal layer. It contains the blood and lymphatic vessels that transport absorbed nutrients to other parts of the body. Sensory and regulatory nerves are also found in the submucosa.

4. Surrounding the submucosa are two layers of smooth muscle oriented at right angles to one another. *Circular muscle* surrounds the circumference of the gut tube, while *longitudinal muscle* runs down its length. Alternating contractions of these muscles produce wavelike churning movements in the stomach and intestines that mix food and move it through the digestive tract.

5. A final layer of fibrous tissue called the *serosa* surrounds and stabilizes the gut tube. The digestive system is also covered and supported by *peritoneum* tissue.

6. Food enters the digestive tract via the mouth, where it is usually chewed and then swallowed. Swallowed food passes through the *pharynx* and into the *esophagus*, a tube that connects the oral cavity with the remainder of the gut. The pharynx is located at the junction between the oral cavity, the trachea (windpipe), and the esophagus. The pharynx directs air and food into their proper pathways.

7. Smooth muscle contractions begin in the esophagus and normally progress toward the anus, creating a wavelike *peristalsis* that pushes food through the digestive tract. The presence of food causes the gut to stretch, which stimulates the muscle contractions of peristalsis. Sometimes the direction of peristalsis is reversed, as when the stomach is too full (e.g., heartburn) or during the violent contractions of the vomiting reflex.

8. Reversed movement of food within the digestive tract is normally prevented by thick rings of smooth muscle called *sphincters*. They are found at the junction of the esophagus and the stomach (esophageal sphincter) and where the stomach attaches to the intestine (pyloric sphincter). The release of feces through the anus is also regulated by sphincter muscles.

9. Chemical digestion of food is initiated in the mouth. The most notable enzyme in this process is amylase, which initiates the breakdown of carbohydrates. The tongue then pushes a bolus of food to the back of the mouth, where a complex set of neural reflexes causes swallowing.

10. Although animals typically consume their food rapidly, digestion is a relatively slow process. Food must therefore be stored in the stomach and metered into the digestive tract at a slow, steady pace. While awaiting release into the intestine, the churning motions of the stomach continue the mechanical breakdown of food.

11. The stomach secretes *hydrochloric acid*, which creates a very low pH that kills ingested microorganisms and is essential for the proper digestion of proteins. Stomach cells located in deep folds of the stomach lining called *gastric pits* secrete the zymogen *pepsinogen*, which is converted into the active endopeptidase *pepsin* by the acid environment of the stomach. This conversion process is known as *autocatalysis*.

12. The combined action of acid, muscular action, and pepsin activity causes the breakdown of ingested foods (especially proteins) in the stomach, exposing more nutrient surface area to enzymatic degradation. The resultant mixture, called *chyme*, is pushed to the bottom of the stomach where it is slowly released into the small intestine through the pyloric sphincter.

13. Little absorption occurs across the cell walls of the stomach, primarily because of the stomach's protective mucus coating. However, a few substances such as alcohol, aspirin, and caffeine can enter the stomach cells; this explains the rapid physiological effects of these drugs on the body.

14. Attached to the stomach is the *small intestine*. Named for its diameter, it is actually a very long, coiled tube with many minute foldings of the mucosal lining that create an enormous absorptive surface area.

15. The small intestine is divided into three sections. The *duodenum* is the initial portion responsible for most digestion. The *jejunum* and the *ileum* make up the bulk of the intestine; they perform 90 percent of the nutrient absorption.

16. The liver and pancreas are two accessory organs that produce digestive enzymes and other secretions used in the small intestine. The liver produces *bile*, a cholesterol derivative used to emulsify fats. Bile is secreted from the liver via the *hepatic duct* and stored in the gall bladder. In response to the presence of fat in the small intestine, bile travels from the gall bladder through the *common bile duct* and is released into the duodenum.

17. Emulsification is necessary because large, hydrophobic clusters of fat molecules are not easily digested. They must first be broken up into very small fat clusters called *micelles* to facilitate the action of *lipases* (fat-digesting enzymes). Bile molecules emulsify fats by inserting their lipophilic ends into the hydrophobic portions of the large fat clusters, leaving their lipophobic (hydrophilic) ends sticking out into the aqueous solution. This prevents the fat molecules from coalescing into large clusters and provides more surface area for the action of lipases.

18. A wide variety of exocrine products of digestive importance are made in the *pancreas*. They are released as zymogens in order to prevent self-digestion of the pancreas, and are activated inside the small intestine. One especially important zymogen is *trypsinogen*, which is activated by an *enterokinase* to become the functional protease *trypsin*. Trypsin then cleaves more trypsinogen molecules through autocatalysis to produce additional trypsin, and also activates other kinds of zymogens.

19. In addition to digestive enzymes, the pancreas releases bicarbonate ions (HCO_3^-), which neutralize the chyme entering the small intestine from the stomach. This is necessary because pancreatic enzymes, unlike the stomach proteases, do not function well at low pH.

20. The small molecules produced during the digestion of nutrient macromolecules eventually find their way to the cell walls of the microvilli lining the small intestine. These cells release dipeptidases and the enzymes maltase, lactase, and sucrase. The individual amino acids and monosaccharides produced during this final digestion are then absorbed across the mucosal layer.

21. Amino acids and monosaccharides are probably absorbed across the mucosal layer by active transport. A *sodium cotransport* system plays a role in this process by providing energy in the form of a sodium concentration difference between the mucosa and the submucosa

to actively transport nutrients through the mucosal membranes.

22. Since they are components of lipids, the fatty acids and monoglycerides produced by fat digestion can dissolve through the lipid membranes of mucosal cells. Inside the cells they are reassembled and combined with other molecules to form water-soluble *chylomicrons*, which are then transported by the lymph and released into the blood.

23. Bile molecules are reused by shuttling back and forth between the microvilli and the gut contents. Eventually they reach the ileum, where many are reabsorbed and transported back to the liver. Some bile loss is considered healthy, however, since it helps remove excess cholesterol from the body.

24. The large intestine, or *colon*, receives undigested food by peristalsis from the ileum. Water and inorganic ions are removed in the colon, producing the semisolid feces. These are stored near the end of the colon and periodically excreted. If the colon removes too much water from the feces, constipation results; too little water removal produces diarrhea.

25. Large numbers of beneficial bacteria such as *E. coli* live within the colon, feeding on undigested food and producing beneficial products such as vitamin K and biotin for the host. Some animals enhance these nutritional benefits by eating their feces, a behavior called *coprophagy*.

26. The *appendix* of humans is a small, blind-ended sac that emerges from the colon. It is a *vestigial* (nonfunctional remnant) of the much larger caecum found in herbivores, which functions in cellulose digestion. Although it still plays a role in immune activity, the human appendix is not essential and is often surgically removed if it becomes infected.

27. Ironically, most herbivores are incapable of producing *cellulases*, the enzymes needed to digest their primary food source, cellulose. These animals therefore harbor vast numbers of microorganisms inside their digestive tracts that do manufacture cellulases.

28. *Ruminant* animals such as cows and sheep have modified digestive anatomies that enhance the activity of microorganisms. The typical mammalian stomach has been replaced in these animals by a four-chambered structure, each part of which is specialized for a particular function in cellulose digestion.

29. The first two chambers are the *rumen* and the *reticulum*. They are packed with anaerobic organisms that carry out fermentation of cellulose to produce fatty acids. Periodically this fermenting mixture (the cud) is regurgitated into the mouth and rechewed before being sent back to the chambers for additional digestion. Large quantities of gaseous CO_2 and CH_4 produced through the fermentation process are expelled by belching.

30. Partially digested materials from the rumen and reticulum are passed on to a third chamber called the *omasum* that is the site of water reabsorption. The food then enters the *abomasum*, or true stomach, where high acidity and proteases kill the microorganisms. The fatty

acids, proteins, and carbohydrates obtained from both the digested plant materials and microorganisms are then absorbed into the body by the intestine.

31. Nonruminant herbivores such as rabbits maintain their own microbial fermentation vats inside the *caecum.* There is, however, no small intestine after this structure, so the nutritionally beneficial materials produced by microbial activity can only be made available for absorption through coprophagy.

Activities *(for answers and explanations, see page 398)*

- Label the tissue layers and gut space indicated in the following schematic cross section of a vertebrate small intestine.

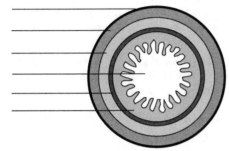

- Explain the functional significance of the large size of the rumen and the reticulum, as compared with the smaller omasum and abomasum, in the digestive tracts of ruminant mammals.

Questions *(for answers and explanations, see page 398)*

1. A blood vessel entering the small intestine from the peritoneum will pass through the _____ before reaching the submucosa.
   a. lumen and serosa
   b. serosa and mucosa
   c. serosa and two smooth muscle layers
   d. two smooth muscle layers and mucosa
   e. two smooth muscle layers only

2. The contraction of smooth muscle is important in which of the following digestive processes?
   a. Swallowing
   b. Peristalsis and sphincter control
   c. Secretion of digestive enzymes
   d. Zymogen formation
   e. All of the above

3. Carbohydrate digestion first begins in the human _____ , but occurs mostly in the _____ .
   a. mouth; stomach
   b. stomach; small intestine
   c.. small intestine; colon
   d. mouth; small intestine
   e. stomach; colon

4. The reabsorption of water within the vertebrate digestive system
   a. takes place mainly in the large intestine.
   b. occurs only in the duodenum.
   c. occurs both in the jejunum and the ileum.
   d. takes place through the gastric pits of the small intestine.
   e. is regulated by the sodium cotransport system.

5. Which of the following statements about the digestive roles of the gall bladder and pancreas is *false*?
   a. The exocrine products of the pancreas empty into the duodenum.
   b. The gall bladder stores carbohydrases for subsequent release into the small intestine.
   c. Trypsinogen is one of the zymogens released from the pancreas.
   d. Large amounts of bicarbonate ions are released from the pancreas to neutralize the chyme entering the small intestine.
   e. The gall bladder is an important element in the process of lipid digestion.

6. The last enzyme to act upon the remnants of an ingested protein before its component amino acids cross the membranes of cells lining the intestinal mucosa is
   a. a carbohydrase.
   b. an enterokinase.
   c. a dipeptidase.
   d. pepsinogen.
   e. amylase.

7. A biochemical test for the presence of chylomicrons within the vertebrate body would find them located mainly in the
   a. stomach and colon.
   b. mucosal cells of the small intestine.
   c. mucosal cells of the small intestine, the lymph fluid, and the blood.
   d. duodenum, jejunum, and ileum.
   e. abomasum and reticulum of ruminant mammals.

8. Which of the following statements about the digestion of cellulose is *false*?
   a. Cellulose can only be digested by highly specialized mammalian herbivores.
   b. Cellulases are the enzymes that hydrolyze cellulose.
   c. Cellulose fermentation occurs in the rumen and reticulum of a bison.
   d. Two byproducts of cellulose digestion in mammals are carbon dioxide and methane gas.
   e. The caecum is used to carry out cellulose digestion in nonruminant, herbivorous mammals.

CONTROL AND REGULATION OF DIGESTION • CONTROL AND REGULATION OF FUEL METABOLISM • NUTRIENT DEFICIENCY DISEASES • TOXIC COMPOUNDS IN FOOD (pages 1057–1063)

Key Concepts

1. Neural reflexes mediated by the central nervous system coordinate important digestive functions like swallowing and salivation. Stimuli such as the touch of food in the mouth or smell trigger these reflex responses.

2. The digestive tract is unique in that it also possesses an intrinsic nervous system that sends signals from one part of the gut to another without the signals passing through the CNS. An example is the *gastrocolic reflex* that initiates bowel movements through direct stimulation of the colon when food load increases in the stomach.

3. In addition to neural control, hormones also influence the activity of digestive systems. The first-ever demonstration of hormonal influences in the body involved *secretin*, which travels from the mucosal lining of the duodenum, through the bloodstream, and causes the release of pancreatic secretions.

4. *Cholecystokinin* is another digestive hormone that is secreted by the intestinal mucosa to stimulate the release of bile from the gall bladder. *Gastrin* is released by the stomach in response to the presence of food. It causes secretions from, and muscular contractions of, the upper stomach walls.

5. The regulation of nutrients for use in energy metabolism is mediated primarily by the liver, which stores large quantities of glycogen and fat and synthesizes important blood proteins. The liver also interconverts carbohydrates with fats and regulates the conversion of amino acids, pyruvate, and lactate into glucose through a process known as *gluconeogenesis*.

6. *Lipoproteins* are made up of a fat core and a hydrophilic protein covering. They help control fat metabolism and transport in the blood. The protein coat not only serves to make the lipoprotein water-soluble, but also helps target the fat to a specific tissue in the body. Chylomicrons are large lipoproteins used in dietary fat and cholesterol transport between absorptive regions of the intestine and other body tissues.

7. The liver also produces several kinds of lipoproteins that differ in their densities. Very-low-density lipoprotein (VLDL), low-density lipoprotein (LDL), and high-density lipoprotein (HDL) are all believed to influence the rate of cholesterol accumulation in the human body. HDL, the so-called "good cholesterol," serves the beneficial role of removing cholesterol from body tissues and returning it to the liver for conversion into bile.

8. During the *absorptive period*, when food is present in the gut and nutrients are actively being absorbed into the circulatory system, the liver converts glucose into glycogen and fat. At this time, body cells store fatty acids in the form of fat, and preferentially use glucose as their metabolic fuel.

9. These processes are reversed during the *postabsorptive period*, when free nutrients are no longer available in the gut or circulatory system. The lone exception to this reversal is nervous tissue, which requires glucose for energy metabolism at all times.

10. The switch between absorptive and postabsorptive energy metabolism is largely controlled by the hormones *glucagon* and *insulin* released from the pancreas. Insulin is secreted in response to elevated blood glucose levels. It facilitates the entry of glucose into cells, stimulating the use of glucose as metabolic fuel and its conversion into fat or glycogen.

11. Glucagon is released when blood glucose falls below normal postabsorptive levels. Its effects are opposite those of insulin, stimulating the conversion of glycogen and fat into glucose, which is then released into the blood.

12. The primary control mechanism in this system is a reduced concentration of insulin in the blood; the presence of glucagon is a secondary stimulus regulating fuel metabolism.

13. Under stressful conditions, postabsorptive blood glucose levels are also influenced by the hormones *cortisol* and *epinephrine*, which are released by the adrenal cortex and the adrenal medulla, respectively. Epinephrine has similar effects on liver cells such as glucagon, stimulating glycogen breakdown and increasing gluconeogenesis. Cortisol inhibits glucose metabolism and promotes the use of fats as metabolic fuel.

14. In summary, the absorptive state is characterized by the channeling of glucose and other fuel molecules toward a variety of storage forms. The postabsorptive state is marked by the conversion of stored fuels into glucose for use by the nervous system, with other tissues switching over to the use of fatty fuels. Several hormones are involved in regulating this process.

15. Deficiency diseases develop when animals experience chronic nutrient shortages. The human disease *kwashiorkor* results from severe protein deficiencies that cause extreme edema, breakdown of the immune system, and the degeneration of metabolic and neural functions.

16. Vitamin deficiencies can also produce disease. The lack of thiamin (vitamin B_1) causes the extreme weakness associated with the disease *beriberi* . A lack of vitamin C can lead to *scurvy*. This debilitating and sometimes fatal disease produces massive tissue breakdown resulting from problems with collagen formation; it was once common among sailors on long oceanic voyages who had no access to fresh fruits or vegetables.

17. *Pellagra* is another vitamin-deficiency disease caused by a lack of nicotinamide; it is characterized by diarrhea, itching skin, and abdominal pain and is often associated with alcoholism. *Rickets* results from a vitamin D deficiency that causes skeletal deformation as bones soften through decreased calcium absorption. *Pernicious anemia*, a blood disease caused by a lack of the animal vitamin B_{12}, can occur in strict vegetarians who do not take vitamin supplements.

18. Deficiencies in mineral nutrition can also lead to disease. For example, iodine is a component of the thyroid hormone thyroxine. If there is a dietary deficiency of iodine, a *goiter* develops as the thyroid gland enlarges to try to compensate for the lack of functional thyroxine.

19. Many nutrients contain natural toxic substances that have evolved mainly to deter predators. Examples include the poisons and hallucinogens in mushrooms, neurotoxins in animals, and plant compounds that influence heart function or modify nervous system function. These compounds include nicotine, opium, and the psychoactive agents in marijuana.

20. In addition to natural toxins, humans regularly add enormous quantities of potentially toxic substances into the environment that can ultimately end up endangering plant and animal life. This problem is studied by scientists in the field of *environmental toxicology*.

21. *Hydrophilic* toxins are less dangerous to organisms because they are accessible to a wide range of degradative enzymes and can be quickly metabolized and detoxified. *Lipophilic* (hydrophobic) toxins are metabolized more slowly and can thus accumulate in body tissues, often reaching very high concentrations.

22. Lipophilic toxins can also *bioaccumulate* because of the concentration of substances that occurs when nutrients move through a food web. Deleterious effects stemming from large *body burdens* of toxins such as the pesticide DDT are now well documented in animals, and have clearly been linked to bioaccumulation by top predators in food webs.

23. A serious problem with many synthetic environmental toxins is that the body lacks natural mechanisms to metabolize them. Metabolic enzymes called *cytochrome P450s* are generalist detoxification enzymes that can handle some synthetic toxins, but many others escape detoxification by P450s.

24. Some synthetic toxins are structurally similar to hormones. They bind to hormone receptors but do not trigger normal hormonal responses in cells. Polychlorinated biphenyls, or *PCBs*, which were once used to insulate electrical transformers, are an example. PCBs are stable, lipophilic, and are now common in the environment. They can have dramatic effects on reproductive cycles, maintenance of body weight, immune system function, and embryonic development in animals.

Activities (for answers and explanations, see page 398)

* Match each of the following hormones involved in the control and regulation of energy metabolism with its best description.

___ Secretin	a. Secreted by mucosal lining; causes release of chemicals from pancreas
___ Glucagon	b. Causes muscular contraction of the stomach wall in response to food
___ Gastrin	c. Adrenal gland hormone that inhibits glucose metabolism
___ Epinephrine	d. Causes the gall bladder to release bile into the small intestine
___ Insulin	e. Has same effects as glucagon under conditions of physiological stress
___ Cholecystokinin	f. Pancreatic hormone that stimulates the conversion of glycogen to glucose
___ Cortisol	g. Released from the pancreas in response to elevated blood glucose levels

* Briefly describe the process of bioaccumulation, and explain how it is a particular problem with respect to toxic substances in the environment.

Questions (for answers and explanations, see page 398)

1. A fundamental difference between the gastrocolic and salivation reflexes in humans is that
   a. one reflex involves only the peripheral nervous system, the other only the central nervous system.
   b. one reflex involves only the intrinsic nervous system of the gut, while the other is mediated by the central nervous system.
   c. one reflex is influenced by hormonal regulation, the other is not.
   d. one reflex produces an appropriate digestive response, the other does not.
   e. none of the above

2. Damage to the mucosal lining of the gut would directly impair the control function of which of the following hormones?
   a. Gastrin only
   b. Secretin and insulin
   c. Gastrin, secretin, and cholecystokinin
   d. Cholecystokinin and glucagon
   e. Cholecystokinin only

3. The rate of gluconeogenesis is affected by which of the following hormones?
   a. Glucagon
   b. Insulin
   c. Epinephrine
   d. Cortisol
   e. All of the above

4. During the absorptive period, when food is present in the gut, _____ is released by the pancreas and stimulates the _____ in the liver.
   a. insulin; conversion of glucose to glycogen and fat
   b. glucagon; conversion of glucose to glycogen and fat
   c. cortisol; conversion of glucose to glycogen and fat
   d. insulin; conversion of glycogen and fat to glucose
   e. glucagon; conversion of glycogen and fat to glucose

5. The postabsorptive period of vertebrate digestion is characterized mainly by
   a. elevated blood glucose levels and the rapid secretion of insulin.
   b. accelerated rates of glycogen storage in liver and muscle cells.
   c. decreased levels of gluconeogenesis.
   d. conversion of stored fuels into glucose, and a switch to most cells using fatty fuels.
   e. an overall state of metabolic quiet.

6. Chylomicrons, VLDL, LDL, and HDL are related to one another in that they are all
   a. proteins.
   b. lipids.
   c. lipoproteins.
   d. digestive hormones.
   e. none of the above

7. All but one of the following diseases is caused by a vitamin deficiency. Select the *exception*.
   a. Rickets
   b. Pernicious anemia
   c. Scurvy
   d. Kwashiorkor
   e. Beriberi

8. The environmental toxins known as PCBs exert their biological influence
   a. by preventing the process of bioaccumulation.
   b. by interfering with the normal function of animal hormones.
   c. only when glucagon is absent from the blood.
   d. if vitamin C is lacking in the diet.
   e. none of the above

Integrative Questions (for answers and explanations, see page 398)

1. A cow, an eagle, a snake, and a human have which of the following digestive characteristics in common?
   a. They are ingestive heterotrophs.
   b. They are predators.
   c. They have a stomach and a small intestine.
   d. They utilize autocatalysis.
   e. All of the above

2. Which of the a following is *not* a type of nutritional problem in humans?
   a. Anorexia nervosa
   b. Malnutrition
   c. Undernourishment
   d. Beriberi
   e. Defecation

3. The human liver carries out all but one of the following functions. Select the *exception*.
   a. Insulin production
   b. Synthesis of plasma proteins
   c. Gluconeogenesis
   d. Glycogen storage
   e. Interconversion of fuel molecules

4. In which of the following animals would you find both a mouth, an anus, and extracellular digestion?
   a. Nematode
   b. Snail
   c. Cockroach
   d. Earthworm
   e. All of the above

48

Salt and Water Balance and Nitrogen Excretion

CHAPTER LEARNING OBJECTIVES—after studying this chapter you should be able to:

❑ Describe the primary roles of excretory systems and the three mechanisms employed in their operation.

❑ Contrast the osmoregulatory problems associated with life in fresh water, life in salt water, and a terrestrial existence.

❑ Differentiate between osmoconformers and the two types of osmoregulators, hypertonic osmoregulators and hypotonic osmoregulators.

❑ Contrast ionic conformers with ionic regulators.

❑ Describe the concepts of osmoregulation illustrated by the brine shrimp *Artemia* living in different environments.

❑ Describe the three major types of nitrogenous waste products that result from the metabolism of the major classes of macromolecules.

❑ Define the terms "ammonotelic," "ureotelic," and "uricotelic," and classify animals relative to these terms.

❑ Describe the structure and functioning of the protonephridial excretory system based on the flame cell seen in flatworms.

❑ Describe the structure and functioning of the metanephridial excretory system seen in most of the annelid worms.

❑ Describe the structure and functioning of Malpighian tubule–based excretory systems as seen in the insects.

❑ Describe the structure of a typical vertebrate nephron and associated circulatory elements, and explain how they participate in filtration, secretion and reabsorption, and urine processing.

❑ Discuss osmoregulatory adaptations seen in the following groups: marine bony fish, freshwater fish, cartilaginous fish, amphibians, reptiles, and birds.

❑ Describe the components of the mammalian excretory system and the major circulatory elements serving it.

❑ Describe the components of the mammalian nephron and the capillary beds associated with it.

❑ Describe the regions of the mammalian kidney and explain what nephron components are found in these regions.

❑ Explain the quantitative relationship between the amount of blood entering the kidney, the amount of filtrate formed, and the amount of urine produced.

❑ Relate the characteristics of the cells lining the different regions of the nephron to the activities occurring in those regions.

❑ Explain how the countercurrent multiplier system in the vertebrate kidney establishes a NaCl concentration gradient in the medulla.

❑ Describe the roles of each of the following nephron components in the production of concentrated urine: proximal convoluted tubule, descending limb of the loop of Henle, ascending limb of the loop of Henle, distal convoluted tubule, collecting duct.

❑ Describe the relationship between the following variables: concentration of urine formed, magnitude of the NaCl concentration gradient in the medulla, length of the loop of Henle.

❑ Discuss the major mechanisms involved in the autoregulation of the glomerular filtration rate, especially the role of renin, angiotensin, and aldosterone.

❑ Discuss the central control mechanisms involving antidiuretic hormone (ADH) that regulate the concentration of the urine and the osmolarity and volume of the blood.

❑ Explain the involvement of aquaporins in determining the water permeability of tissues and the action of ADH.

❑ Explain the osmoregulatory principles involved in the operation of the artificial kidney.

❑ Describe some adaptations of the mammalian kidney to produce very concentrated urine.

EXTRACELLULAR FLUIDS AND WATER BALANCE • DIS-TINGUISHING ENVIRONMENTS AND ANIMALS IN TERMS OF SALT AND WATER • EXCRETING NITROGEN AS AMMONIA, URIC ACID, OR UREA   (pages 1067–1070)

Key Concepts

1. *Excretory systems* regulate the volume, composition, and osmotic potential of body fluids. Excretion of excess and toxic molecules and retention of needed molecules is part of the function of excretory systems.

2. The functioning of excretory organs is closely tied to the environment in which the animal lives, be it marine, freshwater, or terrestrial.

3. Since water cannot be moved by active transport, all excretory mechanisms employ pressure or osmotic gradients to move water.

4. Most excretory organs include a system of *tubules* in which the composition of the extracellular fluid is modified to produce *urine*—the liquid excretory waste product. Extracellular fluid enters the tubules by *filtration* and its composition is modified by active *secretion* and *reabsorption*.

5. The *salinity*, or salt content, of aqueous environments varies continuously from fresh water to tide pools that are saltier than the sea.

6. Most marine invertebrates are *osmoconformers*—the osmotic potential of their body fluids matches that of the environment.

7. Even osmoconformers must become *osmoregulators* in environments with extreme osmotic potentials.

8. In fresh water, excess water passively enters organisms by osmosis and must be excreted. Salts are lost from the body in fresh water and must be conserved. Osmoregulators in fresh water produce large volumes of dilute urine.

9. In salt water, water is lost to the external environment and must be conserved. Osmoregulators in salt water produce small amounts of concentrated urine.

10. A *hypertonic osmoregulator* maintains body fluid that has a more negative osmotic potential (it is saltier) than the environment (i.e., the body fluid is hypertonic to the environment).

11. A *hypotonic osmoregulator* maintains body fluid that has a less negative osmotic potential (it is less salty) than the environment (i.e., the body fluid is hypotonic to the environment).

12. The brine shrimp (*Artemia*) can be either a hypertonic osmoregulator, an osmoconformer, or a hypotonic regulator depending on the osmotic potential of its environment. Its main mechanism of osmoregulation is active transport of NaCl across its gills.

13. If the ionic composition of an osmoconformer's body fluids is like that of its environment, it is called an *ionic conformer*—it does not regulate its ions at all. However, most osmoconformers are *ionic regulators*—active transport is used to absorb or excrete ions so that the ionic composition of the body fluid differs from that of the environment.

14. Most terrestrial animals must conserve water and obtain needed ions as part of their diet. Because plant biomass is low in sodium, herbivores must ingest salt directly. Some marine birds have *nasal salt glands* to rid themselves of excess dietary salt.

15. Unlike carbohydrates and fats, which are metabolized entirely to water and carbon dioxide, the metabolism of proteins produces the nitrogenous waste product *ammonia* in addition to H_2O and CO_2.

16. Ammonia is highly toxic and must be immediately excreted or converted into the nontoxic molecules *urea* and *uric acid* prior to excretion.

17. *Ammonotelic* animals, including most aquatic invertebrates, bony fishes, crocodiles, amphibian tadpoles, etc., excrete ammonia directly by diffusion from the blood through the gill membranes to the environment.

18. Because ammonia must be kept dilute, ammonia excretion requires large amounts of water. Consequently, most terrestrial animals convert ammonia to urea or uric acid.

19. *Ureotelic* animals, such as mammals, amphibians, and cartilaginous fishes, excrete their nitrogenous waste as urea. Urea is highly soluble in water. To conserve water, some vertebrate kidneys can produce quite concentrated urine.

20. Sharks and rays accumulate high concentrations of urea in their tissues as a way of increasing the osmotic potential of their body fluids and limiting osmotic loss of water to their marine environment.

21. *Uricotelic* animals, including insects, reptiles, birds, and some amphibians, excrete their nitrogenous waste as uric acid. Because uric acid is very insoluble in water, it can be excreted with little loss of water.

22. Many species produce more than one nitrogenous waste product. Being mammals, humans mostly excrete urea, but small amounts of ammonia and uric acid are also produced.

23. Some species switch their major excretory products during development. For example, aquatic frog and toad tadpoles excrete ammonia, but as terrestrial adults they switch to producing urea.

Questions (for answers and explanations, see page 399)

1. Which of the following is *not* a feature common to the excretory systems of most animals?
   a. Tubular design
   b. Production of urine
   c. Active transport of water
   d. Filtration
   e. Secretion and reabsorption

2. Select *all* of the following features that would be characteristic of a penguin.
   a. Ammonotelic
   b. Ureotelic
   c. Uricotelic
   d. Presence of nasal salt glands
   e. Accumulation of urea in body tissues

3. Select *all* of the following features that would be characteristic of a freshwater bony fish.

a. Ammonotelic
b. Ureotelic
c. Uricotelic
d. Produces urine more concentrated than its blood
e. Absorbs salt from water across gill membranes
f. Drinks little

4. The osmotic potential of the body fluid of a hypertonic osmoregulator should be _____ negative than the external environment, and this animal would probably live in a _____ environment.
    a. more; freshwater
    b. more; saltwater
    c. less; freshwater
    d. less; saltwater
    e. less; terrestrial

5. Ammonia results directly from the metabolism of
    a. carbohydrates only.
    b. fats only.
    c. proteins only.
    d. nucleic acids only.
    e. proteins and nucleic acids.
    f. all four classes of macromolecules.

Activities (for answers and explanations, see page 399)

- Draw a curve in the following figure to show the relationship between the osmotic potential of the body fluid of the brine shrimp *(Artemia)* and its external environment. Label the portion of the curve where *Artemia* shows hypertonic osmoregulation, osmoconformity, and hypotonic osmoregulation.

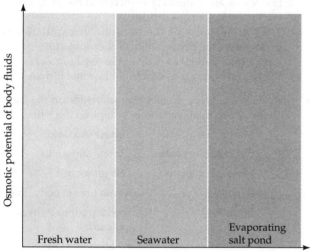

Osmotic potential of environment

- Would you expect *Artemia* to be an ionic conformer or an ionic regulator? Explain.

THE DIVERSE EXCRETORY SYSTEMS OF INVERTEBRATES • THE NEPHRON: THE BASIC UNIT OF VERTEBRATE EXCRETION   (pages 1070–1075)

Key Concepts

1. Marine invertebrates are largely osmoconformers and ionic regulators. They have few specific adaptations for salt and water balance, other than active transport of specific ions, and most excrete ammonia directly to the environment.

2. Flatworms, such as *Planaria*, have excretory systems consisting of a network of tubules that include blind ends formed by ciliated *flame cells* and excretory pores that open to the exterior. Tubules and flame cells together are called *protonephridia.*

3. Body fluid filters into the cavity formed by each flame cell, and as it is moved down the tubule by the beating cilia, active reabsorption of ions occurs so that the urine leaving the excretory pores is hypotonic to the body fluid.

4. The same processes of filtration of the body fluids and tubular reabsorption of ions is well developed in the *metanephridia* of annelid worms, such as the earthworm.

5. Blood pressure within the closed circulatory system causes blood to filter through the capillary membranes. The resulting *coelomic fluid* fills each segment of the worm's body and consists of water, ions, and small molecules; blood cells and large protein molecules are retained in the circulatory system as part of the filtration process.

6. A metanephridium consists of a funnellike ciliated opening, called a *nephrostome*, into which coelomic fluid flows, a long tubular portion where active reabsorption and secretion of molecules occurs, and a *nephridiopore* that dumps the hypotonic urine to the outside of the worm. Each segment of the worm's body has a pair of metanephridia.

7. Terrestrial arthropods, including the insects, have excretory organs called *Malpighian tubules* that are highly efficient at conserving water. Malpighian tubules are a collection of 2 to over 100 tubular outpocketings of the digestive tract that are bathed by coelomic fluid.

8. Cells in the Malpighian tubules absorb uric acid and potassium and sodium ions from the coelomic fluid, creating an osmotic potential gradient between the coelomic fluid and the interior of the tubule. Water moves by osmosis into the tubule. Muscle fibers within the tubules move the contents into the gut cavity.

9. Acidic conditions within the gut precipitate the uric acid, and sodium and potassium are actively pumped back into the body cavity. Water now follows the reabsorbed salts to rejoin the coelomic fluid, and this leaves behind a semisolid mixture of uric acid and undigested food.

10. *Kidneys* are the major salt and water balance organs found in vertebrates. *Nephrons* are the functional units of the kidneys. Each human kidney includes about one million nephrons.

11. Two capillary beds are located between the arteriole and the venule associated with each nephron. The first bed, called the *glomerulus*, is a knot of very permeable vessels enclosed within the cup-shaped end of the tubular part of the nephron called the *Bowman's capsule*.

12. An *afferent arteriole* carries blood to the glomerulus and an *efferent arteriole* transports blood from the glomerulus to the second capillary bed, called the *peritubular capillaries*, which surrounds the renal tubule.

13. The glomerulus and capsule together are called the *renal corpuscle*. Specialized cells of the capsule called *podocytes* have fingerlike projections that wrap around the capillaries of the glomerulus.

14. The blood entering the glomerulus is under high pressure. Water and small molecules are filtered through pores in the cells of the capillaries and the slits of the podocytes and are collected in the Bowman's capsule. Large molecules and blood cells remain behind.

15. Glucose, amino acids, and other needed molecules and ions are reabsorbed from the filtrate as it moves through the *renal tubule*, and are returned to the circulatory system via the peritubular capillaries. Cells of the tubule also secrete unwanted substances into the urine.

16. The earliest vertebrates lived in fresh water and the primary function of the nephron was to eliminate excess water taken on osmotically, while conserving salts.

17. Various adaptations are seen in the different vertebrate groups living in environments where water must be conserved and salt excreted.

18. Marine bony fishes drink seawater, and salt is actively secreted from their gill membranes. Ammonia also diffuses across the gill membranes. Since these animals must conserve water, their kidneys have very few (or no!) glomeruli, and consequently, very little urine is produced.

19. Whereas marine bony fishes are hypotonic to seawater and continuously lose water, the cartilaginous fishes retain urea within the body tissues so that they become isotonic, or even hypertonic, to seawater. Cartilaginous fishes have a salt-secreting *rectal gland* to remove excess NaCl.

20. As with freshwater fishes, amphibians mostly produce copious, dilute urine and conserve salts. Other adaptations for life in dry environments seen in amphibians include secretion of waxes and fats to retard water loss from the skin and burrowing and estivation during long dry periods.

21. Reptiles and birds have evolved some similar adaptations for water conservation, including internal fertilization, shelled eggs, waterproof skin coverings, and uric acid as the nitrogenous excretory product. As with mammals, birds can also produce urine that is hypertonic to their body fluid.

Questions (for answers and explanations, see page 399)

1. Match the excretory system characteristics on the right with the appropriate animal groups on the left.

    _____ Flatworms   *a.* Nephrons
    _____ Annelid worms   *b.* Flame cells
    _____ Insects   *c.* Metanephridia
    _____ Vertebrates   *d.* Protonephridia
                             e. Malpighian tubules

2. Select the following feature that is *not* characteristic of both the metanephridia of earthworms and the nephrons of vertebrates.
    a. Both are tubules.
    b. Selective reabsorption occurs there.
    c. Selective secretion occurs there.
    d. Both produce urine with a composition different than that of the body fluid.
    e. Coelomic fluid enters the excretory organ directly.

3. Which of the following statements about the excretory system of insects is *not* true?
    a. Active transport moves materials from the coelomic fluid into the Malpighian tubules.
    b. The Malpighian tubules can produce a highly concentrated urea solution, allowing insects to inhabit some of Earth's driest habitats.
    c. Reabsorption of salts takes place mostly in the gut.
    d. Water reabsorption is by osmotic movement only.
    e. There are no circulatory vessels associated with the Malpighian tubules.

4. Which of the following statements about the vertebrate nephron is *not* true?
    a. The blood pressure is highest in the afferent arterioles.
    b. Podocytes are intimately involved in blood filtration within the Bowman's capsule.
    c. The permeability of the peritubular capillaries is greater than the permeability of the glomerulus.
    d. Active transport takes place in the renal tubules.
    e. Plasma proteins are absent from the renal filtrate.

5. Match the excretory system characteristics on the right with the appropriate vertebrate groups on the left.

    _____ Marine bony fishes   *a.* Excrete uric acid
    _____ Cartilaginous fishes   *b.* Excrete ammonia
    _____ Most amphibians   *c.* Few glomeruli
    _____ Reptiles   *d.* Retain urea in body fluids
    _____ Birds   *e.* Produce copious, hypotonic urine
                                  f. Waterproof skin covering
                                  g. Produce hypertonic urine
                                  h. Rectal glands

Activity (for answers and explanations, see page 399)

- Label the diagram of a typical vertebrate nephron shown on the next page with the following labels: afferent arteriole, efferent arteriole, glomerulus, Bowman's capsule, podocytes, renal tubule, peritubular capillaries, renal venule, and collecting duct. Circle the portion of the nephron where filtration occurs.

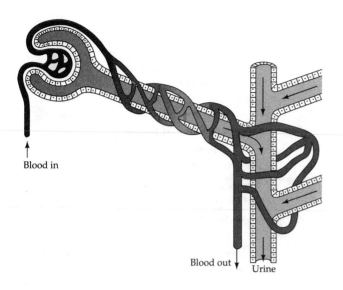

Blood in

Blood out ↓   Urine

THE MAMMALIAN KIDNEY • TREATING KIDNEY FAILURE • CONTROL AND REGULATION OF KIDNEY FUNCTIONS (pages 1075–1082)

Key Concepts

1. In humans, the excretory system consists of two *kidneys*, *ureters* that connect each kidney to a *urinary bladder* where urine is stored, and a *urethra* that conducts the urine to the outside. Two sphincter muscles at the base of the urethra control the timing of urination.

2. Each kidney is supplied with a *renal artery* and *renal vein*. The renal artery branches extensively to form the afferent arterioles that serve each nephron. The peritubular capillaries join to form venules that fuse to become the renal vein.

3. The ureter enters the kidney on its concave side. It divides into several branches that envelop kidney tissue called *renal pyramids*. The collecting ducts to which the nephrons are connected empty into the branches of the ureter at the tips of the renal pyramids.

4. In addition to the Bowman's capsule, the tubular part of the mammalian nephron can be divided into a *proximal convoluted tubule* (the part of the tubule nearest the capsule), a *distal convoluted tubule* (the part of the tubule nearest the collecting duct), and the *loop of Henle*, connecting the two convoluted tubules.

5. The kidney consists of two layers: an outer *cortex* and an inner *medulla*.

6. The cortex includes the glomeruli and the proximal and distal convoluted tubules of the nephrons. Most of the peritubular capillaries are also in the cortex.

7. The loop of Henle is of varying length in different mammalian groups. The *descending limb* dips down into the medulla, shows a hairpin loop, and becomes the *ascending limb* that returns to the cortex and joins the distal convoluted tubule.

8. The collecting tubules start in the cortex and run in parallel with the loops of Henle through the medulla. Branches

of the peritubular capillaries called the *vasa recta* are located in the medulla and surround the loop of Henle.

9. All substances enter the kidney via the renal artery and leave the kidney via the renal vein as blood or via the ureter as urine. In humans, about 180 liters of filtrate enter the glomeruli of the kidneys each day (about 12 percent of the total blood volume moving through the renal arteries), but only about 2 to 3 liters of urine are normally produced. The remaining filtrate is reabsorbed into the blood.

10. Cells of the proximal convoluted tubules are cuboidal and have extensive microvilli and numerous mitochondria. Most of the reabsorption of solutes, including NaCl, glucose, amino acids, and other needed molecules, occurs via active transport mechanisms in the cells of the proximal convoluted tubules.

11. Water osmotically follows the reabsorbed solutes into the cells of the proximal convoluted tubule, and the water and solutes are then returned to the blood in the peritubular capillaries.

12. Although the filtrate in the proximal convoluted tubule changes dramatically in composition due to reabsorption of solutes, the filtrate entering the loop of Henle is isotonic to blood plasma because a proportional share of water is also reabsorbed in the proximal convoluted tubule.

13. The loops of Henle in the mammalian kidney set up a *countercurrent multiplier system* in the medulla that enables mammals to produce hypertonic urine (4 times more concentrated than blood plasma in humans and 12 to 15 times more concentrated in desert mammals such as the kangaroo rat).

14. Cells of the descending limb and the lower part of the ascending limb of the loop of Henle are permeable to water and small molecules. They have few microvilli and mitochondria and perform little active transport.

15. Cells of the upper part of the ascending limb of the loop of Henle and the distal convoluted tubule are impermeable to water and actively transport NaCl out of the tubular fluid (urine) and into the surrounding extracellular fluid of the renal medulla. Consequently, the urine becomes more dilute as it moves up the ascending limb.

16. Because the NaCl pumped out of the ascending limb moves into tubular fluid in the descending limb, a NaCl concentration gradient builds up in the medulla.

17. Cells of the collecting duct are permeable to water but not mineral ions. As urine travels down the collecting duct it encounters extracellular fluid with ever-increasing NaCl concentration. Water moves osmotically out of the collecting duct, concentrating the urine.

18. Water reabsorbed from the collecting ducts returns to the circulatory system via the vasa recta.

19. Kidney failure (acute renal failure) causes retention of salt and water, increase in blood pressure, and a poisonous accumulation of urea *(uremia)* and metabolic acids *(acidosis)*.

20. The *artificial kidney* depends on the process of *dialysis*; blood is separated by a *semipermeable membrane* from an isotonic *dialyzing fluid* of precise composition. In the dialyzing fluid, the concentration of valuable molecules is

the same as their concentration in the blood while the concentration of waste products is zero. Waste products, such as urea and sulfate, move by osmosis from the blood into the dialysis fluid, while other molecules remain in the blood.

21. The kidney shows *autoregulatory adaptations* that maintain the *glomerular filtration rate (GFR)*. The GFR depends on both an adequate blood volume and a high blood pressure.

22. Autoregulatory responses to decreases in either cardiac output or blood pressure include dilation of the afferent arterioles and secretion by the kidney of the enzyme *renin*. Renin catalyzes a reaction that leads to production of the hormone *angiotensin*.

23. The effects of angiotensin leading to an increase in blood pressure include constriction of efferent renal arterioles and peripheral blood vessels. Angiotensin also stimulates thirst and causes the *adrenal medulla* to produce the hormone *aldosterone*. Aldosterone increases sodium reabsorption in the kidney. These effects lead to water retention and an increase in blood volume.

24. A decrease in blood pressure also leads to release of *antidiuretic hormone (ADH,* also called vasopressin) by the posterior pituitary. In response to a reduction in stretch receptor activity in large veins of the body, this hormone is produced by cells in the hypothalamus and delivered to the posterior pituitary via their axons. The hormone increases the water permeability of the collecting duct, more water is reabsorbed, and blood volume and pressure increase.

25. Differences in water permeability of different tissues can be related to the presence of membrane proteins called *aquaporins* that form water channels. ADH has an effect on the expression of the gene for aquaporin-2, a water channel protein specific to collecting tubules.

26. The disease *nephrogenic diabetes insipidus*, characterized by production of copious, dilute (unsweet) urine, is caused by mutations in the gene for aquaporin-2 that can result in an insensitivity to ADH.

27. *Osmosensors* in the hypothalamus detect changes in the osmolarity of the blood. An increase in blood osmolarity stimulates release of more antidiuretic hormone and also causes thirst. Both these effects lead to increased blood volume and a reduction in blood osmolarity.

28. The osmotic concentration of the urine can almost equal the greatest concentration of the NaCl gradient set up by the countercurrent multiplier system. The magnitude of the NaCl concentration gradient in the medulla is dependent on the length of the loop of Henle relative to the size of the kidney. Desert jumping mice have extremely long loops of Henle. Their kidneys are so efficient that they can live on metabolic water alone.

Questions (for answers and explanations, see page 399)

1. Use numbers to order the following structures through which a molecule of urea would pass during excretion in a mammal.

   _____Collecting duct

   _____Urethra

   _____Bladder

   _____Glomerulus

   _____Ureter

2. Complete the following statements with reference to the simplified diagram of a mammalian nephron shown below.

   a. The composition of the filtrate would be most like plasma in the tubule next to letter ____.

   b. The NaCl concentration in the extracellular fluid would be greatest in the area of letter ____.

   c. NaCl is passively diffusing from the extracellular fluid into the tubule in the area of letter ____.

   d. The osmolarity of the filtrate next to letters _____ is similar to the osmolarity of blood plasma.

   e. The urine would be least concentrated in the collecting duct next to letter ____.

3. Next to each of the following kidney components, place an "M" if it is located mostly within the kidney *medulla* and a "C" if it is located mostly within the kidney *cortex*.

   _____Peritubular capillaries

   _____Bowman's capsule

   _____Proximal convoluted tubule

   _____Distal convoluted tubule

   _____Renal pyramids

   _____Vasa recta

   _____Loop of Henle

4. You measure your urine output in 12 hours and discover that you have produced 1 liter. This amount of urine would represent about ___ liters of filtrate leaving Bowman's capsules. This amount of filtrate would have been derived from about ___ liters of blood entering the renal arteries.

   a. 1; 10

   b. 10; 100

   c. 100; 1,000

   d. 1,000; 10,000

   e. 1,000; 100

5. Some desert mammals do not need to drink, but can survive entirely on metabolic water. This results directly from kidneys with
   a. very high glomerular filtration rates.
   b. very long loops of Henle.
   c. collecting ducts that are impermeable to water.
   d. very high rates of active transport of NaCl into the ascending limb of the loop of Henle.
   e. reduced numbers of nephrons.

6. Match all of the characteristics on the right with the kidney areas on the left that have cells with those characteristics.

   _____ Proximal convoluted tubule   *a.* Many microvilli

   _____ Distal convoluted tubule   *b.* Many mitochondria

   _____ Ascending limb   *c.* Reabsorption of glucose and amino acids

   _____ Descending limb   *d.* Active transport of NaCl

   _____ Collecting duct   *e.* Impermeable to water

   f. Tubular fluid concentration through osmosis

7. A mammal is given a transfusion of a hypotonic saline solution that reduces the osmolarity of its blood by 5 percent and increases its blood volume and pressure by 10 percent. Select *all* of the following effects that will result from this treatment.
   a. Increase release of renin by the kidney
   b. Increase in the glomerular filtration rate
   c. Dilation of the afferent arterioles
   d. Reduced release of antidiuretic hormone by the posterior pituitary
   e. Production of copious, dilute urine

8. Which of the following would *not* be characteristic of the dialyzing fluid used in an artificial kidney machine?
   a. Isotonic to blood
   b. Glucose concentration same as blood
   c. Urea concentration initially zero
   d. Sodium concentration initially zero
   e. Same osmotic potential as blood

9. Place a "+" or "−" next to each of the following events to indicate whether it would increase or decrease urine production.

   _____Insertion of aquaporin-2 into cells of collecting ducts

   _____More stretch receptor activity in major arteries

   _____More aldosterone released by adrenal cortex

   _____Constriction of efferent renal arterioles

   _____Decreased permeability of collecting ducts

Activities (for answers and explanations, see page 400)

- Discuss the role that water channels called aquaporins have in the molecular mechanism by which ADH controls the permeability of the collecting ducts of the kidney.

- Label the boxes next to the arrows in the figure below to indicate what materials are moving. Also, of the materials moving, indicate whether active transport or diffusion is involved.

Blood plasma (300 mosm)

Urine

49

Animal Behavior

GENETICALLY DETERMINED BEHAVIOR • HORMONES AND BEHAVIOR   (pages 1086–1092)

Key Concepts

1. Animal behavior is an act or series of acts performed by an animal with respect to another organism or the environment.

2. *Ethologists* are scientists who study animal behavior.

3. *Fixed action patterns* are genetically determined behaviors that can be expressed without prior experience and are not modifiable by learning. Once begun, the series of movements constituting a fixed action pattern is performed from beginning to end.

4. Fixed action patterns are stereotypic (performed the same way each time) and species-specific. Studying fixed action patterns can lead to a better understanding of the mechanisms and genetics of behavior.

5. In a *deprivation experiment*, an animal is reared in isolation from its normal experiences. If an animal can perform a behavior after being reared in isolation, the behavior is a fixed action pattern.

6. Studies done by *Konrad Lorenz* and *Niko Tinbergen* showed that specific sign stimuli or *releasers* are required to elicit fixed action patterns.

7. Releasers are a simple subset of all the sensory information available to the animal. Recognition of the releasers of a fixed action pattern is genetically programmed.

8. Tinbergen and A. C. Perdick showed that bill shape and the presence of a contrasting bill spot color are the releasers that direct the pecking of herring gull chicks at the bill of the parent.

9. *Supernormal releasers* are exaggerations of the normal sign stimuli and lead to a greater response than the normal sign stimuli. Natural selection has favored exaggerated releasers in the courtship behavior of many species, such as the bowerbirds (see Chapter 50).

10. An animal's developmental and physiological condition determines its motivational state. The motivational state determines to which releasers the animal will respond. A motivated animal may search for specific releasers while ignoring others.

11. Fixed action patterns evolve when learning is not possible (many species have nonoverlapping generations) or when mistakes are costly (mate selection, avoidance of predators, capture of dangerous prey).

12. Highly stereotypic, species-specific behavior can also result through learning. Acquisition of song in the white-crowned sparrow is dependent on exposure to the song during the nestling stage.

13. In some species, learning is limited to a rather narrow *critical period*. For example, white-crowned sparrows must hear their species-typical song as nestlings in order to form a song template in their nervous systems. Some bird species, such as the chaffinch, can only learn their species-typical song during a critical period.

14. *Imprinting* is the rapid learning that takes place during a critical period. Examples include song acquisition in birds and the individual recognition of offspring and parents that occurs in many species. Imprinting allows the encoding of complex sensory information in the nervous system.

15. Fixed action patterns are often strongly influenced by hormones because the endocrine system has a direct influence on the developmental and motivational state of the animal.

16. A study of the copulatory *lordosis* posture exhibited by female rats shows the influence of hormones on a fixed action pattern. If the ovaries of a female rat are removed at any age, she will not show lordosis unless injected with a supplement of female sex steroids. Furthermore, if a spayed female is exposed to testosterone early in life, her sexual behavior becomes masculinized. Males castrated early in life and then later exposed to female hormones show lordosis, but exhibit no sexual response when treated with testosterone.

17. These results indicate that both genetic males and females develop female sexual behaviors if testosterone is absent early in life. Testosterone must therefore trigger the development of male fixed action patterns.

18. Further evidence for hormonal influences on behavior come from experiments on songbirds. Males are normally the only sex that sings, and although this song is learned, there is a stereotyped sequence of events that must occur under the influence of testosterone if the song is to *crystallize* and become functional. If a young male who hears his father sing while still a nestling is castrated before he hears himself sing, his song is never properly expressed and does not become fixed into his nervous system.

19. Females do not sing because they are not normally exposed to testosterone. However, they apparently do learn the male's song, because if treated with testosterone they express normal male song patterns. Proper hormonal stimulation thus appears critical for the expression of this important behavior.

20. A recent and unexpected finding is that testosterone causes the song control centers of the bird brain to become larger. Prior to this work it was believed that new brain cells could not be added after birth in vertebrates. It now appears that hormones influence not only the functioning of the nervous system, but also the actual growth of new cells during adult life.

Questions (for answers and explanations, see page 400)

1. Which of the following is *not* an example of a fixed action pattern?
   a. Nut burying behavior in tree squirrels
   b. Behaviors that are expressed in a deprivation experiment
   c. Song in white-crowned sparrows
   d. Web building in orb-weaving spiders
   e. Lordosis

2. Studies done on hand-reared chaffinchs showed that
   a. several critical periods may be involved in song acquisition.
   b. the song of this species is programmed as a fixed action pattern.
   c. imprinting determines only when a song template can be learned.
   d. this species is genetically programmed to recognize what song to learn.
   e. its song would be learned even by individuals held in isolation all their lives.

3. A newborn male rat is castrated and then allowed to mature. When he reaches adulthood, he is given an injection of male sex steroids and exposed to a normal adult female rat exhibiting lordosis. Which of the following will occur?
   a. The male will mount the female and attempt to copulate.
   b. The male will show lordosis.
   c. The male will attempt to copulate, and also show lordosis.
   d. The male will aggressively repel the female.
   e. The male will show no sexual behavior at all toward the female.

4. Based on your knowledge of the releasers directing the pecking behavior of herring gull chicks, which of the following models should elicit the *greatest* response?

5. Select the following example that is *least* likely to be a fixed action pattern.
   a. Courtship display
   b. Predator escape behavior
   c. Predator's prey-search behavior
   d. Behavior triggered by a releaser
   e. Web spinning in garden spiders

6. Select one of the following choices to make a *false* statement. Supernormal releasers
   a. result in a greater response than the normal releasers.
   b. are exaggerations of normal sign stimuli.
   c. would not be effective in a deprivation experiment.
   d. are possible because of the simplicity of most sign stimuli.
   e. are conceptually similar to many of the elaborate displays and features that have evolved in courting animals.

7. Which of the following statements about hormonal influences on bird song is *not* true?
   a. Male songbirds castrated as nestlings do not sing proper song as adults.
   b. Female songbirds can be induced to sing by treatments with male sex hormone.
   c. Testosterone is not necessary for singing in males once they have learned their song.
   d. Females learn their species song, but do not normally express it under natural conditions.
   e. Testosterone stimulates young male songbirds to practice singing during their first spring.

Activities (for answers and explanations, see page 401)

- Discuss the process of song crystallization in birds.

- Explain why the process of cocoon construction is considered to be a fixed action pattern in many spiders.

- You spay a newborn female rat and treat her with testosterone. In each of the following two situations (*a* and *b*), describe what type of sexual behavior you expect to observe and explain what these outcomes suggest about the hormonal control of sexual behavior in rats.
  a. At maturity, the female is injected with female sex steroids.

  b. At maturity, the female is injected with testosterone.

THE GENETICS OF BEHAVIOR • COMMUNICATION (pages 1092–1100)

Key Concepts

1. Genes do not directly cause specific behaviors. Genes code for proteins that influence the development and functioning of the nervous and endocrine systems, which indirectly results in behavior.

2. Hybridization experiments show that the behavior of hybrids is usually intermediate to that of the two parent species, as shown in nest material carrying by the lovebirds *Agapornis roseicollis* and *A. fischeri*.

3. Konrad Lorenz showed that the series of components of the courtship behavior of dabbling ducks differs in the hybrids resulting from crossing two different species. Display components from the two parent species are put together in different combinations in the hybrids or new components not seen in the two parent species are added.

4. The behavior of domesticated animals produced by artificial selection clearly shows the importance of genetics. Hybridization between artificially selected strains of *Drosophila* fruit flies indicates that many aspects of the behavior of this species are indirectly determined by multiple genes that affect the general properties of the nervous system.

5. Mendelian segregation of behavioral traits is rare. *Hygienic behavior* in honeybees seems to be controlled by two genes—the recessive alleles for these genes indirectly lead to the uncapping and larval removal behavior involved in combatting a bacterial disease that kills larvae.

6. Some of the powerful techniques of molecular genetics discussed in Chapter 16 are now being used to study the behavioral functions of specific genes. In the marine mollusk *Aplysia*, a single gene has been discovered that codes for a precursor molecule consisting of nearly 300 amino acids. This precursor is modified into a number of different peptides that function as neural signals to coordinate the egg-laying behavior in this species.

7. Communication is behavior designed to influence the actions of other individuals. Communication involves the exchange of displays or signals between a sender and a receiver.

8. A *display* or *signal* is a behavior, frequently an exaggerated version of a normal behavior, that has evolved to influence other individuals.

9. Communication systems are sometimes exploited by illicit senders or receivers of signals. Predators, for example, can intercept signals and use them to locate potential prey.

10. *Chemosensation* is the most ancient mode of communication and probably evolved from intracellular signaling. Intracellular-signaling molecules have occasionally been directly adapted for use in chemical communication, as in the use of cAMP for aggregation of myxamoebas in cellular slime molds.

11. *Pheromones* are chemical signals exchanged between species members. Characteristics of pheromones are that they can communicate very specific messages, are relatively long lasting (compared to auditory or visual signals), and the pheromone concentration gradient can be useful for locating the organisms producing it.

12. The chemical features and size of the pheromone molecule determine its diffusion coefficient, which, in turn, determines its rate of spread and dissipation. Pheromones with greater diffusion coefficients spread rapidly, but disappear quickly—useful for sex pheromones. Pheromones with smaller diffusion coefficients spread slowly, but also disappear slowly—useful for trail- and territory-marking pheromones.

13. The *visual communication mode* allows an infinite variety of signals that can be produced and propagated rapidly, but detection may be difficult and requires adequate light. Some species have evolved their own light-emitting mechanisms, as in the luciferin/luciferase mechanisms in fireflies.

14. Although few animal species use *auditory communication*, sound can be used when light is limited, can go around objects, and can be transmitted over great distances. However, sound cannot convey as much information as rapidly as visual signals.

15. The physical properties of sound can be varied to make them easy or difficult to locate. A pure tone with a sharp beginning and end is easier to locate than a complex sound with a gradual onset and completion.

16. *Tactile communication* occurs primarily in environments where visual signaling is impossible mainly due to low light levels, such as in the deep sea, in dark burrows or caves, and inside beehives. Chemicals are also frequently exchanged during tactile communication.

17. If a honeybee discovers a food source within 80 meters of the hive, it performs a *round dance* on the vertical combs within the hive's dark interior. Other worker bees detect the round dance tactilely and respond by leaving the hive to search randomly for the food.

18. In the waggle dance of the honeybee, the speed of the waggle run is directly related to the distance from the hive to the food source and the orientation of the waggle run relative to vertical shows the direction to fly to get to the food source relative to the hive–sun direction. The waggle dance is given if the food source is greater than 80 meters from the hive.

19. Certain fish produce electric fields to sense objects in the murky water in which they live. Some fish species show true *electrocommunication*; they can produce electric pulses that vary in frequency in order to convey information about their individual identity, sex, and dominance status.

20. Signals may be elaborated or exaggerated to reduce ambiguity concerning the signaller's identity, emotional or reproductive state, or intentions. Sources for communication signals include *intention movements, autonomic responses,* and *displacement behaviors*.

21. Intention movements are an important source of behavior for visual displays.

22. Autonomic nervous system responses, such as the erection of fur and feathers, have become elaborated into displays in many animals.

23. Displacement behaviors are seemingly irrelevant displays that result from conflict between different motives like attack and escape.

Questions (for answers and explanations, see page 401)

1. Which of the following is *not* a valid conclusion resulting from studies done on hygienic behavior in honeybees?
   a. The inheritance of this behavior could be explained in a simple Mendelian manner.
   b. This was a rare case where a gene resulted directly in the expression of a behavioral trait.
   c. The uncapping and larval removal components of this behavior seemed to be determined by different genes.
   d. Since all of the F_1 bees resulting from a cross between hygienic and nonhygienic strains were nonhygienic, the trait must be determined by recessive alleles.
   e. A backcross of the F_1 to the nonhygienic strain should yield one-quarter hygienic to three-quarter nonhygienic individuals.

2. Studies of which of the following organism types have led most directly toward connecting a specific gene with a specific behavior?
   a. Lovebirds in the genus *Agapornis*
   b. Dabbling ducks
   c. Domestic animals
   d. The marine mollusk *Aplysia*
   e. *Drosophila* fruit flies

3. Which of the following waggle dances should be given by a bee returning from the food source shown in the diagram below?

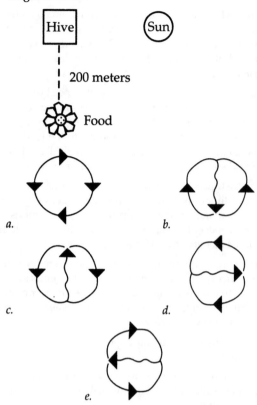

Which of the conditions in the premise statement for this question determines whether a foraging bee returning to the hive will give a round dance or a waggle dance?

How does the search initiated by a round dance differ from the search initiated by a waggle dance?

4. Which communication mode would you expect an animal to use if it lives in a dark, terrestrial environment, at a low population density, and needs a rapid response?
   a. Auditory
   b. Olfactory
   c. Visual
   d. Tactile
   e. Electric

5. Which of the following communication modes is frequently associated with chemical communication?
   a. Auditory
   b. Olfactory
   c. Visual
   d. Tactile
   e. Electric

6. When two dogs interact aggressively, the submissive dog will frequently roll onto its back and urinate. This is an example of
   a. an intention movement.
   b. an autonomic response.
   c. a displacement behavior.
   d. a display.
   e. chemical communication.

Activities (for answers and explanations, see page 401)

- Discuss how the auditory characteristics of a mating call might differ from an alarm call given in response to detection of a nearby predator.

- Discuss how the diffusion coefficients of a sex pheromone and a trail-marking pheromone might differ, and explain these differences.

THE TIMING OF BEHAVIOR: BIOLOGICAL RHYTHMS • FINDING THEIR WAY • HUMAN BEHAVIOR (pages 1100–1108)

Key Concepts

1. Many organisms have evolved *rhythmicity* in response to the 24-hour cycle of environmental conditions caused by the rotation of Earth on its axis.

2. The presence of *circadian rhythms* in the activity of many animals when kept under constant environmental conditions is evidence for the existence of an internal *(endogenous)* clock.

3. A rhythm consists of a series of cycles and the length of a cycle is the *period* of the rhythm. Typically the period of a circadian rhythm is not exactly 24 hours.

4. A point on a cycle is a *phase* of that cycle. Because a circadian rhythm does not usually have a period of 24 hours, under uniform conditions a circadian rhythm will *free-run* with its natural periodicity.

5. *Entrainment* is the process whereby the circadian rhythm is reset by environmental cues so that it is in phase with the 24-hour cycle. During entrainment, a circadian rhythm of more than 24 hours is *phase-advanced*; a circadian rhythm of less than 24 hours is *phase-delayed*.

6. Typically, the natural light–dark cycle entrains a free-running clock to a 24-hour period. Brief exposure to pulses of light or dark every 24 hours can be used to experimentally entrain animals. Within limits, an animal's endogenous clock can also be entrained to cycles less than or greater than 24 hours.

7. Because your internal clock can only be phase-shifted 30 to 60 minutes per day, *jet lag* results if you fly across several time zones.

8. In mammals, the circadian clock is located in the *suprachiasmatic nuclei* (SCN), small groups of cells located in the area of the brain above the place where the two optic nerves join. Removal of the SCN results in a loss of circadian rhythmicity and SCN transplantation can restore it. The mammalian circadian clock is entrained by light cues from the photoreceptors in the eyes.

9. The avian circadian clock resides in the *pineal gland* and is entrained by light cues detected by that gland. The pineal gland is located on the dorsal surface of the brain, between the two hemispheres; it secretes the hormone melatonin.

10. Many organisms have also evolved rhythmicity in response to the year-long cycle of environmental conditions caused by the revolution of Earth around the sun.

11. Changes in *photoperiod*, the daily day–night lengths, is used by many animals to coordinate their activities with seasonal changes. For example, the reproductive cycle in many birds is *photoperiodic*, that is, coordinated by photoperiod.

12. Hibernators and equatorial migrants cannot use photoperiod to coordinate their activities. Many of these species have endogenous annual rhythms called *circannual rhythms*. The period of a circannual rhythm is usually somewhat shorter than 365 days.

13. In *orientation*, the behavior of an animal is directed by an object or stimulus. *Navigation* is long-distance orientation.

14. *Piloting* is orientation with reference to familiar landmarks in the animal's environment. Piloting depends on knowledge of the structure of the environment.

15. *Homing* involves an oriented movement from a specific location to the animal's nest site or burrow. Although homing can be accomplished by piloting, some animals are able to home without using landmarks. Homing pigeons and albatrosses have amazing homing abilities.

16. *Migration* is the seasonal movement, typically between summer breeding grounds and wintering areas, seen in many animals. Bird banding shows that many bird species fly long distances, sometimes over large bodies of water where no landmarks are available.

17. In many bird species, adults leave the breeding grounds at the end of the season before the juveniles. This observation suggests that young birds must be able to navigate on their own and cannot simply follow experienced birds.

18. Navigation through unfamiliar territory can be accomplished by two systems: distance and direction navigation and bicoordinate navigation.

19. *Distance and direction navigation* requires that the animal know the direction to the goal and the distance to travel.

20. *Bicoordinate navigation* requires that the animal know its current map location (latitude and longitude) and the map location of where it wants to go.

21. In many species, if individuals are displaced prior to their normal migration, they migrate in the direction and distance that would be appropriate if they had not been displaced. This observation suggests that these species are using distance and direction navigation.

22. The increased, oriented activity shown by many caged birds during the migratory period is called *migratory restlessness*.

23. If an animal knows the direction to its destination, the duration of the migratory restlessness will cause it to move the correct distance. Once within the approximate location of its destination, piloting by landmarks could be used to complete the journey.

24. In the Northern Hemisphere, the sun rises in the East, is due south at noon, and sets in the West. If an animal knows the time, compass directions can be obtained from the sun's position.

25. Many animals use circadian clocks to determine direction from the sun's position.

26. Clock-shifting experiments cause birds to orient in predictably incorrect directions based on the amount of the phase shift. A six-hour phase shift in circadian clocks causes a 90° error in their orientation.

27. Some nocturnally migrating birds use the North Star (Polaris) to determine the correct migratory direction. In the northern hemisphere, all other stars circle a point located over Earth's axis of rotation. This point is near the North Star.

28. Stephen Emlen showed that birds imprint on the North Star only if they are exposed to a rotating sky as juveniles. Birds raised in a planetarium will imprint on any star about which the sky is made to rotate.

29. Because the location of the North Star is fixed, a circadian clock is not necessary to derive directional information from it. Experienced birds are not confused if the sky in the planetarium does not rotate or rotates at an incorrect rate.

30. Some animals have several, redundant navigation systems. For example, in addition to a sun compass, pigeons can derive directional information from Earth's magnetic field.

31. Sensitivity to the planes of the polarization of light, very low frequency sound, and weather patterns are used by some organisms to navigate.

32. Bicoordinate navigation requires a map sense—the animal must know its geographic location and the latitude and longitude of its destination.

33. Longitude can be deduced from a knowledge of the time of day and the sun's position. The time of sunrise is earlier if you are east of home and later if you are west of home.

34. Latitude can be deduced from a knowledge of the time of day and the sun's position. At a given time, the sun is higher in the sky if you are south of home and lower in the sky if you are north of home.

35. Although there are genetically based limitations on what we can learn and when we can learned based on the organization of our sensory and nervous systems, the most notable aspect of human behavior is its ability to be modified by experience.

36. The genetically determined features of human behavior include some behavioral drives, emotions, and motor patterns, such as facial expressions.

Questions (for answers and explanations, see page 401)

1. The following figure shows the periods of activity (white bars) and inactivity (black bars) of an animal kept under constant conditions of light and temperature. What is the period of this animal's circadian rhythm, and does it need to be phase-advanced or phase-delayed to become entrained to a 24-hour cycle, or is it already entrained?

a. 24 hours, entrained
b. 11 hours, phase-advanced
c. 11 hours, phase-delayed
d. 22 hours, phase-advanced
e. 22 hours, phase-delayed

Note: the following diagram should be consulted in answering the next three questions.

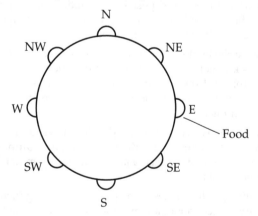

2. A starling is trained to feed from a food bin in the east end of a circular cage from which it can see the sky. At sunrise the cage is covered and a fixed light source is presented above the east end of the cage. If the starling treats the light as the sun, where will it search for food at noon?
a. North
b. East
c. South

d. West
e. Random

3. A starling is trained to feed from a food bin in the east end of a circular cage from which it can see the sky. If a mirror is used to shift the apparent position of the sun at noon from the south to the northwest, where will the starling search for food?
a. North
b. Northwest
c. Southwest
d. Northeast
e. East

4. A starling is trained to feed from a food bin in the east end of a circular cage from which it can see the sky. The bird is then placed into a light-controlled environment for three weeks and phase-delayed by six hours. If the bird is returned to the cage under natural lighting conditions at sunset, where will the bird search for food?
a. North
b. South
c. East
d. West
e. Southeast

5. You are conducting studies on a species of bird that lives in the northern hemisphere and normally migrates to the south for the winter using the North Star as a navigational cue. In a planetarium, caged birds are exposed to a night sky that is shifted six hours ahead of the normal star pattern. In what direction would the birds direct their migratory restlessness?
a. North
b. East
c. South
d. West
e. Random

6. Which one of the following animals would *not* be expected to have an endogenous circannual rhythm?
a. A ground squirrel that hibernates
b. An arctic hare
c. A cave salamander
d. A tropical insect
e. A bird that summers in North America and migrates to South America for the winter.

7. Which of the following statements about circadian rhythms is true?
a. The effect of jet lag on your circadian rhythms is the same if you fly 1,000 miles west or 1,000 miles south.
b. A regular short pulse of light or dark can entrain the free-running circadian rhythms of an animal kept under constant environmental conditions.
c. Circadian rhythms dependent on endogenous clocks are restricted to birds and mammals.
d. An animal's circadian rhythm usually has a period of 24 hours.
e. If Earth were not tilted on its axis, circadian rhythms would not be necessary.

8. Select *all* of the following treatments that would likely eliminate the circadian rhythms of the animal.
   a. Removing the suprachiasmatic nuclei of a mammal
   b. Blinding a mammal
   c. Removing the pineal gland of a bird
   d. Removing the pineal gland of a mammal
   e. Blinding a bird
   f. Placing a bird or mammal under constant darkness

9. In most homing species, homing is accomplished by
   a. piloting.
   b. a time-compensated solar compass.
   c. bicoordinate navigation.
   d. use of the Earth's magnetic field.
   e. random searching.

10. Suppose you are an animal that is capable of bicoordinate navigation. You are displaced from home and you notice that the sun rises later and is higher in the sky than it should be based on your endogenous clock and your knowledge of the sun's position in your home area. You conclude that
    a. you need to travel northeast to get home.
    b. you need to travel southeast to get home.
    c. you need to travel southwest to get home.
    d. you need to travel northwest to get home.
    e. you need more information!

11. Suppose you capture a bird at position *A* on a large island and transport it 200 km due south to position *D* on the other side of the island just before its annual migration starts. The population from which this bird came normally migrates to position *B*. Assuming that the bird is capable of the orientation modes listed below, for each mode predict the position (A, B, C, D, or E) to which the displaced bird should move.

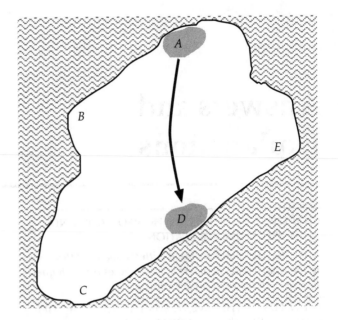

Homing only: _____

Migration by piloting only: _____

Migration by distance and direction navigation only: _____

Migration by bicoordinate navigation only: _____

Answers and Explanations

Chapters 37–49

CHAPTER 37 – PHYSIOLOGY, HOMEOSTASIS, AND TEMPERATURE REGULATION

Note: the page numbers listed with the section titles refer to the study guide; numbers in parentheses for each question refer to the relevant key concepts.

THE NEEDS OF THE ONE VERSUS THE NEEDS OF THE MANY • ORGANS AND ORGAN SYSTEMS (pages 271–273)

1. Choice *e*. All of these processes contribute to the maintenance of appropriate internal body conditions. (1–3)
2. Choice *b*. Eyes are sensory structures adapted for light detection. They relay the information they gather to other parts of the central nervous system. (9)
3. Choice *a*. Endocrine tissue is a type of epithelium specialized for secretion. (5)
4. Choice *d*. Connective tissue includes such things as blood, cartilage, and bone, all of which are characterized by having loose arrays of cells embedded in some kind of an extracellular matrix. (6)
5. Choice *e*. Electrical excitability is a hallmark of the nervous system. The blood circulatory system acts as a conduit for essential nutrients and wastes to all other organ systems. (9, 16)
6. Choice *b*. The excretory system handles removal of nitrogenous wastes and helps to regulate salt and water balance. (9, 12, 17)

- (5–8)
Epithelium:	protective barrier against invasion by microorganisms; prevents water loss; secretes oils and other substances
Connective:	support base for other tissues (especially epithelium and muscle); provides flexibility and strength; blood
Muscle:	controls action of some glands and causes hair to move
Nervous:	various sensory functions (e.g. touch, pain, temperature); controls muscle movements

- (9–18)
  b: Excretory
  h: Circulatory
  f: Skeletal
  g: Muscle
  a: Nervous
  d: Endocrine

c: Digestive
e: Reproductive

CONTROL, REGULATION, AND HOMEOSTASIS • TEMPERATURE AND LIFE • MAINTAINING OPTIMAL BODY TEMPERATURE (pages 273–275)

- The lower and upper set points for this thermostat are 18° and 22°, respectively. As long as the room temperature remains within this range no action is required. If, however, the sensor detects that the room temperature has fallen below 18° or risen above 22°, an appropriate error signal is sent to the thermostat, which turns on either the heater or the air conditioner to counteract the change in room temperature. Once the temperature is back within the set point range, negative feedback shuts down the heating or cooling system. (2–5)
- Acclimatization is a seasonal adjustment in metabolic activity. This is accomplished by metabolic compensation, in which a shift is made between enzymes having different optimal temperatures. (12)

1. Choice *d*. Changing set points in anticipation of environmental changes is feedforward regulation. (3, 7)
2. Choice *c*. The amplification of uterine contractions due to persistent stimulation is indicative of a positive feedback mechanism. (6)
3. Choice *c*. Because it can adjust its metabolic rate to compensate for changes in environmental temperature, the mouse qualifies as an endotherm. (14–15)
4. Choice *d*. Q_{10} = (rate at 30°C)/(rate at 20°C). Therefore, $1.5 = 200/X$. Solving for X yields 133. (9)
5. Choice *e*. Since the lizard obtains its heat mostly from the environment, it is best described as an ectotherm. Because it holds a somewhat constant body temperature through changes in its behavior, it can also be considered a behavioral homeotherm. (13–16)
6. Choice *b*. Anything that would cool the animal, either by preventing more heat from entering the body or by facilitating heat loss, would be an appropriate response in this situation. Since preventing blood flow to the periphery would tend to trap heat within the animal's body, this would be an inappropriate response. (16–17)
7. Choice *e*. Muscle activity produces heat, and the special vascular anatomy of these fish helps conserve heat within the body core. (19–20)

THERMOREGULATION IN ENDOTHERMS • THE VERTEBRATE THERMOSTAT (pages 275–277)

- Your drawing should resemble Figure 37.19 on page 823 in the textbook. (2–4)

- Evaporative cooling is used to dump excess heat above the upper critical temperature. Both panting and sweating require an energy expenditure, thus the increase in metabolic rate. (9)

1. Choice *c*. Shivering is a metabolic heat production mechanism employed when temperature drops below the lower critical limit. (2–5)
2. Choice *b*. The upper critical temperature is the highest body temperature possible before the metabolic rate will increase as a result of physiological efforts to cool the animal. (3–4)
3. Choice *e*. Panting and sweating are the means by which evaporative cooling is achieved; both require ATP and lead to water loss. (9)
4. Choice *b*. Hypothalamic thermal set points change often as with fever, sleep, hibernation, etc. (10–15)
5. Choice *d*. Torpor is a short-term response (a few days at most) to cold or food scarcity, while hibernation usually lasts for an entire climatic season. (18–20)
6. Choice *a*. Pyrogens released in response to pathogenic substances trigger mechanisms that promote fever. (15)
7. Choice *d*. Brown fat metabolism is a physiological response to cold, not a morphological one. (6)

Integrative Questions

1. Choice *e*. All these animals engage in behaviors related to thermoregulation.
2. Choice *b*. Adipose (fat) tissue is a type of connective tissue.
3. Choice *a*. The conditions described suggest that the rodent is in its thermoneutral zone (see textbook Figure 37.19). If the animal were hibernating or in torpor its body temperature would be close to 0°C, and if it had a fever its body temperature would be greater than 37°C.

CHAPTER 38 – ANIMAL HORMONES

Note: the page numbers listed with the section titles refer to the study guide; numbers in parentheses for each question refer to the relevant key concepts.

HORMONES AND THEIR ACTIONS • HORMONAL CONTROL OF MOLTING AND DEVELOPMENT IN INSECTS (pages 278–280)

- (10–13)

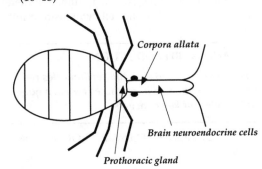

Corpora allata

Brain neuroendocrine cells

Prothoracic gland

- (1, 4, 9, 11–12)
  d: Histamine
  a: Endocrine gland
  b: Target cell
  e: Ecdysone
  c: Juvenile hormone

- The hormonal receptors for this transformation are likely to be widespread in the body of this fish. Since changes in body coloration occur, there are probably receptors in the surface epithelial cells. Since the brain signals the pituitary to release hormones, there are receptors in the pituitary as well. Reproductive tissues are also affected, as are muscle cells that grow to make the fish larger, and so receptors would be expected in these cells as well. Nervous tissue may also be influenced to produce increased aggression, suggesting the presence of receptors in these locations also. (1, 7)

1. Choice *d*. Vasopressin is released from the posterior pituitary (an endocrine gland), travels through the bloodstream (a circulating hormone), and affects target cells in the kidney. (2, 7, 9)
2. Choice *b*. Histamine is produced for use inside the body and acts in the immediate vicinity of the cells where it is produced. (2, 4)
3. Choice *a*. Hormones affect specific target cells because only those cells possess appropriate receptors to recognize the hormone. Exactly what response they stimulate depends on the particular biochemical pathways of that cell. (1, 7)
4. Choice *c*. Even though the bug is fed, the fact that it lacks its supply of brain hormone means that ecdysone is never released from the prothoracic gland. Thus, molting cannot occur. (10–13)
5. Choice *d*. Partial decapitation means that the corpora allata, the source of juvenile hormone, are still present. Since decapitation occurred a week after feeding, there was plenty of time for brain hormone and ecdysone to be released, thereby initiating molt. Since the juvenile hormone concentration is still high, the bug molts into another juvenile instar instead of an adult. (10–13)
6. Choice *e*. The lack of juvenile hormone is what allows larvae to molt into adults. As long as juvenile hormone concentrations remain high, larvae will not become adults and no reproduction can occur. (15–16)

VERTEBRATE ENDOCRINE SYSTEMS (pages 280–283)

1. Choice *c*. The anterior pituitary makes all three of these hormones. (8, 10, 12)
2. Choice *a*. See Figure 38.8 in your textbook for a flow diagram showing the regulation of thyroxine release. (14–15)
3. Choice *d*. By selectively destroying the posterior pituitary, you would be eliminating only the influence of vasopressin on the kidney (the effects of oxytocin, the other hormone of the posterior pituitary, are not particularly important in male dogs). Another approach would be to destroy the hypothalamus, but this would also affect many other physiological systems. (4–5)
4. Choice *c*. Insulin is needed to turn the excess glucose obtained from the meal into stored glycogen. (20)
5. Choice *a*. This situation is likely to produce stress in the rat, which would dramatically increase its levels of cortisol. (25–26)
6. Choice *d*. The child described is genetically a male. The fact that his cells are insensitive to female hormones will not affect his development either in the womb or at the time of puberty. (31, 33)

7. Choice *c*. Puberty is preceded by increased levels of gonadotropin-releasing hormones, secreted by the hypothalamus. This triggers the pituitary to release luteinizing hormone and follicle-stimulating hormone, which stimulate the gonads to produce sex steroids. (34)

8. Choice *a*. Epinephrine is the only hormone listed that is not a steroid. (23–25, 29)

• Your diagram should resemble textbook Figure 38.8 on page 844. (14–15)

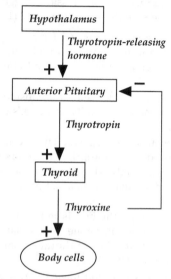

• The mother's ears pick up the baby's cry and this information is rapidly transmitted to her brain, which relays it to the hypothalamus. The neurohormonal connections between the hypothalamus and the posterior pituitary facilitate the rapid release of oxytocin, which travels quickly through the circulatory system to stimulate the release of milk from mammary glands in the breasts. (4, 6)

MECHANISMS OF HORMONE ACTION (pages 283–285)

1. Choice *e*. Just the reverse is true; a single hormone molecule can result in the activation of hundreds or thousands of enzymes. (9)

2. Choice *e*. In the process of amplifying the hormone signal, second messenger systems cause conformational changes in many metabolic enzymes, including protein kinases. (3–7)

3. Choice *d*. Receptor tyrosine kinases are involved in the stimulation of gene expression by growth factors. (16)

4. Choice *d*. The basic signalling principle of hormones is the same for both lipid- and water-soluble varieties. Both involve manufacture by some gland or tissue at a site different from where they have their effect, and they also require receptors on target cells in order to be detected. (1–2)

5. Choice *e*. ATP is used to phosphorylate various molecules during the cascade of reactions in a second messenger system, but is not itself a second messenger. (3–6)

6. Choice *c*. Lipid-soluble hormones cross cell membranes and work directly on gene expression. Their effects thus take longer to develop than the more transient second messenger changes that occur when water-soluble hormones bind to the surface of a cell. Moreover, since they

affect DNA transcription and protein synthesis, the effects of lipid-soluble hormones also tend to last longer. (19)

• (3–6)
The correct sequence is:
1. Hormone
2. Receptor molecule
3. Inactive G protein
4. GTP
5. Active G protein
6. Inactive adenylate cyclase
7. Active adenylate cyclase
8. ATP
9. cAMP
10. Protein kinase
11. Phosphoprotein phosphatase

• (17–19)
a. The water-soluble hormone, by binding with a surface receptor and triggering a rapid and amplifying second messenger signal, will act faster than the lipid-soluble hormone, which must first penetrate the cell and then influence gene transcription.

b. Lipid-soluble hormones all have their main effects on transcription in the nucleus and so are much more likely to effect gene expression. (Note, however, that some water-soluble hormones, such as growth factors, can also influence gene expression).

Integrative Questions

1. Choice *d*. All of these hormones are either peptides, polypeptides, or protein molecules. Thus, they are all made of amino acids.

2. Choice *c*. Prolactin, like oxytocin, is involved in milk production and secretion in mammals. It would thus be found associated with mammary tissue. Since it is a water-soluble hormone, it would be found attached to plasma membrane surface receptors.

3. Choice *a*. Progesterone is a critical hormone for maintaining the condition of the uterus in pregnant women. Calcitonin regulates the deposition and reabsorption rates of calcium from the mother's bones; this process is normally disrupted by the growing infant's need for calcium to build its own bones.

4. Choice *b*. Thyroxine is a lipid-soluble hormone. It therefore needs a receptor within the cell cytoplasm to initiate its effects.

CHAPTER 39 – ANIMAL REPRODUCTION

Note: the page numbers listed with the section titles refer to the study guide; numbers in parentheses for each question refer to the relevant key concepts.

ASEXUAL REPRODUCTION • SEXUAL REPRODUCTION (pages 286–289)

• (14)
2n Primary spermatocytes
n Spermatids
2n Spermatogonia
n Secondary spermatocytes
n Functional sperm

- Since they have fully functional gonads, most organisms that are capable of regeneration also reproduce sexually. Depending on their sex, sea stars release either sperm or eggs to form zygotes, thus benefiting from the genetic diversity created in their offspring. On the other hand, if an adult animal loses an arm to a predator, or more dramatically, if the predator gets most of the body but leaves an arm, the remaining section can regenerate the lost body parts. This "new" individual is genetically identical to the original one because regeneration is an asexual process. (1–4, 6)

1. Choice *b*. Parents who are able to survive long enough to reproduce are probably also fairly well adapted for their habitat. If the environmental conditions remain stable, it is a good strategy to make more offspring that are just like themselves, which is what occurs in asexual reproduction. (1–4, 8)

2. Choice *e*. In parthenogenesis, a mother's egg is stimulated to begin development without sperm penetration. Although a mating act may cause this, there is no union of sperm and egg. Therefore, parthenogenetic offspring have no father. (7–9)

3. Choice *e*. All of these are valid differences between sperm and eggs. (10–11)

4. Choice *b*. The presence of sperm cells indicates that you are examining a male. Although very young males would not have these cells in their immature testes, any male beyond the point of sexual maturity would have them, so the best you could say was that the animal was a reproductive adult. (11, 14)

5. Choice *c*. Dioecious means that the species has two separate sexes, spermatogonia are only made in males, and squid do utilize a spermatophore to transfer sperm from male to female. (13, 17, 23)

6. Choice *a*. Since external fertilization can only take place in an aquatic habitat, there are no terrestrial animals that use it. (20)

7. Choice *a*. Many animals reproduce both by asexual and sexual means. (1–3)

8. Choice *e*. Gametes are necessary for sexual reproduction, as is synchrony between mates. Terrestrial life requires internal fertilization, with the likely presence of an intromittent organ in the male to achieve fertilization. (10, 20, 22)

REPRODUCTIVE SYSTEMS IN HUMANS AND OTHER MAMMALS (pages 289–292)

1. Choice *d*. Located on the surface of each testis, the epididymis stores mature sperm until they are ready to be emitted just prior to ejaculation. (2)

2. Choice *e*. The woman's reproductive tract is acidic to help protect against infection by microorganisms. Unfortunately, this environment is also hostile to the sperm. The prostate gland secretions are alkaline, which helps neutralize the pH of the female's system and permits the sperm to survive longer. (6–7)

3. Choice *b*. Sperm are deposited in the vagina during copulation, travel past the cervix and into the uterus, then up the oviducts where they encounter the egg. (10)

4. Choice *c*. The corpus luteum is the primary source of progesterone. It is formed at the time of ovulation, so progesterone levels are therefore only high enough to sustain endometrial development during the second half of the cycle. (19–22)

5. Choice *e*. The gonadotrophic hormones are follicle-stimulating hormone and luteinizing hormone, both of which are inhibited by the high levels of progesterone and estrogen present during the latter half of the menstrual cycle. They are at their absolute lowest levels between days 20 to 25. (18–20)

6. Choice *d*. Seminal vesicles produce the fructose sugars that power the sperm's mitochondria. (6–7)

7. Choice *a*. Of the choices listed, only the intrauterine device (IUD) prevents implantation of the embryo. All the others are aimed at preventing fertilization. (28–32)

8. Choice *b*. Men can have only a single ejaculation before undergoing a refractory period of at least 20 minutes, during which they cannot obtain the full erection needed for another orgasm. (26, 27)

9. Choice *b*. It is virtually impossible to halt all sperm production in males through chemical means without creating serious medical side effects. (33)

- (16–22)
Progesterone concentration increases during the second half of the menstrual cycle, when the corpus luteum is functional, in order to sustain the endometrial lining of the uterus. If pregnancy does not occur, progesterone levels drop precipitously just prior to menstruation. Follicle-stimulating hormone rises slowly through the first third of the cycle, stimulating the growth of the follicle. When the follicle ruptures, estrogen and progesterone levels rise and inhibit the release of FSH. If pregnancy does not occur, its concentration then begins to rise once again at the end of the cycle. Luteinizing hormone remains fairly constant at low concentrations early in the cycle, then peaks in a dramatic midcycle surge that stimulates ovulation.

- (2, 5, 8–9)

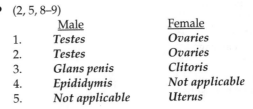

	Male	Female
1.	*Testes*	*Ovaries*
2.	*Testes*	*Ovaries*
3.	*Glans penis*	*Clitoris*
4.	*Epididymis*	*Not applicable*
5.	*Not applicable*	*Uterus*

FROM CONCEPTION TO BIRTH IN HUMANS
(pages 292–294)

- (1–12) The correct order of events is:
  fertilization
  embryo formation
  blastocyst implantation
  initial placental development
  embryo becomes a fetus
  rapid fetal growth
  mature fetal organ systems become functional
  parturition
  afterbirth

1. Choice *e*. (3, 5)
2. Choice *a*. The blastocyst is one of the very first stages in human embryonic development, occurring early in the first trimester. (3)
3. Choice *c*. The corpus luteum's production of hormones is essential for the maintenance of early pregnancy. During the second trimester, however, the placenta becomes fully formed and takes over hormone production, leading to degeneration of the corpus luteum. (6–8)
4. Choice *b*. The development of organs and organ systems, a process called organogenesis, occurs during the latter half of the first trimester. (4)
5. Choice *e*. Labor is brought on and intensified by all of these factors. (10–12)

Integrative Questions

1. Choice *b*. All these animals can reproduce sexually. However, only the chicken lays an external egg.
2. Choice *a*. The corpus luteum forms after ovulation, which does not occur prior to puberty and sexual maturation in women.
3. Choice *e*. Sexual reproduction leads to genetic variability in offspring. Asexual reproduction produces offspring that are genetically identical to the parent.
4. Choice *c*. This is effectively what happens to a woman taking birth control pills.

CHAPTER 40 – ANIMAL DEVELOPMENT

Note: the page numbers listed with the section titles refer to the study guide; numbers in parentheses for each question refer to the relevant key concepts.

FERTILIZATION: CHANGES IN SPERM AND EGG • CLEAVAGE: REPACKAGING THE CYTOPLASM
(pages 295–298)

1. (2, 5, 8, 9, 14, 16, 21–22)
   h: Blastocoel
   i: Fertilization
   d: Acrosome
   b: Zona pellucida
   f: Capacitation
   a: Blastula
   g: Radial cleavage
   e: Blastomeres
   j: Trophoblast
   c: Vegetal hemisphere

2. Choice *c*. During cleavage, the embryo remains about the same size, even though there is a large increase in cell number. Consequently, cells (blastomeres) become smaller and the ratio of nuclear volume to cytoplasmic volume increases. (17)
3. Choice *a*. These are features of mammalian eggs. (8)
4. Choice *d*. A sperm penetrates the egg just after formation of the fertilization cone. (5, 6)
5. Choice *e*. Both the slow block to polyspermy and increases in postfertilization metabolism of the zygote are linked to the release of calcium ions. (12, 13)
6. Choice *b*. Regulated development means that the loss of early blastomeres can be adjusted for to still produce a normal embryo. Monozygotic twins form when a young embryo splits and each set of undifferentiated cells gives rise to a complete embryo. (25, 26)
7. Choice *a*. In the mammalian blastocyst, the trophoblast forms the fetal part of the placenta. A disc-shaped portion of the inner cell mass surrounded by the trophoblast becomes the embryo. (22, 23)

- (14–16)

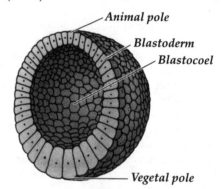

- Your diagram should resemble the first (or second) and third rows of textbook Figure 40.6 on page 889. The difference in cell size of the embryo is due to the large difference in yolk present in the egg, which determines if cleavage will be complete or incomplete. (18, 19)

GASTRULATION: PRODUCING THE BODY PLAN • NEURULATION: INITIATING THE NERVOUS SYSTEM
(pages 298–300)

1. Choice *a*. (19–21)
2. Choice *d*. The gray crescent forms opposite the site of sperm penetration before gastrulation even begins. (9–10; and previous section of this chapter)
3. (6, 9, 16)
   b: Sea urchin
   e: Frog
   a: Bird
4. Choice *c*. All vertebrates have a notochord, which develops as part of the process of neurulation. (19–20)
5. Choice *c*. (5)
6. Choice *a*. In the frog, cells immediately below the blastopore become part of the gut tube. (10)
7. Choice *d*. This structure determines the principal axes of the embryo. (13)

- (2, 6, 7)

Secondary mesenchyme
Ectoderm
Endoderm
Archenteron
Blastocoel
Primary mesenchyme
Blastopore

- (9, 10, 19–21)

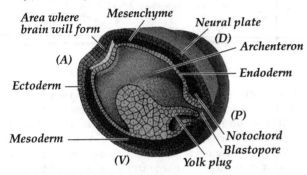

Area where brain will form
Mesenchyme
Neural plate
(D)
Archenteron
(A)
Ectoderm
Endoderm
(P)
Mesoderm
Notochord
Blastopore
(V)
Yolk plug

EXTRAEMBRYONIC MEMBRANES • DEVELOPMENT CONTINUES THROUGHOUT LIFE   (pages 300–302)

1. Choice *a*. Fish reproduce in water and so do not have amniotic eggs. They therefore lack these membranes. (1–6)
2. Choice *c*. (9)
3. Choice *d*. (5)
4. Choice *b*. (3)
5. Choice *e*. All of these are characteristics of animal growth. (12–15)
- The amniotic egg was crucial in the evolution of terrestrial vertebrates because it freed them from a dependancy on aquatic reproduction, thus opening up whole new terrestrial environments for colonization. (1–6)

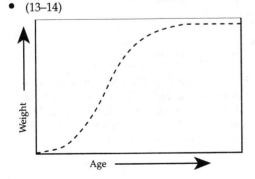

Embryo
Shell
Chorion
Allantois
Yolk Sac

- (13–14)

Weight
Age

- (13–14)

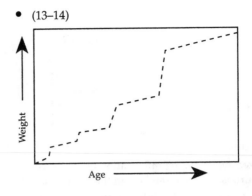

Weight
Age

Integrative Questions

1. Choice *e*. Amniocentesis is a medical procedure for sampling embryonic human cells to check for genetic disorders while the developing child is still in the mother's uterus.
2. 6: Neurulation
   2: Cleavage
   1: Fertilization
   4: Invagination at the blastopore
   3: Blastual formation
   5: Mesoderm formation
3. *d* Trophoblast embryo
   b Spherical blastocoel
   a,b,c,d Anus forms first
   b Vitelline envelope
   c Gray crescent
   a,d Primitive streak
   e Continuous growth curve

CHAPTER 41 – NEURONS AND NERVOUS SYSTEMS

Note: the page numbers listed with the section titles refer to the study guide; numbers in parentheses for each question refer to the relevant key concepts.

NERVOUS SYSTEMS: CELLS AND FUNCTIONS (pages 303–305)

1. Choice *d*. While it is true that chemical communication occurs at the synapse, the portion of the neuron that actually stores and releases the neurotransmitter chemicals is the axon terminal. (13, 14)
2. Choice *c*. Since the axon is typically the longest part of a neuron, this is where your cut would likely be placed. (10–13)
3. Choice *d*. Nerve nets are characteristic of animals with less complex nervous systems and behaviors. (5, 6)
4. Choice *e* All of these features can be used to distinguish between fish and mammal brains. (5–8)
5. Choice *c*. A specialized category of glial cells called astrocytes wrap themselves around blood vessels in the brain tissue to form the blood–brain barrier. (15–18)
6. Choice *a*. Axon terminals are the place where the neurotransmitter is stored and from where it is released into the synaptic cleft. (13, 14)

- (10–13)

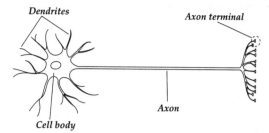

Dendrites receive incoming information from the axons of other neurons located elsewhere in the body. The cell body may also receive incoming information, but its main function is to house the nucleus and carry out metabolic activities necessary for cell function. The axon is the conduction zone of the neuron, sending action potentials to the axon terminal, which synapses with other neurons in the circuit or with some kind of effector cell.

- The nerve net seen in the sea anemone permits relatively simple coordinated movements of tentacles for feeding and predator avoidance. It fits the limited needs of a sessile animal. The squid has many more neurons, distinct nerve tracts, paired nerves, and clusters of neuron cell bodies (ganglia); these greatly speed information transfer and analysis. This allows for much more complex coordination of body systems and behavior than does a simple nerve net. (1, 5, 6)

NEURONS: GENERATING AND CONDUCTING NERVE IMPULSES (pages 305–307)

1. Choice **d**. The interior of the cell is negative with respect to the outside at the resting membrane potential (i.e., it is polarized). Depolarization reduces this polarized state. This requires the influx of positive charge, which occurs as Na^+ ions enter the cell. (15)

2. Choice **b**. If an already polarized cell with a negative membrane potential is hyperpolarized, the potential becomes even more negative. (10)

3. Choice **d**. In many neurons, voltage-gated potassium channels are responsible for the repolarization phase of an action potential (the downward swing in membrane potential). The depolarization phase (upward swing) would not be affected, however, since this results from the action of gated sodium channels. (15, 16)

4. Choice **c**. Sodium activation gates open and inactivation gates close at the onset of an action potential. When the activation gates close again, the inactivation gates remain closed for several milliseconds before reopening. This temporarily prevents the sodium influx necessary to start another action potential. (17)

5. Choice **a**. Invertebrates, such as ants, do not have myelinated axons to increase conduction velocity. They achieve increases in the speed at which action potentials are propagated by increasing the diameter of the axon, not by decreasing it. (20, 21)

6. Choice **e**. All of these characteristics could be true of such an axon. (21–23)

- (1, 2, 13–17)

- The sodium–potassium pump of a neuron establishes ion concentration gradients across the cell membrane, with sodium more concentrated outside the cell and potassium more concentrated within. At rest, more K^+ ion channels are open than Na^+ ion channels, so there is a net loss of positive charge from the inside of the cell due to the diffusion of K^+ ions down their concentration gradient. This creates a relative negative charge inside the cell reinforced by the negative charges of chloride ions and proteins within the cell. This charge difference across the membrane is the resting potential. (4–9)

NEURONS, SYNAPSES, AND COMMUNICATION
• NEURONS IN CIRCUITS (pages 307–311)

- (3, 15–16, 19)
  d: Curare
  e: Acetylcholine
  a: Glycine
  b: Calcium
  c: Atropine

- Acetylcholinesterases are the enzymes that degrade the neurotransmitter acetylcholine in the synaptic cleft. This must occur or the transmitter will keep stimulating receptors in the postsynaptic membrane, which will keep the neurons continuously active, and, if sustained, will lead to spasm and death. By blocking the action of the cholinesterase enzyme, anticholinesterases promote uncontrolled nerve and muscle spasms. (2, 21)

- (23–24)

1. Choice **a**. Neurotransmitters bind to membrane receptors on the postsynaptic cell. This causes specific gated ion channels to open, which may either depolarize or hyperpolarize the cell, depending on whether Na^+ or K^+ ions move through the membrane. (2)

2. Choice *d*. Dendrites and most of the cell body have few gated sodium channels. These channels mediate the production of action potentials by permitting the membrane to rapidly depolarize, a characteristic that is concentrated at the axon hillock and in the axon itself. (7–10)

3. Choice *c*. Atropine is a substance that blocks the inhibitory effects of muscarinic receptors in muscle tissue, especially the heart. (16, 19)

4. Choice *a*. The diagram shows that three different stimuli arrive at the same time and evoke a large EPSP that leads to an action potential, while less than three do not. This is spatial summation. Temporal summation involves the same stimulus arriving in rapid succession. (7–10)

5. Choice *b*. Excitatory postsynaptic potentials make it easier for an action potential to occur, so they must depolarize the postsynaptic membrane. (5, 7)

6. Choice *b*. Active transport is used by some axon terminals as a way to clear the synaptic cleft by gathering up released neurotransmitter chemicals. (22)

7. Choice *e*. Gap junctions are electrical synapses that physically connect neurons via membrane proteins. Electrical signals can thus travel in either direction. (13)

8. Choice *c*. The basic knee-jerk response is controlled by a monosynaptic circuit. Interneurons are only involved in the voluntary or conscious modification of this reflex. (23–27)

9. Choice *e*. (29–30)

Integrative Questions

1. Choice *e*. These would most likely be true of any neuron in the body, including those of the brain.

2. The correct sequence for these events is **c, a, e, d, b.**

3. Choice *c*. Since the action potential started in the middle of the axon, rather than at the axon hillock as typically occurs, there is no previous patch of membrane currently undergoing a refractory period. The action potential is thus free to propagate in both directions.

4. Choice *b*. Hyperpolarizing neuron B by 15 mV would make its membrane potential –65 mV, actually taking it further away from the threshold voltage needed to fire an action potential.

CHAPTER 42 – SENSORY SYSTEMS

Note: the page numbers listed with the section titles refer to the study guide; numbers in parentheses for each question refer to the relevant key concepts.

SENSORS, SENSORY ORGANS, AND TRANSDUCTION
• CHEMOSENSORS: RESPONDING TO SPECIFIC MOLECULES (pages 312–314)

• Receptor potentials are the result of a physical stimulus being transduced into an electrical signal. The stimulus causes changes in the ion permeability of the sensor membrane to yield the receptor potential, the size of which is directly proportional to the intensity of the stimulus. In sensors that produce an action potential directly, the receptor potential is called a generator potential. In other sensors, the receptor potential causes the release of neurotransmitter onto an associated sensory neuron, which then produces the action potential. (6–8)

• A blow to the head is a sufficiently intense physical stimulus to easily cause changes in the ion permeability of sensor membraness associated with vision and hearing. This would facilitate transduction and the generation of action potentials from these sensors. Since the action potentials produced are targeted on specific brain areas that interpret light and sound, the brian would perceive generalized sensations of "stars" and "bells" as the meaning of these action potentials, even though no light or sound stimuli were actually present. (2, 3)

1. Choice *a*. Transduction is the conversion of a physical stimulus (energy) into electrical impulses (action potentials) that the nervous system can use for communication. The only answer that meets this criterion is the first one. (2)

2. Choice *b*. Adaptation occurs when a sensor cell ceases its response to an unchanging, steady-state stimulus, such as the one described in this situation. The action potentials recorded in the sensory nerve are a direct reflection of the activity in the sensor cell. (9)

3. Choice *c*. Human taste sensors are not neurons. Instead, they produce receptor potentials in response to appropriate chemical stimuli. This causes the release of neurotransmitter onto adjacent sensory neurons, which fire the action potentials. (17)

4. Choice *b*. cAMP is the second messenger in this signalling system. It is produced in the presence of a chemical stimulus and causes gated sodium channels to open. This results in an influx of Na^+ ions, causing the membrane potential to become more positive. The cell depolarizes as a result. (13)

5. Choice *d*. When you have nasal congestion, the olfactory epithelium becomes inflamed, and excess mucus covers the olfactory sensors; this makes your sense of smell less efficient. Since the odors perceived while eating strongly influence our sense of taste, a stuffed-up nose causes us to perceive that the food has lost much of its flavor. (12, 20)

6. Choice *e*. Humans actually perceive at least four categories of taste. The one missing from this list is salty. (15–20)

7. Choice *b*. Although technically a type of hormone, since pheromones are used outside the body they have their own name. (11)

8. Choice *e*. All sensors are capable of transduction in some fashion. (2)

MECHANOSENSORS: STIMULI THAT DISTORT MEMBRANES (pages 314–317)

1. Choice *d*. Meissner's corpuscles are extremely sensitive touch receptors in the skin. They achieve transduction through the direct deformation of the sensor cell membrane rather than through the bending of stereocilia, as is the case with hair cells. (3)

2. Choice *b*. The malleus is the first of the three ear ossicles found in the middle ear of vertebrates; it attaches directly to the tympanic membrane. (13)

3. Choice *c*. All mechanosensors (indeed, *all* sensory cells) ultimately have their effects by creating changes in the ion permeability of cell membranes. This either leads directly to the production of an action potential, or indirectly to action potentials by way of receptor potentials. (1, 2)

4. Choice *a*. Statocysts and the vestibular system both rely on hair cells to control balance and body position. Expanded-tip tactile receptors and Pacinian corpuscles are both touch receptors in the skin. (3, 4, 8, 9)

5. Choice *e*. Hair cells are an important class of mechanosensors that do not qualify as neurons because they do not fire action potentials. (6)

6. Choice *b*. The proximal end of the basilar membrane (the end nearest the middle ear) responds only to high-frequency sound. Permanent damage to this region of the cochlea would impact the man's ability to hear high-pitched tones. (17, 18)

7. Choice *c*. Meissner's corpuscles are sensitive to light touch, and so adapt quickly to a continuous unchanging stimulus. (3)

● (4–5, 7–10, 16)

    d: Organ of Corti

    e: Pacinian corpuscle

    b: Golgi tendon organ

    f: Statocyst

    c: Vestibular apparatus

    a: Lateral line

● The semicircular canals are fluid-filled chambers that respond to head movements. As the fluid moves within a canal, it moves an ampulla and bends hair cells, causing the production of action potentials. The vestibular apparatus also detects head position, acceleration, and deceleration. Otoliths rest upon sensory hairs in the chambers of this system, and when the head moves so do the otoliths. This causes hairs to bend and action potentials to be produced. (9–10)

● X will be perceived as a relatively high-pitched sound, Y a mid-range pitch, and Z a low-pitched sound. This is because specific regions of the basilar membrane only vibrate in response to certain frequencies. High-frequency sounds travel least far within the cochlea and so stimulate vibration in the basilar membrane proximal to the oval window. Lower frequency sounds travel farther within the cochlear fluid and therefore cause more distal portions of the basilar membrane to vibrate. (17, 18)

PHOTOSENSORS AND VISUAL SYSTEMS: RESPONDING TO LIGHT • SENSORY WORLDS BEYOND OUR EXPERIENCE (pages 317–320)

1. Choice *e*. The photopigment used by all animals is rhodopsin, which contains the protein opsin. (1–3)

2. Choice *b*. There should be a blue indicator whenever the 11-*cis* isomeric form of retinal is present. This will only occur prior to the breakdown of rhodopsin, that is, before the pigment has been exposed to light. (2, 3)

3. Choice *a*. The only retinal neurons with direct links to the brain are ganglion cells. (23–25)

4. Choice *d*. cGMP is the substance that keeps sodium channels open in photosensors during the resting state. When stimulated by light, a chain of biochemical events causes cGMP concentrations to drop. This lets the sodium channels close, and as a result the membrane hyperpolarizes. (5–7)

5. Choice *c*. Retinula cells are found in the ommatidia of arthropod compound eyes. (10–16)

6. Choice *e*. Each of these cell layers overlies the photosensors of vertebrates, so a photon of light must pass through all the layers before striking the rods or cones. (16, 21)

7. Choice *d*. One of the most important functions of the lens and its associated ciliary muscles is to change the len's shape or position to permit proper focusing of light on the retina. This process is called accommodation. (15)

8. Choice *a*. The tapetum is a reflective layer at the back of the eye that helps nocturnal animals capture the maximum amount of light. Capturing light causes rhodopsin to bleach, not regenerate. (22)

● (12–17)

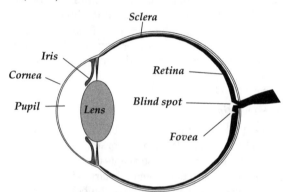

● You are told that this is an off-center receptive field, so the surround will be excited by light and the center will be inhibited. Thus, condition #1 will produce a brief burst of action potentials for the duration of the stimulus. Condition #2 will either produce no response at all from the cell or will cause it to cease the production of any spontaneous action potentials it was generating. Condition #3 will begin with lots of action potentials until the spot hits the center, at which time the production of action potentials will stop. (26–28)

Integrative Questions

1. Choice *c*. Echolocation is used by bats and dolphins to create acoustic images by reflection. Snakes do not make these kinds of high-intensity, high-frequency sounds, nor are they capable of hearing the echoes produced.

2. Choice *b*. Mechanoreception is the transduction mechanism for all of these stimuli.

3. Choice *b*. The vestibular system uses hair cells and otoliths to help maintain balance. After a day at sea, the hair cells have become overstimulated and continue to give the impression of rocking motions.

4. Choice *a*. Since chemosensors are the most ancient and commonplace of all sensor cells, they are what you

would certainly expect to find. All the others listed would probably not be necessary for the animal to survive, although they might be present as well.

CHAPTER 43 – THE MAMMALIAN NERVOUS SYSTEM: STRUCTURE AND HIGHER FUNCTIONS

Note: the page numbers listed with the section titles refer to the study guide; numbers in parentheses for each question refer to the relevant key concepts.

THE NERVOUS SYSTEM: STRUCTURE, FUNCTION, AND INFORMATION FLOW • FUNCTIONAL SUBSYSTEMS OF THE NERVOUS SYSTEM
(pages 321–324)

1. Choice *a*. The autonomic portion of the vertebrate nervous system controls the function of involuntary organs such as glands. (7)
2. Choice *b*. Nerves contain bundles of axons from many different neurons, along with connective tissue that holds the nerve together. Since the axons are coming from the cochlea, it would be the auditory nerve. (4)
3. Choice *c*. The convolutions of the cerebral cortex create the gyri and sulci that significantly increase cortical surface area. This allows the presence of many more neurons than would otherwise be possible. (26)
4. Choice *d* Agnosias are disorders resulting from damage to the temporal lobe that can produce the symptoms described. (28)
5. Choice *e*. All of these structures are located posterior to the telencephalon within the main axis of the vertebrate central nervous system. The impulse must therefore pass through all of them before reaching the telencephalon. (10–14)
6. Choice *d*. The sensory stimulus of pain is triggered by actions in the periphery, but the sensation of pain itself is located in the brain. Interneurons gather information from the afferent elements of simple reflex loops controlling foot movements and send this information through the spinal cord to the brain. (5–7, 19, 20)
7. Choice *b*. The occipital lobe, which is important in controlling vision, is located at the back of the cerebral cortex and would be impacted by such an injury. (29)
8. Choice *a*. The telencephalon is relatively undeveloped in fish but increases in size for the more complex vertebrates. Fish therefore have the smallest ratio of telencephalon size to body size. (13, 33, 34)
9. Choice *c*. The cell bodies of sensory neurons are located in the periphery and send their axons to the CNS. (6)
10. Choice *e*. The telencephalon is most highly developed in mammals and birds. It is therefore the area that controls most of the conscious behavior in an animal like the rat. The behavior of all the other animals listed is regulated primarily by lower brain structures. (13, 33)

- Based on the computer reconstruction shown at the opening of this chapter, Phineas Gage's injury likely involved damage to anterior portions of the cerebral cortex and limbic system. The frontal lobes of the cortex are involved in motor control, as well as planning. The limbic system controls drives and other fundamental emotions such as rage. All these characteristics changed after Gage's accident. (25, 26, 32)

- (9–15, 26)

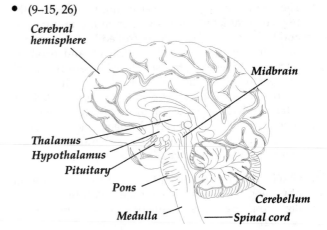

Cerebral hemisphere—site of sensory processing, learning, memory, and consciousness
Thalamus—relay station for sensory information going to telencephalon
Hypothalamus—regulates physiological functions and biological drives
Pituitary—endocrine master gland
Pons and *medulla*—control circulation and breathing functions
Cerebellum—coordinates and refines behavior patterns
Midbrain—processes aspects of visual and auditory information
Spinal cord—conducts and integrates information between brain and periphery

- All vertebrates are fairly similar in the early development of their nervous system. Mammals, and humans in particular, differ from fish by having a well-developed forebrain. Based on the drawings in Figure 43.3, the emphasis on forebrain development starts to emerge in the 35–40 day old human embryo. (9–13)

INFORMATION PROCESSING BY NEURONAL CIRCUITS • UNDERSTANDING HIGHER BRAIN FUNCTIONS IN CELLULAR TERMS
(pages 324–328)

- Voluntary muscle movements are not possible during REM sleep, when dreaming occurs. Thus it is not possible to move the limbs and sleepwalk during this stage of sleep. Rapid contractions of the ocular muscles and their resulting eye twitches are the obvious facial characteristic from which REM sleep derives its name. (16–19)

- Conditioned reflexes, a type of associative learning, occur when a novel stimulus is associated with another stimulus that normally triggers a response. In rabbits, irritation of the eye from a puff of air causes the blinking response. If this normal reflex is paired with a sound stimulus, the sound will soon be sufficient to trigger blinking in the absence of the air puff. If the cerebellar nucleus is lesioned, then conditioning of this reflex is not possible. Thus, the cerebellum must play a key role in controlling the behavior. (24–26)

• The corpus callosum is the structure that permits the two cerebral hemispheres to communicate with one another. In epileptics, uncontrolled bursts of action potentials cross the corpus callosum and produce seizures. By severing this structure the seizures can be prevented. Since the two hemispheres still function internally, however, other behavior remains relatively intact. What suffers are higher brain functions that require connection between the hemispheres, such as certain aspects of language, as demonstrated by Sperry's split-brain studies. (33, 34)

1. Choice *a*. Sympathetic pathways are stimulatory for physiological responses associated with the fight-or-flight response, parasympathetic pathways are inhibitory. (1, 2)

2. Choice *c*. Refer to textbook Figure 43.11 on page 961 for this information. (1–3)

3. Choice *d*. The occipital cortex is the higher brain region controlling vision. (8, 9)

4. Choice *b*. Disparity is the property used to judge distance for binocular vision and results from the fact that each eye views an object from a different angle. Nearby objects are more affected by this angular difference than are distant objects, and specific binocular cells respond best to images with particular disparity. (11, 13)

5. Choice *e*. The axons of sympathetic postganglionic cells terminate on effectors, which include all of the structures listed. (5)

6. Choice *e*. Dreams occur during REM sleep, which is characterized by all of these conditions. (17–19)

7. Choice *e*. Memories do not seem to be located in any one portion of the brain. Rather, they appear to be spread across large regions of the cerebral cortex with many interconnecting pathways between these areas. (28)

8. Choice *b*. Different aspects of language production and use reside in different cortical regions such as Broca's and Wernicke's areas. The brain pathways of patients with aphasias have trouble exchanging information between these areas. (32–35)

9. Choice *a*. Changes in neural activity of specific cortical and thalamic brain nuclei have been linked to transitions between the stages of sleep. (21, 23)

Integrative Questions

1. Choice *e*. Sensory information from vision and touch, as well as motor control to move your arm and mouth, have peripheral elements. The mental recognition of chocolate and the decision to eat it are central processes. The parasympathetic branch of the autonomic nervous system stimulates salivation.

2. Choice *e*. Association areas and long-term memories are present in all these regions of the cerebral cortex.

3. Choice *c*. Fish have the most poorly developed forebrains of all the vertebrates.

4. Choice *a*. The primary visual cortex, located in the occipital lobe at the back of the brain, is where the visual signal of a written word is first decoded via higher brain functions like memory and recognition. The decoded information is then sent to the angular gyrus and on to Wernicke's area.

CHAPTER 44 – EFFECTORS

Note: the page numbers listed with the section titles refer to the study guide; numbers in parentheses for each question refer to the relevant key concepts.

CILIA, FLAGELLA, AND CELL MOVEMENT (pages 329–331)

• The cilia of the respiratory tract move mucus toward the oral cavity. This mucus traps airborne particles and keeps them from possibly damaging the delicate respiratory tissue. Mucus movement is made possible because adjacent microtubules inside the cilia slide past one another and cause the cilia to bend. The power stroke produced by thousands of cilia working in coordinated fashion propels the mucus up and out of the respiratory tract. (3–5, 8, 9)

• (4, 7–10)
  b: Axoneme
  d: Recovery stroke
  e: Myosin
  a: Dynein
  c: Pinocytosis

1. Choice *c*. Energy released from the splitting of a phosphate bond in ATP powers the conformational change of dynein, which causes the microtubules to slide past one another. (8, 9)

2. Choice *b*. Plasmagel is a special category of cytoplasm that influences the amoeboid movement of some cells. It is not a structural component of a cilium. (12)

3. Choice *a*. Although cell movements are one of the most important functions of effectors, they are also involved in other activities such as chemical release and the emission of various forms of energy. (1)

4. Choice *d*. The movement of flagella is controlled mainly by microtubules. (7, 10)

5. Choice *e*. Axoneme movement is produced by the chemical events that cause microtubules to slide past each other. (8, 9)

6. Choice *e*. Since no cross links or spokes are present, when the microtubules slide by one another they elongate the entire axoneme, rather than cause it to bend. (8, 9)

7. Choice *c*. As in the previous question, severing the connections between the cell and the cilium eliminates bending, but does not stop the microtubules from sliding past one another. (8, 9)

MUSCLE CONTRACTION: ACTIN AND MYOSIN INTERACTIONS (pages 331–333)

1. Choice *a*. All digits and limbs carry out voluntary motion and are controlled by skeletal muscles. (9)

2. Choice *d*. Microfilaments are the contractile elements that make up a muscle cell or myofibril. These are then bundled to form whole muscles. (9, 10)

3. Choice *d*. Microfilaments are the basic contractile elements found in all muscle. (2)

4. Choice *b*. Since the A band represents the myosin, its size cannot change. The Z lines are the end points of the sarcomere, where the actin microfilaments attach. As

actin slides past the myosin from both directions toward the middle of the sarcomere, the Z lines approach each other. (12–14)

5. Choice *c*. Tropomyosin is the double-stranded protein that covers the actin binding sites for myosin. Troponin is the other protein to which the calcium ions bind. (15, 21)

6. Choice *c*. The action potential is initiated on the plasma membrane at the neuromuscular junction. It then spreads through the cell via the T-tubule system before reaching the triads that initiate the opening of the Ca^{2+} channels of the sarcoplasmic reticulum. (18–20)

7. Choice *e*. Calcium binding to troponin induces a conformational change that moves the attached tropomyosin strands away from the actin binding sites. (21, 22)

8. Choice *b*. The continuous current from a live wire would cause very rapid volleys of action potentials to all motor units, producing a tetanus response. (26, 27)

9. Choice *e*. Marathon running is an aerobic activity. The muscle best suited for this work is dark muscle, because it has many slow-twitch fibers for endurance and abundant mitochondria to fuel the activity of these fibers. (28, 29)

10. Choice *e*. Both of these characteristics are important for the function of cardiac muscle. (6–8)

- (11–17)

- (11–17)

The sarcomere has shortened because the actin microfilaments have slid past the myosin filaments and toward each other.

SKELETAL SYSTEMS PROVIDE RIGID SUPPORTS • OTHER EFFECTORS IN THE ANIMAL KINGDOM (pages 333–336)

- (7, 15, 22, 28, 29)
  a: Osteoblast
  e: Nematocyst
  b, *d*: Extensor
  d: Exoskeleton
  c, *d*: Chromatophore

- The bones of the vertebral column and the skull (cranium, maxilla, mandible) are axial. All others are part of the appendicular skeleton. (10, 11)

- The elongated limbs indicate the presence of long load arms and short power arms in the levers that make up the legs; these are adaptations for speed. In addition, the sleek aerodynamic body design means light weight, also an adaptation to increase speed. (26)

1. Choice *a*. Skeletal muscle is usually what makes skeletons move, not cardiac or smooth muscle. (2, 5, 10)

2. Choice *d*. By contracting either its circular or longitudinal muscles, a sea anemone can change the shape of its hydrostatic skeleton just as the proper squeezing of a water balloon causes it to elongate or shorten. (2)

3. Choice *e*. The main difference has to do with the relative location of muscles and skeletal elements. The skeleton is outside of the muscles in an exoskeleton, and on the inside for an endoskeleton. (5–10)

4. Choice *c*. The appendicular skeleton includes all bones associated with the appendages. The jaw bone is part of the skull, which belongs to the axial skeleton. (11)

5. Choice *e*. All of these cells are involved in the process of regulating extracellular calcium concentration. (15, 16)

6. Choice *c*. Since ligaments hold bones together, he has probably suffered a ligament tear. (23)

7. Choice *d*. Vertebrae are part of the axial skeleton, and so cannot be a category of appendicular skeleton injuries. (11)

8. Choice *b*. Effectors that release communication chemicals are the most common effector organs not related to movement. (27–32)

9. Choice *a*. The butterfly has an exoskeleton. All the other animals listed have some kind of hydroskeleton, including the snail, which also has an exoskeleton covering its hydroskeletal parts. (2–7)

Integrative Questions

1. Choice *e*. The antagonists in a hydrostatic skeleton are opposing circular and longitudinal muscles; contraction of one elongates the animal, contraction of the other shortens it.

2. Choice *b*. Electric organs are useful in aquatic environments and are known only from certain species of fish.

3. Choice *a*. The fact that flagella create whiplike motions is due simply to their length compared to a that of a typical cilium; the molecular mechanisms for how the movement is produced are the same for both types of effectors.

4. Choice *c*. Only cardiac muscle cells have the unique potassium channels that give this type of muscle its myogenic properties.

CHAPTER 45 – GAS EXCHANGE IN ANIMALS

Note: the page numbers listed with the section titles refer to the study guide; numbers in parentheses for each question refer to the relevant key concepts.

RESPIRATORY GAS EXCHANGE • ADAPTATIONS FOR GAS EXCHANGE (pages 337–340)

- Barometric pressure is lower on top of Mt. Everest because of its higher altitude. Thus, P_{O_2} is also lower on Mt. Everest. Lower P_{O_2} makes diffusion, and therefore respiration, more difficult. (11, 12)

- (9, 17, 23, 24)
  a. Maximize the surface area for exchange between air and water
  b. Use thin membranes for the exchange surfaces
  c. Cause water and air to flow across the exchange surfaces in opposite directions (i.e., use countercurrent flow)
- (23, 24)

The tube with the higher O_2 saturation must be the oxygen source (i.e., the water); the other is therefore the blood. The two flows are antiparallel, establishing a countercurrent exchange that greatly enhances the net diffusion of O_2 from the water into the blood along the entire length of the tube.

1. Choice *b.* Temperature usually has little or no effect on gas exchange for air breathers, but can be of major importance to ectothermic water breathers. (10)
2. Choice *a.* The amount of O_2 that a given volume of water can hold does, in fact, decrease with increasing temperature, which is why ectothermic water breathers can suffer ill effects as the water warms. (10)
3. Choice *e.* CO_2 dissolves easily in water, so getting rid of this gas is not a problem for water breathers. (13)
4. Choice *c.* Large, metabolically active animals cannot survive without specialized respiratory structures to provide adequate gas exchange. (8, 9)
5. Choice *a.* This set of characteristics is common to all respiratory organs. (1, 9, 15, 19, 25)
6. Choice *c.* Spiracles are the abdominal openings to the tracheal system of insects. The others are part of gill or lung systems. (14, 15, 19, 24, 26)
7. Choice *d.* O_2 moves into the fish's mouth, over the gill filaments and lamellae, across the respiratory membranes, and then into the blood capillaries. (21)
8. Choice *d.* The inhalation phase of bird breathing is when the air sacs are filling. (26–28)
9. Choice *b.* Salmon, being fish, use countercurrent exchange for breathing. All the others listed are air breathers, which use tidal mechanisms. (23–25)
10. Choice *e.* The resting tidal volume is the amount of air moved in and out of the lungs during normal breathing. All other lung volumes listed are larger, as indicated in Figure 45.11 of your textbook. (31, 32)

MAMMALIAN LUNGS AND GAS EXCHANGE
(pages 341–342)

- Your diagram should resemble the left-hand drawing in Figure 45.13 of the textbook. Contraction of the diaphragm and intercostal muscles expands the thoracic cavity, creating negative pressure that pulls air into the lungs. This is an active process. (9, 10)
- An oxygen molecule entering the human respiratory system would first pass through the nasal cavity, then travel to the pharynx, larynx, trachea, bronchi, bronchioles, alveoli, and finally the blood capillaries. (1, 2)

1. Choice *c.* Just the reverse is true. Inhalation is the active part of breathing in mammals, dependent upon the contraction of the diaphragm and intercostal muscles. Exhalation is usually passive. (10, 12)
2. Choice *b.* Alveoli are the exchange surfaces in mammalian lungs. All the other structures listed are dead space used to transport air to and from the alveoli. (1–3)
3. Choice *d.* The unidirectional air flow in birds is more efficient than the tidal flow of mammals. (1–4, and previous section)
4. Choice *d.* This mechanism serves a very importance cleansing role for the mammalian respiratory system. (7, 8)
5. Choice *a.* The underdeveloped state of the premature infant lung means that surfactant is not produced in the alveoli and breathing is therefore very difficult. (5, 6)

BLOOD TRANSPORT OF RESPIRATORY GASES
- REGULATING BREATHING   (pages 342–345)

1. Choice *c.* Hemoglobin contains four polypeptide chains. Myoglobin has only a single chain with high oxygen affinity. (2, 3, 9, 10)
2. Choice *e.* All the listed items are commonly found suspended or dissolved in mammalian blood plasma. (2, 16)
3. Choice *d.* A left shift in the oxygen-binding curve means that the pigment operates at lower P_{O_2} values. (7)
4. Choice *c.* Changes in pH due to the accumulation of acidic metabolites resulting from increased cell activity are the basis of the Bohr effect. (12)
5. Choice *e.* By changing the shape of hemoglobin, diphosphoglyceric acid makes it more likely that hemoglobin will release its O_2 cargo. This is an adaptation for increased O_2 availability that occurs during sustained exercise or acclimation to high altitude. (13)
6. Choice *a.* CO_2 is rapidly removed from mammalian lungs by exhalation, thus keeping its concentration low in the alveolar air spaces. (18)
7. Choice *d.* The contraction state of the diaphragm is not directly involved in the *neural* control of ventilation. All the other structures listed either sense, relay, or act upon stimuli and neural information related to ventilation. (20, 21, 24)
8. Choice *a.* Cutting the brain stem below the medulla eliminates neural contact with the respiratory mechanism and could easily result in a complete stoppage of breathing. (20)

- (4, 7, 16, 17, 20)
  e: Carbonic anhydrase
  a: Myoglobin
  b: Oxygen binding
  d: Carboxyhemoglobin
  c: Medulla

- *A* = llama, *B* = adult human, *C* = adult human at high altitude. Llamas and humans make opposite kinds of adjustments in the oxygen-binding properties of their hemoglobin for life at high altitude, although the end result of making more O_2 available to their tissues is the same. The llama's hemoglobin can saturate at very low P_{O_2} values (a left shift in the curve), while the human's hemoglobin changes shape under the influence of diphosphoglyceric acid to more easily release O_2 (a right shift in the curve). (4, 11, 13)

Integrative Questions

1. Choice *b*. In order to avoid drying, external gills can only be present in aquatic animals. Fish, although aquatic, have internal gills.

2. Choice *c*. Birds have increased their respiratory efficiency by using unidirectional air flow rather than tidal respiration.

3. Choice *b*. Diffusion is key to the functioning of all gas exchange systems.

4. Choice *e*. Refer to Figure 45.12 in your textbook for scale.

CHAPTER 46 – CIRCULATORY SYSTEMS

Note: the page numbers listed with the section titles refer to the study guide; numbers in parentheses for each question refer to the relevant key concepts.

CIRCULATORY SYSTEMS: VESSELS, BLOOD, AND A PUMP (pages 346–349)

1. Choice *d*. Butterflies are insects, which have open circulatory systems. They have only a minimal number of short, open-ended transport vessels; capillaries are not present. (8–10)

2. Choice *e*. None of the possible choices are found in both hydras and chickens, which have very different types of circulatory systems. (1–4, 14–17, 26)

3. Choice *a*. One of the advantages of a closed circulatory system is that it can serve as a transport system for large, multicellular animals whose body size is much too big for exchange by simple diffusion alone. Diffusion occurs between individual cells and the blood only after the required materials have traveled through the transport system from another location. (8, 11, 12)

4. Choice *c*. Arteries are large-diameter, high-volume vessels that carry blood away from the heart. They are therefore most likely to carry the greatest amount of blood during any given time period. (16, 17)

5. Choice *b*. The two-chambered fish heart is the evolutionary precursor of the three-chambered hearts of amphibians and reptiles. (18, 23)

6. Choice *d*. The ventricles and arteries are the places where blood pressure is highest. The systemic ventricle of a bird or mammal would probably have the highest pressure of all, since it is a large muscular pump, completely distinct from the pulmonary ventricle, and designed to move blood to all parts of the body except the lungs. (26, 27)

7. Choice *e*. Blood traveling through either the systemic or pulmonary circuits would make its way through all possible types of vessels before returning to the heart. (15–17)

8. Choice *b*. While it is true that blood leaving the heart through the systemic circuit in birds and mammals is oxygenated, the blood returning from the body tissues via the systemic circuit is mostly deoxygenated. (15, 26, 27)

9. Choice *a*. Since there is no oxygen entering the respiratory system when the reptile is not breathing, blood flow to the lungs is greatly reduced until breathing resumes. (24, 25)

- (9, 16–18)
  a: Ventricle
  d: Ostia
  e: Atrium
  f: Venule
  b: Capillaries
  c: Arteriole

- The open circulatory system of an insect consists of a muscular pump (heart) that propels intercellular fluid through dorsal vessels leading to the anterior and posterior ends of the body. Fluid flows out through smaller vessels along the way and then percolates through the body tissues. The fluid pools up in special sinuses on the ventral side of the animal and reenters the heart via valved holes called ostia. An earthworm has a closed system, in which the blood remains inside vessels throughout its journey around the body. Finely branched capillaries take blood to within one cell distance of every portion of the body before returning it to the heart. (8–13)

- (15–17, 26)

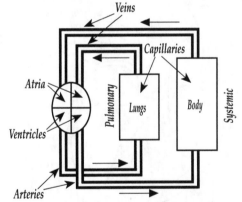

The systemic circuit delivers blood to the body tissues and then returns it to the heart. The pulmonary circuit sends blood to the lungs and then returns it to the heart. Blood always leaves the heart from the ventricles by way of arteries, and returns to the heart at the atria by way of veins. Capillaries could be found both in the lungs and the body tissues.

THE HUMAN HEART: TWO PUMPS IN ONE • THE VASCULAR SYSTEM: ARTERIES, CAPILLARIES, AND VEINS (pages 349–352)

1. Choice *e*. All of the characteristics listed apply to the venae cavae. (3)

2. Choice *d*. The numerator in this ratio is the systolic pressure, which occurs when the ventricle is contracting (i.e., when the atrium is relaxing). The denominator is

the diastolic pressure, which results when the ventricle relaxes. "Normal" values for a young adult human would be about 120/80 mm Hg, so the diastolic pressure appears to be rather low. (5, 7)

3. Choice *a*. Tachycardia refers to a condition of excessively high heartbeat rate. (14)

4. Choice *a*. The conduction of the action potential from SA to AV node is very rapid. The AV node is slow to conduct to the bundle of His. Thus, an electrical stimulus not applied directly to the bundle would still involve a delay. (11–13)

5. Choice *b*. A stroke occurs when the small vessels of the brain become blocked through an embolism and cut off oxygen flow to brain cells. Embolisms and strokes do not occur in the arteries. (30–33)

6. Choice *a*. Capillaries directly link arterioles with venules, not arteries with veins. (16)

7. Choice *c*. Although the blood and lymph are chemically quite similar, the lymph has somewhat less protein and no red blood cells at all. (23)

8. Choice *b*. Stimulation by the parasympathetic branch of the autonomic nervous system tends to slow heartbeat rate, which would not help the movement of venous blood back to the heart. (8–10)

9. Choice *d*. Histamine changes the permeability of capillaries and venules, causing them to lose water to the body tissues. This condition is known as edema. (18–20)

• (1–3)

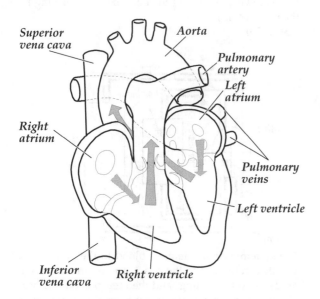

• (1–3)

(1) left ventricle → (2) aortic valve → (3) aorta → (4) systemic arterioles → (5) systemic capillaries → (6) systemic venules → (7) vena cavae → (8) right atrium → (9) atrioventricular valve → (10) right ventricle → (11) pulmonary valve → (12) pulmonary artery → (13) pulmonary arterioles → (14) pulmonary capillaries → (15) pulmonary venules → (16) pulmonary vein → (17) left atrium → (18) atrioventricular valve

• (5)

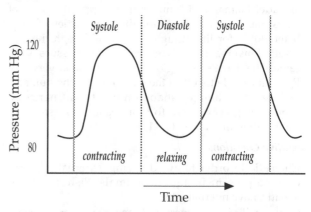

Time

BLOOD: A FLUID TISSUE • CONTROL AND REGULATION OF CIRCULATION   (pages 352–355)

1. Choice *e*. New red blood cells are made by the stem cells of the bone marrow in response to the release of erythropoietin from the kidney. (4)

2. Choice *b*. This description matches that of platelets, another cellular component of the blood. (8)

3. Choice *c*. The activation of platelets causes circulating prothrombin, which is inactive, to convert to its active form, thrombin. This in turn activates fibrinogen, which causes the polymerization of fibrin threads. (9–11)

4. Choice *e*. All of these parameters influence the autoregulatory mechanisms of cardiovascular control. (14)

5. Choice *d*. Angiotensin and vasopressin cause the constriction of arterioles, not venules. (18)

6. Choice *a*. During the diving response of marine mammals, heartbeat rate is slowed and peripheral blood flow is restricted in order to shunt O_2 to vital internal organs. (21)

7. Choice *d*. The precapillary sphincters control the size of the capillaries. If dilated, more blood will flow through the capillary beds. (16)

8. Choice *e*. Information from all of these sources helps the medulla to regulate cardiovascular function. (19, 20)

• (1–13)

   d: Plasma
   c: Erythrocyte
   f: Leukocyte
   e: Platelet
   b: Thrombin
   a: Albumin
   g: Hematocrit

• When mammals dive, they are unable to breathe underwater. Furthermore, it may take some time for them to find food. The diving reflex therefore allows these animals to conserve oxygen by decreasing heart rate and reducing blood flow to all but the most vital internal organs such as the brain, heart, and kidneys. They breathe heavily upon returning to the surface in order to relieve the oxygen debt built up during the dive. (21)

Integrative Questions

1. Choice *c*. Cellular and hormonal blood composition would be basically the same for all mammals. However, the O_2 concentration in veins (vena cavae) and arteries (aorta) would be quite different.

2. **Choice _b._** Clams are mollusks and have an open circulatory system. They do have a pumping heart, but since there are no lungs, pulmonary valves would not be present in this heart. The clam heart pumps interstitial fluid, which lacks red blood cells.

3. **Choice _a._** Hemoglobin is contained in erythrocytes. Leukocytes have immune and other defense functions within the body.

4. **Choice _d._** As blood pools in the venous system, less of it is available to be oxygenated and sent to the brain. If this condition persists for too long, fainting can occur.

CHAPTER 47 – ANIMAL NUTRITION

Note: the page numbers listed with the section titles refer to the study guide; numbers in parentheses for each question refer to the relevant key concepts.

NUTRIENT REQUIREMENTS (pages 356–359)

- (1, 4, 7, 8, 12, 19, 22)
  _f_: Anorexia Nervosa
  _e_: Malnutrition
  _c_: Vitamin C
  _g_: Heterotrophy
  _a_: Digestion
  _d_: Undernourished
  _b_: Macronutrient
- No. Overnourishment results from a caloric intake in excess of metabolic needs, while undernourishment is just the opposite. (7, 9)
- Yes. It is possible to be overnourished and at the same time malnourished since malnutrition is caused by the lack of essential dietary components, regardless of the number of calories being consumed. (9, 10, 12)

1. **Choice _b._** Fat is the mostly easily stored and energetically valuable fuel for a migrating bird. (6)

2. **Choice _d._** By combining linoleic acid with acetyl groups, virtually all of the lipid needs of an animal can be met. (18)

3. **Choice _a._** The chipmunk's daily caloric intake is 25 kcal less than its total metabolic rate, making it slightly undernourished. The lack of thiamin produces a condition of malnutrition. (3, 7, 12, 22)

4. **Choice _c._** Certain amino acids are essential because the animal lacks a biochemical means of converting its available carbon skeletons into these particular amino acids. They must therefore be acquired preformed in the diet. (13, 14)

5. **Choice _e._** (19)

6. **Choice _b._** Fat-soluble vitamins cannot dissolve in the blood or urine (i.e., water). Any excess therefore accumulates in the body tissues. (21–23)

7. **Choice _a._** Carbohydrate reserves are the first to be exhausted during starvation, often disappearing in 1 to 2 days. (4–6)

8. **Choice _a._** (22, 23)

ADAPTATIONS FOR FEEDING • DIGESTION (pages 359–361)

1. **Choice _b._** All five organisms are heterotrophs. Only the mushroom (a saprophyte) is not a predator. The termite and the bison are both herbivores. Of the two carnivores, only the blue whale is a filter feeder. (1, 2)

2. **Choice _e._** Sheep, being animals, ingest their food. Because they prey on plants they are also herbivores. Since the only animals that do not carry out extracellular digestion are the sponges and a few specialized internal parasites, this characteristic also applies. (1–3, 9)

3. **Choice _b._** The absence of any flattened molars or premolars, which function in grinding or chewing, suggests that the toothed whales rip and tear their food and swallow it whole. (5–7)

4. **Choice _e._** Actually, just the opposite is true; the animal prey of carnivores is much more elusive and resistant to being captured than are the plants on which herbivores feed. This requires a wider diversity of adaptations for prey capture in the carnivores. (3, 4)

5. **Choice _c._** The stomach stores food (and performs some digestion too) before passing it on to the intestines. The small intestine (midgut) finishes the digestion and carries out most of the nutrient absorption, while the large intestine (hindgut) reabsorbs water and ions. (15–18)

6. **Choice _a._** Since most of the absorption of nutrients occurs in the intestine, this is the structure that would be most severely impacted. (17)

7. **Choice _d._** Dentine is a component of teeth, which hummingbirds lack. The other materials listed are present in some form in all animal digestive systems. (6)

8. **Choice _e._** The meal described contains carbohydrates, proteins, and fats. Therefore, the enzymes that digest these materials will all be at work in the stomach. (21–22)

9. **Choice _b._** Both of these structures improve absorption, either by increasing surface area or by slowing down the passage of digested food through the intestine. (20)

- The _foregut_ includes the mouth and buccal cavity (food gathering, mechanical breakdown, initial digestion), the gizzard (mechanical breakdown), and the stomach and crop (storage and digestion). Note that not all of these structures are present in all animals.
  The _midgut_ is the small intestine, which performs most of the digestion and virtually all of the nutrient absorption.
  The _hindgut_ includes the large intestine, rectum, and anus. Water and ions are reabsorbed in the large intestine, and the feces are condensed. The muscular rectum stores the feces until they are excreted through the anus. (15–19)

- There are a number of structural features evident in the dentition of these two species that reflect differences in feeding habits. Humans are omnivores, and our teeth are therefore generalized to deal with a variety of food sources. We have canines and incisors to bite and tear meat, but they are not well adapted for use as actual weapons. We also have molars and premolars to grind and chew plant foods. The cat has large canines for subduing and killing its prey, and large incisors for ripping meat from bones. Its premolars and molars are also adapted for meat eating and bone crushing, rather than grinding up plant matter. (5–7)

STRUCTURE AND FUNCTION OF THE VERTEBRATE GUT (pages 361–363)

- (1–5)

Serosa
Longitudinal muscle
Circular muscle
Lumen
Submucosa
Mucosa

- The first two chambers (rumen and reticulum) of the multichambered digestive system in a ruminant mammal act as fermentation vats for cellulose digestion. Since so much plant matter must be ingested in order to meet the nutritional demands of these often very large animals, a large storage chamber is needed. Also, by having a large volume of cellulose material fermenting at any one time, large colonies of microorganisms can be sustained to produce the cellulases needed to make this type of digestion possible. The second set of chambers (omasum and abomasum) perform water reabsorption and protein or carbohydrate digestion functions, neither of which requires the large spaces needed for cellulose digestion. (28–30)

1. Choice c. The peritoneum lies outside the gut tube, so the vessel must first penetrate the serosa and then both layers of smooth muscle before reaching the submucosa. (1–5)

2. Choice b. Peristalsis occurs because of the alternating contraction of circular and longitudinal smooth muscles within the gut. Sphincters are thick rings of smooth muscle that control the movement of materials from one part of the digestive tract to another. (7, 8)

3. Choice d. The action of salivary amylase begins carbohydrate digestion in the mouth, and other carbohydrases complete this process in the small intestine. (6, 15)

4. Choice a. An important function of the colon is retrieving the large amounts of water and ions released into the gut during the digestive process. (24)

5. Choice b. The gall bladder stores the bile that has been produced in the liver. The bile is then released into the small intestine to emulsify fats. (16)

6. Choice c. Just before individual amino acids are absorbed across the membranes of mucosal cells lining the small intestine, dipeptidases cleave the remaining dipeptides created by the earlier action of proteases. (20)

7. Choice c. Chylomicrons are assembled in the mucosal cells from absorbed monoglycerides, fatty acids, and proteins. They are then transported via the lymph and blood to all parts of the body. (22)

8. Choice a. Strictly speaking, cellulose is not digested by mammals. Rather, it is the microorganisms that these mammals house within their gut that manufacture the cellulases needed to digest cellulose. (27–31)

CONTROL AND REGULATION OF DIGESTION • CONTROL AND REGULATION OF FUEL METABOLISM • NUTRIENT DEFICIENCY DISEASES • TOXIC COMPOUNDS IN FOOD (pages 363–366)

- (3–4, 10, 13)
  a: Secretin
  f: Glucagon
  b: Gastrin
  e: Epinephrine
  g: Insulin
  d: Cholecystokinin
  c: Cortisol

- Bioaccumulation results from the structure of food webs. As energy and certain chemical compounds travel up the food web they are either lost (energy) or concentrated (chemicals). Unmetabolized toxic substances such as DDT or PCBs may only be present in low levels in organisms at the bottom of the web, but they become increasingly concentrated (and ever more toxic) at higher levels. (22)

1. Choice b. The gastrocolic response involves neural signals that never leave the intestinal tract. The salivation reflex, however, utilizes the typical peripheral and central nervous system pathways. (1–2)

2. Choice c. Gastrin, secretin, and cholecystokinin are all secreted by or have their effects upon the mucosal lining of some structure within the vertebrate digestive tract. (3–4)

3. Choice e. Gluconeogenesis is the process of producing glucose from other macromolecules (such as stored glycogen or fat). Glucagon and epinephrine increase the rate of gluconeogenesis, while insulin and cortisol reduce the rate. (5, 10, 13)

4. Choice a. The role of insulin is to lower the high levels of circulating blood glucose that occur after a large meal. It does this by stimulating the conversion of glucose into its primary storage form, glycogen. (8, 10)

5. Choice d. During the postabsorptive period, circulating blood glucose levels drop because no new food is available in the gut. Thus, glucose stores are mobilized through the breakdown of glycogen. Most of this glucose is used to maintain the cells of the nervous system (which constantly require glucose), so other body cells switch to using fat as their metabolic fuel. (9)

6. Choice c. Chylomicrons and the various density-variant forms of lipoproteins in humans control fat metabolism and transport in the blood. (6, 7)

7. Choice d. Kwashiorkor is caused by extreme protein deficiency, not a vitamin deficiency as are all the others. (15–17)

8. Choice b. PCBs are structurally similar to certain animal hormones. They bind to hormone receptors and block the normal function of the hormone, thus producing dramatic physiological effects. (24)

Integrative Questions

1. Choice e. All of these animals are ingestive heterotrophs and predators of some type. All vertebrates have complete digestive tracts and thus have a stomach or stomachs and a small intestine. Vertebrates also use autocatalysis to activate digestive enzymes.

2. Choice *e*. Defecation is the normal process of expelling feces from the body.

3. Choice *a*. Insulin is produced in the pancreas.

4. Choice *e*. All animals with a complete digestive tract also have a mouth and an anus, and all four choices have complete digestive tracts. In addition, all animals utilize extracellular digestion.

CHAPTER 48 – SALT AND WATER BALANCE AND NITROGEN EXCRETION

Note: the page numbers listed with the section titles refer to the study guide; numbers in parentheses for each question refer to the relevant key concepts.

EXTRACELLULAR FLUIDS AND WATER BALANCE • DISTINGUISHING ENVIRONMENTS AND ANIMALS IN TERMS OF SALT AND WATER • EXCRETING NITROGEN AS AMMONIA, URIC ACID, OR UREA (pages 367–369)

1. Choice *c*. Water can only move osmotically, it is never actively transported. (1–4)

2. Choices *c, d*. Birds excrete uric acid, and marine species with high salt diets also have nasal salt glands. (14, 16–22)

3. Choices *a, e, f*. Freshwater fishes take on water from their environment and lose salt. Therefore, they produce lots of dilute urine, absorb salts, and excrete ammonia. (8, 16–22)

4. Choice *a*. The osmotic potential of the body fluid of a hypertonic osmoregulator would be more negative than its environment; thus, it would probably be living in fresh water (10, 11)

5. Choice *c*. Only proteins are converted directly into ammonia; nucleic acids are converted first into purines and pyrimidines, then into uric acid and urea. (15–16)

• (5–11)

• *Artemia* is an ionic regulator. Since it lives in a very saline environment it will need to excrete NaCl, but absorb other necessary ions. (12, 13)

THE DIVERSE EXCRETORY SYSTEMS OF INVERTEBRATES • THE NEPHRON: THE BASIC UNIT OF VERTEBRATE EXCRETION (pages 369–371)

1. (2–10)
   Flatworms: *b, d*
   Annelid worms: *c*
   Insects: *e*
   Vertebrates: *a*

2. Choice *e*. In the metanephridium of earthworms, coelomic fluid enters the nephrostome of the excretory organ directly. In the vertebrate kidney, filtration of blood in the glomerulus of the nephron forms the tubular fluid. (4–6, 10–16)

3. Choice *b*. The Malpighian tubules produce a highly concentrated, solid mass of *uric acid*, not a urea solution. (7–9)

4. Choice *c*. Permeability of the glomerulus is greater than that of the peritubular capillaries. (10–16)

5. (17–21)
   Marine bony fishes: *b, c*
   Cartilaginous fishes: *d, h*
   Most amphibians: *b, e*
   Reptiles: *a, f*
   Birds: *a, f, g*

• (10–16)

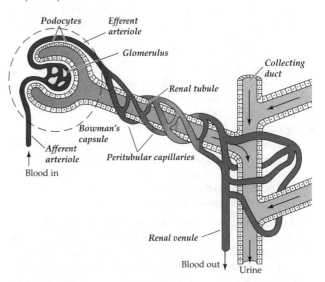

THE MAMMALIAN KIDNEY • TREATING KIDNEY FAILURE • CONTROL AND REGULATION OF KIDNEY FUNCTIONS (pages 371–373)

1. (1)
   2: Collecting duct
   5: Urethra
   4: Bladder
   1: Glomerulus
   3: Ureter

2. See Figure 48.11 on page 1078 in the textbook. (10–19)
   a. A
   b. G
   c. C
   d. A, B, E
   e. F

3. (2–8)

C: Peritubular capillaries

C: Bowman's capsule

C: Proximal convoluted tubule

C: Distal convoluted tubule

M: Renal pyramids

M: Vasa recta

M: Loop of Henle

4. Choice *c*. About 10 to 12 percent of the blood entering the renal arteries is filtered. Of that, only about 1 to 2 percent is not reabsorbed. So 1 liter of urine would represent about 100 liters of filtrate and 100 liters of filtrate would be derived from about 1,000 liters of blood. (9)

5. Choice *b*. The length of the loop of Henle determines the size of the NaCl gradient set up by the countercurrent multiplier system in the kidney. The greater the gradient, the greater the water reabsorption from the urine in the collecting ducts and the more concentrated the resulting urine. (7, 17, 18, 28)

6. (10–18)

Proximal convoluted tubule: *a, b, c, d*

Distal convoluted tubule: *d, e*

Ascending limb: *d, e*

Descending limb: *f*

Collecting duct: *f*

7. Choices *d, e*. Higher blood pressure would lead to increased production of urine. Reduction in antidiuretic hormone release by the posterior pituitary would result in less reabsorption of water from the collecting ducts and help to reduce blood osmolarity by producing a more dilute urine. (21–25)

8. Choice *d*. Since sodium is an important constituent of blood, its concentration in the dialyzing fluid would be fairly great. (20)

9. (20–27)

+ Insertion of aquaporin-2 into cells of collecting ducts

+ More stretch receptor activity in major arteries

− More aldosterone released by adrenal cortex

+ Constriction of efferent renal arterioles

+ Decreased permeability of collecting ducts

• Aquaporins are membrane proteins that form water channels and increase a tissues' water permeability. Aquaporin-2 is specific to the collecting ducts. There is a long-term effect of ADH levels on the expression of the gene for aquaporin-2. In the short term, ADH also influences the insertion of aquaporin-2 into the cell membrane to form the water channels. (25, 26)

• (4–19)

Blood plasma (300 mosm)

Urine

CHAPTER 49 – ANIMAL BEHAVIOR

Note: the page numbers listed with the section titles refer to the study guide; numbers in parentheses for each question refer to the relevant key concepts.

GENETICALLY DETERMINED BEHAVIOR • HORMONES AND BEHAVIOR (pages 374–376)

1. Choice *c*. Bird song in many species is an example of a highly stereotypic, species-specific behavior that is acquired through learning and is not inherited, as are fixed action patterns. (3–5, 12–14)

2. Choice *d*. The chaffinch can only learn its species-typical song during the critical period. (12–14)

3. Choice *e*. If castrated at birth, no androgens will be present to masculinize the rat's nervous system. It will therefore not respond to treatment with male sex steroids later in life. (15–17)

4. Choice *d*. Bill shape and a strong contrast between the color of the bill spot and the bill seem most important. (6–8)

5. Choice *c*. Although prey capture behavior is likely to involve fixed action patterns, search behavior is more flexible and influenced by the predator's past experience. (3–5, 7, 11–14)

6. Choice *c*. Deprivation experiments help to reveal which behaviors are fixed action patterns. Since supernormal releasers are exaggerations of the normal releasers of a fixed action pattern, they would be highly effective in a deprivation experiment. (6–9)

7. Choice *c*. Even though they have previously learned and used their song, male birds still require elevated testosterone levels to stimulate singing behavior. Thus, males normally only sing in the spring when their gonads are producing lots of male sex steroids. (18–20)

• In many birds, males learn by imprinting a song template for their species upon hearing their father sing as nestlings. During their first breeding season, these males are stimulated by increased testosterone levels to practice this song. The song becomes crystallized when their rendition matches the song template they learned. Thereafter, the song is always rendered the same way. (18–20)

• Cocoon construction in many spiders is a series of behaviors that, once begun, is always performed from beginning to end. Cocoon construction is an inherited behavior that does not require learning, is species-specific, and sterotypic. These are all characteristic of fixed action patterns. (3–5)

• (15–17)
   a. The female will show neither lordosis nor the mounting behavior typical of males. As a result of the injection of testosterone after spaying, the nervous system of the female has been masculinized and now she can not respond to female sex steroids.

   b. The female will show the mounting behavior typical of males, but not lordosis. As a result of the injection of testosterone after spaying, the nervous system of the female has been masculinized and she can now respond only to male sex steroids.

THE GENETICS OF BEHAVIOR • COMMUNICATION
(pages 376–378)

1. Choice *b*. Even here the behavior is much too complex to be caused directly by a gene's presence; many intervening steps affecting the sensory and effector systems of the bee are involved in producing this behavior. (1–5)

2. Choice *d*. Egg-laying in *Aplysia* is one of the few cases where we know the effect of a specific gene product on the behavior of an organism. (6)

3. Choice *d*. The food source shown is 90° to the right of the hive–sun direction, so the waggle dance should be 90° to the right of the vertical axis.
Distance between the hive and the food source determines the type of dance. A round dance is given if the dance is about 80 meters or less; a waggle dance, if more than 80 meters.
A random search is initiated by a round dance; a directed search is initiated by a waggle dance. (16–18)

4. Choice *d*. The auditory mode is energy intensive, odors are slow to diffuse, the environment is too dark for visual communication, and the electrical mode requires an aqueous environment. (7–20)

5. Choice *d*. Due to the close proximity of sender and receiver, chemicals are frequently exchanged during tactile communication. (7–20)

6. Choice *c*. A displacement behavior is a seemingly irrelevant display resulting from conflict between different motives. (20–23)

• Since mating calls are designed to advertise the location of the caller, they would be expected to be pure tones with abrupt beginnings and endings. Alarm calls are designed to alert other species individuals to the presence of a predator without disclosing the location of the caller. They would be expected to be complex sounds with gradual onset and completion. (15)

• Since sex pheromones are used to advertise the location of the sender, they should disperse rapidly and dissipate quickly. Sex pheromones are usually small molecules with large diffusion coefficients. A trail- or territory-marking pheromone should last a long time and spread slowly. They tend to be larger molecules with small diffusion coefficients. (11, 12)

THE TIMING OF BEHAVIOR: BIOLOGICAL RHYTHMS
• FINDING THEIR WAY • HUMAN BEHAVIOR
(pages 378–381)

1. Choice *d*. The period or length of the cycle of this circadian rhythm is 22 hours and it would need to be phase-advanced to become entrained to a 24-hour cycle. (1–5)

2. Choice *a*. The bird's sense of time tells it that the sun should be due south at noon and that the feeding dish is 90° to the left (east). Because the artificial sun is stationary at the east, it will go to the north bin. (7, 24–26)

3. Choice *c*. Normally at noon, the east food bin would be 90° to the left. The bird will go to the food bin 90° to the left of the sun's displaced position. (7, 24–26)

4. Choice *b*. At sunset the sun is in the west and the bird would normally go to the food bin 180° to the left. However, since the bird has been phase-delayed by 6 hours, it thinks it is noon, so it searches 90° to its left to find the east food bin, but instead goes to the south food bin. (7, 24–26)

5. Choice *c*. Because the location of the North Star is fixed, birds are not confused if the sky in the planetarium does not rotate or rotates at an incorrect rate. (27–29)

6. Choice *b*. Since hibernators, animals living underground, and equatorial migrants cannot use photoperiod to coordinate their year-long activities, many of these species have circannual rhythms. Only the arctic hare does not fit into one of these categories. (10–12)

7. Choice *b*. All other statements are false. In *a*, jet lag is only a problem if your flight is across time zones, as it would be in traveling 1,000 miles west. In *e*, if Earth were not tilted on its axis, seasonality would be greatly reduced. This would affect the occurrence of *circannual*, but not circadian rhythms. (1–9)

8. Choices *a, c*. Removing the suprachiasmatic nuclei of a mammal or the pineal gland of a bird would eliminate any circadian rhythms. Blinding a mammal (or covering the pineal gland of a bird) or placing mammals or birds in complete darkness only prevents entrainment to a true 24-hour cycle; it does not destroy the endogenous circadian rhythm. (8, 9)

9. Choice *a*. Piloting, orientation using familiar landmarks, is the most basic form of homing. (13–21)

10. Choice *a*. Since the sun rises later than usual, you must have been displaced to the west. Since the sun is higher in the sky than your home location you must have been displaced to the south. Therefore, you must travel to the east and north to get home. (7, 8, 20–22)

11. (14–16, 18–20)
    Homing only: *A*
    Migration by piloting only: *E* (assuming the birds are following the coastline during migration as in the gray whale migration mentioned in the textbook)
    Migration by distance and direction navigation only: *C*
    Migration by bicoordinate navigation only: *B*

50

Behavioral Ecology

CHAPTER LEARNING OBJECTIVES—after studying this chapter you should be able to:

❑ Define "ecology" in terms of the interactions between organisms and the environment.

❑ Define habitat, and explain the choices that organisms make in the process of habitat selection.

❑ Describe the major predictions that foraging theory makes about predator–prey interactions and provide an example.

❑ Explain the assumption underlying the approach to constructing foraging theories called optimality modeling.

❑ Relate the cost–benefit analysis of behavior to the concept of fitness.

❑ Differentiate between the following terms: "energetic cost," "risk cost," "opportunity cost."

❑ Provide several examples showing how social behavior in a species can vary with changes in the environment.

❑ Characterize differences in male and female mating tactics, and relate these differences to their investment in reproduction.

❑ Define "territoriality," describe several different territory types, and provide some examples of each.

❑ Describe the characteristics of a lek, and provide an example.

❑ Explain what is meant by sexual selection, describe the conditions under which it might operate, and provide some examples.

❑ Describe the conditions that promote the formation of a social group and discuss the benefits and costs of group living.

❑ Define, differentiate between, and provide an example of the following types of behavioral acts: "altruistic," "selfish," "cooperative," and "spiteful."

❑ Define the conditions under which altruistic behavior would evolve, and relate altruistic behavior to the concept of inclusive fitness.

❑ Describe the coefficient of relatedness (r) and be able to calculate r for specified conditions.

❑ Describe the model of reciprocal altruism, and explain its importance.

❑ Explain what a natal group is, and describe the importance of this concept in explaining the evolution of social systems in birds and mammals.

❑ Discuss the social system of the scrub jay, and relate it to concepts of altruism, helping, and natal groups.

❑ Explain the term "eusocial," and discuss the strengths and weaknesses of W. D. Hamilton's haplo-diploid hypothesis for the evolution of eusociality.

❑ Describe some differences in parental investment by males and females of various vertebrate groups including any notable exceptions.

❑ Discuss the varied social systems in African hoofed mammals, African weaverbirds, and primates and explain how these systems are related to the environment in which the species lives.

❑ Provide an example of the importance of phylogenetic history in determining the type of society that a species has.

CHOOSING ENVIRONMENTS • COSTS AND BENEFITS OF BEHAVIORS • MATING TACTICS AND ROLES
(pages 1113–1120)

Key Concepts

1. The science of *ecology*, an outgrowth of natural history, is concerned with the interactions among different organisms and the interaction of organisms with their environment.

2. An organism's *environment* includes both physical factors, such as temperature, light intensity, and nutrient availability, and biological factors, such as interactions with members of the same species and other species.

3. Organisms are affected by their environment but their activities can also modify the environment.

4. An organism's *habitat* is the area in the environment where it normally lives.

5. The cues that organisms use in selecting a habitat are usually good predictors of habitat quality. For example, red abalone larvae will only settle where a specific peptide produced by coralline algae is present.

6. *Foraging theories* were developed in order to predict how predators will decide where to forage and what type of prey to take. The first step in constructing a foraging theory is to specify the objective of the forager's behavior, which may be to maximize the rate of energy return, to get specific nutrients or minerals, to minimize its risk while foraging, etc.

7. *Optimality modeling* assumes that through natural selection animals evolve to make the best choices among those that are available to them.

8. Time costs associated with predation include searching for, capturing, handling, and digesting prey. These costs must be included in determining the value of prey of different energy content for predators.

9. A top-ranked prey item yields the greatest energy content per predator time invested. If prey differing in energy value are available to a predator, foraging theory predicts that the predator will always take the top-ranked prey item and will add the prey with the next highest value to its diet only when the density of top-ranked prey is reduced below a certain point. Experiments with *bluegill sunfish* preying on water fleas support these predictions.

10. A *cost–benefit analysis* of behavior assumes that an organism has limited time, energy, and materials available to it.

11. The increased energy required to perform a behavior over that expended in resting is the *energetic cost* of that behavior.

12. The increased risk of injury or death of a behavior over that associated with resting is the *risk cost* of that behavior.

13. When an animal performs a certain behavior, it forfeits the benefits of other behaviors that it could not perform. The forfeited benefits are the *opportunity costs* of the behavior performed.

14. If the costs associated with a behavior exceed the benefits in terms of an animal's survival and reproductive success, the individual is not expected to perform the behavior.

15. Behavioral costs and benefits can be measured indirectly by examining how behavior changes with variation in the environment.

16. Within social systems, individuals must choose with whom to associate, with whom to interact, with whom to mate, and when to leave the group. Mate choice can depend on an individual's phenotypic traits, the resources it controls, or a combination of the two.

17. Because females generally make a greater energy commitment to producing and raising young than males do, they benefit by discriminating in choice of mates; males usually benefit from maximizing the number of females they inseminate and are often less discriminating in choice of mate.

18. The evolution of sex-specific traits that enhance an individual's reproductive success is called *sexual selection*. Mating preferences by individuals of the opposite sex often determine the direction of sexual selection.

19. Usually males evolve traits through sexual selection that improve their competitiveness for females or resources that attract females.

20. That sexual selection was responsible for the remarkably long tails of widowbirds was demonstrated by artificially lengthening and shortening the tails.

21. Male courtship behavior provides information to the female as to the health of the male, his ability to provide for offspring, and the quality of his genes. Some male behavior is designed to insure that it is his sperm that fertilizes the eggs.

22. *Territoriality* is aggressive behavior in the defense of space. The space defended is called a *territory*.

23. Many songbirds have territories that include everything needed by the pair for reproductive success: a mating area, nest site, and an adequate food supply.

24. In some species, such as the red-winged blackbird, a typical territory includes breeding and nesting areas, but not an adequate food supply to rear young. The birds must seek some food outside the territory.

25. Marine colonial birds such as gulls and penguins have territories that include only a nest site and a small area around it.

26. In some species, territories are used only for courtship and mating. Species with this type of territory frequently have a mating system in which one male mates with several females. Females raise the young elsewhere with no help from the father.

27. An area containing many small territories used only for displaying and mating is called a *lek*. Males fight for central territories because females prefer to mate with males in these places.

28. Females have evolved behaviors that focus on the male reproductive signals that are reliable indicators of their fitness. For example, female barn swallows prefer males with long tails. Tail length is a reliable indicator of parasite load in males.

29. Recently, DNA fingerprinting methods have shown that extra-pair copulations are common in some species previously thought to be monogamous.

Questions (for answers and explanations, see page 435)

1. Which of the following is *not* usually included within the domain of ecology?
   a. Interactions between conspecifics
   b. Modification of the environment by organisms
   c. Factors determining the habitat of a species
   d. Modification of the environment by physical processes
   e. Evolution of different social organizations

2. Which of the concepts listed below best describes the following situation: Birds spend some of their time scanning the horizon for predators. While scanning, they cannot be foraging for food.
   a. Foraging theory
   b. Energetic cost
   c. Risk cost
   d. Opportunity cost
   e. Cooperative hunting

3. Which of the following animals would *likely* have a territory that is used exclusively for a nest site?
   a. The song sparrow
   b. The red-winged blackbird
   c. The fruit-eating birds called manakins
   d. A colonial marine bird
   e. A species in which mating is restricted to a lek

4. You perform a series of experiments in which the density of prey species 1 is varied while the density of prey species

2 is kept constant. At each prey species 1 density, you determine how many prey of species 1 and 2 the predator takes. The two curves (*a* and *b*) plotted in the following graph show the outcome of this study.

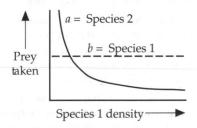

Select *all* of the following statements that correctly interpret these curves.
a. Curve *a* indicates that species 2 is not the preferred prey.
b. Curve *a* indicates that species 2 is the preferred prey.
c. Curve *b* indicates that species 1 is the preferred prey.
d. Curve *b* indicates that species 2 is not the preferred prey.
e. Neither of these curves provides insight to the prey preference of the predator.

5. Which of the following statements is *not* characteristic of the elephant seal social system?
a. The energetic costs of a dominant male are low.
b. The social system is based on polygyny.
c. Most matings involve only a few males.
d. The males show many traits that resulted from sexual selection.
e. Most females are able to mate.

6. Which of the following is *not* a true statement describing the study on nest singing in female red-winged blackbirds described in the textbook?
a. Singing by females attracts predators to the nest.
b. Nest defense by males was greater if the female regularly sang at the nest.
c. The most successful females did not sing, but depended on the male to defend their nest.
d. A model of a crow was used to solicit nest defense by males.
e. An experimental approach can be useful in cost–benefit analysis of behaviors

Activities (for answers and explanations, see page 435)

• Describe the experiments that were done with bluegill sunfish preying on waterfleas and the implications of these studies for foraging theory.

• Describe what is meant by sexual selection and how the sexual selection of tail length in male widowbirds was studied.

SOCIAL BEHAVIOR: COSTS AND BENEFITS • CATEGORIES OF SOCIAL ACTS • THE EVOLUTION OF ANIMAL SOCIETIES (pages 1120–1129)

Key Concepts

1. Social behavior evolves when cooperation among individuals of the same species improves their fitness. Such organized collections of individuals are called *social groups*.

2. Individuals can sometimes obtain nutrients more efficiently when in a group than when alone; *cooperative hunting* is an example of this benefit of group living, as seen in many social carnivores.

3. Studies showing that predators are less able to capture individuals in a group than isolated individuals suggest that protection from predation is a benefit of group living.

4. Increased spread of disease and parasites is a cost of group living. Other costs of group living include increased competition for food and mates, and increased chance of injury from conspecifics.

5. Costs of group living are sometimes related to the size of the individual. For example, in the solitary bee *Centris pallida*, small male bees avoid competition with larger individuals by patrolling for undiscovered females rather than digging them up.

6. An *altruistic act* (behavior) benefits other social group members, but at a cost to the performer.

7. A *selfish act* benefits the performer, but at a cost to other social group members.

8. A *cooperative act* benefits both the performer and other social group members.

9. A *spiteful act* has a cost to both the performer and other social group members.

10. Selfish or cooperative behaviors evolve if the behavior improves the fitness of the performer. That is, its genetic contribution is to the next generation.

11. Altruistic behavior evolves if the behavior improves the *inclusive fitness* of the performer. Inclusive fitness is the sum of individual fitness and kin selection (see Chapter 20).

12. Altruistic behavior is beneficial when the improvement in the reproductive success of relatives exceeds the reduced direct reproductive success of the individual performing the act.

13. The coefficient of relatedness (r) is the probability that two alleles in different individuals are related by descent; $r = \sum(0.5)^k$, where k is the number of different generation links separating the two individuals. The value in parentheses is 0.5 because at each generation link there is a 0.5 probability that the allele will be passed on to an offspring due to meiosis.

14. Species with social systems in which there are classes of individuals that do not breed are called *eusocial*. Most social insects (termites, ants, and some bees and wasps) and the naked mole rat are eusocial.

15. W. D. Hamilton developed the haplo-diploid hypothesis for the evolution of eusociality. In a species in which males are haploid and females diploid, sister workers are identical in terms of paternal genes and share 50% of the maternal genes. The coefficient of relatedness (r) is 0.75 for sisters, but only 0.5 for any offspring a female produced on its own. Thus, a haplo-diploid condition favors a system in which females forgo their own reproduction and help to raise additional sisters.

16. Although there is strong support for the haplo-diploid hypothesis, important exceptions remain. The hypothesis cannot explain the existence of eusocial species with either multiple queens per colony or in which queens mate with more than one male. Also, many eusocial species exist in which both sexes are diploid, as in all termite species and the African naked mole rat.

17. The naked mole rat is the only known eusocial mammal. Its colonies consist of 70 to 80 individuals living in underground tunnel systems constructed by sterile workers. Each colony has a single queen and several kings.

18. Individuals of some species, such as the *white-fronted bee-eater*, are able to determine their genetic relatedness to social group members and preferentially help their close relatives.

19. Most eusocial species are colonial and utilize elaborate nest or burrow systems. The colony-founding success rate is usually low. These conditions favor species in which many individuals defer reproduction and serve as helpers to promote colony success.

20. The model of *reciprocal altruism* states that altruistic behavior can occur between unrelated individuals if the individual performing the altruistic act later benefits from a reciprocal act.

21. To reconstruct the events in the evolution of social systems, we must study the stages of complexity shown among living social species.

22. Parental care increases the chances of an offspring's survival but it may simultaneously limit the parent's ability to have additional offspring and may lower its own survival chances.

23. Females, by definition, contribute larger gametes and usually have a greater investment in care of the young than males.

24. Biparental investment is more prevalent in birds than in mammals because birds do not provide milk as food for their young, an ability limited to female mammals.

25. Parental investment in fishes, if present, consists of guarding eggs and young; males are the primary guarders in most species.

26. Communal breeding systems probably evolved from extended families with overlapping generations.

27. In the *Florida scrub jay*, breeding pairs with helpers (previous offspring) fledge more young than pairs without helpers. Most helpers are offspring from an earlier breeding season. Male helpers may be able to assume control of the territory if their father dies.

28. Most mammal societies that evolved via the extended family route, range from solitary females or female–male pairs to *natal groups* consisting of a breeding pair and offspring, especially females. Most mammalian helpers are female.

29. Social systems are best understood in terms of the costs and benefits accrued by the participants that associate within them.

30. Within a social system, associations constantly change and often depend directly on genetic relatedness.

31. The evolution of a species' social system depends on diet, body size, and the environment in which it lives.

32. The social organization of a species is strongly related to the features of its environment. For example, societal types shown by African weaverbird species vary with the availability of nest sites: monogamous, isolated pairs characterize species living in forests with many available nest sites; polygyny and colonial nesting characterize savanna-dwelling species with limited nest sites.

33. Social organization in African hoofed mammals is correlated with body size: small species with high metabolic rates live solitarily in forests where they eat scattered, high-quality food and hide effectively from predators; large species with lower metabolic rates live in grasslands where they feed on the abundant but lower-quality vegetation and form large, complex social groups (herds) that constantly scan for predators.

34. The smallest primate troops are found in forest-dwelling species and the largest in ground-dwelling species. Females tend to remain with the troop, young males are driven away, and both males and females within the troop may form dominance hierarchies.

35. Phylogenetic history plays a part in determining the type of society that a species has. Studies of lekking in about 40 species of manakins show that once a particular lek type evolved in a lineage, it was maintained among descendant lineages regardless of habitat type.

Questions (for answers and explanations, see page 435)

1. Some birds give a species-typical vocalization called an "alarm call" when they see a predator, although this call may direct the predator toward them. Other members of their species respond to these calls by taking cover. This would be an example of altruistic behavior that is beneficial to the calling bird if
   a. the bird giving the vocalization survives the attack.
   b. all birds survive the attack.
   c. the birds benefiting are members of the same social group.

d. the birds benefiting are offspring of the bird giving the vocalization.

e. the inclusive fitness of the bird giving the vocalization is increased.

2. A dominant individual displaces a subordinate individual from food that the subordinate has discovered. This is an example of a (an)

a. selfish act.

b. cooperative act.

c. altruistic act.

d. spiteful act.

e. nonsocial act.

3. Which of the following choices correctly calculates the coefficient of relatedness to show the probability that an allele in a child is a direct copy of one present in her grandfather?

a. $r = (0.5)^1$

b. $r = (0.5)^2$

c. $r = (0.5)^2 + (0.5)^2$

d. $r = 2(0.5)^4$

e. None of the above

4. Select all of the following conditions that must be *true* for a behavioral act to be considered an example of reciprocal altruism.

a. The individuals involved must be closely related.

b. Both individuals must potentially benefit from the act at some time.

c. The act must have a genetic basis.

d. The act must have evolved by sexual selection.

e. The individuals involved must be unrelated.

5 Select *all* of the characteristics listed below that are shared by most eusocial species.

a. Haplo-diploid sex determination

b. Elaborate nests or burrows

c. Presence of sterile classes

d. Colonies founded by a single female

e. Queens that mate with a single male

6. For each of the two y-axes in the following graph, draw a labeled curve that correctly summarizes observations made on goshawks attacking wood pigeons, as described in the textbook.

7. Which of the following statements describing the social systems in Florida scrub jays is *not* true?

a. It is a eusocial organization.

b. Individuals in the social group are genetically related.

c. The social system consists of natal groups.

d. Only the parent birds breed.

e. Males may benefit from helping more than females benefit.

8. Of the following factors, select the one that *was not* mentioned as being important in the evolution of social organization in African hoofed mammals.

a. Diet

b. Dispersal of food resources

c. Dispersal of breeding territories

d. Metabolic needs

e. Predation

9. Select all of the following factors that can influence the type of social organization typical of a species.

a. Presence of a dominance hierarchy

b. Availability of nest sites

c. Evolutionary history

d. Haplo-diploidy

e. Asymmetry in reproductive investment

Activities (for answers and explanations, see page 436)

• Imagine that you are in a lifeboat after a disaster at sea. Your half-sibling and full cousin remain in the water. In order to maximize your inclusive fitness, which relative should you invite to take the last remaining space in the lifeboat? Explain your answer in a quantitative manner.

• Describe W. D. Hamilton's haplo-diploid hypothesis for the evolution of eusociality. What are some major phenomena not explained by this hypothesis?

• Describe important environmental variables that seem to have guided the evolution of social organization in species of African weaverbirds.

***Integrative Questions (for answers and explanations,
    see page 436)***

1. Select all of the following concepts that apply to elephant seal social organization.
   a. Sexual selection
   b. Eusociality
   c. Polygyny
   d. Type territory not including a food supply
   e. Helpers
   f. Lekking

2. Match the characteristics on the right with the species on the left that show those characteristics. (A species may be matched with more than one characteristic.)

   _____Red-winged blackbirds   *a.* Sterile workers

   _____Manakins   *b.* Multi-purpose territory

   _____Honeybees   *c.* Helpers

   _____Naked mole rats   *d.* Cooperative hunting

   _____Scrub jays   *e.* Lekking

   _____Song sparrows   *f.* Eusociality

   _____African lions   *g.* Territory accommo-
                                     dating several females

   _____Bee-eaters

51

Population Ecology

CHAPTER LEARNING OBJECTIVES—after studying this chapter you should be able to:

❑ Describe the subject matter of population ecology.

❑ List the typical life history stages of organisms.

❑ Describe some important life history trade-offs associated with reproduction.

❑ Describe the concept of reproductive value, and relate it to attempts to extend human life expectancy.

❑ Explain what determines the structure of a population.

❑ Differentiate between unitary and modular organisms.

❑ Define the term "population density," and describe how to estimate it for unitary and modular organisms.

❑ Estimate population density based on the mark and recapture method.

❑ Describe three common spacing patterns within populations and the types of conditions that produce each pattern.

❑ Infer relative birth and death rates from graphs showing the age distribution of a population.

❑ Describe the demographic events that determine population size, and express these ideas as an equation.

❑ Explain what a life table is, and estimate survivorship and death for different age classes from a life table.

❑ Describe the three types of survivorship curves and the types of populations that each characterizes, and provide an example of each.

❑ Describe some differences between how population ecologists study modular and unitary organisms.

❑ Define exponential population growth, recognize it graphically, and express it as an equation.

❑ Discuss the role of the intrinsic rate of increase (r_{max}) in determining the magnitude of exponential growth.

❑ Define logistic population growth, recognize it graphically, and express it as an equation.

❑ Describe population growth when the population is very small, when it is medium-sized, and when it has reached its carrying capacity.

❑ Describe the major density-dependent factors that limit population growth.

❑ Describe the major density-independent factors that limit population growth.

❑ Explain typical variation in the density of a species within its range.

❑ Characterize common population disturbances and the adaptations that organisms have evolved for dealing with disturbances.

❑ Describe a common practice for maximizing the number of individuals that can be harvested from a managed population and the problems associated with overharvesting.

❑ Explain some strategies for managing populations of undesirable species.

❑ Characterize historical trends in human population growth and what factors may limit Earth's carrying capacity for humans.

VARIED LIFE HISTORIES • POPULATION STRUCTURE: PATTERNS IN SPACE   (pages 1132–1137)

Key Concepts

1. A *population* includes all of the individuals of the same species in a certain area.

2. *Population ecology* is the study of population structure and growth and the factors that regulate population growth.

3. The *life history* of a species consists of the stages that a typical individual goes through from birth to death. Life histo-

ries vary greatly and different stages may be specialized for different activities.

4. The life history stages seen in organisms may include a growth stage, a metamorphosis stage, a dispersal stage, a reproductive stage, a postreproductive stage, and a senescence stage.

5. The amount of time an individual spends in each life history stage and how much the stages overlap varies among species. Life history stages involve trade-offs and constraints.

6. Although allocating more food to offspring increases their chances of surviving, it reduces the number of offspring that the parent can produce.

7. Thus, in organisms that provide parental care, there is a trade-off between the number of young produced and the amount of care each can receive.

8. When and how often a species reproduces varies, as does the number of offspring produced each time (known as the *clutch* or *litter size* in animals or *seed crop* in plants).

9. Some organisms reproduce only once in their lifetime; examples include agaves, Pacific salmon, and annual plants.

10. Engaging in reproduction reduces the potential for growth and usually reduces the probability of parental survival.

11. Natural selection favors early reproduction except in cases where it would result in poor juvenile or parental survival; in these cases delaying reproduction may be advantageous.

12. *Reproductive value* is the average number of offspring remaining to be born to individuals of a specific age; reproductive value increases as an individual approaches the age of reproduction, is highest when it first starts to reproduce, and decreases to zero as it becomes postreproductive.

13. Because older individuals have less reproductive value than younger individuals, genes that are expressed only late in life are less affected by natural selection than genes that are expressed earlier.

14. Alleles that delay the phenotypic effects of the deleterious alleles of other genes are favored by natural selection. This leads to a buildup of genetically based problems expressed after reproduction is complete.

15. *Senescence* is the increased probability of dying with age.

16. *Population structure* from an ecological viewpoint consists of the numbers, spacing, and age of population members.

17. *Population density* is the number of population members per unit area, for most terrestrial species, or per unit volume, for aquatic species. Population density can be determined by direct counting or by the mark and recapture method.

18. *Biomass*, the total mass of organisms, is best for expressing population density for species whose individuals differ greatly in size, such as plants, fish, mollusks, and so on.

19. Population members may show *uniform spacing, clumped spacing,* or *random spacing* within the population range.

20. Uniform spacing can result if population members are competing for a uniformly distributed resource, such as light, or if they defend territories.

21. Clumped spacing characterizes populations whose members tend to settle close to their birthplaces, or that live in patchy environments.

22. Random spacing can result if numerous factors interact to determine the locations of individuals.

23. The *age distribution* of a population, the proportion of population members in each age category, is determined by the timing of births and deaths.

24. If both birth and death rates are high, the population will be dominated by young individuals; if both birth and death rates are low, the population age structure will be more uniform.

25. The age structure of a population determines how fast it will grow and utilize resources.

Questions (for answers and explanations, see page 436)

1. From an *ecologist's* point of view, population structure does *not* include the
   a. distribution of genotypes within a population.
   b. population density.
   c. spacing of population members.
   d. age structure of the population.
   e. biomass of the population.

2. The number of seeds that a tree produces increases, but the resources available to the tree remain the same. Put a "+," "−," or "0" before each of the following variables, to express the expected effect of this change on the variable.

   _____ Weight of each seed

   _____ Width of the tree's growth rings

3. A random distribution of population members is usually caused by
   a. settling near one's birthplace.
   b. territoriality.
   c. competition for a uniformly distributed resource.
   d. the interaction of several factors that affect survival.
   e. patchy habitat.

4. In which of the following life history stages are annual plants and salmon *most* similar?
   a. Growth
   b. Dispersal
   c. Reproduction
   d. Senescence
   e. Metamorphosis

5. For a species that breeds only during one season and lives in a disturbed environment where new habitat is constantly being created, it would be best to
   a. produce many, small offspring only once.
   b. produce a few, large offspring only once.
   c. produce many, small offspring more than once.
   d. produce many, large offspring more than once.
   e. produce a few, small offspring only once.

6. You collect a sample of 100 bats that are hibernating in a cave. Each bat is marked and released. Two weeks later you return and collect another sample of 100 bats, 20 of whom are marked. The estimated number of bats in the cave is _____.

Activities (for answers and explanations, see page 436)

- Discuss the three life history trade-offs presented in the textbook.

- Discuss the implications that the concept called "reproductive value" has for hopes of greatly increasing human life expectancy.

POPULATION DYNAMICS: CHANGES OVER TIME • EXPONENTIAL POPULATION GROWTH • POPULATION GROWTH IN LIMITED ENVIRONMENTS
(pages 1137–1142)

Key Concepts

1. The number of population members depends on the *demographic events* of *birth, death, immigration,* and *emigration*. The current population size is the population size in the past, plus individuals added by birth and immigration, minus individuals removed by death and emigration, or
$$N_1 = N_0 + B - D + I - E$$
where N_1 = population size at time 1, N_0 = population size at time 0, B = number that are born, D = number that die, I = number that immigrate, and E = number that emigrate.

2. A group of individuals all born at the same time is a *cohort*.

3. A *life table* shows changes in the number of cohort members alive at different times. From these data, one can calculate the survivorship and death rates at different times.

4. A *survivorship curve* plots the number of survivors of a cohort over time.

5. In a *type I* survivorship curve, nearly all cohort members die at the same time, as in the *Poa annua* example in the text.

6. In a *type II survivorship* curve, survivorship is uniform during life; this curve is typical of most bird populations.

7. In a *type III* survivorship curve, most young die, but survivorship is high for the remaining individuals; this curve is typical for organisms producing large numbers of young with low parental investment.

8. Many organisms have survivorship curves that combine different aspects of curves I, II, or III, and survivorship may differ for males and females of the same species, as in the red deer.

9. Species in which adults are similar in shape and size are called *unitary organisms*. Most mollusks, echinoderms, insects, and vertebrates consist of unitary organisms.

10. Species in which adults differ markedly in shape and size because they consist of modules that are added as the organism grows are called *modular organisms*. Modular organisms include most plants, some protists, fungi, and most colonial animals.

11. The number, size, and shape of the modules of modular organisms are more important to population ecologists than the number of individuals. Differential growth of modules allows an attached modular organism to spread and to compete with neighboring individuals.

12. All populations have the potential to grow *exponentially*, because as a population increases in size, more and more individuals are alive to reproduce.

13. Exponential population growth can be expressed mathematically as
$$\frac{\Delta N}{\Delta t} = (b - d) N$$
where $\Delta N/\Delta t$ is the rate of change in the population size (ΔN = change in number of individuals; Δt = change in time), b is the birth rate, d is the death rate, and N is the population size.

14. The *intrinsic rate of increase* of a species, r_{max}, is the difference between the average birth and death rates under optimal conditions, so the rate of growth for a species under ideal conditions is $\Delta N/\Delta t = r_{max} N$.

15. Species that reproduce only once or only during a short breeding season have ragged growth curves that are best modelled with a different equation:
$$N_{t+1} = \lambda N_1$$
where λ is the *finite rate of increase* (the ratio of population size during the next time period to the population size during the current time period).

16. As populations grow, the environment limits exponential growth by increasing death rates and decreasing birth rates.

17. The environmental *carrying capacity* is the equilibrium number of individuals sustainable by the resources available to the population in a particular environment.

18. As birth rates decrease and death rates increase in a population, its growth curve departs from the J-shaped exponential curve and becomes S-shaped.

19. *Logistic growth*, typical of populations growing in environments with limited resources, is expressed mathematically as
$$\frac{\Delta N}{\Delta t} = r \left(\frac{K - N}{K} \right) N$$
where K is the carrying capacity, $r = (b - d)$, and other symbols are as in the exponential equation.

20. The expression $(K - N)/K$ becomes zero when the population size (N) reaches K; this causes the population growth rate also to be zero.

21. Because of time delays in the effects of the environment on birth and death rates and in the effects that population members have on resources, populations frequently oscillate around the carrying capacity.

22. *Metapopulations* are groups of subpopulations among which some exchange of individuals occurs. This exchange provides a buffering effect that may prevent extinction of subpopulations experiencing a population decline. This process is called the *rescue effect*.

Questions (for answers and explanations, see page 436)

1. Which of the expressions (*a* through *e*) of the exponential growth equation should be *increased* in order for curve 1 to become more like curve 2 in the following graph?

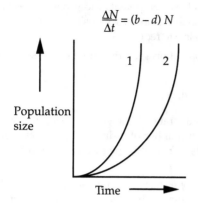

$$\frac{\Delta N}{\Delta t} = (b - d) N$$

a. N
b. d
c. b
d. $(b - d)$
e. $\Delta N/\Delta t$

2. Species A and B have intrinsic rates of increase (r_{max}) of A = 0.25 and B = 0.50. In reference to the graph shown in Question 1, the population growth curve for A should be more like curve _____.

3. In the logistic population growth curve shown below, the rate of growth is greatest at which point? _____

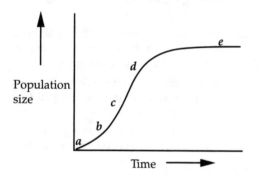

4. At which point in the graph shown above would there be zero population growth ($\Delta N/\Delta t = 0$)? _____

5. Based on the following life table, during what time interval is survivorship greatest?

Age (years)	Number Alive
0	800
1	770
2	550
3	125
4	75
5	0

a. 0–1 years
b. 1–2 years
c. 2–3 years
d. 3–4 years
e. 4–5 years

Activity *(for answers and explanations, see page 436)*

- Relate the concept of a metapopulation to the process known as the rescue effect.

POPULATION REGULATION • HUMANS MANAGE POPULATIONS (pages 1142–1148)

Key Concepts

1. Population regulation is *density-dependent* if per capita birth and death rates vary with the density of the population.

2. A larger population uses up resources faster, attracts more predators, and facilitates the spread of disease; these events will increase the death rate and may lower the birth rate within the population.

3. Population regulation is *density-independent* if birth and death rate are unrelated to population size; weather-related events can act either in a density-independent or density-dependent fashion.

4. The population size at which the birth rate equals the death rate is an equilibrium value toward which the population returns if the birth rate, the death rate, or both are density-dependent.

5. *Disturbances* are short-term events that adversely affect natural populations; disturbances can be physical or biological.

6. Organisms have evolved behavioral and physiological responses to predictable disturbances. Dispersal is a common response to disturbances.

7. *Migration*, a regular movement of individuals of a species between two different areas, is a common response to seasonal changes in habitat quality.

8. Life cycles with resting stages are frequently a strategy for dealing with seasonal changes in habitat quality.

9. For most organisms, population density tends to be higher toward the center of the species range and lower as the periphery is reached.

10. Humans have constantly attempted to regulate the populations of other species, encouraging desirable species and attempting to eradicate undesirable species.

11. A species can be harvested maximally if its population size is kept well below the carrying capacity, because population growth rates are greatest there.

12. Both the age class and the number of individuals harvested can influence sustainable yield greatly; in many fish, removing smaller, prereproductive individuals has less effect than removing the larger, reproductive individuals.

13. Overharvesting occurs when too few reproductive individuals remain after harvesting. Overharvesting is most likely in species with low reproductive rates, as among whales.

14. In managing undesirable species, it is more effective to reduce the carrying capacity by limiting the food supply or introducing additional predators/parasites than to try to directly kill them.

15. The carrying capacity for humans now seems to be determined by Earth's ability to absorb the waste materials of our industrialized societies.

Questions (for answers and explanations, see page 437)

1. The *best* way to decrease the population size of a pest species is to
   a. poison it.
   b. introduce additional predators.
   c. decrease the carrying capacity of the habitat for the species.
   d. add competitors.
   e. sterilize females.

2. Select all of the following that are *not* adaptations to seasonal changes in habitat quality.
   a. Migration
   b. Expansion of the species range
   c. Production of seeds
   d. Hibernation
   e. Food caching

3. Based on the graph shown below, which of the following events is *least likely* to be true of this population?

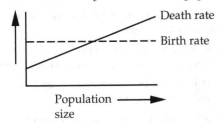

   a. Parasites spread between population members.
   b. Increased competition for food causes some individuals to delay reproduction.
   c. The number of predators in the area varies with population density.
   d. Territorial disputes led to injury and death of some males.
   e. The number of young produced remains stable from year to year.

4. In the graph shown in Question 3, the population size where the curves for the birth and death rates intersect is
   a. the population size with the greatest sustainable yield.
   b. an estimate of the carrying capacity.
   c. an estimate of the intrinsic rate of increase.
   d. the point where density-dependent regulation begins.
   e. the point where density-independent regulation begins.

Activities (for answers and explanations, see page 437)

- Provide an explanation for the observation that population density is usually greatest in the center of the range of a species.

- In the two following figures showing variation in the population density in two species (*A* and *B*), compare and contrast population regulation in these two species. What types of organisms would show these types of regulation?

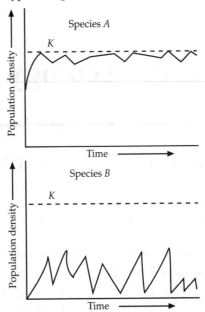

Integrative Questions (for answers and explanations, see page 437)

1. Choose the collection of terms that completes the following sentence: Plants are _____ organisms, they frequently show a _____ spatial distribution, and their population density is most appropriately expressed in terms of _____ .
   a. modular; clumped; biomass
   b. modular; random; individuals per unit area
   c. modular; uniform; biomass
   d. unitary; uniform; biomass
   e. unitary; random; individuals per unit area

2. The following age distribution indicates that

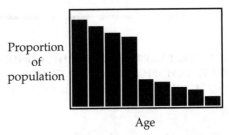

   a. the population's birth rate and death rates are both high.
   b. the population's birth rate and death rates are both low.
   c. the population's birth rate is high but its death rate is low.
   d. the population's birth rate is low but its death rate is high.
   e. the population's survivorship curve is type II.

52

Community Ecology

CHAPTER LEARNING OBJECTIVES—after studying this chapter you should be able to:

- ❑ Define what constitutes an ecological community, and specify the biological questions of concern to community ecologists.
- ❑ Characterize the different types of ecological interactions.
- ❑ Define and relate the terms "resource" and "consumer."
- ❑ Explain what is meant by the term "ecological niche" and how the niche of a species is affected by the presence of other species.
- ❑ Discuss the concept of "limiting resource" and relate it to an organism's niche.
- ❑ Differentiate between intraspecific and interspecific competition and between exploitative and interference competition.
- ❑ Explain how the logistic growth equation can be modified to account for a competing species and what competition coefficients represent.
- ❑ Describe the methods used by G. F. Gause to study competition and explain his principle of competitive exclusion.
- ❑ Discuss the role of environmental heterogeneity in avoiding competitive exclusion.
- ❑ Describe the effects that predators can have on the outcome of competition.
- ❑ Characterize some factors that influence the type of predator–prey interactions that occur.

- ❑ Differentiate between parasites, parasitoids, herbivores, carnivores, and suspension feeders, and discuss the unique features of each type of consumer.
- ❑ Explain the factors that cause predator–prey oscillations or that prevent them from occurring.
- ❑ Discuss the importance of spatial and temporal environmental heterogeneity to predator–prey oscillations.
- ❑ Describe the characteristic features of Batesian and Müllerian mimicry.
- ❑ Discuss some of the important factors that determine the virulence of parasites and the relationships between parasites and vectors.
- ❑ Differentiate between grazers and browsers and discuss common plant defenses to herbivory.
- ❑ Provide some examples of amensalism, commensalism, animal–animal mutualism, and plant–animal mutualism.
- ❑ Describe what is meant by "coevolution" and "diffuse coevolution," and provide some examples.
- ❑ Characterize some important effects that plants have on ecological communities.
- ❑ Define ecological succession and describe the three major types of succession.
- ❑ Characterize some important effects that animals have on ecological communities.
- ❑ Characterize some important effects that microorganisms have on ecological communities.
- ❑ Explain what a keystone species is and provide some examples.

ECOLOGICAL INTERACTIONS • NICHES: THE CONDITIONS UNDER WHICH SPECIES PERSIST • COMPETITION: SEEKING SCARCE RESOURCES (pages 1151–1156)

Key Concepts

1. An *ecological community* consists of organisms of two or more species that live together in the same area and interact.

2. *Community ecology* is the study of the structure of biological communities and the interactions of the organisms living there.

3. In *predator–prey* or *host–parasite interactions*, one organism eats another organism; the consumer is the predator or parasite, the consumed is the prey or host.

4. If two organisms use a common resource whose supply is insufficient to support both, the two organisms are *competitors* and the interaction type is *competition*.

5. If two organisms interact in a mutually beneficial way, the interaction is called *mutualism*.

6. If one participant benefits while the other is unaffected by the interaction, it is called *commensalism*.

7. If one participant suffers from the interaction while the other is unaffected, it is called *amensalism*.

8. A *resource* is any feature of the community that is needed for normal growth of an organism and whose quantity is reduced when it is used; food is a resource, but so are space, nesting material, oxygen in aquatic environments, and so forth

9. The *ecological niche* of a species is a characterization of all the requirements of an organism that determine where it can live.

10. In the absence of any negative influences from other organisms (competitors, predators, parasites, etc.) the distribution of a species usually can be much broader.

11. The difference between a species' potential and actual distributions reflects the influence of other community members on the species and indicates which resources are most important in determining where the species can exist.

12. *Limiting resources* are resources that are present at an inadequate level to meet the demand of consumers.

13. A resource may be limiting in some environments but not in others. For example, although oxygen is required by all organisms, it is typically abundant in terrestrial environments, but may be regularly depleted in aquatic environments.

14. Competition can occur when one species limits the growth of another species by using up its resources; or when one species behaves in a way that directly prevents access of another species to needed resources.

15. *Intraspecific competition*, competition between individuals of the same species, limits the density of a species in a given area.

16. *Interspecific competition*, competition between members of different species, can also reduce the density of another species, but may also exclude a species from an area.

17. The logistic equation can be modified to account for the effects of interspecific competition on the populations:

$$\frac{\Delta N_1}{\Delta t} = r_1 N_1 \frac{(K_1 - N_1 - \alpha N_2)}{K_1}$$

$$\frac{\Delta N_2}{\Delta t} = r_2 N_2 \frac{(K_2 - N_2 - \beta N_1)}{K_2}$$

where N is the number of individuals, r is the intrinsic rate of growth, K is the carrying capacity, and the subscript denotes the species (1 or 2). α and β are competition coefficients that measure the per-individual effect of one species on the growth rate of the other. α is the effect of species 2 on species 1; β is the effect of species 1 on species 2.

18. G. F. Gause conducted the first controlled experiments on interspecific competition in 1934, working primarily with protists such as *Paramecium*.

19. Many of Gause's studies showed *competitive exclusion*, in which one species completely replaced the other species; competitive exclusion could be avoided if environmental heterogeneity was provided.

20. Performing studies of competition between plant species may be easier than with animal species because it is easier to manipulate plant population densities and resource availability.

21. In field experiments, removing a species from an area is a good way to evaluate its impact on another, potentially competing species.

22. Natural environments vary in space and time. Increased environmental complexity allows competing species to co-exist and avoid competitive exclusion.

Questions (for answers and explanations, see page 437)

1. A bird eats the fruit of a plant species. The seeds are not digested and germinate in the bird's excrement at some distance from the parent plant. This is an example of
   a. predation.
   b. parasitism.
   c. competition.
   d. mutualism.
   e. amensalism.
   f. commensalism.

2. Certain birds follow swarms of foraging army ants and prey upon the insects that the ants flush. The relationship between these birds and the ants is an example of
   a. predation.
   b. parasitism.
   c. competition.
   d. mutualism.
   e. amensalism.
   f. commensalism.

3. According to the definition of resource developed in the text, which of the following would *not* be a resource?
   a. Sunlight for a plant species
   b. Oxygen for an aquatic insect
   c. Hydrogen sulfide for a sulfur bacterium
   d. Nesting boxes for a bluebird
   e. Oxygen for a terrestrial animal

4. As shown below, species A and B originally had nonoverlapping ranges. Species A expanded its range and now overlaps with species B in area 2. The two species still live separately in areas 1 and 3.

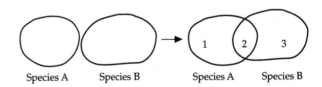

Species A   Species B   Species A   Species B

Assuming that the two species compete for common, limiting resources, which of the following statements is *not* a likely outcome of the range expansion of species A?
   a. The ecological niches of both species would be more constrained in area 2 than in areas 1 and 3.
   b. The range of conditions under which species B can exist will be greater in area 3 than in area 2.
   c. The range of conditions under which species A can exist will be greater in area 1 than in area 2.
   d. The ecological niche of species A in area 1 would be the same as it had been before the range expansion.
   e. The ecological niche of species B in area 2 would be the same as it had been before the range expansion.

5. In many of the studies done by G. F. Gause, two species were grown alone and in the presence of another species. The following graphs show growth of species 1 (left) and species 2 (right), both alone and when in mixed culture.

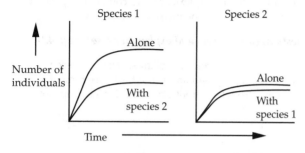

Interpretation of these graphs shows that

a. both species are affected by interspecific competition but species 1 is affected less.

b. both species are affected by interspecific competition but species 2 is affected less.

c. neither species is affected by interspecific competition.

d. both species are affected equally by interspecific competition.

e. competitive exclusion occurred in this study.

Activities (for answers and explanations, see page 437)

• In Question 5 above, what should be the relative sizes of the competition coefficients (α and β) in the equations showing the effect of interspecific competition on population growth?

$$\frac{\Delta N_1}{\Delta t} = r_1 N_1 \frac{(K_1 - N_1 - \alpha N_2)}{K_1}$$

$$\frac{\Delta N_2}{\Delta t} = r_2 N_2 \frac{(K_2 - N_2 - \beta N_1)}{K_2}$$

Explain your answer.

• Describe how you would assess the degree of interspecific competition that exists between two rodent species living in the same habitat and how you would interpret the results.

PREDATOR–PREY INTERACTIONS • BENEFICIAL INTERSPECIFIC INTERACTIONS • TEMPORAL CHANGES IN COMMUNITIES • COEVOLUTION OF INTERACTING SPECIES • KEYSTONE SPECIES: MAJOR INFLUENCES ON COMMUNITY COMPOSITION • INDIRECT EFFECTS OF INTERACTIONS AMONG SPECIES (pages 1156–1170)

Key Concepts

1. Predation can have direct effects on interspecific competition if one of the competing species is preferentially removed by the predator.

2. Both the consistency of prey availability and the relative sizes of predator and prey determine the type of predator–prey interaction.

3. *Herbivores* are predators on plant tissues. Because wood is difficult to digest, few herbivores eat it—most consume other plant parts.

4. *Carnivores* are predators on animals, which they usually capture individually.

5. Prey defend themselves against attack by chemical means, by physical armor, by escape behavior, and by camouflage.

6. *Suspension feeders,* found in many different phyla, use a filtering device to remove the abundant, tiny prey from the medium in which they live; the structural details of the filtering device determine the size range and, in some cases, the type of prey captured.

7. *Parasites* are smaller than their prey *(hosts)* and feed on them internally; *parasitoids* are about the same size as their prey.

8. In some parasite–host interactions, it is the egg-laying female that chooses the host for her offspring, as in many parasitic wasps and flies.

9. As a predator population increases, the death rate of its prey increases and may cause the prey population to decrease.

10. As a prey population decreases, predators may starve or emigrate and their population will decrease.

11. Decrease in a predator population allows its prey to recover, and as the prey population increases, the predator population can also recover.

12. The interactions between prey and predator population sizes cause *oscillations,* during which the prey population always recovers more quickly than the predator population.

13. Environmental heterogeneity over space can suppress predator–prey oscillations if prey are more vulnerable to predators in some patches than in others.

14. Oscillations can also be dampened if the prey is able to reinhabit areas and flourish there until the predator arrives and exterminates it. In this case, the environment is heterogeneous over time.

15. Over the course of time, predators and prey often evolve adaptations that reduce the effect of each on the other.

16. In *Batesian mimicry,* a palatable species, called the *mimic,* evolves an appearance and/or behavior similar to a distasteful or dangerous species, called the *model.*

17. Because palatable mimics "dilute" the effectiveness of the model's adaptation, directional selection causes models to diverge in appearance from mimics.

18. To have a stable Batesian mimicry system, the mimic must evolve toward the model faster than the model evolves away from the mimic.

19. Because Batesian mimicry relies on predator learning, mimics are typically less common than models.

20. *Müllerian mimicry* results from convergence in the appearance of two or more unpalatable species.

21. Müllerian mimicry makes it easier for predators to learn a common appearance; since all species are unpalatable, the relative abundances do not matter.

22. Host–parasite relationships tend to stabilize if the parasite does not kill the host or does not kill it before the parasite can spread to a new host.

23. Parasites that can be transmitted by physical means after the host has died are typically more deadly than those transmitted by biological *vectors* which can pick up the parasite only from living hosts.

24. Plants defend their leaves by making them tough, adorning them with hairs and spines, and by producing *secondary compounds* (see Chapter 33).

25. One group of secondary compounds is toxins, such as nicotine. Some secondary compounds imitate hormones and interfere with insect metamorphosis. Other toxins are unusual amino acids, as in canavanine (see the discussion on page 741 in the textbook). Another group of defensive chemicals, including *tannins*, is difficult for herbivores to digest.

26. Some plants increase their production of secondary chemicals in response to being damaged by herbivores.

27. Plants being damaged by animal trampling or falling leaves and branches from larger, neighboring plants are examples of *amensalism* because one party is harmed, while the other party is unaffected.

28. *Cattle egrets* foraging on insects flushed by hoofed mammals and *epiphytes* growing on large trees are examples of *commensalism* because one party benefits, while the other party is unaffected.

29. Mutualisms are widespread among organisms of virtually every phylum; an important early result of a mutualism was the evolution of the eukaryotic cell from previously free-living prokaryotes.

30. Examples of mutualisms between members of different kingdoms include the association of plant roots and nitrogen-fixing bacteria such as *Rhizobium*, and the association of plant roots and certain fungi to form *mycorrhizae*.

31. Other trans-kingdom mutualisms involve the harboring of photosynthetic protists by corals and some tunicates and the association between termites and cellulose-digesting protists.

32. Some examples of animal–animal mutualisms include aphid "farming" by certain ants, *cleaner* shrimp and small coral reef fish removing parasites from larger fish, and the birds called honeyguides leading honey badgers and humans to bee nests.

33. Some examples of plant–animal mutualisms include the association between *Acacia* trees and ants of the genus *Pseudomyrmex*, and the many associations between plants and their animal pollinators and seed dispersers.

34. *Ecological succession* is the gradual change in the species composition of a community through time.

35. Ecological succession may start in areas without organisms; in many areas, lichens and mosses are important soil-forming, pioneering organisms and the nitrogen-fixing bacteria associated with the roots of many trees help to enrich the soil.

36. Succession can also begin after a disturbance has removed a portion of the original community.

37. Degradative succession is the decomposition of the dead bodies or excrement of organisms.

38. Coevolution results when interactions between two or more species mutually influence their evolution. The clearest examples of coevolution involve only two species, such as the mutualisms described between *figs* and *fig wasps* and between species of *Yucca* and their moth pollinators.

39. *Diffuse coevolution* results from interactions between a diversity of organisms, including predators, parasites, mutualists, and so forth.

40. Diffuse coevolution produces generalized traits that adapt an organism to a variety of species acting as pollinators, seed dispersers, predators, and so forth. Some examples are bat-pollinated flowers that are white and only open at night, nectar guides on flowers best seen under ultraviolet light, and fleshy red fruit for dispersal by birds.

41. *Keystone species* are organisms that have a larger effect on the structure and functioning of an ecological community than would be expected based on their abundance. In certain areas, beaver and moose are keystone species.

42. Plants are the major modifiers of physical conditions in terrestrial environments; plants intercept sunlight and rain, moderate climate (temperature, humidity, and wind), and reduce soil evaporation.

43. Predators are frequently keystone species because they control the population sizes of important prey species that might otherwise dominate the community.

44. Microorganisms have major influences on ecological communities, including nitrogen fixation, mycorrhizal association with plant roots, and degradation and recycling of organic material.

45. Most interactions at the community level are indirect and involve many different species.

Questions (for answers and explanations, see page 437)

1. Which vertical line in the following graph represents the time at which the predator population is *increasing* and the prey population is *decreasing*?

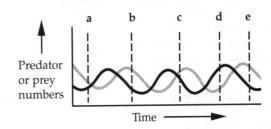

2. Some predator–prey interactions do not show oscillations. Select *all* of the following factors that could be responsible for damping oscillations.

   a. Spatial habitat heterogeneity
   b. Temporal (time-based) environmental heterogeneity
   c. Dispersal of prey
   d. Density-dependent interactions between prey and predator populations
   e. Extinction of prey
   f. Alternate prey species

3. You are studying three butterfly species (1, 2, 3) that have very similar, bright coloration patterns. You collect the following data on the abundance of these species in two different regions, A and B. Based on your data shown below, which of the following conclusions is *most probable*?

Region	Numbers collected for species:		
	1	2	3
A	90	20	120
B	190	35	155

   a. Species 1 is the model, species 2 and 3 are Batesian mimics.
   b. Species 1 is a Batesian mimic, species 2 and 3 are Müllerian mimics.
   c. Species 1 and 3 are Müllerian mimics, and 2 is a Batesian mimic.
   d. All three species are Müllerian mimics.
   e. Species 3 is the model, species 1 and 2 are Batesian mimics.

4. Which of the following statements about parasite–host interactions is *not* true?
   a. Competition with other parasites for the same host can lead to faster killing of the host.
   b. Transference to a new host is improved if the current host lives longer.
   c. Water-borne diseases are less serious to the host than diseases involving biological vectors.
   d. Most parasites with biological vectors do not harm their vectors as much as they harm the primary host.
   e. Hosts act as selective agents on their parasites.

5. A species of orchid has flowers that look and smell like the female of a wasp species. Male wasps of that species attempt to copulate with these flowers and in the process pollinate them. This is an example of
   a. parasitism.
   b. diffuse coevolution.
   c. mutualism.
   d. amensalism.
   e. commensalism

6. If a species of plant that is trampled by an animal eventually evolves sharp spines that prevent trampling, we can say that its association with the animal has changed from
   a. amensalism to competition.
   b. amensalism to commensalism.
   c. commensalism to competition.
   d. parasitism to mutualism.
   e. parasitism to competition.

7. Which of the following is *not* an effect that plants normally have on biological communities?
   a. Moderate climate

   b. Influence which animal species will be present
   c. Increase CO_2 levels
   d. Reduce water evaporation from the soil
   e. Reduce moisture reaching the soil

8. Place numbers before each of the stages of ecological succession to reflect the order of the stages in northern coniferous forests.

   _____   Soil formation by physical processes

   _____   Arrival of willows and alders

   _____   Arrival of lichens

   _____   Arrival of conifers

   _____   Increase in soil nitrogen content

9. Select all of the following that is/are characteristic of a keystone species.
   a. Very abundant
   b. Removal has a great effect on community structure
   c. Herbivore
   d. Large
   e. Uniform distribution

Activities (for answers and explanations, see page 438)

• Discuss four common plant defenses against browsers.

• Discuss the costs and benefits for both the moth and the plant in the Yucca–Yucca moth coevolution described in the textbook.

• You discover two fig wasp nematode parasites. Each nematode parasitizes a different species of fig wasp. Nematode *A* is significantly less virulent than nematode *B*. Discuss life history differences in the fig wasp species that could account for differences in parasite virulence.

53

Ecosystems

CHAPTER LEARNING OBJECTIVES—after studying this chapter you should be able to:

❑ Define what an ecosystem is and provide several examples of ecosystems, ranging from local to global.

❑ Explain what determines the solar energy input to different regions of Earth and the consequences of these differences.

❑ Describe the movement of air masses over mountains and the conditions that create rain shadows.

❑ Describe the global circulation of air in the atmosphere and the consequences of this circulation for Earth's climates.

❑ Explain the effect of the tilt in Earth's axis on the location of the intertropical convergence zone and the seasonal climatic changes this creates.

❑ Describe the general circulation of the world's oceans and its effects on climate and marine primary production.

❑ Differentiate between continental and maritime climates.

❑ Define the term "gross primary production," and explain how it varies with temperature and precipitation in terrestrial ecosystems, and in aquatic ecosystems, with depth.

❑ Relate net primary production to gross primary production.

❑ Explain what a trophic level is, and list the major types of trophic levels.

❑ Relate the concepts of food chain and food web to energy flow in an ecosystem.

❑ Estimate the efficiency of energy transfer between trophic levels and express it as an equation.

❑ Explain how pyramids of energy and biomass are constructed, and know what determines their shapes.

❑ Describe the conditions that might lead to an inverted biomass pyramid for an ecosystem.

❑ Compare energy flow in grasslands and forests.

❑ Explain how integrated pest management is addressing some of the ecological shortcomings of modern agriculture.

❑ Characterize the major compartments of the global ecosystem, and describe the cycling of material through each.

❑ Describe the annual temperature and oxygen cycles in temperate lakes.

❑ Explain what a biogeochemical cycle is, and describe some key generalizations about movement of materials through biogeochemical cycles.

❑ Compare and contrast the features of the following biogeochemical cycles: hydrological cycle, carbon cycle, nitrogen cycle, phosphorus cycle, and sulfur cycle.

❑ Describe lake eutrophication—a local alteration of a biogeochemical cycle.

❑ Describe acid rain—a regional alteration of a biogeochemical cycle.

❑ Describe global warming—a global alteration of a biogeochemical cycle.

❑ Characterize communities that are based on chemoautotrophy and how they differ from Earth's more typical ecosystems.

CLIMATES ON EARTH • ENERGY FLOW THROUGH ECOSYSTEMS   (pages 1174–1182)

Key Concepts

1. An *ecosystem* includes communities of organisms and their physical environment. Ecosystems can be studied at many levels, from local units to a global scale.

2. The goal of ecosystem ecology is to understand the flow of energy and the cycling of materials within ecosystems.

3. Earth's climate patterns are determined by circulation of the atmosphere and ocean currents, which are driven by energy from the sun and Earth's rotation.

4. Although all parts of Earth receive equal amounts of sunlight per year, the amount of heat received varies with the angle of sunlight striking the ground.

5. The rate of heat arrival per unit of Earth's surface area depends primarily on the angle of sunlight. When the sun is low in the sky, more heat energy is absorbed by the atmosphere before it reaches the ground, and upon reaching the ground, the energy is spread over a wider area.

6. Seasonal changes in day length and sun angle increase directly with distance from the equator (increasing latitude), so mean annual temperature decreases with latitude (about 0.4°C for each degree of latitude).

7. As gases expand, they cool, so air temperature decreases with altitude. Since cooler air holds less moisture, as air rises, clouds form and rain often falls.

8. *Rain shadows* form on the leeward sides of mountain ranges, because air loses much of its moisture as rain on the windward side of mountains.

9. Air heated at the equator rises, moves poleward, and descends at latitudes of about 30° north and south, creating hot deserts there, such as those of the Sahara and Australia.

10. Air rising at about 60° north and south latitude descends at the poles to create cold deserts there.

11. Surface air moving latitudinally is deflected to the right in the Northern Hemisphere and to the left in the Southern Hemisphere by the spinning of Earth on its axis.

12. Air descending at 30° north and south latitude becomes the northeast and southeast *trade winds* if it flows back toward the equator or the *westerlies* if it flows toward the poles. The area where the trade winds converge is called the *intertropical convergence zone*.

13. Solar energy flux is greatest where the sun is directly overhead at noon; this region shifts north and south seasonally because of the tilt of Earth's axis.

14. The location of the intertropical convergence zone shifts latitudinally each year with the region of greatest solar energy input, but lags behind it by about a month. Consequently, the intertropical convergence zone shifts north during the northern summer and south during the northern winter (southern summer).

15. Tropical and subtropical rainy seasons usually occur when an area is located in the intertropical convergence zone and ascending air produces rain. The tropical and subtropical dry seasons occur when the intertropical convergence zone has departed and the area is dominated by the trade winds.

16. The *specific heat* of a substance is the amount of heat required to change the temperature of 1 gram of that substance by 1°C. The specific heat of water is 1 cal/g; the specific heat of air or land surfaces is much lower.

17. Ocean currents are driven by prevailing wind, diverted by continents, and affected by Earth's rotation. The result is a general clockwise circulation in the Northern Hemisphere and a counterclockwise circulation in the Southern Hemisphere.

18. Upwelling of deeper, cooler water on the western sides of continents replaces some of the water that has flowed toward the equator and away from the land.

19. If air blowing over water crosses warmer land it retains its moisture; if air blowing over water crosses colder land it releases its moisture.

20. *Continental climates* are characterized by large seasonal variation in temperature; the largest land masses have the greatest seasonal variations in temperature.

21. *Maritime climates* prevail on coasts of continents where sea breezes moderate the climate.

22. Only about 5 percent of solar energy is "fixed" by photosynthesis; the remainder is absorbed by the atmosphere or used by plants to power transpiration.

23. The total amount of energy captured by plants during photosynthesis is called *gross primary production*.

24. Gross primary production is greatest in parts of the world where water is abundant and temperatures are moderate, such as in the tropics.

25. Gross primary production in aquatic ecosystems decreases with depth due to extinction of light with depth, but increases with nutrient availability (greatest in areas of upwelling).

26. *Net primary production* is gross primary production minus the energy used by plants for their own maintenance and biosynthesis.

27. A *trophic level* includes all organisms that obtain their energy from the same general source within an ecosystem.

28. Common trophic level designations include *photosynthesizers*, *herbivores* (organisms that consume photosynthesizers), *primary carnivores* (organisms that consume herbivores), *secondary carnivores* (organisms that consume primary carnivores), *detritivores* (consumers of dead organisms), and *omnivores* (organisms that consume at several levels).

29. Photosynthesizers are also called *primary producers*. Organisms in all other trophic levels are called *consumers*.

30. A *food chain* is a set of linkages showing the energy flow through trophic levels; a *food web* shows the linkages among food chains in an ecosystem.

31. Because much energy is lost as heat at each trophic level, less is available for transfer to the next level. The efficiency of energy transfer between trophic levels is
$$E = P/(P + R)$$
Where E is efficiency, P is net production, and R is respiration.

32. Endotherms, like birds and mammals, have very low efficiencies because of their high metabolism. Herbivores are less efficient than carnivores due to the difficulty in digesting plant material relative to animal material. In general, the efficiency of energy transfer between trophic levels is less than 20%.

33. A *pyramid of energy* is a graph in which the amount of energy captured at each trophic level is plotted from the lowest to the highest levels. Most energy pyramids have a broad primary producer base, with each successive consumer level proportionally smaller.

34. A *pyramid of biomass* plots the biomass at each trophic level. Generally, energy and biomass pyramids from the same ecosystem have a similar shape.

35. Inverted biomass pyramids are sometimes observed in marine ecosystems. In these ecosystems, the primary producers, mostly bacteria and protists, reproduce at such a great rate that the biomass of herbivores may be larger than the biomass of producers.

36. Grasslands can support higher grazing rates by herbivores than forests can because little energy is allocated by grasses to produce supportive, woody tissue.

37. Detritivores are essential for releasing minerals tied up in dead organisms so that they can be used again. High production rates in the tropics are possible, in part, because of the high rates of decomposition due to increased humidity and temperatures there.

Questions (for answers and explanations, see page 438)

1. In comparing an acre of land in Colombia with an acre of land in Michigan, which of the following would *not* differ?
   a. The angle of the sun reaching the ground in the month of July
   b. The solar energy flux in the month of July
   c. The annual solar energy flux
   d. The total hours of daylight per year
   e. The mean annual air temperature

2. The following diagram shows a mountain with a sea breeze blowing as indicated by the arrow. Circle the letter for the area with air that would be *both* relatively warm and dry.

3. In what area of the diagram in question 2 is a process occurring that is similar to the process that occurs in the intertropical convergence zone? _____

4. If the Earth did *not* spin on its axis, from what direction would the northeast trade winds blow?
   a. Northeast
   b. South
   c. North
   d. East
   e. Southwest

5. Ocean circulation patterns are influenced by all of the following *except*
   a. circulation of Earth's atmosphere.
   b. deflection by land masses.
   c. upwelling of deeper water.
   d. rotation of Earth on its axis.
   e. prevailing winds.

6. Differences in the specific heat of water and land have most to do with creating
   a. the intertropical convergence zone.
   b. maritime climates.
   c. continental climates.
   d. the westerlies.
   e. oceanic upwelling.

7. Select all of the following that could *not* be considered ecosystems.
   a. A small pond
   b. A large lake
   c. All the fish in a coral reef
   d. Earth
   e. A pile of dung in a pasture

8. A plant in the dark uses 0.02 ml of O_2 per minute. The same plant in sunlight releases 0.14 ml of O_2 per minute. A correct estimate of its rate of *gross* primary production is
   a. 0.02 ml of O_2 per minute.
   b. 0.12 ml of O_2 per minute.
   c. 0.14 ml of O_2 per minute.
   d. 0.16 ml of O_2 per minute.

9. In the following food web, organism 9 is a:

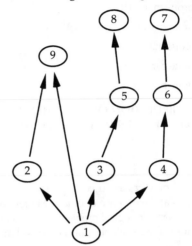

   a. primary producer.
   b. herbivore.
   c. primary carnivore.
   d. secondary carnivore.
   e. omnivore.

10. The food web shown in Question 1 has _____ trophic levels?

11. If the dry biomass of the primary producers of an ecosystem is 1000 g per square meter and the efficiency of transfer between trophic levels is 20%, what dry biomass weight would you expect at the *secondary carnivore* level? Show your answer and calculations below.

12. In the equation for efficiency of energy transfer between trophic levels, $E = P/(P + R)$, what does P represent?

Activities (for answers and explanations, see page 438)

- Where do energy pyramids like the one shown below occur and what are the conditions that create them?

	Dry Weight (g/m^2)
Consumers	4
Producers	21

- Rank the following organism types in terms of their expected efficiency of energy transfer between trophic levels, where 1 is most efficient and 5 is least efficient.

_____ Herbivorous mammal

_____ Invertebrate detritivore

_____ Carnivorous mammal

_____ Invertebrate carnivore

_____ Invertebrate herbivore

AGRICULTURAL MANIPULATION OF ECOSYSTEM PRODUCTIVITY • CYCLES OF MATERIALS IN ECOSYSTEMS (pages 1182–1185)

Key Concepts

1. Modern agricultural practices are used to promote the success of some species at the expense of others. Modern agriculture has created many ecological problems and has flourished largely because of the availability of cheap fossil fuel.

2. *Integrated pest management* (IPM) is an attempt to minimize the adverse effects of agriculture by reducing reliance on chemicals through the application of cultural and biological control methods.

3. Unlike the cycling of energy through the trophic levels of ecosystems, chemical elements are not lost as they cycle. To follow the cycling of chemicals, it is useful to divide the global ecosystem into oceans, fresh water, atmosphere, and land compartments.

4. Oceans receive materials from land as runoff from rivers and exchange material with the atmosphere only at the air–water interface.

5. Most materials that enter the oceans sink and become part of the bottom sediment. Consequently, except for areas of *upwelling,* most of the oceans are nutrient-poor and populations of organisms are sparse.

6. Freshwater ecosystems contain only a small fraction of the total water on Earth; most water resides in the oceans.

7. Because rivers and lakes are much less deep than oceans, elements in sediments are recycled more quickly in the former than in the latter.

8. Unlike oceans, where oxygen levels are fairly uniform throughout, oxygen in many lakes becomes depleted at the bottom due to decomposition.

9. In lakes, mixing is important in redistributing oxygen and nutrients.

10. Water is most dense at 4°C. In most temperate lakes, mixing—*turnover*— brought about by wind occurs in the spring and in the fall when the entire water column is at this temperature.

11. The surface water of lakes is warmed during the spring and a *thermocline* is established, where the temperature changes rapidly with depth. Since water colder than 4°C expands and floats on warmer water, ice forms only on the surface.

12. Deep tropical lakes are more stratified than temperate lakes because the surface water is never cooled to 4°C.

13. The *troposphere*, the lowest layer of the atmosphere, contains 80% of Earth's gases. The troposphere is about 17 km thick in the tropics, but thins to about 10 km at higher latitudes.

14. The *stratosphere* extends from the troposphere to about 50 km above the surface. Materials enter the stratosphere at the intertropical convergence zone and remain there relatively long because stratospheric circulation is largely horizontal. Ozone (O_3) is an important component of this layer and the "ozone hole" develops here.

15. The atmosphere is 78 percent nitrogen, 21 percent oxygen as O_2, and 0.03 percent carbon dioxide, with the remainder consisting of argon, neon, krypton, helium, hydrogen, ozone, and methane.

16. Certain atmospheric components, especially water vapor, carbon dioxide, and ozone, trap a significant proportion of infrared radiation, causing the temperature of Earth to be higher than it would be without an atmosphere.

17. One-fourth of Earth's surface is land. Because lateral movement of material within this compartment is very slow, there is large variation in the distribution of nutrients in terrestrial ecosystems.

18. Exchange of materials between the land and the atmosphere is mainly through the activities of organisms.

19. Exchange of materials between the land and the oceans is via runoff from groundwater into rivers. The uplifting of marine sediments slowly returns materials to the land compartment.

20. Soils, resulting from weathering of parent rock, cover most of the land; aged soils have lost most of their nutrients to the groundwater.

21. Plants play key roles in the development and modification of soils.

Questions (for answers and explanations, see page 438)

1. Next to each of the following features, place an "O" or "F," if it is characteristic of the *ocean* or *freshwater* ecosystem.

_____ Seasonal mixing of materials

_____ Nutrient concentrations low, except locally

_____ Oxygen concentration vertically uniform

_____ Elements buried in bottom sediments for long periods of time

_____ Receives material from land mostly via groundwater

2. Next to each of the following features, place a "T" or "S," if it is characteristic of the *troposphere* or *stratosphere*.

_____ Most water vapor resides here.

_____ Most ozone resides here.

_____ Circulation of gases is mostly horizontal in this layer.

_____ Represents the greatest mass of the total atmosphere.

_____ Circulation of this layer influences ocean currents.

3. Next to each of the following features, place an "O," "F," "A," or "T" if it is characteristic of the *oceans, freshwater, atmospheric,* or *terrestrial* compartments of the global ecosystem (some features may apply to several).

_____ Very slow movement of materials within compartment

_____ Exchange of gases mostly via organisms

_____ Circulation of materials affected by Earth's rotation on its axis

_____ Circulation of materials affected by Earth's revolution around the sun

_____ Organisms mostly in uppermost layer

4. In the following diagram, the temperature of a freshwater lake is plotted against depth.

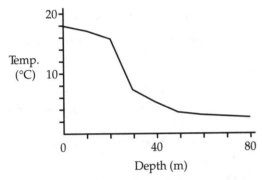

This temperature profile would be most characteristic of the lake during
a. early winter.
b. late winter.
c. spring.
d. summer.
e. fall.

5. Based on data in the graph from Question 4, the thermocline for the lake is located between _____ and _____ meters.

6. Which of the following would *not* likely be a method employed in integrated pest management?
a. Mixed plantings of crop plants
b. Introduction of a natural disease organism or an important pest
c. Aerial spraying of insecticides
d. Use of chemical attractants for controlling pest species
e. Development of pest-resistant strains of crop plants

BIOGEOCHEMICAL CYCLES • HUMAN ALTERATIONS OF BIOGEOCHEMICAL CYCLES • CHEMOSYNTHESIS-POWERED ECOSYSTEMS   (pages 1185–1192)

Key Concepts

1. Earth is a closed system with respect to the essential elements: carbon, hydrogen, oxygen, nitrogen, phosphorus, and sulfur; they circulate between organisms and the environment in what are known as *biogeochemical cycles*.

2. Elements cycle through organisms quickly; the gaseous elements cycle through the nonliving environment most quickly.

3. The *hydrological cycle* refers to the exchange of water between land, the ocean, and the atmosphere. Evaporation (including transpiration within the leaves of plants) and precipitation exchange water between the atmosphere and land/ocean. The movement from land to ocean as runoff equals the amount returned as ocean-derived precipitation.

4. Carbon enters the living world when CO_2 is fixed by a plant during photosynthesis; carbon moves into heterotrophs when they consume autotrophs.

5. Respiration and decomposition return carbon to the atmosphere as CO_2.

6. Carbonate minerals in sedimentary rocks and dissolved carbon in the oceans are the two major reservoirs of carbon in Earth's *carbon cycle*.

7. Fossil fuels, consisting of coal, petroleum, natural gas, and peat, are a reservoir of carbon produced when organisms decayed anaerobically.

8. Increased atmospheric CO_2 levels resulting, in part, from combustion of fossil fuel, may lead to global warming because CO_2 is a greenhouse gas; it is transparent to sunlight but traps infrared heat radiation.

9. Nitrogen (N_2) is biologically inert until it is converted to ammonia by certain bacteria and cyanobacteria, or through industrial fertilizer production.

10. Other types of bacteria are responsible for interconverting nitrogen among its various inorganic forms (ammonia, nitrite, nitrate), and for returning it to the atmosphere as N_2.

11. Plants obtain inorganic forms of nitrogen from the soil (or root nodules) and animals get their organically fixed nitrogen by consuming plants or other animals.

12. Flux through the atmosphere is of minor importance in the *phosphorus cycle*. Phosphorus cycles through organisms quickly in the oceans, is eventually incorporated into sedimentary rock, and returns as dissolved phosphate ion when the rock is uplifted and undergoes weathering.

13. Unlike nitrogen and phosphorus, sulfur is seldom limiting in natural ecosystems.

14. In the *sulfur cycle*, natural venting of SO_2 and H_2S by volcanoes and fumaroles is significant, but *dimethyl sulfide* release by marine algae and terrestrial decomposition of organically fixed sulfur by bacterial fermentation are also important fluxes.

15. Dimethyl sulfide is an important nucleation agent for water drop formation during cloud development. Increased cloud cover due to higher atmospheric levels of dimethyl sulfide may help slow global warming.

16. Because the availability of essential elements in the biogeochemical cycles sometimes limits the metabolism of organisms, human modification of these cycles can have local, regional, and global biological consequences.

17. Because aquatic photosynthesis is frequently limited by phosphorus, *eutrophication*, the addition of phosphorus to fresh water can lead to blooms of algae.

18. The effects of eutrophication include changes in the species of algae present, reduction in oxygen levels due to increased organic decomposition, and the dominance of anaerobic organisms in the ecosystem.

19. Eutrophication of *Lake Erie* was caused by nutrients added to the lake from domestic and industrial wastes and as agricultural runoff.

20. Lakes recover quickly from eutrophication if nutrient inputs are reduced and if the rate of water turnover in the lake is rapid.

21. *Acid precipitation* is an example of a regional disturbance caused by human-produced changes in the nitrogen and sulfur cycles.

22. The low pH of acid precipitation is caused by the presence of sulfuric and nitric acids derived from the burning of fossil fuels.

23. Acid precipitation damages plants directly and may cause the soil to lose calcium and magnesium, which in turn may limit plant growth.

24. Lower pH in lakes may reduce populations of nitrifying bacteria, which leads to accumulation of ammonia and death of fish populations.

25. Perturbation of the carbon cycle, due to increased atmospheric CO_2 levels from combustion of fossil fuels and forest burning, has had an effect on the heat balance of Earth.

26. Atmospheric CO_2 levels are expected to double from pre-Industrial Revolution levels. Possible effects include massive climate change, melting of the Arctic and Antarctic ice caps, flooding of coastal areas, and disruption of agriculture.

27. Unlike most ecosystems on Earth, some novel ecosystems are not dependent on sunlight. These include communities based on chemoautotrophs that can oxidize hydrogen sulfide emitted from deep-sea thermal vents and certain cave communities also dependent on chemoautotrophic production.

Questions *(for answers and explanations, see page 439)*

1. Which of the following statements about biogeochemical cycles is *not* true?
   a. Gaseous elements cycle more quickly than elements without a gaseous phase.
   b. Most elements remain longest in the living portion of their cycle.
   c. You may have some atoms in your body that were originally part of a dinosaur.
   d. Earth is a closed system relative to the essential elements of life.
   e. Biogeochemical cycles all include both organismal and nonliving components.

2. Next to each of the following features, place a "C," "N," "P," or "S" if it is characteristic of the biogeochemical cycles of *carbon, nitrogen, phosphorus,* or *sulfur* (some features may apply to several cycles).

   _____ Major reservoir is atmospheric

   _____ Major reservoir is in sedimentary rocks

   _____ Often in short supply in ecosystems

   _____ Fossil fuel reserve is part of this cycle

   _____ Major human impact on cycle in many ecosystems

   _____ Involved in cloud formation

   _____ Lacks a gaseous phase

   _____ Major inorganic form is directly available to only a small group of bacteria

   _____ Includes a form that is a greenhouse gas

   _____ Most fluxes involve organisms

3. Place numbers next to the following events in the eutrophication of Lake Erie to reflect their correct chronological sequence.

   _____ *Escherichia coli* populations proliferate.

   _____ Phosphorus input from sewage and agricultural runoff increases.

   _____ Respiratory demand from decomposers increase.

   _____ Fish species change.

   _____ Oxygen levels drop in deeper water.

   _____ Algal blooms occur.

4. Which of the following statements is *not* a major concern about our alteration of the carbon cycle?
   a. The mean temperature of the Earth may increase 3–4°C.
   b. High-sulfur fuels are used by power plants because they are less expensive than low-sulfur fuels.
   c. The increase in atmospheric CO_2 exceeds the ability of the oceans to absorb the increase.
   d. The polar ice caps are expected to melt if global warming continues.
   e. CO_2 is a gas that traps infrared radiation.

Activity *(for answer and explanation, see page 439)*

- What factors determine how quickly a lake can recover from eutrophication? Discuss why Lake Erie will never be restored to its historical condition.

54

Biogeography

CHAPTER LEARNING OBJECTIVES—after studying this chapter you should be able to:

❏ Define what biogeography is, and describe the types of questions of concern to biogeographers.

❏ Know the fundamental processes that determine the presence or absence of a species in an area.

❏ Know the important role Alfred Wegener played in the development of biogeography.

❏ Describe the distributions of some contemporary organisms that have been influenced by continental drift.

❏ Characterize the principal concerns and approaches of historical biogeography.

❏ Characterize the principal concerns and approaches of ecological biogeography.

❏ Explain the principle of parsimony and its application in biogeography.

❏ Describe the use of cladistic methods in biogeography and how a taxonomic cladogram can be transformed into an area cladogram.

❏ Differentiate between a vicariant distribution and a dispersal distribution, and give an example of each.

❏ Define what endemic species are and how they can arise.

❏ Characterize the relationship between the degree of endemism and the amount of time that an area has been isolated.

❏ Use cladistic information to make inferences about continental drift or the appearance of barriers.

❏ Describe the major assumptions and predictions of the equilibrium species richness model and be able to apply it.

❏ Explain how island size and distance from the mainland affect the predictions of the equilibrium species richness model.

❏ Describe the variety of evidence that supports the equilibrium species richness model.

❏ Describe what is meant by a habitat island and the applicability of the equilibrium species richness model for studying habitat islands.

❏ Explain how species richness varies latitudinally, with topographic complexity, and in peninsulas and mainland areas.

❏ Define what a biome is, describe how biomes are named, and explain the basis for their latitudinal and altitudinal distribution.

❏ Name and know the locations and boundaries of the major terrestrial biomes.

❏ Name and describe the characteristic features of the major terrestrial biomes.

❏ Characterize the major types of freshwater biomes.

❏ Describe the general features of marine biomes, the names and features of the zones of the ocean, and the names and characteristics of the major marine communities.

❏ Describe the major biogeographic forces that determine the distribution of marine and freshwater species.

WHY ARE SPECIES FOUND WHERE THEY ARE? • HISTORICAL BIOGEOGRAPHY (pages 1195–1201)

Key Concepts

1. *Biogeography* is the study of the present and past distribution of organisms. Biogeographers want to know why specific kinds of organisms live in certain places and, also, why they do not live in other places.

2. A species inhabits an area because it either evolved there or dispersed there from a different area. A species that is absent from an area either evolved in that area but became extinct there, or it evolved in a different area and did not disperse to and survive in that area.

3. Based on a consideration of the interlocking continental shapes and the distributions of plants and animals, in 1912 *Alfred Wegener* proposed that the continents have moved over large distances.

4. Some widespread species alive today evolved on *Pangaea* before that supercontinent broke up at the close of the Mesozoic Era, about 65 million years ago. For example, many groups of insects, freshwater fishes, and frogs with widespread distributions evolved before Pangaea broke up.

5. Originally, biogeographers assumed that the continents were fixed in place and that most biogeographical patterns resulted from *dispersal* of organisms. Early biogeographers also underestimated the age of Earth.

6. Information on the current distributions of species can be applied to a cladogram for the group. These *area cladograms* can then be used to make inferences about continental drift or the appearance of barriers.

7. *Historical biogeography* is concerned with the large-scale distribution of groups and their relationship to evolutionary histories. It concentrates on long time periods and global distribution patterns.

8. *Ecological biogeography* examines the interactions of organisms with one another and the environment to explain their present distribution. It concentrates on recent history and local distribution patterns.

9. Ecological biogeographers attempt to determine to what degree biogeographical patterns can be explained by interspecific interactions such as competition, predation, and mutualism.

10. Three types of data are especially useful to historical biogeographers: phylogenies that show relationships relative to a time scale (perhaps using molecular clock data), fossils that indicate past distributions, and distributions of contemporary organisms.

11. An organism that evolved in an area that is now separated into two subareas by a barrier has a *vicariant distribution*, and the appearance of the barrier is a *vicariant event*.

12. An organism that evolved in one area and crossed a barrier to occupy a second area has a *dispersal distribution*.

13. Species or higher taxa that are found in only one area are *endemic* to that area.

14. Endemic taxa can be taxa of either ancient origin or recent taxa that have not yet dispersed to new areas.

15. The number of endemic taxa in an area is related to the period of time that the area has been separated from other areas; Australia has the most distinct biota (period of separation, 65 million years); North America and Eurasia have the most similar biotas because they frequently were joined during Earth's history.

16. Biogeographers compare the distribution patterns of many different types of organisms currently living in the same area and apply *parsimony* to explain any underlying similarities. The principle of parsimony states that the simplest explanation that can account for known facts should be accepted.

17. Examples of biogeographic histories of groups that have been influenced by continental drift are the southern beeches (genus *Nothofagus*) and the woody plant family Proteaceae.

18. An area cladogram based on mitochondrial DNA for the black-throated green warbler lineage suggests when different taxa were separated by intervals of glaciation in North America.

19. Since the separation of the continents, terrestrial biotas have evolved into distinctive *biogeographic regions*, called the *Nearctic*, the *Neotropical*, the *Australasian*, the *Ethiopian*, the *Oriental*, and the *Palearctic* regions.

20. Plant biogeographers also recognize the southern portion of South America and the Cape Region of Africa as distinct because the floras of these two areas are unique.

21. The biological distinctiveness of the terrestrial biomes are maintained by major water, mountain, or desert barriers.

Questions (for answers and explanations, see page 439)

1. Which of the following types of information would be *least used* by a historical biogeographer?
   a. Experiments on species richness
   b. Fossils showing past distributions of organisms
   c. Present distributions of organisms
   d. Cladistic systematics
   e. Application of the principle of parsimony
   f. Molecular clock data for dating divergences

2. The following diagram shows the present locations of four continents (W, X, Y, Z) and the number of endemic beetle families found on each. Three were part of a supercontinent in the past, the other was always separate.

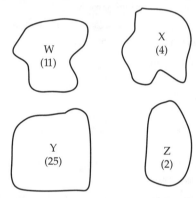

Based on this information, choose the correct conclusions
   a. W was always separate, Y separated from the supercontinent first.
   b. X was always separate, Y separated from the supercontinent first.
   c. Y was always separate, W separated from the supercontinent first.
   d. Y was always separate, Z separated from the supercontinent first.
   e. Z was always separate, X separated from the supercontinent first.

3. The following diagram shows three islands. Islands A and B were connected in the past, C was always separate. A species of land snail is found on all three islands.

Select all of the following that *do not* correctly describe the distribution of this snail relative to the three islands.
   a. Vicariant distribution relative to A and B
   b. Dispersal distribution relative to A and C
   c. Dispersal distribution relative to B and C
   d. Vicariant distribution relative to A and C
   e. Endemic relative to C

4. Which of the following biogeographical regions represents the largest area?
   a. Nearctic
   b. Palearctic
   c. Neotropical
   d. Oriental
   e. Australasian

5. Circle an area on the following graph that would most apply to the spatial and temporal scale of processes studied by ecological biogeographers.

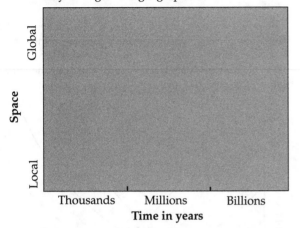

Time in years

Activities (for answers and explanations, see page 439)

• Based on an examination of the following cladogram and information on the distribution of the taxa members on four continents, discuss what can be inferred about the histories of the four continents.

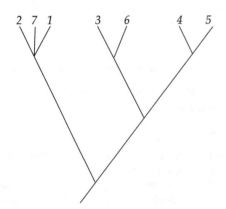

Species	Continent A	B	C
1	✔	✔	✔
2	✔	✔	✔
3	✔	✔	
4		✔	
5		✔	
6	✔	✔	
7	✔	✔	✔

• How does a knowledge of continental drift help explain the past and present distribution of southern beeches (genus *Nothofagus*)?

ECOLOGICAL BIOGEOGRAPHY (pages 1201–1204)

Key Concepts

1. One objective of ecological biogeography is to discover the factors that influence the *species richness* (total number of species) of an area.

2. Arrival of new species to an area increases species richness; extinction of species already there decreases species richness.

3. *Robert MacArthur* and *E. O. Wilson* developed a *species richness equilibrium model* to predict species richness of islands of known size and distance from the mainland.

4. The species present on the mainland that are available to colonize an island make up the *species pool*.

5. As the number of species on an island increases, the rate of arrival of new species decreases until it equals zero when all members of the species pool have colonized the island.

6. As the number of species on an island increases, interspecific competition increases and reduces the resources available for each species.

7. As the number of species on an island increases, predation and parasitism increase as new predators and parasites of species already present arrive.

8. The *extinction rate* of resident species increases with the number of species present on an island.

9. The species number where the extinction rate equals the new arrival rate is the equilibrium number of species that the island can support. The species composition (which species are present) continues to change through time.

10. The model assumes that the arrival and extinction rates are relatively constant at the island's equilibrium number of species.

11. Distance from the mainland affects the predictions of the equilibrium species richness model. Distant islands are harder to find than nearer islands. As a result, the arrival rates of new species on distant islands are less and these islands support fewer species.

12. Island size affects the predictions of the equilibrium species richness model. Extinction rates are higher on smaller islands because they have fewer resources and population sizes are smaller. Arrival rates are lower since smaller islands are harder to find than larger islands.

13. Natural distributions of birds, plants, insects, lizards, and mammals generally support the predictions that species numbers on islands correlate positively with island size and negatively with distance from the source area.

14. On *Krakatoa*, an island on which all life was exterminated in 1883 by a volcanic eruption, 271 plant species and 27 resident bird species recolonized within 50 years. New species continue to arrive, but at a diminishing rate.

15. An experimental test of the species richness equilibrium model involving fumigation of small red mangrove islands in the Florida Keys also confirmed the model. Within a year the fumigated islands had about the same number of species of arthropods.

16. Patches of suitable habitat for a species separated by areas of unsuitable habitat are called *habitat islands*.

17. Although arrival rates are higher for habitat islands than for islands separated by water, the equilibrium species richness model makes the same qualitative predictions for those "islands."

Questions (for answers and explanations, see page 439)

1. According to the equilibrium species richness model, which of the statements about the following graph showing the effect of species number on arrival and extinction rates is *not* true?

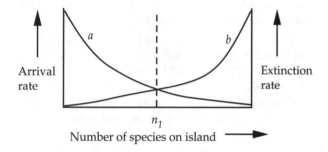

a. *a* is the arrival rate curve.
b. *b* is the extinction rate curve.
c. n_1 is the equilibrium species number.
d. The arrival rate equals the extinction rate at n_1.
e. The extinction rate is zero at n_1 .

2. The species numbers of many islands of different sizes are plotted against their distance from the mainland. Data points for islands of *similar size* were connected to form the five curves shown in this figure.

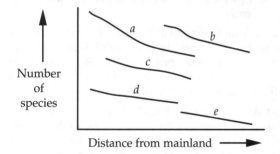

Circle the letter of the curve corresponding to the group of islands with the *smallest* size.

3. A small volcanic island was destroyed by an eruption. The following graph shows recolonization by a number of different organism types.

Identify the correct curve for each organism type.

_____   Birds

_____   Plants

_____   Insects

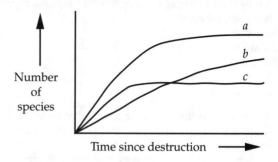

4. The equilibrium species richness model makes predictions about the effects of species number on the rate of extinction. In the following figure, the solid curve shows this relationship for a large island. Which of the curves shows the expected relationship for a *small* island?

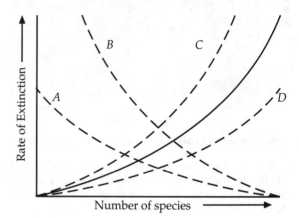

a. Curve *A*
b. Curve *B*
c. Curve *C*
d. Curve *D*
e. It would be the same curve as for the large island

TERRESTRIAL BIOMES • AQUATIC BIOMES (pages 1204–1220)

Key Concepts

1. Organisms occupying similar climatic zones are classified into several major terrestrial ecosystems called *biomes*. Biome names are usually derived from their characteristic vegetation types and location or climate types.

2. The climate, especially how temperature and rainfall vary throughout the year, determines the type of vegetation present in a biome. The dominant plant community of the biome strongly influences what other types of organisms can exist there.

3. Rainy seasons change with latitude in the tropics and subtropics because the intertropical convergence zone shifts latitudinally in a seasonally predictable way.

4. Many biomes are distributed altitudinally in mountains, in addition to their latitudinal distribution.

5. Species richness varies latitudinally with more species occurring in the tropics than at higher latitudes. Species richness is greater in more topographically complex mountainous regions than flat regions and less in peninsulas than adjacent, mainland areas.

6. The *tundra biome* is found in the Arctic and above the treeline in mountains at all latitudes. In the Arctic tundra, most primary production occurs during the short summers and many animals leave during the long winters or are dormant. Because of *permafrost*, tundra soil is very wet although precipitation is low.

7. The *boreal forest* biome is dominated by evergreen, coniferous trees in the Northern Hemisphere, but by beech in the Southern Hemisphere. Long cold winters, short warm summers, and communities with low species numbers characterize this biome. Most plants are wind-pollinated and produce large crops of wind-dispersed seeds that are the major food supply for many animal species.

8. The *temperate deciduous forest* biome has a greater diversity of tree species than the boreal forest, and its species are adapted for maximum photosynthesis during the summers when temperatures are warmer and precipitation is good. Forest structure is more complex, with a well-defined understory, and many tree species rely on animals for pollination and seed dispersal.

9. The *grassland* biome is common in areas with restricted rainfall during a significant portion of the year. Unlike tropical grasslands, temperate grasslands experience large seasonal temperature differences. Grasses store much energy in their root systems and can withstand heavy grazing. In addition to grasses, sedges and forbs are also common plant types.

10. The *cold desert* biome is common in the interiors of large continents and in rain shadow areas. Most rain falls in the winter and the dominant low shrubs grow mostly in spring. Many seed-eating animals live in this biome.

11. The *hot desert* biome is found in a belt 30° north and south of the equator where dry air descends in its circulation through the troposphere and rain falls mostly during the summer. Succulent plants and annuals predominate and the large seed crop supports many insects and rodents.

12. The *chaparral,* or Mediterranean, biome is found in areas with maritime climates where winters are cool and wet and summers hot and dry. Vegetation is dominated by low evergreen shrubs with drought-resistant leaves and bird-dispersed fruit.

13. The *thorn forest* is usually found on the equatorial side of hot desert biomes where a rainy season occurs in the summer. The biota is similar to that of the hot desert biome.

14. The *tropical savanna* biome is common in dry, tropical areas of Africa, Australia, and South America. The savanna is a grassland with scattered trees and large, diverse populations of grazing and browsing mammals.

15. The *tropical deciduous forest* biome replaces thorn forest where the rainy season is longer. Species diversity is great and most trees lose their leaves during the dry season. Much of this biome has been cleared for agriculture because soils are more fertile than in wetter areas.

16. The *tropical evergreen forest* biome occurs where the rainy season predominates and total rainfall exceeds 250 cm per year. This biome has the greatest species diversity and energy flux found on Earth. Its productivity is based on rapid nutrient cycling. Food webs and ecological interrelationships are complex and fragile.

17. Land is a barrier for dispersal of many aquatic organisms. Some aquatic species have flying adults, wind-borne spores or seeds, or can be transported by other organisms.

18. Species richness is great in freshwater biomes despite their small extent. Insects are especially prominent in freshwater biomes. The life cycle of many aquatic insects includes a winged adult disperser stage.

19. Salt water is a barrier to the dispersal of freshwater fish species that are not able to tolerate it. Many of these species are restricted to a single continent or they are derived from an ancient lineage that was widely distributed in Pangaea or Gondwanaland.

20. There are no physical barriers that separate marine biomes, since all of the ocean basins are connected. Environmental limits do restrict the distribution of marine organisms.

21. Horizontal and vertical gradients of light, temperature, pressure, and food supply divide the oceans into zones.

22. The ocean bottom is called the *benthic zone*, with the portion of this zone below the penetration of light named the *abyssal zone*.

23. The open waters of the ocean are the *pelagic zone*. Because nutrients sink, the open ocean zone has limited productivity except in areas where upwelling returns nutrients to the surface.

24. The *littoral zone* is the coastal area between high and low tides. Communities in the littoral zone are more diverse and complex than pelagic communities.

25. Biotas that are adapted for one zone have difficulty dispersing across intervening zones. For example, deep water is a barrier to organisms that are adapted for life in shallow water. Nutrient-rich waters have different biota than nutrient-poor waters.

26. Because of the prevalence of ocean currents, the distribution of most marine organisms cannot be related to continental drift. Distributions of marine organisms can be attributed to more recent vicariant events, such as rapid changes in sea level due to tectonic events or expansion and contraction of continental glaciers.

Questions (for answers and explanations, see page 440)

Questions (for answers and explanations, see page 440)

1. The biome that is maintained by browsers, grazers, or fire is the

    a. tundra.

    b. savanna.

    c. cold desert.

    d. boreal forest.

    e. tropical deciduous forest.

2. Match the letters of the following biomes with the descriptions that follow.

   a. Tundra biome   b. Boreal forest biome
   c. Cold desert biome   d. Chaparral biome
   e. Tropical deciduous forest biome

   _____ Mostly coniferous, wind-pollinated and wind-dispersed tree species

   _____ Leaves lost during dry season; agriculturally desirable land

   _____ Cool winters, hot dry summers; maritime climate

   _____ Frequently found in wind shadows

   _____ Distribution altitudinally or latitudinally determined; permafrost present

3. The following climograph shows yearly variation in rainfall and temperature for four biomes. Select the correct curve for each of the following biomes.

   _____ Tundra

   _____ Tropical evergreen forest biome

   _____ Temperate deciduous forest biome

   _____ Cold desert biome

4. The tropical evergreen forest biome has relatively constant _____ , but _____ varies seasonally; the temperate deciduous forest biome has relatively constant _____ , but _____ varies seasonally.

5. Which of the following comparisons of freshwater and marine biomes is *not* true?
   a. The global volume of the marine biome is much greater than that of the freshwater biome.
   b. There are many more insect species in the freshwater biome than the marine biome.
   c. The number of species of marine organisms is greater than the number of freshwater species.

   d. Freshwater species richness is less than marine species richness in proportion to the relative extent of the two biomes.
   e. About 10 percent of all aquatic species live in freshwater.

6. Rank the following locations from lowest (1) to highest (5) in terms of the expected species diversity.

   _____ Mountainous forest area on a peninsula in the temperate zone

   _____ Lowland forest area on a peninsula in the temperate zone

   _____ Mountainous forest area on a mainland in the temperate zone

   _____ Lowland forest area on a mainland in the temperate zone

   _____ Mountainous forest area on an island in the tropics

7. Next to each of the following features, place "P," "L," "B," or "A," if it is characteristic of the *pelagic, littoral, benthic,* or *abyssal* zones of marine communities (Note: some features may apply to more than one zone).

   _____ The ocean floor below the penetration of light

   _____ Coastal zone affected by wave action

   _____ Organisms in this zone must have adaptations to resist sinking

   _____ Divided into a lighted and dark region

   _____ Globally, the zone occupying the smallest total area

8. Select *all* of the following that might be *true* of a group of freshwater fish species that are widely distributed on several continents.
   a. They are able to tolerate saltwater.
   b. They are a recently derived group.
   c. They have an ancient lineage that was widely distributed in Pangaea and Gondwanaland.
   d. Their distribution was not affected by continental drift.
   e. Their distribution can be explained by movements of glaciers.

9. Which of the following statements about the biogeography of marine biomes is *not* true?
   a. Vertical gradients of light and temperature divide the ocean into different zones.
   b. In interpreting the biogeography of the marine biome, oceanic currents are more important than physical barriers.
   c. Organisms adapted for life in nutrient-poor waters usually flourish in nutrient-rich areas.
   d. Continental drift does not explain the distribution of many marine organisms.
   e. Recent vicariant events, such as changes in sea level, are important in determining the distribution of marine organisms.

55

Conservation Biology

CHAPTER LEARNING OBJECTIVES—after studying this chapter you should be able to:

❑ Describe the objectives of conservation biology, and list the areas of biology that have contributed to its development.

❑ Explain how species–area relationships can be used to estimate extinction rates caused by habitat destruction.

❑ Characterize the factors that increase a species' probability of becoming extinct as its numbers are reduced to a low level.

❑ Define the term "minimum viable population" (MVP).

❑ Describe the process of population viability analysis and its objectives.

❑ Differentiate between demographic stochasticity and genetic stochasticity, and explain how each contributes to determining the minimum viable population.

❑ Provide some examples of extinctions caused by overexploitation.

❑ Provide some examples of extinctions caused by introduced pests, predators, and competitors.

❑ Discuss the effect of modern agriculture on species diversity and the general phenomenon of co-option.

❑ Explain what ecosystem services are, and describe the major ecosystem services.

❑ Provide some examples of extinctions caused by habitat destruction and creation of habitat islands.

❑ Describe the projected magnitude of global warming and its expected effects on communities and individual species.

❑ Define what endemic species are and explain where most endemic species occur.

❑ Explain what types of organisms are likely to be keystone species and why identification of a community's keystone species is important to conservation efforts.

❑ Describe how the community importance of a species is measured and how this concept relates to keystone species.

❑ Discuss the process of habitat fragmentation in terms of size effects and edge effects, and provide some examples of its importance.

❑ Explain the objectives of captive propagation, and provide some examples of its successful application.

❑ Describe the features of a megareserve, and provide an example.

❑ Describe the features of a forest reserve, and provide an example.

❑ Characterize the objectives of the subdiscipline of conservation biology called restoration ecology, and provide an example of its successful application.

ESTIMATING CURRENT RATES OF EXTINCTION • WHY CARE ABOUT SPECIES EXTINCTIONS? • DETERMINING CAUSES OF ENDANGERMENT AND EXTINCTION (pages 1223–1232)

Key Concepts

1. *Conservation biology* is concerned with the causes of species richness and the preservation of genes, species, and communities. Its conceptual basis is in the fields of ecology, population genetics, evolutionary biology, biogeography, and wildlife management.

2. The species–area relationship, that the number of species present increases with the size of an area, can be used to estimate the extinction rate by habitat destruction.

3. Rare species are more susceptible to extirpation (disappearance from an area) due to sudden local disturbances.

4. The *minimum viable population (MVP)* is the estimated population size required for continued species survival and growth.

5. *Demographic stochasticity* is the amount of variation in birth and death rates in a population.

6. Small populations are more at risk of extinction from demographic stochasticity, especially if low birth rates and high death rates coincide.

7. Many species are distributed as metapopulations—collections of subpopulations, with each subpopulation occupying a suitable habitat patch. Extinction of subpopulations and recolonization of unoccupied patches determines the metapopulation structure.

8. *Genetic stochasticity* is the amount of variation in the genetic makeup of a population. Lowering of fitness due to inbreeding is common in small populations.

9. *Population viability analysis (PVA)* estimates the influence of population size on the probability of extinction for a species; PVA must take into account both demographic and genetic stochasticity.

10. Our ability to identify useful natural plant and animal products will be compromised if species richness is reduced by an increase in the global extinction rate.

11. *Ecosystem services* include absorption of CO_2, release of O_2, cleansing of water, regulation of stream flow, recreation, and aesthetics.

12. Many ecosystem services, like aesthetic benefits, cannot be replaced by technology.

13. Species requiring uncommon habitats (caves, saline lakes, etc.) are naturally rare. Secondary carnivores are rare for trophic reasons. Rare species are more likely to become extinct than common species.

14. Overexploitation due to hunting by humans has been a common pattern. Large mammals were exterminated by overhunting in North America and Australia after human colonization of those areas.

15. Accidental or purposeful introduction of pests, predators, and competitors has caused the extinction of many native species. Introduced pigs and black rats have been destructive to many island organisms, such as the tortoises in the Galapagos archipelago. Chestnut blight and Dutch elm disease were inadvertently carried to North America from overseas.

16. Plant–pollinator mutualisms are often highly species-specific on islands where ecological communities are species-poor. Extinction of many species of Hawaiian honeycreepers, sole pollinators of Hawaiian lobelias, have left many lobelias without pollinators.

17. Traditional agricultural ecosystems maintain many more natural species than does modern, high energy input agriculture. Modern agriculture uses chemicals to eliminate competing species.

18. *Co-option* is the diversion of an ecosystem's primary production into species intended for human use; today, about 30 percent of terrestrial production is co-opted.

19. Habitat loss or the creation of habitat islands too small to support dependent species is likely to be the major cause of extinction during the coming decades.

20. Loss and deterioration of habitat results in isolated, small, local populations that are very susceptible to local deleterious effects.

21. Species with large home ranges require large natural patches to survive and reproduce.

22. Because of area–edge relationships, small habitat patches have proportionally greater *edge effects* than larger patches.

23. Because of *exploitation of resources,* some habitats are never able to reach the late stages of succession needed by some species. For example, logging of old-growth forests of the northwestern United States and recutting forests after 60 to 80 years prevents mature forests from being reestablished.

24. Information on the sizes of habitat patches required by different species is important to know before habitats become fragmented.

25. Increased concentrations of greenhouse gases may cause a 2 to 5°C increase in temperature within the next 100 years. This warming trend will occur more rapidly than past global climate changes.

26. The rapidity of global warming, combined with the fragmented habitats of many species, may result in the loss of many species and the creation of new, hotter climates in the lowland tropics.

Questions (for answers and explanations, see page 440)

1. The concepts of conservation biology come mainly from all *but* which of the following fields?
   a. Ecology
   b. Evolutionary biology
   c. Population genetics
   d. Immunology
   e. Wildlife management

2. Which of the following is *not* presently a major cause of the global reduction in biodiversity?
   a. Overexploitation
   b. Global warming
   c. Habitat destruction
   d. Overhunting
   e. Introduction of foreign predators and disease

3. Based on species–area relationships, ecologists predict that
   a. a 90 percent loss of habitat will result in loss of about 9 percent of the species living there.
   b. 50 percent of Earth's species may become extinct in the next 50 years.
   c. most species will be able to adapt by moving into parks.
   d. the area required by most species will need to be reduced.
   e. extinction is inevitable.

4. Co-option is
   a. a result of overhunting.
   b. mostly caused by modern agriculture and forestry.
   c. more likely to affect flightless island bird species.
   d. unimportant in terrestrial ecosystems.
   e. a natural process.

5. Which of the following is *not* true of small populations in comparison with larger populations?
   a. More subject to extinction caused by demographic stochasticity
   b. Show greater genetic heterozygosity
   c. More subject to inbreeding
   d. May be more susceptible to extirpation
   e. Less able to respond to environmental change

6. Global warming may lead to extensive decimation of species because (select *all* correct answers)
   a. the rate of climate change will be more rapid than past periods of climate changes.
   b. the magnitude of climate change will be much greater than past periods of climate change.
   c. habitat fragmentation will make it difficult for some species to shift their ranges.
   d. many plant species may not be able to shift their ranges at the same pace as the northern movement of temperature zones.

e. since little change in plant community composition has occurred in the past, we cannot expect present communities to adapt to climate change.

7. In the following graph, select the curve (*a, b,* or *c*) that *correctly* shows the expected relationship between habitat patch area and the proportion influenced by edge effects.

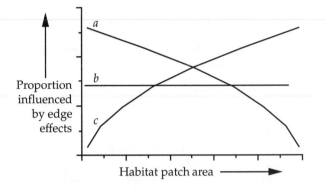

Activities (for answers and explanations, see page 440)

• Discuss the importance of the "rescue effect" for species that are distributed as a metapopulation.

• Aside from the aesthetic benefits of biodiversity, list three economic reasons people should care about species extinctions.

DESIGNING RECOVERY PLANS • ESTABLISHING PRIORITIES FOR CONSERVATION EFFORTS • THE IMPORTANCE OF COMMERCIAL LANDS • RESTORING DEGRADED ECOSYSTEMS • MARKETS AND CONSERVATION • INVESTING TO PRESERVE BIODIVERSITY (pages 1233–1241)

Key Concepts

1. Preservation of habitats is important because most species can survive only in the habitats in which they evolved.

2. Preservation of a key species—so-called "umbrella species"—that requires a large amount of suitable habitat assists the preservation of other species that live in the same community. Examples include elephants, grizzly bears, and the spotted owl.

3. Songbirds in eastern North America are declining because of increased predation pressure from domestic dogs and cats and native predators.

4. Habitat modification may favor certain native species of predators and parasites that can have unforeseen effects on ecological communities. Reduction in songbird populations in North America in recent years may be related to the increase in populations of the brood-parasitic brown-headed cowbird associated with habitat modification.

5. North American song bird populations are also affected by destruction of their wintering habitats in Central and South America.

6. Kirtland's warbler is an endangered species due to heavy brood parasitism by cowbirds and loss of its breeding habitat—8 to 18 year-old stands of jack pine previously maintained by fires.

7. When individuals are reintroduced into an area from which they were extirpated, the genotypes of the colonists must be carefully matched to the habitats they will occupy.

8. Because *keystone species* strongly influence the structure and functioning of an ecological community, it is important that they be identified and preserved.

9. *Community importance* is a measure of the change in a community or ecosystem trait per unit change in the abundance of a species, where a trait is any measurable feature of the community or ecosystem. Keystone species typically have great community importance.

10. Species that serve as an important food source can be keystone species, as are herbivores that affect habitat structure and succession or predators that control the population sizes of important prey species that might otherwise dominate the community.

11. *Captive propagation* allows a species to be maintained in captivity while threats to its existence in the wild can be reduced.

12. Captive propagation was a key factor in the successful reintroduction of the peregrine falcon. Young from captive-bred birds were returned to former breeding sites after reduced levels of DDT in the peregrine's food chain no longer posed a threat to its reproductive biology.

13. The primary purpose of parks and reserves is to set aside areas that are relatively free of human disturbances. Preservation of ecological communities and ecosystems insures survival of its inhabitants.

14. The most species-rich communities are in the tropics, so major conservation efforts focus on this global region.

15. *Endemic species* are found in only one place. Islands, mountainous regions, and areas with many habitat islands are likely to have a high proportion of endemic species. Madagascar is a good example of a center of endemism—nearly all of its birds and mammals are found nowhere else.

16. A region's conservation potential depends on estimates of the number of endemic species present, the amount of habitat that is already protected, and the amount of natural

habitat that is unprotected. High-priority regions have much unprotected habitat available; low-priority regions have little remaining natural habitat available to protect.

17. *Forest reserves* are maintained as forests from which economically valuable products are harvested.

18. A *megareserve* is centered on an undisturbed natural area and is surrounded by an area where nondestructive economic activities are permitted. A buffer zone where some forestry and agriculture occurs encircles the megareserve. The eight megareserves in Costa Rica should maintain 80 percent of the country's biodiversity.

19. *Restoration ecology* applies basic knowledge of the functioning of an ecosystem to restore it to its former state in order to conserve biodiversity. The ability to restore natural ecosystems is still limited by our knowledge of their functioning.

20. Although species are common property resources, in some cases an incentive to conserve species can be gained if individuals are given property rights to the economic benefits derived from a species. Butterfly farming in Papua New Guinea is an example.

21. Although genes, species, and habitats are economic and aesthetic resources, their current and future value is difficult to measure. However, the irreversibility of extinction makes the preservation of biodiversity an urgent issue.

Questions (for answers and explanations, see page 440)

1. Endemism is likely to be common (select *all* correct answers)
   a. on islands.
   b. among keystone species in continental areas with uniform habitat.
   c. for aquatic species in a region with isolated lakes.
   d. wherever habitat islands exist.
   e. in mountainous areas.

2. Which of the following is *not* a keystone species?
   a. Kirtland's warbler
   b. Fig and palm species in some tropical forests
   c. Beaver
   d. Humans
   e. Any species with a large community importance

3. Which of the following is *not* an "umbrella species"?
   a. A species that is endemic
   b. Grizzly bear
   c. African elephant
   d. Spotted owl
   e. A species requiring large areas of a specific habitat

4. Which one of the following factors is *not* probably important in causing the current declines in the populations of many North American songbird species?
   a. Increased predation
   b. Destruction of wintering habitat
   c. Global warming
   d. Increases in populations of brown-headed cowbirds
   e. Decreased habitat patch size

5. Which of the following statements about the successful reintroduction of the peregrine falcon is *not* true?
   a. The peregrine's habitat became more suitable after use of DDT was restricted.
   b. DDT adversely affected the peregrine's reproductive physiology.
   c. Reintroducing peregrines into habitats where they were not at the top of the food chain was critical to the success of the program.
   d. Captive propagation was an important part of the peregrine reintroduction effort.
   e. Peregrine eggs are less fragile today then they were in 1960.

6. Select *all* of the following characteristics that would cause a country to be considered a high-priority region for conservation efforts.
   a. A high degree of endemism
   b. Having much natural habitat remaining
   c. Having little natural habitat already protected
   d. Having most of the available natural habitat in megareserves
   e. Having low species richness

Activities (for answers and explanations, see page 441)

● What is the structure of a megareserve and how does this structure help maintain biodiversity?

● Why is cattle grazing being encouraged as part of the effort to restore the tropical deciduous forest in Guanacaste National Park in Costa Rica?

Answers and
Explanations

Chapters 50–55

CHAPTER 50 – BEHAVIORAL ECOLOGY

Note: the page numbers listed with the section titles refer to the study guide; numbers in parentheses for each question refer to the relevant key concepts.

CHOOSING ENVIRONMENTS • COSTS AND BENEFITS OF BEHAVIORS • MATING TACTICS AND ROLES (pages 403–405)

1. Choice *d*. All except the modification of the environment by physical factors are normally included within the domain of ecology. (1–4)

2. Choice *d*. The forfeited benefits of behaviors that could not be performed as a result of performing a different behavior, like scanning, comprise the *opportunity cost* of the performed behavior. (10–14)

3. Choice *d*. A territory that is used exclusively for nesting and rearing young is characteristic of many colonial marine birds. The song sparrow has an all-purpose territory; the red-winged blackbird has a territory that provides a space for rearing young and some, but not all, food resources; polygamous species subject to strong sexual selection, such as many manakins, have small display territories that are part of a lek. (22–27)

4. Choices *a, c, d*. The number of species 2 taken by the predator increases markedly at low species 1 densities, indicating that the predator includes species 2 in its diet only when insufficient species 1 individuals are available. The number of species 1 taken is independent of species 1 density. (6–9)

5. Choice *a*. The energetic costs of a dominant male are great. It must fight to establish its territory and spend much of its time defending its females from challengers. Elephant seals are polygynous; males with territories have a "harem" of many females. The larger size of the males and their aggressive behavior is the result of sexual selection. (8, 18, 22).

6. Choice *c*. The females that had successfully raised young sang at their nest more than unsuccessful females. The cost of attracting predators to the nest by singing was less than the benefit derived by increased nest defense from neighboring males. (10–15)

• In studies where bluegill sunfish were provided with small, medium, and large waterfleas at three different prey densities, the fish took equal proportions of the three prey sizes at low densities, but mostly large water fleas at high prey densities. This agrees with predictions from foraging theory on how a predator should behave to maximize its energy input. (6–9)

• Sexual selection is the spread of a trait that improves the reproductive success of the individual, as determined by mating preferences of individuals of the opposite sex. By artificially lengthening or shortening the tails of male widowbirds, behavioral ecologists were able to show that increased tail length attracted more females but did not confer an advantage in interactions with other males. (15–18)

SOCIAL BEHAVIOR: COSTS AND BENEFITS • CATEGORIES OF SOCIAL ACTS • THE EVOLUTION OF ANIMAL SOCIETIES (pages 405–408)

1. Choice *e*. Altruistic behavior is beneficial to the performer when the improvement in the reproductive success of relatives exceeds the reduced reproductive success of the individual performing the act. This would be true if the behavior improves the *inclusive fitness* (individual fitness + kin selection) of the performer. (6–12)

2. Choice *a*. A selfish act benefits the performer at the expense of the recipient of the act. (1, 6–9)

3. Choice *b*. Because of segregation of alleles during meiosis, the probability that the allele would be transferred to the child's parent from the grandfather is 0.5. Similarly, the probability that the allele would then be transferred to the child from the parent is 0.5. The probability of the joint occurrence of two or more independent events is the *product* of their separate probabilities. Here, the probability would be 0.5×0.5 or $(0.5)^2 = 0.25$. (13)

4. Choices *b, c, e*. Reciprocal altruism involves cooperative behavior between unrelated individuals, the behavior is reciprocated so all benefit, and the behavior has a genetic basis. (20)

5. Choices *b* and *c*. Only some eusocial species have features a, d, and e. (14–19)

6. Your curves should show a positive relationship between the hawk's distance when spotted and pigeon flock size and a negative relationship between the hawk's attack success and pigeon flock size (see below). (2, 3)

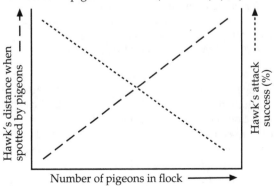

7. Choice *a*. (14, 27).

8. Choice *c*. The distribution of breeding territories was not mentioned as an important factor in the evolution of social organization in African hoofed mammals. (33)

9. Choice *b, c, d, e*. A dominance hierarchy *is* a type of social organization, not a determining factor. (31–35)

- You and a half-sibling would each have a 0.5 probability of inheriting the same allele from your common parent, so the coefficient of relatedness is $r = (0.5)^2 = 0.25$. For full cousins it is $r = 2(0.5)^4 = 0.125$. To increase your inclusiveness fitness you should save the more closely related relative; so, save your half-sibling. (6, 8)

- Because females are more closely related to their sisters ($r = 0.75$) than they are to any offspring they could produce ($r = 0.5$), this theory predicts that in a species with haplo-diploid sex determination, females can increase their inclusive fitness by foregoing reproduction and helping to raise sisters. In these societies, the sex ratio should be skewed in favor of females, and it usually is. The haplo-diploid hypothesis cannot explain the evolution of eusociality in species without this mode of sex determination or in haplo-diploid species with multiple queens per colony or in species in which the queen mates with multiple males. (9, 10)

- Forest-dwelling African weaverbird species are monogamous, eat insects, build inconspicuous nests, and males and females are similar in appearance. Savanna-dwelling species are polygynous, eat seeds, build elaborate nests in colonies in *Acacia* trees, and males are brightly colored. The major environmental variable causing these differences seems to be that nest sites are dispersed randomly in forests, but are highly clumped in savannas. (12)

Integrative Questions

1. Choices *a, c, d*.

2. Red-winged blackbirds: *g*
   Manakins: *e*
   Honeybees: *a, f*
   Naked mole rat: *a, f*
   Scrub jay: *c*
   Song sparrow: *b*
   African lions: *d*
   Bee-eaters: *c*

CHAPTER 51 – POPULATION ECOLOGY

Note: the page numbers listed with the section titles refer to the study guide; numbers in parentheses for each question refer to the relevant key concepts.

VARIED LIFE HISTORIES • POPULATION STRUCTURE: PATTERNS IN SPACE (pages 409–411)

1. Choice *a*. Population structure includes numbers, spacing, and age distribution of population members, but not the distribution of genotypes. Biomass is best used as a measure of density for some organisms like plants. (16–19)

2. If the resources available to the plant are fixed, increasing the number of seeds produced would result in a decrease in both variables. (3–8)

3. Choice *d*. A uniform distribution normally results from competition for a uniformly distributed resource or from territoriality. Settling near one's birthplace or a patchy habitat can cause a clumped distribution. The interaction of several factors may lead to a random distribution. (19–22)

4. Choice *c*. Both salmon and annual plants reproduce and then die, so they are most similar in the reproductive stage of their life histories. (3–5, 9)

5. Choice *a*. Since new, unoccupied habitat is constantly being created in disturbed areas, a good strategy is to produce large numbers of smaller offspring. (5–11)

6. The marked individuals were 20% of the second sample. If you assume that the original 100 marked bats redistributed themselves randomly within the whole population, then 100 is 20% of the total population. So, the total population is $100/0.2 = 500$ individuals. (17)

- Given fixed resources, there is an inverse relationship between the number of offspring produced and the amount of energy allocated to each. Also, there is an inverse relationship between the number of offspring and the amount of care each can receive. A final trade-off is that energy expended in reproduction is not available for growth and maintenance of the parent, so reproduction usually reduces the probability of parental survival. (5–8)

- Reproductive value is the number of offspring remaining to be born to an individual of a particular age. Reproductive value decreases with age and is zero for postreproductive individuals. Natural selection cannot act on traits that are expressed after reproduction is complete. Thus, many of the diseases afflicting postreproductive segments of populations in developed countries (cancer, cardiovascular disease, etc.) are beyond the reach of natural selection. (12–15)

POPULATION DYNAMICS: CHANGES OVER TIME • EXPONENTIAL POPULATION GROWTH • POPULATION GROWTH IN LIMITED ENVIRONMENTS (pages 411–412)

1. Choice *b*. Both curves show exponential growth, but the rate of exponential growth for curve 2 is less than the rate for curve 1. Since the rate of growth is determined by the expression $(b - d)$, increasing d, the death rate, would cause curve 1 to be more like curve 2. (12–14)

2. Curve **2**. Since $r_{max} = (b - d)$, you would expect the growth curve for species A to correspond to the curve with the smaller rate of exponential growth. (12–14)

3. Choice *c*. The curve showing the relationship between population size and time is steepest at point *c*, so the growth rate would be greatest at that point. (12–20)

4. Choice *e*. The growth rate at *e* would be zero. This is the point called the carrying capacity, the point where the birth and death rates are equal. (12–21)

5. Choice *a*. The decrease in consecutive age classes is least between 0 and 1 year, so the survivorship is greatest during that interval. (2–8)

- The rescue effect occurs if a subpopulation that has undergone a large decline is saved from extinction by the immigration of individuals from other subpopulations within the metapopulation. (22)

POPULATION REGULATION • HUMANS MANAGE POPULATIONS (pages 412–413)

1. Choice *c*. The best way to control a pest species is to reduce the carrying capacity of the habitat for the species. (14)

2. Choice *b*. Expansion of a species' range is a long-term population change and not a response to seasonal changes in habitat quality. (6–9)

3. Choice *b*. The graph indicates that the birth rate is independent of population size, so event *b* would not be true. The graph shows that the death rate is density-dependent. (1–4)

4. Choice *b*. The carrying capacity is the equilibrium population size when the birth rate equals the death rate The birth rate equals the death rate at the intersection of the two curves. (1–4)

- The environmental conditions needed by the species are usually best towards the center of the range and become less optimum towards the periphery of the range. (9)
- Species *A* shows population regulation dominated by limitations on the environmental carrying capacity. Bird populations are frequently regulated in this manner. Species *B*'s population is maintained well below the carrying capacity and its fluctuations probably are caused by environmental disturbances. Small organisms like insects typically show this type of regulation. (2–6)

Integrative Question

1. Choice *a*. Plants are *modular* organisms, they frequently show a *clumped* spatial distribution, and their population density is most appropriately expressed in terms of *biomass*.

2. Choice *a*. When a population's birth rate and death rate are both high, the age distribution is dominated by young individuals. A type II survivorship curve results if survivorship is uniform throughout life. The age distribution of such a population would have a constant decline in proportions of the population in all age categories.

CHAPTER 52 – COMMUNITY ECOLOGY

Note: the page numbers listed with the section titles refer to the study guide; numbers in parentheses for each question refer to the relevant key concepts.

ECOLOGICAL INTERACTIONS • NICHES: THE CONDITIONS UNDER WHICH SPECIES PERSIST • COMPETITION: SEEKING SCARCE RESOURCES (pages 414–416)

1. Choice *d*. Both species have benefited; the bird dispersed and provided fertilizer for the plant's seed, the plant provided food for the bird. (3–7)

2. Choices *c* and *f*. Both ants and birds are competing for the same food supply; the birds also benefit from the ants' activity in flushing the insects, but the ants might not be affected. (3–7)

3. Choice *e*. A resource must be needed by an organism, but also must be present in limited supply (use of a resource depletes it). Atmospheric oxygen is so abundant that terrestrial animals never reduce its concentration. (8, 12, 13)

4. Choice *e*. If the species compete for limited resources, the ecological niches of *both* species will be more constrained in areas where the ranges overlap and will differ from their niches before the range expansion. (8–14)

5. Choice *b*. Both species grow less well when in the presence of the other; the effect is greater for species A. (14–19)

- The graphs show that the population growth of species 1 is more affected by the presence of species 2 than the reverse. Since the competition coefficients quantify the per capita effect of one species on the growth of the other, α, the effect of species 2 on species 1, should be *larger* than β, the effect of species 1 on species 2. (15–19)
- A removal experiment could be done in which one or the other species is removed in two similar areas. One then looks for changes in the abundance of the remaining species. The species with the largest increase in population density was most affected by competition with the removed species. (15, 16, 21, 22)

PREDATOR–PREY INTERACTIONS • BENEFICIAL INTERSPECIFIC INTERACTIONS • TEMPORAL CHANGES IN COMMUNITIES • COEVOLUTION OF INTERACTING SPECIES • KEYSTONE SPECIES: MAJOR INFLUENCES ON COMMUNITY COMPOSITION • INDIRECT EFFECTS OF INTERACTIONS AMONG SPECIES (pages 416–418)

1. Choice *c*. The key to answering this question is realizing that prey cycles always lead predator cycles, so the black line indicates the prey and the gray line the predator. (9–12)

2. Choices *a, b, c, e*. Density-dependent interactions between predator and prey populations are what normally lead to oscillations. (9–14)

3. Choice *c*. Although it is true that all three could be Müllerian mimics, the observation that species 2 is much less abundant than the other two species suggests that it may be a Batesian mimic. (15–21)

4. Choice *c*. Diseases spread by nonliving vectors, like water, are more serious than diseases spread by a living vector because such disease organisms can still be dispersed after the host dies. (22, 23)

5. Choice *a*. This is an example of parasitism since the wasp is harmed by wasting its time courting a flower instead of a female wasp. The flower benefits from pollination. (27–33)

6. Choice *a*. Initially, the plant is harmed but the animal is unaffected (amensalism). After the evolution of sharp spines, the two species are competitors for space. (27–33)

7. Choice *c*. Most plants decrease CO_2 levels by incorporating it within their biomass. (42)

8. (34–36)
 1 Soil formation by physical processes
 3 Arrival of willows and alders
 2 Arrival of lichens
 5 Arrival of conifers
 4 Increase in soil nitrogen content

9. Choice *b*. By definition, removal of a keystone species has a larger effect on the community than would be expected based on its abundance. (41–44)

- Plants can produce thorns, hairs, or spines, or construct tough leaves. Leaves with a lot of tannin are difficult for herbivores to digest. Many plants produce secondary products that are toxic or mimic insect hormones that disrupt normal development. (24–26)
- The moth has given up the opportunity to feed from other species of plants and the Yucca has given up the opportunity to use other pollinators. Both benefit from being able to specialize at exploiting the other as an exclusive resource. (38–40)

- Nematode *A* probably parasitizes a wasp species in which there is only a single foundress per fig; the wasp species parasitized by nematode *B* probably has multiple foundresses per fig. If all of the offspring are killed by nematodes of species *A*, the nematodes perish as well, so this species evolves to be less virulent. Nematode *B* may have access to additional wasp broods, so it can evolve to be more virulent. (7, 38)

CHAPTER 53 – ECOSYSTEMS

Note: the page numbers listed with the section titles refer to the study guide; numbers in parentheses for each question refer to the relevant key concepts.

CLIMATES ON EARTH • ENERGY FLOW THROUGH ECOSYSTEMS (pages 419–422)

1. Choice *d*. All areas on Earth receive equal hours of daylight per year. The seasonal distribution of those hours, the solar energy input, and the sun angle do vary latitudinally. (4–6, 13)

2. Choice *e*. The air would have lost most of its moisture while rising on the windward side of the mountain and, in descending to area *e*, would have warmed again. (7, 8)

3. Choice *b*. In the equatorial convergence zone, warm air rises and loses much of its moisture. These same events occur when moist air rises over a mountain. (9–12)

4. Choice *c*. If Earth did not spin on its axis, air flowing south toward the equator would not be deflected to the right. So the northeast trade winds would blow directly from the north instead of from the northeast. (10–12)

5. Choice *c*. The upwelling of deeper water, especially on the western side of continents, is an effect of the pattern of ocean current circulation and not a cause of the pattern. (17, 18)

6. Choice *c*. Because of the high specific heat of water, a large body of water moderates the climate of adjacent land masses, creating a maritime climate. (16, 19–21)

7. Choice *c*. An ecosystem includes the environment *and* the biological community living there. A coral reef ecosystem would include much more than just the fish. (1, 2)

8. Choice *d*. The amount of production used in maintenance and biosynthesis is added to net primary production to determine gross primary production. Photosynthetic release of O_2 in the light is an estimate of net production; O_2 use in the dark is an estimate of maintenance and biosynthesis costs. You would add 0.02 ml to 0.14 ml to get an estimate of gross primary production of 0.16 ml per minute. (23–26)

9. Choice *e*. Since organism 9 eats from both the primary producer level (1) and the herbivore level (2), it would be an omnivore. (27–30)

10. *5* trophic levels. The levels are primary producer (1), herbivore (2, 3, 4), primary carnivore (5, 6), secondary carnivore (7, 8), and omnivore (9). (27–30)

11. *8 grams*. The flow of energy is from primary producers to herbivores to primary carnivores to secondary carnivores. If each of these three energy transfers is only 20% efficient, then the biomass of secondary carnivores should be:
$$1000 \text{ g} \times 0.2 = 200 \text{ g}; \quad 200 \text{ g} \times 0.2 = 40 \text{ g}; \quad 40 \text{ g} \times 0.2 = 8 \text{ g}$$
(27–31)

12. *Net production*. (31)
- In marine ecosystems, the primary producers are usually unicellular phytoplankton. Phytoplankton populations multiply so rapidly and are cropped so efficiently by herbivores (zooplankton) that they can sometimes support a larger herbivore biomass than their own. (33–35)
- (32)
  5: Herbivorous mammal
  1: Invertebrate detritivore
  4: Carnivorous mammal
  2: Invertebrate carnivore
  3: Invertebrate herbivore

AGRICULTURAL MANIPULATION OF ECOSYSTEM PRODUCTIVITY • CYCLES OF MATERIALS IN ECOSYSTEMS (pages 422–423)

1. (4–11)
   F: Seasonal mixing of materials
   O: Nutrient concentrations low, except locally
   O: Oxygen concentration vertically uniform
   O: Elements buried in bottom sediments for long periods of time
   F: Receives material from land mostly via groundwater

2. (13–16)
   T: Most water vapor resides here.
   S: Most ozone resides here.
   S: Circulation of gases is mostly horizontal in this layer.
   T: Represents the greatest mass of the total atmosphere.
   T: Circulation of this layer influences ocean currents.

3. (3–21)
   T: Very slow movement of materials within compartment
   T: Exchange of gases mostly via organisms
   O, A: Circulation of materials affected by Earth's rotation on its axis
   O, A, F: Circulation of materials affected by Earth's revolution around the sun.
   O, F, T: Organisms mostly in uppermost layer.

4. Choice *d*. The steep thermocline that occurs between 20 and 30 meters would typically only be established by mid- to late summer. (9–12)

5. *20* and *30* meters. This is the depth over which the temperature drops most quickly. (9–12)

6. Choice *c*. Indiscriminate aerial spraying that kills both useful and harmful insects alike is not normally employed in integrated pest management. (1, 2)

BIOGEOCHEMICAL CYCLES • HUMAN ALTERATIONS OF BIOGEOCHEMICAL CYCLES • CHEMOSYNTHESIS-POWERED ECOSYSTEMS (pages 423–424)

1. Choice *b*. Most elements cycle through organisms more quickly than they cycle through the nonliving world. (1, 2)

2. (3–15)
   N: Major reservoir is atmospheric
   C, P: Major reservoir is in sedimentary rocks
   N, P: Often in short supply in ecosystems
   C: Fossil fuel reserve is part of this cycle
   C, P: Major human impact on these cycles
   S: Involved in cloud formation
   P: Lacks a gaseous phase
   N: Major inorganic form is directly available to only a small group of bacteria
   C: Includes a form that is a greenhouse gas
   N: Most fluxes involve organisms

3. (17–20)

3: Escherichia coli populations proliferate (this probably occurred concomitant with algal blooms, since both resulted from increased input of nutrients and sewage).

1: Phosphorus input from sewage and agricultural runoff increases.

4: Respiratory demands from decomposers increase.

6: Fish species change.

5: Oxygen levels drop in deeper water.

2: Algal blooms occur.

4. Choice *b*. The use of high sulfur coal has more to do with acid precipitation than it does with global warming. (21–25)

• The rate at which a lake can recover from eutrophication depends on the rate of turnover of its water. Lake Erie cannot be returned to its pre-eutrophication condition because irreversible changes have occurred. For example, in Erie, several key species became extinct and other new species were introduced. (20)

CHAPTER 54 – BIOGEOGRAPHY

Note: the page numbers listed with the section titles refer to the study guide; numbers in parentheses for each question refer to the relevant key concepts.

WHY ARE SPECIES FOUND WHERE THEY ARE? • HISTORICAL BIOGEOGRAPHY (pages 425–427)

1. Choice *a*. Experimental results to explain the distribution and abundance of present-day species would be more typically used by an ecological biogeographer. (5–11)

2. Choice *c*. Because the number of endemic taxa in an area is related to the period of time that the area has been separated from other areas, we would expect that continent Y had always been separate and continent W had been separated first from the other two. (14–16)

3. Choices *d, e*. A vicariant distribution requires that the species once had a continuous distribution in what are now separate areas. The species is not endemic on island C since it is also found on islands A and B. (12–14)

4. Choice *b*. (19, 20)

5. Ecological biogeography is mostly concerned with a local spatial scale and events occurring within the past few thousand years. (8, 9)

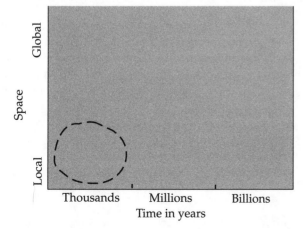

• Species 2, 7, and 1 represent the clade that first separated from the others. That each species in this clade is found on all these continents suggests that the three continents were once joined. The next clade includes species 3 and 6, which are found on continents A and B. This suggests that the original super continent split to form two continents. One became continent C. Species 3 and 6 evolved on the other continent, which then split to form continents A and B. Species 4 and 5, in the final clade, evolved on continent B after the split. (17, 18)

• Southern beeches are today found in South America, Australia, Tasmania, New Zealand, but not Africa or India. They once existed in Antarctia but became extinct there. Their distribution suggests that southern beeches were distributed across Gondwanaland after Africa and India drifted away. (3, 4)

ECOLOGICAL BIOGEOGRAPHY (pages 427–428)

1. Choice *e*. The extinction rate is nonzero and equals the arrival rate at n_1. Thus, n_1 is the equilibrium species number on the island. (3–10)

2. Choice *e*. Each curve corresponds to a group of similar-sized islands whose species number is plotted against their distance from the mainland. For each curve, species number decreases with distance from the mainland and the curve that is lowest relative to the *y* axis would be the group of smallest islands. (11–13)

3. More mobile groups like insects and birds should reach an equilibrium species number on the island more quickly than the plants, and we would expect more insect species than bird species. (16)

c: Birds

b: Plants

a: Insects

4. Choice *c*. With less space available, populations of the different species would be smaller. This would subject them to higher extinction rates than would be expected on larger islands. (16)

TERRESTRIAL BIOMES • AQUATIC BIOMES (pages 428–430)

1. Choice *b*. (6–16)

2. (6–16)

b: Mostly coniferous, wind-pollinated and wind-dispersed tree species

e: Leaves lost during dry season; agriculturally desirable land

d: Cool winters, hot dry summers; maritime climate

c: Frequently found in wind shadows

a: Distribution altitudinally or latitudinally determined; permafrost present

3. (6–16)

a: Tundra

d: Tropical evergreen forest biome

c: Temperate deciduous forest biome

b: Cold desert biome

4. The tropical evergreen forest biome has relatively constant *temperature*, but *rainfall* varies seasonally; the temperate deciduous forest biome has relatively constant *rainfall*, but *temperature* varies seasonally. (6–16)

5. Choice *d*. Even though the freshwater biome occupies a very small global area, it has about 10 percent of all aquatic species, which by proportion gives it a very large species richness.

6. (5)

   2: Mountainous forest area on a peninsula in the temperate zone

   1: Lowland forest area on a peninsula in the temperate zone

   4: Mountainous forest area on a mainland in the temperate zone

   3: Lowland forest area on a mainland in the temperate zone

   5: Mountainous forest area on an island in the tropics

7. (22–24)

   A: The ocean floor below the penetration of light

   L: Coastal zone affected by wave action

   P: Organisms in this zone must have adaptations to resist sinking

   P, B: Divided into a lighted and dark region

   L: Globally, the zone occupying the smallest total area

8. Choices *a, c.* They were either able to cross oceans to colonize new continents or they had ancestors existing on Pangaea and Gondwanaland before they split up. (19)

9. Choice *c.* Usually organisms adapted for one zone do not do well in a different zone. (20–26)

CHAPTER 55 – CONSERVATION BIOLOGY

Note: the page numbers listed with the section titles refer to the study guide; numbers in parentheses for each question refer to the relevant key concepts.

ESTIMATING CURRENT RATES OF EXTINCTION • WHY CARE ABOUT SPECIES EXTINCTION? • DETERMINING CAUSES OF ENDANGERMENT AND EXTINCTION (pages 431–433)

1. Choice *d.* Although immunology is sometimes a useful tool in obtaining information on the genetic stochasticity of a population, its importance is less than the other fields listed. (1)

2. Choice *b.* Global warming is predicted to be a major cause of extinction in the future, but is not yet having an impact on populations. (14–26)

3. Choice *b.* Based on estimates of current and future disappearance of tropical forests, perhaps 50 percent of Earth's species will become extinct in the next 50 years. This loss is not inevitable if we change our co-option of these ecosystems. (2, 18)

4. Choice *b.* As mentioned in the text, some 30 percent of terrestrial habitats on Earth have already been co-opted, mostly by agriculture and forestry. (18)

5. Choice *b.* Because small populations are more subject to inbreeding than larger populations, they generally show less genetic heterozygosity. (3–9)

6. Choices *a, c, d.* Although the expected magnitude of climate change due to global warming may be similar to past climate changes, the rate of warming will be greater. This, combined with habitat fragmentation, will make it impossible for many species to extend their ranges at the same rate as the northward movement of the temperature zones. (25, 26)

7. Choice *a.* Since the edge of a habitat patch equals the circumference of the patch, it is proportionally greater for smaller habitat patches. So, the relationship between edge effect and habitat patch area is an inverse relationship, as shown by curve *a.* (21, 22)

- Species that are distributed as a metapopulation consist of many smaller subpopulations each occupying a patch of suitable habitat. If there is a significant immigration rate between subpopulations, any single subpopulation is less likely to become extinct. The rescue effect is the prevention of subpopulation extinction by occasional immigrants from nearby patches. (5–7)

- A list of economic reasons people should care about species extinctions would include the importance of natural products, ecosystem services, and nature tourism. (10–12)

DESIGNING RECOVERY PLANS • ESTABLISHING PRIORITIES FOR CONSERVATION EFFORTS • THE IMPORTANCE OF COMMERCIAL LANDS • RESTORING DEGRADED ECOSYSTEMS • MARKETS AND CONSERVATION • INVESTING TO PRESERVE BIODIVERSITY (pages 433–434)

1. Choices *a, c, d, e.* Endemic species are common wherever habitat islands exist. Habitat islands would not be common in continental areas with uniform habitat. (14, 15)

2. Choice *a.* A keystone species has a large effect on the structure and functioning of an ecological community. This is not true of Kirtland's warbler, but it is true of the other species listed. (8–10)

3. Choice *a.* The defining characteristic of an "umbrella species" is that it requires large amounts of a specific habitat. (1, 2)

4. Choice *c.* Although rapid climate change may be a major concern in the future, at present it is not considered the cause for the decline in North American songbird populations. All of the other factors mentioned have been implicated in their decline. (3–6)

5. Choice *c.* Peregrines were reintroduced into the same habitats that they had disappeared from due to the effects of DDT on eggshell thickness. Reintroduction took place after restricted use of DDT had made those habitats safe for peregrines. (11, 12)

6. Choices *a, b, c.* (14–16)

- A megareserve is typically a large area that includes a multipurpose region surounding an undisturbed natural area. Megareserves are managed for both biodiversity and economic activities. This structure allows regulated use of a natural area for such things as drinking water, hydroelectric production, eco-tourism, forest products, and so forth, while preserving its biodiversity. (17, 18)

- Domestic livestock help to keep the grasses cropped down so they cannot serve as fuel for extensive fires that damaged adjacent forests. Cattle are also good seed dispensers for some tree species. Cattle grazing will be discontinued after woody succession has reached a point where fire is no longer a danger. (19)